Medical – Chapter 18	Scientific – Chapter 19	
Obstacle detection for the blind	Satellite detection Space vehicle navigation and flight control Horizon sensors Sun followers for in orientation Studies of the optical of the horizon	1. Search, track
Measurement of skin temperature Early detection of cancer Monitor healing of wounds and onset of infection, without removing bandages Remote biosensors Studies of skin heating and temperature sensation	Measurement of lunar, planetary, and stellar temperatures Remote sensing of weather conditions Study of heat transfer in plants Measurement of the earth's heat balance	2. Radiometry
Detection and monitoring of air pollution Determination of carbon dioxide in the blood and in expelled air	Determination of the constituents of earth and planetary atmospheres Detection of vegetation or life on other planets Terrain analysis Monitor spacecraft atmospheres Zero-G liquid level gauge Measurement of magnetic fields	3. Spectroradiometry
Early detection and identification of cancer Determination of the optimum site for an amputation Localization of the placental site Studies of the efficiency of Arctic clothing Early diagnosis of incipient stroke	Earth resource surveys Locate and map the gulf stream Detect forest fires from satellites Study volcanoes Detect and study water pollution Locate crevasses Sea-ice reconnaissance Petroleum exploration	4. Thermal imaging
Measurement of pupillary diameter Location of blockage in a vein Monitoring eye movements Study the nocturnal habits of animals Examination of the eye through corneal opacities Monitoring healing processes	Detection of forgeries Determine thickness of epitaxial films Determination of the surface constituents of the moon and the planets Gem identification Analysis of water quality Detection of diseased crops	5. Reflected flux
Ranging and obstacle detection for the blind Heat therapy	Space communications Understand the mechanism of animal communication Peripheral input for computers Study the nocturnal habits of animals Terrain illumination for night photography	6. Cooperative source

RICHARD D. HUDSON, JR., Senior Scientist, Infrared Laboratories, Hughes Aircraft Company, served as consultant to the Director, assisted in the establishment of foreign joint-venture companies, and was responsible for programs to upgrade the technical competence of the engineering staff—an assignment that led to his present position as Manager of the Hughes Professional Development Center.

Mr. Hudson received BS and MS degrees in Optics from the Institute of Optics of The University of Rochester. He is Executive Secretary of The IRIS Specialty Group Program and was Associate Chairman of its Infrared Standards Group. He has published widely in the fields of infrared, night vision, traffic safety, and engineering education.

Infrared System Engineering

WILEY SERIES IN PURE AND APPLIED OPTICS

Advisory Editor

STANLEY S. BALLARD, University of Florida

Lasers, BELA A. LENGYEL

Ultraviolet Radiation, second edition, LEWIS R. KOLLER

Introduction to Laser Physics, BELA A. LENGYEL

Laser Receivers, MONTE ROSS

The Middle Ultraviolet: Its Science and Technology, A.E.S. GREEN, *Editor*

Optical Lasers in Electronics, EARL L. STEELE

Applied Optics, A Guide to Optical System Design/Volume 1, LEO LEVI

Laser Parameter Measurements Handbook, HARRY G. HEARD

Gas Lasers, ARNOLD L. BLOOM

Advanced Optical Techniques, A.C.S. VAN HEEL, *Editor*

Infrared System Engineering, RICHARD D. HUDSON, JR.

Infrared System Engineering

RICHARD D. HUDSON, JR.
SENIOR SCIENTIST
INFRARED LABORATORIES
HUGHES AIRCRAFT COMPANY

WILEY – INTERSCIENCE

John Wiley & Sons, New York/London/Sydney/Toronto

Copyright © 1969 by John Wiley & Sons, Inc.

All rights reserved. No part of this book may
be reproduced by any means, nor transmitted,
nor translated into a machine language without
the written permission of the publisher.

10 9 8 7 6 5 4 3 2 1

Library of Congress Catalog Card Number: 68-8715
SBN 471 41850 1
Printed in the United States of America

To my wife BETTY

Foreword

Modern infrared technology was born during World War II. In its early years it was nurtured chiefly by military funds and was directed toward military applications. Thus technical details were protected by military security and were not released for general civilian use; that is, the literature of infrared technology was largely classified and was unavailable to the scientific and engineering fraternities. The security barrier was gradually lowered, and in the late 1950's and early 1960's books on infrared techniques began to appear. Concurrently, the technical journals printed more and more infrared articles, and professional societies scheduled papers on infrared physics and technology.

The present book is the latest, but probably not the last, on this general subject. It is distinguished from the others by its completeness in coverage, its down-to-earth attitude toward engineering details, its system point-of-view, and its very comprehensive bibliography. Just as life involves a series of compromises, or tradeoffs, so does engineering design. Here the tradeoffs in infrared systems are clearly identified and evaluated.

The author has bravely, but not brazenly, tackled a very big job in covering such an extensive field. Ordinarily this would be too great a task for a single author, but, as stated in the Preface, Mr. Hudson has been assisted by many specialists in his home organization and has profited by teaching the material several times to Hughes engineers.

The first 14 chapters speak for themselves—they cover the broad fields of infrared physics, technology, and system engineering. I invite the reader's attention to the extensive listing of applications of infrared techniques in Chapters 15 to 19. Here is an excellent, exhaustive bibliography that includes industrial, medical, and scientific entries as well as the perhaps better-known military applications. These chapters constitute an exclusive and especially valuable feature of the book. The four appendices should not be overlooked and the end papers will prove useful.

I am both happy and proud to have this fine addition to the Wiley Series in Pure and Applied Optics.

STANLEY S. BALLARD

Gainesville, Florida
September 1968

Preface

This book is written for those who design, build, test, or use infrared equipment to solve problems that occur in the military, industrial, medical, and scientific fields. Unlike previous books of this nature, it stresses the engineering approach rather than the scientific approach with its inevitable mathematical rigor. It is written for those who work with infrared equipment rather than for those who merely think or write about such equipment.

The Infrared Laboratories (now the Electro-Optical Laboratories) of the Hughes Aircraft Company comprise the largest organization in the free world (and probably in the entire world) devoted to the design, development, and production of sophisticated military infrared equipment. As Senior Scientist in these Laboratories it has been my privilege to associate with top-flight engineering designers, to observe their methods of design, and to follow many of the systems that they design through their full life cycles. One of my assignments during this time was the responsibility for continual upgrading of the technical competence of our engineering staff. This book evolved from a series of courses and seminars that I conducted for this purpose. It began as a set of lecture notes (of which some 1200 copies have been distributed within Hughes) and gradually developed into the present book. This series of courses offered me the opportunity to determine how well the material presented met the needs of more than 450 engineers. Their comments and reactions have been a great help in deciding what topics the book should cover and how they should be presented.

As its title implies, I have written this book for system engineers and for those who aspire to do system engineering. The development of today's complex systems requires the efforts of a team of specialists skilled in a variety of engineering discipline. At the center of this team is the system engineer. To be successful he must understand something of each discipline and must coordinate and guide the efforts of each member of the project team. This book provides the system engineer with the important facts about the various disciplines that are needed in the development of an infrared system, it discusses the terminology peculiar to each, and, perhaps most important of all, it tries to point out the limitations of each of these disciplines. In short, it provides the system engineer with the means of expressing himself in terms the specialist will understand, and it gives him sufficient insight to make an intelligent assessment of the efforts of each specialist. In doing this the book also provides a perspective from which members of the project team can assess their own contributions and better appreciate the problems faced by their colleagues.

The book opens with a history of the development of the infrared portion of the electromagnetic spectrum, probes the system engineering process, and then examines the characteristics of the successful system engineer. The next eleven chapters delve deeply into the engineering aspects of the elements that comprise the infrared system. These chapters cover such topics as the radiation characteristics of typical targets, the transmission of the atmosphere, optical systems and optical materials, optical modulation, infrared detectors and their associated coolers, signal processing, and displays. Chapter 13 returns to the system viewpoint and examines the functional relationships between the various system elements and the effects of their interactions when assembled into a system. Analytical expressions are developed for estimating the performance of various types of infrared equipment and the basis is laid for tradeoff studies, which are an essential part of the system engineering process. Finally, in Chapter 14, the reader is invited to peer over the system engineer's shoulder to watch the development of an infrared search system for use in commercial jet transports. It is unusual in a book of this sort to devote an entire chapter to the discussion of a single design problem, but it appears to be the only way to give the reader an accurate understanding of how an engineering team really does converge on the optimum system design.

Part II of this book is devoted to an examination of the application of infrared techniques to the solution of military, industrial, medical, and scientific problems. This is not the usual stereotyped collection of how-to-do-it ideas interlarded with photos and descriptions taken from manufacturers' literature. Instead, annotated references to the *published literature* of the world are used. Such an extensive collection of references has not been published elsewhere. Taken in their entirety, these references literally chronicle the world's progress in the application of infrared techniques. The references have been grouped according to a carefully prepared classification scheme that is shown in Figure 15.1 and repeated in the end papers. In many instances virtually all of the open literature is listed that is relevant to a particular application. This provides a rarely encountered historical perspective and allows the reader to trace the efforts of his predecessors and to examine in detail how various techniques have been applied to the solution of particular problems. The annotations summarize the contents of each reference, describe the equipment that was used and how well it worked, and indicate the significant results that are reported.

I have emphasized that the references are to the *published* literature — a very important point. The literature of the infrared is weighed down with a huge bulk of unpublished literature, much of which is classified. All too often authors cite this unpublished literature knowing full well that many of their readers have no means of access to it. With a very few well-justified exceptions, I have refused to include any citations to this unpublished literature. My references can be obtained through any well-stocked public library. You, the reader, will experience none of the frustrations that arise when a newly found reference cites an internal report of the XYZ Company or, worse yet, an unpublished communication between authors.

In addition to the nearly 1400 annotated references in Part II, there are more than 700 carefully selected (unannotated) references in Part I pertaining to the various elements that form the building blocks of an infrared system. A large number of these references are from literature published outside of the United States. This underscores the fact that no nation has a monopoly on infrared techniques and it helps broaden the reader's perspective of the field. Among the countries whose literature is cited are Australia, Czechoslovakia, England, France, Germany, Holland, Hungary, Japan, Poland, Rumania, and the Soviet Union.

Appendix 4 is a guide to the unpublished and the classified literature of the infrared. Those readers having the necessary credentials can use the material in this appendix to guide their entry into this special branch of the literature.

Another unusual feature of this book is the extent to which the patent literature has been used. Although much of the information pertaining to the infrared is classified, it is not always easy to know where to draw the line between material that is classified and that which is not. By definition, patents are in the public domain and contain no classified or proprietary information. Many of the patent citations serve a dual function — they contain valuable information and they clearly establish that this information is not classified. An example of such usage is found in Chapter 6, Optical Modulation, where two-thirds of the references cited are patents. Since the information contained in these patents is not available elsewhere in the published literature, such a detailed discussion of reticles could not have been written without reference to these patents.

I have chosen to use the units of measurement that appear to be most prevalent in each of the fields concerned. Hence the consistency that would have resulted from the use of the International System of Units (the modern version of the Metric System) is lacking, but it does not seem fair to burden readers with a continuing need to make unfamiliar conversions. Most U.S. engineers can readily visualize a detection range stated in miles, but if it is given in kilometers many will have to convert it to miles before they can fully appreciate it. If the sequence 107, 58, 97 centimeters is mentioned, few of you will even ponder its meaning, but if it is restated as 42, 23, 38 inches nearly all will instantly recognize the dimensions every Hollywood starlet covets. Those who need assistance with conversions will find it in Appendix 3.

How does one properly acknowledge all the help received in the preparation of a book? I cannot possibly hope to list all those who deserve it, since virtually every friend and acquaintance of my professional life has contributed in some way to the thoughts and ideas expressed here. Regrettably, I must limit acknowledgement to a selected few chosen from those who contributed most directly (and, perhaps, most recently) to the final synthesis, this book.

Dr. Warren E. Mathews, (the then) Director of the Hughes Infrared Laboratories, gave me the initial organization-wide technical upgrading assignment, suggested that I conduct classes and prepare a set of lecture notes, and gave me the encouragement and the opportunity to turn them into this book. May everyone have such a friend!

Dr. Stanley S. Ballard, advisory editor of the Wiley Series in Pure and Applied Optics, is a valued friend of long standing. He has painstakingly read the manuscript and offered countless valuable suggestions for its improvement. In 1956 Dr. Ballard co-authored a RAND Corporation report (R-297) entitled *Fundamentals of Infrared for Military Applications*. Those who know him will remember how he referred to it as "What Every Young Engineer Should Know About The Infrared." I have long admired this report and the practical philosophy that it contains. It has undoubtedly had a strong influence on the development of my own book.

More than 20 years ago Dr. R. Bowling Barnes, one of the true pioneers in the application of infrared techniques, gave me my first opportunity to explore the infrared. More recently he has been extremely helpful in bringing to my attention many of the references to medical applications that are cited in Chapter 18.

Many others have reviewed portions of the manuscript and contributed ideas and encouragement. These include D. E. Bode, J. S. Buller, H. F. Carl, K. W. Cowans, W. A. Craven, Jr., R. H. Frels, R. H. Genoud,

N. B. Hammond, A. S. Jerrems, R. C. McCormack, B. Shube, and R. M. Talley.

Mrs. Joyce Kachin and Mrs. Shirley Collins patiently typed the several versions of the manuscript and its many changes. Mrs. Hazel Whipple and Mrs. Betty Mathewes handled the voluminous correspondence that a book entails and assisted with the classes that served as a proving ground for the text.

The Hughes Aircraft Company and its subsidiary The Santa Barbara Research Center have been most kind in supporting my efforts and in providing me with a multitude of services that I might not otherwise have enjoyed.

And, finally, I voice a special note of thanks to my family for the support they gave and the sacrifices they made in order than I might realize my dream of writing a book of this kind.

RICHARD D. HUDSON, JR.

Woodland Hills, California
September 1968

Contents

PART I THE ELEMENTS OF THE INFRARED SYSTEM

Chapter 1 Introduction to Infrared System Engineering — 3

1.1 The Development of the Infrared Portion of the Spectrum — 3
1.2 The Market for Infrared Devices — 9
1.3 System Engineering — 10
1.4 The System Engineer — 13
1.5 The Infrared System and the Organization of This Book — 14
1.6 The Literature of the Infrared — 16
1.7 The Symbols and Abbreviations Used in This Book — 17

Chapter 2 Infrared Radiation — 20

2.1 The Electromagnetic Spectrum — 20
2.2 Terminology Used in the Measurement of Radiant Energy — 23
2.3 The Measurement of Radiant Flux — 30
2.4 Thermal Radiation — 33
    Thermal Radiation Laws — 35
2.5 Emissivity and Kirchhoff's Law — 39
2.6 Selective Radiators — 46
    Absorption Spectra of Gases — 47
    Absorption Spectra of Liquids and Solids — 50
    Molecular Emission Spectra — 50
2.7 Aids for Radiation Calculations — 53
    Radiation Slide Rules — 53
    Charts and Monographs — 57
    Tables of Blackbody Functions — 60
2.8 Other Blackbody Relationships — 60
    Efficiency of Radiation Production — 60
    Radiation Contrast — 63

Chapter 3 Sources of Infrared Radiation   67

- 3.1 Blackbody-Type Sources   67
  - Theoretical Principles   68
  - Construction of a Blackbody-Type Source   72
- 3.2 Standards for Sources of Radiant Energy   78
- 3.3 General-Purpose Sources of Infrared   82
  - The Nernst Glower   82
  - The Globar   83
  - The Carbon Arc   83
  - The Tungsten Lamp   83
  - The Xenon Arc Lamp   84
  - The Laser   84
  - The Sun   84
- 3.4 Targets   85
  - The Turbojet Engine   85
  - The Turbofan Engine   89
  - The Boeing 707 Jet Transport   90
  - Afterburning   96
  - The Ramjet   98
  - The Rocket Engine   99
  - Aerodynamic Heating   100
  - Personnel   103
  - Surface Vehicles   103
  - Stars and Planets   104
- 3.5 Backgrounds   104
  - The Earth   106
  - The Sky   108
  - Outer Space   109
  - Stars and Planets   109

Chapter 4 Transmission of Infrared Radiation Through the Earth's Atmosphere   114

- 4.1 The Earth's Atmosphere   116
- 4.2 Water Vapor   119
- 4.3 Carbon Dioxide   126
- 4.4 Other Infrared-Absorbing Gases   127
- 4.5 Field Measurements of Atmospheric Transmission   129
- 4.6 Laboratory and Analytical Methods of Predicting Atmospheric Transmission   136
- 4.7 Tables of Atmospheric Transmission Data   142
- 4.8 Scattering Effects in the Atmosphere   159

CONTENTS xvii

4.9	Transmission Through Rain	163
4.10	Atmospheric Scintillation	165

Chapter 5   Optics 171

5.1	Refraction and Reflection	171
5.2	Describing an Optical System	175
5.3	Factors Affecting Image Quality	182
	Diffraction	183
	Aberrations	186
5.4	Typical Optical Systems for the Infrared	195
	Reflective Optics	195
	Refractive Optics	200
	Miscellaneous Considerations in the Choice of Optics	201
5.5	Auxiliary Optics	202
5.6	Methods of Generating Scan Patterns	206
5.7	Optical Materials for the Infrared	209
5.8	Antireflection Coatings	218
5.9	High-Reflection Coatings	220
5.10	Optical Filters	222
5.11	Collimators	224

Chapter 6   Optical Modulation 235

6.1	Optical Filtering for Background Discrimination	236
6.2	The Use of Reticles for Background Suppression	237
6.3	The Use of Reticles to Provide Directional Information	240
	Rotating Reticles	241
	Stationary Reticles	250
	Two-Color Reticles	254
6.4	Tracking Systems Without Reticles	255
6.5	Comments on Reticle Design	256
6.6	Fabrication of Reticles	258

Chapter 7   Introduction to Detectors 264

7.1	How the Performance of a Detector is Described	266
7.2	Thermal Detectors	271
	The Thermocouple	271
	The Thermopile	272
	The Bolometer	273
	The Pneumatic or Golay Detector	274
	The Calorimetric Detector	274
	Problems of Blackening Thermal Detectors	274
7.3	Photon or Quantum Detectors	275

	The Photoelectric Detector	276
	The Photoconductive Detector	278
	The Photovoltaic or p-n Junction Detector	284
	The Photoelectromagnetic Detector	286
	Spectral Response of Photon Detectors	286
	Fabrication of Photon Detectors	287
7.4	The Comparison of Detectors	287
7.5	Optically Immersed Detectors	289
7.6	Imaging Detectors	296
	Infrared Film	296
	The Image Converter	296
	The Vidicon	297
	The Photothermionic Image Converter	299
	The Evaporograph	299
	The Infrared-Sensitive Phosphor	299

Chapter 8 Noise   304

8.1	Types of Noise	305
	Johnson or Thermal Noise	306
	Shot Noise	308
	Partition Noise	309
	$1/f$ Noise	309
	Generation-Recombination Noise	309
	Radiation or Photon Noise	310
	Temperature Noise	310
	Summary — Noise in Detectors	310
8.2	Equivalent Noise Bandwidth	311
8.3	The Statistical Description of Noise	313
8.4	Meters for the Measurement of Noise	316
	Peak-Responding Meter	317
	Rms-Responding Meter	317
	Average-Responding Meter	317
8.5	Noise Figure	318

Chapter 9 The Measurement of Detector Characteristics   321

9.1	Quantities To Be Measured	321
9.2	The Basic Detector Test Set	322
9.3	Use of the Basic Detector Test Set	330
	Measurement of Detector Area	330
	Determining the Operating Point of a Detector	330
	Determining the Operating Point for a Detector that Requires Bias	334

CONTENTS   xix

		Determining the Operating Point for a Self-Generating Detector	336
		Calibrating the Amplification of the Test Set	337
		Measurement of Frequency Response	339
		Measurement of the Detector Noise Spectrum	339
		Calculation of the Various Figures of Merit	340
	9.4	The Measurement of Spectral Response	340
	9.5	The Measurement of Time Constant	345
	9.6	The Measurement of Detector Response Contours	346

Chapter 10 Modern Detectors and the Ultimate Limits on Their Performance — 348

	10.1	Background-Limited Photon Detectors	348
	10.2	Limitations on the Performance of Thermal Detectors	357
	10.3	Considerations in the Selection of a Detector	358
	10.4	Engineering Data on Selected Detectors	363

Chapter 11 Techniques for Cooling Detectors — 373

	11.1	Packaging Cooled Detectors	374
	11.2	Low-Temperature Coolants	376
	11.3	Open-Cycle Refrigerators	377
		Liquid-Transfer Refrigerators	377
		Joule-Thomson Refrigerators	380
		Solid-Refrigerant Coolers	384
		Radiative-Transfer Coolers	385
		Comparison of Typical Open-Cycle Refrigerators	386
	11.4	Closed-Cycle Refrigerators	386
		Joule-Thomson (Closed-Cycle) Refrigerators	387
		Claude Refrigerators	388
		Stirling Refrigerators	389
		Refrigerators Using Other Refrigeration Cycles	390
		Comparison of Typical Closed-Cycle Refrigerators	391
	11.5	Solid-State Refrigerators	391
		Thermoelectric Refrigerators	391
		Thermomagnetic Refrigerators	394
	11.6	Integrating the Detector and Refrigerator	394

Chapter 12 Signal Processing and Displays — 398

	12.1	General Considerations	399
	12.2	Preamplifiers	400
		Preamplifiers Using Vacuum Tubes	401
		Preamplifiers Using Transistors	405

Preamplifiers Using Microelectronics 405
12.3 Additional Considerations in Signal Processing 408
12.4 Multiple-Channel Systems 410
12.5 Displays 411

Chapter 13 The Analysis of Infrared Systems 417

13.1 The Generalized Range Equation 417
     Tradeoff Analysis 420
13.2 The Generalized Range Equation for a Background-Limited Detector 421
13.3 The Range Equation for Specific Types of Systems . 423
     Search Systems 424
     Tracking Systems that Use Reticles 426
     Tracking Systems that Use Pulse Position Modulation 428
13.4 Line-Scan Thermal Mapping Systems 428
13.5 Radiometry 432
13.6 The Specification of System Performance 435

Chapter 14 The Design of an Infrared Search System 438

14.1 Preliminary Studies 438
14.2 System Synthesis and Analysis 441
14.3 Tradeoff Studies and Final System Design 447

PART II THE APPLICATIONS OF INFRARED TECHNIQUES

Chapter 15 An Introduction to the Applications of Infrared Techniques 455

15.1 The Applications of Infrared Techniques 458
15.2 Miscellaneous References 459

Chapter 16 Military Applications of Infrared Techniques 464

16.0 General 466
16.1 Search, Track, and Ranging Applications 469
     16.1.1 Search Systems 469
     16.1.2 Track Systems 475
     16.1.3 Search and Track Systems 477
     16.1.4 Weapon Guidance 479
     16.1.5 Navigation and Flight Control Systems . . . 486
     16.1.6 Ranging Systems 489
16.2 Radiometric Applications 490
     16.2.1 Measurement of Flux 490

CONTENTS xxi

16.3	Spectroradiometric Applications	491
	16.3.1 Target and Background Signatures	491
	16.3.2 Miscellaneous	493
16.4	Thermal Imaging Applications	494
	16.4.1 Reconnaissance	494
16.5	Applications Involving Reflected Flux	497
	16.5.1 Applications of Image Converter Tubes	498
	16.5.2 Infrared Photography	502
16.6	Applications Involving a Cooperative Source	502
	16.6.1 Terrestrial Communications	503
	16.6.2 Ranging	506
	16.6.3 Infrared Countermeasures	507
	16.6.4 Command Guidance	508

Chapter 17 Industrial Applications of Infrared Techniques 511

17.1	Search, Track, and Ranging Applications	511
	17.1.1 Search Systems	512
17.2	Radiometric Applications	513
	17.2.1 Measurement of Temperature	513
	17.2.2 Position Sensing	522
17.3	Spectroradiometric Applications	523
	17.3.1 Measurement of Temperature	523
	17.3.2 Miscellaneous	524
17.4	Thermal Imaging Applications	527
	17.4.1 Nondestructive Test and Inspection	528
17.5	Applications Involving Reflected Flux	531
	17.5.1 Applications of Image Converter Tubes	531
	17.5.2 Infrared Photography	532
	17.5.3 Miscellaneous	532
17.6	Applications Involving a Cooperative Source	533
	17.6.1 Intrusion Detection	533
	17.6.2 Miscellaneous	534

Chapter 18 Medical Applications of Infrared Techniques 537

18.1	Search, Track, and Ranging Applications	538
	18.1.1 Obstacle Detection (Passive)	538
18.2	Radiometric Applications	539
	18.2.1 Measurement of Temperature	539
18.3	Spectroradiometric Applications	541
	18.3.1 Miscellaneous	541
18.4	Thermal Imaging Applications	541
	18.4.1 Diagnostic Assistance	542

18.5	Applications Involving Reflected Flux		547
	18.5.1	Applications of Image Converter Tubes	547
	18.5.2	Infrared Photography	548
	18.5.3	Miscellaneous	549
18.6	Applications Involving a Cooperative Source		550
	18.6.1	Obstacle Detection (Active)	550

Chapter 19 Scientific Applications of Infrared Techniques — 551

19.1	Search, Track, and Ranging Applications		552
	19.1.1	Search and Track Systems	553
	19.1.2	Navigation and Flight Control Systems	554
19.2	Radiometric Applications		563
	19.2.1	Measurement of Temperature	563
	19.2.2	Measurement of Flux	575
	19.2.3	World Weather Watch	579
19.3	Spectroradiometric Applications		581
	19.3.1	Remote Sensing of the Earth and its Atmosphere	581
	19.3.2	Remote Sensing of Astronomical Bodies	586
	19.3.3	Instrumentation and Miscellaneous Applications	594
19.4	Thermal Imaging Applications		599
	19.4.1	Earth Resource Surveys	599
	19.4.2	Meteorological Applications	603
	19.4.3	Lunar and Planetary Studies	605
	19.4.4	Miscellaneous	606
19.5	Applications Involving Reflected Flux		607
	19.5.1	Infrared Photography	607
	19.5.2	Reflectance Properties of Materials	612
19.6	Applications Involving a Cooperative Source		613
	19.6.1	Space Communications	613
	19.6.2	Miscellaneous	614

Appendix 1	**The Symbols and Abbreviations Used in This Book**	**615**
	a. Simple English Letter Symbols	615
	b. Simple Greek Letter Symbols	617
	c. Special and Composite Symbols	617
	d. Selected Abbreviations	618
Appendix 2	**Symbols and Nomenclature for Radiometry and Photometry**	**619**
Appendix 3	**Conversion Factors**	**623**
Appendix 4	**The Unpublished Literature of the Infrared**	**626**
Index		**631**

Tables

Number	Legend
2.1	Subdivisions of the Infrared
2.2	Recommended Radiometric Terminology
2.3	Emissivity (Total Normal) of Various Common Materials
2.4	Solar Absorptance α and Low-Temperature (300°K) Emissivity ϵ for Spacecraft Materials (adapted from data in [29, 31–33]
2.5	Some Convenient Relationships for Blackbody Spectral Distribution Curves (for λ in μ, T in °K, and W in W cm^{-2})
3.1	Characteristics of the Boeing 707 Intercontinental Jet Transports (courtesy of The Boeing Company, Airplane Division, Renton, Washington, and Pratt and Whitney Aircraft, Division of United Aircraft Corp., East Hartford, Conn.)
3.2	Summary of Calculated Radiometric Quantities from Example 3.4.1 for the Boeing 707 Intercontinental Jet Transports (altitude 35,000 ft, Mach 0.8, maximum cruise thrust)
3.3	Summary of Plume Calculations in Example 3.4.2 for the Boeing 707 Intercontinental Jet Transports (altitude 35,000 ft, Mach 0·8, maximum cruise thrust)
4.1	Designated Regions of the Earth's Atmosphere
4.2	Composition of the Earth's Atmosphere (Dry)
4.3	Mass of Water Vapor in Saturated Air, g m^{-3}

Number	Legend
4.4	Effective Transmittance of a Vertical Path Through the Atmosphere for Solar Radiation in the 16- to 24-μ region (adapted from Adel[41])
4.5	Window Limits and Constants for Use with equation 4-7 (Adapted from Elder and Strong[43])
4.6	Constants for use in equations 4-14 and 4-15 (adapted from Howard, Burch, and Williams[46])
4.7	Spectral Transmittance of Water Vapor for a Horizontal Path at Sea Level (from Passman and Larmore[65], courtesy of The Rand Corporation)
4.8	Spectral Transmittance of Carbon Dioxide for a Horizontal Path at Sea Level (from Passman and Larmore[65], courtesy of The Rand Corporation)
4.9	Spectral Transmittance of Water Vapor for a Horizontal Path at Sea Level
4.10	Spectral Transmittance of Carbon Dioxide for a Horizontal Path at Sea Level
4.11	Values of the Altitude Correction Factor $(P/P_0)^k$ for Reduction to Equivalent Sea-Level Path (adapted from Passman and Larmore[65], courtesy of The Rand Corporation)
4.12	Effective Transmittance of the 9·6 μ Absorption Band of Ozone
4.13	Calculated Transmittance Through 6 kft of Haze
4.14	Calculation of the Scattering Coefficient for Rain: Cloudburst Conditions, Rainfall Rate 4 in. (100 mm) per Hour (rainfall data from Laws and Parsons [90])
4.15	Transmittance of a 6 kft Path Through Rain
5.1	Values of b for Calculating the Increase in Blur Circle Diameter Due to Chromatic Aberration (use with equation 5-27)
5.2	Characteristics of Lenses Designed for Use in the Infrared Portion of the Spectrum
5.3	Linear Expansion Coefficient for Common Mounting Materials
5.4	Materials for Low Reflection Coatings (adapted from Smakula et al.[64])

Number	Legend
5.5	Reflectance of Evaporated Metal Films (adapted from Bennett et al.[92, 94], and Hass[93])
5.6	Estimate of the Probable Error in the Calculated Value of the Irradiance in the Beam from Double-Mirror Collimators
6.1	A Comparison of Several Types of Reticles for Tracking Systems
7.1	Some Important Characteristics of Infrared Detectors (data adapted from publications of various detector manufacturers)
8.1	Comparison of Equivalent Noise and 3 dB Bandwidths for Amplifiers with Different Interstage Coupling Networks (adapted from Lawson and Uhlenbeck [7])
8.2	Percent of Time a Given Peak Value Is Exceeded by Gaussian and by Rayleigh Noise
9.1	Figures of Merit Used to Describe the Performance of a Detector
9.2	Quantities that Might Be Reported on a Detector Data Sheet (adapted from [1, 2])
9.3	Values of the Rms Conversion Factor for Various Chopper Geometries (courtesy of Santa Barbara Research Center)
11.1	Physical Properties of Low-Temperature Coolants
11.2	Approximate Inversion Temperatures of Commonly Used Gases
11.3	Rate of Gas Flow to Produce 1 W of Refrigeration with a Joule-Thomson Cryostat
11.4	Attainable Temperatures and Required Storage Pressures for Solid Refrigerants (adapted from data in Gross and Weinstein [20])
11.5	Weight of Solid Refrigerant and Cylindrical Storage Container Needed to Cool a 0.1 W Heat Load for One Year. Length of Container Is Equal to Its Diameter and the Temperature of Its Outer Shell

Number	Legend
	is 300°K (adapted from data in Gross and Weinstein [20])
11.6	Characteristics of Typical Open-Cycle Refrigerators (data adapted from Gross and Weinstein [20] and from publications of the Hughes Aircraft Co. and of Santa Barbara Research Center)
11.7	Characteristics of Commercially Available Closed-Cycle Refrigerators (data adapted from publications of the Hughes Aircraft Co. and of the Santa Barbara Research Center)
13.1	Values of Differential Radiance ΔN for the Three Atmospheric Windows and for a Temperature of 300°K
14.1	Characteristics of the Preliminary Design for an Infrared Search System
14.2	Characteristics of the Final Design for an Infrared Search System
14.3	Estimated Weight, Volume, and Power Consumption of the Infrared Search System (final design)
16.1	Missiles that Are Thought to Employ Infrared Guidance (data adapted from 16.1.4 [9, 19, 33, 41, 44, 46, 47, 49, 52, 53, and 55])
16.2	Typical Characteristics of Night-Viewing Equipment that Uses Infrared Image Converters (data adapted from 16.5.1 [2, 4, 19, 27, 31, and 34])
16.3	Tactical Missiles that Are Thought to Use Infrared as an Element of Their Command Guidance (data adapted from 16.1.4 [47,49,55] and 16.6.4 [14 and 15])
19.1	Equilibrium Temperature for a Spherical Satellite (50 cm radius) in an Orbit 300 Miles Above the Earth's Surface

Infrared System Engineering

1

The Elements of the Infrared System

1

Introduction to Infrared System Engineering

"... There are rays coming from the sun ... invested with a high power of heating bodies, but with none of illuminating objects.... The maximum of the heating power is vested among the invisible rays.... It may be pardonable if I digress for a moment and remark that the foregoing researches ought to lead us on to others..."[1, 2].

Thus wrote Sir William Herschel early in 1800, in a series of papers in which he was the first to reveal the existence of what we today call the infrared portion of the spectrum[1, 2, 3]. By this time, Herschel, Royal Astronomer to King George III of England, was already famous for his discovery of Uranus. During the first few months of 1800 he sought a means of protecting his eyes while observing the sun, and this search led him to the discovery of the "invisible rays." Despite his prophetic remark concerning further researches, Herschel virtually ignored his discovery. He referred to it again in 1801 in two papers but made no mention of it thereafter. With the rise of infrared to a key role in modern technology, it is, indeed, a sad but curious commentary that the definitive article on Herschel in the *Encyclopedia Brittanica*[4] does not credit him with the discovery of the infrared, nor does it even hint that he worked in this area.

1.1 THE DEVELOPMENT OF THE INFRARED PORTION OF THE SPECTRUM

When we speak of the infrared, we mean that portion of the electromagnetic spectrum that lies between visible light on the one hand and microwaves on the other. Expressed quantitatively, it is the region that extends from a wavelength of 0.75 to 1000 μ. In the century and a half since Herschel's first hesitant steps into the region, many devoted workers have patiently contributed the bits and pieces necessary to transform a chance discovery into a solid engineering discipline. All of us owe a debt of gratitude to these early workers, and we can only hope that those who follow us will view our efforts as favorably. It is fortunate that a top-flight historian of science, E. Scott Barr, has published a series of papers that illuminate the efforts of many of these early workers[5 – 8]. Since it stresses twentieth-century developments, the paper by Arnquist[9] is an excellent supplement to those of Barr. Some of this history will be summarized here, because this author believes that most new concepts are more readily understood and accepted if something is known of their origins.

Herschel referred to the new portion of the spectrum by such names as "the invisible rays," "the thermometrical spectrum," "the rays that occasion heat," and "dark heat." Herschel did not use the term "infrared," and it does not seem to have appeared in the literature until the 1880's. It is vexing to this author, and to others[5] that we have been unable to identify and give credit to the originator of the term "infrared." This author is about ready to accept the unsatisfying explanation that since this term has an obvious Latin root,[1] it may have entered the language without any particular sponsorship.

In the course of his observations, Herschel tried colored glass filters as a means of reducing the brightness of the solar image. He noted that even though they gave similar reductions in brightness, some filters passed very little heat whereas others passed so much heat that observation had to be limited to a few seconds to prevent permanent damage to his eyes. Therefore he decided to examine various filters systematically to find one that gave the desired reduction in brightness as well as the maximum reduction in heat. His first step was to study the heating effect of sunlight. Using a technique developed by Newton nearly 100 years before, Herschel formed a spectrum by passing sunlight through a glass prism. The various colors of the spectrum fell on a table that served as a convenient support for his radiation detector, a sensitive mercury-in-glass thermometer. By noting the temperature increase that occurred when the blackened bulb of the thermometer was placed in the spectrum, he measured the heating effect of the different colors. Starting at the blue

---

[1]*Infra* — a prefix from the Latin meaning below or beneath. Hence the infrared region is the region below the red. The Latin root is readily recognizable in the equivalent word in other languages, for example, *infrarouge* in French, *infrarot* in German, and *infrakrasnye* in Russian. The German language still recognizes an alternative term *ultrarot* that is equivalent to the term "ultra red" found in some of the English literature of the 1880's. Most dictionaries recognize "infrared" as an adjective, but it is quite commonly used as a noun by workers in the field. As used in the United States, the term has undergone an evolution from "infra red," to "infra-red," and now finally to "infrared."

end of the spectrum, Herschel found that the heating effect increased as he moved the thermometer toward the red. Up to this point, his experiment was not unique; Landriani had performed a similar one before 1777 and several others had repeated it[5]. However, Herschel was the first to realize that there must be a point at which the heating effect reaches a maximum and that measurements limited to the visible portion of the spectrum failed to locate this point. Moving the thermometer into the dark portion beyond the red end of the spectrum, where his eye could perceive no light, he found that the heating effect not only persisted but continued to increase. The maximum, when he found it, lay considerably beyond the red end of the spectrum, that is, in the region that we now call the infrared.

Using the same apparatus, Herschel measured the radiation from fires, candles, and kitchen stoves. On the basis of these measurements, he posed questions concerning the similarity of light and heat that were to remain unanswered for many years. By comparing the temperature increases with and without a filter placed in front of the thermometer, Herschel was able to make the first crude measurements of the extent to which a filter transmits the various colors of light. His investigations covered nearly 50 samples of such diverse materials as colored glass, natural crystals, water, wine, gin, brandy, paper, and muslin. These measurements led to his selection of a filter consisting of a glass cell filled with ink-tinted water.

In retrospect, perhaps Herschel's most important contribution was the remarkably firm foundation that he provided for later workers: (1) he showed that there was a satisfactory detector available for investigating the new region of the spectrum; (2) he raised provocative questions concerning the basic similarity of light and heat; (3) he showed that there are significant differences in the way various materials transmit light and heat. For the next 100 years, many workers painstakingly followed the leads he provided. Few applications for their discoveries appeared, however, until the early part of the twentieth century. From our present vantage point, it is clear that the nineteenth century represents a period of component development that made possible today's applications of infrared techniques to the solution of military, scientific, industrial, and medical problems.

Thermometers remained unchallenged as radiation detectors until about 1830. With Herschel's thermometer, it took as long as 16 min to make an observation, and his readings were recorded to the nearest 0.5°. Later workers developed smaller, faster-reacting thermometers equipped with microscopes to permit the reading of 0.1° increments. In 1829 Nobili made the first thermocouple; it was an improved electrical thermometer based on the thermoelectric effect discovered by Seebeck in 1821. In 1833 Melloni constructed a thermopile by connecting a number of thermocouples in series. It was at least 40 times more sensitive than the best thermometer available and could detect the heat from a person at a distance of 30 ft[7].

In 1840 Herschel's son John, who was a famous astronomer in his own right, developed a radiation-detection process based on the differential evaporation of a thin film of oil to form a "heat picture." The same process was resurrected and improved by Czerny in 1929, and it subsequently formed the basis for the modern-day Evaporograph described in Chapter 7. In 1843 Becquerel found that certain materials phosphoresced when exposed to infrared radiation. Such phosphors are still used in relatively simple detection and communication systems. At about the same time, Becquerel was able to show that the photographic process could be extended a short distance into the infrared. In 1883, using especially sensitized photographic plates, Abney detected a wavelength of 1.3μ, a limit that exists to this day.

During the 1880's, several highly sensitive new detectors were developed. Most notable was the Langley bolometer[8], which was about 30 times more sensitive than Melloni's thermopile. In 1901 Langley and Abbot reported their development of an improved bolometer that could detect the heat from a cow at a distance of $\frac{1}{4}$ mile.

In 1917 Case developed the thallous sulfide detector, the first use of the photoconductive effect in the infrared. Unlike the earlier thermometers, thermocouples, and bolometers that utilize the heating effect of the incident radiation, the photoconductive detector uses a direct interaction between the photons in the incident radiation and the electronic structure of the detector material. Such a detector was more sensitive than any previously available, and it also responded more rapidly. During World War II, workers in Germany made noteworthy contributions to the development of photoconductive detectors, and they were the first to demonstrate the increase in sensitivity that could be obtained by cooling such detectors. Since World War II the development of photoconductive, or photon, detectors has proceeded so rapidly that they are now available for use in any part of the infrared region.

We know today that infrared, or heat radiation, and visible light are both forms of electromagnetic radiation and that they differ from each other in wavelength and frequency. The question of their similarity was of interest to Herschel, and it is evident from his papers that he changed his opinion several times during the course of his researches. It appears that Ampere clearly verified this similarity in 1835, but dissenting opinions are found in the literature as late as 1850. The difficulties faced by these early workers can perhaps be attributed to their inability to measure wavelength. In the late 1600's it was proposed that light was a form of wave motion, but it was not until 1804 that Young made the first measurement of the wavelength of visible light. Progress toward the accurate measurement of wavelength in the infrared was very slow. In 1847 Foucault and Fizeau made measurements extending to $1.5\,\mu$. In 1880 Desains and Pierre Curie succeeded in measuring wavelengths as long as $7\,\mu$. In 1897 Rubens measured out to $20\,\mu$, and working with Nichols in 1898, he extended this to more than $150\,\mu$.

Despite his crude equipment, Herschel showed that there were significant differences in the transparency of optical materials in the infrared. He also demonstrated that the newly discovered heat rays were reflected in exactly the same way as visible light. He therefore concluded that reflective optics, that is, plane or curved mirrors, would be particularly useful for investigations of the new portion of the spectrum. This conclusion is still valid. Melloni made a more detailed study of the transparency of optical materials and published the first of a series of papers on the subject in 1883. He showed that conventional optical glass has only a limited transparency in the infrared. His most important discovery was that rock salt is remarkably transparent to infrared. He made lenses and prisms of rock salt and used them in developing techniques that ultimately formed the basis of modern analytical infrared spectroscopy. Through the years that followed, many other workers contributed to the growing literature of optical materials until, today, information is readily available on the characteristics of well over 100 optical materials having a useful transparency in some part of the infrared.

Melloni made quantitative measurements of solar radiation, and in 1839 he noted variations that he correctly attributed to the absorption of solar radiation by water vapor in the earth's atmosphere. He originated the technique, still extensively used, in which the sun serves as a source for the measurement of the transmission of the atmosphere. Langley continued these investigations, made detailed maps of the various absorption bands, and correctly identified most of the absorbers. These measurements, reported between 1883 and 1900, extended to beyond $5\,\mu$. By 1917 Fowle had extended them to $13\,\mu$ and by 1942 Adel reached $24\,\mu$. By the early 1950's, the gap between $24\,\mu$ and the microwave region at $1000\,\mu$ had been bridged by several workers.

The literature of the early 1900's shows an increasing interest in the application of infrared to the solution of numerous problems. This period is dominated by the work of Coblentz[10, 11] who in 1903, started a forty-year career that included pioneering applications of infrared spectroscopy and precision radiometry, the establishment and maintenance of national standards of thermal radiation, the radiometric measurement of stellar and planetary temperatures, the application of infrared sources for therapeutic purposes, the development of heat-absorbing glasses for protecting the eyes, and the debunking of the claims of scientific quacks (for example, that infrared devices could promote the growth of hair on bald heads). The patent literature shows that many inventors "discovered" the infrared between 1910 and 1920. Patents issued during this period cover devices for the detection of ships, aircraft, personnel, artillery pieces, and icebergs, as well as devices for secure communications, intrusion detection, the remote sensing of temperature, and the guidance of aerial torpedoes.

During World War I both sides had research programs devoted to the military applications of infrared, and a few experimental communication devices underwent limited field evaluation. An infrared search system that was developed during this period could detect aircraft at a distance of 1 mile and people at a distance of 1000 ft.

The period between World Wars I and II is marked by the development of photon detectors and image converters and by the emergence of infrared spectroscopy as one of the key analytical techniques available to chemists. Since a discussion of spectroscopy is not within the scope of this book, the reader with further interest in this subject should consult [12-14]. The image converter, developed on the eve of World War II, was of tremendous interest to the military because it enabled man to see in the dark. In essence, it converts an image formed initially at infrared wavelengths into one that is visible to the human eye. Because few military targets produce a significant amount of radiation in the wavelength region around $1\,\mu$ where the image converter responds, it is necessary to provide a source of illumination. Such a source may consist of a tungsten lamp combined with a filter to block any visible light. A by-product of the wartime development of these sources is the sealed-beam headlamp found on virtually all modern automobiles. A system using an image converter is an *active* system in the same sense that radar is, that is, the system must first illuminate the target and then detect the reflected radiation. In contrast, *passive* systems emit no radiation but only sense the radiation emitted naturally by the target.

During World War II numerous infrared devices were proposed and investigated by both sides. However, a close study of this period shows that relatively little infrared equipment reached production status, and of those devices that did, nearly all were active. The Germans made effective use of an infrared communication system, the Lichtsprecher, in the African desert during the major tank battles from 1941 to 1943. They did what is probably the first infrared-system engineering job by integrating image converters into fire control systems for tanks. These systems, which were used on the eastern front in 1944, proved to be remarkably effective in nighttime battles. Apparently these successes did not impress the German High Command[15] since such devices were never used on the western front. The United States developed infrared communication systems for naval use and simple viewers for detecting the sources required by active systems. The best known U.S. development, the sniperscope, consisted of an image converter and an illuminator mounted on a carbine. It enabled a soldier to fire accurately in complete darkness at targets as far away as 75 yd. The sniperscope was first used in combat in April 1945 during the invasion of Okinawa. Both the United States and Germany were making progress in the development of night-driving systems utilizing image converters, but few such systems saw field use before the end of the war. Influenced by the German successes, the Japanese had plans for the production of similar active devices, but the war ended before they could be put into effect.

Despite their limited deployment during World War II, infrared devices showed sufficient merit to justify a strong postwar development supported by military funding. The period is remarkable for the rapid development of new and improved detectors and infrared-transparent optical materials and their application to a host of military problems. Subsequently, many of the same techniques were applied to the solution of industrial, scientific, and medical problems. In the late 1950's the release of information on the Sidewinder and Falcon heat-seeking infrared-guided missiles caught the public fancy, and subsequent applications of infrared techniques to the attitude stabilization of space vehicles, measurement of planetary temperatures, nondestructive testing, and the early detection of cancer have been eagerly reported in the popular news media.

1.2 THE MARKET FOR INFRARED DEVICES

As infrared matured into a recognized technology, the annual sales of infrared devices assumed significant proportions. The exact dollar value in the United States today is open to considerable speculation, but it is possible to indicate some general limits on the figure. Most estimates assume that sales to the military constitute about three-fourths of the total infrared market. Because of security restrictions and the limited detail given in published budgets, many of the military funds ultimately spent on infrared devices are difficult to identify. The remaining one-fourth of the market, the nonmilitary portion, consists of equipment for such applications as analytical infrared spectroscopy, process control, intrusion detection, fire monitoring or warning, and medical diagnosis.

The most comprehensive study of the total infrared market is that made in 1959 by Sanford Research Institute[16]. They estimate that the value of the market was $100 million in 1958. They offered several extrapolations showing that the value of the market should have reached $1 billion sometime between 1961 and 1964. A consulting firm, Robert Manley Associates, concluded that the value of the infrared market was about $250 million in 1962[17, 18]. This firm thought the figure would reach $500 million in 1965. Based on his own studies, this author favors a figure somewhat lower than either of those given above—$125 million in 1960, increasing to $350 million by 1968. A study limited to the nonmilitary market concluded that its value was about $15 million in 1959[19]. It should be noted, however, that one manufacturer of infrared hot-box detectors expected that his sales to U.S. railroads would reach almost $20 million by 1961. Regardless of which estimate we choose to believe, it is obvious that Herschel's discovery has developed into an important and lucrative industry.

1.3 SYSTEM ENGINEERING

The increasing rate at which scientific discoveries are made has created a demand for new ways to speed their application to the fulfillment of human needs. One result of this demand is the gradual disappearance of the lone inventor; he has been replaced by teams organized to follow a general discipline that Kelly[20] has appropriately termed "organized creative technology."

Organized creative technology encompasses the entire gamut of techniques that can be applied to transform a basic discovery into a useful and manufacturable item. The newest and hence the most glamorous of these techniques is system engineering. It offers a methodology specifically designed for finding the shortest, most efficient path through the transformation cycle. Opinions regarding the value of system engineering range all the way from the fervent belief that it is nothing less than sheer magic, to the conviction that it is nothing more than the logical application of common-sense engineering. The purpose of this book is neither to join this controversy nor to add to it, but instead to show how some of the methods of system engineering can assist in guiding the development of infrared systems.

A *system* is a set of elements whose interrelated functions are coordinated so as to serve some human purpose. This definition, which has purposely been made as broad as possible, becomes more specific when we define the elements, their functions, or the purpose of the system. Some systems, such as weapon or communication systems, are *physical*; others, such as accounting or business procedures, are *abstract*. It is well to recognize that the set of elements that is one man's system may be no more than a component or a subsystem to another man. The optical designer thinks of his creations as optical systems; the missile designer may have no more than a passing interest in the optical, mechanical, and electronic components that form the guidance subsystem of his missile; the air defense officer may consider a missile to be just another component carried aboard the interceptors that form one of the subsystems in his far-flung air defense system.

System engineering, as a term, defies precise definition in a single sentence.[1] In the very simplest sense, system engineering is the discipline that offers an orderly approach to the design of systems and, in particular, to systems that are so complex that no one individual can possibly understand all the pertinent details. A definition more in harmony with the objectives of this book is an operational definition that stresses the pattern of actions characteristic of the system engineering process. Although no two systems are ever designed in exactly the same way, a study of successful designs shows a common pattern that consists of some, or all, of the following phases:

1. Identification studies.
2. Problem definition.
3. Determination of performance specifications.
4. System synthesis.
5. System analysis.
6. System design.
7. Evaluation of system performance.
8. Sustaining engineering.

Identification studies provide background information to assist management in planning for future programs. They provide a broad survey of an entire area of technology, such as communications, battlefield surveillance, or power distribution, with emphasis on present and future needs and on means of achieving them. The ultimate objective is, of course, to identify an existing or future need that the organization is uniquely qualified to fill. This phase may not always be part of the system engineering process; it is not uncommon for it to be handled as a market analysis function or for a customer to recognize his own needs and ask others for assistance in meeting them.

---

[1] In his excellent book on the subject, Hall[21] devotes an entire chapter to the definition of system engineering. Even after filling 18 pages, he still seems to be somewhat unhappy with the result.

Problem definition commences when a need has been identified and management has decided that further study is in the best interests of the organization. It starts with a reexamination of the pertinent information developed during the identification studies and the collection of the further information needed for a clear statement of the problem. The real requirements of the customer must be determined and criteria established for measuring the value of meeting, or partially meeting, each requirement. The functions to be performed by the system must be identified not only for the existing environment but for the best estimate of the future environment as well. System constraints, such as cost, weight, volume, reliability, and operational complexity, must be determined and criteria established to measure the relative importance of each.

The *determination of performance specifications* is the logical complement to the problem definition phase. This phase results in a detailed description of exactly what the system is required to do. These performance specifications, often called *design objectives*, provide the system designer with quantitative measures for system inputs and outputs and the various system constraints. In addition, they serve as an idealized conceptual framework against which the system engineer can continually compare the various design alternatives that will evolve during the later design phases.

System synthesis is the formulation of a system model or concept capable of meeting the previously determined performance specifications. It is customary to develop several alternative system concepts, since, at this stage, it is still too early to tell whether a particular concept is entirely suitable.

The alternative system concepts are critically evaluated during the *system analysis* phase. The ultimate intent is to select the system concept, or perhaps a combination of several concepts, that appears to provide the most promising means of attaining the system objectives. Using various analytical models, the system analyst attempts to predict the performance of the systems that result from each of the system concepts. These results can be readily compared with the performance specifications to provide a quantitative evaluation of over-all system performance. Since it is usually possible to take a first look at costs during the system analysis phase, it is here that the system engineer has his first indication of the relationship between cost and the achievement of certain system objectives.

The *system design* process includes the steps necessary to translate the optimum system concept into physically realizable hardware. It is the first place in the system engineering process in which there is a real concern with the nuts-and-bolts aspect of hardware design. The design process rarely proceeds in a perfectly straightforward manner; instead, it is more likely to cycle through a preliminary design and then back to system analysis in order to compare the over-all performance with that described in the performance specifications. This cycle, often called *tradeoff studies*, results ultimately in a design that satisfactorily meets the system objectives and that can be successfully implemented within the cost and delivery limitations imposed by the customer. One of the most critical elements of the system design phase is knowing when to stop the design effort; this is a decision that is usually directly translatable into dollars. It is very likely that this decision depends more on art than on science, on the application of seasoned engineering experience backed up by the system analysis process.

The *evaluation of system performance* begins with the emergence of the first hardware from the design phase. It seeks to provide the final evaluation of the system operating in its normal working environment. The results of this phase provide the customer with the assurance that he has in fact received what he has paid for. The system engineer is vitally interested in these results because any shortcomings that they reveal can be brought to the attention of the designers for early correction. The lessons learned during this phase provide the basis for much of the intangible expertise that is the real stock-in-trade of the expert system engineering organization.

The *sustaining engineering* phase starts with the final installation of the system and extends throughout its useful life. Its purpose is to provide a continuous flow of information back to the design organizations concerning system performance, design weaknesses, reliability, failure mechanisms, and unusual system capabilities not previously recognized. System objectives rarely remain fixed throughout the life of a system. It is more likely that they will broaden in scope, and thus the system engineer must be ever alert for the need to upgrade and modify the system to provide for these broader objectives.

Readers interested in further discussions of the system engineering process can consult [21–24].

1.4 THE SYSTEM ENGINEER

After contemplating the elements of the system engineering process, we might well ask: "Just who is this jack-of-all-trades, the system engineer, who oversees the entire process?" Jack-of-all-trades he is, but he is usually master of at least one, the speciality he developed in depth before moving into system engineering. System engineers are rarely developed by the academic process; more often they drift into this field from the ranks of successful engineers after discovering that their interests lie in over-all concepts and functional relationships rather than in a narrow area of specialization. As Affel[22] has observed, the system engineer has shown a desire to grow broad as well as deep.

If there is a single distinguishing characteristic of the successful system engineer it is that he is an accomplished communicator. The evolution of almost any system requires the efforts of a team of specialists skilled in a wide variety of engineering disciplines. The system engineer must coordinate and guide the efforts of these specialists in order to achieve a system that meets all of the goals set for it. He is, in effect, the master planner who provides the conceptual framework within which the other team members must operate. As such, he does not have to be a specialist in all of the disciplines involved, but he must speak the language peculiar to each and must understand what can reasonably be expected from each. He has a knack for recognizing what it is that is really important to know about each discipline, and he does not involve himself with the details that are better left to the specialist. Being, as he is, at the center of the process, the system engineer must provide a clear communication channel between the specialists on his team; his failure to do so will almost certainly result in the team failing to provide a system that fulfills the required objectives.

It is the intent of this book to provide the system engineer with the important facts about the various disciplines needed in the development of a successful infrared system. This book stresses the language peculiar to each discipline, develops the important facts about each and, perhaps most important of all, tries to point out the limitations of each discipline. In short, it provides the system engineer with the means of expressing himself in terms that the specialist will understand and, equally important, gives him sufficient insight to permit him to make intelligent assessments of the efforts of each specialist.

1.5 THE INFRARED SYSTEM AND THE ORGANIZATION OF THIS BOOK

The elements of an infrared system are shown in block diagram form in Figure 1.1. The target is the object of interest, usually the real reason for the existence of the system. For the purposes of this book it is assumed that the target radiates energy somewhere in the infrared portion of the spectrum. The system may be designed to detect the presence of the target, to track it as it moves, to glean information leading to its identity, or to measure its temperature. If the radiation from the target passes through any portion of the earth's atmosphere, it will be attenuated

INTRODUCTION TO INFRARED SYSTEM ENGINEERING

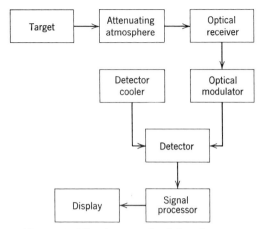

Figure 1.1 The elements of an infrared system.

because the atmosphere is not perfectly transparent. The optical receiver, which is closely analagous to a radar antenna, collects some of the radiation from the target and delivers it to a detector which converts it into an electrical signal. Before reaching the detector, the radiation may pass through an optical modulator where it is coded with information concerning the direction to the target or information to assist in the differentiation of the target from unwanted details in the background. Since some detectors must be cooled, one of the system elements may be a means of providing such cooling. The electrical signal from the detector passes to the processor where it is amplified and the coded target information is extracted. The final step is the use of this information to automatically control some process or to display the information for interpretation by a human observer.

Since it succinctly describes an infrared system, Figure 1.1 also provides a convenient framework for the organization of this book. Each of the elements is discussed in detail in one or more chapters: the generation of infrared radiation and the radiation characteristics of targets are discussed in Chapters 2 and 3, the attenuation of radiation by the earth's atmosphere is covered in Chapter 4, optics and optical materials are considered in Chapter 5, optical modulation processes are described in Chapter 6, detectors are discussed in Chapters 7 through 10, methods of cooling detectors are found in Chapter 11, and signal processing and displays are covered in Chapter 12. Chapter 13 marks a return to the system viewpoint. It emphasizes the examination of the functional relationships between the various elements of the system and the effects of their interactions. Analytical expressions are developed from which the performance of various types of infrared systems can be predicted, and the foundation is laid for tradeoff studies, which are an essential part of the system engineering process. In Chapter 14 the reader is invited to peer over the system engineer's shoulder to watch the evolution of an infrared search system for use in commercial jet transports. The last five chapters form Part 2 of the book; they examine the application of infrared techniques to the solution of military, industrial, medical and scientific problems.

1.6 THE LITERATURE OF THE INFRARED

The literature that pertains to the infrared can be divided into two categories, *published* and *unpublished*. In more practical terms, these might well be called the readily available and the not so readily available literature. This author has strong feelings about the distinction between the two and is particularly delighted with the way in which Goody has so succinctly described the situation:

"... This subject is weighed down by a huge, unpublished literature, partly classified. The United States is perhaps the principal source through its project reports, but other countries also contribute to this grey literature. I believe that it should not be recognized in the same sense as publication in a book or commercially available journal, so that authors may be encouraged to submit their papers to scrutiny by their colleagues, or lose recognition. I have therefore attempted to avoid reference to unpublished material but, in a few cases where important data are not elsewhere available, I have been prepared to abandon my principles..." (from R. M. Goody, *Atmospheric Radiation*, New York: Oxford at the Clarendon Press, 1964, by permission).

This book contains a carefully selected and unusually extensive set of references. With a few exceptions,[1] all refer exclusively to the published literature. In short, the reader interested in pursuing these references will find that they are available in any well-stocked public library. In the United States most of the published literature on the infrared is found in the pages of *Applied Optics, The Journal of the Optical Society of America*, and *Infrared Physics*. Those with an interest in the applications of infrared techinques will find much of value in *Aviation Week, Electronics*, and *Technology Week*.

A significant number of the references are to literature published outside of the United States. In some cases, indicated by the inclusion of a CFSTI, OTS, or SLA number at the end of the reference, English-

[1]References 65 and 66 in Chapter 4 are two of these exceptions. They are good examples of extremely valuable work that has, unfortunately, never been published.

INTRODUCTION TO INFRARED SYSTEM ENGINEERING   17

language translations are available from the Clearinghouse for Federal Scientific and Technical Information. For further details concerning this service, refer to Appendix 4.

An unusual feature of the references is the extensive citation of the U.S. patent literature. It is regrettable that so few engineers acquire an adequate regard for the value of patents and the patent literature. At the very least, patents represent a tremendous fund of detailed technical knowledge that is remarkably well catalogued so that information retrieval is quite simple.[1] A patent, of course, represents a contract between an inventor and his government in which the inventor is granted certain exclusive rights for a limited period of time. The inventor earns these rights by a full public disclosure of his invention.[2] Thus, by definition, the patent literature is in the public domain and contains no proprietary or classified information. For this reason it is cited extensively in this book to establish clearly that certain topics are not classified. An example of such usage is found in Chapter 6, where two-thirds of the references are to the patent literature. Since the information contained in these patents does not exist elsewhere in the published literature, such a detailed discussion of reticles could not have been written without reference to these patents.

For the benefit of those who wish to assemble a reference library to supplement the present volume, a selected reading list is given at the end of this chapter. It consists of review papers, special issues of journals and books that are so broad in their coverage that they are not necessarily referenced at any other point in this book. They are arranged chronologically, so that they may be used as a historical guide to the development of the infrared field over the last three or four decades.

The reader having the credentials necessary for access to the unpublished literature will find that it, too, is extensive. In Appendix 4 an attempt is made to describe the principal sources of this literature and the way that a qualified reader can approach them.

1.7 THE SYMBOLS AND ABBREVIATIONS USED IN THIS BOOK

The letter symbols and abbreviations used in this book follow the recommendations of the Optical Society of America, the Institute of Electrical and Electronic Engineers, and the United States of America Standards Institute, altered in some cases by the limited number of char-

---

[1]The U.S. Patent Office has an extensive numerical classification system. Of particular interest is Class 88 pertaining to optical devices, Class 250 pertaining to devices for the detection of radiant energy, and Classes 201 and 338 pertaining to detectors.
[2]Copies of any U.S. patent are available from the Patent Office for $0.50 each.

acters available in the English and Greek alphabets. A list of letter symbols and the quantities they represent is important to the reader in helping him thread his way through a book. However, this author feels a strong antipathy to placing such a list in the front of a book, for then a reader is forced to wade through the list in order to discover the first chapter. In this book the list of letter symbols and abbreviations has been placed at the back, in Appendix 1.

REFERENCES

[1] W. Herschel, "Investigation of the Powers of the prismatic Colours to heat and illuminate Objects; with Remarks, that prove the different Refrangibility of radiant Heat. To which is added, an Inquiry into the Method of viewing the Sun advantageously, with Telescopes of large Apertures and high magnifying Powers," *Phil. Trans. Roy. Soc. London*, **90**, 255 (1800).
W. Herschel, "Experiments on the Refrangibility of the invisible Rays of the Sun," *Phil. Trans. Roy. Soc. London*, **90**, 284 (1800).
[3] W. Herschel, "Experiments on the solar, and on the terrestrial Rays that occasion Heat; with a comparative View of the Laws to which Light and Heat, or rather the Rays which occasion them, are subject, in order to determine whether they are the same, or different," *Phil. Trans. Roy. Soc. London*, **90**, 293, 437 (1800).
[4] A. Pogo, "Herschel, Sir William," *Encyclopedia Britannica*, **11**, 520 (1963).
[5] E. S. Barr, "Historical Survey of the Early Development of the Infrared Spectral Region," *Am. J. Phys.*, **28**, 42 (1960).
[6] E. S. Barr, "The Infrared Pioneers—I. Sir William Herschel," *Infrared Phys.*, **I**, 1 (1961).
[7] E. S. Barr, "The Infrared Pioneers—II. Macedonia Melloni," *Infrared Phys.* **2**, 67 (1962).
[8] E. S. Barr, "The Infrared Pioneers—III. Samuel Pierpont Langley," *Infrared Phys.*, **3**, 195 (1963).
[9] W. N. Arnquist, "Survey of Early Infared Developments," *Proc. Inst. Radio Engrs.*, **47**, 1420 (1959).
[10] "List of Scientific Publications of W. W. Coblentz," *J. Opt. Soc. Am.*, **36**, 62 (1946).
[11] "Coblentz Commemorative Issue," *Appl. Opt.*, *Vol. 2* (November 1963).
[12] N. Jones, "Infrared Spectroscopy," *Intl. Sci. Tech.*, January 1965, p. 35.
[13] W. Brugel, *An Introduction to Infrared Spectroscopy*, New York: Wiley, 1962.
[14] R. P. Bauman, *Absorption Spectroscopy*, New York: Wiley, 1962.
[15] H. Guderian, *Erinnerungen Eines Soldaten*, Nuernberg, Germany: Kurt Vowinkel, 1950. An abridged translation is available in the United States: *Panzer Leader*, New York: Ballantine, 1957.
[16] "Infrared," SRI Long Range Planning Reports, Stanford Research Institute, Menlo Park, Cal., 1959.
[17] "Infrared," *Control Engrg.* **10**, 20 (February 1963).
[18] "Infrared Products Market Seen Reaching $500 Million by 1965," *Weekly Report*, Electronic Industries Association, Washington, D.C., Vol. 18, December 3, 1962.
[19] "Infrared Spots New Markets," *Electronics*, **32**, 32, April 3, 1959.
[20] M. J. Kelly, "The Bell Telephone Laboratories—An Example of an Institute of Creative Technology," *Proc. Roy. Soc. London*, **203A**, 287 (1950).
[21] A. D. Hall, *A Methodology for Systems Engineering*, Princeton, N.J.: Van Nostrand, 1962, Chapter 1.

[22] H. A. Affel, Jr., "System Engineering," *Intl. Sci. Tech.*, November 1964, p. 18.
[23] H. W. Chestnut, *Systems Engineering Tools*, New York: Wiley, 1965.
[24] R. E. Machol, W. P. Tanner, Jr., and S. N. Alexander, *System Engineering Handbook*, New York: McGraw-Hill, 1965.

A SELECTED READING LIST ON INFRARED TECHNIQUES AND THEIR APPLICATIONS (ARRANGED CHRONOLOGICALLY)

Papers

V. Z. Williams, "Infrared Instrumentation and Techniques," *Rev. Sci. Instr.* **19**, 135 (1948).
C. R. Brown et al., "Infrared; a Bibliography," Library of Congress, Technical Information Division, Washington, D.C., December 1954. (Covers the published literature from 1935 to 1951. Additional reports in this series, covering the unpublished literature, are listed in Appendix 4.)
G. B. B. M. Sutherland, "Infrared Radiation," *J. Inst. Elec. Engrs.* (London), **105B**, 306 (1958).
S. S. Ballard, ed., "Special Issue on Infrared Physics and Technology," *Proc. Inst. Radio Engrs.* Vol. 47 (September 1959).
S. S. Ballard and W. L. Wolfe, ed., "Special Issue on Infrared," *Appl. Opt.*, Vol. 1 (September 1962).
J. King et al., "Infrared," *Intl. Sci. Tech.*, April 1963, p. 26.

Books

J. Lecomte, *Le Spectre Infrarouge*, Paris: University of Paris Press, 1928.
C. Schaefer and F. Matossi, *Das Ultrarot Spektrum*, Berlin: Springer, 1930.
W. E. Forsythe, *Measurement of Radiant Energy*, New York: McGraw-Hill, 1937.
R. B. Barnes et al., *Infrared Spectroscopy—Industrial Applications and Bibliography*, New York: Reinhold, 1944.
W. Clark, *Photography by Infrared*, 2d ed., New York: Wiley, 1946.
J. Lecomte, *Le Rayonnement Infrarouge*, Paris: Gauthier-Villars, 1948.
W. Brugel, *Physik und Technik Der Ultrarotstrahlung*, Hanover: Vincent, 1951.
J. A. Sanderson, "Emission, Transmission, and Detection of Infrared," Chapter 5 in A. S. Locke, ed., *Guidance*, Princeton, N.J.: Van Nostrand, 1955.
I. A. Margolin and N. P. Rumyantsev, *Fundamentals of Infrared Technology*, 2d ed., Moscow: Voenizdat, 1957 (SLA: 60-13034).
R. A. Smith, F. E. Jones, and R. P. Chasmar, *The Detection and Measurement of Infrared Radiation*, London: Oxford University Press, 1957.
G. K. T. Conn and D. G. Avery, *Infrared Methods*, New York: Academic, 1960.
H. L. Hackforth, *Infrared Radiation*, New York: McGraw-Hill, 1960.
T. R. Harrison, *Radiation Pyrometry and Its Underlying Principles of Radiant Heat Transfer*, New York: Wiley, 1960.
M. R. Holter et al., *Fundamentals of Infrared Technology*, New York: Macmillan, 1962.
P. W. Kruse, L. D. McGlaughlin, and R. B. McQuistan, *Elements of Infrared Technology*, New York: Wiley, 1962.
J. A. Jamieson et al., *Infrared Physics and Engineering*, New York: McGraw-Hill, 1963.
L. Z. Kriksunov and I. F. Usol'tsev, *Infrared Equipment for Missile Homing*, Moscow: Voenizdat, 1963 (OTS: 64-31416).
W. L. Wolfe, ed., *Handbook of Military Infrared Technology*, Office of Naval Research, Department of the Navy, Washington, D.C., 1965.

2

Infrared Radiation

This chapter presents the basic material needed to understand the physical basis of infrared radiation. After establishing the relationship between the infrared region and the rest of the electromagnetic spectrum, the terminology used to characterize radiant energy and the instrumentation used to measure it are described. The important differences between the radiation from solids and that from gases are noted, together with their implications for the infrared system engineer. The equations used to describe thermal radiation are presented, along with a discussion of the various tables, nomograms, and slide rules that are available for simplifying radiation calculations.

2.1 THE ELECTROMAGNETIC SPECTRUM

In everyday life we encounter many different types of radiation. Seemingly different forms, such as sunlight, heat, radio waves, and X-rays to name only a few, are inherently similar in nature and can be conveniently grouped under a single classification called electromagnetic radiation. It is common practice to describe these radiations by their position in the *electromagnetic spectrum*—an arrangement of the various radiations by wavelength or frequency. A portion of the electromagnetic spectrum is indicated in Figure 2.1. All of the radiations obey similar laws of reflection, refraction, diffraction, and polarization. The velocity of propagation, popularly called the "velocity of light," is the same for all. They differ from one another only in wavelength and frequency.

The portion of the spectrum that includes the infrared is depicted in greater detail in the lower part of Figure 2.1. It shows that the infrared region is bounded on the short-wavelength side by visible light and on the long-wavelength side by microwaves. Since heated objects radiate energy in the infrared, it is often referred to as the heat region of the spectrum. It is convenient to subdivide the infrared into the four parts shown in Figure 2.1 and detailed in Table 2.1. The specific limits and nomen-

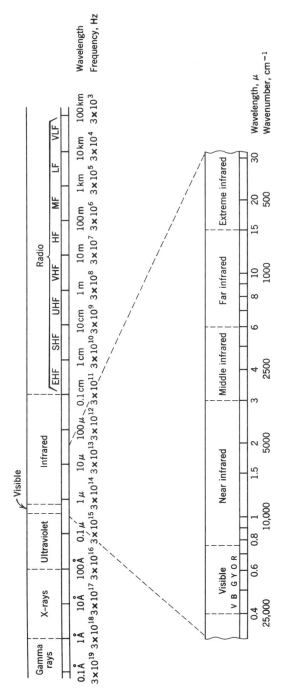

Figure 2.1 The electromagnetic spectrum.

clature vary with authors; those chosen here are compatible with a standard terminology suggested for the ultraviolet by Ballard[1].

TABLE 2.1   SUBDIVISIONS OF THE INFRARED

Designation	Abbreviation	Limits, Microns
Near infrared	NIR	0.75 to 3
Middle infrared	MIR	3 to 6
Far infrared	FIR	6 to 15
Extreme infrared	XIR	15 to 1000

The logic on which these choices are based will become apparent after Chapter 4 has been read. Each of the first three subdivisions includes regions in which the earth's atmosphere is relatively transparent; in the last subdivision the atmosphere is essentially opaque for paths more than a few meters long.

Since all of the types of radiation shown in Figure 2.1 are considered to be a form of wave motion, they must obey the general equation

$$\lambda \nu = c, \qquad (2\text{-}1)$$

where λ is the wavelength, ν the frequency, and c the velocity. *Wavelength* is the distance, measured in the direction of propagation, between any two successive points on a wave having the same phase. It is customary to use the micron (μ) (which some authors call the micrometer) as the unit of wavelength in the infrared. We may occasionally encounter the millimicron (mμ), the nanometer (nm), or the angstrom (Å) as units for wavelength. These are related as follows:

$$1\mu = 10^{-4}\text{ cm} = 10^{-6}\text{ m} = 3.937 \times 10^{-5}\text{in}.$$
$$1\mu = 10^3\text{ m}\mu = 10^3\text{nm} = 10^4\text{Å}$$

Many of the relationships of quantum physics governing the emission and absorption of radiation are simplified by the use of frequency rather than wavelength. *Frequency* is the number of waves passing a point per unit time; the units, cycles per second, are called the hertz (Hz). As shown in Figure 2.1, frequencies in the infrared are on the order of 10^{13} to 10^{14} Hz. There are no methods available for the direct measurement of such high frequencies; instead, wavelength is measured and converted to frequency by equation 2-1. The accuracy with which frequency can be determined in this way is limited to about one part in 10^6 because of uncertainties in the value of c[2]. Wavelength can be measured interferometrically to an accuracy of about one part in 10^7. To preserve this increased accuracy we can use the *wavenumber* (σ), the number of waves per unit distance. If the distance is measured in centimeters, the units are reciprocal centimeters (cm^{-1}), as shown in Figure 2.1. Despite the fact that they are derived from wavelengths, wavenumbers are often confused with frequencies. It is incorrect to refer to a wavenumber of 5000 cm^{-1} as a frequency of 5000 wavenumbers.

2.2 TERMINOLOGY USED IN THE MEASUREMENT OF RADIANT ENERGY

It is unfortunate that the concepts basic to the measurement of light evolved many years before it was known that light is not a separate entity but is instead merely the radiation in one narrow region of the electromagnetic spectrum. *Photometry*, the measurement of light, implicitly involves the visual sensations produced by light in the consciousness of an observer. Thus the methods of photometry are psychophysical rather than physical, and photometry cannot be classed along with such familiar physical measurements as those of mass, length, or time.

More pertinent to the infrared region are the methods of *radiometry*, the measurement of radiant energy, which are based on a system of physical measurements. In principle, radiometers absorb some of the radiant energy from a source and convert it to another form, such as electrical, thermal, or chemical energy. To date, there is no known way in which radiant energy can be measured directly; it must always be converted to some other form. Conversion devices, which are called *detectors*, include the radiation thermocouple, bolometer, photographic plate, photocell, and photon detector, to name just a few.

Since it provides a physical measurement of radiant energy, radiometry is equally applicable throughout the electromagnetic spectrum. References in the literature to microwave, infrared, ultraviolet, and X-ray radiometry all pertain to the measurement of the energy contained in the incident radiation. Because of its dependence on the characteristics of the human eye, photometry is, by contrast, applicable over only a narrow portion of the electromagnetic spectrum[3].

As in many other branches of engineering, the terminology and symbols of radiometry undergo periodic, often violent, changes. It is believed that those used in this book reflect the current preferences of the majority of the U.S. members of the infrared community. They are based on the report of the Committee on Colorimetry of the Optical Society of America[4], from which their usage can be traced back to the mid-1930's. More recently, the Working Group on Infrared Backgrounds (WGIRB), a semiofficial industry and government group, has reaffirmed their usage[5]. Several of the international standards organizations have chosen a different set of symbols and are urging member societies to adopt them. Their recommendations are summarized in Appendix 2, but the author does not favor their usage. The terminology used in this book is summarized in Table 2.2, together with the symbols, the definitions, and the units in which they are expressed.

The energy transferred by electromagnetic waves is called *radiant energy(U)* and is measured in joules. This term is used to describe the entire amount of energy radiated from a source in a given time interval. It also describes the energy received by an accumulative or integrating type of detector, such as the photographic plate.

Most detectors used in infrared equipment respond to the time rate of transfer of radiant energy rather than to the total amount of energy transferred. *Radiant flux(P)* is the measure of the time rate of transfer of radiant energy and is given in watts; a watt is numerically equal to one joule per second. An equally acceptable equivalent term, preferred by some authors, is *radiant power*.

Three terms — radiant emittance, radiant intensity, and radiance — may be used to describe the radiant flux from a source. They are usually determined by radiometric measurements made at a distance from the source. If no correction is made for attenuation, scattering, or emission by the atmosphere between the source and radiometer, the measured values are referred to as *apparent* values. If corrections for these effects are applied, the details of the correction should be clearly stated.

The radiant flux emitted per unit area of a source is the *radiant emittance (W)*. It is the limiting value of the expression $W = \partial P/\partial A$, where P is the radiant flux emitted by a source element having an area A. One way of measuring the flux is to place the source in a device that collects flux from all directions with equal efficiency. After a correction is made for the efficiency of the collector, the flux collected is divided by the area of the source to obtain the radiant emittance. For an inaccessible source, such as the sun, we would operate as follows:

1. Measure the radiant flux at the detector.
2. Divide this value by the solid angle subtended by the receiver at the sun.
3. Correct for attenuation, and so forth, along the line of sight.
4. Multiply by 4π.
5. Divide this product by the area of the sun.

Step 4 yields a value for the total radiant flux from the sun based on the assumption that the flux it emits is the same in all directions; dividing

INFRARED RADIATION

TABLE 2.2 RECOMMENDED RADIOMETRIC TERMINOLOGY

Symbol	Term	Description	Unit
U	Radiant energy	Energy transferred by electromagnetic waves	Joule
u	Radiant energy density	Radiant energy per unit volume	Joule cm^{-3}
P	Radiant flux	Rate of transfer of radiant energy	Watt (W)
W	Radiant emittance	Radiant flux emitted per unit area of a source	W cm^{-2}
Q	Radiant photon emittance	Number of photons emitted per second per unit area	Photon sec^{-1} cm^{-2}
J	Radiant intensity	Radiant flux per unit solid angle	W sr^{-1}
N	Radiance	Radiant flux per unit solid angle per unit area	W cm^{-2} sr^{-1}
H	Irradiance	Radiant flux incident per unit area	W cm^{-2}
P_λ	Spectral radiant flux	Radiant flux per unit wavelength interval at a particular wavelength	W μ^{-1}
W_λ	Spectral radiant emittance	Radiant emittance per unit wavelength interval at a particular wavelength	W cm^{-2} μ^{-1}
Q_λ	Spectral radiant photon emittance	Radiant photon emittance per unit wavelength interval at a particular wavelength	Photon sec^{-1} cm^{-2} μ^{-1}
J_λ	Spectral radiant intensity	Radiant intensity per unit wavelength interval at a particular wavelength	W sr^{-1} μ^{-1}
N_λ	Spectral radiance	Radiance per unit wavelength interval at a particular wavelength	W cm^{-2} sr^{-1} μ^{-1}
H_λ	Spectral irradiance	Irradiance per unit wavelength interval at a particular wavelength	W cm^{-2} μ^{-1}
ϵ	(Radiant) emissivity	Ratio of radiant emittance of a source to that of a blackbody at the same temperature	(Numeric)
α	(Radiant) absorptance	Ratio of absorbed radiant flux to incident radiant flux	(Numeric)
ρ	(Radiant) reflectance	Ratio of reflected radiant flux to incident radiant flux	(Numeric)
τ	(Radiant) transmittance	Ratio of transmitted radiant flux to incident radiant flux	(Numeric)

this value by the area of the sun (step 5) gives the radiant emittance. If step 3 is eliminated, the result is the apparent radiant emittance.

Before discussing radiant intensity and radiance, it is necessary to differentiate between *point sources* and *extended sources*. A true point source is, of course, not physically realizable, but it can be closely approximated by a star, a small source at a great distance, or by an optical device called a collimator. What is important is not the physical size of the source but the angle that it subtends at the detector. The same source can at different times be classed as either point or extended. The tailpipe of a jet aircraft at a distance of 10 miles is effectively a point source, whereas at a distance of 10 ft it represents an extended source. If the radiometer consists simply of a detector without optics, it is acceptable to consider as point sources any sources at a distance greater than ten times the largest dimension of the source[6]. If the radiometer uses optics, the basic criterion is the relationship between the size of the detector and the size of the image of the source. If the image is smaller than the detector, the source is considered to be a point source; if the image is larger than the detector, the source is considered an extended one.[1] It should come as no surprise that we must make a distinction between types of source and use separate terms to describe the radiation from each. For example, in our everyday experience we refer to the intensity of a star (a point source) and to the brightness of the sky (an extended source).

The radiant flux emitted per unit solid angle is called the *radiant intensity* (J). It is the limiting value of the expression $J = \partial P/\partial \omega$, where ω is the solid angle[2] subtended by the detector at the source. Radiant intensity is used to describe a point source. The radiant intensity is found by measuring the radiant flux and dividing it by the solid angle subtended by the detector at the source. Since most sources do not radiate uniformly in all directions, it is important when measuring the radiant flux to have the detector centered about the direction of interest.

If the source is an extended one, such as the sky, it is not possible to define the solid angle subtended by the detector at the source. This difficulty is resolved by describing an extended source by the *radiance* (N), the radiant flux per unit solid angle per unit area of source. It is

---

[1] In Chapter 5 it is shown that one of the terms used to describe an optical device is its field of view. In these terms, the definitions given here are rephrased to say that extended sources fill the field of view while point sources do not.

[2] A solid angle can be defined in terms of the portion of the space within a sphere that is bounded by a conical surface having its vertex at the center of the sphere. It is measured in steradians (sr), and is the ratio of the area of the spherical surface intercepted by the cone to the square of the radius of the sphere. Since the total area of a sphere is $4\pi r^2$, there are 4π sr in a sphere.

the limiting value of the expression $N = \partial^2 P/\partial A \partial \omega$. In order to measure radiance it is necessary to use masks or optical means to limit the measurement to a small area of the extended source. Under these conditions the radiant flux is measured and divided by both the area of the source and the angle subtended by the detector at the source. In effect, we measure the radiant intensity of a small, but known, portion of the extended source. Indeed, a comparison of the two definitions shows that radiance is radiant intensity per unit area of source. Since the flux is always measured from a distance, the exact area involved in the measurement, and its orientation, may be difficult to determine. However, if the source area is defined as the area projected on a plane located at the source and normal to the direction of measurement, the result is a quantity that can be measured with ease, even from a distance.

The *radiant photon emittance (Q)* is the number of photons emitted per second per unit area. Since some types of detectors respond to the number of photons rather than to the radiant flux in the incident beam, it is sometimes convenient to be able to express the radiometric quantities in terms of photons.

The terms introduced thus far have been used to describe the flux emitted by sources. *Irradiance (H)* is the radiant flux incident on a surface of unit area. The units in which it is measured, W cm^{-2}, are the same as those used for radiant emittance; both describe an areal density, one arriving at a surface, the other leaving a source. With a point source, if the effects of any intervening atmosphere can be ignored, the irradiance at a distance d from the source is

$$H = \frac{J}{d^2}. \qquad (2\text{-}2)$$

The error in assuming that the source is a point is less than 1 per cent, if the distance is at least ten times the largest dimension of the source[6]. If the source is an extended one, the irradiance must be found by an integration[7].

Radiant emittance, radiant photon emittance, radiant intensity, radiance, and irradiance refer to the flux contained in a particular solid angle or passing through a particular area. Thus these quantities are differential with respect to solid angle or area. Each is associated with a corresponding spectral quantity in which the radiant flux is that within a small wavelength interval centered about a particular wavelength. Thus the spectral quantities are differential with respect to wavelength. As an example, *spectral radiant flux* (P_λ) is the radiant flux per unit wavelength interval evaluated at a particular wavelength; it is the limiting value of the expression $P_\lambda = \partial P/\partial \lambda$. Use of the subscript λ to indicate a differential with respect to wavelength has had wide acceptance and is followed in this book. Accordingly, a quantity X whose value is a function of wavelength is written as $X(\lambda)$ and not X_λ.

In order to find the radiant flux between wavelengths λ_1 and λ_2 it is necessary to integrate the expression

$$P = \int_{\lambda_1}^{\lambda_2} P_\lambda d\lambda. \qquad (2\text{-}3)$$

If the limits extend from zero to infinity, the result of the integration is the *total* radiant flux. This particular connotation of the term "total," in the sense of integrated over-all wavelengths, is used throughout this book. Any other choice of limits will, of course, cover a smaller wavelength interval, and the result of the integration is called the *effective* radiant flux. Here the term "effective" denotes a measurement in a limited wavelength interval. The concept of effective flux becomes useful in discussing a detector or sensor that responds only to the flux within a limited wavelength interval.

The last four terms in Table 2.2 are all numerics representing the ratio between two quantities, and having values between zero and unity. The modifier "radiant" is used with these quantities only when it is necessary to distinguish radiometric from photometric quantities, a distinction we shall have little occasion to use in this book. If the reader examines the general rules by which physical units are named, he finds that (1) the suffix *-ance* is used to refer to the properties of a particular body, and not to the intrinsic properties of the material of which the body is composed; (2) the suffix *-ivity* is used to refer to the intrinsic properties of a material measured under standard conditions; (3) the suffix *-sion* is used to refer to a process and not to a property of a particular material. For consistency, with these conventions the term "emissivity" in Table 2.2 should be replaced by "emittance". However, few workers in the infrared field make the distinction between the properties of a particular sample and the intrinsic properties of the material it consists of. Of equal importance, "emissivity" avoids the possible confusion of "emittance" and "radiant emittance," and the undesirability of using the same word to mean two entirely different things. Following the convention for the use of *-sion* would require that "radiation" be used to refer solely to the process and not as a synonym "for radiant energy". Since virtually all workers in the field use these two terms interchangeably, this practice is followed here.

The descriptions of reflectance and transmittance given in Table 2.2 are deceptively simple. Each requires a measurement of the incident radiant flux and of the flux reflected or transmitted by the sample. If the surface is a *specular reflector*, that is, a mirror, it is relatively easy

INFRARED RADIATION

to collect and measure all of the reflected flux. If, however, the surface is a *diffuse reflector*, the reflected flux is spread over a wide solid angle and it may be difficult to ensure that all of it is collected and measured. For a perfectly diffuse reflector the reflected flux per unit solid angle is proportional to the cosine of the angle between the direction of interest and the normal to the surface, a relationship known as *Lambert's cosine law*. Such surfaces are often called *Lambertian surfaces*. The cosine distribution can also be used to describe the angular distribution of flux from a *perfectly diffuse source*. For such a source Lambert's cosine law states that the radiant intensity is proportional to the cosine of the angle measured from the normal to the surface of the source. Since the projected area of the source also varies with the cosine of the same angle, the radiance of a perfectly diffuse source is independent of the viewing angle. A simple parallel exists; the sensation of brightness we perceive when viewing a luminous surface is a function of the luminance of the surface (luminance, a photometric term, corresponds to radiance). For a perfectly diffuse source the luminance is the same in all directions and the surface appears to be equally bright from whatever angle it is viewed. Therefore a luminous sphere, such as the sun, appears to be a uniformly bright disk despite the fact that the central portion is viewed normally and the edges are seen tangentially.

Although Lambert's cosine law is an idealized concept, many sources follow it quite closely. The blackbody, which will be discussed in detail, follows it exactly. Most materials that are electrical insulators follow it until the viewing angle exceeds 60 deg from the normal to the surface; materials that are electrical conductors show somewhat greater departures, but until the viewing angle exceeds 50 deg the cosine law is useful in engineering calculations[8]. For a perfectly diffuse source radiating into a hemisphere, the following relationships among radiant emittance (W), radiant intensity (J), radiance (N), and source area (A) hold:

$$W = \pi N = \frac{\pi J}{A} \tag{2-4}$$

$$J = \frac{WA}{\pi} = NA \tag{2-5}$$

$$N = \frac{W}{\pi} = \frac{J}{A}. \tag{2-6}$$

Note that in (2-6) the relationship between radiance and radiant emittance is $N = W/\pi$, a consequence of Lambert's cosine law, and not $N = W/2\pi$, (as we might reason from the fact that there are 2π sr in a hemisphere).

Of all the mistakes a newcomer to radiometry may make, confusion over this factor of 2 is an odds-on favourite.

2.3 THE MEASUREMENT OF RADIANT FLUX

A *radiometer* is a device for measuring radiant flux over a broad spectral interval. A *spectroradiometer* is a device for measuring the spectral radiant flux within a small spectral interval. Thus radiometry provides broad-band measurements, while spectroradiometry is used for narrow-band measurements.

The basic elements of a radiometer are shown in Figure 2.2. Some of the radiant flux from the source is collected by the optics and focused onto the detector; the detector produces an electrical signal that is proportional to the flux input. Since such measurements are invariably made at a distance from the source, the radiometer responds to the irradiance, the areal density of the flux, at its input (the optics). Thus irradiance is the fundamental quantity involved in all radiometric measurements; other quantities, such as radiant emittance, radiant intensity, and radiance, are calculated from the measured value of irradiance.

In the radiometer shown in Figure 2.2, the spectral interval over which the measurement is made is controlled by the spectral response of the detector and the transparency of the optics. If it is assumed that the detector responds equally to all wavelengths and that the optics transmit all wavelengths without absorption, then the output indication will be proportional to the total irradiance at the optics. Radiometers can be constructed that come quite close to achieving these conditions. By using a thermal detector and mirror optics, we can obtain a nearly equal response to wavelengths extending from 2 to 40 μ. If desired, an optical filter can be placed in front of the detector to limit the response of the radiometer to any desired smaller spectral interval.

The elements of a spectroradiometer are shown in Figure 2.3. It

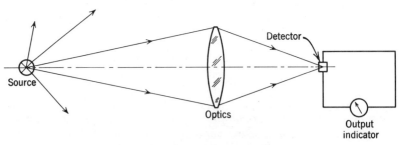

Figure 2.2 Elements of a radiometer.

INFRARED RADIATION

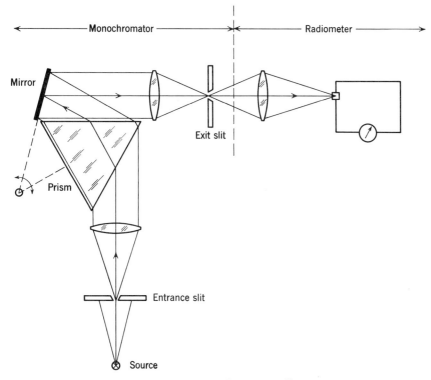

Figure 2.3 Elements of a spectroradiometer.

consists of two major parts: a *monochromator* to provide radiant flux of a narrow band of wavelengths, and a radiometer to measure this flux. The flux from the source is dispersed, or spread into a spectrum, by a prism. A small portion of this flux passes through the exit slit to the radiometer. Rotation of the prism-and-mirror combination varies the wavelength of the flux passing through the exit slit; the width of this slit determines the spectral interval passed by the monochromator. The monochromator and source are seen to be analogous to the signal generator used in electronics. The prism and exit slit are used to select a particular wavelength; in the signal generator a particular frequency is selected by varying a resonant circuit. Note, however, that the signal generator inherently provides a single-frequency signal, whereas the signal from the monochromator must always consist of a narrow band of wavelengths. Perhaps a somewhat closer analogy is a broad-band noise generator followed by a variable-bandpass filter.

A spectroradiometer measures the *spectral distribution* of radiant flux, that is, the variation in flux as a function of wavelength. Such measurements are of basic importance to a system designer because they indicate what portion of the spectrum is likely to be most favorable for detection of the flux from a particular source. If a spectroradiometer is used to investigate the spectral distribution of a variety of sources, it soon becomes apparent that there are two quite different types of sources. If the source is a heated solid or liquid, the spectral distribution curve is continuous and shows a single maximum at a wavelength that varies with the temperature of the source. Such sources are called *thermal radiators*. If the source is a flame or an electrical discharge in a gas, the spectral distribution curve is not continuous, but instead the flux is concentrated in narrow spectral intervals. With a high-resolution monochromator these intervals may appear to be extremely narrow, sharply defined lines, and the distribution is called a *line spectrum*. Alternatively, the spectrum may consist of bands of narrow lines, and in this case it is called a *band spectrum*. Sources giving line or band spectra are called *selective radiators*. At first glance the wavelengths at which the flux is concentrated appear to be located randomly throughout the spectrum. A

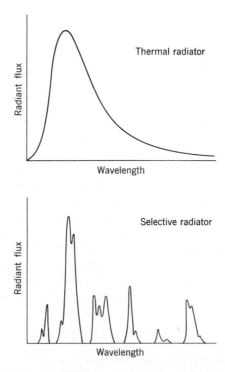

Figure 2.4 Spectral distribution of thermal and selective radiators.

more detailed examination shows that they are characteristic of the particular type of atom or molecule that is radiating. Hence these wavelengths constitute a unique signature of the source. A line spectrum is characteristic of atoms, and a band spectrum is characteristic of molecules. Figure 2.4 shows typical spectral distribution curves for thermal and selective radiators, the latter with poor resolution; that is, lines and band structure are not shown in detail.

Some of the selective radiators that an infrared system designer is likely to encounter include the stream of hot gases from the exhaust of a jet engine or rocket, the shock-excited layer surrounding a reentry body, and the gas-discharge sources used in communication systems. Typical thermal radiators include the hot metal of a jet engine or rocket tailpipe, aerodynamically heated surfaces, motor vehicles, personnel, terrain, space vehicles, and astronomical bodies. Since, from the standpoint of an infrared system designer, thermal radiators are by far the most prevalent, the laws governing these sources are discussed first.

2.4 THERMAL RADIATION

One of the major problems facing physicists during the second half of the nineteenth century was to explain the energy distribution in the spectrum of a thermal radiator. In 1860 Kirchhoff[9] introduced his famous law that states, in effect, that good absorbers are also good radiators. This law is one of the keystones in the theory of radiation transfer. Kirchhoff also proposed the term *blackbody* to describe a body that absorbs all of the incident radiant energy and that, as a consequence of his law, must also be the most efficient radiator. A blackbody, then, provides a standard of comparison; it is the ultimate thermal radiator with which we can compare any other source. In 1879 Stefan concluded from his experimental measurements that the total amount of energy radiated by a blackbody is proportional to the fourth power of its absolute temperature. In 1884 Boltzmann reached the same conclusion by the application of thermodynamic relationships; the result has become known as the *Stefan–Boltzmann law*. In 1894 Wien published the *displacement law* that gives the general form of the equation for the spectral distribution of the radiation from a blackbody. Unfortunately, his attempt to find the specific form of this equation agreed with the experimental data only at short wavelengths and at low temperatures. However, his displacement law did yield useful corollaries relating temperature to the wavelength at which the maximum amount of energy is radiated. In 1900 Rayleigh, basing his argument on the concepts of classical physics, derived an expression that fits the experimental data at long wavelengths and at high temperatures.[1] Unfortunately, this expression predicted that the energy increases without limit as the wavelength decreases, a prediction that earned it the dubious distinction of being called the ultraviolet catastrophe.

Planck observed that the Rayleigh–Jeans law seemed valid at long wavelengths and that the Wien law seemed equally valid at short wavelengths. He sought a means of interpolating between the two laws to find an expression that would be correct at all wavelengths. On October 19, 1900, he presented his radiation formula in a paper read before a meeting of the German Physical Society. Five days later, Rubens and Kurlbaum published a comparison of Planck's formula with their experimental measurements and concluded that the two were in excellent agreement. At their urging, Planck undertook a theoretical derivation of his formula.

Since it was well known by that time that an oscillating electric charge radiates electromagnetic energy, it was customary to consider that a thermal source consisted of a large number of such oscillators. Assuming that a wide range of oscillator frequencies were present and that a suitable distribution of amplitudes for the various frequencies could be found, it was thought that the result would give an expression for the spectral energy distribution of blackbody radiation. Unfortunately, the result of this line of reasoning is the Rayleigh–Jeans law, already shown to be in error. Planck concluded that the laws of classical physics were inadequate to describe processes taking place at the atomic level. As stated previously, there was no restriction placed on the amplitudes of the oscillators. Planck agreed that the oscillators could be allowed all possible frequencies, but he introduced the idea that their amplitudes, and hence their energies, could increase only in discrete steps, differing by the quantity $h\nu$, now called a *quantum* of energy. Planck called h the "quantum of action"; later experience has shown that it is a universal constant, and it is today known as Planck's constant. Thus Planck introduced the key assumption that energy is quantized and appears in discrete packages of size proportional to the frequency of the oscillator.

On December 14, 1900, Planck announced the derivation of his radiation law. Thus in less than two months the inadequate theories of classical physics had been bolstered by the new concepts of quantum physics. As history relates, these concepts gradually gained acceptance that culminated in the formal discipline of quantum mechanics. Indeed, we can today regard the mechanics of classical physics as merely a special case of quantum mechanics that is successful in the engineering realm but is inadequate to describe processes at the atomic level.

[1] In a letter to *Nature,* Jeans pointed out a numerical error in Rayleigh's expression. The corrected expression has since become known as the Rayleigh–Jeans law.

Thermal Radiation Laws

Planck's law describes the spectral distribution of the radiation from a blackbody as

$$W_\lambda = \frac{2\pi hc^2}{\lambda^5} \frac{1}{e^{ch/\lambda kT} - 1}, \tag{2-7}$$

which is usually written as

$$W_\lambda = \frac{c_1}{\lambda^5} \frac{1}{e^{c_2/\lambda T} - 1}, \tag{2-8}$$

where

W_λ = spectral radiant emittance, W cm^{-2} μ^{-1}
λ = wavelength, μ
h = Planck's constant = $(6.6256 \pm 0.0005) \times 10^{-34}$ W sec^2
T = absolute temperature, °K
c = velocity of light = $(2.997925 \pm 0.000003) \times 10^{10}$ cm sec^{-1}
$c_1 = 2\pi hc^2$ = first radiation constant
    = $(3.7415 \pm 0.0003) \times 10^4$ W cm^{-2} μ^4
$c_2 = ch/k$ = second radiation constant
    = $(1.43879 \pm 0.00019) \times 10^4 \mu$°K
k = Boltzmann's constant = $(1.38054 \pm 0.00018) \times 10^{-23}$ W sec °K^{-1}.

The literature shows differing values for the constants appearing in the radiation laws because the values of most physical constants are continually refined as improved measurement techniques become available. An excellent account of the process is given in the book by Cohen, Crowe, and DuMond[10]. The values given above are those recommended in July 1963 by the Committee on Fundamental Constants of the National Academy of Sciences—National Research Council[11]. The estimated error given for each constant is equal to three standard deviations and is to be applied to the final digits in the constant. The manner in which the constants enter each equation is shown so that the reader can, if he wishes, enter new values, should future changes occur.

The spectral radiant emittance of a blackbody at temperatures ranging from 500°K to 900°K[1] is shown in Figure 2.5. This is an interesting range

---

[1]The relationships among the common temperature scales, °C(Celsius or Centigrade), °K(Kelvin), °F(Fahrenheit), and °R(Rankine) are

°C = (°F − 32)5/9   °F = 9/5°C + 32
°K = °C + 273.16   °R = °F + 459.69.

The Centrigrade and Fahrenheit scales coincide at −40. It is convenient to assume that room temperature is 300°K.

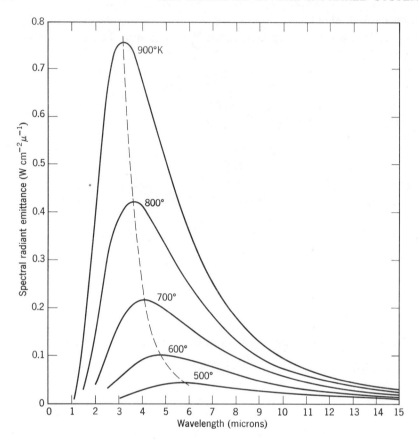

Figure 2.5 Spectral radiant emittance of a blackbody at various temperatures.

because it includes the temperature of the hot metal tailpipes of turbojet aircraft. Several characteristics of the radiation from a blackbody are evident from these curves. The total radiant emittance, which is proportional to the area under the curves, increases rapidly with temperature. The wavelength of maximum spectral radiant emittance shifts toward shorter wavelengths as the temperature increases. The individual curves never cross one another; hence the higher the temperature, the higher the spectral radiant emittance at all wavelengths.

We sometimes see Planck's law written so as to give the spectral radiant energy density within an enclosure; expressed in this way, it is $4/c$ times the value given by (2-7) and (2-8). The form given here is more useful in practice; it expresses the flux within a spectral interval $1\,\mu$ wide radiated into a hemisphere by a blackbody having an area of

1 cm². The significance of the 1 μ spectral interval may still be confusing, however. In the curves in Figure 2.5, the value of the ordinate at any point is proportional to the radiant emittance at that wavelength; the absolute value depends on the width of the spectral interval used in the calculations. In the monochromator shown in Figure 2.3, the width of the exit slit determines the spectral interval. If all other quantities remain constant, the value indicated by the meter will be a direct function of the width of the exit slit. Thus a value for spectral radiant emittance is meaningless unless the spectral interval is also specified. It is customary to refer all radiometric measurements and calculations to a spectral interval equal to the unit in which wavelength is measured.

Time is a quantity that apparently does not appear in Planck's law, and this point may be puzzling. In Table 2.2 it can be seen that the rate of transfer of radiant energy is measured in watts. A watt is numerically equal to one joule per second. Thus the unit of time is implicit in, but obscured by, this choice of the watt.

Integrating Planck's law over wavelength limits extending from zero to infinity gives an expression for the radiant emittance, the flux radiated into a hemisphere above a blackbody 1 cm² in area. This is commonly known as the *Stefan–Boltzmann law*,

$$W = \frac{2\pi^5 k^4}{15 c^2 h^3} T^4 = \sigma T^4, \tag{2-9}$$

where

W = radiant emittance, W cm^{-2}
σ = Stefan–Boltzmann constant
  = $(5.6697 \pm 0.0029) \times 10^{-12}$ W cm^{-2} °K^{-4}.

In Figure 2.5 the rapid increase in radiant emittance with increasing temperature is evident; from the Stefan–Boltzmann law, this increase is proportional to the fourth power of the absolute temperature. Thus relatively small changes in temperature can cause large changes in radiant emittance.

Differentiating Planck's law and solving for the maximum gives *Wien's displacement law*,

$$\lambda_m T = a, \tag{2-10}$$

where

λ_m = wavelength of maximum spectral radiant emittance
$a = 2897.8 \pm 0.4 \mu$ °K.

Thus the wavelength at which the maximum spectral radiant emittance occurs varies inversely with the absolute temperature. The dashed curve in Figure 2.5 is the locus of these maxima.

An alternative form of Wien's displacement law gives the maximum value of the spectral radiant emittance as

$$W_{\lambda_m} = 21.20144 \frac{c_1}{c_2^5} T^5 = bT^5, \quad (2\text{-}11)$$

where

W_{λ_m} = maximum spectral radiant emittance, W cm$^{-2}\mu^{-1}$
$b = 1.2862 \times 10^{-15}$ W cm$^{-2}\mu^{-1}$ °K^{-5}.

Hence the value of the maximum spectral radiant emittance from a blackbody is proportional to the fifth power of its absolute temperature.

The radiation laws can also be written in terms of photons[12], a form that will be useful in discussing the performance of photon detectors. If Planck's expression, (2-7), is divided by hc/λ, which is the energy associated with one photon, the result is the spectral radiant photon emittance

$$Q = \frac{2\pi c}{\lambda^4} \frac{1}{e^{ch/\lambda kT} - 1} \quad (2\text{-}12)$$

$$= \frac{c_1'}{\lambda^4} \frac{1}{e^{c_2/\lambda T} - 1}, \quad (2\text{-}13)$$

where

Q_λ = spectral radiant photon emittance, photon sec^{-1} cm$^{-2}\mu^{-1}$
$c_1' = 2\pi c = 1.88365 \times 10^{23}$ sec^{-1} cm$^{-2}\mu^3$.

The Stefan–Boltzmann law, (2-9), becomes

$$Q = \frac{c_1'}{c_2^3} \frac{2\pi^3}{25.79436} = \sigma' T^3, \quad (2\text{-}14)$$

where

Q = radiant photon emittance, photon sec^{-1} cm^{-2}
$\sigma' = 1.52041 \times 10^{11}$ sec^{-1} cm^{-2} °K^{-3}.

Therefore the rate at which photons are emitted from a blackbody varies as the third power of its absolute temperature, rather than as the fourth-power relationship observed with radiant flux.

The Wien displacement law, (2-10), becomes

$$\lambda_m' T = \frac{c_2}{3.92069} = a', \quad (2\text{-}15)$$

INFRARED RADIATION

where
λ'_m = wavelength of maximum spectral radiant photon emittance
$a' = 3669.73\,\mu\,°K$.

Thus the displacement law has the same form for both flux and photons, but the wavelength at which the maximum occurs is about 25 per cent greater for photons. The alternative form of Wien's law, (2-11), becomes

$$Q'_{\lambda_m} = 4.77984 \frac{c'_1}{c_2^4} T^4 = b'T^4, \qquad (2\text{-}16)$$

where
Q'_{λ_m} = maximum spectral radiant photon emittance, photon sec^{-1} cm^{-2}μ^{-1}
$b' = 2.10098 \times 10^7$ sec^{-1} cm^{-2}μ^{-1} °K^{-4}.

Another pair of useful relationships is

$$\text{energy per photon} = \frac{hc}{\lambda} = \frac{1.9863 \times 10^{-19}}{\lambda} \text{ W sec}, \qquad (2\text{-}17)$$

where λ is in microns. The reciprocal of this expression gives the number of photons per second per watt:

$$1 \text{ watt} = 5.0345 \times 10^{18} \lambda \text{ photon sec}^{-1}. \qquad (2\text{-}18)$$

Although the radiation laws are written here in terms of wavelength, they can also be written in terms of wavenumber or of frequency. Since these forms are of value principally to the spectroscopist, they will not be given here; they can be found in refs. 13 and 14.

In Section 2.7 the various calculating aids that are available to simplify the handling of the radiation formulas are discussed.

2.5 EMISSIVITY AND KIRCHHOFF'S LAW

The formulas in the previous section describe the radiation from a blackbody. A factor can be added to them so that they can also be applied to sources that are not blackbodies. This factor, called the *emissivity* ϵ, is given by the ratio of the radiant emittance W' of the source to the radiant emittance of a blackbody at the same temperature:

$$\epsilon = \frac{W'}{W}. \qquad (2\text{-}19)$$

Thus emissivity is a numeric whose value lies between the limits of zero for a nonradiating source and unity for a blackbody. It is a convenient measure of the degree to which a source approximates a blackbody.

Emissivity is a function of the type of material and its surface finish and it can vary with wavelength and with the temperature of the material. A more general expression in terms of the spectral emissivity $\epsilon(\lambda)$ is

$$\epsilon = \frac{\int_0^\infty \epsilon(\lambda) W_\lambda d\lambda}{\int_0^\infty W_\lambda d\lambda} = \frac{1}{\sigma T^4} \int_0^\infty \epsilon(\lambda) W_\lambda d\lambda. \qquad (2\text{-}20)$$

Three types of sources can be distinguished by the way that the spectral emissivity varies:
1. A blackbody or Planckian radiator, for which $\epsilon(\lambda) = \epsilon = 1$.
2. A *graybody*, for which $\epsilon(\lambda) = \epsilon =$ constant (but less than unity).
3. A *selective radiator*, for which $\epsilon(\lambda)$ varies with wavelength.

Figure 2.6 shows the spectral emissivity and spectral radiant emittance for each type of source. A blackbody, which was defined as the ultimate thermal radiator, radiates more flux, either total or in an arbitrary spectral interval, than any other type of source at the same temperature. Thus the spectral distribution curve of a blackbody provides the limiting envelope for the other types of sources. A graybody, for which the emissivity is a constant fraction of that for a blackbody, is a particularly useful concept because such sources as jet tailpipes, aerodynamically heated surfaces, unpowered space vehicles, personnel, and terrestrial and space backgrounds can be represented as graybodies with an accuracy sufficient for most engineering calculations. As shown in Figure 2.6, a selective radiator can sometimes be considered to be a graybody over a limited spectral interval, thus simplifying calculations.

When radiant energy is incident on a surface, three processes can occur: a fraction of the incident energy α may be absorbed, a fraction ρ may be reflected, and a fraction τ may be transmitted. Since energy must be conserved, the following relationship can be written:

$$\alpha + \rho + \tau = 1. \qquad (2\text{-}21)$$

By definition, a blackbody absorbs all of the incident radiant energy so that $\alpha = 1$ and $\rho = \tau = 0$. While studying the radiation transfer process, Kirchhoff[9] observed that at a given temperature the ratio of radiant emittance to absorptance is a constant for all materials and that it is equal to the radiant emittance of a blackbody at that temperature. Known as *Kirchhoff's law*, it can be stated as

$$\frac{W'}{\alpha} = W. \qquad (2\text{-}22)$$

This law is often paraphrased as "good absorbers are good emitters,"

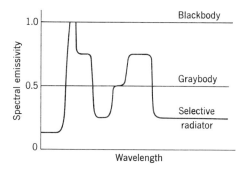

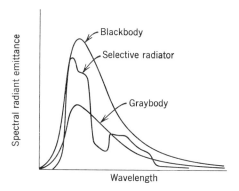

Figure 2.6 Spectral emissivity and spectral radiant emittance of three types of radiators.

an obvious relationship when (2-22) is rewritten as $W' = \alpha W$. Combining (2-9) and (2-22) gives

$$\frac{\epsilon \sigma T^4}{\alpha} = \sigma T^4. \qquad (2\text{-}23)$$

From this it follows that

$$\epsilon = \alpha. \qquad (2\text{-}24)$$

Thus the emissivity of any material at a given temperature is numerically equal to its absorptance at that temperature. Since an opaque material does not transmit energy, $(\alpha + \rho)$ equals 1 and

$$\epsilon = (1 - \rho). \qquad (2\text{-}25)$$

This is a particularly convenient relationship, since it is often easier to measure reflectance than to measure emissivity directly [15].

Since emissivity can vary with the direction of measurement, particularly for polished metals, it is necessary to define several types of emissivity. *Hemispherical emissivity* ϵ_h is defined by (2-20) and gives the emissivity of a source radiating into a hemisphere. This type of emissivity is important in calculating the amount of heat transferred by radiation. *Directional emissivity* ϵ_θ is the emissivity measured in a small solid angle at an angle θ from the normal to the radiating surface. The particular case in which the angle θ is zero is called the *normal emissivity* ϵ_n. Each type may be either a total (implying that it is measured over all wavelengths) or a spectral quantity. Since most infrared systems respond to the flux contained in a small solid angle at a definite direction from the source, ϵ_θ and ϵ_n are of particular interest. Fortunately the differences between ϵ_h, ϵ_θ and ϵ_n are usually small, and they can be ignored except for polished metals, for which the hemispherical emissivity is about 20 per cent greater than the normal emissivity.[1] Since one rarely encounters sources made from polished metals, the symbol ϵ will be used to represent emissivity and subscripts will be used only if it is necessary to distinguish between hemispherical and normal emissivity.

Table 2.3 gives values of the emissivity (total normal) of a wide variety of materials; additional data can be found in [17-21]. Measurement techniques, their theoretical basis, and data on the emissivity of terrains and of materials exposed to extreme environmental conditions are given in [22-28].

For metals, emissivity is low, but it increases with temperature and may increase tenfold or more with the formation of an oxide layer on the surface. For nonmetals, emissivity is high, usually more than 0.8, and it decreases with increasing temperature. The radiation from a metal or other opaque material originates within a few microns of the surface; hence emissivity is a function of the surface state of a material rather than of its bulk properties. For this reason, the emissivity of a coated or painted surface is characteristic of the coating rather than of the underlying surface; notice in Table 2.3 the effect of an oil layer on a nickel base. We can expect to find differences in the published values of emissivity for the same material; such differences are almost certainly due to variations in the condition of the surface of the samples. For instance, Table 2.3 shows that polishing sheet aluminum can halve its emissivity. Barnes et al. [22] report that all of their polished metal samples became tarnished

---

[1] For a diffuse radiator, that is, one that obeys Lambert's cosine law, the three types of emissivity are equal. For electrical insulators ϵ_h/ϵ_n varies from 0.95 to 1.05; an average value is about 0.98. For these materials ϵ_n and ϵ_θ are essentially equal until θ exceeds 65 or 70 deg. For electrical conductors, ϵ_h/ϵ_n varies from 1.05 to 1.33; an average value is 1.20 for most polished metals [16]. Differences between ϵ_n and ϵ_θ become significant when θ exceeds 45 deg.

INFRARED RADIATION

TABLE 2.3 EMISSIVITY (TOTAL NORMAL) OF VARIOUS COMMON MATERIALS

Material	Temperature (°C)	Emissivity
Metals and their Oxides		
Aluminium:		
polished sheet	100	0.05
sheet as received	100	0.09
anodized sheet, chromic acid process	100	0.55
vacuum deposited	20	0.04
Brass:		
highly polished	100	0.03
rubbed with 80-grit emery	20	0.20
oxidized	100	0.61
Copper:		
polished	100	0.05
heavily oxidized	20	0.78
Gold: highly polished	100	0.02
Iron:		
cast, polished	40	0.21
cast, oxidized	100	0.64
sheet, heavily rusted	20	0.69
Magnesium: polished	20	0.07
Nickel:		
electroplated, polished	20	0.05
electroplated, no polish	20	0.11
oxidized	200	0.37
Silver: polished	100	0.03
Stainless Steel:		
type 18-8, buffed	20	0.16
type 18-8, oxidized at 800°C	60	0.85
Steel:		
polished	100	0.07
oxidized	200	0.79
Tin: commercial tin-plated sheet iron	100	0.07
Other Materials		
Brick: red common	20	0.93
Carbon:		
candle soot	20	0.95
graphite, filed surface	20	0.98
Concrete	20	0.92
Glass: polished plate	20	0.94
Lacquer:		
white	100	0.92
matte black	100	0.97

TABLE 2.3 (*Continued*)

Material	Temperature (°C)	Emissivity
Oil, lubricating (thin film on nickel base):		
nickel base alone	20	0.05
film thickness of 0.001, 0.002, 0.005 in.	20	0.27, 0.46, 0.72
thick coating	20	0.82
Paint, oil: average of 16 colors	100	0.94
Paper: white bond	20	0.93
Plaster: rough coat	20	0.91
Sand	20	0.90
Skin, human	32	0.98
Soil:		
dry	20	0.92
saturated with water	20	0.95
Water:		
distilled	20	0.96
ice, smooth	−10	0.96
frost crystals	−10	0.98
snow	−10	0.85
Wood: planed oak	20	0.90

at temperatures of less than 320°C and underwent marked changes in emissivity. In addition, fingerprints, dust, dirt, and surface scratches caused by careless handling can change the measured value of emissivity.

We must resist the temptation to estimate the emissivity of a material on the basis of its visual appearance. A good illustration of this point is furnished by snow; as shown in Table 2.3, it has an emissivity of 0.85. To the eye, snow is an excellent diffuse reflector, and from Kirchhoff's law we might be tempted to guess that its emissivity is very low. The maximum spectral radiant emittance of a blackbody at the temperature of snow occurs at 10.5 μ and 98 per cent of its radiant energy lies between 3 and 70 μ. Hence a visual estimate is meaningless because the human eye, which is sensitive to wavelengths near 0.5 μ, cannot possibly sense conditions at 10 μ. If the eye did respond at 10 μ, then snow would, indeed, look quite black and we could readily guess that it should be a good emitter. Sunlight has the approximate spectral characteristics of a 6000°K blackbody; its maximum spectral radiant emittance occurs at 0.5 μ, and 98 per cent of its radiant energy lies between 0.15 and 3 μ. Thus snow illuminated by sunlight absorbs radiant energy in the 0.5 μ region and reradiates the energy in the 10 μ region.

For a body in space, such as a space vehicle or a satellite, or, to a good approximation, vehicles and aircraft on the ground, the only means by

INFRARED RADIATION

which energy absorbed from the sun can be transferred away is radiation. Thus for thermal control of such vehicles, it is important to know for their skin materials the absorptance for solar radiation α and the emissivity for low-temperature (300°K) radiation ϵ. It can be shown[29, 30] that the equilibrium temperature of a passive satellite, that is, one that dissipates no power internally, depends only on the value of α/ϵ. High values of α/ϵ result in a "hot" satellite and low values result in a "cold" satellite. For an active satellite, other factors must be taken into account, but the value of α/ϵ remains the most important one in establishing the equilibrium temperature.

Table 2.4 gives values of α, ϵ, and α/ϵ for several materials that appear to be useful for the skins of space vehicles[29–33]. The values of emissivity in Table 2.4 are not necessarily in agreement with those in Table 2.3 because of differences in the preparation of samples. However, the data in the two tables can be used to supplement one another. A comparison of the values of α/ϵ for polished aluminum and for white titanium dioxide paint shows why many aircraft are painted to reduce their internal temperatures while they are parked on the ground under a hot sun.

TABLE 2.4 SOLAR ABSORPTANCE α AND LOW-TEMPERATURE (300°K) EMISSIVITY ϵ FOR SPACECRAFT MATERIALS (ADAPTED FROM DATA IN REFERENCES 29,31–33)

Material	α	ϵ	α/ϵ
Aluminum:			
polished and degreased	0.387	0.027	14.35
foil, dull side, crinkled and smoothed	0.223	0.030	7.43
foil, shiny side	0.192	0.036	5.33
sandblasted	0.42	0.21	2.00
oxide, flame sprayed, 0.001 in. thick	0.422	0.765	0.55
anodized	0.15	0.77	0.19
Fiberglass:	0.85	0.75	1.13
Gold: plated on stainless steel and polished	0.301	0.028	10.77
Magnesium: polished	0.30	0.07	4.3
Paints:			
Aquadag, 4 coats on copper	0.782	0.490	1.60
aluminum	0.54	0.45	1.2
Microbond, 4 coats on magnesium	0.936	0.844	1.11
TiO_2, gray	0.87	0.87	1.00
TiO_2, white	0.19	0.94	0.20
Rokide A	0.15	0.77	0.20
Stainless steel:type 18-8, sandblasted	0.78	0.44	1.77

2.6 SELECTIVE RADIATORS

In our discussion of the spectroradiometer we pointed out that emission spectra may be either continuous or discontinuous. The continuous type is, of course, the thermal radiator discussed in section 2.4. With the discontinuous type, called a selective radiator, the radiant energy is found in narrow spectral intervals, called spectral lines because of their appearance in a photograph of the spectrum, or in groups of closely spaced lines called bands. Line spectra are observed in the visible and ultraviolet and, to a lesser extent, in the near infrared. The wavelengths of the lines are characteristic of the particular radiating atom. Band spectra are observed in all spectral regions, but especially in the infrared, and are characteristic of molecules. A brief review of the origin of line spectra will provide an introduction to the more complicated band or molecular spectra.

In 1906 Rutherford proposed that an atom consisted of one or more negatively charged electrons rotating about a positively charged nucleus. A familiar principle of mechanics states that a body moving in a circular path is accelerated toward the center of the circle; in addition, classical electromagnetic theory indicates that an accelerated charge must radiate energy. Thus, if no external energy is supplied to the atomic system, classical physics predicts that the electron will spiral into the nucleus and radiate energy in a continuous band of frequencies as it goes. Unfortunately for this theory, neither prediction is true in practice. In 1913 Bohr suggested that certain laws of classical physics were not valid at the atomic level, much as Planck had done in deriving his blackbody law. The hydrogen atom with its single electron is the simplest atom available, and it provided a starting point for an explanation of atomic spectra. To account for the stability of the atom, Bohr suggested that the electron of the hydrogen atom could remain indefinitely in certain orbits without radiating energy. The only possible orbits were those in which the angular momentum of the electron was an integral multiple n of the quantity $h/2\pi$, where h is Planck's constant. The value of the angular momentum for an orbit determined its *energy level*. Finally, Bohr suggested that the electron could jump only between these specific orbits and that the energy difference would be gained or lost by the absorption or emission of a photon. The energy of the emitted photon is $\Delta E = h\nu$, where ΔE is the difference between the two energy levels and ν is the frequency. In its normal state, in which $n = 1$, the hydrogen atom has its lowest energy and the electron is in an orbit close to the nucleus. If the atom acquires energy, as, for example, by collision with other atoms in an electrical discharge, the electron will jump to an outer orbit;

it is said to have been raised to an excited state. Since an excited state is an unstable one, the electron soon falls back to a lower energy level and emits a photon in the process. Not all possible transitions between the energy levels of the atom occur; so-called selection rules have been evolved to describe the permitted transitions.

The Bohr model of the atom succeeded in explaining the observed spectrum of hydrogen and provided a pleasing physical picture of the process. It was gradually extended into the so-called Vector Model, to explain the spectra of atoms having more than one electron. However, it remained for quantum mechanics, at least partially, to explain the spectra of more complex atoms and molecules. The definite Bohr orbits were replaced by probability distributions describing the possible locations of the electrons. The definite orbit was not essential to the Bohr model; only the energy associated with each orbit was of importance. Quantum mechanics retains the concept of definite values for the energy states, whereas the selection rules, introduced into the Bohr theory to give agreement with observation, follow logically from the mathematical treatment.

Absorption Spectra of Gases

Thus far the discussion has dealt with the emission spectrum of an atom. The same principles can be applied to explain its absorption spectrum. If radiation from a blackbody passes through a gas, certain wavelengths are absorbed; these wavelengths are the same as those observed in the emission spectrum of the gas. In an absorption spectrum the atom absorbs, from the incident blackbody radiation, a photon having the energy needed to raise it to an excited state. We may wonder why these wavelengths are missing from the beam, for the excited atom is unstable and will shortly return to its stable state with the emission of a photon. The answer is that since the photon can be emitted in any direction, very few will have the same direction as the original radiation. Thus these wavelengths appear to be missing from the beam.

Spectra observed in the infrared are mostly due to transitions between the energy levels of molecules rather than atoms. The spectra appear as diffuse bands, although if examined with sufficient resolution each is seen to consist of many sharp lines. The energy of a molecule is of four types: electronic, translational, rotational, and vibrational. Large energies are required to produce transitions between the electronic states in a molecule. The resulting spectra are found in the ultraviolet, visible, and near infrared. Since changes in translational energies have a negligible effect on molecular energy levels, they will be disregarded. The *rotational spectrum* of a molecule results from changes in its rotational energy. Since the energies of the various rotational states are quite small, the frequencies are low and they occur in the extreme infrared. The *vibrational spectrum* results from changes in the vibrational energy of a molecule. The energy differences between vibrational states are many times greater than those between rotational states, the frequencies are correspondingly higher, and the spectra are observed in the region from about 2 to 30 μ, a region of prime interest to the infrared system designer.

For the production of absorption spectra, an interaction must occur between the incident radiation and the molecular system. Infrared spectra are found only when the resulting vibration (or rotation) of the molecule produces a change in its electric dipole moment.[1] For this reason, there are no vibrational spectra from the symmetric vibrations of the molecules of nitrogen, oxygen, and argon, the three most abundant gases in the earth's atmosphere.

Water vapor and carbon dioxide are responsible for most of the absorption of radiant energy by the earth's atmosphere. Carbon dioxide is a polyatomic molecule; its three atoms lie along a straight line with the two oxygen atoms equidistant from the carbon atom. Such a molecule has nine degrees of freedom; that is, nine numbers are needed to describe the motions of its atoms: three for translation, two for rotation, and four for vibration. The four vibrational modes, two of which have the same frequency, are responsible for the fundamental-vibration spectra of carbon dioxide, as shown in Figure 2.7. Since the symmetrical-stretching mode is symmetrical about the carbon atom, there is no change in the dipole moment and no absorption occurs. In the other modes the two oxygen atoms move equal distances parallel to one another and the carbon atom moves in the opposite direction as to maintain a fixed center of gravity for the molecule. Both vibrations result in a change in the dipole moment. Absorption is observed at 4.25μ for the antisymmetrical stretch and at 15μ for the bending mode. The fourth vibrational mode lies at right angles to that shown for the bending mode, so that the atoms oscillate in and out of the plane of the paper.

The fundamental vibration modes of the water vapor molecule are also shown in Figure 2.7, from which it is seen that absorption occurs at 2.74, 2.66, and 6.27μ. In molecules having no symmetry, every fundamental vibration causes a change in the dipole moment and a corresponding absorption. In symmetrical molecules, one or more of the fundamental vibrations occur without changing the dipole moment and no absorption is found, as, for example, with the symmetric-stretch mode of carbon dioxide.

[1]The electric dipole moment is a measure of the separation between the positive and negative charges in a molecule. If a positive charge $+e$ and a negative charge $-e$ are a distance d apart, the dipole moment is given by the product ed.

INFRARED RADIATION

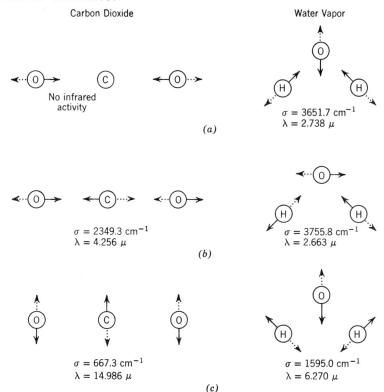

Figure 2.7 Vibrational modes of carbon dioxide and water vapor molecules. (*a*) Symmetrical-stretch mode; (*b*) antisymmetrical-stretch mode; (*c*) bending or degenerate mode.

The vibrational spectra observed in practice are almost always far more complex than the discussion up to this point would indicate. Interactions between neighboring molecules distort the form of the fundamental vibrations and cause the appearance of *overtone bands* at multiples of the fundamental frequency and *combination bands* at various sums or differences of two or more fundamental frequencies. The vibrational frequencies are hundreds or thousands of times higher than those of rotation. The frequency of a vibrational transition thus differs very little from that of the corresponding transition, in which both vibrational and rotational states change. The result is a *vibration-rotation band*, one in which the line due to the fundamental is surrounded by a series of closely spaced lines caused by the superposition of a series of rotational frequencies on the much higher vibrational frequency. The $15\,\mu$ absorption band of carbon dioxide, shown in Figure 2.8, is an excellent example of the complexity of a vibration-rotation band[34].

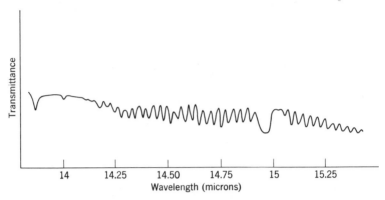

Figure 2.8 The 15 μ vibration-rotation band of carbon dioxide (adapted from Oetjen, Kao, and Randall[34]).

Absorption Spectra of Liquids and Solids

The previous discussion of absorption spectra was simplified by limiting it to gases. The distinguishing characteristic of a gas in comparison with a liquid or solid is its much lower density. The effects of neighboring molecules are slight, and it is reasonable to assume that each molecule in a gas is independent of any other. As the pressure of a gas increases, the number of molecular collisions and the interaction between neighboring molecules both increase; the numerous lines in the vibration-rotation spectrum are successively broadened, smeared out, and finally disappear entirely. Figure 2.9 shows the effect of pressure, temperature, and change of state on the 1.75 μ overtone band of hydrogen chloride[35]. (Individual curves have been displaced vertically for clarity. The transmittance scale applies to all curves, but it must be shifted for each. The left end of each curve is a point of 100 per cent transmittance.) The vibration-rotation structure disappears with increasing gas pressure, but there is no shift in the wavelength of the absorption band. As is shown by the lower curves, rotational spectra are not observed for the liquid state; all that remains is the broadened vibrational band that has been shifted slightly toward longer wavelengths. Further decrease in the temperature of the liquid, including freezing, tends to narrow the band slightly and to continue to shift it toward longer wavelengths.

Molecular Emission Spectra

For atomic spectra, absorption occurs at the same wavelengths as does emission. To a very close approximation, the same is true of molecular spectra. Such a result is, of course, to be expected from Kirchhoff's law

INFRARED RADIATION

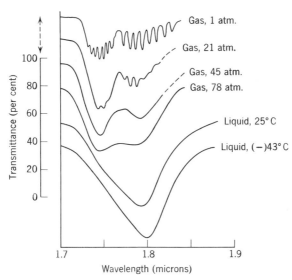

Figure 2.9 Effect of pressure, temperature, and change of state on the 1.75 μ overtone band of hydrogen chloride (adapted from West[35]).

that good absorbers are good emitters, if we apply this law to specific wavelengths rather than to the entire spectrum.

Most combustion processes result in the formation of carbon dioxide and water vapor; thus a knowledge of their spectra is likely to be directly applicable to the prediction of the radiation from rocket and turbojet exhausts. The emission spectrum of a Bunsen flame burning natural gas is shown in Figure 2.10. A strong emission band occurs at 4.4μ with a weaker but somewhat broader one at 2.7μ. Numerous other weak bands have been observed to beyond 25μ. The 2.7μ band is formed by the superposition of several bands of water vapor and carbon dioxide, and the 4.4μ band is due to carbon dioxide. In a flame the emission band is broader than the corresponding absorption band and it is shifted toward longer wavelengths by the higher temperature and pressure in the flame. Gaydon [37], for example, gives the wavelength of the maximum of the emission band as 4.403μ for an ordinary Bunsen flame, 4.388μ for a flame at 1000°C, and 4.344 μ for a 600°C flame. The spectra of many different types of flames are remarkably similar; all show a strong emission band between 4 and 5μ and a weaker band just below 3μ. In certain fuels hydrogen chloride is formed during combustion. Its principal emission lies in a series of bands near 3.5μ, a region in which most flames show no emission bands.

What is the spectral radiance in an emission band, is there a limit to

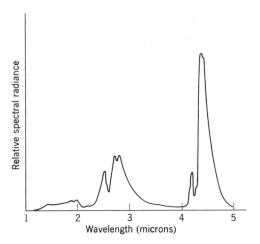

Figure 2.10 Infrared emission from a Bunsen flame (adapted from Plyler[36]).

the value it may have, and how is the limit calculated? The answer, which really should be no surprise, is that the blackbody laws set the limit. The spectral radiance in an emission band cannot exceed the spectral radiance of a blackbody at the same wavelength and temperature. If the spectral emissivity of a flame is known for a particular set of conditions, the radiation laws can be used to calculate its spectral radiance. Many workers have determined the spectral emissivity of gases[38–41].

The conditions under which blackbody radiation is observed from a gas have been examined by Finkelnburg[42] He finds that there are two conditions that must be fulfilled, an infinite thickness of gas and temperature equilibrium within the volume of gas. Monochromatic (of a single wavelength) radiation is absorbed gradually as it passes through a volume of gas. Thus the absorption by a volume of gas depends not only on its temperature and pressure but also on the size and shape of the volume. By Kirchhoff's law, emissivity will show the same dependence (at the same temperature). Only for an infinite thickness of gas will the absorption be complete so that the emissivity is equal to unity. If the gas absorbs strongly at the same wavelength as that of the incident radiation, the thickness of gas required for complete absorption may be quite moderate and the gas will radiate as a blackbody at that wavelength. For the Bunsen flame shown in Figure 2.10, a flame thickness of about 1 m is needed to ensure that its spectral radiance at 4.4 μ is equal to that of a blackbody.

The radiation from a volume of gas does not obey Lambert's cosine law. When a volume of gas is confined between two parallel planes, its radiance is least in the direction normal to these planes, because the thickness of

INFRARED RADIATION 53

gas is a minimum in this direction. Therefore the relationship $W = \pi N$ holds for a volume of gas only under those conditions necessary to make its emissivity equal to unity.

The treatment of molecular spectra given here is both short and non-analytical; it emphasizes the physical factors involved. In addition to the references already given, the reader desiring further information can consult Herzberg[43], the classical reference in the field. For a less rigorous treatment the reader is referred to Brugel[44], and for a short but comprehensive summary, to Harrison, Lord, and Loofbourow[45].

2.7 AIDS FOR RADIATION CALCULATIONS

Calculations involving Planck's law in the form given in (2-8) soon become tedious. Fortunately there are many ways in which such calculations can be simplified. They can be grouped into three general categories: (1) radiation slide rules, (2) charts and nomographs, and (3) tables of blackbody functions.

Radiation Slide Rules

Although many individuals probably realized the convenience of a slide rule for solving the radiation formulas, the first "reduction to hardware" was made in 1943 by Czerny at Frankfurt University[46]. Czerny constructed a small number of these rules by hand and they later became the prototype for the rule produced in Germany by Aristo.

In 1947 a group led by Makowski and Verra at the Admiralty Research Laboratory developed a nomograph to simplify radiation calculations. Later, inspired by one of Czerny's slide rules, this group decided to extend its work to include the development of an improved radiation slide rule capable of an accuracy of 1 per cent[47]. A commercial model, known as the Thornton Radiation Slide Rule, has been available since 1953 and is shown in Figure 2.12.[1] It covers the temperature range of 100 to 10,000°K. The scales on one side of the slide are in terms of energy, and those on the other side give the corresponding photon quantities.

Perhaps the best-known radiation slide rule in the United States is the one designed in 1948 by A. H. Canada[48] and shown in Figure 2.11.

---

[1] Designated the F5100 Radiation Slide Rule, it is manufactured by A. G. Thornton, Ltd., Wythenshawe, Manchester 22, England. It is available in the United States from Jarrel-Ash Co., 590 Lincoln St., Waltham, Massachusetts, or International Scientific and Precision Instrument Co., 910 Seventeenth St., N.W., Washington, D.C. 20206. A new model, featuring an improved construction and a new type of case, was introduced early in 1965. A reprint of reference 47 serves as an instruction manual.

Through the years this has become known as the *GE slide rule*[1] and it is virtually a badge of the infrared fraternity. Since it is available for a very nominal price, every worker in the infrared field should have one. Canada characterized his rule as being suitable for order-of-magnitude calculations; despite this, probably 90 per cent of the radiation calculations made in the United States during the last decade were made with this rule. Constructed of heavy cardboard and plastic, it covers the temperature range from 100 to 10,000°K.

A radiation slide rule based on the original Czerny design is available in Germany.[2] This rule is simple to use, has a minimum number of scales, and covers the temperature range from 100 to 10,000°K.

The Autonetics Blackbody Photon Calculator computes six closely allied photon functions.[3] Its temperature scale extends down to 2°K, making it particularly useful for calculations involving the blackbody flux from the very low temperature backgrounds found in space.

In addition to these commercially available slide rules, several do-it-yourself rules are described in the literature [49, 50].

The following examples are presented to show some of the wide variety of calculations that can be made with a radiation slide rule.

EXAMPLE 2.7.1. With the GE radiation slide rule find, for an 873°K blackbody, (a) the radiant emittance, (b), the effective radiant emittance between 2 and 6 μ, and (c) the radiance.

Referring to Figure 2.11, align 873°K on the slide with the °K index at ①.

(a) Read at ② the radiant emittance, 3.2 W cm^{-2}.
(b) Read at ③ that 66.5 per cent of the flux lies below 6 μ; likewise at ④ read that 3.5 per cent lies below 2 μ. Thus 66.5 − 3.5 or 63 per cent of the radiant emittance lies between 2 and 6 μ and the effective radiant emittance is 0.63 × 3.2 or 1.98 W cm^{-2}.

---

[1]Designated the GEN-15C, it is available from the General Electric Company, Advertising and Sales Promotion Department, One River Road, Schenectady 5, New York.

[2]Designated the Rechenstab für Temperaturstrahlung, Model 922, it is available from Dennert and Pape Aristo Werke, Juliusstrasse 10, Hamburg, Germany. Users are cautioned that the scale marked "watt cm^{-2}" is not the radiant emittance, as might be assumed, but is the radiance in watt cm^{-2}sr^{-1}. As (2-6) shows, these two quantities differ by a factor of π. Apparently this confusing scale marking is justified on the basis that dimensionless units, such as the steradian, need not be shown.

[3]Reproductions of the Blackbody Photon Calculator can be obtained by inquiry through the Electro-Optical Laboratory, Autonetics Division, North American Aviation, Inc., 3370 Miraloma Ave., Anaheim, California, 92083.

INFRARED RADIATION 55

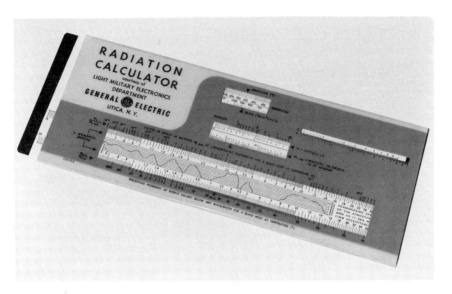

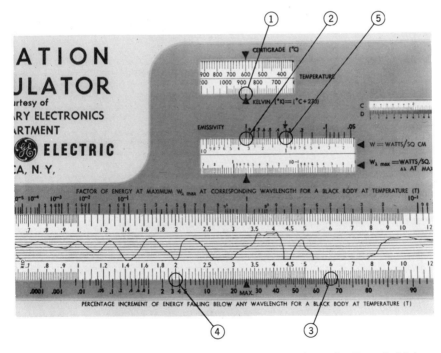

Figure 2.11 Use of the General Electric Radiation Slide Rule to solve Example 2.7.1.

THE ELEMENTS OF THE INFRARED SYSTEM

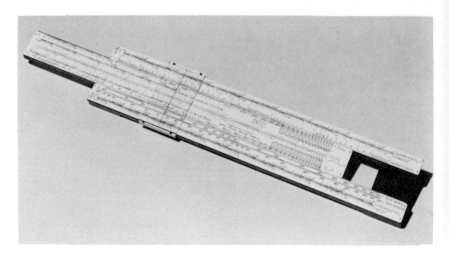

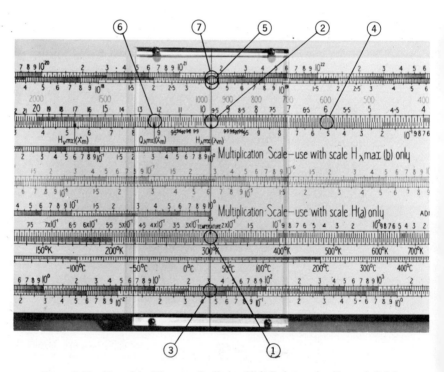

Figure 2.12 Use of the Thornton Radiation Slide Rule to solve Example 2.7.2.

(c) From (2-6), $N = W/\pi = 3.2/\pi = 1.02$ W cm^{-2}sr^{-1}. This value can be read directly from the slide rule by using the special $1/\pi$ index at ⑤.

EXAMPLE 2.7.2. With the Thornton radiation slide rule find, for a 300°K blackbody, (a) the wavelength and the value of the maximum spectral radiant emittance, (b) the spectral radiant emittance at 6 μ, (c) the radiant photon emittance, and (d) the wavelength and the value of the maximum spectral radiant photon emittance.

Referring to Figure 2.12, align the temperature index on the slide with 300°K at ①.

(a) Read at ② the wavelength of the maximum spectral radiant emittance, 9.68 μ. Read at ③ its value, 3.1×10^{-3} W cm$^{-2}\mu^{-1}$. (This scale reads W cm^{-2}cm^{-1} and should be multiplied by 10^{-4} for the units used here.)
(b) Read at ④ that the spectral radiant emittance at 6 μ is 51.6 per cent of that at the peak or $0.516 \times 3.1 \times 10^{-3} = 1.6 \times 10^{-3}$ W cm$^{-2}\mu^{-1}$.
(c) Read at ⑤ the radiant photon emittance, 4.1×10^{18} photon sec^{-1}cm^{-2}.
(d) Read at ⑥ the wavelength of the maximum spectral radiant photon emittance, 12.25 μ. Read at ⑦ its value, 1.7×10^{17} photon sec^{-1} cm$^{-2}\mu^{-1}$. (As in (a) above the values read from this scale must be multiplied by 10^{-4} to convert to a wavelength interval in microns.)

Charts and Nomographs

The choice of scales for plotting blackbody spectral distribution curves affects their shape. Figure 2.13 shows the effect of various combinations of linear and logarithmic scales on a family of blackbody curves covering the temperature range from 500 to 900°K. Notice that curves for different temperatures have the same shape only on the log-log plot. This is also the only plot on which the locus of the successive maxima, the Wien function, is a straight line. These unique properties of the log-log plot have been used by Moon[51] as the basis of a simple yet effective calculator. If the straight line representing the Wien function in the log-log plot of Figure 2.13d is marked with a temperature scale, a single blackbody curve on a transparent overlay can be slid along this line until the peak of the curve coincides with the proper temperature. Having established the position of the curve in this way, we can read the values of wavelength and spectral radiant emittance directly from the scales.

A blackbody spectral distribution curve for one temperature can be used for any other temperature by applying simple scaling laws. If the new temperature is n times the old temperature, the new wavelength scale

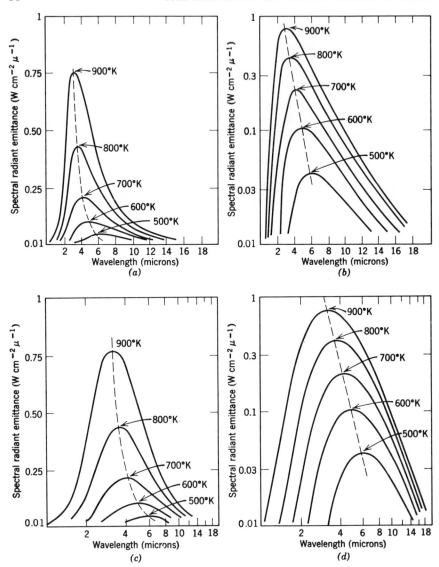

Figure 2.13 A family of blackbody spectral distribution curves plotted with (a) linear-linear, (b) log-linear, (c) linear-log, and (d) log-log scales.

is $1/n$ times the old scale and the new radiant emittance scale is n^5 times the old scale.

Several convenient relationships for blackbody spectral distribution curves are given in Table 2.5. Memorizing the few constants involved

INFRARED RADIATION 59

TABLE 2.5 SOME CONVENIENT RELATIONSHIPS FOR BLACKBODY SPECTRAL DISTRIBUTION CURVES (FOR λ IN μ, T IN °K, AND W IN W cm^{-2})

Peak wavelength	$\lambda_m T = 2898$
0.25 W lies between 0 and λ_m	
0.75 W lies between λ_m and ∞	
Half-power (3 dB) wavelengths	$\lambda' T = 1780$
	$\lambda'' T = 5270$
0.04 W lies between 0 and λ'	
0.67 W lies between λ' and λ''	
0.29 W lies between λ'' and ∞	
Median wavelength	$\lambda''' T = 4110$
0.50 W lies between 0 and λ'''	

allows us to quickly visualize the blackbody curve for any temperature, without recourse to slide rules, printed curves, or tables.

An examination of Planck's equation, (2-8), shows that the spectral radiant emittance is a function of the product of λ and T. Therefore, if the spectral distribution curve is plotted against λT instead of λ, one curve

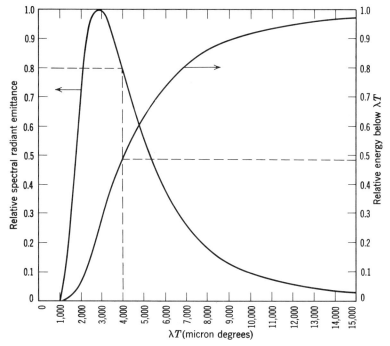

Figure 2.14 Universal blackbody curve.

will suffice for all temperatures. Such a universal curve is shown in Figure 2.14. The other curve in Figure 2.14 gives the integrated radiant emittance between zero and any desired λT as a fraction of the total radiant emittance. Thus at a value of $\lambda T = 4000\,\mu°K$, the value of the spectral radiant emittance is 0.8 times its maximum value; this is true for any combination of λ and T for which the product is 4000, for example, $\lambda = 8\,\mu$ and $T = 500°K$, $\lambda = 4\,\mu$ and $T = 1000°K$, and so on. Similarly, 48 per cent of the radiant emittance lies below this point.

Tables of Blackbody Functions

Many tables of blackbody functions are available in the literature; making no claim for completeness, Pivovonsky and Nagel[52] have listed a total of 48 such tables in their references. The values assumed for the radiation constants c_1 and c_2 differ in the various tables. A user who is interested in extreme accuracy may find it necessary to correct the tabulated values for the most recent values of c_1 and c_2. Unfortunately, many of the available tables do not detail the method for deriving and applying such corrections. For some of the more useful tables that include instructions for taking account of new values of c_1 and c_2, see (52–56).

2.8 OTHER BLACKBODY RELATIONSHIPS

Efficiency of Radiation Production

The approach in discussing the radiation laws has been essentially that of the physicist; the emphasis has been on how much radiant energy is produced and where in the spectrum it may be found. In an extremely interesting paper on blackbodies, Benford[57] has pointed out that an engineer interested in the efficiency with which thermal radiation is produced may write the radiation formulas in a different form. Most infrared systems are designed to operate against noncooperative targets, such as aircraft, missiles, vehicles, and personnel. Consider, however, a system to assist in the rendezvous of two cooperating vehicles, such as two spacecraft or an aircraft and its refueling tanker. Such a system might consist of infrared equipment carried on one vehicle that searches for and tracks a beacon carried on another vehicle. One of the system design constraints will be the limited power available for operating the beacon. The problem is to get the maximum spectral radiant emittance from the beacon at the desired wavelength without exceeding the power limitation. Assume for the moment that the search system operates at a single wavelength and that the efficiency with which the input power is converted to radiant flux is constant over the range of temperatures considered for the beacon. Reasoning intuitively, we would probably

INFRARED RADIATION

guess that the beacon temperature should be chosen by the Wien relationship so that the peak of the spectral distribution curve coincided with the operating wavelength. This solution, although quite good, can be significantly improved.

Benford defined the efficiency with which radiation is produced at a particular wavelength as the ratio of the spectral to the total radiant emittance:

$$\text{Efficiency} = \frac{W_\lambda}{W} = \frac{c_1}{\lambda^5} \frac{1}{e^{c_2/\lambda T} - 1} \frac{1}{\sigma T^4}. \qquad (2\text{-}26)$$

Unfortunately the efficiency shows no simple dependence on the temperature, since temperature appears in the denominator as both an exponent and a factor. However, taking the derivative of (2-26) with respect to temperature and solving for the maximum gives

$$\lambda_e T_e = 3625\mu°\text{K}. \qquad (2\text{-}27)$$

This shows that for a given wavelength λ_e there is a particular temperature T_e at which radiation is produced most efficiently *for a fixed value of flux radiated by the source*. Benford calls this the engineering maximum to distinguish it from the physical maximum given by Wien's law. The relationship between these maxima (the subscript m denotes the physical or Wien relation) is

$$T_e = 1.276\, T_m. \qquad (2\text{-}28)$$

Thus at a particular wavelength the temperature to give an engineering maximum is 27.6 per cent higher than that for a physical maximum. Benford shows that the spectral radiant emittance is 11.6 per cent greater for the temperature corresponding to the engineering maximum.

Benford's example is an interesting illustration of the application of these concepts; it deals with the use of heat in treating body tissues. The skin is relatively transparent at 1.1 μ, but the total flux incident on the skin is limited by the discomfort from heating effects. Thus the problem is to find the source temperature that produces the maximum flux at 1.1 μ without exceeding the skin's total flux limitation. According to Wien's displacement law, a temperature of 2630°K gives the maximum spectral radiant emittance at 1.1 μ. The engineering maximum of 3360°K gives 11.6 per cent more radiant flux at 1.1 μ for the same total flux incident on the skin. (Increasing the temperature of the source requires that its area be reduced so as to maintain a constant total radiant emittance.) Curves of spectral radiant emittance for these and several other temperatures are shown in Figure 2.15. (These curves appear different from those given earlier in this chapter because they have been adjusted for constant

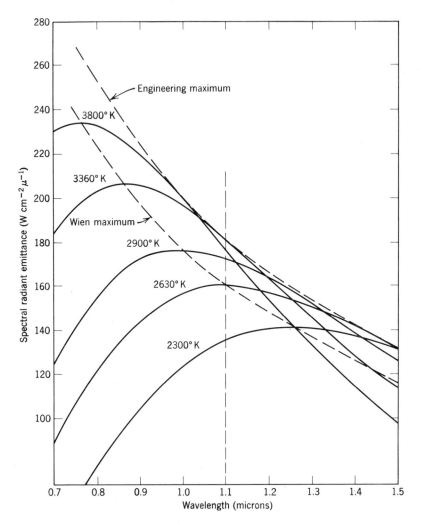

Figure 2.15 Determination of the source temperature to give the maximum radiation at 1.1 μ for a constant value of the total radiant emittance. See text for further details.

radiant emittance rather than for constant source area.) The envelope enclosing the curves is the locus of the engineering maximum; clearly this locus is higher than that of the Wien maximum. This discussion was simplified by limiting it to a single wavelength; from Figure 2.15 it is evident that the same general conclusions hold for a band of wavelengths.

Worthing[58] has shown that the concept of an engineering maximum is equally applicable to photons.

Radiation Contrast

When the temperatures of a target and its background are nearly the same, detection becomes very difficult. This situation occurs, for instance, with thermal mappers viewing targets against an earth background. The level of the radiant flux from personnel, vehicles, structures, and roads is not very different from that of the terrain, trees, rocks, or water in the background. Thus the radiation contrast between the target and the background is likely to be very low.

Is it possible to maximize the radiation contrast by a proper choice of the spectral bandpass for the system? In order to answer this question it is necessary to know the way radiant emittance changes with temperature as a function of wavelength. This can be found by taking the partial derivative of the Planck function with respect to temperature. The result is tedious to evaluate without a computer. Pivovonsky and Nagel[52] show that its value is given by

$$\frac{\partial W_\lambda}{\partial T} = W_\lambda \frac{\psi}{T}, \tag{2-29}$$

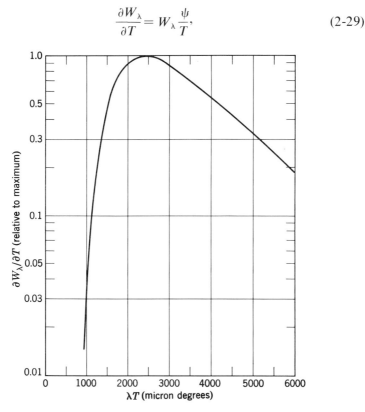

Figure 2.16 Rate of change of spectral radiant emittance, $\partial W_\lambda/\partial T$.

where ψ is tabulated in their Table 3. Relative values of (2-29) are shown in Figure 2.16 plotted against λT. Alternatively, Sanderson[59] has shown that

$$\frac{\partial W_\lambda}{\partial T} \approx W_\lambda \frac{c_2}{\lambda T^2}, \qquad (2\text{-}30)$$

where c_2 is the second radiation constant. The error resulting from the use of (2-30) is less than 1 per cent for $\lambda T < 3100\,\mu°\text{K}$ and less than 10 per cent for $\lambda T < 6200\,\mu°\text{K}$. Since the principal application of this derivative is to thermal mappers viewing targets at a temperature of 300°K, Sanderson's approximation is usually quite acceptable.

The wavelength for which the rate of change of spectral radiant emittance is a maximum for a given blackbody temperature is

$$\lambda_c T = 2411, \qquad (2\text{-}31)$$

where λ_c is in microns and T is in °K[53]. Thus λ_c, the wavelength of maximum radiation contrast, follows a relationship similar to the Wien displacement law and the two are related by

$$\lambda_c = 0.832\lambda_m. \qquad (2\text{-}32)$$

For a temperature of 300°K, λ_c is equal to 8 μ. In the absence of other considerations this would be the most desirable operating wavelength for a thermal mapper viewing terrestrial objects. It is shown in Chapter 4 that the earth's atmosphere is not transparent at all wavelengths. There are three principal spectral intervals, called atmospheric windows, extending from 2 to 2.5 μ, 3 to 5 μ, and 8 to 13 μ, in which the transmission is high. From Figure 2.16 it is evident that λ_c is considerably higher within the 8 to 13 μ window than it is within the other two.

It is well to note that variations in emissivity between the various elements in a scene may contribute as much to the radiation contrast as do temperature differences. Gill[60] has computed the relative effect of variations in the emissivity.

REFERENCES

[1] L. R. Koller, *Ultraviolet Radiation*, 2d ed. New York: Wiley, 1965, p. 5.
[2] A. G. McNish, "The Speed of Light," *IRE Trans. Instrumentation*, **1-11**, 138 (1962).
[3] L. M. Biberman, "Apples, Oranges, and Unlumens," *Appl. Opt.*, **6**, 1127 (1967).
[4] Committee on Colorimetry, *The Science of Color*, New York: Crowell, 1953, Chapters 1, 6, 7.
[5] G. Kelton et al., "Infrared Target and Background Radiometric Measurements — Concepts, Units, and Techniques," *Infrared Phys.* **3**, 139 (1963).
[6] D. J. Lovell, "The Concept of Radiation Measurements," *Am. J. Phys.* **21**, 459 (1953).

[7] P. Moon, *The Scientific Basis of Illuminating Engineering*, New York: Dover, 1961, Chapter 9. (First published by McGraw-Hill, New York, 1936.)
[8] M. Jakob, *Heat Transfer*, New York: Wiley, 1949, Vol. 1, p. 41.
[9] M. Planck, *Theory of Heat Radiation*, New York: Dover, 1959. (A reprint of the 1910 edition.)
[10] E. R. Cohen, K. M. Crowe, and J. W. M. DuMond, *Fundamental Constants of Physics*, New York: Interscience, 1957.
[11] "New Values for the Physical Constants" *Phys. Today*, **17**, 48 (February 1964).
[12] A. G. Worthing, "Radiation Laws Describing the Emission of Photons by Blackbodies," *J. Opt. Soc. Am.*, **29**, 97 (1939).
[13] R. A. Smith, F. E. Jones, and R. P. Chasmar, *The Detection and Measurement of Infrared Radiation*, London: Oxford University Press, 1957, Chapter 2.
[14] M. Eppley and A. R. Karoli, "Use of Wavenumber in Radiation Formulas," *J. Opt. Soc. Am.*, **43**, 957 (1953).
[15] F. Nicodemus, "Directional Reflectance and Emissivity of an Opaque Surface," *Appl. Opt.*, **4**, 767 (1965).
[16] Jakob, *op. cit.* (reference 8), p. 52
[17] W. H. McAdams, *Heat Transmission*, 3d ed., New York: McGraw-Hill, 1954, p. 472.
[18] C. L. Mantell, ed., *Engineering Materials Handbook*, New York: McGraw-Hill, 1958, pp. 35-10.
[19] D. E. Gray, ed., *American Institute of Physics Handbook*, 2d ed., New York: McGraw-Hill, 1963, pp. 6-157.
[20] A. Goldsmith, T. E. Waterman, and H. J. Hirschhorn, *Handbook of Thermophysical Properties of Materials*, New York: Macmillan, 1961.
[21] G. G. Gubareff, J. E. Janssen, and R. H. Torborg, *Thermal Radiation Properties Summary*, Minneapolis, Minn.: Minneapolis-Honeywell Regulator Co., 1960.
[22] B. T. Barnes, W. E. Forsythe, and E. Q. Adams, "The Total Emissivity of Various Materials at 100–500°C," *J. Opt. Soc. Am.*, **37**, 804 (1947).
[23] D. Weber, "Spectral Emissivity of Solids in the Infrared at Low Temperatures," *J. Opt. Soc. Am.*, **49**, 815 (1959).
[24] D. Weber, "Low Temperature, Directional, Spectral Emissivity of Translucent Solids," *J. Opt. Soc. Am.*, **50**, 808 (1960).
[25] H. O. McMahon, "Thermal Radiation From Partially Transparent Reflecting Bodies," *J. Opt. Soc. Am.*, **40**, 376 (1950).
[26] K. J. K. Buettner and C. D. Kern, "Infrared Emissivity of the Sahara from Tiros Data," *Science*, **142**, 671 (1963).
[27] F. J. Clauss, ed., *Surface Effects on Spacecraft Materials*, New York: Wiley, 1960.
[28] H. H. Blau, and H. Fischer, eds., *Radiative Transfer from Solid Materials*, New York: Macmillan, 1962.
[29] G. B. Haller, "Problems Concerning the Thermal Design of Explorer Satellites," *IRE Trans. Mil. Electronics*, **MIL-4**, 98, (1960).
[30] G. Hass, L. F. Drummeter, Jr., and M. Schach, "Temperature Stabilization of Highly Reflecting Spherical Satellites," *J. Opt. Soc. Am.*, **49**, 918 (1959).
[31] R. E. Graumer, "Materials for Solar-Energy Systems," *Space/Aeronautics*, **36**, 60 (October 1961).
[32] A. L. Johnson, "Spacecraft Radiators," *Space/Aeronautics*, **37**, 76 (January 1962).
[33] A. L. Alexander, "Thermal Control in Space Vehicles," *Science*, **143**, 654 (1964).
[34] R. A. Oetjen, C. L. Kao, and H. M. Randall, "The Infrared Prism Spectrograph as a Precision Instrument," *Rev. Sci. Instr.*, **13**, 515 (1942).

[35] W. West, "The Influence of Temperature and Pressure on the Infrared Absorption Spectrum of Gaseous and Liquid Hydrogen Chloride up to the Critical State," *J. Chem. Phys.*, **7**, 795 (1939).
[36] E. K. Plyler, "Infrared Radiation from a Bunsen Flame," *J. Research Natl. Bur. Standards*, **40**, 113 (1948).
[37] A. G. Gaydon, *The Spectroscopy of Flames*, New York: Wiley, 1957, p. 169.
[38] E. E. Bell, P. B. Burnside, and F. P. Dickey, "Spectral Radiance of Some Flames and their Temperature Determination," *J. Opt. Soc. Am.*, **50**, 1286 (1960).
[39] H. J. Babrov, P. M. Henry, and R. H. Tourin, "Methods of Predicting Infrared Radiance of Flames by Extrapolation from Laboratory Measurements," *J. Chem. Phys.*, **37**, 581 (1962).
[40] S. S. Penner, "The Emission of Radiation from Diatomic Gases. I. Approximate Calculations," *J. Appl. Phys.*, **21**, 685 (1950).
[41] R. H. Tourin, "Quantitative Infrared Spectroradiometry of Liquid Propellant Flames," *Appl. Opt.*, **2**, 1133 (1963).
[42] W. Finkelnburg, "Conditions for Blackbody Radiation of Gases," *J. Opt. Soc. Am.*, **39**, 185 (1949).
[43] G. Herzberg, *Infrared and Raman Spectra of Polyatomic Molecules*, Princeton, N.J.: Van Nostrand, 1945.
[44] W. Brugel, *An Introduction to Infrared Spectroscopy*, New York: Wiley, 1962, Chapters 1-7.
[45] G. R. Harrison, R. C. Lord, and J. R. Loofbourow, *Practical Spectroscopy*, Englewood Cliffs, N. J.: Prentice-Hall, 1948, Chapters 10,11.
[46] M. Z. Czerny, "A Slide Rule for Integration of the Planck Radiation Function," *Phyzik Z*, **9**, 205 (1944).
[47] M. W. Makowski, "A Slide Rule for Radiation Calculations," *Rev. Sci. Instr.*, **20**, 876 (1949).
[48] A. H. Canada, "Simplified Calculation of Blackbody Radiation," *General Electric Review*, December 1948, p. 50.
[49] A. M. Crooker, and W. L. Ross, "A Note on Blackbody Radiation," *Can. J. Phys.*, **33**, 257 (1955).
[50] R. S. Knox, "Direct Reading Planck Distribution Slide Rule," *J. Opt. Soc. Am.*, **46**, 879 (1956).
[51] Moon, *op. cit.* (reference 7), section 5.04.
[52] M. Pivovonsky and M. R. Nagel, *Tables of Blackbody Radiation Functions*, New York: Macmillan, 1961.
[53] M. Gzerny and A. Walther, *Tables of the Fractional Functions for the Planck Radiation Law*, Berlin: Springer, 1961.
[54] Gray, *op. cit.* (reference 19), pp. 6-154.
[55] W. E. Forsythe, ed., *Smithsonian Physical Tables*, 9th ed., Washington, D.C.: Smithsonian Institute, 1954, p. 79.
[56] A. N. Lowan and G. Blanch, "Tables of Planck's Radiation and Photon Functions," *J. Opt. Soc. Am.*, **30**, 70 (1940).
[57] F. Benford, "Laws and Corollaries of the Blackbody," *J. Opt. Soc. Am.*, **29**, 92 (1939).
[58] A. G. Worthing, "New λT Relations for Blackbody Radiation," *J. Opt. Soc. Am.*, **29**, 101 (1939).
[59] J. A. Sanderson, "Emission, Transmission, and Detection of the Infrared," Chapter 5 in A. S. Locke, ed., *Guidance*, Princeton, N.J.: Van Nostrand, 1955.
[60] T. P. Gill, "Some Problems in Low-Temperature Pyrometry," *J. Opt. Soc. Am.*, **47**, 1000 (1957).

3

Sources of Infrared Radiation

This chapter emphasizes practical details concerning radiation from sources that are of interest to an infrared system designer. Considered first is the design of blackbody-type sources for use in calibration and the peculiar problems arising from the lack of a traceable national standard of blackbody radiation. Sources useful in the laboratory, and others such as turbojets, rockets, vehicles, and personnel, which are often the targets for infrared systems, are examined to see what is known about their radiation characteristics. When such information is not available, it is shown how useful engineering estimates can be made by properly applying the radiation laws. Finally, sources that often interfere with target detection are described.

3.1 BLACKBODY-TYPE SOURCES

Blackbody-type sources are widely used for the absolute calibration of infrared equipment. Before discussing the various theoretical and practical factors to be considered in their design and use, a plea should be made for proper terminology. A blackbody represents a theoretical concept; that is, it is an ideal thermal radiator to which all others can be compared. Thus, by its very definition, we cannot hope to build a blackbody. The often-encountered statement concerning a "blackbody with an emissivity of 0.9" is inherently contradictory because by definition a blackbody has an emissivity of unity. The blackbody-type sources used for calibration purposes have an emissivity somewhat less than unity (and probably independent of wavelength) and should thus be called graybodies or, perhaps, blackbody simulators. Because virtually all workers in the infrared field refer to such calibration sources as blackbodies, that usage will be (reluctantly) followed here.

Theoretical Principles

In 1860 Kirchoff stated the conditions that must be met in constructing a blackbody. He pointed out that the radiation within an isothermal enclosure is blackbody radiation; therefore if a small hole is cut through the wall of the enclosure, the radiation leaving this hole should closely simulate that from a blackbody. He further noted that neither the geometrical form of the enclosure nor the material of which it is constructed affect the result. What is important is that the enclosure be truly isothermal and that the area of the hole be very much less than that of the internal surface of the enclosure.

The practical realization of Kirchhoff's ideas has been the subject of many studies. The difficulty in applying his ideas lies in determining the relationship between the relative size of the hole and the accuracy with which the emergent radiation simulates that from a blackbody. The accuracy of simulation is called the *effective emissivity*. It is a function of the size of the hole, the shape and constructional material of the resulting cavity, and the extent to which the cavity departs from a true isothermal condition.

The best-known analysis of the blackbody design problem is that of Gouffé[1]. Although the validity of some of his approximations has been questioned in subsequent literature, it is still the most widely quoted reference on the subject. Most of the papers describing the construction of highly precise blackbodies for use as calibration standards use the method of Gouffé to calculate their effective emissivity [30, 33–35]. Since the original reference may be difficult to obtain, the Gouffé method will be summarized and a nomograph given for the convenient solution of the resulting equations. Any one who must design blackbodies for highly accurate calibration purposes will wish to supplement this material with a study of some of the additional references to be mentioned later.

On the assumption that the walls are diffuse reflectors, Gouffé finds that the effective emissivity of a cavity is

$$\epsilon' = \frac{\epsilon(1+k)}{\epsilon(1-A/S)+A/S}, \qquad (3\text{-}1)$$

where

ϵ' = effective emissivity of the cavity
ϵ = emissivity of the cavity walls
A = area of the opening through which radiation leaves the cavity, cm^2
S = total surface area of the cavity, including that of the opening, cm^2
$k = (1-\epsilon)(A/S - A/S_0)$

SOURCES OF INFRARED RADIATION

S_0 = surface area of a sphere whose diameter is equal to the depth of the cavity (measured from the plane of the opening to the deepest point of the cavity).

It is convenient to write (3-1) as

$$\epsilon' = \epsilon_o(1+k). \tag{3-2}$$

The numerical value of k is small; for a spherical cavity k is equal to zero and the effective emissivity is equal to ϵ_o.

Three typical cavity configurations are shown in Figure 3.1. They are more easily characterized by the depth of the cavity L and the diameter of the opening $2r$ than by the areas used in (3-1). The values of L and $2r$ are identical in each of the configurations shown. Note that for a sphere, L is not equal to the diameter since it is measured normal to the plane of the opening to the deepest point of the cavity. Formulas and tables for determining the values of A/S and $(A/S - A/S_0)$ for various values of L/r are given in the original reference[1]. Rather than make these calculations, it is simpler to use the nomogram shown in Figure 3.2; directions for using it are given in the figure. In the example shown, the effective emissivity of a conical cavity with an L/r of 6 and a wall emissivity of 0.9 is found to be 0.995. The *cavity effect* is clearly evident; the effective emissivity of a cavity always exceeds that of its surfaces. As the emissivity of the surface decreases, the cavity effect becomes increasingly evident. If the cone shown in the example in Figure 3.2 is constructed from metal having an emissivity of 0.1, its effective emissivity will be 0.53. For the same

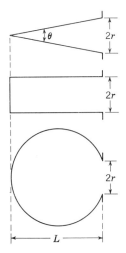

Figure 3.1 Typical cavity configurations for blackbody-type sources.

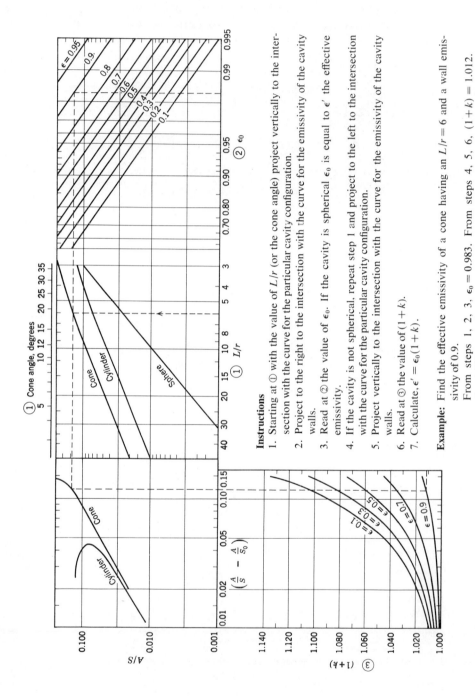

Instructions

1. Starting at ① with the value of L/r (or the cone angle) project vertically to the intersection with the curve for the particular cavity configuration.
2. Project to the right to the intersection with the curve for the emissivity of the cavity walls.
3. Read at ② the value of ϵ_0. If the cavity is spherical ϵ_0 is equal to ϵ' the effective emissivity.
4. If the cavity is not spherical, repeat step 1 and project to the left to the intersection with the curve for the particular cavity configuration.
5. Project vertically to the intersection with the curve for the emissivity of the cavity walls.
6. Read at ③ the value of $(1+k)$.
7. Calculate, $\epsilon' = \epsilon_0(1+k)$.

Example: Find the effective emissivity of a cone having an $L/r = 6$ and a wall emissivity of 0.9.

From steps 1, 2, 3, $\epsilon_0 = 0.983$. From steps 4, 5, 6, $(1+k) = 1.012$.
$\epsilon' = 0.983 \times 1.012 = 0.995$.

material this value could be raised to 0.93 by increasing the value of L/r to 40. The cavity effect explains why the emissivity of a surface marred by pits and scratches is greater than that of the same material in a well-polished condition.

Another important result from the analysis of Gouffé is that *for a given value of L/r, the cavity with the largest surface area has the highest effective emissivity.*

What of the later papers, some of which are not in agreement with Gouffé? The formulation of the problem is so complex that the resulting equations can be solved only by introducing simplifying assumptions. Since different authors have made different assumptions, a detailed comparison of their results is difficult and often requires further assumptions on the reader's part. Williams[2] has compared the various theories proposed up to 1960; his paper is a logical starting point for anyone wishing to pursue the subject in greater detail than that given here. Gouffé assumed that the cavity walls were perfectly diffuse reflectors, that is, that they obeyed Lambert's cosine law. DeVos[3] considered the effect of walls having diffuse as well as nondiffuse reflection characteristics. His results for the diffuse case agree closely with those of Gouffé. However, he found that the effective emissivity of the cavity decreases as its walls become less diffuse. For a particular spherical cavity for which the value of L/r was equal to 10, the effective emissivity was calculated to be 0.992 for diffusely reflecting walls and 0.894 for smooth walls. In a widely quoted but unpublished paper Edwards[4] extended DeVos' results to a conical cavity. He concluded that for a given value of L/r the effective emissivity of a cone exceeds that of a cylinder, a conclusion that is apparently incorrect.

A series of papers by Sparrow et al.[5-7] contain an analysis of conical, cylindrical, and spherical cavities with diffusely reflecting walls. A computer was used in solving the equations given in these papers, and it is claimed that the analysis is far more rigorous than that of earlier investigators. The calculated values of effective emissivity are lower than those calculated by Gouffé's method. As an example, with a conical cavity for which the value of L/r is 8 and the emissivity of the walls is 0.9, Sparrow's equations give an effective emissivity of 0.97 whereas those of Gouffé give 0.997. However, both Sparrow and Gouffé agree that for a given value of L/r the effective emissivity of a cylindrical cavity is higher than that of a conical cavity. An experimental investigation by Kelly and Moore [8] of the effective emissivity of shallow cylindrical cavities ($L/r = 0.5$ to 2) with diffusely reflecting walls showed good agreement with the values calculated by the equations of Gouffé and Sparrow. In this region the two methods appear to agree within 1 per cent.

Campanaro and Ricolfi[9, 10] have given new expressions for the effective emissivity of cylindrical, conical, and spherical cavities. They conclude that the method of Gouffé gives the correct value for a spherical cavity but is somewhat in error for cylindrical and conical cavities. They also conclude that DeVos' expressions are in error for all three configurations.

For most commercially available blackbody-type sources and those built in the laboratory, the values of L/r are usually greater than 6 and the wall emissivities usually exceed 0.85. Under these conditions *we cannot depend on the present theories to predict the effective emissivities of such sources to an accuracy of better than ± 1 per cent*. As a result, it is not presently possible to have a primary standard blackbody-type source whose radiating characteristics are known to an accuracy commensurate with those of other primary physical standards.

Construction of a Blackbody-Type Source

Despite the lack of a completely satisfactory theory for predicting the effective emissivity of a cavity, the situation is not as bleak as it might appear. One interpretation of the various theories is that it is difficult to make a poor blackbody-type source, that is, one for which the effective emissivity is less than 0.95. The statement is not as facetious as it sounds; look again at Figure 3.2 and notice the restrictions we must adopt in order to make the effective emissivity of a cavity less than 0.95. For instance, the conical cavity shown in the example in Figure 3.2 has an effective emissivity greater than 0.95 as long as the emissivity of its walls exceeds 0.6. The blackbody is unique among theoretical concepts because of the ease with which it can be so nearly realized. Although the present theories are not entirely satisfactory, they fail us only in the last 1 or 2 per cent. Contrast this situation with the problems of realizing some of the other well-known theoretical concepts, such as a noiseless receiver, a frictionless plane, and an ideal heat engine.

Most blackbody-type sources used for the calibration of infrared equipment are of the cavity type, have an opening of 0.5 in. or less, and operate in the temperature range from 400 to 1300°K. The design problems associated with such sources include the choice of a cavity configuration, the means of achieving an isothermal condition in the cavity, the provision of a high wall emissivity, and the means for ensuring that the cavity is maintained at a known and stable temperature. Although not intended as a treatise on the design of blackbodies, the following discussion of these problems should be of value to any potential user.

The choice of a cavity configuration usually involves consideration of a cone, sphere, or cylinder. It is clear from Figure 3.2 that for a given value of L/r a spherical cavity has the highest effective emissivity. Unfortunately, it is a difficult shape to fabricate, and it is also hard to heat it uniformly. A cylindrical cavity is easy to fabricate, but it too is difficult to heat uniformly, particularly across the closed end. A conical cavity seems to be the best compromise in terms of a shape that is easy to fabricate and one that can be heated reasonably uniformly.

The large mass of material containing the cavity is called the *core*. Ideally, the core material should have a high thermal conductivity so as to minimize temperature gradients, a good resistance to surface oxidation and subsequent scaling at high temperatures, and a high emissivity. Unfortunately, there are few materials that can satisfy all of these requirements, and it is necessary to accept some compromises. For temperatures up to 1400°K the core is usually made of metal; above 1400°K, graphite or a ceramic is used. Among the metals that could be used, copper appears, at first glance, to be an ideal choice because of its high thermal conductivity. Unfortunately, the oxide layer formed on the surface of copper by heating is unstable and continually scales off at temperatures above 600°K. Since the stainless steels in the 18-8 series have good thermal conductivity and their oxidized surfaces are highly stable, they are an almost universal choice for core materials up to temperatures of 1400°K.

The core is usually heated by a nichrome wire that is wound around it. To improve the uniformity of the cavity temperature, the designer can vary the outer contour of the core so that the cross-sectional area of the metal at any point is constant. If the heater winding is uniform, each turn has a constant volume of metal to heat. Alternatively, an arbitrary outer contour can be used with a nonuniform heater winding adjusted so that each turn heats a constant volume of metal. The greatest heat loss occurs near the open end of a cavity and it is generally recommended that the number of heater turns be increased in this area so as to increase the heat input[11]. The reentrant conical cavity, shown in Figure 3.3, is often used because it is less susceptible to excessive cooling at its open end and it probably has a higher effective emissivity than does a simple conical cavity. Still another precaution is to place a thermally isolated limiting aperture in front of the opening in the cavity. A typical

Figure 3.3 Reentrant conical cavity.

conical cavity may have an opening of 0.5 in. diameter and a limiting aperture of diameter no larger than 0.2 in.

Various steps can be taken to increase the emissivity of the cavity walls. A rough-machined finish should be specified, and no attempt should be made to smooth or polish it. An exception to this is a polished surface that is subsequently liquid-honed to make it a good diffuse reflector. For the 18-8 series of stainless steel, heating to 300°C causes the surface to tarnish and increases its emissivity to 0.5. Treating the surface with chromic and sulfuric acid results in an emissivity of 0.6. Heating the surface to 800°C will form a stable oxide film having an emissivity of 0.85[12]. If the operating temperature is no higher than 100°C, the cavity may be coated with a black enamel, such as Sicon black,[1] to give a wall emissivity of 0.93[13].

A platinum resistance thermometer is often used to sense the temperature of the core. For the most accurate control of the core temperature an electronic proportional controller should be used. However, many applications do not require such high accuracy, and one of the inexpensive meter relays or time-proportioning relay controllers can be used. Ideally, the platinum resistance thermometer should sense the cavity temperature rather than that of the core. Achieving this is very difficult (but not nearly as difficult as proving that it has been accomplished). Much of the difficulty can be avoided by adjusting the set point of the temperature controller with reference to a precisely calibrated thermocouple inserted into the cavity (but not so as to touch the walls). Before setting the temperature controller in this way, it is wise to cement the platinum resistance thermometer into the core so that it cannot move and spoil the setting of the controller.

A complete blackbody-type source that can be built in almost any well-equipped laboratory is shown in Figure 3.4. The type 18-8 stainless steel core is about 4 in. long and has a conical cavity for which the value of L/r is equal to 8 (an included cone angle of about 15°). The constant-pitch nichrome heater winding is insulated from the core by a thin sheet of asbestos. If one values his happy disposition, he should be sure to anneal the nichrome wire prior to winding the core (passing enough current through the wire to bring it to a red heat will anneal it). The heater requires an input power of about 125 W in order to maintain the cavity at a temperature of 800°K. Under these conditions it is estimated that the temperature variations in the cavity do not exceed 5°C. Most of this variation is near the open end of the core and is effectively eliminated by the limiting aperture placed in front of the source. If the emissivity of the

---

[1]Manufactured by Midland Industrial Finishes, Waukegan, Illinois.

SOURCES OF INFRARED RADIATION

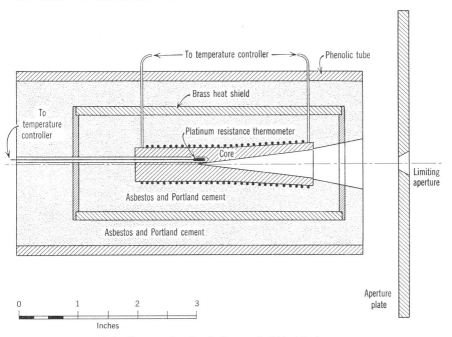

Figure 3.4 Construction details for a typical blackbody-type source.

cavity walls is assumed to be 0.85, the effective emissivity of the cavity is 0.995 (from Figure 3.2).

It is of interest to calculate the variation of radiant emittance caused by small changes in the cavity temperature. At an assumed temperature of 800°K and an effective emissivity of 0.99, the radiant emittance is given by the Stefan-Boltzmann law (2-9) as

$$W = \epsilon\sigma T^4$$
$$= 0.99 \times 5.67 \times 10^{-12} \, (800)^4$$
$$= 2.30 \text{ W cm}^{-2}. \quad (3\text{-}3)$$

To determine the change in radiant emittance resulting from a 1°C change in cavity temperature, we could use (3-3) again with a new temperature of 801°K. It is much simpler, however, to differentiate (3-3):

$$\frac{dW}{dT} = 4\epsilon\sigma T^3$$
$$= 0.0115 \text{ W cm}^{-2}\,°\text{K}^{-1}. \quad (3\text{-}4)$$

The change in radiant emittance is $0.0115/2.30 = 0.5$ per cent for each degree of change in cavity temperature (at 800°K).[1] Thus, for precision calibration of infrared devices, the temperature of a blackbody-type source must be known and controlled to within a fraction of a degree.

In Figure 3.4 a limiting aperture is shown in front of the cavtiy. It is the area of this aperture that is to be used in calculating the radiant emittance from the source. In calculating irradiance, the distance is measured from this aperture, not from some point of the cavity. Gouffé[1] and De Vos[3] discuss the effect of a limiting aperture on effective emissivity, but their treatment does not apply to the thermally isolated type of aperture shown here. For convenience, most blackbody-type sources used for calibration contain a series of apertures so arranged that they can be easily changed. A toolmaker's microscope with a traveling stage can be used to measure the diameter of each aperture in order that its area may be calculated. An optical comparator, which projects an enlarged image of the aperture onto a calibrated screen, is also a convenient means of measuring this area. With experience one finds that apertures made by conventional drilling techniques are rarely round. When the apertures are not round, it is far easier to determine their area with an optical comparator than with a toolmaker's microscope. Experienced workers have found that it is impractical to use apertures having a diameter less than 0.010 in. Particles of dust and dirt lodged in apertures smaller than this can seriously change their area, special techniques are needed to clean them without damage, and it is difficult to measure their area accurately. The accuracy of the calculated areas should be checked radiometrically. This can be done by placing a radiometer at a convenient distance from the source and recording the signal from each aperture. Provided that the radiometer response is linear, a plot of signal versus aperture area should be a straight line. The deviation of individual points from this straight line is a strong indication that an error has been made in determining the area of that aperture.

In Chapter 2 it was noted that a blackbody is a Lambertian, or perfectly diffuse, radiator; as a consequence, its radiant emittance varies with the

---

[1]This result can be estimated from a radiation slide rule in the following way: Since the scales cannot be read for increments of 1°, it is necessary to use a larger increment, such as 50° and to divide the final answer by 50; for example,

$$\text{for } T = 775°K \quad W = 2.04 \text{ W cm}^{-2}$$
$$\text{for } T = 800°K \quad W = 2.32 \text{ W cm}^{-2}$$
$$\text{for } T = 825°K \quad W = 2.62 \text{ W cm}^{-2}.$$

(Since a ratio will be taken, there is no need to multiply these values by the emissivity.) The change in W is 0.58 W cm^{-2} for a 50°K change in the temperature, or an average change of 0.0116 W cm^{-2} °K^{-1}, a value essentially in agreement with that calculated above.

cosine of the viewing angle (measured from the normal). A blackbody-type source such as that shown in Figure 3.4 follows the cosine relationship for no more than 5 or 10 deg from the axis. When such a source is used with an optical system that collects flux over a larger angle than this, the simple relationships among W, N, and J (2-4) no longer hold. The magnitude of the resulting error is difficult to calculate; the angular distribution of the flux from the source must be measured and an integration based on the methods given in reference 14 must be carried out. Many workers simply ignore the error, but for the most accurate calibration work its effect must be considered.

Blackbody-type sources in wide variety are commercially available. Some idea of this variety is evident from Figure 3.5, in which the sources shown have aperture diameters ranging from 0.080 to 3 in., maximum temperatures as high as 1300°K, and weights ranging from a few ounces to over 300 lb. The source with a rectangular aperture is not a cavity type; it has a flat-plate radiator 4.5 in. square that supposedly has an emissivity of 0.9. Similar flat-plate sources are available in sizes up to 12 in. square.

Numerous blackbody-type sources have been described in the literature. However, very few of the descriptions are detailed enough to allow the source to be duplicated. Sources operating at temperatures below 1200°K are described in refs. 15, 16, and 28. Sources operating at temperatures above 1200°K are described in refs. 17–21, 24, 26, 30, 33, and 34. Sources having apertures with areas larger than 10 in^2. are described in refs. 13 and 22.

Figure 3.5 Typical blackbody-type sources (courtesy of Infrared Industries, Inc., Santa Barbara, California).

3.2 STANDARDS FOR SOURCES OF RADIANT ENERGY

In the United States the establishment and maintenance of the primary physical standards for such quantities as length, mass, time, temperature, current, and so on are the responsibility of the National Bureau of Standards (NBS). In turn, most groups making physical measurements either own or have access to local standards that are periodically sent to NBS for comparison with the national standards. Local calibrations made with reference to such NBS-certified standards are said to be "traceable to NBS." The maintenance of traceability therefore serves to link the nation's industrial complex together through the common bond of these national standards. Furthermore, most contracts from the U.S. Department of Defense that involve deliverable hardware require that the contractor maintain traceability of his measurement standards to NBS. Analogous procedures are followed in most other heavily industrialized countries, and many of their national standards are linked by traceability to international standards.

The infrared industry as we know it today depends heavily on the use of blackbody-type sources for calibrating the response of infrared devices. Unfortunately, since NBS (like the corresponding organizations in other countries) does not have a primary standard blackbody-type source, it is impossible to establish traceability in this important area. NBS will not accept such sources for calibration. When queried, NBS will only suggest that the user should purchase an NBS-calibrated thermocouple and use it to measure the temperature of the blackbody cavity. Following such a procedure makes it possible to arrive at a calibration of the radiant emittance (or the radiance) with an accuracy of about 5 per cent. There are those who believe that such accuracy is sufficient for present-day needs, but this author is not one of them. The design and manufacture of infrared equipment are rapidly becoming recognized engineering disciplines rather than a black art practiced by only the chosen few. The inference that such equipment is so crude and poorly understood that there is no need to calibrate with an accuracy of more than 5 per cent simply ignores the true state of the art. Many manufacturers are producing infrared equipment and measuring its response with a *precision* of 1 per cent or less by reference to a locally designated standard blackbody-type source. This may be either a commercially available source or one designed and constructed in their own shops. On the other hand, the *absolute accuracy* of these measurements probably ranges from 3 to 10 per cent. Each organization is forced to establish its own standards, and when disagreements arise between organizations, there is no possible recourse to a common traceability or to a third party to referee the dispute.

SOURCES OF INFRARED RADIATION

Workers in the field of standards use the general rule of thumb that the accuracy of a primary standard should be ten times that desired in the calibrations that are made from it. From the discussion in the previous section, it is evident that presently available theories will permit the construction of a blackbody-type source for which the radiant emittance can be specified with an accuracy of about 1 per cent. This author believes that it is foolish to wait until more accurate sources can be built before establishing a national standard. There is an immediate need to be able to certify calibrations of blackbody-type sources to within 1 per cent accuracy. Thus there is already a need for calibration accuracies equal to the best accuracies now attainable with carefully designed and constructed sources and the ideal 10-to-1 ratio between the two is simply not achievable in the near future. The establishment of a 1 per cent primary standard at NBS and a source-calibration service referenced to it would give the infrared community the traceability it now needs and must have in the future. If we must wait until it is possible to construct a 0.1 per cent primary standard, most of us will no longer be in a position to care.

The inability to show traceability for blackbody-type sources poses an awkward problem for many government contractors. It arises when the cognizant government inspector reports that he has verified the traceability of local standards of length, time, and so on and now wishes to verify the traceability of the blackbody sources used for acceptance testing of infrared equipment. It is rarely sufficient to simply state that traceability cannot be established. One must, instead, be prepared to follow an unconventional approach and attempt to convince the inspector that Planck's law does, in fact, accurately describe the radiant emittance from a blackbody and that the particular blackbody-type source in use is a satisfactory approximation to a blackbody. If the inspector will accept these arguments, the problem is reduced to measurements of the temperature of the cavity and the area of the limiting aperture, measurements that are quite familiar to all inspectors. A failure to gain the inspector's acceptance of the key arguments foreshadows an unpleasant time for all. Those who have been through the experience just described know that it is not nearly as funny as it sounds. It is hoped that the material in this section will be useful to any reader unfortunate enough to have to face this problem.

In 1961 it seemed that the House Appropriations Committee had taken an effective step toward reducing the problem. At that time they approved the establishment by the U.S. Air Force of a maintenance and calibration facility that was to include a capability for infrared calibration. In approving the funds, the Committee stipulated that the facility was to be available to all branches of the Department of Defense and, further, they made it quite clear that the establishment of duplicative facilities elsewhere would be discouraged. This action appeared to be of prime importance because it would, in effect, provide a traceability for blackbody-type sources throughout the Department of Defense and its contractors. This ideal situation has not yet been achieved, and, as a result, each branch of the service provides its own facility for the calibration of blackbody-type sources.[1] This arrangement solves only a part of the problem and only for those who are contractors for the Department of Defense. It provides the promise of a calibration service for blackbody-type sources to be used in acceptance testing and in so doing establishes a referee who can be called upon to resolve disagreements. These facilities have no formal arrangements for intercomparing their standard sources, and thus any traceability they provide is limited to their own branch of the service and its contractors.

In early 1967 it appeared that the Department of Defense was making a concerted effort to use the facilities of NBS to provide a solution to this problem. The key to this new attack is the development of blackbodies in which the core is heated by immersion in a bath of molten metal held at the temperature of the freezing point of the metal. Since the freezing points of most metals are known to a small fraction of a degree, this arrangement offers a convenient means of holding the blackbody core at a precisely known temperature. Tentative plans call for NBS to manufacture a number of these blackbody-type sources for the various Department of Defense calibration laboratories. In addition, NBS is to designate one of its sources as a primary standard and provide a traceable calibration service for the molten metal sources. If all of this can, indeed, be achieved, it will represent a major breakthrough in the infrared calibration art.

The National Bureau of Standards has developed four types of incandescent lamps for use as standards: (1) the Standard of Thermal Radiation, (2) the Standard of Spectral Radiance, (3) the New Standard of Spectral Irradiance, and (4) the New Standard of Total Irradiance.[2] Since they are the only standard sources available in the United States and they do have

---

[1] 2802D Inertial Guidance and Calibration Group, Newark Air Force Station, Newark, Ohio.
Bureau of Naval Weapons Representative, 1675 W. Fifth, P.O. Box 1011, Pomona, California.
Frankford Arsenal, Bridge and Tacoma Streets, Philadelphia, Pennsylvania.

[2] The first two types are now available only from The Eppley Laboratory, Inc., Newport, Rhode Island. The last two types are available from the National Bureau of Standards, Washington, D. C. 20234.

some limited usefulness for the calibration of infrared equipment, it is appropriate to examine their characteristics.

The Standard of Thermal Radiation was developed in 1913 by Coblentz [23] for those who wanted to measure visible light in absolute physical units. It is a seasoned 115 V carbon-filament lamp calibrated in terms of the irradiance produced at a distance of 2 m. The original lamps were calibrated against a blackbody-type source operated at temperatures between 1270 and 1420°K [24, 25]. The estimated experimental errors in these measurements are 0.5 per cent. Several of these lamps have been checked against similar standards maintained by national laboratories in Germany and England, and an agreement within about 0.2 per cent between standards was noted [26]. Gillham [27] states that the standards of the National Physical Laboratory in England differ by 0.5 per cent from those of NBS. Bedford [28] of the National Research Council in Canada has described the construction and calibration of a blackbody-type source intended for use as a standard. After a careful analysis of the errors involved, he concludes that the stated values of radiant emittance from his source are accurate to within 0.3 per cent. When compared with a group of four NBS lamps, variations as large as 1.7 per cent were noted; the average variation was 0.3 per cent. Stair and Johnston [29] report that the normal range of laboratory humidities affects the certified calibration of these lamps by less than 0.3 per cent for the 2 m path normally used. Since the temperature of the filament is about 2000°K, nearly 85 per cent of its radiant emergy is at wavelengths shorter than the 4μ transmission limit of the glass bulb. The small amount of energy observed at longer wavelengths is from the heated glass envelope and not from the filament.

The Standard of Spectral Radiance is a lamp with a tungsten ribbon filament and was first available in 1960 [30]. Values of spectral radiance are given at intervals of 0.05 μ from 0.250 to 0.750 μ and at intervals of 0.1 μ from 0.80 to 2.6 μ. The New Standard of Spectral Irradiance was introduced in 1963 and it is a 200 W quartz-iodine lamp with a coiled-coil tungsten filament [31]. Values of the spectral irradiance are given for the interval from 0.25 to 2.6 μ. The estimated uncertainties in the calibration of both lamps range from 3 per cent at the long wavelengths to 8 per cent at the short wavelengths [26]. The New Standard of Total Irradiance was introduced in 1966 [32]. It is a tungsten-filament lamp and is available in three sizes (100, 500, and 1000 W). These lamps are calibrated against a blackbody-type source together with a quartz plate that limits the flux from the blackbody to the region below 4.5 μ. The new standards appear to be in close agreement with the original Coblentz lamps. The two blackbody-type sources used in the calibration of these lamps appear to be the unofficial standards of NBS[30, 33–35]. It is interesting to note that the methods of Gouffé were used in calculating the effective emissivity of these two sources.

The discussion to this point has been based on the assumption that the best way to make a standard source is to make one that is essentially indistinguishable from a blackbody. Another approach, which has thus far enjoyed little popularity, is to forego the idea of a standard source in favor of an absolute or self-calibrating detector[27, 36, 37]. An example of an absolute detector is a blackened thermopile with fine heater wires attached. Incident radiant energy is absorbed, raising the temperature of the thermopile, and generating a thermal voltage proportional to the temperature rise. The incident energy is then measured by shielding the thermopile and passing a current through its heater wires so as to duplicate the heating produced by the incident energy. Thus the response of an absolute detector can be determined without recourse to any sort of standard source. Such detectors have been used to measure the irradiance from high-quality blackbodies, and the results indicate differences of less than 1 per cent between the calculated values and those given by the absolute detectors[27, 37].

3.3 GENERAL-PURPOSE SOURCES OF INFRARED

Several sources are available for use in system checkout and alignment, spectrometers, communications devices, and solar simulators. In general, the characteristics of these sources are not as well known as those of a blackbody, but their intended applications do not require such detailed knowledge. In addition, such sources are often relatively inexpensive, readily portable, and simple to use.

The Nernst Glower

A Nernst Glower is often found in infrared spectrometers that are used to measure the transmittance, reflectance, or absorptance of various materials. It consists of a relatively fragile cylinder made by sintering a mixture of zirconium, yttrium, thorium, and certain oxides. When cold, it does not conduct, but when it is heated to 400°C by a flame or self-contained tungsten filament, it becomes conductive and can be further heated by passing an electrical current through it. For an average glower, that is, one about 3 cm long and 0.15 cm in diameter, an input of 0.5 amp at 20 V is required after the initial heating. Under these conditions the *effective temperature*[1] of the glower is about 2100°K. Because of the

---

[1]The effective temperature of a non-blackbody source is defined in terms of the temperature of a blackbody having the same spectral radiance, at a particular wavelength, as that of the source.

large negative temperature coefficient of resistance, a current-limiting ballast is needed. The emissivity of the glower varies somewhat with wavelength and has an average value of about 0.6 from 2 to 15 μ.

The Globar

Another source often used with infrared spectrometers is the Globar. It is a rod of silicon carbide, typically 5 to 10 cm long and 0.5 cm in diameter. It is heated to an operating temperature of about 1500°K by an input of 3 to 5 amp at 50 V. Unlike the Nernst Glower, the Globar needs no separate heater since the heating current is passed directly through the silicon carbide rod. The emissivity varies somewhat with wavelength and has an average value of about 0.8 from 2 to 15 μ [38, 39].

The Carbon Arc

A low-intensity carbon arc has been used as a spectrometer source when a greater radiance than that of the Globar or Nernst Glower was needed[40]. A source temperature of about 3900°K is reached. A fivefold decrease in emissivity occurs as the wavelength increases from 2 to 10 μ.

The high-intensity carbon arc, which operates at 5800 to 6000°K, is used in solar simulators. The arc current is three to four times greater than that of the low-intensity arc and the operating life of the electrodes is proportionately less.

The Tungsten Lamp

Tungsten lamps are used as sources, but only for the near infrared since their glass envelopes do not transmit radiant energy beyond 4 μ. Filament temperatures as high as 3300°K can be obtained. The average emissivity of a tungsten filament at 2800°K is about 0.23 from 2 to 3 μ [41,42]. Tungsten lamps provide a solution, although not always a satisfactory one, to the problem of finding a suitable source for field calibration of near-infrared equipment. Since the radiant emittance changes rapidly with changes in filament current, it is imperative that this current be closely monitored during measurements.

Tungsten lamps are surprisingly inefficient sources of visible light. Ten per cent of the input power to a typical 100 W household lamp is radiated beyond the bulb as visible light, 70 per cent is radiated in the near infrared, and 20 per cent is absorbed by the gas in the lamp and by its glass envelope. The glass envelope can readily reach a temperature of 150°C. As a result, equipment operating in the intermediate- and the far-infrared may receive strong signals from tungsten lamps. It is important to note that the signals are from the heated envelope and not the filament, since the spectral distribution is quite different for the two.

The Xenon Arc Lamp

The xenon arc lamp has been used in near-infrared communication systems. Its particular advantage is the ease with which the output can be modulated by varying the current supplied to the lamp[43, 44]. Most of the energy from the xenon arc is radiated in the visible and ultraviolet, but there is a useful output in the near infrared, extending to a wavelength of about 1.5 μ.

The Laser

The laser, an acronym for "light amplification of stimulated emission of radiation," represents an entirely new family of quantum electronic devices[45-48]. Lasers provide coherent sources of extremely high radiance in the portion of the spectrum extending from the ultraviolet to microwaves. The first operating laser was demonstrated in 1960 by Maiman of the Hughes Research Laboratories. Since then, a tremendous amount of developmental effort has been expended on lasers and it is very probable that the full potential of the device is, as yet, only dimly seen.

Probably the first application of the laser in the infrared portion of the spectrum will be for communication systems that can exploit the laser's coherence, high radiance, and the ease with which it can be modulated. Other applications are certain to arise in the future, but at the present we can do no more than speculate on what they may be.

The Sun

For convenience in calculations it is often assumed that the sun radiates as a 5900°K Blackbody. Careful measurements, however, show that for accurate calculations, no single effective temperature can be assumed for the sun since the value appears to decrease with increasing wavelength. Murcray *et al.*[18] have made the most recent measurements and they report an effective solar temperature of 5626°K at 4 μ, 5270°K at 5 μ, and 5036°K at 11.1 μ.[1] Moon[49] has published a set of standard solar radiation curves that are very useful for engineering calculations.

The solar constant, which is the irradiance from the sun measured outside the earth's atmosphere at the mean solar (earth-to-sun) distance, has been measured almost constantly since 1900. Since these measurements are made from the surface of the earth, they must be corrected for atmospheric absorption and scattering. The currently accepted value[50] of the solar constant is 0.140 W cm^{-2} (or, as it is usually stated, 2.00 gm

---

[1]Since the peak of the solar spectral radiance curve is at a wavelength of 0.48 μ, the Wien relationship predicts an effective solar temperature of 6040°K.

SOURCES OF INFRARED RADIATION 85

cal cm^{-2} min^{-1}).[1] The irradiance at the surface of the earth is about two-thirds of this value, or 0.09 W cm^{-2}. Since many infrared systems are designed to detect targets that produce an irradiance of 10^{-10} W cm^{-2} or less, an inadvertent look at the sun may seriously overload or even permanently damage these systems.

3.4 TARGETS

As used here, the term "target" refers collectively to those objects that infrared systems are designed to detect. Although specific details on the radiating characteristics of some targets are classified by the military, reasonably accurate estimates can often be made by applying the radiation laws and certain other information readily available in the open literature.

The Turbojet Engine

The turbojet engine is an important target and, quite conveniently, one from which there is considerable radiant energy because of the large quantity of heat developed in the combustion process. Measurements of the radiation from military turbojets are highly classified; similar measurements of civilian turbojets are, by some inscrutable logic, often given an equally high security classification. Fortunately, it is not too difficult to make order-of-magnitude calculations of the radiation by using only temperature and dimensional information that is readily available in the open literature. To the reader in search of such information, the text by Zucrow[52] is recommended as a comprehensive treatment of the design principles of aircraft and missile propulsion units. From this source, typical values for the factors contributing to the radiation are readily available. The article in the *Encyclopedia Britannica*[53] on jet propulsion is an excellent summary of the principles involved; it, too, gives data on the operating characteristics of typical turbojet engines. The aircraft and missile specifications published annually by *Aviation Week*[54] provide a considerable amount of helpful information, as do the "pilot reports" that appear periodically in the same magazine[55]. Finally, the usefulness of the inexpensive plastic models available in hobby stores should not be overlooked. Their dimensioning is usually excellent and their scale factors are often given in the direction sheet; if not, these factors can be determined with the aid of the specifications given in ref. 54.

The elements of a typical turbojet engine are shown in Figure 3.6. Such

---

[1]Stair and Waters[51] published measurements in early 1967 of the spectral irradiance in the 0.31 to 0.53 μ region that indicate a value of 0.136 W cm^{-2} for the solar constant. Since these measurements were taken on Mauna Loa at an altitude of 11,150 ft, there was relatively little difficulty in correcting them for atmospheric absorption.

86 THE ELEMENTS OF THE INFRARED SYSTEM

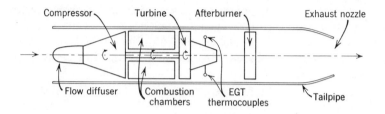

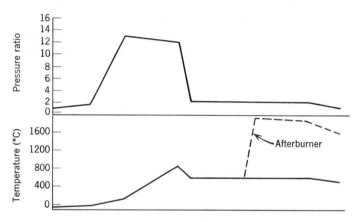

Figure 3.6 The turbojet engine.

an engine consists of a compressor, combustion chambers, turbine, exhaust nozzle and, in some cases, an afterburner. Typical values of temperature and pressure are shown in the lower part of this figure for a turbojet operating at subsonic velocities, both with and without an afterburner. Air to the engine passes a flow diffuser and enters the compressor, which typically has a compression ratio of 10 to 15. As it enters the combustion chambers, the compressed air is mixed with fuel and burns at nearly constant pressure. The hot combustion products pass through the turbine, which extracts enough power from them to run the compressor; the pressure ratio drops to about 2 and the temperature falls about 150°C. Finally, the gases expand to ambient pressure through a nozzle at the end of the tailpipe and produce a high-velocity exhaust stream. There are two sources of radiation from a turbojet engine: the hot metal tailpipe and the stream of hot exhaust gases, often known as the *plume*. For a non-afterburning engine viewed from the rear, the radiation from the tailpipe is far greater than that from the plume. When an afterburner is used, however, the plume becomes the dominant source.

As Figure 3.6 shows, the tailpipe is, in effect, a cylindrical cavity that is

SOURCES OF INFRARED RADIATION

heated by the exhaust gases. In the terminology of Gouffé, the ratio of length to radius (L/r) is in the range of 3 to 8. The tailpipe can therefore be considered a blackbody-type source, and the radiation from it can be calculated from its temperature and nozzle area.

The combined heat and stress limitations of turbine-blade materials require that the temperature of the gases entering the turbine be limited to a maximum of about 900°C. Rather than monitor this temperature directly, it is somewhat easier to monitor the temperature of the gases as they leave the turbine. This quantity, called *exhaust gas temperature* or *EGT*, is one of the most important criteria of engine performance and is always prominently displayed on the pilot's instrument panel. It is measured by a series of thermocouples located a foot or two downstream of the final turbine stage. The limitation on turbine inlet temperature can then be considered as a limitation on EGT. With current engines, an EGT as high as 700°C can be used for short intervals, such as during takeoff; the maximum value that can be sustained over a long flight is from 500 to 600°C, and it may fall to as low as 350 or 400°C during low-speed flight.

As shown in Figure 3.6, the temperature of the gas stream remains nearly constant from the turbine outlet to the exhaust nozzle. The heat-transfer conditions are such that the temperature of the tailpipe wall approaches that of the gas stream. For engineering calculations it can be assumed that temperature of the radiating cavity formed by the tailpipe is equal to the EGT. To be conservative, it is further assumed that the effective emissivity of the tailpipe is 0.9. This assumption tends to compensate for variations in emissivity of the wall, low values of L/r, and small differences between EGT and the actual temperature of the wall. To summarize: *for engineering calculations, a turbojet engine can be considered as a graybody with an emissivity of 0.9, a temperature equal to the EGT, and an area equal to that of the exhaust nozzle.*

Since the principal combustion products in the exhaust plume are carbon dioxide and water vapor, the discussion of molecular emission spectra in section 2.6 is directly applicable to an understanding of plume radiation. The spectrum of a Bunsen flame, shown in Figure 2.10, provides a qualitative estimate of the spectral distribution of the plume radiation. It should be noted that the spectral radiance is higher at $4.4\,\mu$ than it is at $2.8\,\mu$, the ratio being about 3 in Figure 2.10. Measurements of various flames show that this ratio can vary from 2.5 to 10, depending on the fuel used[56]. Thus from the standpoint of detection the emission band at $4.4\,\mu$ is likely to be more useful than that at $2.8\,\mu$ (as will be evident later, it is also more favorable because of reduced interference from sunlight and better atmospheric transmission). The $4.4\,\mu$ emission band of carbon dioxide[57, 58] is shown in more detail in Figure 3.7. The doublet absorption band at 4.25 μ is due to absorption by carbon dioxide in the atmosphere and in the cool layer of gases surrounding the flame. The dashed curve shows the shape of the emission band when corrected for this absorption. Both the pressure broadening of the emission band and its shift toward longer wavelengths are evident when compared with the absorption band, as is done in Figure 3.7; because of the shift to longer wavelengths, only a small fraction of the radiated energy is reabsorbed by the atmosphere. The dotted rectangle is a simple analytical approximation that will be used in Example 3.4.2 for estimating the radiance of the plume.

The temperature of the gas after expansion through the exhaust nozzle is

$$T_2 = T_1 \left(\frac{P_2}{P_1}\right)^{\frac{\gamma-1}{\gamma}} \tag{3-5}$$

where

T_2 = temperature of gas after expansion through exhaust nozzle, °K
T_1 = temperature of gas in tailpipe (essentially EGT), °K
P_2 = pressure of gas after expansion, atmospheres
P_1 = pressure of gas in tailpipe, atmospheres
γ = ratio of specific heats of the gas for constant pressure and constant volume. For combustion products $\gamma = 1.3$.

The ratio of P_2/P_1 for present-day turbojets in subsonic flight is about 0.5.

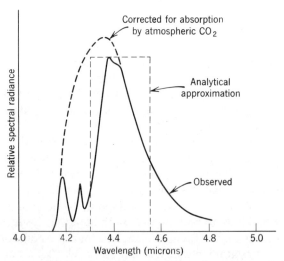

Figure 3.7 The 4.4 μ emission band of carbon dioxide (adapted from Plyler[57]).

SOURCES OF INFRARED RADIATION 89

If expansion to ambient pressure is assumed, (3-5) becomes

$$T_2 = 0.85\, T_1. \qquad (3\text{-}6)$$

Thus the absolute temperature of the plume at the nozzle is about 15 per cent lower than the EGT.

By applying the conclusions of section 2.6, it is evident that the plume radiance depends on the number and temperature of the gas molecules in the exhaust stream. These values depend, in turn, on fuel consumption, which is a function of aircraft flight altitude and throttle setting.. Table 3.1 shows that the fuel consumption of a Pratt and Whitney JT4A turbojet engine is 10,814 lb hr^{-1} at sea level, and 4819 lb hr^{-1} at an altitude of 35,000 ft. Thus the radiance of its plume at an altitude of 35,000 ft is about one-half of the value at sea level.

The Turbofan Engine

Under optimum conditions, the thrust of a jet engine is a function of the mass flow through the engine and the relative velocity of the exhaust gases:

$$F = \dot{m}(v_e - v), \qquad (3\text{-}7)$$

where F is the thrust, $\dot{m}$ the mass flow through the engine, v_e the rearward velocity of the exhaust gases, and v the forward velocity of the engine. Since fuel accounts for only about 2 per cent of the mass flow, $\dot{m}$ is approximately equal to the mass of air passing through the engine. A turbofan engine trades increased mass flow for decreased exhaust velocity. It can be thought of as a turbojet equipped with a fan that draws in considerably more air than a turbojet normally requires. The excess air bypasses the turbojet and is exhausted to the rear so as to produce additional thrust. Thus the thrust of the turbofan engine comes from two sources, the conventional turbojet section and the large mass of excess air from the fan.

How does the radiation from a turbofan compare with that from a turbojet? In most cases it is less, a conclusion that will bring no joy to the heart of an infrared system designer worrying about long-range detection. In essence, it is simply a consequence of the lower EGT of a turbofan engine. The size and temperature distribution of the plume from a turbofan is quite different than that from a turbojet. With a forward fan, that is, one in which the fan is in front of the compressor, the excess air is expelled concentrically about the engine and forms a cold sheath around the plume. As a result the plume is much smaller than that from a conventional turbojet. In an aft-fan engine, that is, one in which the fan is located

90   THE ELEMENTS OF THE INFRARED SYSTEM

downstream from the turbine, some of the excess air is mixed with the hot exhaust gases in the tailpipe so that both plume and tailpipe temperatures are reduced.

The turbofan engine is highly efficient, particularly in subsonic flight. Because of the lower velocity of its exhaust gases it generates less noise than a turbojet does. A turbofan is therefore a very attractive engine for use in commercial transports.

The Boeing 707 Jet Transport

The Boeing 707 series of commercial jet transports are famous the world over. The pertinent characteristics of two models are shown in Table 3.1; the airframes are essentially identical, but one model is equipped with turbojet and the other with turbofan engines. The available thrust is about the same for either type of engine. As we shall see, there are significant differences in the radiation from these two models as well as in the size and shape of their exhaust plumes.

Figure 3.8 shows the Boeing 707-320B with turbofan engines. At the aspect angle shown, only two tailpipes are visible; the other two are obscured by the wing. Thus it is evident why a polar plot of the radiation from a multiengine aircraft shows abrupt changes that occur when individual engines are obscured by the wings, fuselage, or tail. Cutaway

TABLE 3.1 CHARACTERISTICS OF THE BOEING 707 INTERCONTINENTAL JET TRANSPORTS[a]

Model	707-320	707-320B
Engines	4-P&WA JT4A-9	4-P&WA JT3D-3
Type of engine	Turbojet	Turbofan
Maximum rated thrust, lb (per engine)	16,800	18,000
Area of engine exhaust nozzle, cm^2	3,660	3,502
Exhaust gas temperatures:		
Maximum allowable takeoff, °C	635	555
Maximum continuous thrust, °C	515	490
Maximum cruise thrust, °C	485	445
Fuel flow (per engine):		
Sea level, Mach 0.4, lb hr^{-1}	10,814	9,068
35,000 ft, Mach 0.8, lb hr^{-1}	4,819	3,962
Spacing between engines		
Inner engines, ft	66	66
Outer engines, ft	104	104
Maximum speed, mph	585	592

[a]Courtesy of the Boeing Company, Airplane Division, Renton, Washington, and Pratt and Whitney Aircraft, Division of United Aircraft Corp., East Hartford, Connecticut.

SOURCES OF INFRARED RADIATION 91

Figure 3.8 The Boeing 707-320B Intercontinental Jet Transport equipped with turbofan engines (courtesy of the Boeing Company, Airplane Division, Renton, Washington).

views of the Pratt and Whitney JT4A turbojet and JT3D turbofan engines are shown in Figure 3.9. Note the size of the forward fan on the JT3D and the way in which excess air is expelled concentrically about the engine so as to form a cold sheath about the exhaust plume. The tailpipes, which may be almost as long as the engine, are not shown in this figure. The exhaust temperature contours for each type of engine are shown in Figure 3.10 for maximum thrust at sea level. The effect of the fan air in reducing the diameter of the plume is clearly evident, as is the fact that the length of the plume is about the same as that of the aircraft. Therefore it is not surprising that the plume, or a portion of it, is visible from almost any aspect angle.

Since the Boeing 707 jet transports are so widely known, it is appropriate to use them in several examples to show how their radiation characteristics can be calculated.

EXAMPLE 3.4.1. Using the data in Table 3.1 for the Boeing 707-320 turbojet, compute (a) the radiance in the plane of the exhaust nozzle, (b) the radiant intensity of a single engine, (c) the radiant intensity of the aircraft, and (d) the effective radiant intensity in the $3.2-4.8\,\mu$ region.

92   THE ELEMENTS OF THE INFRARED SYSTEM

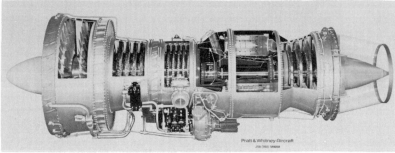

Figure 3.9 The Pratt and Whitney JT4A turbojet and JT3D turbofan engines (courtesy of Pratt and Whitney Aircraft, Division of United Aircraft Corporation, East Hartford, Connecticut).

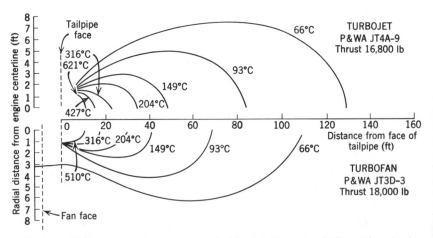

Figure 3.10 Exhaust temperature contours for the turbojet and turbofan engines used on the Boeing 707 (courtesy of Pratt and Whitney Aircraft, Division of United Aircraft Corporation, East Hartford, Connecticut).

SOURCES OF INFRARED RADIATION 93

Assume that the engine operates at maximum cruise thrust, and consider only the thermal radiation from the hot tailpipe.

(a) By assuming that the tailpipe has an effective emissivity of 0.9 and a temperature equal to the EGT, the radiance is found from (2-6) and (2-9):

$$N = \frac{\epsilon \sigma T^4}{\pi} = \left(\frac{0.9}{\pi}\right)(5.67 \times 10^{-12})(485 + 273)^4$$

$$= 0.54 \text{ W cm}^{-2} \text{ sr}^{-1}.$$

(b) The radiant intensity of a single engine can be calculated from (2-5) as
$$J = NA = (0.54)(3660) = 1975 \text{ W sr}^{-1}.$$

(c) If the aircraft is at such a distance that the individual engines cannot be resolved, that is, if all of the engines are simultaneously within the field of view of the infrared sensor, the radiant intensity of the aircraft is the value calculated in (b) multiplied by the number of engines, or 7900 W sr^{-1}. If, as an example, the field of view is 5 mrad (0.29 deg), then the minimum distance at which all four engines fall within the field is view is

$$R = \frac{104}{0.005} = 20{,}800 \text{ ft}.$$

(d) To find the effective radiant intensity in the 3.2–4.8 μ region, use a radiation slide rule to find that 26.6 per cent of the radiant flux lies in this region; thus the effective radiant intensity of each engine is $(1975)(0.266) = 525$ W sr^{-1} or 2100 W sr^{-1} for the entire aircraft.

Table 3.2 summarizes the values calculated in the preceding example as well as the corresponding values calculated for the turbofan version. It is assumed that both aircraft are flying at an altitude of 35,000 ft at a speed of Mach[1] 0.8. It is evident from Table 3.2 that under these conditions, the radiant intensity of the turbofan version is only about three-fourths that of the turbojet. Since the detection range of infrared sensors is proportional to the square root of the radiant intensity, the detection range for the turbofan is about 85 per cent of the range for the turbojet.

It was stated earlier that in subsonic flight the radiant flux from the plume is relatively small compared with that from the tailpipe. The next example will show how that conclusion was reached.

[1]Mach number expresses the ratio of the speed of a body to the speed of sound in the medium through which it passes. Mach numbers less than 1 are called subsonic, those greater than 1 are called supersonic.

TABLE 3.2 SUMMARY OF CALCULATED RADIOMETRIC QUANTITIES FROM EXAMPLE 3.4.1 FOR THE BOEING 707 INTERCONTINENTAL JET TRANSPORTS
(Altitude 35,000 ft, Mach 0.8, maximum cruise thrust)

	707-320 (Turbojet)	707-320B (Turbofan)
For a single engine		
Radiance, W cm^{-2} sr^{-1}	0.54	0.43
Radiant intensity, W sr^{-1}	1975	1505
For the entire aircraft (individual engines not resolved)		
Radiant intensity, W sr^{-1}	7900	6020
Effective radiant intensity, (3.2–4.8 μ), W sr^{-1}	2100	1520

EXAMPLE 3.4.2. Estimate the radiance of the plume for the two models of the Boeing 707 and compare these values with those previously calculated for the thermal radiation from the tailpipe. Assume that the engines operate at maximum cruise thrust.

The temperature of the exhaust gases as they leave the tailpipe can be calculated from (3-6). For the turbojet, this temperature is 370°C (643°K). If only the 4.4 μ emission band of carbon dioxide shown in Figure 3.7 is considered, the radiance can be found by an integration:

$$N = \frac{1}{\pi} \int_{4\mu}^{5\mu} \epsilon(\lambda) \, W_\lambda d\lambda.$$

where $\epsilon(\lambda)$ is the spectral emissivity of the plume gases and W_λ is the spectral radiant emittance of a blackbody, both at a temperature of 643°K.

Evaluation of this integral can be simplified by the use of a radiation slide rule and the assumption that the emission band can be approximated by the dotted rectangular curve shown in Figure 3.7. In effect, this procedure assigns a constant value to $\epsilon(\lambda)$, assumed here to be 0.5, between 4.3 and 4.55 μ, and a value of zero outside this interval. With this change, the integral becomes

$$N = \frac{0.5}{\pi} \int_{4.3\mu}^{4.55\mu} W_\lambda d\lambda.$$

The value of the integral, 3.49×10^{-2} W cm^{-2}, is read from the radiation slide rule and the radiance is found to be 5.6×10^{-3} W cm^{-2} sr^{-1}. Assum-

SOURCES OF INFRARED RADIATION

ing a sensor with a spectral bandpass extending from 3.2 to 4.8 μ, the effective radiance of the hot tailpipe is 0.14 W cm^{-2} sr^{-1} (by combining the results in parts (a) and (d) of Example 3.4.1). The ratio of effective tailpipe radiance to plume radiance for this sensor is $0.14/5.6 \times 10^{-3} = 25$.

The results of the calculations in Example 3.4.2 and those of a similar one for the turbofan are summarized in Table 3.3. It is evident that under full power (but without afterburner) the radiance of the tailpipe is about 25 times that of the plume, even when the value for the plume is calculated for its hottest point, that is, just after the gases leave the exhaust nozzle. The exhaust temperature contours in Figure 3.10 show how rapidly plume temperature decreases with increasing distance from the tailpipe.

When the aircraft is several miles away, so that its engines are no longer resolved, radiant intensity is a more appropriate measure than is radiance. The calculation of the radiant intensity of the plume is extremely difficult since both temperature and emissivity vary in a complex manner throughout its volume. In essence, we must evaluate the integral of Example 3.4.2 over a set of temperature contour curves such as those given in Figure 3.10. Carrying out such a calculation shows that the radiant intensity of the plume averages about 10 per cent of that of the tailpipe (for a system responding between 3.2 and 4.8 μ). Thus it is common in the non-afterburning case to ignore the plume in calculating the maximum detection range. Of course, if the aircraft is viewed from the forward hemisphere or from an aspect angle at which the tailpipes are not visible, the plume is the only source of radiation available.

TABLE 3.3 SUMMARY OF PLUME CALCULATIONS IN EXAMPLE 3.4.2 FOR THE BOEING 707 INTERCONTINENTAL JET TRANSPORTS
(Altitude 35,000 ft, Mach 0.8, maximum cruise thrust)

	707-320 (Turbojet)	707-320B (Turbofan)
For a single engine:		
Temperature of exhaust gases leaving tailpipe, °C	370	337
Radiance of plume, (4.4 μ band only), W cm^{-2} sr^{-1}	0.56×10^{-2}	0.42×10^{-2}
Effective radiance of tailpipe (3.2–4.8 μ) W cm^{-2} sr^{-1} (from Example 3.4.1)	14.4×10^{-2}	10.9×10^{-2}
Ratio of effective tailpipe radiance to plume radiance	25	26

Present day infrared equipment can easily detect and track the Boeing 707 at ranges in excess of 20 miles (32.2 km). Ignoring any absorption by the atmosphere, the irradiance is $7900/(32.2 \times 10^5)^2 = 7.6 \times 10^{-10}$ W cm^{-2} (using the radiant intensity calculated in Example 3.4.1c, equation 2-2, and the distance in centimeters). Such low values of irradiance are difficult to translate into readily comprehensible physical terms. The irradiance at the earth's surface from the sun is, for example, about 0.1 W cm^{-2}, that is, it is about eight orders of magnitude greater than that calculated for the aircraft. Perhaps the following analogy will give the reader a better understanding of the quantities involved. Suppose that we wish to heat water with the radiant energy from the Boeing 707 (while it is at a distance of 20 miles). Assume that 1 mL of water is to be heated and that it is contained in a transparent cubical cell, 1 cm on a side. If the water is exposed for 1 sec to the irradiance calculated above, the temperature of the water will rise about 1.8×10^{-10}°C, a change that could not be measured by conventional thermometry. At this rate of heating, it would require about 175 years to increase the temperature of the water 1°C.

Afterburning

When mixed with the exact amount of air to give complete combustion (stoichiometric mixture), a typical jet fuel, such as JP-4, has a combustion temperature of 2700°C. Since this far exceeds the allowable turbine inlet temperature of 900°C, it must be reduced by mixing considerable excess air with the fuel. As a result, only about one-third of the available oxygen is consumed in the combustion chambers; the rest can be used up by burning additional fuel in the tailpipe. This process, called *afterburning*, results in an increase in thrust; at takeoff, this increase may be 40 per cent, while at Mach 1 it may be as great as 100 per cent. During afterburning, the rate at which fuel is consumed increases drastically, a factor of 5 being typical.

Since afterburning occurs near the end of the tailpipe and no turbine is involved, much higher gas temperatures are allowable; present-day engines have a limit of about 2000°C. Thus when an afterburner is turned on, the temperature and size of the plume increase appreciably. Figure 3.11 shows the exhaust temperature contours of the Pratt and Whitney JT4A turbojet engines at full takeoff thrust with and without an afterburner. Since the afterburner is downstream from the EGT thermocouples, as shown in Figure 3.6, they do not indicate the increase in gas temperature caused by afterburning.

The methods presented for estimating the temperature of the exhaust plume are adequate for engineering calculations up to aircraft speeds of Mach 1.5. At higher aircraft speeds, account must be taken of the *ram*

SOURCES OF INFRARED RADIATION 97

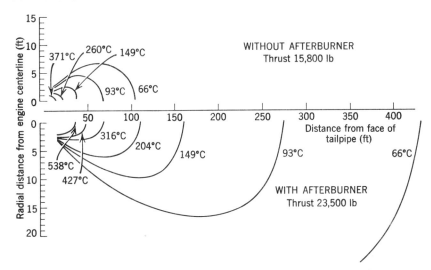

Figure 3.11 Exhaust temperature contours for the Pratt and Whitney JT4A turbojet engines at maximum sea level thrust with and without afterburner (courtesy of Pratt and Whitney Aircraft, Division of United Aircraft Corporation, East Hartford, Connecticut).

effect, the increase, caused by the motion of the aircraft, in the pressure and temperature of the air delivered to the compressor. The ram effect increases rapidly with speed; the pressure ratio at the compressor inlet is about 1.8 at Mach 1, 7 at Mach 2, and 30 at Mach 3. Compressor performance is poor at such high-inlet pressures and limits the pressure ratio in the combustion chambers to about 80. Under these conditions the pressure ratio of the gases leaving the turbine is 30, or about the same as that existing at the inlet to the compressor because of the ram effect. When the net pressure increase across an engine — that is, the difference in pressure measured between the turbine outlet and the compressor inlet — approaches unity, the engine efficiency becomes very low. In addition, as the speed increases, the temperature of the ram air eventually exceeds that which the compressor blades can withstand and the engine is *inlet-temperature limited*, a condition that occurs at speeds of about Mach 2.5.

Equation 3-5 shows that the plume temperature decreases with an increase in the pressure ratio across the exhaust nozzle. For a pressure ratio of 30, such as occurs at Mach 3, the exhaust gas temperature is approximately one-half of what it is at subsonic speeds. A calculation similar to that in Example 3.4.2 shows that this decrease in temperature results in a fivefold decrease in plume radiance. This gives the unexpected conclusion that the temperature and radiance of the plume decrease at high flight speeds. At about Mach 3.5, the plume from the afterburner (an afterburner is assumed simply because it is necessary in order to reach this high speed) is no hotter than that from the same engine operated at maximum power without an afterburner. As a result, at high supersonic speeds the plume radiation from afterburning turbojets becomes small in comparison with that from the hot tailpipe. Fortunately for an infrared system engineer, it is at these same speeds that the aerodynamic heating of the aircraft surfaces provides a significant source of radiation.

Afterburners can be used with either turbojets or turbofans; the advantages and limitations that result are approximately the same with either type of engine. More detailed information on afterburners is given in ref. 52.

The series of examples in this section will be concluded by one showing the method of calculating the radiance of the plume from an afterburning turbojet.

EXAMPLE 3.4.3. In those turbojet engines of the Pratt and Whitney JT4A series that are equipped with afterburners, the temperature of the gases leaving the afterburner (but still in the tailpipe) is limited to about 1850°C (2123°K). Assuming subsonic flight, calculate the radiance of the plume and compare it with the results for the non-afterburning case in Example 3.4.2.

For subsonic flight, equation 3-6 is valid, so that the gas temperature just outside of the nozzle is $(0.85)(2123) = 1804°K$. Use the same method as that used in Example 3.4.2, but because of the higher temperature assume that the emissivity of the gases is 0.8 between 4.3 and 4.55 μ. For these conditions the value of the integral is 1.14 W cm^{-2} and the radiance is 0.29 W cm^{-2} sr^{-1}.

In Example 3.4.2 the plume radiance without afterburner was 5.6×10^{-3} W cm^{-2} sr^{-1}; thus the afterburner caused a 50-fold increase in plume radiance. This new plume radiance is also greater than the effective radiance of the tailpipe calculated in Example 3.4.1. Accordingly, if we integrate over the entire plume in order to estimate its radiant intensity, as would be done if the plume were several miles from the sensor, the radiant intensity will be several times that of the hot tailpipe (given by the product of its radiance and area).

The Ramjet

The operating cycle of a ramjet engine is basically similar to that of a turbojet. There is, however, no compressor ahead of the combustion chambers; advantage is taken of the ram effect to provide compression of the air by the moving engine. At Mach 2 the air entering the engine has

SOURCES OF INFRARED RADIATION 99

already been compressed by a factor of 7. For this reason there is no need for a compressor at this speed or at higher ones. Unfortunately, no thrust is produced at zero velocity and auxiliarly means must be provided to initially accelerate the engine to supersonic speeds and start the ramjet cycle. The efficiency of the ramjet increases rapidly beyond Mach 2.5, a speed at which the efficiency of the afterburning turbojet begins to decrease. Thus the turboramjet appears to be a very efficient combination. At lower speeds the engine operates as an afterburning turbojet. As the speed approaches Mach 2, the inlet air can be bypassed around the turbojet and fed directly into the afterburner, which then operates as a ramjet.

In a ramjet there are no moving parts that are simultaneously subjected to high temperatures and high dynamic loads, as there are in the turbojet. For this reason, combustion temperatures as high as 2000°C can be used. A major problem for the ramjet designer is to find a means of cooling the tailpipe and exhaust nozzle in order that these high combustion temperatures may be sustained for the duration of the flight. Ram air is the principal source of cooling, although the fuel can also be used to absorb some heat from the tailpipe before injection into the combustion chamber. Since the temperature of the ram air increases at high speeds, it eventually becomes useless for cooling, making Mach 4 to 5 the maximum speed at which a ramjet can operate.

The principles described for calculating the radiation from a turbojet can, in general, be applied to a ramjet. Because of the high pressure ratios involved, the exhaust gases from the ramjet become progressively cooler at higher speeds, as they do with the turbojet. The radiation from the hot tailpipe of a ramjet remains relatively constant with altitude and speed. A reasonable estimate of the tailpipe temperature is about 1600°C.

The Rocket Engine

The final member of the thermal-jet engine family is the rocket. It differs from the other engines described, however, in that it requires no atmospheric air because it carries its own supply of both fuel and oxidizer. For this reason the thrust of a rocket is practically independent of speed and environment; a rocket can even operate in a vacuum.

Basically, a rocket engine consists of a combustion chamber and an exhaust nozzle; in a liquid-propellant rocket there must also be means for propellant injection. A variety of propellants are mentioned in the literature that have combustion temperatures ranging from 600 to 4500°C [59]. As with a turbojet, we can estimate the thermal radiation from the hot exhaust nozzle by considering that it radiates as a graybody with an emissivity of 0.9, that it has a temperature equal to that of the exhaust gases, and that it has an area equal to that measured at the exhaust plane of the nozzle. The area of the nozzle is a difficult value to find in the literature; as a last resort, try scaling one of the plastic models available in hobby stores.

Pressures in the combustion chamber range from 15 to 100 atmospheres, depending on the particular propellant and engine. Equation 3-5 can be used to estimate the temperature of the gases as they leave the exhaust nozzle. For a liquid-fueled rocket burning kerosene and liquid oxygen (typical values for use in (3-5) are $T_1 = 3520°K$, $P_2/P_1 = 0.05$, $\gamma = 1.25$), the calculated temperature of the exhaust gases is 1940°K (1667°C). Rosenberg et al.[60] give spectral measurements of the exhaust from an unnamed rocket burning kerosene and liquid oxygen. They report that the radiation from the undisturbed exhaust cone is fairly continuous and approximately matches that of a 2000°K blackbody, a figure lending some credence to the above calculation.

As widely circulated photographs of the firing of the apogee motor of the Hughes Syncom communications satellite clearly show, the plume may become extremely large at the high altitudes reached by a rocket[61]. This motor, which was fired when the satellite was 22,548 miles above the earth, burned for 19.9 sec; the exhaust plume was then 66 miles long and 33 miles wide. Because of its high temperature and large size, the plume is probably the principal source of radiation from a rocket. Seymour[62] discusses the possibility of observing missile plumes with image tubes. He estimates that such plumes can be represented by a 2000°K blackbody source having a radiant intensity of 10^6 W sr^{-1}, and that the radiation from actual missiles is within an order of magnitude either side of this figure. Fontenot[63] describes procedures for calculating the thermal radiation from plumes, but gives no experimental verifications.

Aerodynamic Heating

An object moving at high speed through the atmosphere becomes heated. At speeds above Mach 2, the resulting high temperatures produce sufficient radiation to be of interest to the infrared system designer. This happens to be the same speed regime in which ram air compression starts to reduce the temperature of the exhaust gases from an afterburning turbojet.

When air flows over a body, there is a region near the surface, called the *boundary layer*, in which the flow is affected by its proximity to the surface. The flow in the boundary layer may be either laminar or turbulent. In *laminar flow* the air passes smoothly across the surface. In *turbulent flow* the air suffers a rapid churning or mixing between layers that are at different distances from the surface. In general, the flow over the forward

SOURCES OF INFRARED RADIATION

portion of a body is laminar, but it often becomes turbulent toward the rear of the body.

Any point on a body where the air stream comes to a complete rest is called a *stagnation point*. At such a point, the kinetic energy of the moving stream of air is converted into potential energy in the form of a high temperature and pressure. This temperature, called the *stagnation temperature*, is given by

$$T_s = T_0 \left[1 + r\left(\frac{\gamma - 1}{2}\right) M^2 \right], \qquad (3\text{-}8)$$

where

T_s = stagnation temperature, °K
T_0 = temperature of surrounding atmosphere, °K
r = recovery factor; its value depends on conditions in boundary layer (see text)
γ = 1.4, the ratio of the specific heats of air at constant pressure and at constant volume
M = Mach number.

This relationship is valid up to Mach 10, where dissociation of the air molecules begins. If the value of the recovery factor is equal to unity, (3-8) gives the temperature of the air next to the surface. The transfer of this heat into the body is a complex process strongly dependent on the flow conditions in the boundary layer[64]. If the flow is laminar, the surface reaches the temperature given by (3-8) with the value of r taken as 0.82. For turbulent flow a value of 0.87 should be used.[1]

For laminar flow and stratospheric flight (above an altitude of 37,000 ft), the temperature of an aerodynamically heated surface is found by rewriting (3-8) as

$$T_{\text{lam}} = 216.7(1 + 0.164 M^2). \qquad (3\text{-}9)$$

This equation is plotted in Figure 3.12 (note that the temperature is given in °C). The scale on the right relates the wavelength of peak spectral radiant emittance to the temperature scale on the left. It is helpful to recall that 25 per cent of the radiated energy lies on the short-wavelength side of this maximum. These temperatures are equilibrium values. The time required to reach equilibrium depends on the emissivity and thermal

---

[1] Aerodynamic theory shows that the value of r is given by the square root of the Prandtl number for laminar flow and by the cube root for turbulent flow[64]. The value of the Prandtl number varies slightly with temperature; for our purposes, this variation can be ignored.

conductivity of the surface material, its recent thermal history, and the thermal characteristics of the underlying structure. Surfaces having a high emissivity experience the smallest increases in temperature simply because they can more easily radiate away the heat absorbed from the boundary layer. Also shown in Figure 3.12 are temperatures observed during flights of the X-15[65], the XB-70A[66], and calculated for supersonic transports[67, 68]. That the values observed for the X-15 are lower than those indicated by the curve can be attributed to two causes: all flights lasted less than 10 min and hence thermal equilibrium was probably not achieved; the skin of the X-15 is made of blackened inconel having a high emissivity that tends to reduce its temperature.

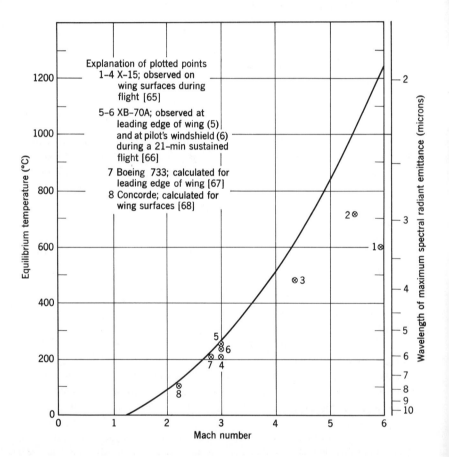

Figure 3.12 Equilibrium surface temperature caused by Aerodynamic Heating (for altitudes above 37,000 ft and laminar flow).

SOURCES OF INFRARED RADIATION 103

Space vehicles reentering the earth's atmosphere convert an enormous amount of kinetic energy into heat, resulting in surface temperatures of 2000°C and more[69].

Personnel

The emissivity of human skin is very high, averaging 0.99 at wavelengths longer than 4μ[70]. It is interesting to note that the value is independent of skin color; our skins are all equally black beyond 2μ[71]. Thus human skin is another excellent example of the inadvisability of estimating the emissivity of a surface on the basis of its visual appearance.

Skin temperature is a complex function of the radiation exchange between the skin and its surroundings. When human skin is exposed to severe cold, its temperature can drop to as low as 0°C. In a normal indoor environment, with an air temperature of 21°C, the temperature of the exposed skin of the face and hands is about 32°C. In order to calculate the radiation from a nude human body, it is necessary to know the radiating area of the body. For analytical purposes, Wissler[72] represents the average male by an assemblage of cylinders having a surface area of 1.86 m² (20 ft²). On the assumption that the skin is a perfectly diffuse radiator, the effective radiating area is equal to the projected area of the body, or about 0.6 m². With a skin temperature of 32°C, the radiant intensity of an average nude male (assuming him to be a point target) is 93.5 W sr^{-1}. At a distance of 1000 ft (if atmospheric absorption is ignored), he produces an irradiance of 10^{-7}Wcm^{-2}. About 32 per cent of this energy lies in the 8 to 13 μ region and only 1 per cent in the 3.2 to 4.8 μ region. The presence of clothing reduces these values since both the temperature and the emissivity of clothing are lower than those of the exposed skin.

Surface Vehicles

Surface vehicles may radiate sufficent energy to be of interest as targets. The paint used on such vehicles usually has an emissivity of 0.85 or greater; weathering and natural deterioration of the paint as well as accumulations of dust and dirt tend to increase the emissivity. Because of their high temperature, exhaust pipes and mufflers may radiate several times as much energy as the rest of the vehicle does. In recent years, designers have learned the importance of keeping mufflers and exhaust pipes well hidden beneath vehicles to limit their detection by infrared systems.

Stars and Planets

Most of the brighter stars are best detected by systems working in the visible or near-infrared portion of the spectrum. Figure 3.13, adapted from the extensive data of Ramsey[73] shows the spectral irradiance at the top of the earth's atmosphere from some of the brighter stars. There are 19 stars that produce a peak spectral irradiance in excess of 10^{-12} W cm$^{-2}\mu^{-1}$. Additional data on stars are given in refs. 74 and 75. Figure 3.14 shows the spectral irradiance from the moon and planets. It includes only self-emitted thermal radiation and not reflected sunlight. Information on the number of stars giving an irradiance above a certain level is found in Figure 3.17.

3.5 BACKGROUNDS

Targets are highly likely to appear in front of some sort of background that will complicate the detection process. Of particular interest are backgrounds such as the earth, the sky, outer space, and the stars and planets.

There is little agreement on the most effective means of describing a background. We might, for example, use an infrared or thermal picture in which the brightness (more properly, the luminance) at any point in the picture is related to the radiance at that point in the scene. Such pictures

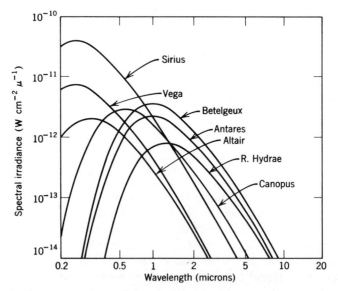

Figure 3.13 Spectral irradiance at the top of the earth's atmosphere from selected stars (adapted from Ramsey[73]).

SOURCES OF INFRARED RADIATION 105

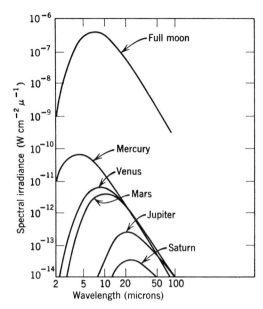

Figure 3.14 Spectral irradiance at the top of the earth's atmosphere from the moon and planets (adapted from Ramsey[73]).

are relatively easy to interpret, but it is difficult to obtain a good differentiation of fine detail with a short exposure time, an accurate correlation between brightness and radiance, and coverage of wide range of temperatures on a single picture. Alternatively, we could present a series of single-line scans across the scene, showing scene radiance as a function of azimuth angle. This type of presentation has been used by Ginsburg et al.[76] to show the radiance of typical urban scenes as well as their variation with time. The disadvantage of this method is the difficulty in correlating the line-scan information with the actual details in the scene. If a suffcent number of line scans are available, it may be valuable to plot a series of isoradiance contours superimposed on an ordinary photograph of the scene.

In recent years considerable effort has been devoted to applying statistical methods to the description of backgrounds. The problem is somewhat analogous to that faced by an electrical engineer in trying to describe by statistical means a recording of electrical noise as a function of time. A common solution is to use the power spectrum of the noise, a plot of the mean-square noise voltage per cycle versus frequency. Since a line scan of a scene is analagous to the record of the noise, we could plot the square of the radiance against wavenumber. No name has been coined for the square of the radiance; it is generally referred to as the Wiener spectrum. Further discussion is beyond the scope of this book; the reader will find detailed information in refs. 77–81.

Because of emission from the atmosphere along the line of sight, the background may actually be three-dimensional rather than two-dimensional. In those wavelength regions where there is little absorption by the atmosphere, dust and haze may attenuate the flux from a distant background and scatter additional flux into the line-of-sight path. At wavelengths at which there is strong absorption by the atmosphere, the line of sight extends only a few feet from the instrument; the radiance of the background is, in reality, just the radiance of this short atmospheric path. Jones has suggested that this radiance generated along the line of sight be called the sterisent [82].

The Earth

It is relatively easy to predict the gross features of the spectral radiance from typical terrains. During the daytime the radiation from the surface of the earth is a combination of reflected and scattered sunlight and thermal emission from the earth itself. The maximum spectral radiance from the sun occurs at about $0.5\,\mu$; that of the earth, radiating as a graybody at 280°K, occurs at about $10\,\mu$. Thus the spectral distribution shows a double peak; that for the short wavelength is due to sunlight and that for the long wavelength is due to thermal radiation from the earth. The minimum between the two peaks occurs at about $3.5\,\mu$. At night, when the solar component is missing, the spectral distribution becomes that of a graybody at the ambient temperature of the earth. Figure 3.15 shows the spectral radiance of snow, grass, soil, and white sand. There is remarkably little difference between these materials; each shows the double-peaked distribution with an intervening minimum in the 3 to $4\,\mu$ region. Comparison of each of these curves with one for a 35°C blackbody shows that these materials radiate as graybodies with an emissivity in excess of 0.9. The curve for white sand shows a dip in the spectral radiance near $9\,\mu$. This is a *restrahlen* or *residual ray* effect and is caused by an enhanced reflectance at a resonant frequency that is characteristic of the particular material. This increase in reflectance is accompanied by a decrease in the emissivity and the result is a reduction in the radiance at the restrahlen wavelength. The gypsum sand shown in Figure 3.15 has a strong restrahlen at $8.75\,\mu$, which makes it possible to identify such a sand at a distance, perhaps even at interplanetary distances. Since the atmosphere is quite transparent at $8.75\,\mu$, it is a poor radiator at this wavelength. The sum of the emission from the sand plus the atmospheric radiation reflected from the sand is not sufficient

SOURCES OF INFRARED RADIATION

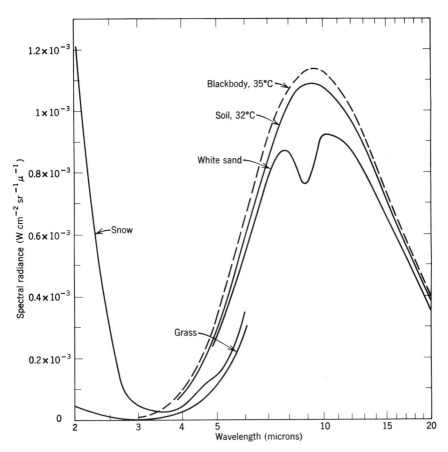

Figure 3.15 Spectral radiance of typical terrain materials as observed during the daytime (adapted from [76], [83], [84]).

to produce a radiance in this spectral region that is equivalent to that of a blackbody at the ambient temperature.

Murcray et al.[85] have reported on the radiance of the earth when viewed from a high altitude. Their equipment used filters to isolate various wavelength bands extending from 1 to 35 μ. It was carried by a balloon to an altitude of about 85,000 feet. Their data is presented in the form of isoradiance plots of the surface of the earth as viewed from the balloon. Block and Zachor report measurements of the spectral radiance of the earth that were made from a satellite[86].

The Sky

Curves of spectral radiance of the sky show features that are grossly similar to those of the terrain, that is, scattered sunlight below 3 μ and thermal radiation beyond. The emissivity of a path through the atmosphere depends on the amount of water vapor, carbon dioxide, and ozone along the path. Thus both the temperature of the atmosphere and the elevation angle of the line of sight must be known in order to calculate sky radiance. Figure 3.16 shows the spectral radiance of a clear night sky and its change with elevation angle [87]. At low elevation angles the path through the atmosphere is very long and the radiance is essentially the same as that from a blackbody at the temperature of the lower atmosphere (8°C in Figure 3.16). At higher elevation angles the path through the atmosphere is shorter and at those wavelengths where the absorption is low the emissivity becomes low. However, in the 6.3 μ water vapor band and the 15 μ carbon dioxide band, absorption is so high that even over a short path the emissivity is essentially equal to unity. The emission at 9.6 μ is due to ozone. These curves are for a night sky. They will be similar during the day except for the addition of scattered sunlight below 3 μ.

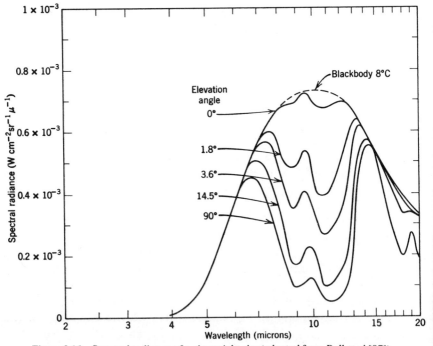

Figure 3.16 Spectral radiance of a clear night sky (adapted from Bell et al.[87]).

SOURCES OF INFRARED RADIATION 109

Sloan et al.[88] have studied the influence that air temperature and humidity at the surface of the earth have on the spectral radiance of a clear sky. Their empirical relationships simplify the calculation of spectral radiance of the sky in the 4 to 15 μ region. Additional measurements of sky radiance may be found in refs. 89–92.

The spectral radiance curve of an overcast sky matches that of a blackbody; little or none of the structure shown between 8 and 14 μ in Figure 3.16 is evident in such a curve. Most overcasts occur at relatively low altitudes, a few hundred to a few thousand feet, and their temperatures are usually within a few degrees of the air temperature at the surface of the earth.

Outer Space

Infrared systems operating outside of the earth's atmosphere will view a background of cold outer space. To a first approximation, the temperature of this background is absolute zero. More accurate calculations [93] that integrate the radiance of the space background over all of outer space (including all of the stars) indicate that the effective temperature is about 3.5°K.

Stars and Planets

When considered as backgrounds, stars and planets are of interest because they may be mistaken for targets. Figure 3.17 shows the number

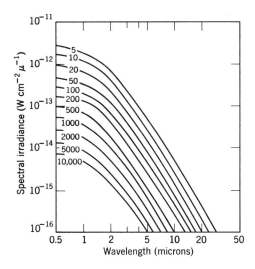

Figure 3.17 The number of stars giving a spectral irradiance at the top of the earth's atmosphere greater than a selected value (adapted from Ramsey[73]).

of stars that will produce a spectral irradiance at the top of the earth's atmosphere in excess of any selected value[73]. If for instance, a particular infrared sensor will respond to a spectral irradiance of 10^{-14} W cm^{-2} μ^{-1} at 2 μ, there are about 1000 stars that it will see. If this system had the same response, but at 5 μ, there are only about 50 stars that it would have to contend with. Detailed information on individual stars and planets is given in Figures 3.13 and 3.14.

REFERENCES

[1] A. Gouffé, "Corrections d'ouverture des corps-noir artificiels compte tenu des diffusions multiples internes," *Rev. optique*, **24**, 1 (1945)
[2] C. S. Williams, "Discussion of the Theories of Cavity-Type Sources of Radiant Energy," *J. Opt. Soc. Am.*, **51**, 865 (1961).
[3] J. C. DeVos, "Evaluation of the Quality of a Blackbody," *Physica*, **20**, 669 (1954).
[4] D. F. Edwards, "The Emissivity of a Conical Blackbody," University of Michigan, Engineering Research Institute Report 2144-105-T, 1956.
[5] E. M. Sparrow, L. U. Albers, and E. R. G. Eckert, "Thermal Radiation Characteristics of Cylindrical Enclosures," *J. Heat Transfer*, **C84**, 73 (1962).
[6] E. M. Sparrow and V. K. Jonnson, "Absorption and Emission Characteristics of Diffuse Spherical Enclosures," *J. Heat Transfer*, **C84**, 188 (1962).
[7] E. M. Sparrow and V. K. Jonnson, "Radiant Emission Characteristics of Diffuse Conical Cavities," *J. Opt. Soc. Am.*, **53**, 816 (1963).
[8] F. J. Kelly and D. G. Moore, "A Test of Analytical Expressions for the Thermal Emissivity of Shallow Cylindrical Cavities," *Appl. Opt.*, **4**, 31 (1965).
[9] P. Campanaro and T. Ricolfi, "Effective Emissivity of a Spherical Cavity," *Appl. Opt.*, **5**, 929 (1966).
[10] P. Campanaro and T. Ricolfi, "New Determination of the Total Normal Emissivity of Cylindrical and Conical Cavities," *J. Opt. Soc. Am.*, **57**, 48 (1967).
[11] E. M. Sparrow, "Radiant Emission Characteristics of Nonisothermal Cylindrical Cavities," *Appl. Opt.*, **4**, 41 (1965).
[12] N. W. Snyder, J. T. Gier, and R. V. Dunkle, "Total Normal Emissivity Measurements on Aircraft Materials between 100 and 800°F," *Trans. Am. Soc. Mech. Engrs,*, **77**, 1011 (1955).
[13] A. La Rocca and G. J. Zissis, "Field Sources of Blackbody Radiation," *Rev. Sci. Instr.*, **30**, 200 (1959).
[14] D. J. Lovell, "The Concept of Radiation Measurements," *Am. J. Phys.*, **21**, 459 (1953).
[15] A. Glaser and A. Ross, "A Calibrated Infrared Signal Generator," *Proc. Nat. Electronics Conf.*, October 1962, p. 228,
[16] A. R. Karoli and J. R. Hickey, "Absolute Calibration Source for Laboratory and Satellite Infrared Spectrometers," *J. Opt. Soc. Am.*, **56**, 1444 (1966).
[17] F. S. Simmons, A. G. De Bell, and Q. S. Anderson, "A 2000°C Slit-Aperture Blackbody Source," *Rev. Sci. Instr.*, **32**, 1265 (1961).
[18] F. H. Murcray, D. G. Murcray, and W. J. Williams, "The Spectral Radiance of the Sun from 4 μ to 5 μ," *Appl. Opt.*, **3**, 1373 (1964).
[19] A. J. Metzler and J. R. Branstetter, "Fast Response, Blackbody Furnace for Temperatures up to 3000°K," *Rev. Sci. Instr.*, **34**, 1216 (1963).
[20] T. J. Quinn, "A Practical Blackbody Cavity for the Calibration of Radiation Pyrometers," *J. Sci. Instr.* **44**, 221 (1967).

[21] H. Y. Yamada, "A High-Temperature Blackbody Radiation Source," *Appl. Opt.,* **6**, 357 (1967).
[22] M. M. Benarie, "Optical Pyrometry below Red Heat," *J. Opt. Soc. Am.*, **47**, 1005 (1957).
[23] W. W. Coblentz, "Measurements of Standards of Radiation in Absolute Value," *Bull. Bur. Standards*, **11**, 87 (1914).
[24] W. W. Coblentz, "Constants of Spectral Radiation of a Uniformly Heated Enclosure, or So-called Blackbody." *Bur. Standards Bull.*, **10**, 1 (1914).
[25] W. W. Coblentz, "Present Status of the Constants and Verification of the Laws of Thermal Radiation of a Uniformly Heated Enclosure," *Bur. Standards Sci. Papers,* **17**, 8 (1921).
[26] H. H. Blau, Jr., and H. Fischer, eds., *Radiative Transfer from Solid Materials,* New York: Macmillan, 1962; see A. G. Maki, "Thermal Radiation Standards and Measurements of the Radiometry Section at the National Bureau of Standards," p. 135.
[27] E. G. Gillham, "Recent Investigations in Absolute Radiometry," *Proc. Roy. Soc.*, **269A**, 249 (1962).
[28] R. E. Bedford, "A Low Temperature Standard of Total Radiation," *Can. J. Phys.*, **38**, 1256 (1960).
[29] R. Stair and R. G. Johnston, "Effects of Recent Knowledge of Atomic Constants and of Humidity on the Calibrations of the National Bureau of Standards Thermal Radiation Standards," *J. Research Natl. Bur. Standards*, **53**, 211 (1954).
[30] R. Stair, R. G. Johnston, and E. W. Halbach, "Standard of Spectral Radiance for the Region of 0.25 to 2.6 Microns," *J. Research Natl. Bur. Standards,* **64A**, 291 (1960).
[31] R. Stair, W. E. Schneider, and J. K. Jackson, "A New Standard of Spectral Irradiance," *Appl. Opt.*, **2**, 1151 (1963).
[32] R. Stair, W. E. Schneider, and W. B. Fussell, "The New Tungsten-Filament Lamp Standards of Total Irradiance," *Appl. Opt.*, **6**, 101 (1967).
[33] G. T. Lalos, R. J. Corruccini, and H. P. Broida, "Design and Construction of a Blackbody and its Use in the Calibration of a Grating Spectroradiometer," *Rev. Sci. Instr.*, **29**, 505 (1958).
[34] Blau and Fisher, *op. cit.* (Reference 26), see J. C. Richmond, "Physical Standards of Emittance and Reflectance," p. 142.
[35] A. G. Maki, R. Stair, and R. G. Johnston, "Apparatus for the Measurement of the Normal Spectral Emissivity in the Infrared," *J. Research Natl. Bur. Standards*, **64C**, 99 (1960).
[36] W. W. Coblentz and W. B. Emerson, "Studies of Instruments for Measuring Radiant Energy in Absolute Value: An Absolute Thermopile," *Bull. Bur. Standards*, **12**, 503 (1916).
[37] M. Eppley and A. R. Karoli, "Absolute Radiometry Based on a Change in Electrical Resistance," *J. Opt. Soc. Am.*, **47**, 748 (1957).
[38] S. Silverman, "The Emissivity of Globar," *J. Opt. Soc. Am.*, **38**, 989 (1948).
[39] J. C. Morris, "Comments on the Measurement of Emittance of the Globar Radiation Source," *J. Opt. Soc. Am.*, **51**, 798 (1961).
[40] C. S. Rupert and J. Strong, "The Carbon Arc as an Infrared Source," *J. Opt. Soc. Am.*, **40**, 455 (1950).
[41] W. E. Forsythe and E. Q. Adams, "Radiating Characteristics of Tungsten and Tungsten Lamps," *J. Opt. Soc. Am.*, **35**, 108 (1945).
[42] R. D. Larrabee, "Spectral Emissivity of Tungsten," *J. Opt. Soc. Am.*, **49**, 619 (1959).
[43] N. C. Beese, "Light Sources for Optical Communication," *Infrared Phys.*, **1**, 5 (1961).
[44] J. K. Buckley, "Xenon Arc Lamps for Modulation," *Illum. Engr.*, **58**, 365 (1963).

[45] A. L. Schawlow, "Advances in Optical Masers," *Sci. Am.*, **209**, 34 (1963).
[46] B. Lax. "Semiconductor Lasers," *Science,* **141**, 1247 (1963).
[47] C. H. Townes, "Production of Coherent Radiation by Atoms and Molecules," *IEEE Spectrum,* **2**, 30 (August 1965).
[48] D. R, Herriott and I. P. Kaminow, eds., "Special Joint Issue on Optical Electronics," *Appl. Opt.*, Vol. 5, October 1966; *Proc. IEEE*, Vol. 54, October 1966.
[49] P. Moon, "Proposed Standard Solar Radiation Curves for Engineering Use," *J. Franklin Inst.*, **230**, 583 (1940).
[50] F. S. Johnson, "The Solar Constant," *J. Meteorol*, **11**, 431 (1954).
[51] R. Stair and W. R. Waters, "The Solar Constant Based on New Spectral Irradiance Data from 3100 to 5300Å," *J. Opt. Soc. Am.*, **57**, 575 (1967).
[52] M. J. Zucrow, *Aircraft and Missile Propulsion*, New York: Wiley, 1958, Vol. 2, chapters 8, 10.
[53] C. R. Soderberg. "Jet Propulsion," *Encyclopedia Britannica*, **13**, 27 (1963).
[54] "Specifications," *Aviation Week*, **88**, 179 (March 18, 1968).
[55] "Pilot Reports" appearing at irregular intervals in *Aviation Week:* "Boeing 720," **73**, 56 (August 15, 1960); "Boeing 720B," **74**, 67 (March 20, 1961); "Convair 880-M," **74**, 48 (May 29, 1961); "Convair 990," **76**, 69 (February 26, 1962); "Douglas DC-8," **72**, 80 (April 4, 1960); "F-104G," **78**, 72 (February 11, 1963); "U-2," **79**, 72 (August 12, 1963).
[56] A. G. Gaydon, *The Spectroscopy of Flames*, New York: Wiley, 1957, p.169.
[57] E. K. Plyler, "Infrared Radiation from a Bunsen Flame," *J. Res. Natl. Bur. Standards*, **40**, 113 (1948).
[58] J. A. Curcio and D. V. Estes Buttrey, "Transmission of Infrared Radiation from Flames by CO_2," *Appl. Opt.*, **5**, 231 (1966).
[59] H. H. Koelle, ed., *Handbook of Astronautical Engineering*, New York: McGraw-Hill, 1961, Chapters 19, 20.
[60] N. W. Rosenberg, W. M. Hamilton, and D. J. Lovell, "Rocket Exhaust Radiation Measurements in the Upper Atmosphere," *Appl. Opt.*, **1**, 115 (1962).
[61] "South African Station's Photo Sequence Shows Firing of Syncom's Apogee Motor," *Aviation Week*, **79**, 30 (August 19, 1963).
[62] M. E. Seymour, "Observing Missile Plumes with Image Tubes," *Electronics*, **34**, 70 (October 20, 1961).
[63] J. E. Fontenot, Jr., "Thermal Radiation from Solid Rocket Plumes at High Altitude," *AIAA J.*, **3**, 970 (1965).
[64] B. E. Gatewood, *Thermal Stresses*, New York: McGraw-Hill, 1957, Chapter 2.
[65] Various articles on the X-15 program, *Aviation Week:* **75**, 52 (November 20, 1961); **75**, 60 (November 27, 1961); **77**, 35 (August 13, 1962); **78**, 38 (June 10, 1963).
[66] C. M. Plattner, "XB-70A Flight Research—Part 2," *Aviation Week*, **84**, 60 (June 13, 1966).
[67] C. M. Plattner, "Variable-Sweep Wing Keynotes Boeing 733 SST Proposal," *Aviation Week*, **80**, 36 (May 4, 1964).
[68] "Size, Speed, Safety of SST are Debated," *Aviation Week*, **80**, 29 (June 1, 1964).
[69] J. V. Becker, "Re-entry from Space," *Sci. Am.*, **204**, 49 (1961).
[70] J. D. Hardy, "The Radiation of Heat from the Human Body," *J. Clin. Invest.*, **13**, 593 (1934).
[71] J. D. Hardy, ed., *Temperature—Its Measurement and Control in Science and Industry*, New York: Reinhold, 1963; see E. Hendler, J. D. Hardy, and D. Murgatroyd, "Skin Heating and Temperature Sensation Produced by Infrared and Microwave Irradiation," p. 211.

SOURCES OF INFRARED RADIATION 113

[72] Hardy, *op. cit.* (reference 71), see E. H. Wissler, "An Analysis of Factors Affecting Temperature Levels in the Nude Human, " p. 603.
[73] R. C. Ramsey, "Spectral Irradiance from Stars and Planets, " *Appl. Opt.*, **1**, 465 (1962).
[74] R. G. Walker and A. D'Agati, "Infrared Stellar Irradiance," *Appl. Opt.* **3**, 1289 (1964).
[75] J. N. Hanson, "An Approximation of the Integrated Infrared Starlight at 2.2 and 3.6 μ over the Sky," *AIAA J.*, **3**, 395 (1965).
[76] N. Ginsburg, W. R. Frederickson, and R. Paulson, "Measurements with a Spectral Radiometer," *J. Opt. Soc. Am.*, **50**, 1176 (1960).
[77] G. Kelton et al., "Infrared Target and Background Radiometric Measurements— Concepts, Units, and Techniques," *Infrared Phys.*, **3**, 139 (1963).
[78] R. C. Jones, "New Method of Describing and Measuring the Granularity of Photographic Materials," *J. Opt. Soc. Am.*, **45**, 799 (1955).
[79] J. A. Jamieson, "Inference of Two-Dimensional Wiener Spectra from One-Dimensional Measurements," *Infrared Phys.*, **1**, 133 (1961).
[80] H. G. Eldering, "Method for the Complete Description of Infrared Sky Backgrounds," *J. Opt. Soc. Am.*, **51**, 1424 (1961).
[81] L. J. Free, "Background Noise Measurements at the Sea Horizon," *J. Opt. Soc. Am.*, **49**, 1007 (1959).
[82] R. C. Jones, "Terminology in Photometry and Radiometry," *J. Opt. Soc. Am.*, **53**, 1314 (1963).
[83] L. Eisner et al., "Spectral Radiance of Sky and Terrain at Wavelengths between 1 and 20 Microns. III. Terrain Measurements," *J. Opt. Soc. Am.*, **52**, 201 (1962).
[84] E. E. Bell and L. Eisner, "Infrared Radiation from the White Sands National Monument, New Mexico," *J. Opt. Soc. Am.*, **46**, 303 (1956).
[85] D. G. Murcray et al., "Optical Measurements from High Altitude Balloons," *Appl. Opt.*, **1**, 121 (1962).
[86] L. C. Block and A. S. Zachor, "Inflight Satellite Measurements of Infrared Spectral Radiance of the Earth," *Appl. Opt.*, **3**, 209 (1964).
[87] E. E. Bell et al., "Spectral Radiance of Sky and Terrain at Wavelengths between 1 and 20 Microns. II. Sky Measurements," *J. Opt. Soc. Am.*, **50**, 1313 (1960).
[88] R. Sloan, J. H. Shaw, and D. Williams, "Thermal Radiation From the Atmosphere," *J. Opt. Soc. Am.*, **46**, 543 (1956).
[89] H. E. Bennett, J. M. Bennett, and M. R. Nagel, "Distribution of Infrared Radiance Over a Clear Sky," *J. Opt. Soc. Am.*, **50**, 100 (1960). An explanatory note appeared in the same journal at **52**, 1305 (1962).
[90] G. A. Wilkins and J. A. Hoyem, "Terrestrial Night Horizon and Sky. 4.3 μ Radiance Data Generated by the HITAB-TRIS Experiments," *J. Opt. Soc. Am.*, **54**, 1409 (1964).
[91] S. D. Drell and M. A. Ruderman, "Infrared Radiation from the Atmosphere Resulting from High Altitude Explosion X-Rays," *Infrared Phys.*, **2**, 189 (1962).
[92] L. D. Gray and R. A. McClatchay, "Calculations of Atmospheric Radiation from 4.2 μ to 5 μ," *Appl. Opt.*, **4**, 1625 (1965).
[93] T. Dunham, "Stellar Spectroscopy," *Proc. Am. Phil. Soc.*, **81**, 277 (1939).

4

Transmission of Infrared Radiation Through the Earth's Atmosphere

From the standpoint of the designer, it is unfortunate that most infrared systems must view their targets through the earth's atmosphere. Before it reaches the infrared sensor, the radiant flux from the target is selectively absorbed by several of the atmospheric gases, is scattered away from the line of sight by small particles suspended in the atmosphere and, at times, is modulated by rapid variations in some atmospheric property.

The general process by which radiant flux is attenuated in passing through the atmosphere is called *extinction*.[1] The transmittance of a path through the atmosphere can be expressed as

$$\tau = e^{-\sigma x}, \qquad (4\text{-}1)$$

where σ is called the *extinction coefficient* and x is the path length. Under most conditions, more than one process contributes to extinction, so that

$$\sigma = a + \gamma, \qquad (4\text{-}2)$$

where a, an *absorption coefficient*, accounts for absorption by the gaseous molecules of the atmosphere, and γ, a *scattering coefficient*, accounts for scattering by gaseous molecules, haze, and fog. Both a and γ can be expected to vary with wavelength.

In the infrared portion of the spectrum, the absorption process poses a far more serious problem than does the scattering process. The spectral transmittance measured over a 6000 ft horizontal path at sea level is shown in Figure 4.1. The molecule responsible for each absorption band, either water vapor, carbon dioxide, or ozone, is shown in the lower part of the figure. The curve can be characterized by several regions of high

---

[1]This usage of the term "extinction" is preferred by workers in the field of meteorological optics. Most physicists and chemists use the term in a more restricted sense.

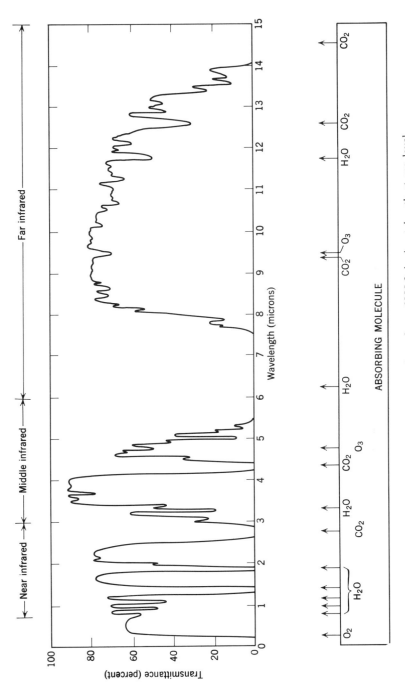

Figure 4.1 Transmittance of the atmosphere for a 6000 ft horizontal path at sea level containing 17 mm of precipitable water (adapted from Gebbie et al.[1]).

transmission called *atmospheric windows* separated by intervening regions of high absorption. The subdivisions of the infrared given in Table 2.1 are also shown; the reason for choosing them should be obvious, since each includes at least one atmospheric window.

Fogs and clouds are strong scatterers and are, in effect, opaque to infrared radiation. As a consequence, infrared systems cannot be considered to have a true all-weather capability. On the other hand, transmission through rain is surprisingly good and should not be overlooked in tradeoff studies of the usefulness of radar and infrared sensors.

It will be our object in this chapter to examine the phenomena affecting the transmission of the atmosphere and to provide methods and supporting data for calculating the effects of these phenomena on system performance.

4.1 THE EARTH'S ATMOSPHERE

For purposes of description it is convenient to divide the earth's atmosphere into four concentric shells, each such region being characterized by a particular manner in which the temperature varies with increasing altitude. The regions are summarized in Table 4.1. The top boundary of

TABLE 4.1 DESIGNATED REGIONS OF THE EARTH'S ATMOSPHERE

Atmospheric Region	Extending From(miles)	To(miles)	Temperature Variation with Increasing Altitude
Troposphere	0	7	Decreases, $2°C$ kft^{-1}
Stratosphere	7	15	Remains constant
Mesosphere	15	50	Increases from 15 to 30 miles, then decreases
Thermosphere	50	~ 5000	Increases from 70 to 150 miles, then remains constant (probably) for several earth radii

each region is named by replacing the suffix *-sphere* with *-pause*. Thus the *tropopause* is located at an altitude of 7 miles (the exact height is a function of latitude and season; at times it may be as low as 4 miles or as high as 11 miles). The region below the stratopause, often called the *lower atmosphere,* is of primary interest because it contains most of the elements detrimental to the transmission of radiation, that is, absorbing molecules, dust particles, fog, rain, snow, and clouds.

Although the various atmospheric properties (pressure, temperature, density, and so on) vary in a complex manner with time and location, the limits are quite well known and it has been possible to establish a series

TRANSMISSION OF INFRARED RADIATION 117

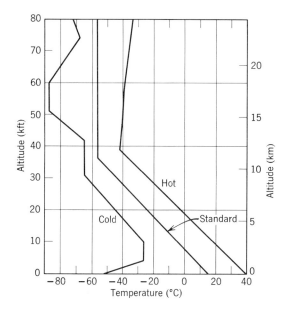

Figure 4.2 Temperature distributions for several model atmospheres (from data in reference 2).

of model atmospheres for aeronautical and scientific purposes[2]. Figure 4.2 shows the variation in temperature with altitude for three such models, a standard, a hot, and a cold atmosphere. Such temperature profiles are useful when estimating the amount of water vapor at a given altitude or calculating the effects of aerodynamic heating.

The atmospheric pressure at a given altitude is just that required to support the weight of the air above it; thus the pressure must decrease as altitude increases. The change in altitude to give a tenfold reduction in pressure is called the *decimal scale height*; for the standard atmosphere this value is about 10 miles. Since molecular absorption bands are broadened by increased pressure, this broadening is most noticeable at sea level.

The gases in the dry atmosphere are shown in Table 4.2. Since their relative proportions are nearly constant up to altitudes in excess of 50 miles, they are often called the *permanent constituents* of the atmosphere. The three most abundant gases, nitrogen, oxygen, and argon, all have symmetrical molecules and produce no absorption in the infrared, a most fortunate situation for those of us who earn our living working with this part of the spectrum. Carbon dioxide, the fourth gas on the list, does,

TABLE 4.2 COMPOSITION OF THE EARTH'S ATMOSPHERE (DRY)

Constituent	Chemical Formula	Percentage by Volume	Absorbs Between 2 and 15 μ
Nitrogen	N_2	78·084	No
Oxygen	O_2	20·946	No
Argon	A	0·934	No
Carbon dioxide	CO_2	0·032	Yes
Neon	Ne	$1·818 \times 10^{-3}$	No
Helium	He	$5·24 \times 10^{-4}$	No
Methane	CH_4	$2·0 \times 10^{-4}$	Yes
Krypton	Kr	$1·14 \times 10^{-4}$	No
Nitrous oxide	N_2O	$5·0 \times 10^{-5}$	Yes
Hydrogen	H_2	$5·0 \times 10^{-5}$	No
Xenon	Xe	$9·0 \times 10^{-6}$	No

however, produce strong absorption in the vicinity of 2.7, 4.3, and 15 μ. Methane and nitrous oxide also absorb in the infrared, but these effects are rarely observed since they are usually present in such small quantities. Near cities and industrial complexes, atmospheric pollutants such as ammonia, carbon monoxide, hydrogen sulfide, and sulfur dioxide are sometimes detectable by their infrared absorption.

The atmosphere contains several other gases, called the *variable constituents*; their amounts vary with temperature, altitude, and location. Chief among them is waver vapor, which may constitute as much as 2 per cent of a very humid atmosphere at sea level. It is evident from Figure 4.1 that most of the atmospheric absorption at sea level is caused by water vapor. The amount of water vapor in the air decreases rapidly with altitude; it is generally assumed that above 40,000 ft the amount present is negligible. Another variable constituent, ozone, is seldom observed at sea level. The amount of ozone increases with altitude, passes through a maximum at about 14 miles and then decreases at higher altitudes. Thus absorption by ozone need be considered only when the line of sight extends into space or from a spaceborne system to the surface of the earth.

In addition to absorption by the gases of the atmosphere, radiant flux may be scattered by the gas molecules or by a variety of particles usually present in the atmosphere. These particles include salt from ocean spray, fine dust blown from the surface of the earth, carbon particles resulting from combustion, and the water droplets and ice crystals that form fogs, clouds, snow, and rain.

4.2 WATER VAPOR

Among the atmospheric constituents, water is the only one that is observed in more than one physical state. It occurs as a solid in snow and ice crystals, as a liquid in rain and clouds, and as a vapor in the atmosphere. As a vapor, it is invisible to the eye but it is readily detected by most infrared systems.

Hygrometry is the branch of physics that is concerned with measuring the amount of water in the atmosphere. An understanding of some of its definitions and of the methods of measurement will make it easier to comprehend the application of meteorological data to the calculation of atmospheric transmittance[3]. *Absolute humidity* is the mass of water vapor contained in a unit volume of air; it is usually stated in grams per cubic meter (g m^{-3}). The amount of water vapor that a sample of air can hold is a function of the temperature of the air. Air holding the maximum possible amount of water vapor is said to be *saturated*. Table 4.3 shows

TABLE 4.3 MASS OF WATER VAPOR IN SATURATED AIR, g m^{-3}

Temperature (°C)	0	1	2	3	4	5	6	7	8	9
−20	0.89	0.81	0.74	0.67	0.61	0.56				
−10	2.15	1.98	1.81	1.66	1.52	1.40	1.28	1.18	1.08	0.98
−0	4.84	4.47	4.13	3.81	3.52	3.24	2.99	2.75	2.54	2.34
0	4.84	5.18	5.54	5.92	6.33	6.76	7.22	7.70	8.22	8.76
10	9.33	9.94	10.57	11.25	11.96	12.71	13.50	14.34	15.22	16.14
20	17.22	18.14	19.22	20.36	21.55	22.80	24.11	25.49	27.00	28.45
30	30.04	31.70	33.45	35.28	37.19	39.19				

the mass of water vapor contained in saturated air for temperatures from −25°C to +35°C. *Relative humidity* is the ratio of the amount of water vapor in a sample of air to the amount that would be required to saturate this sample at the existing temperature. If the relative humidity and temperature are both known, the absolute humidity can be found by multiplying the appropriate value from Table 4.3 by the relative humidity. *Dew point* is the temperature at which a given sample of air becomes saturated. *Mixing ratio* is the mass of water vapor per unit mass of dry air; it is usually given in grams per kilogram (g kg^{-1}). At sea level in the tropics the mixing ratio may be as high as 20 g kg^{-1}. The mixing ratio is a particularly useful specification because it is a function of water vapor content only and does not vary with temperature.

Many techniques are used to measure the water vapor content of the air. Among these are the following.

Evaporative Method. Two thermometers are mounted adjacent to each other with the bulb of one kept wet by a tight-fitting wick dipped in water. Evaporation from this wick lowers the temperature of the wet bulb by an amount that varies inversely with the relative humidity. Tables and simple calculators are available for converting the observed wet and dry bulb temperatures into values of relative humidity.

Chemical Method. A known volume of air is passed through a chemical drying agent, such as silica gel, which is then weighed to determine the amount of water absorbed.

Hygroscopic Method. The length of human hair varies by about 2.5 per cent as the relative humidity changes from 0 to 100 per cent. With a suitable mechanical linkage this change in the length of a hair can be used to drive a pointer over a scale calibrated in terms of relative humidity; its operation is unsatisfactory at temperatures below 0°C. Natural blond female hair makes the best sensing element—one more reason why system engineers prefer blondes!

Electrical Method. The electrical resistance of a strip of lithium chloride is a function of the amount of moisture to which it is exposed. The standard strip found in most radiosondes is about 4 in. long and 2 in. wide. With electrodes placed on the long edges, the resistance varies from about 10 megohms at 15 per cent relative humidity to about 5 kilohms at 100 per cent relative humidity. The time constant is about 1 min. Such strips are useful at temperatures above −30°C; hence water vapor measurements from radiosondes are unreliable at altitudes above 25,000 ft[4].

Dewpoint or Frostpoint Method. A polished metal surface is arranged so that as it is cooled a photocell can detect the instant at which moisture condenses on the surface and a thermocouple can measure the temperature at the same moment. The designations "dewpoint" and "frostpoint" are used interchangeably. If the temperature of the surface is above 0°C, the condensate is water and the temperature is called the dewpoint. If the temperature of the surface is below 0°C, frost forms and the temperature is called the frostpoint. Automatic frostpoint hygrometers have been operated successfully at altitudes up to 50,000 ft, where the frostpoint averages −80°C.

Spectroscopic Method. This method, first proposed by Fowle in 1912, measures the transmittance of an air sample at two different wavelengths. These wavelengths are chosen so that one is at the peak of a water vapor absorption band and the other is one at which no absorption occurs[5, 6]. The ratio of the transmittances at the two wavelengths is a function of the water vapor content of the sample. This method appears to be the most sensitive available and is useful up to altitudes of 90,000 ft.

The absorption of radiation is a function of the number of absorbing molecules in the path. Thus there is need of a quantity to express the amount of water vapor contained along the line of sight. This quantity is called the *precipitable water* and is a measure of the depth of the layer of water that would be formed if all of the water vapor along the line of sight was condensed in a container having the same cross-sectional area as the line of sight. For example, consider an imaginary cylinder whose diameter is equal to that of the system optics and that is long enough to extend from the optics to the target. If the cylinder is placed on end and all of the water vapor is condensed into a container having the same diameter as the cylinder, its depth is a measure of the water vapor content of the path. Notice that is is not necessary to know the area of the cylinder; if the area is increased, more water is condensed but it is spread over a larger area and its depth remains constant.

Figure 4.3 is a convenient means of converting values of relative or absolute humidity to precipitable water content. The precipitable water is given in units of millimeters per thousand feet (mm kft^{-1}), or millimeters per kilometer (mm km^{-1}). In the example given in Figure 4.3, if the temperature is 25°C and the relative humidity is 60 per cent, the absolute humidity is 13.7 g m^{-3} and the precipitable water content is 4.2 mm kft^{-1}. Thus, for a path 10,000 ft long, the total precipitable water is 42 mm. This quantity can also be calculated from the data in Table 4.3, as follows. For a cubic meter of air having an absolute humidity of 1 g m^{-3}, the condensed water would spread over a square 1 m on a side (an area of 10^4 cm^2) and would be 10^{-4} cm (10^{-3} mm) deep.[1] Therefore

$$w' = 10^{-3}\rho, \qquad (4\text{-}3)$$

where w' is the precipitable water expressed in mm m^{-1} of path length and ρ is the absolute humidity expressed in g m^{-3}. The value of ρ can be found by multiplying the appropriate value from Table 4.3 by the relative humidity. For a path length of 1 km,

$$w' = \rho, \qquad (4\text{-}4)$$

where w' is expressed in mm km^{-1}. Similarly, for a path length of 1 kft,

$$w' = 0.3048\rho, \qquad (4\text{-}5)$$

where w' is in mm kft^{-1}.

---

[1] At 4°C, 1 g of water occupies a volume of 1 cm^3. Although this value varies slightly with temperature, this variation can be neglected over the temperature range found in the atmosphere.

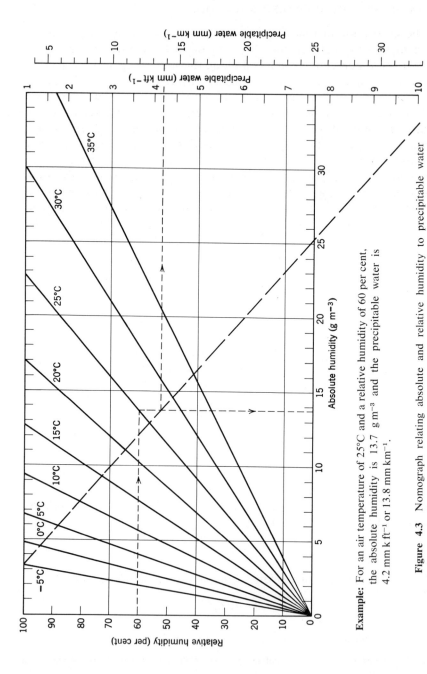

Example: For an air temperature of 25°C and a relative humidity of 60 per cent, the absolute humidity is 13.7 g m^{-3} and the precipitable water is 4.2 mm k ft^{-1} or 13.8 mm km^{-1}.

Figure 4.3 Nomograph relating absolute and relative humidity to precipitable water content at various temperatures.

The water vapor in the air at a particular location shows wide fluctuations that may occur in periods as short as an hour. For well-controlled laboratory tests, this variation presents no problem, because the temperature and humidity along the line of sight can be monitored and the precipitable water calculated by one of the methods given above. For most field situations the exact conditions along the line of sight are unknown and we must use data from other sources. The U.S. Weather Bureau maintains a large network of stations that make radiosonde observations of the atmosphere. Some of these stations make their observations as frequently as every six hours. If such a station is located nearby, it may be useful to obtain the temperature and humidity data taken during the period covered by your tests. If it is necessary to compile statistical data on the weather conditions in a particular area, the National Weather Records Center, Asheville, North Carolina, is an excellent source of raw data. Lacking other data, the curves of Figure 4.4 can be used; they show typical values of precipitable water in a horizontal path as a function of altitude and climatic area. It is evident from these curves that almost all of the water vapor is contained in the troposphere (that is, at altitudes below 35 kft).

There is no general agreement on the exact distribution of water vapor at high altitudes, and we find that there are two equally popular descriptive models. The *dry model* assumes that the mixing ratio decreases steadily with increasing altitude until the tropopause is reached and that it then remains constant at higher altitudes[7]. The *wet model* assumes the same decrease in the mixing ratio up to the tropopause but then supposes that it increases slowly at higher altitudes[8]. Figure 4.5 shows estimates and actual measurements of the water content along a vertical path through the atmosphere. The Gutnick model is based on an analysis of data from all over the world[9]. The other data shown were measured in the United States, Bolivia, England, and Russia[6, 10, 11]. The data from these various sources are in quite good agreement up to an altitude of 30 kft. Above this, there are differences that have led to the wet and dry models. The Gutnick data from Figure 4.5 are redrawn in Figure 4.6 to show the water content of shorter (6.5 kft, 2 km) vertical paths. For slant paths making an angle Z with the vertical, called the *zenith angle,* the water content can be estimated by multiplying the values in Figures 4.5 and 4.6 by secant Z. If the zenith angle is greater than 60 deg, a better estimate of the water content results from using the horizontal path data in Figure 4.4 and multiplying it by secant (90-Z).

It is important that the reader should not confuse the absorption by a given thickness of precipitable water with the absorption by the same thickness of liquid water. A layer of liquid water 10 mm thick is virtually

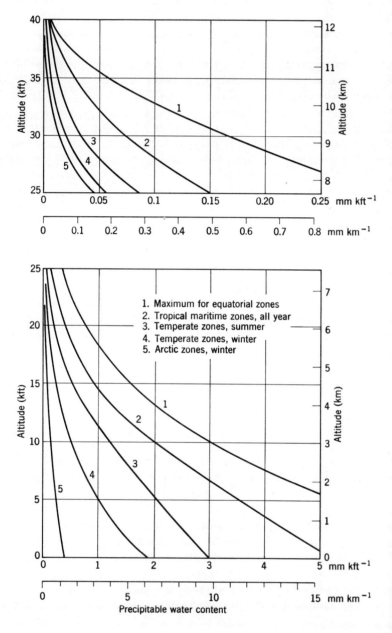

Figure 4.4 Typical values of atmospheric water vapor content along a horizontal path.

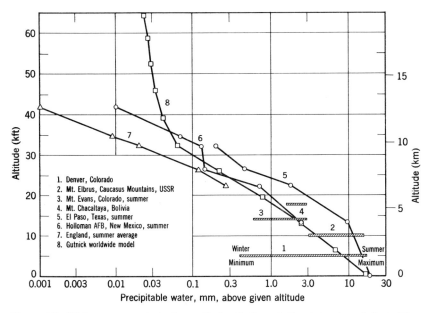

Figure 4.5 Water vapor content of a vertical path through the atmosphere (adapted from [6] [9–11]).

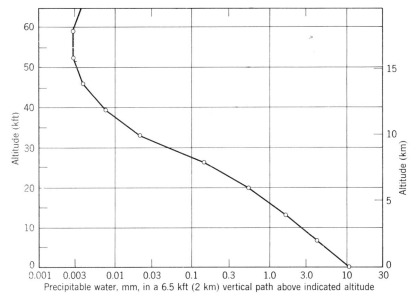

Figure 4.6 Water vapor content of the atmosphere for a 6.5 kft vertical path (adapted from Gutnick[9]).

opaque at wavelengths beyond 1.5 μ[12, 13]. By contrast, the transmittance (within any of the atmospheric windows) of a path containing 10 mm of precipitable water exceeds 60 per cent.

4.3 CARBON DIOXIDE

Carbon dioxide, one of the permanent constituents, makes up about 0.032 per cent of the atmosphere [14]. For the permanent constituents the mixing ratio remains constant up to an altitude of about 30 miles. Thus the decimal scale height for carbon dioxide is the same as that for the atmosphere; a decrease by a factor of 10 for each 10-miles (16 km) increase in altitude. As Figure 4.5 shows, atmospheric water vapor has a decimal scale height of about 3 miles (5 km). Thus, as the altitude increases, the absorption by water vapor decreases much more rapidly than does the absorption by carbon dioxide. As an example, Figure 4.7 shows the atmospheric transmittance in the 2.7 μ region as a function of altitude [15]. These curves were obtained with a balloon-borne spectrometer using the sun as a source. The lowest curve, made at an altitude of 3.9 kft, is dominated by a broad water vapor absorption band that completely obscures bands due to carbon dioxide in this same region. At an altitude of 20 kft the atmospheric water vapor content has decreased to the point that the carbon dioxide bands are barely discernible. At an

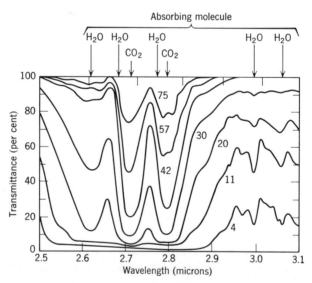

Figure 4.7 Atmospheric transmittance in the 2.7 μ region at various altitudes (in kft) (adapted from Murcray et al. [15]).

altitude of 30 kft the bands due to water vapor have practically disappeared and only those due to carbon dioxide remain. Judging from their intensity in the curve for an altitude of 75 kft, these carbon dioxide bands should be evident up to an altitude of about 150 kft.

The number of gas molecules in the line of sight is expressed in atmosphere centimeters (atm cm), an equivalent path length at standard temperature and pressure that has the same number of gas molecules as does the line of sight. Its value can be calculated from

$$w = \left(\frac{px}{76}\right)\left(\frac{273}{273+T}\right), \quad (4\text{-}6)$$

where w is the absorber content in atm cm, p is the partial pressure of the absorbing gas in cm, x is the length of the absorbing path in cm, and T is the temperature of the gas in °C. The second term, a temperature correction, is usually quite small and may often be ignored. At sea level the value of $p/76$ for carbon dioxide is 3.2×10^{-4}. Hence, for a horizontal path at sea level and a temperature of 273°C, each kilometer of path contains 32 atm cm of carbon dioxide (alternatively, each kilofoot contains 9.8 atm cm).

If all of the atmosphere was converted into a homogeneous layer having constant properties, it would have a thickness of 7.99×10^5 cm[16]. Thus the equivalent thickness of the atmosphere is 7.99×10^5 atm cm. Since carbon dioxide constitutes 0.032 per cent of the atmosphere, the equivalent thickness of the carbon dioxide in a vertical path through the atmosphere is 256 atm cm. As discussed in the previous section, for slant paths we can multiply the vertical-path values by the secant of the zenith angle.

4.4 OTHER INFRARED-ABSORBING GASES

High in the atmosphere the ultraviolet radiation from the sun causes the dissociation of molecular oxygen (O_2) into the atomic form (O). Under certain conditions this atom can recombine with a molecule of oxygen to form a triatomic molecule known as ozone (O_3). At extremely high altitudes no recombination occurs; at low altitudes the intensity of the ultraviolet is reduced and it cannot produce dissociation. Measurements such as those in Figure 4.8 show that the maximum concentration of ozone occurs at an altitude of about 75 kft[17, 18]. Other measurements have shown that the concentration varies with latitude as well as season of the year. In Figure 4.8 the ozone content is given as the number of millimeters of ozone at standard temperature and pressure contained

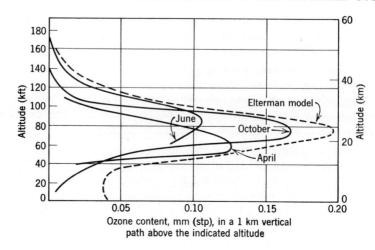

Figure 4.8 Vertical distribution of atmospheric ozone (adapted from [17] and [18]).

along a 1 km vertical path above a given altitude. Integration of these curves gives the effective thickness of atmospheric ozone; values range from 0.1 to 0.4 atm cm with the average being about 0.35 atm cm[18].

The principal infrared absorption band of ozone occurs at 9.6 μ[19]. At sea level its other bands at 4.75 and 14 μ are usually obscured by much stronger bands of carbon dioxide and water vapor. The 9.6 μ band is barely detectable along a 10-mile horizontal path at sea level. Since the concentration of ozone at sea level is about 3 parts per 100 million, such a detection gives some idea of the sensitivity of infrared spectroscopic techniques[20]. The absorption bands of ozone are evident in solar spectra because the line of sight passes through the region of high ozone concentration.

Table 4.3 shows that both methane and nitrous oxide absorb in the infrared. Near industrial complexes, high concentrations of carbon monoxide, ammonia, hydrogen sulfide, and sulfur dioxide sometimes give detectable absorptions. It is unusual for the amounts of these gases to exceed 1 atm cm. The resulting weak absorption is usually obscured by the much stronger bands of carbon dioxide and water vapor. Hence we rarely need to worry about the effect of these miscellaneous gases on atmospheric transmission. The reader faced with unusual concentrations of these gases should consult refs. 21–24, as well as the references they cite.

4.5 FIELD MEASUREMENTS OF ATMOSPHERIC TRANSMISSION

Since it is extremely difficult to derive an analytical expression for the transmission of the atmosphere, it is not surprising that early workers concentrated their efforts on field measurements made over typical paths and under a variety of weather conditions. Many of these measurements are of great value to a system designer who must estimate the performance of his equipment under field conditions.

Most of the early measurements were made by astronomers who needed to know the characteristics of the atmosphere through which they made their observations. This work has been summarized by Goldberg in an excellent review article[25]. The basic technique was to use the sun as a source viewed through the atmosphere. The resulting measurements showed the solar emission curve with the various atmospheric absorption bands (*telluric bands*) superimposed. The separation of solar and telluric effects is simple, at least in principle; the absorption within the telluric bands will change during the day as the zenith angle of the sun varies, whereas effects originating at the sun will remain constant. In the visible and near-infrared region about half of the observed absorption lines and bands are due to the sun; at longer wavelengths most are telluric.

In order to support his measurements of the solar constant, Langley mapped the solar spectrum to 2.8μ in 1881 and extended the measurements to 5.5μ by 1900[26]. His solar spectrum is shown in Figure 4.9. The atmospheric windows, particularly those between 2 and 2.5μ and 3 and 5μ, are clearly evident. Langley was able to identify most of the prominent absorption bands by comparing them with laboratory spectra of known gases. The bands of methane and nitrous oxide evident in his spectra have, however, only recently been identified.

An observant reader, aware that Wien's law predicts a maximum at 0.48μ for a solar temperature of 5900°K, may question the accuracy of the peak found at 1.4μ in Langley's spectrum. A spectrometer of the

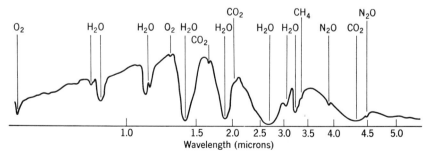

Figure 4.9 The solar spectrum to 5.5 μ (adapted from Langley[26]).

type used for these measurements has the same elements as the spectroradiometer shown in Figure 2.3 and measures the transmittance of a sample by placing it between the source and the entrance slit. In Langley's measurements the sun served as a source and the intervening atmosphere as an absorption cell. His spectrometer used a prism to disperse (spread into a spectrum) the solar radiation, and he then measured the flux passing through an exit slit of fixed width. Since the dispersion of a prism varies with wavelength, the use of a fixed-width exit slit means that the spectral bandpass of such a spectrometer will vary with wavelength. The blackbody spectral distribution curves shown in Chapter 2 are drawn on the basis of the flux contained in a spectral interval of constant width. Langley's curve shows the flux in a constantly varying spectral interval; hence the differences between the two. The use of a variable exit slit, programmed to give a constant spectral bandpass, would shift the apparent peak to $0.48\,\mu$, where we are accustomed to finding it. Barr presents an interesting discussion of early attempts to explain this shift of the spectral peak [27].

By 1917 Fowle[28] had extended transmission measurements of the atmosphere to $13\,\mu$; in 1942 Adel[29] reported measurements to $24\,\mu$. Since that time several workers have extended the measurements to $1000\,\mu$. These measurements, which were made with relatively low resolution, allow one to estimate atmospheric transmittance within any desired wavelength interval *under the conditions prevailing at the time of measurement*. A curve based on data of this type is shown in Figure 4.10.[1] The envelope, a 5900°K blackbody, represents the solar spectral irradiance at the earth's surface in the absence of any atmosphere. The lower curves show the effect of an atmosphere containing 13.7 mm of precipitable water. The effective irradiance within each of the atmospheric windows is also shown. In many applications, sunlight reflected from the background may limit the detection capability of an infrared system; the reduction of such solar effects by the use of a longer-wavelength window is clearly evident. The sum of the effective irradiances for each of the windows is about $903 \times 10^{-4}\,\mathrm{W\,cm^{-2}}$. Since the value of the solar constant is $1400 \times 10^{-4}\,\mathrm{W\,cm^{-2}}$, the atmosphere transmits about 65 per cent of the solar radiation.

More recent measurements of solar spectra have been made with much higher resolution[30–32]. Although they are of great scientific value in the identification of atmospheric constituents, these measurements are not intended to be used in determining the transmittance of the

[1] A similar curve has been in my possession for many years. Although its origin is by now obscure, I believe it was prepared by a group at Eastman Kodak Co. under the direction of E. D. McAlister and that the basic form of presentation was suggested by J. F. Aikins.

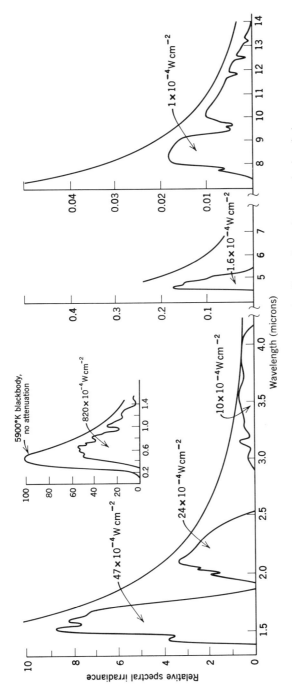

Figure 4.10 A typical solar spectrum and the effective irradiance in each atmospheric window.

atmosphere. Interesting measurements of solar spectra have also been made from high-altitude platforms[10, 33–37].

There are several reports of measurements of the transmittance along horizontal paths near sea level[1, 38–40]. The best-known of these is that of Gebbie et al.[1]; Figure 4.1 shows their measurements of the spectral transmittance at sea level along the east coast of Scotland. The measurements are presented for a path length of 6 kft (1.8 km) with a precipitable water content of 17 mm. For comparison, the transmittance measured in the visible, at a wavelength of 0.61 μ, was 60 per cent. This was used as a control wavelength in a later analysis of the losses caused by scattering along the path.

Taylor and Yates[38] report measurements at sea level in the Chesapeake Bay area over path lengths of 1 kft (0.3 km), 3.4 miles (5.5 km), and 10.1 miles (16.2 km). Their transmission curves are shown in Figure 4.11. Precipitable water contents were 1.1, 13.7, and 52 mm for the respective paths. A spectrometer having relatively high resolution was used in these measurements. Since the curves have not been smoothed, they give a good idea of the complexity of the atmospheric absorption spectrum. The curves presented by Zuev et al.[40] are for path lengths of 3 m, or less, and are useful for estimating transmission losses along the short paths found in the laboratory.

In 1942 Adel[29] reported the existence of a semitransparent window between 16 and 24 μ. The many bands of carbon dioxide, water vapor, and nitrous oxide that are found in this window limit its transmittance. Table 4.4 gives Adel's estimates of the effective transmittance of this window for a vertical path through the atmosphere[41, 42]. This window is obviously of little use to the system designer but it is important in studies of the heat balance of the earth. Farmer and Key[11], working at

TABLE 4.4 EFFECTIVE TRANSMITTANCE OF A VERTICAL PATH THROUGH THE ATMOSPHERE FOR SOLAR RADIATION IN THE 16 TO 24 μ REGION[a]

Spectral Interval (μ)	Precipitable Water (mm)	Effective Transmittance (%)
16–19	10	3.3
16–19	1	14.3
16–22	1	12.6

[a]Adapted from Adel[41].

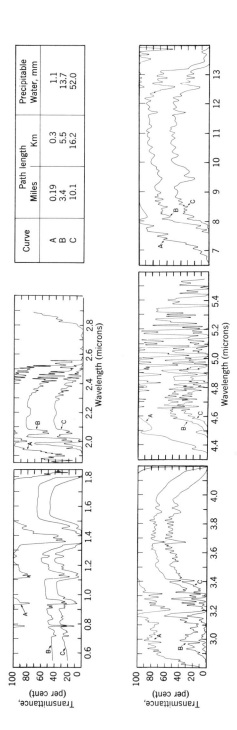

Figure 4.11 Transmittance along several sea-level paths (adapted from Taylor and Yates [38]).

an altitude of 17 kft (5.2 km) on Mt. Chacaltaya, Bolivia, made measurements of the solar spectrum from 7 to 400 μ. Despite the small water content (1.8 mm) above this altitude, they were unable to detect any solar radiation between 40 and 400 μ, except within a small window at 345 μ that showed a peak transmittance of 0.05 per cent[1].

Field measurements are difficult to make, costly, time consuming, and cannot possibly include all likely combinations of weather. Thus it is logical that some workers should have become interested in the possibility of analytical or laboratory simulations of field conditions. Before discussing these efforts, let us review the type of information that the system engineer would like to have about the transmission of the atmosphere. Until the mid-1940's, virtually all infrared detectors were of the thermal type responding from a few tenths to a few tens of microns, so that they spanned all of the atmospheric windows. Sparked by wartime development, a host of new and improved detectors became available in the years following World War II. These detectors were more sensitive than those previously available but, of more fundamental importance, their responses were limited to a much narrower wavelength interval, often only a few microns wide. As designers gained experience with them, it became evident that, in most cases, the response of the detector should be matched to one of the atmospheric windows. This can be done by placing an optical filter over the detector. The reasoning followed is quite simple. At wavelengths where the atmospheric absorption is high, the detector receives no flux from the target, but it does receive flux radiated from the atmosphere. This is evidenced by an increase in the noise output of the detector. The net effect is a degradation in system performance because the increased noise, representing an unwanted signal, is not compensated by a corresponding increase in the signal from the target. As a result, the system designer is vitally interested in the amount of flux *transmitted* through the atmospheric windows. Field measurements readily give this kind of information. Using the curves prepared by Taylor and Yates (Figure 4.11), we can readily study the effect of varying the spectral bandpass of a system; planimetering the area under a given curve, between various wavelength limits, soon shows the point where a further extension of the limits results in a negligible increase in effective transmittance.

Workers involved in laboratory and analytical studies have, for very justifiable reasons, studied the amount of flux *absorbed* by the atmosphere. From the standpoint of a system designer, this emphasis is in the

[1]These measurements should help squelch the perennial rumor that the Russians have discovered a marvelous new window at 47 μ (or some other wavelength) that will make all present equipment obsolete.

wrong place since it is directed at the absorption bands rather than the transmission windows where his system must work. Because most of the investigations of atmospheric transmission since the early 1950's have emphasized laboratory or analytical studies, the system designer has had available only a relatively few new results that are directly applicable to his problems. Quite obviously, however, there are people other than system designers who have benefited from these studies. The greater understanding of the absorption process has been of inestimable value in many other fields.

A paper published in 1953 by Elder and Strong[43] provides a useful transition from the experimental to the analytical approach. They examined the available field measurements to see whether these measurements could be represented by some sort of analytical expression. In order to do this they divided the spectrum into eight windows so that each window extended between the centers of adjacent absorption bands. They were able to show that most of the measurements could be fitted by an equation of the type

$$\tau = -c \log w + t_0, \qquad (4\text{-}7)$$

where τ is the effective transmittance of the window (limits defined in Table 4.5), w is the precipitable water content of the path in mm, and c and t_0 are constants given in Table 4.5. Calculated values of the transmittance through several of the Elder and Strong windows are shown in Figure 4.12 as a function of the water vapor content of the path.

Many factors must be considered in the selection of the exact spectral bandpass for a specific system; these factors are discussed in Chapters 6, 10, 13, and 14. At this point they can be summarized by the statement

TABLE 4.5 WINDOW LIMITS AND CONSTANTS FOR USE WITH EQUATION 4-7[a]

Window	Wavelength (μ)	c	t_0	τ equals 100% if w is less than (mm)
I	0.70–0.92	15.1	106.3	0.26
II	0.92–1.1	16.5	106.3	0.24
III	1.1–1.4	17.1	96.3	0.058
IV	1.4–1.9	13.1	81.0	0.036
V	1.9–2.7	13.1	72.5	0.008
VI	2.7–4.3	12.5	72.3	0.006
VII	4.3–5.9	21.2	51.2	0.005
VIII	5.9–14	(not treated)		

[a]Adapted from Elder and Strong[43].

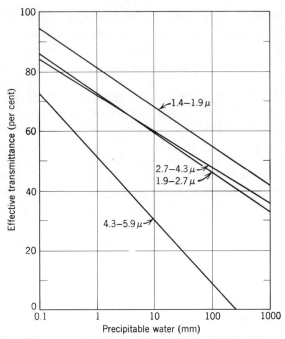

Figure 4.12 Effective transmittance of various atmospheric windows (adapted from Elder and Strong[43]).

that most infrared systems designed to operate within the earth's atmosphere will use one of three spectral bandpasses, extending from approximately 2.0 to 2.5 μ, 3.2 to 4.8 μ, or 8 to 13 μ. An examination of Figure 4.1 or 4.11 shows that these three bandpasses lie in the centers of their respective atmospheric windows and are sufficiently narrow that they include only a minimum of the surrounding absorption bands. Note that the windows adopted by Elder and Strong are a very poor match for these desired bandpasses. Since their windows are wider and have been chosen to include half of the absorption band on each side, Elder and Strong's predicted transmittance values are lower than those that will be observed in the field with equipment using the narrower bandpasses mentioned above. Thus the results of Elder and Strong, although in an extremely useful form, are of limited value to the system designer.

4.6 LABORATORY AND ANALYTICAL METHODS OF PREDICTING ATMOSPHERIC TRANSMISSION

When radiant energy interacts with matter, the energy may be reflected, scattered, or absorbed. For the moment, our only interest is in the absorption process. If a parallel beam of monochromatic flux enters an absorbing layer, the flux at any point within the layer is given by

$$P = P_0 e^{-ax}, \qquad (4\text{-}8)$$

where P is the flux remaining after the beam has traversed a path length of x cm, P_0 is the flux entering the absorbing layer, and a is the absorption coefficient in cm^{-1}. This relationship, originally stated by Bouguer, was later rediscovered by Lambert and is commonly called the *Bouguer-Lambert* law. It states that layers of equal thickness will absorb equal fractions of the flux passing through them. Thus, if one-half of the flux is absorbed after passing through 1 cm of absorber, passage through a second centimeter will again remove one-half of it and leave one-fourth of the original flux. The absorption coefficient usually varies with wavelength (or with wavenumber). It is convenient to remember the physical significance of the absorption coefficient by noting from (4-8) that a is the reciprocal of the length of the absorbing path necessary to reduce the incident flux to $1/e$, or 37 per cent, of its intial value. Beer noted that the absorption coefficient is proportional to the concentration of the absorber in a liquid or to the pressure in a gas, a result commonly known as *Beer's law*.

If the absorber is a solid or a liquid, some flux will be lost by reflection from the surfaces. The absorption laws can only be used to describe the conditions within the body of the absorber. Thus the *internal spectral transmittance*

$$\tau'(\lambda) = \frac{P}{P_0} = e^{-a(\lambda)x} \qquad (4\text{-}9)$$

is a function of the absorption characteristics of the layer and does not include any losses due to reflection at the surfaces. The absorptance of the layer is

$$A(\lambda) = 1 - \tau'(\lambda) = 1 - e^{-a(\lambda)x} \qquad (4\text{-}10)$$

or, in terms of wavenumber,

$$A(\nu) = 1 - \tau'(\nu) = 1 - e^{-a(\nu)x}. \qquad (4\text{-}11)$$

The various atmospheric transmission curves shown earlier were measured with instruments having limited spectral resolution. Figure 4.13 shows portions of the absorption bands of water vapor and carbon dioxide measured with high-resolution spectrometers[44]. These bands consist of many closely spaced absorption lines; the spacing and amplitude

138   THE ELEMENTS OF THE INFRARED SYSTEM

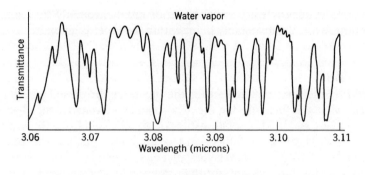

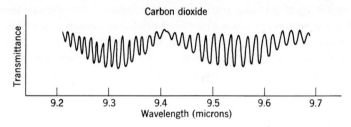

Figure 4.13 High-resolution spectra of water vapor and carbon dioxide (carbon dioxide adapted from Barker and Adel[44]).

appear to be much more random for water vapor than for carbon dioxide. Close examination shows that many lines are just barely resolved, even under the high resolution employed here. Comparison with the earlier spectra in this chapter shows that the true shape of an absorption band is a function of the resolution, or spectral bandpass, of the spectrometer used to record the spectrum.

The truth of the earlier statement that the absorption coefficient varies with wavelength should be obvious from Figure 4.13. How can the absorption law be applied in such a situation? Basically there are two approaches. The first attempts to smooth out the fine detail and examine the effect of the entire absorption band. This has been called the *in toto* approach. The second method attempts to derive analytical models to represent the spectra and then to apply the absorption law to these models. It is called the *spectral model* approach.

Returning now to the in toto approach, the discussion of the absorption law assumed monochromatic radiation. Any physically realizable spectrometer, no matter how high its resolution, has a finite, rather than a zero, spectral bandpass, and thus its output is not truly monochromatic. Theory shows, and indeed, it should be obvious from Figure 4.13, that the absorption coefficient can change rapidly within even a very small spectral interval. The result is a variation of the absorption coefficient within the bandpass of the spectrometer. As a result, the absorptance calculated from (4-10) or (4-11) depends on the characteristics of the spectrometer used to measure the absorption coefficient. Since the bandpass of most spectrometers is determined by the width of the exit slit, truly monochromatic radiation requires a zero-width exit slit, a rather useless accessory for the practical experimenter. The distribution of flux across the exit slit can be described by a slit function. There have been many studies of the relationship between the observed shape of an absorption band and this function. Nielsen et al.[45] have examined the conditions under which the absorptance calculated from (4-10) and (4-11) is independent of the form of the slit function. They show that for an entire absorption band lying between the wavenumber limits ν_1 and ν_2,

$$\int_{\nu_1}^{\nu_2} A'(\nu)d\nu = \int_{\nu_1}^{\nu_2} A(\nu)d\nu = A, \qquad (4\text{-}12)$$

where $A'(\nu)$ is the value of the spectral absorptance measured with a spectrometer having an arbitrary slit function. Physically the integral on the left represents the area under the curve of absorptance versus wavenumber recorded by the spectrometer. Its value is easily found with a planimeter. The integral on the right is the true absorptance, the area under the absorptance curve measured with an ideal spectrometer, and is called the *total absorptance*. The limits ν_1 and ν_2 should, in theory, extend from zero to infinity. As a practical matter, they are chosen by inspecting the individual curves to see that there is no appreciable absorptions outside of the interval $(\nu_2 - \nu_1)$.

The total absorptance has the character of an equivalent bandwidth, a wave number interval over which $A(\nu)$ is equal to unity. Thus the objective of smoothing out the fine detail in the absorption band is accomplished; in its place there is an equivalent band having a simple rectangular shape with the absorption being complete within the band and zero outside. As a result, the use of the in toto method forces us to abandon our interest in transmission windows in favor of absorption bands.

The idea of an equivalent bandwidth is interesting, but alternatively we can calculate an *average absorptance* $\bar{A}$ for the wavenumber interval used in the integration

$$\bar{A} = \frac{\int A(\nu)d\nu}{(\nu_2 - \nu_1)}. \qquad (4\text{-}13)$$

Physically this means that of the flux within the interval $(\nu_2 - \nu_1)$, the fraction $\bar{A}$ will be absorbed.

The heart of the experimental apparatus used by Howard, Burch, and Williams[46] in applying the in toto approach was a multiple-traversal absorption cell of the type first described by White[47]. By an ingenious arrangement of three mirrors, the flux is made to pass back and forth through the cell before it finally emerges and enters a spectrometer. This particular cell was 72 ft long and was usable with up to 68 traversals, resulting in an absorption path length of nearly 1 mile (1.6 km). With this cell it was possible to study the effect of path length, concentration, temperature, and partial pressure on the absorption characteristics of the atmospheric gases. A synthetic atmosphere was made by admitting known amounts of absorber into the cell along with an infrared-transparent gas (nitrogen) to establish the total pressure desired. In this way, pressure-broadening effects could be realistically simulated.

Thousands of absorption spectra were obtained under a wide variety of conditions. The records, which were published in the original reports, have long been one of the most important sources of this type of data. Analysis of these showed that the variation in total absorptance did not follow the logarithmic form used by Elder and Strong for all concentrations of absorber. Instead, two regions were found: a *weak band* region for low absorber concentrations, where the total absorptance was proportional to the square root of the amount of absorber in the path, and a *strong band* region in which the total absorptance was proportional to the logarithm of the amount of absorber in the path. When both pressure and concentration were varied, the total absorptance followed expressions of the type

$$\text{(weak band)} \quad A = \int_{v_1}^{v_2} A(v)dv = cw^{1/2}(P+p)^q \qquad (4\text{-}14)$$

$$\text{(strong band)} \quad A = \int_{v_1}^{v_2} A(v)dv = C + D \log w + Q \log (P+p), \qquad (4\text{-}15)$$

where c, q, C, D, and Q are constants determined empirically for each band and their values are given in Table 4.6, w is the absorber concentration, and P and p are the total and partial pressures. By use of these equations we can calculate the total absorptance for virtually any conceivable set of atmospheric conditions. Burch *et al.*[48–51] have extended the earlier work to cover carbon dioxide, water vapor, nitrous oxide, carbon monoxide, and methane. But, as with the data of Elder and Strong, it is extremely difficult to interpret any of these findings in terms of the transmission windows where our systems operate.

Turning now to the spectral model approach, suppose that an absorption band, such as one of those shown in Figure 4.13, is divided into a series of intervals each of which is a fraction of a wavenumber wide. If the value of the absorption coefficient is known for each interval, we can,

TABLE 4.6 CONSTANTS FOR USE IN EQUATIONS 4-14 AND 4-15[a]

Band μ	Band Limits (cm^{-1})	Weak Band $A = cw^{1/2}(P+p)^q$		Transition[b] (cm^{-1})	Strong Band $A = C + D \log w + Q \log(P+p)$		
		c	q		C	D	Q
Water Vapor							
6.3	1150–2050	356	0.30	160	302	218	157
3.2	2800–3340	40.2	0.30	500	—	—	—
2.7	3340–4400	316	0·32	200	337	246	150
1.87	4800–5900	152	0.30	275	127	232	144
Carbon Dioxide							
15	550–800	3.16	0.44	50	−68	55	47
5.2	1870–1980	0.024	0.40	30	—	—	—
4.8	1980–2160	0.12	0.37	60	—	—	—
4.3	2160–2500	—	—	50	27.5	34	31.5
2.7	3480–3800	3.15	0.43	50	−137	77	68
2.0	4750–5200	0.492	0.39	80	−536	138	114

[a] Adapted from Howard, Burch, and Williams [46].
[b] Use weak band expression for values of A less than value given in the transition column and the strong band expression for values of A that are greater.
w is expressed in precipitable cm of H_2O and atm cm of CO_2.
All logarithms are to the base 10.
P and p are the total and partial pressures, respectively, expressed in mm Hg. For atmospheric CO_2, $p = 32 \times 10^{-4} P$; therefore $(P+p) \approx P$. For water vapor under average conditions, $p = 0.01\ P$; even under the highest humidities it is rare for p to exceed $0.05\ P$; thus $(P+p) \approx P$.

in theory, apply the absorption law to each interval and calculate the absorption throughout the entire band. Unfortunately there is no way to accurately measure the absorption coefficient over the necessarily small wavenumber interval. Instead, we would have to identify and calculate the effect of each molecular transition giving rise to an absorption line, taking into account the exact shape and strength of each line and the effects of any interactions. The number of calculations involved becomes astronomical, and for many years the only apparent solutions was to use simplified models for the calculation. Because of the recent widespread availability of computers, it has for the first time been possible to consider performing the entire calculation.

There is an extensive body of literature on the subject of spectral models. Since it concentrates on the characteristics of the absorption bands, the system designer finds it subject to the same limitations as the data of Elder and Strong and those of Howard, Burch, and Williams. The reader wishing to pursue the subject further will find refs. 52–60 helpful, as are several review papers [61, 62]. Of particular interest are the tables published by Wyatt, Stull, and Plass [63, 64]. These are the result of a computer calculation and represent one of the few cases in which the spectral model approach yields results applicable to the transmission windows.

4.7 TABLES OF ATMOSPHERIC TRANSMISSION DATA

Some workers have presented their experimentally determined transmission data in convenient tables. Among the most useful are those of Passman and Larmore [65]. They examined the available measurements and derived a set of tables showing the spectral transmittance at 0.1 μ intervals from 0.3 to 7 μ for a wide range of absorber concentrations. These are reproduced on the following pages; Table 4.7 shows the spectral transmittance of water vapor for amounts varying from 0.1 to 1000 precipitable mm, and Table 4.8 shows the spectral transmittance of carbon dioxide for path lengths of 0.1 to 1000 km (328 to 3.28×10^6 ft), thus eliminating the need to calculate the number of atmosphere centimeters. It is important to note that although a spectral model was used, it was modified to fit the experimental data and served primarily as a means of scaling the data to other absorber concentrations. The particular spectral model used, that suggested by Elsasser [52], assumes that the variation of transmittance with absorber concentration can be described by an error-function relationship.

The author has always found these tables to be extremely useful in his calculations. His only criticism is that they stop short of the important 8 to 13 μ window. Several years ago, the author extended these tables from 7 to 14 μ. The results are given in Table 4.9 for water vapor and Table 4.10 for carbon dioxide. The format is the same as that used by Passman and Larmore, but the range of absorber concentrations is slightly less. These tables are no contribution to the art of spectral models, but they do accurately represent field measurements in the 7 to 14 μ region.

The value of the transmittance at any wavelength is the product of the individual transmittances read from the tables for water vapor and carbon dioxide,

$$\tau = (\tau_{H_2O})(\tau_{CO_2}). \quad (4\text{-}16)$$

These tables apply only to a horizontal path at sea level. At higher altitudes the absorption lines and bands become narrower because of the reduced pressure, and we would expect the transmittance over a fixed path length to increase; that it does increase is clearly shown in Figure

4.7. The accompanying decrease in temperature also causes a slight increase in the transmittance. Since the temperature effect is very small, it is usually ignored. Passman and Larmore have shown that if a rather simple correction is made, these tables can be used for higher altitudes. They show that the spectral transmittance of a path having a length x at an altitude h is equal to that of an *equivalent sea level path* having a length x_0 given by

$$x_0 = x\left(\frac{P}{P_0}\right)^k, \qquad (4\text{-}17)$$

where P/P_0 is the ratio of atmospheric pressure at altitude h to that at sea level. The exponent k has a value of 0.5 for water vapor and 1.5 for carbon dioxide. Table 4.11 gives values of the correction factor $(P/P_0)^k$ for altitudes up to 100 kft (30.5 km). The concept of an equivalent sea-level path is a useful one in transmittance calculations; it is simply the path length at sea level that has the same transmittance as a longer path at a higher altitude.

A series of atmospheric transmittance curves calculated from these tables is shown in Figure 4.14. The curves were calculated for an altitude of 15 kft (4.6 km) and an assumed precipitable water content of 0.75 mm kft^{-1}. The path lengths shown in Figure 4.14 arise from the use of the water vapor contents heading the columns of Tables 4.7 and 4.9. Examination of Figure 4.14 shows the effect of increasing the path length on the shape of the transmission windows. In the center of the window, for example at 2.2 or 3.8 μ, there is very little change as the length of the path is increased. The obvious change occurs at the sides of the windows; as the path becomes longer, the windows become narrower. Thus a system designed to work at short ranges only might well utilize a wider spectral bandpass than would a similar system designed to work at very long ranges.

For a preliminary analysis of system performance, the designer needs to know the effective atmospheric transmittance for the spectral bandpass of his system. If curves such as those in Figure 4.14 have been prepared, the effective transmittance can be measured with a planimeter or calculated by numerical integration. Figure 4.15 shows the effective transmittance along a horizontal path at sea level for three typical spectral bandpasses, as a function of the water content of the path. These curves were derived from a variety of experimental measurements and may be used with confidence for a preliminary analysis of system performance.

Although a few other tables of atmospheric transmission have been prepared, most of them either give limited coverage or are not readily available. Using the sun as a source, Gates and Harrop[36] measured the

Water Vapor 0.3–$2.4\ \mu$

TABLE 4.7 SPECTRAL TRANSMITTANCE OF WATER VAPOR FOR A HORIZONTAL PATH AT SEA LEVEL[a]

Wavelength (μ)	Water Vapor Content (Precipitable Millimeters)												
	0.1	0.2	0.5	1	2	5	10	20	50	100	200	500	1000
0.3	0.980	0.972	0.955	0.937	0.911	0.860	0.802	0.723	0.574	0.428	0.263	0.076	0.012
0.4	0.980	0.972	0.955	0.937	0.911	0.860	0.802	0.723	0.574	0.428	0.263	0.076	0.012
0.5	0.986	0.980	0.968	0.956	0.937	0.901	0.861	0.804	0.695	0.579	0.433	0.215	0.079
0.6	0.990	0.986	0.977	0.968	0.955	0.929	0.900	0.860	0.779	0.692	0.575	0.375	0.210
0.7	0.991	0.987	0.980	0.972	0.960	0.937	0.910	0.873	0.800	0.722	0.615	0.425	0.260
0.8	0.989	0.984	0.975	0.965	0.950	0.922	0.891	0.845	0.758	0.663	0.539	0.330	0.168
0.9	0.965	0.951	0.922	0.890	0.844	0.757	0.661	0.535	0.326	0.165	0.050	0.002	0
1.0	0.990	0.986	0.977	0.968	0.955	0.929	0.900	0.860	0.779	0.692	0.575	0.375	0.210
1.1	0.970	0.958	0.932	0.905	0.866	0.790	0.707	0.595	0.406	0.235	0.093	0.008	0
1.2	0.980	0.972	0.955	0.937	0.911	0.860	0.802	0.723	0.574	0.428	0.263	0.076	0.012
1.3	0.726	0.611	0.432	0.268	0.116	0.013	0	0	0	0	0	0	0
1.4	0.930	0.902	0.844	0.782	0.695	0.536	0.381	0.216	0.064	0.005	0	0	0
1.5	0.997	0.994	0.991	0.988	0.982	0.972	0.960	0.944	0.911	0.874	0.823	0.724	0.616
1.6	0.998	0.997	0.996	0.994	0.991	0.986	0.980	0.972	0.956	0.937	0.911	0.860	0.802
1.7	0.998	0.997	0.996	0.994	0.991	0.986	0.980	0.972	0.956	0.937	0.911	0.860	0.802
1.8	0.792	0.707	0.555	0.406	0.239	0.062	0.008	0	0	0	0	0	0
1.9	0.960	0.943	0.911	0.874	0.822	0.723	0.617	0.478	0.262	0.113	0.024	0	0
2.0	0.985	0.979	0.966	0.953	0.933	0.894	0.851	0.790	0.674	0.552	0.401	0.184	0.006
2.1	0.997	0.994	0.991	0.988	0.982	0.972	0.960	0.944	0.911	0.874	0.823	0.724	0.616
2.2	0.998	0.997	0.996	0.994	0.991	0.986	0.980	0.972	0.956	0.937	0.911	0.860	0.802
2.3	0.997	0.994	0.991	0.988	0.982	0.972	0.960	0.944	0.911	0.874	0.823	0.724	0.616
2.4	0.980	0.972	0.955	0.937	0.911	0.860	0.802	0.723	0.574	0.428	0.263	0.076	0.012

[a] From Passman and Larmore [65], courtesy of The Rand Corporation.

Carbon Dioxide 0.3–2.4 μ

TABLE 4.8 SPECTRAL TRANSMITTANCE OF CARBON DIOXIDE FOR A HORIZONTAL PATH AT SEA LEVEL[a]

Wavelength (μ)	Path Length (Kilometers)											
	0.1	0.2	0.5	1	2	5	10	20	50	100	200	500	1000
0.3	1	1	1	1	1	1	1	1	1	1	1	1	1
0.4	1	1	1	1	1	1	1	1	1	1	1	1	1
0.5	1	1	1	1	1	1	1	1	1	1	1	1	1
0.6	1	1	1	1	1	1	1	1	1	1	1	1	1
0.7	1	1	1	1	1	1	1	1	1	1	1	1	1
0.8	1	1	1	1	1	1	1	1	1	1	1	1	1
0.9	1	1	1	1	1	1	1	1	1	1	1	1	1
1.0	1	1	1	1	1	1	1	1	1	1	1	1	1
1.1	1	1	1	1	1	1	1	1	1	1	1	1	1
1.2	1	1	1	1	1	1	1	1	1	1	1	1	1
1.3	1	1	1	0.999	0.999	0.999	0.998	0.997	0.996	0.994	0.992	0.987	0.982
1.4	0.996	0.995	0.992	0.988	0.984	0.975	0.964	0.949	0.919	0.885	0.838	0.747	0.649
1.5	0.999	0.999	0.998	0.998	0.997	0.995	0.993	0.990	0.984	0.976	0.967	0.949	0.927
1.6	0.996	0.995	0.992	0.988	0.984	0.975	0.964	0.949	0.919	0.885	0.838	0.747	0.649
1.7	1	1	1	0.999	0.999	0.999	0.998	0.997	0.996	0.994	0.992	0.987	0.982
1.8	1	1	1	1	1	1	1	1	1	1	1	1	1
1.9	1	1	1	0.999	0.999	0.999	0.998	0.997	0.996	0.994	0.992	0.987	0.982
2.0	0.978	0.969	0.951	0.931	0.903	0.847	0.785	0.699	0.541	0.387	0.221	0.053	0.006
2.1	0.998	0.997	0.996	0.994	0.992	0.987	0.982	0.974	0.959	0.942	0.919	0.872	0.820
2.2	1	1	1	1	1	1	1	1	1	1	1	1	1
2.3	1	1	1	1	1	1	1	1	1	1	1	1	1
2.4	1	1	1	1	1	1	1	1	1	1	1	1	1

[a] From Passman and Larmore[65], courtesy of The Rand Corporation.

Water Vapor 2.5–4.9 μ

TABLE 4.7 (Continued)

Wavelength (μ)	Water Vapor Content (Precipitable Millimeters)											
	0.1	0.2	0.5	1	2	5	10	20	50	100	200	500	1000
2.5	0.930	0.902	0.844	0.782	0.695	0.536	0.381	0.216	0.064	0.005	0	0	0
2.6	0.617	0.479	0.261	0.110	0.002	0	0	0	0	0	0	0	0
2.7	0.361	0.196	0.040	0.004	0	0	0	0	0	0	0	0	0
2.8	0.453	0.289	0.092	0.017	0.001	0	0	0	0	0	0	0	0
2.9	0.689	0.571	0.369	0.205	0.073	0.005	0	0	0	0	0	0	0
3.0	0.851	0.790	0.673	0.552	0.401	0.184	0.060	0.008	0	0	0	0	0
3.1	0.900	0.860	0.779	0.692	0.574	0.375	0.210	0.076	0.005	0	0	0	0
3.2	0.925	0.894	0.833	0.766	0.674	0.506	0.347	0.184	0.035	0.003	0	0	0
3.3	0.950	0.930	0.888	0.843	0.779	0.658	0.531	0.377	0.161	0.048	0.005	0	0
3.4	0.973	0.962	0.939	0.914	0.880	0.811	0.735	0.633	0.448	0.285	0.130	0.017	0.001
3.5	0.988	0.983	0.973	0.962	0.946	0.915	0.881	0.832	0.736	0.635	0.502	0.287	0.133
3.6	0.994	0.992	0.987	0.982	0.973	0.958	0.947	0.916	0.866	0.812	0.738	0.596	0.452
3.7	0.997	0.994	0.991	0.988	0.982	0.972	0.960	0.944	0.911	0.874	0.823	0.724	0.616
3.8	0.998	0.997	0.995	0.994	0.991	0.986	0.980	0.972	0.956	0.937	0.911	0.860	0.802
3.9	0.998	0.997	0.995	0.994	0.991	0.986	0.980	0.972	0.956	0.937	0.911	0.860	0.802
4.0	0.997	0.995	0.993	0.990	0.987	0.977	0.970	0.960	0.930	0.900	0.870	0.790	0.700
4.1	0.977	0.994	9.991	0.988	0.982	0.972	0.960	0.944	0.911	0.874	0.823	0.724	0.616
4.2	0.994	0.992	0.987	0.982	0.973	0.958	0.947	0.916	0.866	0.812	0.738	0.596	0.452
4.3	0.991	0.984	0.975	0.972	0.950	0.937	0.910	0.873	0.800	0.722	0.615	0.425	0.260
4.4	0.980	0.972	0.955	0.937	0.911	0.860	0.802	0.723	0.574	0.428	0.263	0.076	0.012
4.5	0.970	0.958	0.932	0.905	0.866	0.790	0.707	0.595	0.400	0.235	0.093	0.008	0
4.6	0.960	0.943	0.911	0.874	0.822	0.723	0.617	0.478	0.262	0.113	0.024	0	0
4.7	0.950	0.930	0.888	0.843	0.779	0.658	0.531	0.377	0.161	0.048	0.005	0	0
4.8	0.940	0.915	0.866	0.812	0.736	0.595	0.452	0.289	0.117	0.018	0.001	0	0
4.9	0.930	0.902	0.844	0.782	0.695	0.536	0.381	0.216	0.064	0.005	0	0	0

Wavelength	Path Length (Kilometers)										
(μ)	0.1	0.2	0.5	1	2	5	10	20	50	100	200	500	1000
2.5	1	1	1	1	1	1	1	1	1	1	1	1	1
2.6	1	1	1	1	1	1	1	1	1	1	1	1	1
2.7	0.799	0.718	0.569	0.419	0.253	0.071	0.011	0	0	0	0	0	0
2.8	0.871	0.804	0.695	0.578	0.432	0.215	0.079	0.013	0	0	0	0	0
2.9	0.997	0.995	0.993	0.990	0.985	0.977	0.968	0.954	0.927	0.898	0.855	0.772	0.683
3.0	1	1	1	1	1	1	1	1	1	1	1	1	1
3.1	1	1	1	1	1	1	1	1	1	1	1	1	1
3.2	1	1	1	1	1	1	1	1	1	1	1	1	1
3.3	1	1	1	1	1	1	1	1	1	1	1	1	1
3.4	1	1	1	1	1	1	1	1	1	1	1	1	1
3.5	1	1	1	1	1	1	1	1	1	1	1	1	1
3.6	1	1	1	1	1	1	1	1	1	1	1	1	1
3.7	1	1	1	1	1	1	1	1	1	1	1	1	1
3.8	1	1	1	1	1	1	1	1	1	1	1	1	1
3.9	1	1	1	1	1	1	1	1	1	1	1	1	1
4.0	0.998	0.997	0.996	0.994	0.991	0.986	0.980	0.971	0.955	0.937	0.911	0.859	0.802
4.1	0.983	0.975	0.961	0.944	0.921	0.876	0.825	0.755	0.622	0.485	0.322	0.118	0.027
4.2	0.673	0.551	0.445	0.182	0.059	0.003	0	0	0	0	0	0	0
4.3	0.098	0.016	0	0	0	0	0	0	0	0	0	0	0
4.4	0.481	0.319	0.115	0.026	0.002	0	0	0	0	0	0	0	0
4.5	0.957	0.949	0.903	0.863	0.807	0.699	0.585	0.439	0.222	0.084	0.014	0	0
4.6	0.995	0.993	0.989	0.985	0.978	0.966	0.951	0.931	0.891	0.845	0.783	0.663	0.539
4.7	0.995	0.993	0.989	0.985	0.978	0.966	0.951	0.931	0.891	0.845	0.783	0.663	0.539
4.8	0.976	0.966	0.945	0.922	0.891	0.828	0.759	0.664	0.492	0.331	0.169	0.030	0.002
4.9	0.975	0.964	0.943	0.920	0.886	0.822	0.750	0.652	0.468	0.313	0.153	0.024	0.001

From Passman and Larmore[65], courtesy of The Rand Corporation.

Water Vapor 5.0–6.9 μ

TABLE 4.7 (Continued)

Wavelength (μ)	Water Vapor Content (Precipitable Millimeters)											
	0.1	0.2	0.5	1	2	5	10	20	50	100	200	500	1000
5.0	0.915	0.880	0.811	0.736	0.634	0.451	0.286	0.132	0.017	0	0	0	0
5.1	0.885	0.839	0.747	0.649	0.519	0.308	0.149	0.041	0.001	0	0	0	0
5.2	0.846	0.784	0.664	0.539	0.385	0.169	0.052	0.006	0	0	0	0	0
5.3	0.792	0.707	0.555	0.406	0.239	0.062	0.008	0	0	0	0	0	0
5.4	0.726	0.611	0.432	0.268	0.116	0.013	0	0	0	0	0	0	0
5.5	0.617	0.479	0.261	0.110	0.035	0	0	0	0	0	0	0	0
5.6	0.491	0.331	0.121	0.029	0.002	0	0	0	0	0	0	0	0
5.7	0.361	0.196	0.040	0.004	0	0	0	0	0	0	0	0	0
5.8	0.141	0.044	0.001	0	0	0	0	0	0	0	0	0	0
5.9	0.141	0.044	0.001	0	0	0	0	0	0	0	0	0	0
6.0	0.180	0.058	0.003	0	0	0	0	0	0	0	0	0	0
6.1	0.260	0.112	0.012	0	0	0	0	0	0	0	0	0	0
6.2	0.652	0.524	0.313	0.153	0.043	0.001	0	0	0	0	0	0	0
6.3	0.552	0.401	0.182	0.060	0.008	0	0	0	0	0	0	0	0
6.4	0.317	0.157	0.025	0.002	0	0	0	0	0	0	0	0	0
6.5	0.164	0.049	0.002	0	0	0	0	0	0	0	0	0	0
6.6	0.138	0.042	0.001	0	0	0	0	0	0	0	0	0	0
6.7	0.322	0.162	0.037	0.002	0	0	0	0	0	0	0	0	0
6.8	0.361	0.196	0.040	0.004	0	0	0	0	0	0	0	0	0
6.9	0.416	0.250	0.068	0.010	0	0	0	0	0	0	0	0	0

From Passman and Larmore[65], courtesy of The Rand Corporation.

Carbon Dioxide 5.0–6.9 μ

TABLE 4.8 (*Continued*)

Wavelength (μ)	Path Length (Kilometers)											
	0.1	0.2	0.5	1	2	5	10	20	50	100	200	500	1000
5.0	0.999	0.998	0.997	0.995	0.994	0.990	0.986	0.979	0.968	0.954	0.935	0.897	0.855
5.1	1	0.999	0.999	0.998	0.998	0.996	0.994	0.992	0.988	0.984	0.976	0.961	0.946
5.2	0.986	0.980	0.968	0.955	0.936	0.899	0.857	0.799	0.687	0.569	0.420	0.203	0.072
5.3	0.997	0.995	0.993	0.989	0.984	0.976	0.966	0.951	0.923	0.891	0.846	0.760	0.666
5.4	1	1	1	1	1	1	1	1	1	1	1	1	1
5.5	1	1	1	1	1	1	1	1	1	1	1	1	1
5.6	1	1	1	1	1	1	1	1	1	1	1	1	1
5.7	1	1	1	1	1	1	1	1	1	1	1	1	1
5.8	1	1	1	1	1	1	1	1	1	1	1	1	1
5.9	1	1	1	1	1	1	1	1	1	1	1	1	1
6.0	1	1	1	1	1	1	1	1	1	1	1	1	1
6.1	1	1	1	1	1	1	1	1	1	1	1	1	1
6.2	1	1	1	1	1	1	1	1	1	1	1	1	1
6.3	1	1	1	1	1	1	1	1	1	1	1	1	1
6.4	1	1	1	1	1	1	1	1	1	1	1	1	1
6.5	1	1	1	1	1	1	1	1	1	1	1	1	1
6.6	1	1	1	1	1	1	1	1	1	1	1	1	1
6.7	1	1	1	1	1	1	1	1	1	1	1	1	1
6.8	1	1	1	1	1	1	1	1	1	1	1	1	1
6.9	1	1	1	1	1	1	1	1	1	1	1	1	1

From Passman and Larmore [65], courtesy of The Rand Corporation.

Water Vapor 7.0–9.4 μ

TABLE 4.9 SPECTRAL TRANSMITTANCE OF WATER VAPOR FOR A HORIZONTAL PATH AT SEA LEVEL

Wavelength (μ)	Water Vapor Content (Precipitable Millimeters)									
	0.2	0.5	1	2	5	10	20	50	100	200
7.0	0.569	0.245	0.060	0.004	0	0	0	0	0	0
7.1	0.716	0.433	0.188	0.035	0	0	0	0	0	0
7.2	0.782	0.540	0.292	0.085	0.002	0	0	0	0	0
7.3	0.849	0.664	0.441	0.194	0.017	0	0	0	0	0
7.4	0.922	0.817	0.666	0.444	0.132	0.018	0	0	0	0
7.5	0.947	0.874	0.762	0.582	0.258	0.066	0	0	0	0
7.6	0.922	0.817	0.666	0.444	0.132	0.018	0	0	0	0
7.7	0.978	0.944	0.884	0.796	0.564	0.328	0.102	0.003	0	0
7.8	0.974	0.937	0.878	0.771	0.523	0.273	0.074	0.002	0	0
7.9	0.982	0.959	0.920	0.842	0.658	0.433	0.187	0.015	0	0
8.0	0.990	0.975	0.951	0.904	0.777	0.603	0.365	0.080	0.006	0
8.1	0.994	0.986	0.972	0.945	0.869	0.754	0.568	0.244	0.059	0.003
8.2	0.993	0.982	0.964	0.930	0.834	0.696	0.484	0.163	0.027	0
8.3	0.995	0.988	0.976	0.953	0.887	0.786	0.618	0.300	0.090	0.008
8.4	0.995	0.987	0.975	0.950	0.880	0.774	0.599	0.278	0.077	0.006
8.5	0.994	0.986	0.972	0.944	0.866	0.750	0.562	0.237	0.056	0.003
8.6	0.996	0.992	0.982	0.965	0.915	0.837	0.702	0.411	0.169	0.029
8.7	0.996	0.992	0.983	0.966	0.916	0.839	0.704	0.416	0.173	0.030
8.8	0.997	0.993	0.983	0.966	0.917	0.841	0.707	0.421	0.177	0.031
8.9	0.997	0.992	0.983	0.966	0.918	0.843	0.709	0.425	0.180	0.032
9.0	0.997	0.992	0.984	0.968	0.921	0.848	0.719	0.440	0.193	0.037
9.1	0.997	0.992	0.985	0.970	0.926	0.858	0.735	0.464	0.215	0.046
9.2	0.997	0.993	0.985	0.971	0.929	0.863	0.744	0.478	0.228	0.052
9.3	0.997	0.993	0.986	0.972	0.930	0.867	0.750	0.489	0.239	0.057
9.4	0.997	0.993	0.986	0.973	0.933	0.870	0.756	0.498	0.248	0.061

Carbon Dioxide 7.0–9.4 μ

TABLE 4.10 SPECTRAL TRANSMITTANCE OF CARBON DIOXIDE FOR A HORIZONTAL PATH AT SEA LEVEL

Wavelength (μ)	Path Length (Kilometers)									
	0.2	0.5	1	2	5	10	20	50	100	200
7.0	1	1	1	1	1	1	1	1	1	1
7.1	1	1	1	1	1	1	1	1	1	1
7.2	1	1	1	1	1	1	1	1	1	1
7.3	1	1	1	1	1	1	1	1	1	1
7.4	1	1	1	1	1	1	1	1	1	1
7.5	1	1	1	1	1	1	1	1	1	1
7.6	1	1	1	1	1	1	1	1	1	1
7.7	1	1	1	1	1	1	1	1	1	1
7.8	1	1	1	1	1	1	1	1	1	1
7.9	1	1	1	1	1	1	1	1	1	1
8.0	1	1	1	1	1	1	1	1	1	1
8.1	1	1	1	1	1	1	1	1	1	1
8.2	1	1	1	1	1	1	1	1	1	1
8.3	1	1	1	1	1	1	1	1	1	1
8.4	1	1	1	1	1	1	1	1	1	1
8.5	1	1	1	1	1	1	1	1	1	1
8.6	1	1	1	1	1	1	1	1	1	1
8.7	1	1	1	1	1	1	1	1	1	1
8.8	1	1	1	1	1	1	1	1	1	1
8.9	1	1	1	1	1	1	1	1	1	1
9.0	1	1	1	1	1	1	1	1	1	1
9.1	1	1	0.999	0.999	0.998	0.995	0.991	0.978	0.955	0.914
9.2	1	1	0.999	0.998	0.995	0.991	0.982	0.955	0.913	0.834
9.3	0.999	0.997	0.995	0.990	0.975	0.951	0.904	0.776	0.605	0.363
9.4	0.993	0.982	0.965	0.931	0.837	0.700	0.491	0.168	0.028	0.001

Water Vapor 9.5–11.9 μ

TABLE 4.9 (Continued)

Wavelength (μ)	Water Vapor Content (Precipitable Millimeters)								
	0.2	0.5	1	2	5	10	20	50	100	200
9.5	0.997	0.993	0.987	0.973	0.934	0.873	0.762	0.507	0.257	0.066
9.6	0.997	0.993	0.987	0.974	0.936	0.876	0.766	0.516	0.265	0.070
9.7	0.997	0.993	0.987	0.974	0.937	0.878	0.770	0.521	0.270	0.073
9.8	0.997	0.994	0.987	0.975	0.938	0.880	0.773	0.526	0.277	0.077
9.9	0.997	0.994	0.987	0.975	0.939	0.882	0.777	0.532	0.283	0.080
10.0	0.998	0.994	0.988	0.975	0.940	0.883	0.780	0.538	0.289	0.083
10.1	0.998	0.994	0.988	0.975	0.940	0.883	0.780	0.538	0.289	0.083
10.2	0.998	0.994	0.988	0.975	0.940	0.883	0.780	0.538	0.289	0.083
10.3	0.998	0.994	0.988	0.976	0.940	0.884	0.781	0.540	0.292	0.085
10.4	0.998	0.994	0.988	0.976	0.941	0.885	0.782	0.542	0.294	0.086
10.5	0.998	0.994	0.988	0.976	0.941	0.886	0.784	0.544	0.295	0.087
10.6	0.998	0.994	0.988	0.976	0.942	0.887	0.786	0.548	0.300	0.089
10.7	0.998	0.994	0.988	0.976	0.942	0.887	0.787	0.550	0.302	0.091
10.8	0.998	0.994	0.988	0.976	0.941	0.886	0.784	0.544	0.295	0.087
10.9	0.998	0.994	0.988	0.976	0.940	0.884	0.781	0.540	0.292	0.085
11.0	0.998	0.994	0.988	0.975	0.940	0.883	0.779	0.536	0.287	0.082
11.1	0.998	0.994	0.987	0.975	0.939	0.882	0.777	0.532	0.283	0.080
11.2	0.997	0.993	0.986	0.972	0.931	0.867	0.750	0.487	0.237	0.056
11.3	0.997	0.992	0.985	0.970	0.927	0.859	0.738	0.467	0.218	0.048
11.4	0.997	0.993	0.986	0.971	0.930	0.865	0.748	0.485	0.235	0.055
11.5	0.997	0.993	0.986	0.972	0.932	0.868	0.753	0.493	0.243	0.059
11.6	0.997	0.993	0.987	0.974	0.935	0.875	0.765	0.513	0.262	0.069
11.7	0.996	0.990	0.980	0.961	0.906	0.820	0.673	0.372	0.138	0.019
11.8	0.997	0.992	0.982	0.969	0.925	0.863	0.733	0.460	0.212	0.045
11.9	0.997	0.993	0.986	0.972	0.932	0.869	0.755	0.495	0.245	0.060

Carbon Dioxide 9.5–11.9 μ

TABLE 4.10 (Continued)

Wavelength (μ)	Path Length (Kilometers)								
	0.2	0.5	1	2	5	10	20	50	100	200
9.5	0.993	0.983	0.967	0.935	0.842	0.715	0.512	0.187	0.035	0.001
9.6	0.996	0.990	0.980	0.961	0.906	0.821	0.675	0.363	0.140	0.029
9.7	0.995	0.986	0.973	0.947	0.873	0.761	0.580	0.256	0.065	0.004
9.8	0.997	0.992	0.984	0.969	0.924	0.858	0.730	0.455	0.206	0.043
9.9	0.998	0.995	0.989	0.979	0.948	0.897	0.811	0.585	0.342	0.123
10.0	1	1	0.999	0.997	0.994	0.989	0.978	0.945	0.892	0.797
10.1	1	0.999	0.998	0.996	0.990	0.980	0.960	0.902	0.814	0.663
10.2	0.997	0.994	0.988	0.977	0.943	0.890	0.792	0.558	0.312	0.097
10.3	0.997	0.994	0.987	0.975	0.939	0.881	0.777	0.532	0.283	0.080
10.4	1	1	0.999	0.998	0.995	0.991	0.982	0.955	0.913	0.834
10.5	1	1	0.999	0.998	0.998	0.995	0.991	0.978	0.955	0.914
10.6	1	1	0.999	0.999	0.998	0.995	0.991	0.978	0.955	0.914
10.7	1	1	1	0.999	0.999	0.997	0.995	0.986	0.973	0.947
10.8	1	1	0.999	0.998	0.998	0.995	0.991	0.978	0.955	0.914
10.9	1	0.999	0.999	0.997	0.993	0.986	0.973	0.934	0.872	0.761
11.0	1	0.999	0.999	0.997	0.993	0.986	0.973	0.934	0.872	0.761
11.1	1	0.999	0.998	0.997	0.992	0.984	0.969	0.923	0.855	0.726
11.2	1	0.999	0.998	0.995	0.989	0.978	0.955	0.892	0.796	0.633
11.3	0.999	0.999	0.997	0.994	0.985	0.971	0.942	0.862	0.742	0.552
11.4	0.999	0.998	0.997	0.993	0.983	0.966	0.934	0.842	0.709	0.503
11.5	0.999	0.998	0.996	0.992	0.980	0.960	0.921	0.814	0.661	0.438
11.6	0.999	0.998	0.995	0.991	0.977	0.955	0.912	0.794	0.632	0.399
11.7	0.999	0.998	0.995	0.991	0.977	0.955	0.912	0.794	0.632	0.399
11.8	0.999	0.998	0.997	0.993	0.983	0.966	0.934	0.842	0.709	0.503
11.9	1	0.999	0.998	0.995	0.989	0.978	0.955	0.892	0.796	0.633

Water Vapor 12.0–13.9 μ

TABLE 4.9 (Continued)

Wavelength (μ)	Water Vapor Content (Precipitable Millimeters)								
	0.2	0.5	1	2	5	10	20	50	100	200
12.0	0.997	0.993	0.987	0.974	0.937	0.878	0.770	0.521	0.270	0.073
12.1	0.997	0.994	0.987	0.975	0.938	0.880	0.773	0.526	0.277	0.077
12.2	0.997	0.994	0.987	0.975	0.938	0.880	0.775	0.528	0.279	0.078
12.3	0.997	0.993	0.987	0.974	0.937	0.878	0.770	0.521	0.270	0.073
12.4	0.997	0.993	0.987	0.974	0.935	0.874	0.764	0.511	0.261	0.068
12.5	0.997	0.993	0.986	0.973	0.933	0.871	0.759	0.502	0.252	0.063
12.6	0.997	0.993	0.986	0.972	0.931	0.868	0.752	0.491	0.241	0.058
12.7	0.997	0.993	0.985	0.971	0.929	0.863	0.744	0.478	0.228	0.052
12.8	0.997	0.992	0.985	0.970	0.926	0.858	0.736	0.466	0.217	0.047
12.9	0.997	0.992	0.984	0.969	0.924	0.853	0.728	0.452	0.204	0.041
13.0	0.997	0.992	0.984	0.967	0.921	0.846	0.718	0.437	0.191	0.036
13.1	0.996	0.991	0.983	0.966	0.918	0.843	0.709	0.424	0.180	0.032
13.2	0.996	0.991	0.982	0.965	0.915	0.837	0.701	0.411	0.169	0.028
13.3	0.996	0.991	0.982	0.964	0.912	0.831	0.690	0.397	0.153	0.025
13.4	0.996	0.990	0.981	0.962	0.908	0.825	0.681	0.382	0.146	0.021
13.5	0.996	0.990	0.980	0.961	0.905	0.819	0.670	0.368	0.136	0.019
13.6	0.996	0.990	0.979	0.959	0.902	0.813	0.661	0.355	0.126	0.016
13.7	0.996	0.989	0.979	0.958	0.898	0.807	0.651	0.342	0.117	0.014
13.8	0.996	0.989	0.978	0.956	0.894	0.800	0.640	0.328	0.107	0.011
13.9	0.995	0.988	0.977	0.955	0.891	0.793	0.629	0.313	0.098	0.010

Carbon Dioxide 12.0–13.9 μ

TABLE 4.10 (*Continued*)

Wavelength (μ)	Path Length (Kilometers)								
	0.2	0.5	1	2	5	10	20	50	100	200
12.0	1	1	0.999	0.999	0.997	0.993	0.986	0.966	0.934	0.872
12.1	1	1	0.999	0.998	0.998	0.995	0.991	0.978	0.955	0.914
12.2	1	1	0.999	0.998	0.998	0.995	0.991	0.978	0.955	0.914
12.3	0.998	0.995	0.990	0.981	0.952	0.907	0.823	0.614	0.376	0.142
12.4	0.994	0.985	0.970	0.941	0.859	0.738	0.545	0.218	0.048	0.002
12.5	0.987	0.968	0.936	0.877	0.719	0.517	0.268	0.037	0.001	0
12.6	0.980	0.950	0.903	0.815	0.599	0.358	0.129	0.006	0	0
12.7	0.996	0.989	0.979	0.959	0.899	0.809	0.654	0.346	0.120	0.015
12.8	0.990	0.974	0.949	0.901	0.770	0.592	0.351	0.072	0.005	0
12.9	0.985	0.962	0.925	0.856	0.677	0.458	0.210	0.020	0	0
13.0	0.991	0.977	0.955	0.912	0.794	0.630	0.397	0.099	0.010	0
13.1	0.990	0.974	0.949	0.900	0.768	0.592	0.348	0.071	0.005	0
13.2	0.978	0.946	0.895	0.801	0.575	0.330	0.109	0.004	0	0
13.3	0.952	0.884	0.782	0.611	0.292	0.085	0.007	0	0	0
13.4	0.935	0.846	0.715	0.512	0.187	0.035	0.001	0	0	0
13.5	0.901	0.767	0.593	0.352	0.070	0.005	0	0	0	0
13.6	0.901	0.792	0.627	0.351	0.097	0.009	0	0	0	0
13.7	0.916	0.803	0.644	0.415	0.110	0.012	0	0	0	0
13.8	0.858	0.681	0.464	0.215	0.021	0	0	0	0	0
13.9	0.778	0.534	0.286	0.082	0.002	0	0	0	0	0

TABLE 4.11 VALUES OF THE ALTITUDE CORRECTION FACTOR $(P/P_0)^k$ FOR REDUCTION TO EQUIVALENT SEA-LEVEL PATH[a]

Altitude		Altitude Correction Factor	
km	kft	Water Vapor	Carbon Dioxide
0.305	1	0.981	0.940
0.610	2	0.961	0.833
0.915	3	0.942	0.840
1.22	4	0.923	0.774
1.52	5	0.904	0.743
1.83	6	0.886	0.699
2.14	7	0.869	0.660
2.44	8	0.852	0.620
2.74	9	0.835	0.580
3.05	10	0.819	0.548
3.81	12.5	0.790	0.494
4.57	15	0.739	0.404
5.34	17.5	0.714	0.364
6.10	20	0.670	0.299
6.86	22.5	0.643	0.266
7.62	25	0.609	0.226
9.15	30	0.552	0.168
10.7	35	0.486	0.115
12.2	40	0.441	0.085
15.2	50	0.348	0.042
18.3	60	0.272	0.020
21.4	70	0.214	0.010
24.4	80	0.167	0.005
27.4	90	0.134	0.002
30.5	100	0.105	0.001

[a] Adapted from Passman and Larmore[65], Courtesy of The Rand Corporation.

transmittance of the atmosphere above Denver, Colorado. Their data cover the range from 1 to 12.5 μ, and they also give constants for use with various approximation formulas derived from spectral models. The tables published by Wyatt, Stull, and Plass[63, 64] represent the result of a computer calculation of a detailed spectral model. Their tables are arranged similarly to those of Passman and Larmore. The use of equal wavenumber intervals results in rather limited coverage at the longer wavelengths. Since the table for water vapor extends from 1 to 9.5 μ, these tables cannot be used for calculations involving the important 8 to

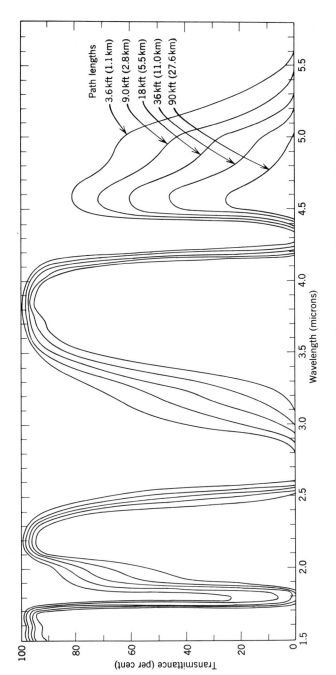

Figure 4.14 Calculated atmospheric transmittance—horizontal path, altitude 15 kft (4.6 km), precipitable water content 0.75 mm kft^{-1}.

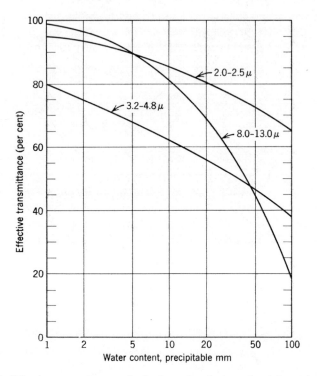

Figure 4.15 Effective transmittance of a horizontal path at sea level for typical spectral bandpasses.

13 μ window. Altshuler[66] analyzed various experimental measurements and presented his results in a most convenient series of curves. It is unfortunate that this important work has never been published in the formal literature (some authors cite an Altshuler reference in *J. Opt. Soc. Am.*; unfortunately it is only a short abstract).

Table 4.12 gives information about the effective transmittance of the 9.6 μ absorption band of ozone. It assumes that the band can be represented as a rectangular-shaped band extending from 9.4 to 9.8 μ with no absorption outside of these limits. Figure 4.8 shows the vertical distribution of ozone in the atmosphere.

The calculation of transmittance over a slant path is far more difficult than that over a horizontal path. The problem is that the atmospheric pressure, and hence the shape of the absorption bands, changes continuously along the path and any method of correcting for this change is extremely complicated. An approximate method, using Tables 4.7 through 4.10, is to divide the path into a number of segments of equal

TRANSMISSION OF INFRARED RADIATION 159

TABLE 4.12 EFFECTIVE
TRANSMITTANCE OF THE 9.6 μ
ABSORPTION BAND OF OZONE

Amount of Ozone (atm cm)	Effective Transmittance (9.4–9.8 μ)
0.1	0.53
0.2	0.47
0.5	0.37
1	0.30

increase in altitude, apply the altitude correction factor at the midpoint of each interval, and assume that the line of sight is horizontal within each segment. The spectral transmittance for the entire slant path is found by multiplying together the values calculated for each horizontal segment. Various other approximations to the slant-path problem can be found in refs. 56, 67, and 68. Howard and Garing[62] have shown that there is no technically satisfying solution at present, by recounting how a number of workers, each using his favorite method, calculated the transmittance of the same slant path; their answers showed a twofold spread in values.

4.8 SCATTERING EFFECTS IN THE ATMOSPHERE

The extinction coefficient in (4-2) includes the effects of scattering by the molecules, hazes, and fogs in the atmosphere. Such processes are of major concern in the visible portion of the spectrum, where they can be observed by the effect they have on visibility. With the exception of scattering by fogs, they are of minor importance in the infrared. Unfortunately there is much misinformation in the literature concerning the importance of scattering effects in the infrared; authors generally proclaim that "infrared penetrates the atmosphere better than visible light." Under certain limited conditions this may be true, but it is far from a general rule.

The principle investigations of scattering processes have been done by those with an interest in meteorological optics; among the most important references are the books by Middleton[69] and van de Hulst[70, 71], and the paper by Deirmendjian[72]. It is important to understand the terminology used by these workers so that their results can be applied to our problems. *Visibility*, as used by the U.S. Weather Bureau, refers to the maximum distance at which prominent objects, such as mountains, buildings, or towers, can be seen and *identified* with the unaided eye. At night, "moderately intense" lights become the objects. Since it requires identification of the object and does not specify the contrast between the object and its background, such a definition is highly subjective. An *aerosol* is a system of dispersed particles suspended in a gas; the term is often applied to the atmosphere and the suspended particles it contains. *Haze* refers to the small particles dispersed throughout the atmospheric aerosol. Haze particles consist of tiny salt crystals, exceedingly fine dust, or products of combustion, and have radii ranging up to 0.5 μ. In regions of high humidity, moisture can condense on these particles and they may grow quite large. When such condensation occurs, the particle is said to act as a *condensation nucleus*. Theory and experiment have shown that the presence of such nuclei is essential before condensation can take place in the atmosphere. By far the most important nucleus is the salt particle, since it is naturally hygroscopic. *Fog* is formed when the condensation nuclei grow into water droplets or ice crystals with radii exceeding 1 μ. A *cloud* is formed in the same way; the two are distinguished by the convention that fogs contact the ground and clouds do not. By international agreement, fog limits the visibility to less than 1 km (0.62 miles), whereas in a *mist* the visibility is greater than 1 km. The precipitation that reaches the earth's surface in the form of liquid drops is called *rain*. These drops usually have a minimum radius of 0.25 mm. *Smog* is defined as a fog contaminated with industrial pollutants. By popular usage it is applied to atmospheric conditions that are irritating or unpleasant to human beings.

In an atmosphere containing only scatterers (no absorbers), the spectral transmittance over a path of length x is

$$\tau = e^{-\gamma x}, \quad (4\text{-}18)$$

where γ is the scattering coefficient. If the atmosphere contains n water droplets per cubic centimeter, each having a radius r, the value of the scattering coefficient is

$$\gamma = \pi n K r^2, \quad (4\text{-}19)$$

where K is the *scattering area ratio*, a measure of the efficiency with which a droplet scatters radiant energy. Its value for spherical water drops is given in Figure 4.16 where it is plotted as a function of r/λ; the ratio of droplet radius to wavelength[73]. The physical interpretation of Figure 4.16 is that K changes rapidly in the early stages of droplet growth and shows a strong dependence on wavelength. Hence the scattering by small droplets ($r \ll \lambda$) changes rapidly with wavelength. The

TRANSMISSION OF INFRARED RADIATION

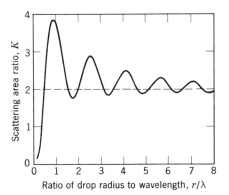

Figure 4.16 Scattering area ratio for spherical water drops (adapted from Houghton and Chalker[73]).

value of K reaches a maximum of about 3.8 when the radius is equal to the wavelength, and this size droplet is the most efficient scatterer. For further droplet growth, the value of K oscillates slightly and finally assumes a value of 2, with virtually no dependence on wavelength.

Measurements of the droplets in fogs show that their radii range from 0.5 to 80 μ and that the peak of their size distribution curve usually occurs between 5 and 15 μ [74-76]. Therefore, fog particles are comparable in size to infrared wavelengths, the ratio of r/λ is approximately unity, and the value of the scattering area ratio is near its maximum. For a fog containing 200 droplets per cubic centimeter, each assumed to have a radius of 5 μ, (4-18) predicts that the spectral transmittance at 4 μ will be only a fraction of a percent for a path 328 ft (0.1 km) long. Changing the wavelength by a factor of 10, to 40 μ, while keeping the other conditions constant, increases the transmittance to over 90 per cent. This is a hollow victory, however, since strong absorption by water vapor bands makes the atmosphere opaque in the 40 μ region for path lengths longer than a few feet. Thus the transmission of fogs, in either the visible or the infrared portion of the spectrum, is extremely poor for any reasonable path length, and little improvement results from a shift to longer wavelengths. The same conclusion applies to clouds. We see here one of the hard facts of life in the infrared: *as long as they are constrained to work within the atmosphere, infrared systems can never qualify as true all-weather systems.* Experimental measurements of the transmission of fogs and clouds clearly support this conclusion[76-82]. Some idea of the latitudinal distribution of cloud cover can be gained from examination of photographs returned by weather satellites[83].

Since typical haze particles rarely exceed a radius of 0.5 μ, they are small with respect to infrared wavelengths, and reduced scattering is noted at the longer wavelengths. This effect is sometimes shown dramatically in photographs of distant landscapes made with infrared film. Such film responds to wavelengths out to about $0.85\,\mu$, where the calculated transmittance of a light haze is about twice what it is at $0.5\,\mu$ in the visible. Thus in the presence of haze the photographer may often be justified in claiming better atmospheric penetration with infrared film[84]. However, when he tries to extend the claim to fogs he is wrong (but, judging from the literature on the subject, he is not alone in his error).

Another way of attacking the scattering problem is to look for a general relationship between the scattering coefficient and the wavelength. One that has been used by many workers is

$$\gamma \sim \lambda^{-\psi}. \tag{4-20}$$

For particles that are small with respect to the wavelength, ψ is equal to 4 and the process is called *Rayleigh scattering*. Such scattering would be present even in a completely clean atmosphere, because the gas molecules themselves would scatter. This strong scattering of the shorter wavelengths is the cause of the blue color of the sky and the red of the setting sun. Its effect is negligibly small for wavelengths longer than $1\,\mu$.

For larger particles the value of ψ in (4-20) approaches zero, and the process is known as *Mie scattering*. When ψ is equal to zero, the scattering is independent of wavelength and there is no hope of any improvement by using longer wavelengths. For most fogs and clouds the value of ψ is approximately zero; indeed, it is this nonselective scattering that makes them appear white. The effect of a change in ψ is easily demonstrated. The smoke drifting upward from a cigarette contains extremely small particles, so that ψ is about equal to 4, and the light scattered from the smoke is very blue. In exhaled smoke, the particles have agglomerated, ψ is approximately equal to zero, and the light scattered from the smoke appears to be white.

There is a relatively simple way to calculate the transmission through haze, based on an observation of the visibility. In viewing a series of objects, we notice that the contrast between them and their background decreases with distance. Eventually the contrast becomes so slight that the object can no longer be perceived; for the human eye this occurs when the contrast drops to about 2 per cent; that is, the luminance of the object differs from that of the background by less than 2 per cent. The *visual range V* is defined as the distance at which the contrast between an object and its background is reduced to 2 per cent. The visual range and the scattering coefficient are related by

$$V = \frac{3.92}{\gamma}. \qquad (4\text{-}21)$$

The choice of units for V and γ is immaterial as long as these units are consistent; thus if V is expressed in feet, then γ must be expressed in (feet)$^{-1}$, and so on. The measurements of Gebbie, shown in Figure 4.1, were made over a 6000 ft path at a time when the spectral transmittance at 0.61 μ was 60 per cent. It is obvious from the increasing transmittance in the peaks of the windows out to 4 μ that there was haze along the measurement path. The transmittance of this haze will be calculated at 2.2, 3.8, and 11 μ, wavelengths near the peak of each window where there is a minimum of absorption. From (4-18) the value of the scattering coefficient γ is found to be 0.085 kft^{-1} and the visual range (observed at 0.61 μ) must have been about 46 kft (14 km). Most workers[69, 85, 86] agree that the value of ψ in (4-20) is about 1.3 for most terrestrial hazes, although there is one report [87] that places the value of ψ at about 0.7. Calculated values of the transmittance for Gebbie's 6000 ft path are shown in Table 4.13 for the two values of ψ. Inspection of the table shows that the transmittance is increased by about 10 per cent by shifting from 2.2 to 11 μ. The difference between panchromatic film (0.5 μ) and infrared film (0.85 μ) is seen to be nearer 50 per cent. For additional data, see reference 88.

4.9 TRANSMISSION THROUGH RAIN

Experience in the field shows that the presence of rain along the line of sight, unlike clouds and fogs, will reduce the performance of most infrared systems, but in many cases they will still be able to perform their intended mission[81]. Since the raindrops are many times larger than the wavelength, the value of ψ is approximately zero and there is no wavelength dependent scattering from rain in the infrared.

Middleton[89] discusses the effect of rain on visual range and shows that the scattering coefficent can be calculated from

$$\gamma = 1.25 \times 10^{-6} \frac{z}{r^3}, \qquad (4\text{-}22)$$

where z is the rainfall rate in centimeters (of depth) per second and r is the radius of the drop, also in centimeters. Under normal conditions,

TABLE 4.13 CALCULATED TRANSMITTANCE THROUGH 6 kft OF HAZE

Wavelength (μ)	Transmittance $\psi=0.7$	$\psi=1.3$
0.61	0.60	0.60
2.2	0.81	0.91
3.8	0.87	0.95
11	0.93	0.99
0.5 (panchromatic film)	—	0.52
0.85 (infrared film)	—	0.72

a raindrop can grow until its radius is about 3 mm; the drop than becomes unstable and breaks up. We cannot assume an average or equivalent drop size for (4-22); instead a statistical distribution must be used. The details of such a calculation are shown in Table 4.14. The rainfall data are taken from Laws and Parsons[90] and are representative of a cloudburst with a rainfall rate of 4 in. (100 mm) per hour. The second column shows the number of drops of each size striking a horizontal area of

TABLE 4.14 CALCULATION OF THE SCATTERING COEFFICIENT FOR RAIN: CLOUDBURST CONDITIONS, RAINFALL RATE 4 in. (100 mm) PER HOUR[a]

Drop Radius (cm)	Number of Drops per cm^2 in 100 sec	Rainfall Rate (cm sec^{-1})	Scattering Coefficient (cm^{-1})
0.025	43	2.78×10^{-5}	2.24×10^{-6}
0.05	21.4	1.11×10^{-4}	1.11×10^{-6}
0.075	14.3	2.50×10^{-4}	7.43×10^{-7}
0.10	9.3	3.89×10^{-4}	4.87×10^{-7}
0.125	5.8	4.72×10^{-4}	$3.02 - 10^{-7}$
0.150	3.6	5.00×10^{-4}	1.865×10^{-7}
0.175	1.8	4.07×10^{-4}	9.59×10^{-8}
0.200	0.75	2.50×10^{-4}	3.91×10^{-8}
0.225	0.35	1.67×10^{-4}	1.83×10^{-8}
0.250	0.13	8.36×10^{-5}	6.76×10^{-9}
0.275	0.064	5.56×10^{-5}	3.34×10^{-9}
0.300	0.024	2.78×10^{-5}	1.29×10^{-9}
0.325	0.019	2.78×10^{-5}	1.02×10^{-9}
	$\Sigma = 100.54$		$\Sigma = 5.234 \times 10^{-6}$

[a] Rainfall data from Laws and Parsons[90].

TRANSMISSION OF INFRARED RADIATION 165

1 cm² in 100 sec (100 sec is used in order to give numbers of a convenient magnitude). There is very nearly 1 drop per square centimeter per second, and about 80 per cent of these drops lie in the first three size intervals. From the last column it is evident that the small drops contribute most heavily to the value of the scattering coefficient. In addition, Middleton shows that the scattering coefficient for rain depends only on the number of drops falling on a unit horizontal area per second. This is, at first glance, a most surprising result but it is clearly borne out in the calculations.

With the scattering coefficient that was calculated in Table 4.14, for a cloudburst, the transmittance over a 6 kft (1.8 km) path is about 38 per cent. Values of the transmittance for other rainfall rates are given in Table 4.15. These values are useful for making order-of-magnitude estimates. Any losses by absorption along the path must, of course, be added.

TABLE 4.15 TRANSMITTANCE OF A
6 kft PATH THROUGH RAIN

Condition	Rainfall Rate (cm hr^{-1})	Transmittance of a 6 kft (1.8 km) path
Light rain	0.25	0.88
Medium rain	1.25	0.74
Heavy rain	2.5	0.65
Cloudburst	10.0	0.38

For example, the precipitable water content of a 6 kft path at sea level is about 18 mm (from Figure 4.4). For a spectral bandpass of 3.2 to 4.8 μ, Figure 4.15 shows that the effective transmittance under these conditions is about 55 per cent. A heavy rain along the path will reduce this value to 36 per cent (0.65 × 0.55).

4.10 ATMOSPHERIC SCINTILLATION

We are all familiar with the twinkling or shimmering of distant objects during the day and of distant lights or stars at night. This phenomenon is called *scintillation*. It is easy to conceive of situations in which scintillation may limit the effectiveness of visual systems, but it is not often recognized that infrared systems can also be affected. When a beam of light passes through regions in which the temperature varies, such as the air above a road heated by sunlight, it is diverted slightly from its original direction. Since such regions of heated air are unstable, the deviation of the beam is a random, time-varying quantity. Thus we expect scintillation to be most pronounced when the line of sight passes close to the surface of the earth.

Two major effects are observed, either singly or in combination; these are the variation in the irradiance from a distant source and the variation in the apparent direction of the source. In many infrared systems a reticle is used to modulate the radiant flux before it arrives at the detector. If scintillation by the atmosphere has modulated the flux before it reaches the reticle, the resulting double modulation can introduce serious errors in the signal processing. Since the variation in irradiance caused by scintillation is, in effect, an amplitude modulation, it would appear unwise to choose a system using amplitude modulation for any application in which the line of sight passes within a few feet of the surface of the earth. Most of the measurements of scintillation have been made in the visible, but they can be used as a guide to the magnitude of the effect to be expected in the infrared [91-96]. One such series of measurements [97] shows that the irradiance from a tungsten lamp at a distance of 3.9 kft had a 65 per cent modulation at a frequency of 9 Hz.

The second effect caused by scintillation, the variation in the apparent direction of a distant target, may limit the accuracy that can be achieved with an angle-tracking system. Once again, only a few measurements have been made in the infrared and we must assume that measurements made in the visible are representative. Riggs et al. [98] set up a camera in a railway tunnel and photographed targets at various distances outside the entrance. For comparison, one target was located in the tunnel on the supposition that the air in the tunnel was at a nearly uniform temperature and would therefore produce little optical disturbance. The line of sight was 30 in. above the ground. Deviations as large as 9 arc sec were observed in the apparent direction of these targets on sunny days. The instantaneous deviations of points separated by more than 5 arc min were found to be uncorrelated, an observation that gives some indication of the size of the atmospheric structure causing the scintillation.

REFERENCES

[1] H. A. Gebbie et al., "Atmospheric Transmission in the 1 to 14 μ Region," *Proc. Roy. Soc.*, **A206**, 87 (1951).

[2] "U. S. Standard Atmosphere, 1962," NAS 1.2: At 6/962, U.S. Government Printing Office, Washington D.C., 1963.

[3] I. J. Spiro, R. C. Jones, and D. Q. Wark, "Symbols, Units and Nomenclature for Atmospheric Transmission," *Soc. Phot. Instr. Engrs. J.*, **4**, 116 (1966).

[4] M. Ference, Jr., "Instruments and Techniques for Meteorological Measurements," p. 1207 in T. F. Malone, ed., *Compendium of Meteorology*, Boston: American Meteorological Society, 1951.

[5] L. W. Foskett et al., "Apparatus for Absorption Spectra Analysis," U. S. Patent No. 2,775,160 December 25, 1956.
[6] K. Y. Kondratyev et al., "Atmospheric Optics Investigations on Mt. Elbrus," *Appl. Opt.*, **4**, 1069 (1965).
[7] M. Gutnick, "How Dry is the Sky," *J. Geophys. Research*, **66**, 2867 (1961).
[8] F. Stauffer and J. Strong, "High Altitude Atmospheric Transmission Measurements," *Appl. Opt.*, **1**, 129 (1962).
[9] M. Gutnick, "Mean Atmospheric Moisture Profiles to 31 km for Middle Latitudes," *Appl. Opt.*, **1**, 670 (1962).
[10] D. M. Gates et al., "Near Infrared Solar Radiation Measurements by Balloon to an Altitude of 100,000 Feet," *J. Opt. Soc. Am.*, **48**, 1010 (1958).
[11] C. B. Farmer and P. J. Key "A Study of the Solar Spectrum from 7 μ to 400 μ," *Appl. Opt.*, **4**, 1051 (1965).
[12] J. A. Curcio and C. C. Petty, "Near Infrared Absorption Spectrum of Liquid Water," *J. Opt. Soc. Am.*, **41**, 302 (1951).
[13] E. K. Plyler and N. Aquista, "IR Absorption of Liquid Water from 2 to 42 μ," *J. Opt. Soc. Am.*, **44**, 505 (1954).
[14] F. Hageman et al., "Stratospheric Carbon-14, Carbon Dioxide, and Tritium," *Science*, **130**, 542 (1959).
[15] D. G. Murcray, F. H. Murcray, and W. J. Williams, "Variation of the Infrared Solar Spectrum between 2800 and 1500 cm^{-1} with Altitude," *J. Opt. Soc. Am.*, **54**, 23 (1964).
[16] H. C. Willett, *Descriptive Meteorology*, New York: Academic Press, 1944, p. 12.
[17] F. S. Johnson et al., "Direct Measurements of the Vertical Distribution of Atmospheric Ozone to 70 km Altitude," *J. Geophys. Research*, **57**, 157 (1952).
[18] L. Elterman, "A Representative Vertical Ozone Distribution for Atmospheric Transmission Studies," *Appl. Opt.*, **3**, 640 (1964).
[19] M. Migeotte, L. Neven, and E. Vigroux, "Fine Structure of the Absorption Band of Ozone at 9.6 μ, Observed in Solar and Laboratory Spectra," *Physica*, **18**, 982 (1952).
[20] D. C. Burch, "Infrared Evidence for the Presence of Ozone in the Lower Atmosphere," *J. Opt. Soc. Am.*, **46**, 360 (1956).
[21] R. L. Bowman and J. H. Shaw, "The Abundance of Nitrous Oxide, Methane, and Carbon Monoxide in Ground-Level Air," *Appl. Opt.*, **2**, 176 (1963).
[22] D. H. Rank et al., "Abundance of N_2O in the Atmosphere," *J. Opt. Soc. Am.*, **52**, 858 (1962).
[23] U. Fink, D. H. Rank, and T. A. Wiggins, "Abundance of Methane in the Earth's Atmosphere," *J. Opt. Soc. Am.*, **54**, 472 (1964).
[24] S. L. Locke and L. Herzberg, "The Absorption Due to Carbon Monoxide in the Infrared Solar Spectrum," *Can. J. Phys.*, **31**, 504 (1953).
[25] L. Goldberg, "The Absorption Spectrum of the Atmosphere," Chapter 9 in G. Kuiper, ed., *The Earth as a Planet*, Chicago: University of Chicago Press, 1954.
[26] S. P. Langley, "Measurements of the Solar Spectrum to 5.5 Microns," *Ann. Astrophys. Obs. Smithsonian Inst.*, **1**, 1 (1900).
[27] E. S. Barr, "Historical Survey of the Early Development of the Infrared Spectral Region," *Am. J. Phys.*, **28**, 42 (1960).
[28] F. E. Fowle, "Extended Measurements of the Solar Spectrum," *Smithsonian Misc. Coll.*, **68**, 1 (1917).
[29] A. Adel, "New Measurements of the Solar Spectrum Extending to 24 Microns," *Astrophys. J.*, **96**, 239 (1942).
[30] J. H. Shaw et al., "A Grating Map of the Solar Spectrum from 3.0 to 5.2 Microns," *Astrophys. J.*, **113**, 268 (1951).

[31] J. H. Shaw, M. L. Oxholm, and M. H. Claassen, "The Solar Spectrum from 7 to 13 Microns," *Astrophys. J.*, **116**, 554 (1952).
[32] M. Migeotte, L. Neven, and J. Swenson, "The Solar Spectrum from 2.8 to 23.7 Microns," Geophysics Research Directorate, Air Force Cambridge Research Center, Bedford, Massachusetts, 1957.
[33] J. Yarnell and R. M. Goody, "Infrared Solar Spectroscopy in a High Altitude Aircraft," *J. Sci. Inst.*, **29**, 353 (1952).
[34] D. G. Murcray et al., "Atmospheric Absorptions in the Near Infrared at High Altitudes," *J. Opt. Soc. Am.*, **50**, 107 (1960).
[35] D. G. Murcray et al., "Study of 1.4, 1.9, and 6.3 Micron Water Vapor Bands at High Altitudes," *J. Opt. Soc. Am.*, **51**, 186 (1961).
[36] D. M. Gates and W. J. Harrop, "Infrared Transmission of the Atmosphere to Solar Radiation," *Appl. Opt.*, **2**, 887 (1963).
[37] C. B. Farmer and S. J. Todd, "Absolute Solar Spectra 3.5–5.5 Microns. I. Experimental Spectra for Altitude Range 0–15 km," *Appl. Opt.*, **3**, 453 (1964). "II. Theoretical Spectra for Altitude Range 15–30 km," **3**, 459 (1964).
[38] J. H. Taylor and H. W. Yates, "Atmospheric Transmission in the Infrared," *J. Opt. Soc. Am.*, **47**, 223 (1957).
[39] A. Arnulf and J. Bricard, "Transmission of Infrared Radiation in the Atmosphere," *Acta Electronica*, **5**, 409 (1961) (SLA: TT-65-10822).
[40] V. E. Zuev, M. E. Elyashberg, and G. A. Safonova, "Transmittance of Thin Atmospheric Layers in the 1–13 Micron Range," *Izv. Vysshikh Ucheb. Zavednii Fiz.*, **5**, 77 (1960) (OTS 61-28701).
[41] A. Adel, "An Estimate of the Transparency of the Atmospheric Window 16 to 24 Microns," *J. Opt. Soc. Am.*, **37**, 769 (1947).
[42] A. Adel, "The Atmospheric Windows at 6.3 μ and 16 to 24 μ," *Infrared Phys.*, **2**, 31 (1962).
[43] T. Elder and J. Strong, "The Infrared Transmission of Atmospheric Windows," *J. Franklin Inst.*, **255**, 189 (1953).
[44] E. F. Barker and A. Adel, "Resolution of the Two Difference Bands of Carbon Dioxide Near 10 Microns," *Phys. Rev.*, **44**, 185 (1933).
[45] J. R. Nielsen, V. Thornton, and E. B. Dale, "The Absorption Laws for Gases in the Infrared," *Rev. Modern Phys.*, **16**, 307 (1944).
[46] J. N. Howard, D. E. Burch, and D. Williams, "Infrared Transmission of Synthetic Atmospheres," *J. Opt. Soc. Am.*, **46**, 186, 237, 242, 334, 452 (1956).
[47] J. U. White, "Long Optical Paths of Large Aperture," *J. Opt. Soc. Am.*, **32**, 285 (1942).
[48] D. E. Burch and D. Williams, "Total Absorptance by Nitrous Oxide Bands in the Infrared," *Appl. Opt.*, **1**, 473 (1962).
[49] D. E. Burch and D. Williams, "Total Absorptance of Carbon Monoxide and Methane," *Appl. Opt.*, **1**, 587 (1962).
[50] D. E. Burch, D. A. Gryvnak, and D. Williams, "Total Absorptance of Carbon Dioxide in the Infrared," *Appl. Opt.*, **1**, 759 (1962).
[51] D. E. Burch, W. L. France, and D. Williams, "Total Absorptance of Water Vapor in the Near Infrared," *Appl. Opt.*, **2**, 585 (1963).
[52] W. M. Elsasser, *Heat Transfer by Infrared Radiation in the Atmosphere*, Harvard Meteorological Studies, No. 6, Cambridge, Mass.: Harvard University Press, 1942.
[53] G. N. Plass, "Models for Spectral Band Absorption," *J. Opt. Soc. Am.*, **48**, 690 (1958).
[54] G. N. Plass, "Useful Representations for Measurements of Spectral Band Absorption," *J. Opt. Soc. Am.* **50**, 868 (1960).
[55] P. J. Wyatt, V. R. Stull, and G. N. Plass, "Quasi-Random Model of Band Absorption," *J. Opt. Soc. Am.*, **52**, 1209 (1962).

[56] G. N. Plass, "Spectral Band Absorptance for Atmospheric Slant Paths," *Appl. Opt.*, 2, 515 (1963).
[57] D. E. Burch and D. Williams, "Tests of Theoretical Absorption Band Model Approximations," *Appl. Opt.*, 3, 55 (1964).
[58] "Line Parameters and Computed Spectra for Water Vapor Bands at 2.7 μ" Monograph 71, National Bureau of Standards, Washington, D.C.: U.S. Government Printing Office, 1964.
[59] R. F. Calfee and W. S. Benedict, "Carbon Dioxide Spectral Line Positions and Intensities Calculated for the 2.05 and 2.7 micron Regions," National Bureau of Standards, Washington, D.C.: U.S. Government Printing Office, 1966.
[60] D. E. Burch, D. A. Gryvnak, and R. B. Patty, "Absorption of Infrared Radiation by CO_2 and H_2O. Experimental Techniques," *J. Opt. Soc. Am.*, 57, 885 (1967).
[61] J. N. Howard, "The Transmission of the Atmosphere in the Infrared," *Proc. Inst. Radio Engrs.*, 47, 1451 (1959).
[62] J. N. Howard and J. S. Garing, "The Transmission of the Atmosphere in the Infrared — A Review," *Infrared Phys.*, 2, 155 (1962).
[63] P. J. Wyatt, V. R. Stull, and G. N. Plass, "The Infrared Transmittance of Water Vapor," *Appl. Opt.*, 3, 229 (1964).
[64] V. R. Stull, P. J. Wyatt, and G. N. Plass, "The Infrared Transmittance of Carbon Dioxide," *Appl. Opt.*, 3, 243 (1964).
[65] S. Passman and L. Larmore, "Atmospheric Transmission," Rand Paper P-897, Santa Monica, Cal.: The Rand Corporation, 11 July 1956.
[66] T. L. Altshuler, "A Procedure for Calculation of Atmospheric Transmission of Infrared," Report R57ELC 15, Ithaca, N.Y.: General Electric Co., 1 May 1957.
[67] A. E. S. Green and M. Griggs, "Infrared Transmission Through The Atmosphere," *Appl. Opt.*, 2, 561 (1963).
[68] R. F. Calfee and D. M. Gates, "Calculated Slant-Path Absorption and Distribution of Atmospheric Water Vapor," *Appl. Opt.*, 5, 287 (1966).
[69] W. E. K. Middleton, *Vision Through the Atmosphere*, Toronto: University of Toronto Press, 1952.
[70] H. C. van de Hulst, *Light Scattering by Small Particles*, New York: Wiley, 1957.
[71] H. C. van de Hulst, "Scattering in the Atmospheres of the Earth and Planets," in G. P. Kuiper, ed., *The Atmospheres of the Earth and Planets*, Chicago: University of Chicago Press, 1952, p. 49.
[72] D. Deirmendjian, "Scattering and Polarization Properties of Water Clouds and Hazes in the Visible and Infrared," *Appl. Opt.*, 3, 187 (1964).
[73] H. G. Houghton and W. R. Walker, "The Scattering Cross Section of Water Drops in Visible Light," *J. Opt. Soc. Am.*, 39, 955 (1949).
[74] M. Neiburger and M. G. Wurtele, "On the Nature and Size of Particles in Haze, Fog, and Stratus of the Los Angeles Region," *Chem. Revs.*, 44, 321 (1949).
[75] A. E. Mikirov, "Measurement of the Size Spectrum of Cloud and Fog Particles," *Izv. Akad. Nauk SSSR Ser. Geofiz.*, 512, no 4 (1957) (OTS AD-608971).
[76] A. Arnulf et al., "Transmission by Haze and Fog in the Spectral Region 0.35 to 10 Microns," *J. Opt. Soc. Am.*, 47, 491 (1957).
[77] J. A. Sanderson, "The Transmission of Infrared Light by Fog," *J. Opt. Soc. Am.*, 30, 405 (1940).
[78] J. A. Sanderson, "The Attenuation of Infrared Light by Fog," *Phys. Rev.*, 57, 1060, 1940.
[79] E. I. Bocharov, "The Attenuation of Infrared Radiation by Water Fogs," *Izv. Akad. Nauk SSSR Ser. Geofiz.*, 791, no 6 (1958) (SLA: TT 63-24291).

[80] S. W. Kurnick, R. N. Zitter, and D. B. Williams, "Attenuation of Infrared Radiation by Fogs," *J. Opt. Soc. Am.*, **50**, 578 (1960).
[81] M. E. Seymour, "In Fog and Rain—Sight, Infrared, or Radar?," *Electronics*, **33**, 64 (January 29, 1960).
[82] D. M. Gates and C. C. Shaw, "Infrared Transmission of Clouds," *J. Opt. Soc. Am.*, **50**, 876 (1960).
[83] A. Arking, "Latitudinal Distribution of Cloud Cover from Tiros III Photographs," *Science*, **143**, 569 (1964).
[84] N. M. Mohler, "Photographic Penetration of Haze," *J. Opt. Soc. Am.*, **26**, 219 (1936).
[85] J. A. Curcio, "Evaluation of Atmospheric Aerosol Particle Size Distribution from Scattering Measurements in the Visible and Infrared," *J. Opt. Soc. Am.*, **51**, 548 (1961).
[86] L. F. Elterman, "Parameters for Attenuation in the Atmospheric Windows for Fifteen Wavelengths," *Appl. Opt.*, **3**, 745 (1964).
[87] M. G. Gibbons, "Wavelength Dependence of the Scattering Coefficient for Infrared Radiation in Natural Haze," *J. Opt. Soc. Am.*, **48**, 172 (1958).
[88] R. G. Eldridge, "A Comparison of Computed and Experimental Spectral Transmissions through Haze," *Appl. Opt.*, **6**, 929 (1967).
[89] Middleton, *op. cit.* (reference 69), p. 121.
[90] J. O. Laws and D. A. Parsons, "The Relation of Raindrop Size to Intensity," *Trans. Am. Geophys. Union*, **24**, 452 (1943).
[91] M. A. Ellison, "The Effects of Scintillations on Telescopic Images," in Z. Kopal, ed., *Proceedings of a Symposium on Astronomical Optics and Related Subjects*, New York: Interscience, 1956.
[92] I. Goldstein, P. A. Miles, and A. Chabot, "Heterodyne Measurements of Light Propagation Through Atmospheric Turbulence," *Proc. IEEE*, **53**, 1172 (1965).
[93] H. R. Carlon, "The Apparent Dependence of Terrestrial Scintillation Intensity upon Atmospheric Humidity," *Appl. Opt.*, **4**, 1089 (1965).
[94] B. N. Edwards and R. R. Steen, "Effects of Atmospheric Turbulence in the Transmission of Visible and Near Infrared Radiation," *Appl. Opt.*, **4**, 311 (1965).
[95] V. I. Tatarski, *Wave Propagation in a Turbulent Medium*, New York: McGraw-Hill, 1961.
[96] J. I. Davis, "Consideration of Atmospheric Turbulence in Laser Systems Design." *Appl. Opt.*, **5**, 139 (1966).
[97] Middleton, *op. cit.* (ref. 69), p. 81.
[98] L. A. Riggs et al., "Photographic Measurements of Atmospheric Boil," *J. Opt. Soc. Am.*, **37**, 415 (1947).

5

Optics

Optical design is a highly complex subject, and the services of an experienced optical designer are essential to the project team responsible for the development of a new infrared system. The system engineer should, however, understand the problems faced by the optical designer and the methods available for their solution. In addition he should clearly understand the limitations that the optics will impose on the performance of his system. Providing this information is the objective of this chapter.

5.1 REFRACTION AND REFLECTION

The velocity of light[1] given in section 2.4 is its velocity in empty space, that is, in a perfect vacuum; its velocity in any other medium is less. For a particular material and wavelength, the ratio of the velocity of light in a vacuum to that in the material is called its *index of refraction*. A graph of the index of refraction plotted as a function of wavelength is called a *dispersion curve*. Ordinary glass has an index of refraction of 1.5; thus the velocity of light in glass is two-thirds of what it is in a vacuum. Optical materials useful in the infrared have indices ranging from about 1.3 to 4. The index of air is about 1.00029, and the correction to vacuum conditions is very small and need only be applied to ultraprecise wavelength determinations. In this book we have no need for this degree of accuracy, and will therefore assume that wavelengths measured in air are equivalent to those measured in a vacuum.

A convenient way to visualize the radiant flux from a source is as a series of concentric, ever-expanding, spherical waves; the ripples from a pebble dropped into a pool provide a useful two-dimensional analogy. When a portion of the wave front passes through an optical system, its curvature is changed. Thus we could describe an optical system by the change it causes in the curvature of the wave front. Since such changes

---
[1]Light is used here in the purely objective or physical sense to refer to electromagnetic waves or photons.

172   THE ELEMENTS OF THE INFRARED SYSTEM

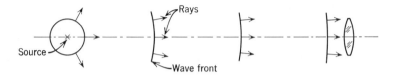

Figure 5.1   Wave fronts and rays at various distances from a source.

are not easily visualized, the idea of rays has been introduced. A *ray* is simply a normal to the wave front and points in the direction in which the wave is moving. When the radiant flux is considered as a stream of photons, the ray (quite conveniently) represents the path followed by a photon.

One of the key assumptions in describing the performance of an optical system is the use of a "distant source" or a "source at infinity." These terms describe a condition in which the wave front entering the optics can be considered as a plane rather than as a portion of a sphere. Alternatively they mean that all of the rays entering the optics are parallel to one another. This effect is shown in Figure 5.1, where several wave fronts and rays are moving toward a lens. If we look only at the portion of the wave front that the lens accepts, it is evident that the curvature of this portion decreases with distance from the source and that the rays tend to become parallel to one another. Obviously, only when the source is at infinity is this portion of wave front truly plane and the rays parallel to one another. The collimator is a simple optical means of producing a source at infinity; it is widely used for testing infrared equipment and will be discussed more fully in section 5.11.

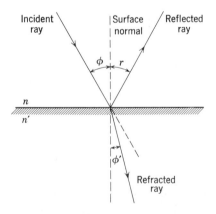

Figure 5.2   Reflected and refracted rays.

OPTICS 173

When a ray meets the interface between two materials having different indices of refraction, it is split into two rays; as shown in Figure 5.2, one ray is *reflected* and the other is *refracted* into the second medium. *Snell's law* or the *Law of Refraction* describes the refracted ray:

$$n \sin\phi = n' \sin\phi', \qquad (5\text{-}1)$$

where n and n' are the indices of refraction of the two media, ϕ is the *angle of incidence*, and ϕ' the *angle of refraction*.[1] Thus the product of the index of refraction and the sine of the angle the ray makes with the normal to the surface remain invariant across the interface. The reflected ray is described by the *Law of Reflection*:

$$r = \phi. \qquad (5\text{-}2)$$

This simply means that the angle of reflection equals the angle of incidence. In addition the incident, reflected, and refracted rays are always coplanar (sometimes called the Second Law of Reflection).

With a *refractive optical system*, that is, one employing lenses, the reflected ray represents energy lost to the system and it is important to know the magnitude of this loss. An exact calculation requires knowledge of the angle of incidence, the index of refraction, and the polarization of the incident flux. However, for the special case of an unpolarized ray incident from air ($n = 1$) and an angle of incidence of zero ($\phi = 0$), the fraction of the flux reflected by a single plane surface is

$$\rho_s = \left(\frac{n-1}{n+1}\right)^2, \qquad (5\text{-}3)$$

where ρ_s is the reflectance of the surface. The fraction of the flux transmitted by the surface is

$$\tau = 1 - \rho_s = 1 - \left(\frac{n-1}{n+1}\right)^2. \qquad (5\text{-}4)$$

In a plate having plane surfaces parallel to one another, some of this flux is absorbed, some emerges from the second surface, and some is reflected back and forth between the surfaces of the plate. Considering all these effects, the transmittance of a parallel-sided plate is

$$\tau = \frac{(1-\rho_s)^2 e^{-ax}}{1 - \rho_s^2 e^{-2ax}}, \qquad (5\text{-}5)$$

[1] By general convention, angles are measured with respect to the normal at the point of incidence, quantities in the incident media are unprimed, and figures are drawn so that rays are incident from the left.

where ρ_s is given by (5-3), a is the absorption coefficient, and x is the thickness of the material. There are several ways to simplify this expression. In many applications we can select an optical material having no absorption bands close to the spectral bandpass of the system. Under these conditions the absorption coefficient is very small in the spectral region of interest and the transmittance is approximately

$$\tau = \frac{(1-\rho_s)^2}{1-\rho_s^2} = \frac{2n}{n^2+1}. \qquad (5\text{-}6)$$

Thus the transmittance of a germanium plate ($n=4$) is about 0.47. If the index of refraction is less than 1.9 or if reflection-reducing coatings have been applied so that the value of ρ_s is less than 0.1, the second term in the denominator of (5-5) can be ignored and the transmittance is given with an error of less than 1 per cent by

$$\tau = (1-\rho_s)^2 e^{-ax}. \qquad (5\text{-}7)$$

Under these conditions the absorption coefficient is given with an error of less than 1 per cent by

$$a = \frac{1}{x} \ln \frac{(1-\rho_s)^2}{\tau}. \qquad (5\text{-}8)$$

Although (5-3) through (5-8) are derived for a ray incident normal to the surface, they can be used without appreciable error for angles of incidence up to 45 deg.

Published data on the transmittance of optical materials are usually given for an arbitrary sample thickness, often 2 mm. At times it may be necessary to estimate the transmittance for some other thickness; Example 5.1 shows how such an estimate can be made.

EXAMPLE 5.1 A sample of a (fictitious) material 2 mm thick is reported to have a spectral transmittance of 0.840 and an index of refraction of 1.75. Estimate the spectral transmittance of a sample 10 mm thick.

From (5-3) the value of ρ_s is 0.0744. From (5-8),

$$a = \frac{1}{2} \ln \frac{(1-0.0744)^2}{0.840} = 0.00985 \text{ mm}^{-1}.$$

For the new thickness of 10 mm, from (5-7),

$$\tau = (1-0.0744)^2 e^{-(0.00985)(10)} = 0.776.$$

The log log scales on many slide rules are particularly convenient for calculations of this type.

OPTICS

The experimental determination of the absorption coefficient is quite simple if two samples of the material are available, each of a different thickness. If the measured spectral transmittances are τ_1 and τ_2 for sample thicknesses of x_1 and x_2 ($x_1 > x_2$), manipulation of (5-7) gives

$$a = \frac{1}{(x_1 - x_2)} \ln \frac{\tau_2}{\tau_1}. \tag{5-9}$$

Notice that it is not necessary to know the index of refraction or to calculate the reflectance term in order to determine the value of the absorption coefficient by this method.

Let us examine more closely the relationship between the incident and refracted rays by rewriting (5-1) as

$$\sin \phi' = \frac{n}{n'} \sin \phi. \tag{5-10}$$

In Figure 5.2, when the incident ray is in the medium having the lower index of refraction, $n/n' < 1$ and $\sin \phi' < \sin \phi$. Thus, even if the incident ray is at grazing incidence ($\phi \sim 90°$), there will still be a refracted ray. If the incident ray is in the medium having the higher index of refraction, $\sin \phi' > \sin \phi$, and there will not be a refracted ray when the incident angle exceeds a certain value. This occurs when $\sin \phi'$ is equal to unity and the refracted ray is tangent to the interface. For greater angles of incidence there is no refracted ray, only a reflected one. This effect is called *total internal reflection* and the angle of incidence at which it occurs is called the *critical angle* ϕ_c where

$$\sin \phi_c = \frac{n'}{n}. \tag{5-11}$$

If the refracted ray is in air, this reduces to

$$\sin \phi_c = \frac{1}{n}. \tag{5-12}$$

Thus, for a ray passing from germanium into air, the critical angle is about 14.5°, and all rays at higher angles of incidence are reflected back into the germanium.

5.2 DESCRIBING AN OPTICAL SYSTEM

To the system engineer the purpose of the optics in his infrared system is to collect radiant flux and deliver it to the detector. Thus the optics are quite analogous to a radar antenna used to receive echoes from a target. He usually knows what field of view the optics must cover, the spectral region over which they will be used, and he has a rough idea of the space into which they must fit. The optics become, in effect, another "black box" that he must integrate into his system.

Reduced to its basic elements, any optical system consists of one or more reflecting or refracting elements. All elements are considered to be centered; that is, the centers of curvature of each of the surfaces all lie on the same straight line, called the *optical axis*. Any departure from this condition because of poor manufacturing or careless mounting impairs the performance of the optics.

Using various simplified formulas, the optical designer lays out a preliminary design and then examines it in detail to see how well it meets the specifications. His principal tool is *ray tracing*, that is, examining the paths followed by rays passing through the optics. This can be done by applying Snell's law, or the Law of Reflection, at each surface that the ray encounters. Such a procedure is inherently very accurate, and it is customary to retain figures to the fifth or sixth decimal place. Modern electronic computers can make such calculations at the rate of a few seconds per ray for a complex system.

Since Snell's law involves the sine of various angles, the ray-tracing equation can be simplified by replacing the sine by its series expansion

$$\operatorname{Sin} \phi = \phi - \frac{\phi^3}{3!} + \frac{\phi^5}{5!} - \frac{\phi^7}{7!} + \cdots \qquad (5\text{-}13)$$

For preliminary evaluation, one often assumes that all terms beyond the first can be neglected. The result is *first-order* or *paraxial* equations. The *paraxial rays* lie very close to the optical axis so that the sines of their angles may be replaced by the angles expressed in radians with negligible error. Better approximations are obtained if the first two terms in the series are included (called *third-order theory*) or if the first three terms are included (called *fifth-order theory*).

A simple thin lens and concave mirror are shown in Figure 5.3. Rays from an object are brought to a focus at the point F'. An image of the object is formed on the *focal plane*, a plane that is normal to the optical axis and passes through F'. Higher-order theory shows that the focal plane may be a slightly curved surface rather than a plane. In a well-designed system the departure from a plane is small and is usually ignored. Evidence for this is seen in photographic cameras, where great pains are taken to hold the film flat. If the rays incident on the lens or mirror are parallel to the optical axis (from an axial object at an infinite distance), the focus at F' is on the axis and is called the *focal point*. The distance from the center of a lens (assumed to have negligible thickness)

OPTICS 177

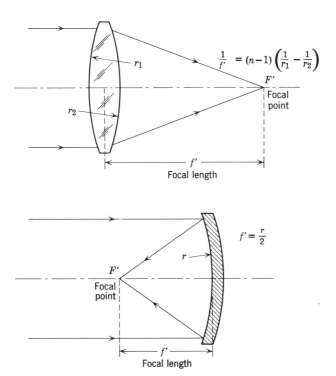

Figure 5.3 Focal length and focal point of a thin lens and of a concave mirror.

to the focal point is called the *focal length* and is the most important parameter used to describe a lens. Likewise, for a concave mirror, the focal length is the distance from the vertex of the mirror surface to the focal point. The formulas in Figure 5.3 show how the focal length can be computed. For a spherical mirror, focal length is equal to half the radius of curvature of the reflecting surface. For a thin lens, it is a function of the index of refraction and the curvature of the surfaces of the lens (the sign convention may be found in [1]).

In defining focal length we referenced our measurement to the center of a thin lens. Such a procedure becomes meaningless for a multielement lens, such as that shown in Figure 5.4. An incident ray that is initially parallel to the optical axis is refracted by the various surfaces and finally passes through the focal point. It would be convenient to replace the various elements by a single fictitious refracting surface that causes a deviation of the ray equivalent to that caused by the actual optics. If the incident and final rays are extended, as in Figure 5.4, in order to intersect one another, they give the location of this equivalent refracting surface.

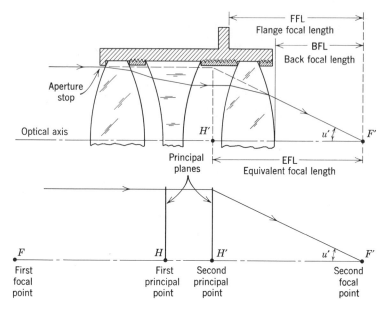

Figure 5.4 Focal length of a multielement lens.

In simple paraxial theory this surface is a plane normal to the optical axis, called the *principal plane*; in higher-order theory, it is spherical with its center at the focal point. In both cases the intersection of the equivalent refracting surface with the optical axis is called a *principal point*. The location of the principal point is a unique characteristic of an optical system and is a convenient point of reference from which to measure focal length. This is called the *equivalent focal length* (EFL); it is the distance between the principal point and the focal point. Referring to the way in which focal length was defined for a thin lens, it is evident that this definition assumes that the principal point lies at the center of the lens.

By tracing another ray through the lens but in the opposite direction, we can locate a second pair of focal and principal points. The equivalent focal length associated with each pair will be identical if the lens is used in air. By convention the pair of points to the left is called first focal point, and so on, and this pair is designated by unprimed quantities. Those points to the right are called second focal point, and so on, and are designated by primed quantities. The lower part of Figure 5.4 shows an equivalent representation of the multielement lens. The same principle can be applied to a multielement reflective system by considering that the principal plane represents an equivalent reflecting surface.

Since the principal and focal points rarely coincide with one of the surfaces in a system, a means of referencing them to some convenient point is needed. One way to do this is to give the distance of the focal point from the vertex of the last optical surface, the *back focal length* (BFL). This is usually unsatisfactory because it is difficult to locate the true vertex of a surface (unless it happens to be a plane). Since any multielement optical system will be assembled into a mounting, it is convenient to give the location of the focal point with respect to a flange on this mounting. This dimension is called the *flange focal length* (FFL). It is unfortunate that these terms are often used interchangeably, because they are not directly equivalent.

In considering the amount of radiant flux collected by an optical system, it is important to know the diameter of the largest bundle of rays that can pass through the optics without obstruction. The physical object that limits this bundle is called the *aperture stop*. Should there be optical elements in front of the aperture stop, the image of the stop that they form is called the *entrance pupil*. In a photographic lens, for instance, the iris diaphragm is the aperture stop and its image as seen from the front of the lens, is the entrance pupil.

There are two important ways of describing the amount of radiant flux collected by an optical system. The first uses the f/number of the optics

$$f/\text{no} = \frac{f}{D}, \tag{5-14}$$

where f is the equivalent focal length and D is the diameter of the aperture stop or entrance pupil. The f/number concept is familiar to most photographers, who usually refer to it as the speed of the optics. It is unfortunate that the f/number is an inverse quantity; that is, the smaller the f/number the greater the radiant flux collected and the higher the speed of the optics.

An alternative means of describing flux collection is the *numerical aperture*, which is given by

$$NA = n' \sin u', \tag{5-15}$$

where n' is the index of refraction of the medium between the final optical surface and the second focal point, and u' is the half-angle of the cone of rays converging at the focal point (see Figure 5.4). The relationship between numerical aperture and f/number is

$$NA = \frac{1}{2(f/\text{no})}. \tag{5-16}$$

If the optics are used in air (so that $n' = 1$), the maximum possible value of the numerical aperture is 1 and the corresponding value of the f/-

number is 1/2.[1] Physically, this means that the cone of rays reaching the second focal point has an included angle of 180°.

If there are a number of objects at various distances along the optical axis, each is brought to a focus at a different distance from the optics. The image of an infinitely distant object lies closest to the optics, and images of closer objects lie at progressively greater distances. It is essential that optical modulators, and sometimes detectors, be accurately located at the image plane. If the system is to be used over a range of object distances, it may be necessary to provide a mechanical device for changing the focus. Although it is possible to accomplish this change by moving the modulator or detector, it is usually easier to move the optics. The amount of focusing motion required can be found from

$$xx' = f^2, \tag{5-17}$$

where x is the object distance *measured from the first focal point*[2] and x' is the image distance *measured from the second focal point*. Thus, for object distances ranging from infinity to 10 m (32.8 ft) and optics having an equivalent focal length of 10 cm, the image will shift 0.1 cm from the second focal point. To accommodate this shift the focusing mechanism must provide a motion of 0.1 cm (0.040 in.).

The bundle of rays converging to a focus has a minimum cross-sectional area at the focus. However, for a short distance on either side of the focus the cross-sectional area is very nearly the same. Thus for a given object distance there is a small range of axial distances over which the focus appears to be equally sharp. Called the *depth of focus*, this range is given by

$$d = 4\lambda (f/no)^2. \tag{5-18}$$

Where d will be in the same units as those selected for the wavelength; the micron is appropriate. For $f/2$ optics and a wavelength of 4μ, the depth of focus is 64 μ (0.0026 in.). This is then the range of uncertainty in the location of the image plane.

Many infrared systems must work against targets at various distances. Since a focusing mechanism is expensive and it may be difficult to provide

---

[1]This conclusion will be disputed by those who reason from simple paraxial theory that f/number as defined in (5-14) is inversely proportional to tan u'. From the exact higher-order theory, the principal plane is actually a spherical surface with its center at the focal point and f/number is thus inversely proportional to sin u'. Hence, in the limit, f/number approaches $\frac{1}{2}$ and not zero.

[2]If the optics are drawn with the rays incident from the left, and object distances measured to the left of the first focal point and image distances measured to the right of the second focal point are considered to be positive quantities. Other sign conventions may require a minus sign in (5-17).

a means of actuation, it is of interest to know over what range of object distances a typical system will work without refocusing. The change in object distance that shifts the image by a distance equal to the depth of focus is called the *depth of field*. If it is assumed that the optics are accurately focused for an infinitely distant object so that the image lies at the center of the depth-of-focus interval, the near limit of the depth of field can be found by combining (5-17) and (5-18):

$$x = \frac{f^2}{2\lambda (f/no)^2} = \frac{D^2}{2\lambda}.$$ (5-19)

Thus optics with an aperture stop 5 cm in diameter and working at a wavelength of $4\,\mu$ would require no refocusing as long as the object distance exceeded 318 m (1015 ft). What is the effect on system performance if the object distance is somewhat less than this value? Probably very little! Defocusing can reduce the efficiency of the optical modulator and it can reduce the electrical signal from the detector if the size of the image grows larger than the size of the detector. However, neither of these effects will be of much consequence since they occur at short target distances where the signal-to-noise ratio of the system is very high. Probably system performance would not be appreciably impaired until the target distance decreased to about one-fourth of the distance given by (5-19). If the system is a seeker for a guided missile, it will by this time be so close to its target that further system degradation will have no effect on the outcome of the encounter. Note that the value given by (5-19) can be halved if the system is initially focused so that the image plane lies at one edge of the depth-of-focus interval; in this way the entire interval is available to accommodate the image shift.

One of the specifications for an optical system is its *field of view*, that is, the angular measure of the volume of space within which the system can respond to the presence of a target. A system designed to search for a target usually has a small *instantaneous field of view* that is moved systematically by optical or mechanical means so as to cover a much larger *search field*. The instantaneous field of view is determined by a *field stop* placed in the image plane. It can consist of a physical aperture, an optical modulator, or a detector. In the camera, a particularly simple example, the field of view is determined by the size of the film. A particularly convenient optical relationship states that in any optical system the object and its image subtend equal angles from their corresponding principal points. Thus the linear size of the field stop is

$$x = f\beta.$$ (5-20)

where β is the instantaneous field of view expressed in radians. When

β exceeds 0.175 (10°), it should be replaced by $\tan \beta$.[1] When using, for example, an optical system having an equivalent focal length of 100 mm, if it is desired to use the detector as a field stop to limit the field of view to 10 milliradians[2] (0.57°), the diameter of the detector will be $100 \times 10^{-2} = 1$ mm.

5.3 FACTORS AFFECTING IMAGE QUALITY

In the preceding section the discussion dealt with perfect optical systems that could be described in terms of focal lengths, principal points, and f/numbers. Further, it was assumed that all rays from an object point that are transmitted by the optics are brought together in a corresponding image point. Such an assumption is useful in the preliminary stages of a system design, but more detailed information on the size of the image and its energy distribution is required in order to design the optical modulator and the associated signal processing circuitry.

Suppose that we use a microscope to view the image of a point source formed by a lens. Since a point source exists only in the world of mathematics, we must simulate one by using a star, a collimator, or an illuminated pinhole placed as far away as possible. Examination of the image shows that it is a bright, somewhat diffuse disk and it is usually called a *blur circle*. If the microscope is moved along the optical axis of the lens, a point will be found at which the diameter of the blur circle is a minimum. This is the smallest image that this lens can form and it is called the *circle of least confusion*. It coincides with the plane of best focus and, depending upon the aberrations present in the lens, may be slightly displaced from the focal point located by ray tracing. The image of an extended object is formed by the superposition of blur circles from each point in the object. As a result the outlines of the image may be softened or blurred and detail within the image may be lost.

[1] We shall often find it convenient in this book to replace the sine or the tangent of a small angle by the angle in radians. Farwell[2] has shown that up to 10° the error in using β in place of $\tan \beta$ is less than 1 per cent, and less than 0.5 per cent for $\sin \beta$.

[2] An unfortunate confusion exists in the literature between the Navy mil and the artillery mil, both of which have long been used to express angular measurements in fire control work, and the milliradian, which many writers carelessly refer to as a mil. The artillery mil is the angle subtended at the center of a circle by 1/6400th of its circumference. The Navy mil is the angle whose tangent is 0.001. The milliradian is the angle subtended by an arc whose length is 1/1000 that of its radius. Approximate numerical values are: 1 milliradian = 1. 000 000 333 Navy mil = 1.018 593 artillery mil. Thus the milliradian and the Navy mil are essentially equal, but the artillery mil differs from both by about 2 per cent. The milliradian (mrad) is used in this book.

OPTICS

Two processes contribute to the size of the blur circle; these are *diffraction*, which is a consequence of the wave nature of radiant energy, and *aberrations*, which depend on the geometrical arrangement of the optical surfaces and on the dispersion of the optical materials. Aberrations can be controlled by the optical designer; indeed, his main task in any design is to reduce them to less than the specified values. Diffraction, on the other hand, is a physical limitation over which he has no control. Even in the absence of aberrations, diffraction still causes a point to be imaged as a blur circle. Such optics are said to be diffraction limited; they represent the ultimate in optical performance.

Diffraction

Diffraction, which is an interaction between a train of waves and an obstacle, is not peculiar to optics and is often observed in acoustics and in the operation of microwave antennas. In optical systems, diffraction occurs at the edges of the optical elements and at the diaphragms used to limit the beam. The image of a point source formed by diffraction-limited optics appears as a bright central disk surrounded by several alternately bright and dark rings. The distribution of radiant flux in the diffraction image is shown in Figure 5.5. The central disk contains 84 per cent of the radiant flux, and the rest is in the surrounding rings. Since Airy was one of the first to analyze the diffraction process, the central disk is usually called the *Airy disk*. The angular diameter of this disk, which is considered to be equal to the diameter of the first dark ring, is

$$\delta = \frac{0.244\,\lambda}{D}, \qquad (5\text{-}21)$$

where δ is expressed in milliradians, λ is in microns, and the aperture diameter D is in centimeters. Figure 5.6 shows the angular diameter of the Airy disk as a function of wavelength for aperture diameters from 1.25 to 100 cm. It shows, for example, that at a wavelength of $4\,\mu$, an

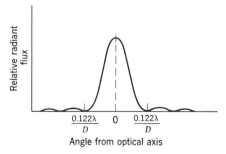

Figure 5.5 Distribution of radiant flux in the diffraction image.

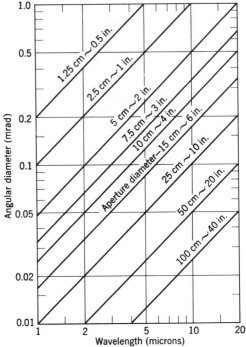

Figure 5.6 Angular diameter of the Airy disk.

optical system 5 cm (~2 in.) in diameter has an Airy disk with an angular diameter of 0.2 mrad.

The linear diameter of the Airy disk can be found by multiplying the angular diameter by the equivalent focal length of the optics:

$$d = (2.44 \lambda)(f/no), \tag{5-22}$$

where d and λ are both expressed in microns. Thus if the optics described above operate at $f/2$, they will have an Airy disk that is 20 μ in diameter. It appears that since the practical lower limit on the size of a detector is about 50 μ, the Airy disk formed by these optics will be much smaller than the smallest detector available. At longer wavelengths the situation may be different; if $f/2$ optics are used at a wavelength of 20 μ, as in a spaceborne tracker of cold objects, the diameter of the Airy disk will be 100 μ and the designer must be sure that the detector is large enough to receive it.

The reader is cautioned that the estimates of image sizes given here and on the following pages are intended to serve only as guides. They must always be interpreted by sound engineering judgment in terms of the specific application. In most infrared systems the image is scanned either by the detector or by the optical modulator. The resulting electrical signal is then a complex function of the size of the image, its flux distribution, the size of the scanning aperture, and the way the image is scanned. Thus the system engineer must consider more than just the size of the image formed by the optics. Consider for a moment some of the factors that must be reviewed when matching a detector to the diffraction image shown in Figure 5.5. Detectors generate electrical noise that ultimately limits the minimum amount of radiant flux they can detect. This noise is proportional to the square root of the detector area; thus we generally try to keep detectors small. If the detector is made large enough to include both the Airy disk and the first two rings surrounding it, the noise contributed by the outer areas of the detector will be far greater than the increase in signal resulting from the radiant flux in the two rings. The criterion used for (5-21) and Figure 5.6 is that the detector should be just large enough to extend to the center of the first dark ring. The detector will then receive 84 per cent of the radiant flux from the image. It can be argued that the detector could be even smaller, perhaps matching the 50 per cent points of the flux distribution in Figure 5.5. This is an exceedingly difficult point, and any answer can apply only to a particular set of conditions. It is the author's opinion that the criterion used for (5-21) represents sound design. This criterion also leaves room for the degradation in imagery that is bound to occur during fabrication, assembly, and use of the system. Dirt or dust on the optics, manufacturing variations, defocusing because of temperature changes, and atmospheric turbulence all contribute to this degradation. When this occurs, the diffraction image broadens out and its peak amplitude decreases. As the image degrades, the signal from a detector matched to the 50 per cent points decreases much more rapidly than that from a detector extending out to the first dark ring. An excellent paper by Scott[3] discusses the engineering aspects of optics for infrared systems; his estimates of image size are about one-half the values given here. Since the subject is such a difficult one, it is regrettable that there was not sufficient space in the paper for him to elaborate on the basis for his assumptions.

The image of two objects consists of two diffraction patterns. If the objects are very close together, the images may overlap and become indistinguishable as separate entities. Lord Rayleigh suggested that the two images could be recognized as separate, or *resolved*, if the center of one Airy disk coincided with the first dark ring of the other. This is called the *Rayleigh criterion* and is a good means of estimating the resolution capability of any diffraction-limited optical system. Thus the minimum angular separation at which two point sources can just be resolved is

$$\alpha = \frac{0.122 \lambda}{D}, \qquad (5\text{-}23)$$

where α is the angle (in mrad) subtended by the two sources at the first principal point, λ is in microns, and D is in centimeters. The previously discussed optics ($f/2$, $D = 5$ cm, $\lambda = 4\,\mu$) should be able to just resolve two points subtending an angle of 0.1 mrad. Note that these values can be read from Figure 5.6 by dividing the angular values given there by 2. Equation 5-23 shows that for any diffraction-limited system, whether it be optics or a radar antenna, the angular resolution is proportional to λ/D. Expressed another way, the ability to resolve objects with a small angular separation is directly proportional to the number of wavelengths in the receiving aperture. Here, then, is one of the fundamental advantages of infrared (or optical) equipment. Since the apertures of such equipment are thousands of times larger than the wavelength, the angular resolution capability is very great. For the $f/2$ lens mentioned previously, the aperture is 12,500 wavelengths wide; to achieve the same angular resolution with a 10 cm (wavelength) radar would require an antenna with a diameter of 1.25×10^5 cm (4100 ft). When radar and infrared search sets are used to track the same group of aircraft, it is common to find that the radar operator cannot determine the number of aircraft in the group, whereas the infrared operator can easily determine both the number of aircraft and the number of engines on each.

In a well-corrected optical system the designer has worked hard to reduce the aberrations so that the imagery will be limited by diffraction. It has been noted that many infrared applications do not need such excellent correction. Recognizing this, the system engineer should insist that the optics not be over designed; correcting each design to the diffraction limit may be satisfying to the optical designer, but it can play havoc with the project budget.

Aberrations

Third-order theory predicts seven types of aberrations. Two of these, called *chromatic aberrations*, are caused by variation in the index of refraction of the lens material with wavelength. The rest, called *monochromatic aberrations*, occur even though only a single wavelength is involved. These seven aberrations can be briefly described as follows:

A. Monochromatic Aberrations
  1. Spherical — Rays from a common axial point which pass through the optics at different distances from the optical axis are not brought to a common focus.

2. Coma—The image of an off-axis object is no longer symmetrical but becomes an enlarged, comet-shaped blur.
3. Astigmatism—The image of an off axis point becomes a pair of lines that are at right angles to each other. The lines lie at different distances from the optics, and the smallest blur circle lies somewhere between them.
4. Curvature of Field—The image of a plane object lies on a curved rather than on a plane surface.
5. Distortion—Straight lines, except those passing through the center of the field, are imaged as curved lines.

B. Chromatic Aberrations
1. Longitudinal—A variation in the position of the focal point as a function of wavelength.
2. Lateral—The size of an image formed by the optics varies as a function of wavelength.

In most infrared systems the optics are usually designed to cover a relatively small instantaneous field of view, and the detector or optical modulator is usually located close to the optical axis. For these reasons those aberrations that occur on the optical axis or near it are of primary importance. These include spherical aberration, coma and, for refractive optics, the two chromatic aberrations. Only these four will be discussed here. Note again that aberrations are a consequence of the laws of reflection and refraction by spherical surfaces; they are not due to faulty fabrication of the optics.

The spherical aberration of a simple lens and of a spherical mirror is shown in Figure 5.7. Rays parallel to the optical axis are not brought to the same focus. In general, the marginal rays (those passing near the edge of the element) are brought to a focus closer to the element than are the paraxial rays (those passing near the optical axis). The optical designer describes the amount of spherical aberration in terms of the axial separation between the marginal and paraxial foci. For our purposes, a more meaningful description is the angular diameter of the smallest blur circle δ_s, the circle of least confusion. For a spherical mirror, this diameter is

$$\delta_s = \frac{15.6}{(f/\text{no})^3}, \qquad (5\text{-}24)$$

where δ_s is in milliradians. The error resulting from the use of (5-24) is only 3 per cent at $f/1$ and becomes even less for higher f/numbers [4]. Equation 5-24 should not be used below $f/1$. For simple lenses, the blur circle due to spherical aberration is always larger than for a spherical mirror. For a given focal length, there is an optimum combination of radii

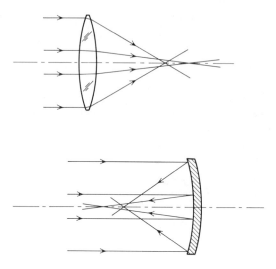

Figure 5.7 Spherical aberration.

that minimizes spherical aberration. Figure 5.8 shows the angular diameter of the smallest blur circle for a spherical mirror and for simple lenses made of fused silica, arsenic trisulfide, and silicon that are designed for minimum spherical aberration. For a blur circle 1 mrad in diameter, an $f/2.5$ spherical mirror, an $f/3$ silicon lens, or an $f/6.5$ quartz lens can be used. For futher details on simple lenses, see the paper by Scott[3]; for reasons already stated, our (5-24) differs from his (3).

The dashed line below $f/1$ shows the rapid increase in the size of the blur circle as the f/number approaches its maximum theoretical value of $f/0.5$.

Spherical aberration cannot be eliminated by using a spherical mirror alone. It can, however, be virtually eliminated by adding a correcting element, such as those used in the Bouwers, Maksutov, or Schmidt systems that are mentioned later. The paraboloid[1] of revolution, an aspheric reflecting surface, is free from spherical aberration, and images formed on the optical axis are limited only by diffraction. Spherical aberration cannot be eliminated from a single thin lens unless one or both

---

[1] A note on proper terminology. The term "parabolic" describes a plane curve. A mirror having a radial cross section that is parabolic is said to have a paraboloidal surface and is often called a paraboloid. Optical designers, perhaps bearing in mind the shape of the cross section, use the word "parabolic." Similarly, they use "hyperbolic" rather than "hyperboloidal" and "elliptical" rather than "ellipsoidal." A spherical mirror, however, is never referred to as circular[5].

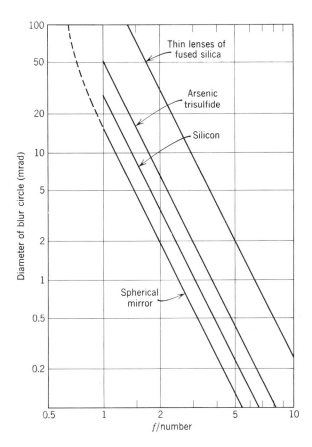

Figure 5.8 Angular diameter of the smallest on-axis blur circle (due to spherical aberration) produced by a spherical mirror and by thin lenses (data for lenses adapted from Scott[3]).

surfaces of the lens are made aspheric. Several thin lenses can be combined to reduce spherical aberration below the diffraction limit.

When optical systems form images away from the optical axis, additional aberrations occur that increase the size of the blur circle. The most important of these is coma. When the incident rays are not parallel to the optical axis, the focus and the magnification are different for rays passing through various zones (ring-shaped areas concentric with the optical axis) of the optics. The behavior of such rays and the appearance of the *comatic image* are shown in Figure 5.9. The rays passing through each zone focus as a circle; the largest circle is formed by the rays passing through the marginal zone. These circles overlap, and the image appears as a flare-like pattern resembling the tail of a comet. In the absence of

THE ELEMENTS OF THE INFRARED SYSTEM

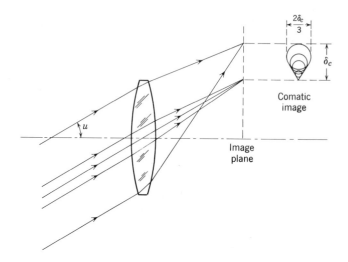

Figure 5.9 Formation of the comatic image.

other aberrations, most of the radiant flux is concentrated near the pointed end of the comatic image; the presence of other aberrations tends to shift more of the flux away from the pointed end.

A single spherical surface has no unique optical axis. If the aperture stop is placed at the center of curvature, any bundle of rays passing through the stop defines its own optical axis and the image is formed on that axis. As a result, a spherical mirror with the aperture stop placed at its center of curvature has no coma. Such an arrangement is rarely useful, however, because of the long structure needed to support the aperture stop. The coma of a parabolic mirror is independent of the location of the aperture stop.

Since the comatic image lacks circular symmetry, it is difficult to describe its size by a single number. A conservative approximation is the angular subtense δ_c shown in Figure 5.9. For a parabolic mirror, this subtense is given by

$$\delta_c = \frac{0.1875 \, u}{(f/\text{no})^2}, \tag{5-25}$$

where u is the field angle, that is, the angle between the incident rays and the optical axis. The units of δ_c will be the same as those chosen for u; the milliradian is a convenient choice. Figure 5.10 shows the angular subtense of the comatic image formed by a parabolic mirror as a function of the field angle. With an $f/2$ paraboloid, if the subtense of the image must be kept to less than 1 mrad, the field angle must not exceed 1.2 deg.

OPTICS

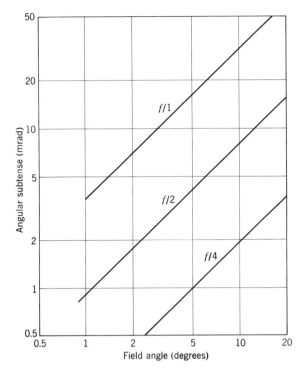

Figure 5.10 Angular subtense of the comatic image produced by a parabolic mirror.

Figure 5.10 can also be used to estimate the subtense of the comatic image formed by a spherical mirror with the aperture stop at the rim of the mirror (a condition commonly met in practice). For small field angles, spherical aberration is the principal contributor to the size of the blur circle; for larger field angles, it is coma. Combining (5-24) and (5-25) gives the field angle or crossover point for which the angular subtense of the comatic image equals the diameter of the on-axis blur circle due to spherical aberration:

$$u = \frac{4.8}{f/\text{no}}.\qquad (5\text{-}26)$$

Here, for convenience, u is given in degrees. Thus we can, for example, approximate the image of a point object formed by an $f/2$ spherical mirror (aperture stop at its rim) by the size of the blur circle given in Figure 5.8 for field angles up to 2.4 deg; for larger field angles, use the size of the comatic image given in Figure 5.10. It would appear that this approximation may be quite crude near the crossover point; actually the principal change in this region is in the distribution of radiant flux within the image, not in its size.

When a high-quality parabolic mirror is used, it is surprising how rapidly coma causes deterioration of the diffraction image. Though few will ever be privileged to use the 200 in. mirror at Mt. Palomar, the figures given by Bowen[6] on the imagery of this $f/3$ paraboloid are fascinating. Under laboratory conditions, coma is evident in images lying only 1 mm off the optical axis, a field angle of about 13 sec. Under average observing conditions, the turbulence of the atmosphere increases the size of the image and coma does not become evident until the images are 5 mm away from the optical axis, a field angle of about 1 min. A Ross corrector lens placed just ahead of the image plane can be used to correct the coma so that images 75 mm from the optical axis are essentially coma free.

The other monochromatic aberrations are of little concern in typical infrared systems. With $f/3$ or faster optics, astigmatism is not noticeable until the field angle exceeds about 15 deg. Curvature of field, which is usually inseparable from astigmatism, becomes a problem only at the larger field angles. If distortion is present in an optical system, the image is not a faithful representation of the object. Rarely is this of concern with an infrared system.

If refractive optical systems are used, chromatic aberrations become important. The formula for the focal length of a thin lens given in Figure 5.3 includes a term that is dependent on the index of refraction of the lens material. Since dispersion—the change in the index of refraction with wavelength—occurs with all optical materials, it follows that the focal length of any simple lens will vary with wavelength. This variation, called *longitudinal chromatic aberration*, is shown in Figure 5.11 (for simplicity, it is assumed that the lens is used with visible light). No sharply defined image is formed, and we must adopt a compromise best focus, as is done for spherical aberration.

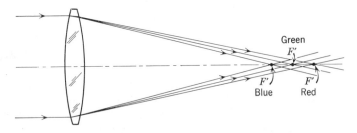

Figure 5.11 Longitudinal chromatic aberration.

A combination of two thin lenses made of different types of glass can be designed to have a common focus for two different wavelengths; in this way longitudinal chromatic aberration can be eliminated at these two wavelengths. Such a lens, which is called an *achromatic doublet*, is shown in Figure 5.12 (again, visible light is assumed for simplicity). The red and blue rays have been brought to a common focus. Unfortunately, since the principal points for the two rays do not coincide, the focal lengths for these two wavelengths are different. Since it is proportional to focal length, image size will vary with the wavelength, an effect called *lateral chromatic aberration*.

There are many low- and medium-resolution applications for which a single thin lens will suffice. If the radii are chosen so as to minimize spherical aberration, the angular size of the blur circle is shown in Figure 5.8 for several infrared optical materials. Chromatic aberration increases the size of the blur circle in direct proportion to the spectral bandpass, and its effect can be calculated from

$$\delta_{ch} = \frac{b}{f/no}. \quad (5\text{-}27)$$

Here δ_{ch} is in milliradians and is to be added to the value of δ_s found from Figure 5.8. Values of b are given in Table 5.1 for fused silica, silicon, and arsenic trisulfide and for several typical spectral bandpasses. Let us estimate the size of the blur circle for a single $f/2$ silicon lens with a spectral bandpass from 3 to 5 μ. From Figure 5.8 the value of δ_s is 3.5 mrad, from (5-27) the value of δ_{ch} is 2 mrad, and the angular diameter of the blur circle is about 5.5 mrad. Moving away from the optical axis, coma increases the size of the blur circle. One can estimate this increase by using the methods already described in discussing the comatic image. A conservative design would approximate the size of the blur circle for field angles exceeding the crossover point (5-26) as the sum of the

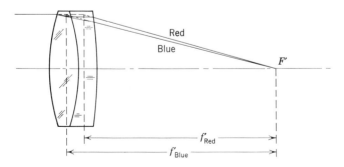

Figure 5.12 Lateral chromatic aberration.

TABLE 5.1 VALUES OF b FOR CALCULATING THE INCREASE IN BLUR CIRCLE DIAMETER DUE TO CHROMATIC ABERRATION (USE WITH EQUATION 5-27)

Material	Wavelength Range (μ)	b
Fused silica	2–2.6	20
Silicon	2–2.6	6
	3–5	4
	8–12	0.5
	2–12	14
Arsenic trisulfide	2–2.6	2
	3–5	4
	8–12	12
	2–12	26

values found for the comatic image (5-25) and the chromatic image (5-27). Thus, with the $f/2$ silicon lens at a field angle of 10 deg, the diameter of the blur circle would be about 10 mrad.

Since optics provides a means for the transfer of information, it is not surprising that optical designers now use much of the theory originally developed for electrical communications. The *modulation transfer function* describes the ability of a system element, or of an entire system, to transfer information. It is of particular value because, at least in theory, the transfer functions of each system element may be multiplied together to give the modulation transfer function for the entire system. Such conclusions are, of course, not new, but the application to optics is new. It is common practice to determine the frequency response of a high-fidelity sound system by multiplying transfer functions, that is, the individual frequency responses of the microphone, amplifier, and speaker. Transfer function will not solve all system problems, but they have proved useful in some applications. The most serious difficulty in their use is the determination of the proper transfer function. Details of the use of these techniques are beyond the scope of this book; the interested reader will find that refs. 7–11 contain excellent summaries.

A simple application of transfer function theory has proved of value in specifying the quality of completed optical systems. It consists of adding to the manufacturing specification a requirement that the axial image of a point source formed by the system must contain x per cent of the radiant flux within a diameter of y microns. Such measurements can be made by using an infrared detector and a series of pinholes of different sizes to measure the distribution of flux as a function of pinhole diameter. The result is often called the *expanding pinhole function*; it cannot be converted to the modulation transfer function. This method is particularly valuable because it tends to reveal manufacturing defects, such as poor centering of elements, dust or dirt on them, and deformation of elements in mountings, any one of which might go unnoticed with other test procedures.

The seven aberrations discussed cannot be simultaneously eliminated or even minimized in a single thin lens. Yet when several elements are used, the aberrations of one element can be balanced against those of another. In general, the better the imagery required, the greater the number of elements needed to achieve it. Regardless of how many elements are used, the designer must concentrate on eliminating those aberrations that will be most detrimental to the intended use of the system. In most infrared systems that have to cover just a small instantaneous field of view, only spherical aberration, coma, and the chromatic aberrations are of concern. In a system covering a wide field of view, astigmatism, curvature of field, and distortion must also be reduced. It is not surprising, therefore, that optical systems designed to cover only a small field of view produce better images than those designed for a wide field of view. The problems and techniques peculiar to modern optical design are described in great detail in refs. 12–14.

5.4 TYPICAL OPTICAL SYSTEMS FOR THE INFRARED

Having examined the limitations of simple lenses and mirrors, let us now consider some of the more complicated optical systems used in the infrared. Until recently most of these systems used mirrors rather than lenses because there were relatively few optical materials transparent in the infrared. This situation no longer exists, and one is as likely to find a lens as a mirror system used in a new design.

Reflective Optics

Most of the mirror systems have evolved from the classical reflective types developed by astronomers. Since the focus of a spherical or parabolic mirror lies in the direction of the incoming rays, some of them must be blocked in order to place a detector at the focus. This is the *prime focus* and is rarely used except in large astronomical telescopes. Newton suggested that a flat *secondary mirror* be so placed as to bring the focus to the side of the telescope (see Figure 5.13*a*). This is a con-

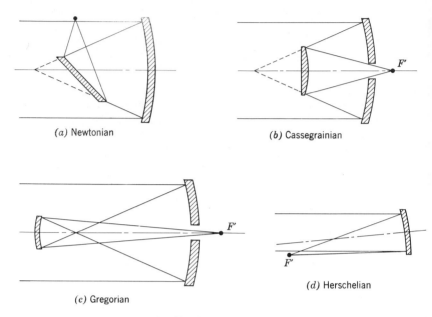

Figure 5.13 Classical reflective optical systems.

venient location for the detector since it minimizes blockage of the incoming rays. The Cassegrainian system (Figure 5.13b) uses a convex secondary mirror placed inside the prime focus; it redirects the rays through a hole in the primary mirror to a new focus. The Gregorian system (Figure 5.13c) is similar to the Cassegrainian, but it uses a concave secondary mirror placed outside the prime focus. In both of these cases the combination of the two mirrors has an *effective focal length* longer than that of the primary mirror. The Herschelian (Figure 5.13d) is occasionally used when it is desirable to minimize the number of optical surfaces in a system. A large-f/number primary mirror is tilted slightly in order to place the prime focus off to the side of the incoming bundle of rays. The large f/number is required to keep coma within reasonable bounds.

The fraction of the area of the entrance aperture that is blocked by secondary mirrors, detectors, or their supports is called the *obscuration*. Typical values vary markedly with optical arrangement and f/number. In general, the obscuration is least with the Newtonian and greatest with the Gregorian system. Note that the diameter of the secondary mirror can be one-half that of the primary mirror, and yet give an obscuration of only 25 per cent. When there is obscuration, the f/number cannot be calculated by the simple expression of (5-14). For these cases we use the *effective f/number*:

$$(f/\text{no})_{\text{eff}} = \frac{f_{\text{eff}}}{D_p}\left(\frac{1}{1-(D_{\text{obs}}/D_p)^2}\right)^{1/2}, \qquad (5\text{-}28)$$

where f_{eff} is the effective focal length of the over-all system (not the primary mirror), D_p is the diameter of the entrance aperture (often the primary mirror), and D_{obs} is the diameter of the secondary mirror (or other obstruction). If there is no obscuration, (5-28) reduces to (5-14).

With any of these arrangements the positions of the primary and secondary mirrors must be accurately maintained in relation to one another. This is generally done by using a tube or open ribbed structure from which both mirrors are supported. The primary mirror is usually mounted directly in one end of the tube. The secondary mirror is often mounted in a cell supported by three of four slender arms fastened to the tube. Appropriately, such a mount is called a *spider*. It is convenient to define the length of the mounting tube, the *tube length*, as being equal to the distance between the primary and secondary mirrors. It is usually desirable to keep the tube length short in order to minimize the weight and inertia of the tube. To show more clearly the variation of the tube length, Figure 5.14 shows Newtonian, Gregorian, and Cassegrainian systems drawn so that the equivalent focal lengths are equal. The superiority of the Cassegrain in producing a long focal length with a short tube

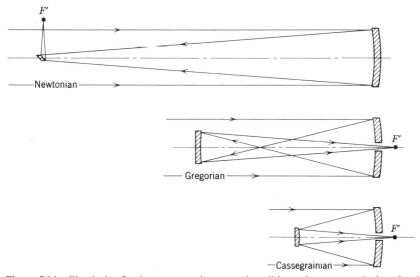

Figure 5.14 Classical reflective systems drawn so that all have the same equivalent focal length and primary mirror diameter.

length is clearly evident. Thus it is no surprise that this system, or variations of it, is the most popular for use in the infrared.

A central obstruction, such as a secondary mirror, decreases the size of the Airy disk and apparently increases the angular resolution. Ritchey[15] gives experimental data showing that a central obstruction one-half the diameter of the primary mirror reduces the diameter of the Airy disk to 81 per cent of its diameter without the obstruction. Unfortunately, the reduction in the size of the Airy disk is accompanied by an increase in the radiant flux in the outer rings of the diffraction pattern. Thus any increase in resolution caused by the presence of the central obstruction will be a complex function of the target, the background, and the characteristics of the optical modulator[16].

In its classical form the Cassegrainian consists of a paraboloidal primary and a convex hyperboloidal secondary mirror. There is no spherical aberration, and the hyperboloid is used by the designer to reduce the coma. In the Cassegrainian (or Gregorian) the effective focal length is A (the *amplification ratio*) times the focal length of the primary mirror. Jones[17] has shown the important result that coma in the Cassegrainian (or Gregorian) system is reduced to $1/A^2$ that of the primary mirror. An $f/2$ system can be made in which the size of the comatic image does not exceed 1 mrad for field angles of up to 2.5 deg. Since a convex hyperboloid is expensive to make, a variation called the Dall–Kirkham is often used; it consists of a spherical secondary and a higher-order aspheric (nonspherical) primary mirror. It is much cheaper to manufacture, and its imagery on axis is essentially identical to that of the classical form. Unfortunately, coma is about four times as bad; an $f/2$ Dall–Kirkham must be limited to a field angle of 0.6 deg if the size of the comatic image is not to exceed 1 mrad. Further design information on the Cassegrainian is given in refs. 18 and 19.

When it is necessary to achieve good image quality over a wide field of view, a fine choice is a *catadioptric system*, one combining a primary mirror with a thin correcting lens. The most famous of these systems is that devised by Schmidt (see Figure 5.15a). It uses a spherical primary mirror with an aspheric correcting lens placed at its center of curvature. The shape of the corrector is chosen to eliminate the spherical aberration of the primary mirror. Note in this figure how the various rays are deviated by the corrector so that they are brought to a common focus by the mirror. If desired, a secondary mirror can be used to bring the image plane to a convenient point behind the primary. Schmidt systems are unsurpassed in providing excellent imagery over a wide field; an $f/1$ system covering a 25 deg field of view with an image size less than 1 mrad can be achieved. Two possibly minor disadvantages are evident; the tube

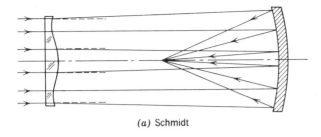

(a) Schmidt

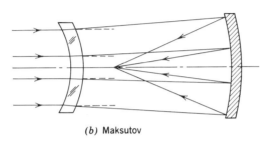

(b) Maksutov

Figure 5.15 Schmidt and Maksutov systems.

length is equal to twice the focal length of the primary mirror, and the image plane is curved (with a radius equal to the equivalent focal length of the system).

Since the aspheric curves on the Schmidt corrector lens are costly to manufacture, a simplified version has been designed by Maksutov[20, 21] and Bouwers[4]. Both men did their work independently during World War II and were unable to publish their results immediately; as a result, priority of discovery has never been established. In the United States, it is customary to refer to these as Maksutov systems. As shown in Figure 5.15b, they consist of a primary mirror with a meniscus corrector lens. All surfaces are spherical, a distinct advantage in simplifying manufacture. The quality of the imagery is comparable to that obtained with the Schmidt except for systems operating at $f/1$ or faster; in addition, the tube length is considerably shorter in the Maksutov system. It is easy to add a secondary mirror by applying a reflecting coating to the central portion of the second surface of the corrector. Dubner[22] has described the application of systems of this type to infrared missile seekers.

The *Mangin Mirror* is a single element that acts, in effect, like a Maksutov or Schmidt system, although the imagery it produces is very much poorer. The reflecting coating is placed on the back or second surface of the mirror. An incoming ray enters and is refracted at the first surface, is reflected from the second surface, and is refracted again as it leaves the first surface. Since the two surfaces of the mirror are spherical, manufacturing costs are quite low. Rosin and Amon[23] describe design procedures for Mangin mirrors.

Refractive Optics

Because of the wide variety of infrared-transparent optical materials now available, multielement refractive optical systems can be designed with imagery equal to the best obtainable with mirrors. Herzberger and Salzberg[24] have discussed the problems of lens design in the infrared and give data on the index of refraction for 14 suitable optical materials. Their paper is an excellent starting point for anyone contemplating such design work. Many lenses designed for the infrared have been described in the literature. The characteristics of some of these lenses are given in Table 5.2. The column labeled "Blur Circle" gives the diameter of the blur circle and the angular field over which the blur circle does not exceed

TABLE 5.2 CHARACTERISTICS OF LENSES DESIGNED FOR USE IN THE INFRARED PORTION OF THE SPECTRUM

Lens Type and Materials	f/number	Useful Spectral Bandpass (μ)	Blur Circle Diam (mrad)	Field (deg)	Reference
Singlet, arsenic trisulfide	3.5	2–3	2	2	25
Singlet, IRTRAN-2	1	2–4.5	14.4	—	26
Singlet, IRTRAN-2	1	2–4.5	4.1	—	26
Doublet, silicon and germanium	1	2.2–2.6	—	—	27
Doublet, arsenic trisulfide	2	2–5	—	—	28
Doublet, germanium	2	8–12	<1	2	29
Doublet, silicon	1	3.5	<1.5	7.5	30
Doublet, silicon	1	3.5	0.4	10	31
Doublet, IRTRAN-1 and IRTRAN-2	3.6	3.0–5.5	—	—	32
Triplet, germanium and IRTRAN-2	0.75	8–12	Diffraction limited	12	33
Triplet, silicon and arsenic trisulfide	2	2.7–11	—	—	34
Triplet, silicon and sapphire	1	3.5	0.3	10	31

this size. Unfortunately, some of the references do not indicate whether the value they quote for the field is a half angle or a total angle; that is, does the field extend 5 deg from the axis (half angle) or is it a 10 deg field (total angle)?

Occasionally it is necessary to measure the transmission of a lens. The method suggested by Olson[35] is particularly convenient.

Miscellaneous Considerations in the Choice of Optics

Several factors influence the cost of an optical system. Spherical curves are the simplest and cheapest to produce. Most optical shops have had enough experience with paraboloids so that for large f/numbers their cost is not too much more than a spherical mirror of equal diameter and f/number. For f/numbers less than $f/3$, the paraboloid becomes increasingly more expensive and may cost several times as much as an equivalent spherical mirror. For the same f/number, the cost of a mirror increases approximately in proportion to the square of the increase in its diameter. Other aspheric curves can be produced only by the more experienced optical shops and they are quite expensive. On lenses, most optical shops prefer to make a concave aspheric surface rather than one that is convex.

Relatively little information has appeared in the literature concerning optical workshop procedures, probably because many of these procedures are still an art rather than a science. Reference 36 contains a very good summary of current optical shop practices and [37] describes measurement techniques used in these shops. Standards are available for specifying surface defects on optical elements[38].

Since the constructional material of a mirror serves only to support the reflecting surface, there are no requirements on its homogeneity or optical properties. If the system must operate over a wide range of temperatures, it is customary to use mirrors made of Pyrex or fused silica in order to reduce thermal deformations of the optical surface. The time that it takes a mirror to reach thermal equilibrium after a sudden change in the temperature of its surroundings varies as the square of its thickness. Barnes[39] has made calculations that show quite clearly how temperature-induced deformation of mirrors affect image quality.

The flexure of a circular disk supported only at its edge is proportional to the fourth power of its diameter and inversely proportional to the square of its thickness. In view of these conflicting requirements, a convenient rule of thumb is to make the thickness of the mirror one-sixth of its diameter. For very large diameters, the weight of the mirror may become prohibitive. Since some of the space programs require very large mirrors, much effort has been devoted to the development of lightweight mirrors of high optical quality. One possible solution is to make the bulk of the mirror of a honeycomb construction rather than solid[40]. Another solution is to make mirrors of metals, such as beryllium, rather than of glass[41]. Beryllium provides the required lightness, dimensional stability, and stiffness, but its crystalline nature prevents the attainment of a satisfactory optical polish. For this reason, it is customary to apply a very thin coating of Kanigen, an amorphous nickel compound, prior to final polishing. After the final polishing, a coating of aluminium or gold is applied to the mirror in order to provide the maximum reflectance. Plans have been announced [42] to carry various types of mirrors into space and monitor them for possible deterioration of their reflecting surfaces. The mounting of large mirrors requires very careful mechanical design in order to preserve the optical quality of the mirror[40, 43].

The maximum diameter achievable with refractive systems is considerably less than that possible with reflective systems. The difficulties in producing a piece of optical material increase rapidly as the diameter increases. The maximum size may be limited by the equipment available to the manufacturer or by the basic physics of the production process. Figure 5.21 (in section 5.7) indicates the maximum diameter that can be realized with some of the more popular infrared optical materials. Flexure of a reflecting surface by mechanical forces causes an immediate deterioration of the image quality. A lens is relatively insensitive to flexure, because the effects of it on the two surfaces nearly cancel one another. Since the index of refraction varies with temperature, the imagery may deteriorate more rapidly with a lens than with a mirror. After considering all factors it is probably reasonable to say that refractive systems of more than 8 in. diameter will rarely be used in the infrared.

5.5 AUXILIARY OPTICS

Unless the detector is located at the image plane of the primary optics, additional optical elements may be necessary. Figure 5.16 shows an

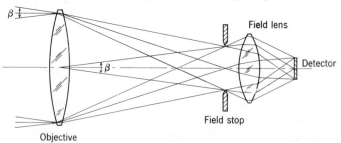

Figure 5.16 Function of the field lens.

objective with a field stop subtending an angle β. The exact details of this stop are unimportant; it can be a physical aperture or a portion of an optical modulator. All of the radiant flux from the target collected by the objective passes through this stop. For maximum system efficiency the detector must be large enough to accept all of this flux. Since most detectors are protected by a window and the optical modulator may rotate at high speed, the detector must be placed a few millimeters back of the field stop. As the detector is moved back from the field stop, it must be made larger in order to accept all of the radiant flux. No increase in signal results, but the noise increases with the square root of detector area.

A convenient solution to this problem is to use a *field lens* behind the field stop, as shown in Figure 5.16. This lens forms an image of the entrance aperture of the system, collects all of the radiant flux passing through the field stop, and redirects it through this image. By using a field lens having a short focal length, the image of the entrance aperture can be made smaller than the field stop. Thus the signal from a detector placed at this image will be the same as that from a larger detector placed at the field stop (ignoring transmission losses in the field lens), but the noise will be less. The diameter D' of the detector when a field lens is used is

$$D' = D \frac{(\text{NA})}{(\text{NA})_f}, \qquad (5\text{-}29)$$

where D is the diameter of the detector if placed at the field stop without a field lens, (NA) is the numerical aperture of the objective (or primary optics), and $(\text{NA})_f$ is the numerical aperture of the field lens. The field lens does not have to produce a high quality image; its function is to condense all of the radiant flux passing through the field stop onto the detector. For this reason it is often called a *condenser*. Field lenses, using two or three elements, can be made with a numerical aperture of 0.7. With an $f/2$ objective, such a condenser reduces the diameter of the detector by a factor of 3. Theoretically the signal-to-noise ratio would increase threefold, but because of transmission losses in the condenser the actual increase would be nearer a factor of 2.

Since the field lens images the entrance aperture and not the target onto the detector, the distribution of radiant flux across the detector is uniform. This uniformity is important because the response of most infrared detectors varies from point to point across their sensitive area. If the flux distribution across the detector is not uniform, spurious signals may be generated.

It is often feasible to replace the field lens with a simple *light pipe* or *conical condenser*, that is, with a portion of a hollow cone placed so that its axis coincides with the optical axis of the system. Since the large end of this cone is located immediately behind the aperture stop, it collects all of the radiant flux at this point and reflects it to the detector at the small end of the cone. The design of conical condensers has been discussed by Williamson[44]. In some cases insufficient attention has been given to the fabrication process, and designers have mistakenly concluded that the efficiency of these condensers is too low for practical use. This is an unfortunate conclusion, because a well-made conical condenser will function as efficiently as a field lens and may be cheaper to manufacture in quantity production. In order to realize this performance, the walls of the cone must have an extremely fine surface finish and the highest possible reflectance. Surface finish should be specified as 2 μin. (rms). A reflecting layer of gold should be applied to the walls of the cone by evaporation in a vacuum; it is desirable to polish this gold layer after application. Obviously any dirt or fingerprints on the gold surface will seriously impair the efficiency of the condenser. Williamson points out how easy it is to make a conical condenser having a numerical aperture of unity. Thus with an $f/2$ objective one can use such a condenser to achieve a fourfold reduction in the diameter of the detector. This results in a theoretical increase of 4 in signal-to-noise ratio; assuming 75 per cent efficiency for the condenser, the actual increase is 3.

In infrared systems that search or track, the optics are mounted in gimbals so as to be free to look in any direction. The detector, which is usually fixed in position, is placed at the *gimbal center*, the center of rotation of the gimbal structure. It is desirable that the protective window in front of the system have a constant optical effect, independent of the direction in which the optics look through the window. The only shape satisfying this condition is that one in which the surfaces of the window are concentric about the gimbal center. Such a window is a portion of a spherical shell and is called an *irdome*. For some applications a more suitable aerodynamic shape is preferable, but the concentric spherical form is the only one having the required negligible effect on the quality of the optical image. Since the irdome is a part of the optical system, tests of the objective without the irdome will give erroneous indications of the optical quality of the system.

There are many possibilities for the application of fiber optics in infrared systems. A ray entering a slender fiber of some suitably transparent materials travels the length of this fiber in a series of reflections from the side walls. As long as the angle of incidence exceeds the critical angle, total internal reflection occurs. The ray may be reflected several hundred times per inch of fiber length; yet so perfect is the process of total internal reflection that only a fraction of 1 per cent of this flux will be lost. The theory and application of fiber optics were first discussed in the patent literature prior to 1930. The fibers available at that time were made of glass or plastic and were a millimeter or more in diameter. More recently, techniques have been developed for making fibers as small as $2\,\mu$ in diameter[45]. These fibers resemble a dielectric waveguide, and the analytical techniques developed by workers in the microwave field are directly applicable. As the diameter of the fiber approaches the wavelength to be transmitted, an increasing fraction of the flux is transmitted along the outer surface of the fiber. To avoid the problems this introduces, it is customary to use fibers with a diameter several times larger than the wavelength.

A careful study of the internal reflection process shows that during reflection the ray penetrates a short distance, perhaps a wavelength, into the second medium. As a result, when many fibers are gathered into a bundle, it is inevitable that some of the flux will be transferred from one fiber to another where the two are in contact. Such a transfer is analogous to cross talk in a communications system. To prevent it, the fibers are given a thin cladding of some material having an index of refraction less than that of the fibers.

If the ray is to remain in the fiber, it must not strike the walls at less than the critical angle. This requirement can be expressed in terms of the numerical aperture of the fiber as

$$\mathrm{NA} = \sin u = (n_1^2 - n_2^2)^{1/2}, \qquad (5\text{-}30)$$

where n_1 and n_2 are the indices of refraction of the fiber and of the cladding, respectively, and u is the maximum angle an incident ray can make with the fiber axis and still be transmitted. Fibers made of arsenic trisulfide are the only infared-transmitting fibers generally available. Typical fibers are $50\,\mu$ (0.002 in.) in diameter and are clad with an arsenic sulfur compound having a thickness of about 10 per cent of the core diameter. The approximate values of the index of refraction at a wavelength of $4\,\mu$ are 2.411 for the fiber and 2.358 for the cladding. From (5-30), the numerical aperture of such fibers is nearly equal to 0.5. Standel and Hendrickson[46] have reported the results of an extensive engineering investigation of these fibers. They state that the internal transmittance of bundles of fibers is 0.90 per in. of bundle length over the wavelength interval from 3.2 to $5.5\,\mu$. Thus a 6 in. bundle has an internal transmittance of only $(0.90)^6$ or 0.53. This somewhat discouraging result is not satisfactorily explained, since the internal transmittance of bulk arsenic trisulfide appears to be more than 0.95 per in. Thus the system engineer may find that in practice the losses are prohibitive if such bundles are more than a few inches long.

Since the bundles of fibers are flexible, they can collect flux at a point on a moving structure and conduct it to a detector at a fixed location. As an example, a detector packaged with a refrigerator can be larger and heavier than it is feasible to place on a gimbaled structure. By use of a fiber bundle the detector/refrigerator package can be placed at a distance from the gimbal. Standel and Hendrickson[46] describe a simple test device for flexing bundles through an angle of 100 deg ($\pm$50 deg from center) at a rate of 700 deg sec^{-1}. With properly designed terminations for the fibers, bundles were flexed 1.5 million times without breakage of individual fibers.

5.6 METHODS OF GENERATING SCAN PATTERNS

Some infrared systems must seek a target by scanning a large *search field*. In essence, we start with a small instantaneous field of view and then find a way to move it so as to scan the search field completely. The *frame time* is the time required for one complete scan of the search field. Most of these systems generate a rectangular *raster*, that is, a line-by-line scan of the search field formed in much the same way that a television picture is formed. The gimbaled optics can be moved in the desired scan pattern by a servomotor. Angular scanning rates as high as 250 deg sec^{-1} can be achieved. Means for mechanically generating a conical scan are described in [47].

If mechanical means are not satisfactory for generating scan patterns, optical means can often be used. They offer the advantages of higher angular scan rates, better scan linearity, a variety of scan patterns, and reduced power consumption since the mass to be driven is usually much smaller[48, 49]. A means of generating a raster scan by a plane mirror placed in front of the optical system is shown in Figure 5.17. Continuous rotation of the mirror about the vertical axis provides a full 360 deg coverage in azimuth, and rotation about a horizontal axis provides elevation coverage. The scan element, which is the projection of the instantaneous field of view on the object plane, appears to spiral upward in the hemisphere above the system. Since the projected height of the mirror must be equal to the diameter of the optics, large elevation angles are impractical. Two-mirror systems can be devised for scanning the entire hemisphere; one such system has been described by Hicks[50]. If the scan mirror is placed near the image plane, it can be very small and can be driven by electromagnetic or piezoelectric elements[51–53].

Another scanning means, the rotating eyeball, is shown in Figure 5.18.

OPTICS

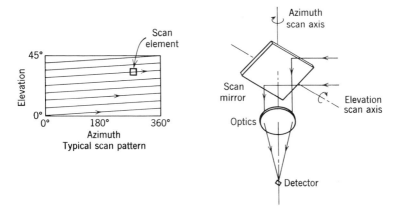

Figure 5.17 Use of an external mirror to generate a raster scan.

Four lenses rotate about a single fixed detector[54, 55]. A shield placed around the detector ensures that it "sees" only one set of optics at a time and limits the azimuthal coverage to 60 deg. If each optical axis lies in the plane of the paper, each lens scans in turn the same scan line. To provide elevation coverage, the entire system can be rocked in and out of the plane of the paper. Alternatively, the optical axes can be displaced vertically to give a four-line (or four-bar) raster.

A pair of thin prisms placed in front of the optics as shown in Figure 5.19 can be rotated to generate a variety of scan patterns. If the prisms rotate in *opposite* directions and if their angular velocities are equal, a linear scan results; if these velocities are not equal, a rosette scan is obtained. Similarly, if they rotate in the *same* direction and if their angular velocities are equal, a circular scan results; if their velocities are unequal, a spiral scan is obtained. Rosell[56] gives a detailed analysis of the spiral scan case. A single rotating prism can be used to provide a circular scan having a fixed diameter.

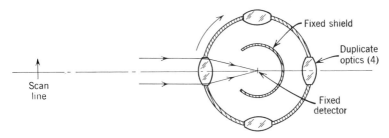

Figure 5.18 Use of rotating optics to generate a raster scan.

208   THE ELEMENTS OF THE INFRARED SYSTEM

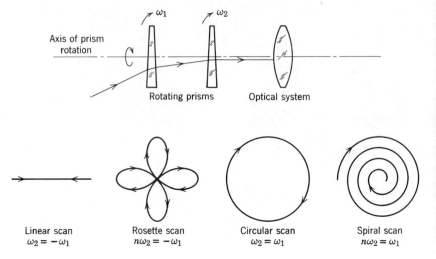

Figure 5.19   Scan generation with rotating prisms.

The efficiency with which the search field is covered can be increased by using multielement detectors, as shown in Figure 5.20. With the linear array, a single scan generates several lines in a raster pattern, one for each element in the array. By using a *mosaic*, that is, a two-dimensional array of detectors, it may be possible to cover the search field without any mechanical or optical scanning motion. Because no

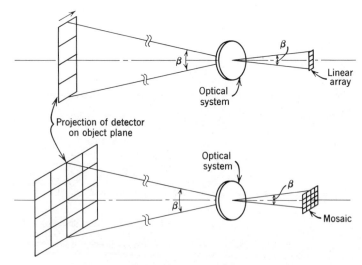

Figure 5.20   Search systems using multielement detectors.

scanning is involved, the entire search field is observed at all times. This is a great advantage in detecting a target that may appear only briefly and at an unknown time and location within the field. With any of the techniques previously described, a target that appears only briefly will be missed if the scan element is in some other part of the search field at that time. Systems of this type have been described in detail by Dubner[57].

Although it simplifies the scanning system, the use of a linear array or mosaic increases the complexity of the electronics. It has not yet proved practical to apply conventional sampling or time-multiplexing methods at the low signal levels typical of infrared detectors. Instead, each detector element must have its own preamplifier to bring the signal level up to the point where it is practical to use sampling. Multielement detectors are both difficult to manufacture and expensive. Reliability is a particularly serious problem, since failure of a single element may render the entire array or mosaic useless for the intended application.

With a linear array it is usually necessary to use a mechanical rather than an optical method of scanning. With a scanning mirror the spatial orientation of the scan element rotates throughout the scan cycle. With a single detector this is no problem, but an array becomes useless. Consider a system similar to that shown in Figure 5.17 with its single detector replaced by a linear array of detectors. Assume that this array is to scan a portion of the hemisphere above the system and that the scan element is vertical when the system is directed to the north. As the line of sight moves around to the east, the scan element rotates into a horizontal position; only a single scan line is generated and the advantage of the linear array is lost. When the line of sight reaches the south, the scan element has again become vertical, but it is inverted from the orientation it had when it was directed to the north. The top detector element has now become the bottom one. If the scanning is limited to no more than ± 30 deg in azimuth, the rotational effect can usually be ignored. For greater coverage in azimuth, mechanical scanning must be used. In principle, we could use some sort of optical derotation device or could rotate the detector at the same speed as the mirror to eliminate this effect. In practice, however, these schemes are too complex to be feasible.

5.7 OPTICAL MATERIALS FOR THE INFRARED

Early workers in the infrared field rarely had to spend much time in choosing between refractive or reflective optics for their systems. Since so few infrared-transparent materials were available, the choice was almost invariably in favor of reflective optics. As a result of a favorable wartime experience with infrared equipment, the military has supported the development of many new optical materials for the infrared. Better methods of producing large pieces of synthetic materials have been devised, a number of infrared-transmitting glasses have been developed, and high-purity semiconducting materials have become available in large sizes. As a result, a designer today has a great deal more freedom in his choice of optics.

There are perhaps a hundred optical materials that transmit in some part of the infrared. The system engineer concerned with equipment that must operate in the field finds that most of these materials are of little use to him because of their undesirable physical properties. Many of these materials are useful only in a laboratory environment; they are soft, easily scratched, readily attacked by the moisture in the atmosphere, and very susceptible to fracture by mechanical shock. Others are either too expensive or are available only in very small pieces; diamonds, a girl's best friend, have excellent transmission in the infrared, but few system engineers will find them a practical optical material. Readers interested in all of these materials will find that the report by Ballard, McCarthy, and Wolfe[58] is the most complete compilation available. Its contents have been well summarized by Wolfe and Ballard[59]. Summaries by other authors can be found in [24] and [60–66]. Many papers are available that deal with one or more materials in detail; for a sampling of these see [67–76].

The author sees no justification for describing here an endless list of optical materials, many of which are of no practical value in the design of infrared systems. Instead the intention is to present a list of preferred materials. The criterion for selection is simple; these are the materials that experienced designers are using in today's and tomorrow's systems. The result is a list of 16 materials proved to be suitable by actual usage. The omission of a particular material from the preferred list is not to be interpreted as a permanent indictment; however, any designer considering the use of such a material should be careful to find out why it has not been used more widely.

Before considering these materials, let us see what physical properties are important. The day when only the transmission and index of refraction of an optical material were significant has long since passed. Before selecting a material, the following properties should be examined in terms of the intended system application.

1. Spectral transmittance and its variation with temperature.
2. Index of refraction and its variation with temperature.

OPTICS
211

3. Hardness.
4. Resistance to surface attack by liquids.
5. Density.
6. Thermal conductivity.
7. Thermal expansion.
8. Specific heat.
9. Elastic moduli.
10. Softening and melting temperatures.
11. RF properties.

The first two items are, of course, crucial in any application. Hardness and resistance to surface attack must be considered for protective windows and irdomes. A knowledge of the thermal coefficients and the elastic moduli is important when the system will be exposed to high transient heat loads and in the design of mountings to prevent thermally induced stresses from causing the optics to fracture. Since a few of the materials soften at unusually low temperatures, they should not be used in certain applications. Occasionally there is interest in dual-mode systems, that is, those combining infrared and radar and using a common protective dome (irradome); it is then necessary to know the RF properties of the material as well.

The preferred optical materials for the infrared are shown in Figure 5.21. To the left are listed the 16 materials with an indication of the spectral range over which they have useful transmission. This range is defined as the wavelength interval for which a sample 2 mm (0.080 in.) thick transmits 10 per cent or more. If one must use any of these materials near their transmission limit, the curves of Figures 5.22 through 5.26 should be consulted. Some of these materials cannot be made in large pieces. Hence the second column shows the diameter of the largest flat window that can be made from each material. The maximum thickness will be from 10 to 20 per cent of the maximum diameter. Hemispherical irdomes are limited to a diameter about two-thirds of that shown in the table. The density of the materials may be an important consideration if the designer is faced with a tight limit on system weight. The value of the index of refraction is an average one for the useful spectral range for each material. It is included here for estimating surface reflection losses in conjunction with (5-3). If the system is to experience large variations in temperature, knowledge of the linear coefficient of expansion will be important. For convenience, these coefficients are given in Table 5.3 for the common mounting materials. When the coefficients of expansion for the mounting and the optics are different and large temperature differences are anticipated, special design steps must be taken. What is

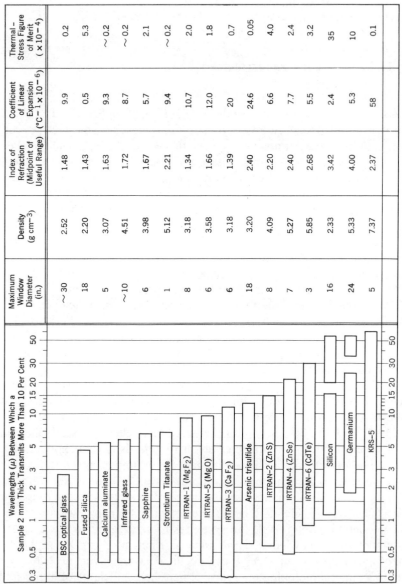

Figure 5.21 Preferred optical materials for the infrared.

OPTICS

TABLE 5.3 LINEAR EXPANSION
COEFFICIENT FOR COMMON
MOUNTING MATERIALS

Material	Linear Expansion Coefficient ($°C^{-1} \times 10^{-6}$)
Aluminum	23.9
Beryllium	12.4
Brass	30.2
Cast iron	10.4
Magnesium	26.0
Steel	13.3
Stainless steel	
300 series	9.3
400 series	16.2
Titanium	8.5

desired is a mounting that is sufficiently rigid to maintain alignment of the optics and yet still has sufficient compliance to accomodate the relative motions between the two materials. Scott[40] and Deterding[77] have described several useful techniques for this purpose that involve the use of metallic springs, plastic pads, and special potting compounds.

The final column of Figure 5.21 gives a thermal-stress figure of merit for predicting which materials may be useful under exposure to high heat loads such as those experienced during supersonic flight. Unfortunately there is no uniquely satisfactory means of describing the ability of a material to withstand thermal shock. Although it is based on simple reasoning, the figure of merit given here has proved very useful in practice. It should be considered as a guide only; any specific application will require a detailed study based on a suitable estimate of the actual conditions.

Aerodynamic heating was discussed in section 3.3, and an estimate was given of the range of temperatures to be expected. An exposed irdome or window will be heated rapidly on its outer surface. Since the temperature of the inner surface does not heat as rapidly (unless the material has an infinite thermal conductivity), there will be a thermal gradient between the two surfaces. As a result, thermal stresses develop that may ultimately cause the material to fracture. The temperature gradient between the two surfaces is approximately inversely proportional to the thermal conductivity k. If the gradient is uniform, the resulting stress F is

$$F = \frac{\alpha Y \Delta T}{2}, \qquad (5-31)$$

where α is the linear expansion coefficient, Y is Young's modulus of elasticity, and ΔT is the temperature difference between the two surfaces. If ΔT is approximated by $1/k$, the stress can be expressed as

$$F = \frac{\alpha Y}{k}. \tag{5-32}$$

(The factor of 2 has been dropped since the only interest here is in comparative values.) The thermal-stress figure of merit M_f can be defined as the reciprocal of the stress

$$M_f = \frac{1}{F} = \frac{k}{\alpha Y}. \tag{5-33}$$

Thus the larger the value of M_f, the more likely it is that the material can resist fracture due to thermal stresses.

A few additional words are in order concerning special characteristics of some of the preferred materials. Six are members of a remarkable family of IRTRAN ⓣ materials developed by the Eastman Kodak Co. Each represents a familiar material (shown in parentheses following the IRTRAN entries in Figure 5.21) produced by a hot-pressing process. For precise work, one must not assume that material produced by other processes has optical properties that are identical to those of the corresponding IRTRAN; differences of several per cent may be encountered. As an example, Olsen and McBride[75] have compared single-crystal magnesium fluoride with IRTRAN-1.

BSC (borosilicate crown) optical glass is used for applications involving the near infrared. It is included here mainly to emphasize the fact that conventional optical glasses do not transmit beyond 2.7 μ.

Fused silica is made by several manufacturers under a variety of trade names. Some types contain a rather broad water absorption band that affects transmission from about 2.7 to 3.4 μ.

Calcium aluminate is available from several sources, but not all of it is of the same quality because of a strong O–H absorption band extending from about 2.8 to 3.8 μ. At least one manufacturer has been able to eliminate this absorption, as is evidenced by the transmittance curve in Figure 5.22. Prospective users would do well to inquire about this characteristic before making a purchase of this material. Polished surfaces of calcium aluminate are rapidly attacked by water and atmospheric moisture. If it is not protected, the polish can be destroyed in a few hours. Fortunately, antireflection coatings give excellent surface protection.

A number of special glasses have been developed for use in the infrared. The data shown in Figure 5.21, while specifically for Corning 9752 glass, are typical of several other types with transmission extending to 5 or 6 μ [67-70].

Sapphire (synthetic)[72] is the hardest material listed in Figure 5.21. Not all optical shops have sufficient experience to make high-quality optical elements from it. The shine applied to sapphire in the jewelry trade is an unacceptable substitute for a true optical polish.

Strontium titanate is available only in small pieces. It is used for field lenses, immersion lenses, and windows for detectors.

Arsenic trisulfide is a most versatile material when properly employed [71]. With the exception of KRS-5, it is the softest material listed and is therefore unsuitable for any unprotected application. Because of its low thermal conductivity and high expansion coefficient, it has very poor resistance to thermal shock. A hand-held plate of this material will often shatter when trust into a stream of cold tap water. Despite these problems, arsenic trisulfide has proved very useful as one of the components in high-quality multielement lenses.

Silicon and germanium have been widely used [73, 74]. Smaller pieces are cut from single crystals; larger pieces are polycrystalline. There are small differences in the mechanical properties, but no differences have been observed in the transmission of either the single or the polycrystalline types. Both materials have secondary transmission bands extending out to 50 μ or beyond. There is some indication of small-angle scattering in the polycrystalline materials, and some workers believe this prevents achieving diffraction-limited performance. At least one supplier claims to have eliminated the problem; however, *caveat emptor*! The transmission of both silicon and germanium decreases drastically at high temperature. Practical limits are about 150°C for germanium and 325°C for silicon; at higher temperatures the loss in transmission is intolerable.

KRS-5 (thallium bromide-iodide) is valuable because it transmits to very long wavelengths. It is often used as a window for long-wavelength detectors and for bolometers responding over the entire infrared spectrum. It is the softest material listed in the table and has both a high density and a high coefficient of expansion. It should be handled with caution during polishing because thallium salts are toxic.

Because infrared equipment has proved so successful in space applications, it has become necessary to know more about the transmission of infrared optical materials at both low and high temperatures [78-81]. The primary reference on the effects of the space environment on materials is [82]. Becker[83] has provided a similar treatment that pertains only to optical materials.

Collyer[84] discusses the problems of the optical workshop in processing infrared-transmitting optical materials. He also gives a list of

216   THE ELEMENTS OF THE INFRARED SYSTEM

preferred optical materials, all of which are included in the list given in Figure 5.21.

It is known that some infrared-transmitting optical materials fluoresce in the near infrared when stimulated by visible light. Rogers et al.[85] examined 23 materials and found 13 that fluoresced.

Transmittance curves for the preferred optical materials are shown in Figures 5.22 through 5.26. These curves are for well-polished samples

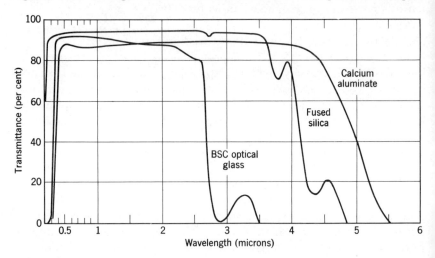

Figure 5.22 Transmittance of preferred optical materials. Thickness 2 mm (0.080 in.)

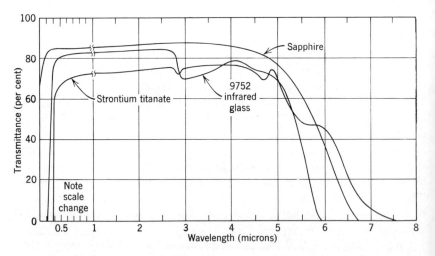

Figure 5.23 Transmittance of preferred optical materials. Thickness 2 mm (0.080 in.).

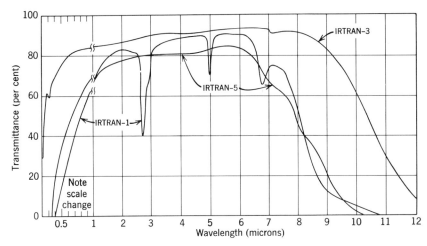

Figure 5.24 Transmittance of preferred optical materials. Thickness 2 mm (0.080 in.). (Data courtesy of Eastman Kodak Co.)

without antireflection coatings. All curves are for a sample thickness of 2 mm (0.080 in.). These curves have been extended to very short wavelengths in order to show the transmission characteristics of each material in the visible portion of the spectrum and in its short-wavelength-cutoff region. Most data sources do not show the transmission of infrared optical materials in the visible, a situation that is particularly frustrating when

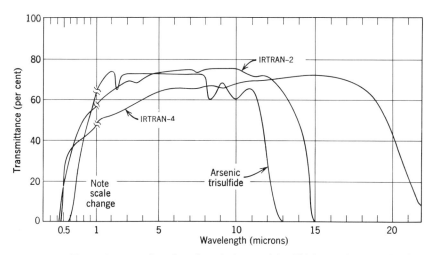

Figure 5.25 Transmittance of preferred optical materials. Thickness 2 mm (0.080 in.). (Data on IRTRAN courtesy of Eastman Kodak Co.).

217

218   THE ELEMENTS OF THE INFRARED SYSTEM

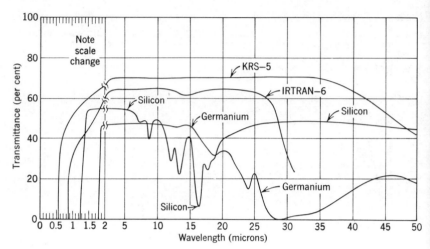

Figure 5.26 Transmittance of preferred optical materials. Thickness 2 mm (0.080 in.). (Data on IRTRAN courtesy of Eastman Kodak Co.)

one is considering the choice of optical materials for multisensor systems.

Very little information is available outside of the classified literature on the spectral emissivity of these preferred materials. We can make an estimate accurate to perhaps 20 per cent by remembering that (2-21) and (2-24) show that

$$\epsilon = 1 - (\rho + \tau). \tag{5-34}$$

This relationship is strictly true only at thermal equilibrium; under other conditions it is useful for estimates in the absence of better data. Since the spectral transmittance curves of Figures 5.22 through 5.26 include losses by both reflection and absorption, they are, to a first approximation, plots of the quantity $(\rho + \tau)$ in (5-34). All that one has to do is subtract the value of the spectral transmittance from unity and plot the result to obtain the approximate spectral emissivity curve. McMahon[86] treats in detail the theory of the emissivity of partially transparent materials. Measured data on the emissivity of IRTRAN-1,2 and 3 are given in [87] and data on IRTRAN-1,2,4, and 5, sapphire, silicon, and several black paints are given in [88].

5.8 ANTIREFLECTION COATINGS

The index of refraction of most of the preferred optical materials is high enough that significant amounts of the incident radiant flux are lost by reflection from the surfaces. A thin film or coating can be applied to the surface by vacuum evaporation in order to eliminate completely the reflection at a given wavelength. The reduction attainable over a band of wavelengths, such as an atmospheric window, although not complete, can still be very impressive.

An antireflection coating for optical materials used in air must meet two criteria; its index of refraction must be equal to the square root of the index of the optical material to be coated, and its optical thickness must be equal to one-fourth of the wavelength at which minimum reflection is to occur. *Optical thickness* is the product of the index of refraction and the physical thickness of the coating. Since optical thickness changes with the angle of incidence, an antireflection coating will not be equally efficient for all of the rays in a converging bundle. Fortunately the effect is negligible for angles less than 30 deg (corresponding to $f/1$ optics).

Materials that have been used for antireflection coatings are shown in Table 5.4. The useful spectral range may be limited by absorption bands in the coating material or by the difficulty of making a stable layer

TABLE 5.4 MATERIALS FOR LOW REFLECTION COATINGS[a]

Material	Useful Spectral Range (μ)	Average Index of Refraction
Cryolite (Al F_3 · NaF)	0.2–10	1.34
Magnesium fluoride (MgF_2)[b]	0.12–5	1.35
Thorium fluoride (Th F_4)	0.2–10	1.45
Cerium fluoride (CeF_3)	0.3–5	1.62
Silicon monoxide (Si O)[b]	0.4–8	1.45–1.90[c]
Zirconium dioxide (Zr O_2)	0.3–7	2.10
Zinc sulfide (Zn S)[b]	0.4–15	2.15
Cerium dioxide (Ce O_2)	0.4–5	2.20
Titanium dioxide (Ti O_2)	0.4–7	2.30–2.80[c]

[a]Adapted from Smakula et al. [64].
[b]Most often used because of their excellent durability.
[c]Index controlled by manner of deposition.

thick enough to meet the quarter-wavelength criterion. A convenient rule of thumb is that a 10 per cent deviation in the optimum index of refraction for the coating (one that meets the square-root relationship given above) results in a decrease of only 1 per cent in the efficiency of the coating. Thus the transmittance of a silicon plate at the design wavelength can be made to exceed 98 per cent by applying coatings that range from 1.65 to 2.05 in index. Uncoated, the transmittance is about 54 per cent. Of the materials listed in Table 5.4, magnesium fluoride, silicon monoxide, and zinc sulfide have the greatest resistance to attack by chemicals and moisture, and they are hard enough that special cleaning methods are not required. Consequently they are logical choices for coating irdomes and windows exposed to the field environment. Since the use of an antireflection coating is one of the cheapest ways to increase system sensitivity, it is axiomatic that any transmitting material having an index of refraction greater than 1.6 should have such a coating applied to it.

Cox and Hass[89] have reported the results of coating silicon and germanium with a variety of materials. In a later paper[90], they report on one-, two-, and three-layer coatings that are even more efficient over wider optical bandwidths. Figure 5.27 shows the transmittance of a germanium plate with several different coatings. From these curves it is evident that the transmittance is more than 90 per cent over a wavelength interval the limits of which are in the ratio of 1.5 to 1, 2.7 to 1, and 3 to 1 for the single-, double-, and triple-layer coatings, respectively, Figure 5.28 shows the transmittance of a germanium plate coated for peak transmittance at 9.8 μ. The transmittance is 86 per cent or more over the range of 8 to 12 μ; compare this with the 47 per cent transmittance of the uncoated sample. One manufacturer has advertised[91] a proprietary coating that results in a germanium plate having a transmittance in excess of 90 per cent between 2 and 12 μ.

5.9 HIGH-REFLECTION COATINGS

For many years chemically deposited silver was the traditional material for coating mirrors. Freshly applied films had a high reflectance but

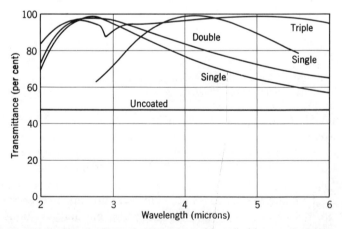

Figure 5.27 Transmittance of a germanium plate with and without antireflection coatings, 2 to 6 μ (partially adapted from Cox et al.[90]).

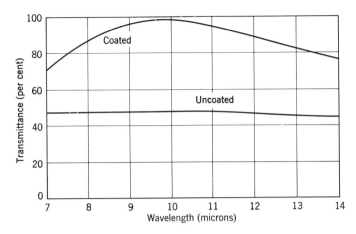

Figure 5.28 Transmittance of a germanium plate with and without antireflection coatings, 7 to 14 μ (adapted from Cox et al.[90]).

they tarnished rapidly when exposed to air. Since the introduction during World War II of the techniques for evaporating metal films, virtually all mirrors have been coated by this means.

The reflectance of most metals increases at longer wavelengths. Table 5.5 shows the spectral reflectance of various evaporated metal

TABLE 5.5 REFLECTANCE OF EVAPORATED METAL FILMS[a]

Wavelength (μ)	Reflectance (per cent)				
	Aluminum	Silver	Gold	Copper	Rhodium
0.5	90.4	97.7	47.7	60.0	77.4
1.0	93.2	98.9	98.2	98.5	85.0
3.0	97.3	98.9	98.3	98.6	92.5
5.0	97.7	98.9	98.3	98.7	94.5
8.0	98.0	98.9	98.4	98.7	95.2
10.0	98.1	98.9	98.4	98.8	96.0

[a]Adapted from Bennett et al. [92] [94] and Hass. [93].

films. The characteristics of aluminum, the most commonly used material, have been carefully studied by Bennett et al.[92]. Occasionally the slightly higher reflectance of gold warrants using it despite its higher cost[94].

If a mirror is exposed to a dusty environment, it is wise to protect its surface with a thin evaporated coating of magnesium fluoride or silicon monoxide. Bennett et al.[95] report measurements to a wavelength of 35 μ on aluminized mirrors with these coatings. Between 3 and 15 μ, the reflectance of the coated mirrors was a fraction of a per cent less than without the coatings. Beyond 15 μ the coatings produced no measurable effect. Such coatings should be used with caution at wavelengths of less than 3 μ since they are likely to absorb in this region.

Methods for the measurement of spectral reflectance are given in refs.[96–101]. Sanderson[102] describes a reflectometer and presents measurements of the diffuse spectral reflectance of various paints from 0.76 to 2.5 μ. A monochromator is the key element in most of the reflectometers described above, as well as in devices for measuring the spectral transmittance of optical materials and the spectral response of detectors. Convenient and accurate means for calibrating the wavelength scales of monochromators are given in refs. 103–105.

5.10 OPTICAL FILTERS

With the development of better photoconductive detectors it has become increasingly important to match the spectral bandpass of the system to one of the atmospheric transmission windows. The principal means of accomplishing this is to use optical filters. Two types are available. The *absorption filter* depends for its effectiveness on the absorbing characteristics of various dyes, plastics, and optical materials. The *reflection* or *interference filter* uses interference effects to reflect rather than to absorb unwanted wavelengths.

The simplest way to describe any filter is by a curve showing its spectral transmittance. The relatively new interference filters can be made to have almost any desired spectral transmittance characteristics. Thus it is desirable to have a more quantitative descriptive terminology for specifying filters. No standard exists for this purpose; until one does, the terminology suggested by the Bausch and Lomb Optical Company[106] will serve admirably.

A *bandpass filter* transmits a band of wavelengths sharply bounded by extended regions of low transmittance.

The *spectral bandwidth* describes the wavelength interval transmitted by the filter in terms of the center wavelength and the half-width.

The *center wavelength* λ_0 is the wavelength (in microns) at the center of the spectral bandwidth.

The *half-width* is the wavelength interval (in microns) over which the transmittance exceeds one-half the peak transmittance of the filter.

It is often expressed as a decimal fraction of the center wavelength; a bandpass filter having a half-width that is 25 per cent of the center wavelength is specified as 0.25 λ_0.

The *peak transmittance* is the maximum transmittance within the spectral bandwidth of the filter. It is expressed as a percentage of the transmittance of the uncoated substrate at the same wavelength.

The *substrate* is the optical material on which the filter is deposited.

A *long-wavelength pass filter* transmits all wavelengths longer than a specified cut-on wavelength.

A *short-wavelength pass filter* transmits all wavelengths shorter than a specified cut-off wavelength.

The *cut-on wavelength* or *cutoff wavelength* λ_c is the wavelength at which the transmittance is 5 per cent of the peak transmittance.

The *slope* specifies the rate at which the transmittance increases between the cut-on or cut-off wavelength and the wavelength λ'_c where the transmittance is 80 per cent of the peak transmittance. It is given by the wavelength difference between λ'_c and λ_c expressed as a decimal fraction of λ_c.

Absorption filters are rarely used as bandpass filters, because their spectral bandwidths are usually very wide. However, they are often used as short-wavelength or long-wavelength pass filters. For example, to reject all wavelengths shorter than 1.8 μ, it is only necessary to include a piece of germanium at some point in the optical path. Glass and fused silica are often used as short-wavelength pass filters. One of the few absorption filters whose characteristics can be modified to meet the wishes of the designer is the Kodak Far Infrared Filter[107]. It is made by applying a layer of silver sulfide on a substrate of silver chloride. The cut-on wavelength can be located anywhere between 1 and 5 μ. Unfortunately the slope is very poor, 0.7λ_c being a typical value. For comparison, an interference filter can be made that has a slope of 0.05 λ_c. Plastic absorption filters for the 1 to 6 μ region have been described by Blout et al.[108]; Cox et al.[90] describe filters of various semiconductors. Because the absorption filter absorbs the radiant flux lying outside its spectral bandwidth, it may become warm and perhaps even shatter. For this reason, a system employing such a filter should never be allowed to look directly at the sun.

Interference filters are made by the vacuum deposition of several layers of dielectric material onto a suitable substrate; the index of refraction and the thickness of each layer must be precisely controlled[109, 110]. The design equations are quite complex, but most manufacturers of filters have their computers programmed so that data for the construction of almost any interference filter can be provided in a few minutes of computer time. As many as 100 layers may be used for filters whose optical characteristics must be rigidly defined.

For the region from 1 to 13 μ, bandpass filters are available with half-widths from 0.02 to 0.5 λ_0 and peak transmittances exceeding 80 per cent. Half-widths of 0.01 λ_0 are available for the 1 to 5 μ region, but their peak transmittance is limited to 35 to 40 per cent. From 15 to 30 μ, half-widths of 0.04 λ_0 are available with peak transmittances of 80 per cent.

Long- and short-wavelength pass filters can be made with λ_c placed at any wavelength between 1 and 30 μ and with a slope as steep as 0.05 λ_c. Peak transmittance is about 80 per cent from 1 to 10 μ, approximately 65 per cent from 10 to 20 μ, and about 45 per cent beyond 20 μ. By applying a short-wavelength pass filter to one side of a substrate and a long-wavelength pass filter to the other side, it is possible to make a bandpass filter with a half-width of many microns.

A word of caution is in order: interference filters often contain secondary bandpasses that are separated from their primary transmitting regions. Manufacturers of filters can limit the transmittance of any of these secondary regions to less than 1 per cent[111]. Since the reduction of these secondary regions increases the cost of the filter, the manufacturer may not automatically do it. Thus the designer must be careful to specify in his procurement specification the extent of any spectral region to be blocked by the filter. Many an engineer has spent days puzzling over the erratic performance of a system that is supposedly insensitive to solar radiation, only to eventually find a filter that leaks at 0.5 μ.

The radiant flux that is not transmitted through an interference filter is reflected and does not increase the temperature of the filter. A system equipped with such a filter can look at the sun indefinitely without damage to the filter (although the detector may suffer either short-term or permanent damage due to the high flux level within the system bandpass.

5.11 COLLIMATORS

The collimator is probably the most important piece of test equipment available to the infrared system engineer. It provides a precisely known irradiance and simulates an infinitely distant source. Its principal use is in the measurement of the sensitivity of infrared systems, but it is also useful in the alignment of optics and the measurement of their focal length. Reduced to its essentials, a collimator consists of optics, either refractive or reflective, and a source of radiant energy located at the focal point of the optics. Rays entering an optical system parallel to the optical axis are brought to a focus at the second focal point. Conversely, if a point source is placed at the second focal point, its rays, after passing through the optics, emerge parallel to the optical axis. It is convenient to visualize the resulting bundle of parallel rays as being much like the beam from a searchlight. Since a point source can only be approximated, it is assumed in the following discussion that the diameter of the source is very small with respect to the focal length of the optics. The effect of this approximation will be discussed later.

The sensitivity of an infrared system can be determined by measuring the signal-to-noise ratio at the output of the system as a function of the irradiance at the optics. The simplest method of providing a known irradiance is to place a source at some convenient distance. From (2-2), (2-5), and (2-9), the irradiance from a blackbody is

$$H = \frac{\epsilon \sigma T^4 A_s}{\pi d^2}, \quad (5\text{-}35)$$

where A_s is the area of the source or its limiting aperture, expressed in square centimeters. Since the system test procedure usually specifies the temperature of the source, the irradiance can be varied either by changing the area of the source or its distance from the system. Assume that the temperature of the blackbody is 800°K (a typical value for tests of this sort) and its emissivity is unity. For the reasons already stated in section 3.1, a practical minimum diameter for the source aperture is 0.010 in. (0.025 cm). Using these values, the irradiance at a distance of 20 ft (6.5 m) from the source is 8.5×10^{-10} W cm^{-2}. Since many infrared systems will respond to an irradiance as low as 10^{-14} W cm^{-2}, it is evident that the source used in this way provides an irradiance that is far greater than needed. Short of building a longer laboratory there is little that can be done to provide the desired reduction of irradiance.

An alternate approach is to try the single-mirror collimator shown in Figure 5.29. The irradiance in the beam from this collimator is

$$H = \frac{\epsilon \sigma T^4 A_s}{\pi} \frac{\rho_1}{f_1^2}, \quad (5\text{-}36)$$

where ρ_1 is the reflectance of the collimating mirror and f_1 is its focal length expressed in centimeters. Since the distance between the source and the collimating mirror is equal to the focal length of the mirror, and ρ_1 is nearly unity, (5-36) is essentially equal to (5-35). As a result, the single-mirror collimator offers little hope of decreasing the irradiance beyond that already achieved with the distant source.

Although there are good and sufficient reasons for establishing a minimum diameter for the source aperture, there is no reason why its

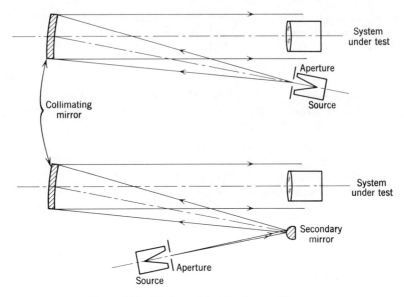

Figure 5.29 Single- and double-mirror collimators.

size cannot be reduced by optical means. This is exactly what is done with the two-mirror collimator shown in Figure 5.29[1]. A small convex secondary mirror forms a greatly reduced image of the source aperture. This image serves as a new source for the collimating mirror. The irradiance in the beam from a double-mirror collimator is

$$H = \frac{\epsilon \sigma T^4 A_s \rho_1 \rho_2}{\pi} \left(\frac{f_2}{f_1 d'} \right)^2, \quad (5\text{-}37)$$

where ρ_2 is the reflectance of the secondary mirror, f_2 is its focal length, and d' is the distance from the source aperture to the vertex of the secondary mirror. Both f_2 and d' are expressed in centimeters. The secondary mirror can be quite small and should have a very short focal length. It is often convenient to use a bearing ball for this mirror although it is desirable to apply an evaporated aluminum coating to increase its reflectance. The quantity $(f_2/d')^2$ expresses the optical reduction in the area of the source; values of 10^{-3} to 10^{-4} can easily be achieved. Defining d' as the distance to the vertex of the secondary mirror is a convenient approximation. It should actually be measured to the focal point, which lies behind the reflecting surface of a convex mirror. However, if d' is

---

[1]The collimating mirrors in Figure 5.29 are tilted so that the source does not lie on the optical axis. The coma introduced by this arrangement can be intolerable if a high-resolution optical system is to be tested. For such instances, special off-axis paraboloids are available.

equal to $10f_2$, the error in irradiance from the use of this approximation is only 1 per cent and it decreases rapidly for larger values of d' [112].

Let us design a two-mirror collimator to fit on the top of a standard 6 ft (1.8 m) laboratory bench. A collimator mirror with a focal length of 5 ft (1.53 m) can be conveniently accommodated by such a bench. Assume a secondary mirror with a focal length of 0.5 cm placed 30 cm from the source. Assume also T is 800°K; ϵ is unity; ρ_1 and ρ_2 are both equal to 0.95; and the diameter of the source aperture is 0.025 cm. The calculated irradiance in the beam of this collimator is 4×10^{-12} W cm^{-2}. The result is a collimator that fits conveniently on a standard laboratory bench and produces an irradiance several orders of magnitude less than that achievable with the single-mirror collimator or distant source.

Notice that the distance between the system under test and the collimating mirror does not appear in (5-36) and (5-37). The irradiance is independent of the distance from the collimating mirror. This is one of the prime advantages of the collimator; the system under test can be placed at any convenient point in the beam and one need only ensure that the beam fills the entrance aperture. It is important to examine the conditions for which this statement is correct. With a true point source, the irradiance in the beam is, in fact, independent of the distance from the collimating mirror (as long as aberration-free optics are used and tests are conducted in an absorption-free atmosphere). With a point source, the diameter of the beam from the collimator is constant throughout its length.

What if the fictitious point source is replaced with a physically realizable one, that is, one having a definable size? As before, the rays from a single point of the source will leave the collimator in a parallel bundle. However, the bundles from various points of the source will not be parallel to one another, and the diameter of the beam from the collimator, which includes all of these bundles, is no longer constant. Figure 5.30 shows a simplified single-mirror collimator.[1] The limiting rays are shown for bundles coming from the center and from the edges of the source. The beam from the collimator diverges through an angle β, which is equal to the angle subtended by the source at a distance equal to the focal length of the collimating mirror. Typically, β is very small, usually no more than a fraction of a milliradian. The irradiance in the beam is independent of the distance from the collimating mirror at all points lying within the shaded region in Figure 5.30. The *crossover point* lies at the intersection of the rays that limit the region of constant irradiance.

[1]The size of the source has been exaggerated in order to show the angles involved. Also, there is a blockage of the center of the beam. With the collimators of Figure 5.29, no such blockage occurs.

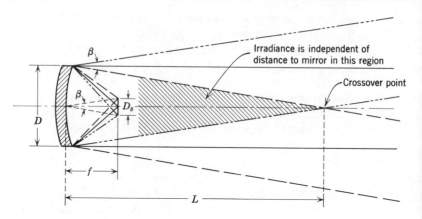

Figure 5.30 Determination of the region in which the irradiance in the beam from a collimator is independent of the distance from the collimator.

Beyond the crossover point the irradiance in the beam varies inversely as the square of the distance from the collimating mirror (this is the region considered in discussions of searchlight optics [113]).

For a single-mirror collimator the distance L from the mirror to the crossover point (assuming β is small) is

$$L = \frac{D}{\beta} = \frac{Df}{D_s}, \qquad (5\text{-}38)$$

where D is the diameter of the collimating mirror, f is its focal length, and D_s is the diameter of the source aperture. This expression can be used for a double-mirror collimator by assuming that D and f are, as before, the diameter and focal length of the collimating mirror and that D_s is the diameter of the image of the source formed by the secondary mirror. If it is assumed that the two-mirror collimator we just designed has a 12-in. diameter primary mirror, the distance to the crossover point is equal to 3.5×10^5 ft. Obviously, with this particular collimator the distance to the crossover point is of academic interest only. Even if the diameter of the source is increased a thousand times (increasing the irradiance a million times), the distance to the crossover point is still 350 ft and it is very unlikely that the collimator would be placed that far from the system under test.

It is considered good design practice to make the diameter of the collimating mirror at least 10 per cent greater than that of the optics to be tested. With this precaution, alignment of the optics and the collimator is not critical and one can be sure that the beam completely fills the optics.

OPTICS

It is of interest to estimate the probable error in the calculated value of the irradiance in the beam of a double-mirror collimator. Two collimators are considered; one type represents a unit constructed and measured as carefully as possible for use as a laboratory standard, the other type represents an average unit designed for daily operation on a production line. Table 5.6 shows the quantities that must be determined

TABLE 5.6 ESTIMATE OF THE PROBABLE ERROR IN THE CALCULATED VALUE OF THE IRRADIANCE IN THE BEAM FROM DOUBLE-MIRROR COLLIMATORS

Quantity	Laboratory standard collimator		Collimator for daily use on a production line	
	Probable Error in Quantity	Contribution to Probable Error in the Irradiance	Probable Error in Quantity	Contribution to Probable Error in the Irradiance
T	±2 deg	±1%	±5 deg	±2.5%
ϵ	±1%	±1%	±1%	±1%
A_s	±0.5%	±0.5%	±1%	±1%
ρ_1	±1%	±1%	±3%	±3%
ρ_2	±1%	±1%	±3%	±3%
f_1	±0.5%	±1%	±0.5%	±1%
f_2	±1%	±2%	±1%	±2%
d'	±0.5%	±1%	±1.5%	±3%
Probable error in the calculated value of the irradiance		±3.2%	—	±6.3%

in order to calculate the irradiance and gives estimates of the probable error in each quantity. From the simple theory of errors, the probable error in the calculated irradiance from the laboratory-standard collimator is ±3.2 per cent while it is ±6.3 per cent for the collimator used daily on a production line. Maintenance of such small probable errors requires continuous diligence by responsible personnel.

A word about terminology. To *collimate* is to render parallel; when we collimate a source, it is adjusted in relation to the focal point of the optics, so that all rays from a given point of the source are parallel to one another after leaving the optics. Any departure from this condition of parallelism, which may be caused by defocusing, aberrations, or local deformations of the optics, is called *decollimation*. The fact that the source is not a true point is indicated by specifying β the angular subtense of the source when used with the particular collimating optics. This usage differs from that of some authors who make no distinction between decollimation and the effect of the angular subtense of the source. It is important that the collimator does not cause an appreciable increase in the size of the blur circle of the optics under test and upset a careful match between the blur circle and a detector or the openings in an optical modulator. Such a mismatch can be avoided by ensuring that the angular subtense of the source does not exceed one-fifth the angular subtense of the blur circle of the optics under test.

REFERENCES

[1] F. W. Sears, *Optics*, Reading, Mass.: Addison-Wesley, 1958.
[2] H. W. Farwell, "When is the Sine of an Angle Equal to the Angle," *Am. J. Phys.*, 17, 449 (1949).
[3] R. M. Scott, "Optics for Infrared Systems," *Proc. Inst. Radio Engrs.*, 47, 1530 (1959).
[4] A. Bouwers, *Achievements in Optics*, New York: Elsevier, 1946.
[5] G. Z. Dimitroff and J. G. Baker, *Telescopes and Accessories*, Philadelphia: Blakiston, 1945.
[6] I. S. Bowen, "Optical Problems at Palomar Observatory," *J. Opt. Soc. Am.*, 42, 795 (1952).
[7] R. M. Scott, "The Practical Applications of Modulation Transfer Functions," *J. Soc. Phot. Instr. Engrs.*, 2, 132 (1964).
[8] F. D. Smith, "Optical Image Evaluation and the Transfer Function," *Appl. Opt.*, 2, 335 (1963).
[9] G. C. Higgins, "Methods for Engineering Photographic Systems," *Appl. Opt.*, 3, 1 (1964).
[10] F. H. Perrin, "Manipulation and Significance of Sine Wave Response Functions," *J. Soc. Motion Picture Television Engrs.*, 69, 239 (1960).
[11] R. L. Lamberts, G. C. Higgins, and R. N. Wolfe, "Measurements and Analysis of the Distribution of Energy in Optical Images," *J. Opt. Soc. Am.*, 48, 487 (1958).
[12] R. E. Hopkins, "Re-Evaluation of the Problem of Optical Design," *J. Opt. Soc. Am.*, 52, 1218 (1962).
[13] D. P. Feder, "Automatic Optical Design". *Appl. Opt.*, 2, 1209 (1963).
[14] B. Brixner, "Automatic Lens Design for Nonexperts," *Appl. Opt.*, 2, 1281 (1963).
[15] G. W. Ritchey, "The Modern Photographic Telescope and the New Astronomical Photography. Part IV," *J. Roy. Astron. Soc. Canada*, 23, 15 (1929).
[16] H. E. Dall, "Diffraction Effects Due to Axial Obstructions in Telescopes," *J. Brit. Astron. Assoc.*, 48, 163 (1938).
[17] R. T. Jones, "Coma of Modified Gregorian and Cassegrainian Mirror Systems," *J. Opt. Soc. Am.*, 44, 630 (1954).
[18] P. R. Yoder, Jr., F. B. Patrick, and A. E. Gee, "Analysis of Cassegrainian-Type Telescopic Systems," *J. Opt. Soc. Am.*, 43, 1200 (1953).
[19] A. T. Young, "Design of Cassegrain Light Shields," *Appl. Opt.*, 6, 1063 (1967).
[20] D. D. Maksutov, "New Catadioptric Meniscus Systems," *J. Opt. Soc. Am.*, 34, 270 (1944).
[21] D. D. Maksutov, *Technologie der Astronomischen Optik*, Berlin: VEB Verlag Technik, 1954.
[22] H. Dubner, "Optical Design for Infrared Missile Seekers," *Proc. Inst. Radio Engrs.*, 47, 1537 (1959).

[23] S. Rosin and M. Amon, "Color-Corrected Mangin Mirror," *Appl. Opt.*, **6**, 963 (1967).
[24] M. Herzberger and C. D. Salzberg, "Refractive Indices of Infrared Optical Materials and Color-Correction of Infrared Lenses," *J. Opt. Soc. Am.*, **52**, 420 (1962).
[25] D. S. Cary, "Optical System for Infrared Target Tracking Apparatus," U.S. Patent No. 3,002,092, September 26, 1961.
[26] "Infrared Advantageously Reflected or Refracted," *Sci. Am.*, **209**, 51 (October 1963).
[27] R. G. Treuting, "An Achromatic Doublet of Silicon and Germanium," *J. Opt. Soc. Am.*, **41**, 454 (1951).
[28] E. H. Thielens, "Infrared Achromat Lens," U.S. Patent No. 2,865,253, December 23, 1958.
[29] R. Gelles, "A General-Purpose Infrared Lens," *Appl. Opt.*, **1**, 78 (1962).
[30] A. E. Murray, "Fast Refractive Infrared Optical Systems," *Infrared Phys.*, **2**, 37 (1962).
[31] A. E. Murray and M. J. Herzberger, "Infrared Lens," U.S. Patent No. 3,217,596, November 16, 1965.
[32] A. E. Murray, "An All-Irtran Doublet Objective," *Appl. Opt.*, **4**, 254 (1965).
[33] T. P. Vogl, "An $f/0.75$ Diffraction Limited Triplet for the 8-12 Micron Atmospheric Window," *J. Soc. Phot. Instr. Engrs.*, **1**, 218 (1963).
[34] J. R. Snyder, "Infrared Optical Objective," U.S. Patent No. 3,160,700, December 8, 1964.
[35] V. F. Olson, "Measurement of Transmission Through a Lens," *Appl. Opt.*, **6**, 1140 (1967).
[36] J. W. Larmer and E. Goldstein, "Some Comments Upon Current Optical Shop Practices," *Appl. Opt.*, **5**, 677 (1966).
[37] E. J. Tew, Jr., "Measurement Techniques Used in the Optics Workshop," *Appl. Opt.*, **5**, 695 (1966).
[38] J. H. McLeod and W. T. Sherwood, "A Proposed Method of Specifying Appearance Defects of Optical Parts," *J. Opt. Soc. Am.*, **35**, 136 (1945).
[39] W. P. Barnes, Jr., "Some Effects of Aerospace Thermal Environments on High-Acuity Optical Systems," *Appl. Opt.*, **5**, 701 (1966).
[40] R. M. Scott, "Optical Engineering," *Appl. Opt.*, **1**, 387 (1962).
[41] W. P. Barnes, Jr., "Considerations in the Use of Beryllium for Mirrors," *Appl. Opt.*, **5**, 1883 (1966).
[42] G. S. Hunter, "ATS-C Satellite to Carry Device to Measure Mirror Deterioration," *Aviation Week*, **86**, 73 (March 13, 1967).
[43] D. Chin, "Optical Mirror-Mount Design and Philosophy," *Appl. Opt.*, **3**, 895 (1964).
[44] D. E. Williamson, "Cone Channel Condenser Optics," *J. Opt. Soc. Am.*, **42**, 712 (1952).
[45] N. S. Kapany and R. J. Simms, "Fiber Optics. XI. Performance in the Infrared Region," *J. Opt. Soc. Am.*, **55**, 963 (1965).
[46] R. R. Standel and R. E. Hendrickson, "Infrared Fiber Optics Techniques," *Infrared Phys.* **3**, 223 (1963).
[47] H. Blackstone, "Apparatus for Producing a Conical Scan of Automatically Varying Apex Angle," U.S. Patent No. 2,855,521, October 7, 1958 (Filed September 4, 1952).
[48] M. R. Holter and W. L. Wolfe, "Optical-Mechanical Scanning Techniques," *Proc. Inst. Radio Engrs.*, **47**, 1547 (1959).
[49] W. B. McKnight and D. E. Holter, "Optical Scanning System," U.S. Patent No. 3,054,899, September 18, 1962.

[50] H. F. Hicks, Jr., "Hemispherical Scanning System," U.S. Patent No. 3,023,662, March 6, 1962.
[51] W. B. McKnight, L. N. McClusky, and G. H. Widenhofer, "Scanning Mirror Assembly," U.S. Patent No. 3,020,414, February 6, 1962.
[52] W. B. McKnight, L. N. McClusky, and R. D. Cleven, "Nutational Scanning Mirror," U.S. Patent No. 3,071,036, January 1, 1963.
[53] A. H. Rosenthal, "Scanning System for Photosensitive Light Tracking Device," U.S. Patent No. 3,135,869, June 2, 1964.
[54] "Infrared — The New Horizon," *Electronic Equipment Eng.*, **9**, 44 (May 1961).
[55] H. H. Koelle, ed., *Handbook of Astronautical Engineering*, New York: McGraw-Hill, 1961, pp.13–87.
[56] F. A. Rosell, "Prism Scanner," *J. Opt. Soc. Am.*, **50**, 521 (1960).
[57] H. A. Dubner, "Multi-Cell IR Search System," U.S. Patent No. 3,041,460, June 19, 1959.
[58] S. S. Ballard, K. A. McCarthy, and W. L. Wolfe, "Optical Materials for Infrared Instrumentation," Report 2389-11-S, The University of Michigan, Willow Run Laboratories, Ann Arbor, Mich., Janurary 1959. A supplement, Report 2389-11-S1, April 1961, is also available. Both reports should be ordered from DDC.
[59] W. L. Wolfe and S. S. Ballard, "Optical Materials, Films, and Filters for Infrared Instrumentation," *Proc. Inst. Radio Engrs.*, **47**, 1540 (1959).
[60] S. S. Ballard and J. S. Browder, "Thermal Expansion and Other Physical Properties of the Newer Infrared-Transmitting Optical Materials," *Appl. Opt*, **5**, 1873 (1966).
[61] S. S. Ballard, "Infrared Optical Materials, Old and New," *Japan. J. App. Phys.*, **4**, Supplement 1, 23 (1965).
[62] E. M. Voronkova et al., *Optical Materials for Infrared Technology*, USSR Academy of Sciences, 1965.
[63] D. E. McCarthy, "The Reflection and Transmission of Infrared Materials: 1. Spectra from 2-50 microns," *Appl. Opt.*, **2**, 591 (1963); "2. Bibliography," **2**, 596 (1963); "3. Spectra from 2μ to 50μ," **4**, 317 (1965); "4. Bibliography," **4**, 507 (1965).
[64] A. Smakula, J. Kalnajs, and M. J. Redman, "Optical Materials and Their Preparation," *Appl. Opt.*, **3**, 323 (1964).
[65] D. L. Stierwalt, J. B. Bernstein, and D. D. Kirk, "Measurement of the Infrared Spectral Absorptance of Optical Materials," *Appl. Opt.*, **2**, 1169 (1963).
[66] P. Billard, "Materials Utilizable in Optical Systems Working in the Infrared," *Acta Electronica*, **6**, 7 (1962) (CFSTI: TT-64-11809).
[67] A. R. Hilton, C. E. Jones, and M. Brau, "New High Temperature Infrared Transmitting Glasses," Part I, *Infrared Phys.*, **3**, 69 (1963); Part II, **4**, 213 (1964); Part III, **6**, 183 (1966).
[68] A. R. Hilton, "Nonoxide Chalcogenide Glasses as Infrared Optical Materials," *Appl. Opt.*, **5**, 1877 (1966).
[69] J. A. Muir and R. J. Cashman, "GeSeTe — A New Infrared Transmitting Chalcogenide Glass," *J. Opt. Soc. Am.*, **57**, 1 (1967).
[70] G. W. Cleek, J. J. Villa, and C. H. Hahner, "Refractive Indices and Transmittances of Several Optical Glasses in the Infrared," *J. Opt. Soc. Am.*, **49**, 1090 (1959).
[71] R. Frerichs, "New Optical Glasses with Good Transparency in the Infrared," *J. Opt. Soc. Am.*, **43**, 1153 (1953).
[72] I. H. Malitson, "Refraction and Dispersion of Synthetic Sapphire," *J. Opt. Soc. Am.*, **52**, 1377 (1962).
[73] C. D. Salzberg and J. J. Villa, "Infrared Refractive Indexes of Silicon, Germanium, and Modified Selenium Glass," *J. Opt. Soc. Am.*, **47**, 244 (1957).

[74] R. C. Lord, "Far Infrared Transmission of Silicon and Germanium," *Phys. Rev.*, **85**, 140 (1952).
[75] A. L. Olsen and W. R. McBride, "Transmittance of Single-Crystal Magnesium Fluoride and IRTRAN-1 in the 0.2–15 μ Range," *J. Opt. Soc. Am.*, **53**, 1003 (1963).
[76] L. S. Ladd, "Cadmium Telluride Infrared Transmitting Material," *Infrared Phys.*, **6**, 145 (1966).
[77] L. G. Deterding, "High Precision, Strain-Free Mounting of Large Lens Elements," *Appl. Opt.*, **1**, 403 (1962).
[78] G. Linsteadt, "Infrared Transmittance of Optical Materials at Low Temperatures," *Appl. Opt.*, **3**, 1453 (1964).
[79] C. D. Salzberg, "Infrared Transmittance of Strontium Titanate from Room Temperature to (−) 180°C," *J. Opt. Soc. Am.*, **51**, 1149 (1961).
[80] D. T. Gillespie, A. L. Olsen, and L. W. Nichols, "Transmittance of Optical Materials at High Temperatures in the 1 to 12 μ Range," *Appl. Opt.*, **4**, 1488 (1965).
[81] U. P. Oppenhein and A. Goldman, "Infrared Spectral Transmittance of MgO and BaF_2 Crystals between 27° and 1000°C," *J. Opt. Soc. Am.*, **54**, 127 (1964).
[82] L. Jaffe and J. B. Rittenhouse, "Behavior of Materials in Space Environments," *ARS J.*, **32**, 320 (1962).
[83] R. A. Becker, "Optical Material Problems of Interplanetary Space," *Appl. Opt.*, **6**, 955 (1967).
[84] P. W. Collyer, "Selection and Processing of Infrared Materials," *Appl. Opt.*, **5**, 765 (1966).
[85] A. K. Rogers, A. L. Olsen, and L. W. Nichols, "Measurements of Infrared Fluorescence of Certain Optical Materials as Stimulated by Visible Light," *Appl. Opt.*, **5**, 1899 (1966).
[86] H. O. McMahon, "Thermal Radiation from Partially Transparent Reflecting Bodies," *J. Opt. Soc. Am.*, **40**, 376 (1950).
[87] S. E. Hatch, "Emittance Measurements on Infrared Windows Exhibiting Wavelength Dependent Diffuse Transmittance," *Appl. Opt.*, **1**, 595 (1962).
[88] D. L. Stierwalt, "Infrared Spectral Emittance Measurements of Optical Materials," *Appl. Opt.*, **5**, 1911 (1966).
[89] J. T. Cox and G. Hass, "Antireflection Coatings for Germanium and Silicon in the Infrared," *J. Opt. Soc. Am.*, **48**, 677 (1958).
[90] J. T. Cox, G. Hass, and G. F. Jacobus, "Infrared Filters of Antireflected Si, Ge, InAs, and InSb," *J. Opt. Soc. Am.*, **51**, 714 (1961).
[91] "Another Advanced State-of-the-Art Infrared Coating from OCLI," *Phys. Today*, **19**, 98 (December 1966).
[92] H. E. Bennett, J. M. Bennett, and E. J. Ashley, "Infrared Reflectance of Evaporated Aluminum Films," *J. Opt. Soc. Am.*, **52**, 1245 (1962).
[93] G. Hass, "Filmed Surfaces for Reflecting Optics," *J. Opt. Soc. Am.*, **45**, 945 (1955).
[94] J. M. Bennett and E. J. Ashley, "Infrared Reflectance and Emittance of Silver and Gold Evaporated in Ultrahigh Vacuum," *Appl. Opt.*, **4**, 221 (1965).
[95] H. E. Bennett, J. M. Bennett, and E. J. Ashley, "The Effect of Protective Coatings of MgF_2 and SiO on the Reflectance of Aluminized Mirrors," *Appl. Opt.*, **2**, 156 (1963).
[96] H. E. Bennett and W. F. Koehler, "Precision Measurement of Absolute Specular Reflectance with Minimized Systematic Errors," *J. Opt. Soc. Am.*, **50**, 1 (1960).
[97] J. E. Shaw and W. R. Blevin, "Instrument for the Absolute Measurement of Direct Spectral Reflectances at Normal Incidence," *J. Opt. Soc. Am.*, **54**, 334 (1964).
[98] J. U. White, "New Method for Measuring Diffuse Reflectance in the Infrared," *J. Opt. Soc. Am.*, **54**, 1332 (1964).

[99] R. T. Neher and D. K. Edwards, "Far Infrared Reflectometer for Imperfectly Diffuse Specimens," *Appl. Opt.*, **4**, 775 (1965).
[100] J. M. Davies and W. Zagieboylo, "An Integrating Sphere System for Measuring Average Reflectance and Transmittance," *Appl. Opt.*, **4**, 167 (1965)
[101] W. R. Blevin and W. J. Brown, "An Infrared Reflectometer with a Spheroidal Mirror," *J. Sci. Instr.*, **42**, 385 (1965).
[102] J. A. Sanderson, "The Diffuse Spectral Reflectance of Paints in the Near Infrared," *J. Opt. Soc. Am.*, **37**, 771 (1947).
[103] A. R. Downie et al., "The Calibration of Infrared Prism Spectrometers," *J. Opt. Soc. Am.*, **43**, 941 (1953).
[104] E. K. Plyler et al., "Vibration-Rotation Structure in Absorption Bands for the Calibration of Spectrometers from 2 to 16 Microns," *J. Research Natl. Bur. Standards*, **64A**, 29 (1960).
[105] K. N. Rao, C. J. Humphreys, and D. H. Rank, *Wavelength Standards in the Infrared*, New York: Academic, 1966.
[106] "Near Infrared Transmission Filters," Infrared Progress Report No. 3, Bausch and Lomb Optical Co., Rochester, New York, September 8, 1958.
[107] M. L. Sugarman, Jr., F. C. Bennett, Jr., and G. W. Hammar, "Methods of Manufacturing Infrared Transmitting Filters," U.S. Patents No. 3,063,861 and 3,063,862, November 13, 1962, (filed February 17, 1950).
[108] E. R. Blout, R. S. Corley, and P. I. Snow, "Infrared Transmitting Filters. II. The Region 1–6 μ," *J. Opt. Soc. Am.*, **40**, 415 (1950).
[109] J. Grant, E. Michel, and A. Thielen, "Recent Developments in Infrared Narrow Band Pass Filters," *Infrared Phys.*, **2**, 123 (1962).
[110] I. H. Blifford, Jr., "Factors affecting the Performance of Commercial Interference Filters," *Appl. Opt.*, **5**, 105 (1966).
[111] J. S. Seeley and S. D. Smith, "High Performance Blocking Filters for the Region 1 μ to 20 μ," *Appl. Opt.*, **5**, 81 (1966).
[112] R. Stair, W. B. Fussell, and W. E. Schneider, "A Standard for Extremely Low Values of Spectral Irradiance," *Appl. Opt.*, **4**, 85 (1965).
[113] Benford, F., "The Projection of Light," *J. Opt. Soc. Am.* **35**, 149 (1945).

6
Optical Modulation

An *optical modulator* is used to provide directional information for tracking and to suppress unwanted signals from backgrounds. The optical modulator can assume many forms, but basically each can be described as a pattern of alternately clear and opaque areas carried on a suitably transparent substrate. It is common practice to call the optical modulator a *reticle* or a *chopper*; occasionally it is referred to as an *episcotister*, a designation usually restricted to scholarly literature. Reticle patterns for infrared systems range from very simple patterns for converting a dc to an ac system, through patterns used to discriminate against unwanted backgrounds, to patterns that code the radiant flux with information about the direction of a target. In this chapter we shall examine some of these patterns, the characteristics of the signals they generate, and the situations in which they can be used.

Reticles are a relatively recent addition to the infrared scene. The first recorded instance of their use was in 1928 by A. H. Pfund[1]. He used a pendulum having a period of 1.5 sec to alternately expose and obscure a thermopile detector or, in other words, to *chop* the flux incident on the detector at intervals of 0.75 sec. The thermopile was connected to a D'Arsonval galvanometer tuned to a period of 1.5 sec. In effect, the source and detector were tuned to a specific frequency that could readily be distinguished from spurious signals such as noise, that occurred at random. The literature of the 1930's contains many references to radiometers and spectrometers using this basic technique. As the use of such choppers became more widespread and vacuum tube amplifiers replaced galvanometers, it was recognized that the addition of a chopper permitted the substitution of an ac amplifier for the older dc amplifier that was usually plagued by problems of thermal drift. German workers during World War II conceived the use of reticles in tracking and guidance systems. Postwar revelation of these systems showed that some had been nearly ready for production. In 1950 Clark[2] described a sun seeker using a simple reticle. This is apparently the first mention in the open literature of such a use for reticles.

It is unfortunate that the general subject of optical modulation is so heavily obscured by the curtain of military classification. A search of the unclassified literature reveals that there are pitifully few papers dealing with the design, analysis, and comparative performance of reticles. It is all too apparent that most authors simply take the easy way out and automatically classify any paper dealing with this subject. The absurdity of this approach is clearly evident from a study of the patent literature, in which the basic principles of reticle design and performance have been described in detail during the last three or four decades. For example, an application filed in 1934 resulted in a patent being issued to Zahl[3] in 1946 on the art of locating objects by their heat radiation. Even today, the teachings of this patent provide a remarkably lucid description of the problems involved in infrared search systems and the means of solving them. Much of the material in this chapter is taken from the patent literature in order to permit a detailed coverage of modern reticles and to forestall the inevitable cries of those who mistakenly think that the entire subject is classified.

6.1 OPTICAL FILTERING FOR BACKGROUND DISCRIMINATION

Whenever the spectral distribution of the flux from a target and its background are different, an optical filter is an inexpensive means of providing some rejection of the unwanted signals from the background. Although it is not a modulator in the strictest sense of the word, a filter is used with a reticle in almost every infrared system. Its primary purpose is to define the spectral bandpass of the system, but at the same time it can also effectively supplement the background rejection capabilities of the reticle. Subject, of course, to other restrictions, we attempt to choose a filter having a high transmittance for the flux from the target and a low transmittance for the flux from the background.

To understand the principles involved, consider the factors that influence the choice of a spectral bandpass for a system that is to detect jet aircraft in the presence of reflected sunlight and thermal radiation from the surface of the earth. The apparent spectral radiance of typical terrain is shown in Figure 3.15. This curve has a double peak, one at short wavelengths, which is due to reflected sunlight, and one at long wavelengths, which is due to thermal emission. Because the rather broad minimum between these two peaks occurs at about 3.5 μ, one is led to conclude that a system operating in this spectral region should have minimum interference from sunlit terrestrial backgrounds. The spectral distribution of the flux from a turbojet has its maximum in the 3.5 to 4 μ region so that the ratio of target-to-background flux has a broad max-

OPTICAL MODULATION   237

imum around 4 μ. The atmospheric transmittance curve in Figure 4.1 shows that this maximum lies within the 3.2 to 4.8 μ window—a particularly convenient act of nature, to provide a window so well placed for a very common detection task. The theory of optimum spectral filtering has been described by Eldering[4].

When the temperatures of the target and of its backgound are nearly the same, spectral filtering is of little help in discriminating one from the other and detection is dependent on there being an adequate radiation contrast, as discussed in Section 2.8. The effect of contrast can be readily appreciated from the following experiment. A vehicle, such as a truck, is parked in an open field so that it can be viewed by an appropriate infrared system. When observations extend over a period of 24 hours, marked variations are found in the contrast between the vehicle and its background. In the afternoon the vehicle has been thoroughly heated by the sun and is warmer than the background, so that the contrast is positive. Because of its large thermal capacity, the vehicle cools more slowly than does the background during the early evening hours and the contrast is even greater than that observed during the afternoon. As the night progresses, the vehicle cools more rapidly and eventually becomes colder than the background, causing the contrast to pass through zero and become negative. After sunrise the background warms more rapidly than does the vehicle and the negative contrast is enhanced. By midmorning the heating of the vehicle by the sun is sufficient to again cause a period of zero contrast before the value finally becomes positive. The point to remember is that *for many targets and background combinations, there are two intervals in any 24-hour period during which the target cannot be detected because there is insufficient radiation contrast between it and its background.* Such effects, often called *washout*, have been observed with vehicles, structures, and roads[5]. No amount of spectral filtering or optical modulation can eliminate the possibility of such a washout; one must either wait for it to pass or employ some other means of detection.

6.2 THE USE OF RETICLES FOR BACKGROUND SUPPRESSION

The increase in target-to-background ratio from spectral filtering is rarely sufficient to render system operation independent of background conditions. For a system operating in the 2 to 2.5 μ atmospheric window, for instance, the irradiance on the detector due to sunlight reflected from clouds in the background may be 10^4 to 10^5 times that due to a distant turbojet target. Fortunately, reticles can provide this magnitude of background suppression.

238   THE ELEMENTS OF THE INFRARED SYSTEM

The use of a reticle to increase the detectability of a particular target in the presence of extraneous background detail is called *space filtering*. Most targets of interest have the common characteristic that they are much smaller in angular extent than are objects in the background; a turbojet in front of a sunlit cloud, a ship against the sea, and a vehicle against terrain are typical examples. For such detection tasks, space filtering is used to enhance the signal from objects of small angular extent and to suppress signals from objects subtending large angles.

A simple example of space filtering by a rotating reticle is shown in Figure 6.1. The reticle pattern consists of a series of fan-shaped segments, alternately transparent and opaque. The reticle is placed at the image plane of the optics, and its center is coincident with the optical axis. The target and a (background) cloud illuminated by sunlight would normally be imaged on the reticle. They are shown to the side in Figure 6.1 in order that the action of the reticle can be seen more readily. As the reticle rotates at high speed about the center of the pattern, imagine that it moves slowly to the right so that it passes across the images of the target and the cloud. As the reticle passes across the target image, the image is chopped, that is, it is alternately transmitted and blocked by the

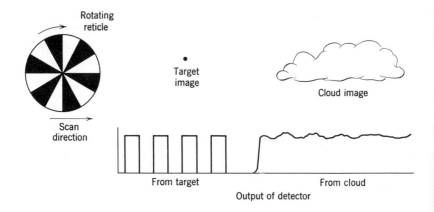

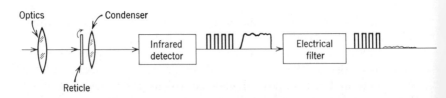

Figure 6.1   Space filtering by a rotating reticle.

OPTICAL MODULATION   239

elements of the pattern. Since the openings in the reticle are of approximately the same size as the image of the target, the electrical signal from the detector is a series of pulses at the chopping frequency f_c,

$$f_c = nf_r, \qquad (6\text{-}1)$$

where n is the number of pairs of clear and opaque segments in the reticle and f_r is its rotational frequency expressed in revolutions per second.[1] As the reticle passes across the relatively large cloud image, the image covers several segments of the reticle pattern at any given instant of time. As a consequence, the irradiance on the detector is increased but there is very little chopping action on the cloud image. With the target and cloud both imaged on the reticle, the output of the detector consists of a large dc signal with a small ripple from the cloud and a pulsed signal from the target. When these are amplified and passed through an electrical filter with its passband centered at the chopping frequency, only the ac signals remain and the effect of the cloud is suppressed.

In practice the suppression is rarely as complete as indicated in this simplified example. Since most cloud edges are irregularly shaped, they will undergo some chopping action. This is responsible for the ripple on the otherwise constant signal from the cloud in Figure 6.1. As the range to the target increases, the chopped signal from the target eventually becomes less than that from the cloud edges and the system becomes *background limited*. Most of the early infrared systems working in the 2 to 2.5 μ atmospheric window were background limited in the presence of sunlit clouds. Even the most advanced reticle techniques were not efficient enough to remove this limitation except at short target ranges. As new detectors became available and it was possible to use the 3.2 to 4.8 μ window in order to escape more of the reflected solar radiation, sunlit clouds ceased to be a problem for most infrared systems.

The reticle is, in effect, a selective modulator that is most efficient against point-source targets. For the highest modulation efficiency the openings in the reticle should match the size of the target image. In practical designs they may be up to three times as large as the size of the image. The data on the diameter of blur circles given in Section 5.3 is useful for estimating the size of the reticle openings. If, for instance, the optics have a focal length of 10 cm and a blur circle that is 1 mrad

---

[1]In Figure 6.1 and those to follow, the chopped target signal is shown as a series of square pulses. The reader will realize that the exact shape of the pulses depends on the relative sizes of the image and the openings in the reticle, as well as the manner in which the signal is modified by subsequent signal processing circuitry.

in diameter, the reticle openings should be approximately 0.01 cm (0.004 in.). Thus typical reticles have extremely fine patterns and require great care in their manufacture. Since the size of the blur circle varies with the distance from the optical axis, the design of the reticle usually entails some compromises. One effective solution is to use a checkerboard pattern[6] similar to that shown in Figure 6.13. The size of the openings increases toward the edge of the reticle so that they are kept approximately matched to the increased size of the off-axis blur circle. The chopping frequency generated by such a reticle is a function of the distance from the axis, an effect that is sometimes useful in tracking systems[32].

The mathematical analysis of space filtering is difficult; it has been treated definitively by Aroyan[7] and Biberman[8]. Their work provides an essential foundation for anyone interested in the analytical techniques involved.

If there is an irradiance gradient in a large-area image formed on the reticle, it will cause a signal in the output of the detector. Since the radiance of the sky is not uniform, the simple system shown in Figure 6.1 will generate a small signal, even in the absence of a target. A frequency analysis of such spurious signals shows that they consist principally of low frequencies centered around the rotation frequency of the reticle. Thus a bandpass filter centered around the chopping frequency will eliminate most spurious signals of this type. Whitney[9] has described reticle designs that, in theory, completely suppress spurious signals from sky backgrounds and, in addition, make possible the elimination of the bandpass filter. Rabinow[10] gives a particularly lucid description of the use of reticles to suppress unwanted background signals in automatic headlight dimmer systems. Though not an application for infrared, the description of the principles involved is excellent.

6.3 THE USE OF RETICLES TO PROVIDE DIRECTIONAL INFORMATION

Many infrared systems are designed to detect and track targets of interest. In this category are seekers for guided missiles, star trackers for navigational purposes, and fire control systems[11]. In such systems the reticle is used to modulate the incident flux with information that can be used to determine the direction of the target. Since backgrounds are likely to be a problem with such systems, the reticle must provide good background rejection together with the directional information. Reticles developing amplitude (AM), frequency (FM), and pulse (PM) modulation are found in contemporary infrared equipment[11]. The reader wishing

OPTICAL MODULATION 241

to review modulation theory will find that the introductory chapters by Black[12] are excellent.

Rotating Reticles

The two-sector reticle shown in Figure 6.2 is one of the simplest ways to provide directional information[13–16]. This reticle, which has one transparent and one opaque sector, rotates about the optical axis of the system. At (*a*) the image of a target is shown slightly to the left of the center of the reticle. The detector output is shown to the right; it is a train of pulses at the chopping frequency, which, in this case, is identical with the rotational frequency of the reticle. At (*b*) the target is shown with the same radial displacement, but the angular displacement has been increased. The output of the detector is similar to that obtained before except that the pulses have been displaced along the time axis. The relative phase shifts δ_1 and δ_2 of the pulses are seen to be proportional to the angular displacement of the target. It is necessary to provide a phase reference so that the phase shifts can be measured. One of the simplest phase-reference generators is a small magnet fastened to the

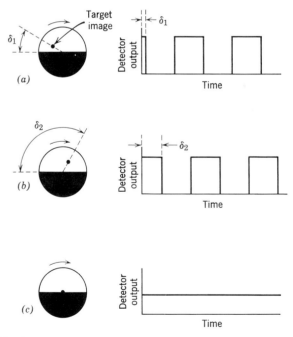

Figure 6.2 Simple two-segment reticle for generating target directional information (adapted from Carbonara et al.[15]).

periphery of a reticle with one or more pickup or *pip coils* mounted on a fixed frame near the path of the magnet [16]. Each time that the magnet passes the coil, a sharp pulse or *pip* is generated that can be used as a stable reference from which phase can be measured.

When the system of Figure 6.2 is pointed directly at the target, so that there is no pointing error, the target image lies at the center of the reticle and there is no chopping action, as shown at (*c*). As a result there is no output signal generated with zero pointing error—an unfortunate condition, since it is indistinguishable from that found when there is no target in the field of view. The reticle generates a chopping frequency analagous to the *carrier frequency* in a communications system. The carrier is phase modulated with information concerning the direction to the target. At zero pointing error the carrier disappears, taking with it the phase information. One way to prevent this is to use a double modulation scheme, such as that shown in Figure 6.3, so that a carrier is present whenever a target is in the field of view [16, 17]. A second reticle, with a

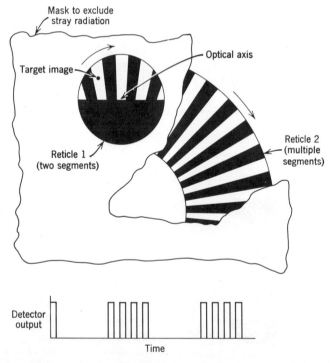

Figure 6.3 Double modulation arrangement to prevent loss of carrier with zero pointing error (adapted from Robert and Deslaudes [16] and Chitayet [17]).

OPTICAL MODULATION 243

large number of segments, is placed immediately behind the two-segment reticle. It produces a high-frequency carrier regardless of the position of the target within the field of view. The two-segment reticle pulse modulates the carrier, and the phase of the pulses contains the information on target direction. The carrier and the modulation frequencies thus generated can be separated after amplification by passing them through filters tuned to the respective frequencies. Since the amplitude of the carrier signal is proportional to the irradiance from the target, it can be used for radiometric purposes, for automatic gain control, or for operating a device to indicate the presence of a target. Since the opening in the two-segment reticle is relatively large, its background rejection capability is poor. The addition of the second reticle gives a combination having quite good background rejection.

A design combining the merits of the reticles shown in Figures 6.1 and 6.2 was developed by Biberman and Estey[18]. Using the simple fan-bladed chopper shown in Figure 6.1, they made a systematic study of the signals it produced when scanning various types of sky backgrounds. They found strong signals from radiance gradients at the rotational frequency of the reticle and at its first few harmonics. The signals generated by clouds were similar but somewhat richer in harmonics. In no case did they find background signals beyond the eighth harmonic. By contrast, the signals from distant small targets, such as aircraft, were found to have strong harmonics to well beyond the twentieth.

On the basis of these observations, Bibermen and Estey designed the "rising sun" reticle, shown in Figure 6.4, so that it would generate a carrier frequency at least eight times higher than the reticle rotation frequency. Their reticle consists of two semicircular sectors. One contains alternately transparent and opaque fan-shaped segments for target sensing and background suppression. The other sector is semitransparent

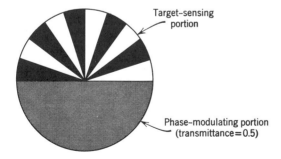

Figure 6.4 Reticle with good rejection of sky backgrounds (adapted from Biberman and Estey[18]).

and has a transmittance of 0.5. It provides the phase modulation that indicates target direction. The carrier frequency f_c generated by this reticle is

$$f_c = Knf_r, \qquad (6\text{-}2)$$

where n is the number of pairs of clear and opaque segments in the target-sensing portion of the reticle, f_r the reticle rotational frequency, and K the reciprocal of the fraction of the total area of the reticle occupied by the target-sensing portion. In Figure 6.4 the value of K is 2 and n is 5 so that the carrier frequency is 10 times the reticle rotational frequency.

Biberman and Estey were the first to recognise that in a reticle of this type the transmittance of the phasing sector should be equal to 0.5. Since the average transmittance of the target-sensing portion is also 0.5, the reticle is balanced; that is, the transmittance of the two sectors is identical for images of large area. Figure 6.5 shows the detector outputs

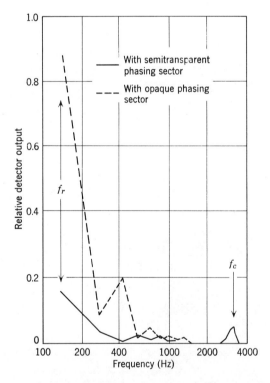

Figure 6.5 Comparison of the frequency spectrum of signals generated by reticles having semitransparent and opaque phasing sectors (adapted from Biberman and Estey[18]).

OPTICAL MODULATION 245

caused by a sky background and by a distant aircraft, with two different reticles. In one reticle the phasing section was opaque; in the second reticle it had a transmittance of 0.5. It is evident from Figure 6.5 that the maximum amplitude of the background signal from the reticle with the opaque phasing sector is five or six times that from the reticle with the semitransparent phasing sector. On the other hand, the target signal (at the carrier frequency) has about the same amplitude with either reticle. Thus the change from an opaque to a semitransparent phasing sector increases the ratio of target-to-background signals by a factor of 5 or 6. An amplifier having a narrow bandpass centered about the carrier frequency will further reject the background signals. Note, however, that this design, like that shown in Figure 6.2, generates no carrier for zero pointing error.

The "rising sun" reticle generates an amplitude modulation as the target image moves out from its center that is due to the changing relationship between the size of the blur circle and the size of the openings in the reticle. Thus the amplitude of the modulation indicates the radial coordinate of the target position and the phase of the modulation indicates the angular coordinate. The phase discriminator designed for use with this reticle is described in ref. 19.

When the reticle pattern consists of segments bounded by straight lines, there is a tendency for the reticle to generate a larger signal when chopping a line image that when chopping a point image. This is a particularly undesirable characteristic if the horizon should appear in the background. Davis[20] has described an improved pattern for the target-sensing sector in the Biberman and Estey reticle. As shown in Figure 6.6, it

Figure 6.6 Reticle pattern having improved rejection of straight-line background elements, such as the horizon (adapted from Davis[20]).

consists of a series of zigzag elements bounded by portions of curved lines spiraling outward from the center of the reticle. The reticle is balanced since the average transmittance of either portion is 0.5. Davis claims that this pattern increases the rejection of spurious signals from cloud and horizon backgrounds. However, his performance curve does not support this conclusion since it is identical to that given by Biberman and Estey to support the claims for their reticle.[1] Aroyan and Cushner[21] have described the usefulness of involute patterns for the rejection of signals from straight-line sources.

Another reticle pattern that generates modulation indicative of both the angular and radial coordinates of target position[22] is shown in Figure 6.7. In the target-sensing portion the pattern is formed from curvilinear strips that originate along a diameter of the reticle and have equal arcuate

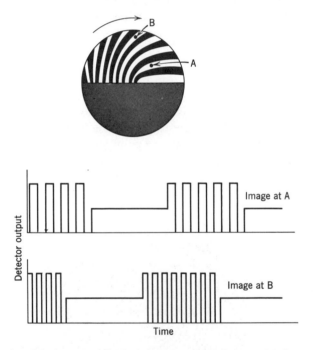

Figure 6.7 A reticle for generating both frequency and phase modulation (adapted from Lovell[22]).

[1]Biberman and Estey support their claims by reference to Figure 6 of their patent. Davis, in turn, supports his claims by reference to Figure 6 of his patent. A careful comparison shows that the two figures are, in fact, identical. Although filed two years apart, both patent applications were prepared by the same attorney, thus offering a possible explanation for the duplication. The information in these identical figures has been adapted for our Figure 6.5.

widths at equal distances from the center of the reticle. The carrier frequency generated by this pattern is a function of the radial position of the image. In Figure 6.7 two images are shown on the reticle. For the image at A, the carrier frequency is 10 cycles per reticle revolution; for the image at B the frequency is 16 cycles per reticle revolution. The output of the detector is shown in the lower part of the figure for both image positions. As before, the phasing sector pulse modulates the carrier so that the phase of the pulses indicates the angular position of the target. In addition, the carrier is frequency modulated to indicate the radial position of the target.

The basic fan-bladed pattern can be modified as shown in Figure 6.8 so as to generate a frequency-modulated carrier[23, 24]. The angular width of the individual elements of the pattern varies as a sinusoidal function of the azimuth angle around the reticle. Other reticles that generate frequency modulation are described in refs. 25–28.

Two reticles, placed one behind the other, can be used to scan a circular field of view[29, 30]. Usually one reticle carries a single spiral slit and is rotated in front of a fixed reticle carrying one or more straight slits.

Figure 6.9 shows a reticle that generates pulse width and phase modulation. The pattern consists of triangular-shaped elements with the bases of the triangles alternately toward or away from the center of the reticle. An image on the optical axis is chopped into a series of pulses. If the image moves vertically from the optical axis, the pulse width varies in proportion to the distance from the axis. If the image moves horizontally, it causes a change in the phase of the pulses. Unlike the previous reticles, the modulation generated by this reticle is proportional to the cartesian rather than to the polar coordinates of target position. The equally spaced slots around the periphery of the reticle are used to generate a phase reference. A photodiode viewing a small source, such as a pilot lamp, through these slots is an effective phase-reference generator. Merlen[31] has described

Figure 6.8 A reticle for generating a frequency-modulated carrier.

248 THE ELEMENTS OF THE INFRARED SYSTEM

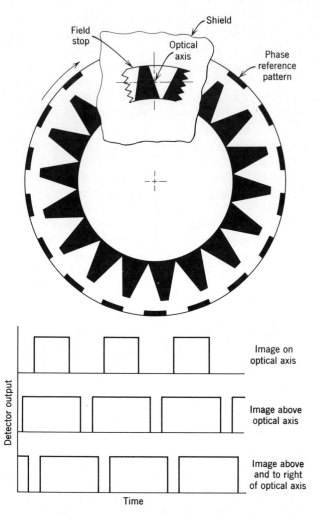

Figure 6.9 A reticle for generating both pulse-width and phase modulation (adapted from Merlen[31]).

a similar reticle in which the triangular elements contain much smaller fan-shaped segments. The number of segments in alternate triangular elements is in the ratio of 2 to 1. The output of the detector consists of bursts of alternately high- and low-frequency pulses. The smaller chopping segments improve the background rejection capabilities. A shield in front of the reticle carries an aperture that forms the field stop for the optics. Merlen claims improved background rejection by the use of sawtooth edges on the field stop.

A reticle that divides the field of view into a number of concentric zones and generates a different carrier frequency for each zone is shown in Figure 6.10, along with its rather unusual detector array [32, 33]. The detectors are made in the form of 10 rings concentric about the center of a flat disk. A narrow space between rings ensures the electrical isolation of each detector. The reticle, which is just in front of the detectors, has an opaque phasing portion and a target-sensing portion consisting of 10 concentric rings. The number of pairs of segments in each ring are in the ratio of 20 to 22 to 24, and so on, with the smallest number being in the innermost ring. The radius of each ring is chosen so that all rings have the same area.

If the reticle rotation rate is chosen so that the chopping frequency of the inner ring is, for instance, 1000 Hz, the second ring will generate 1100 Hz, the third 1200 Hz, and so on. Thus the chopping frequency is a function of the radial position of the image. All detectors are connected in parallel to a single amplifier. After amplification, a bank of 10 bandpass filters, each tuned to one of the frequencies generated by the reticle, is used to separate the signals. This arrangement can be used to track several targets simultaneously because no confusion exists unless more than one target occupies the same ring of the reticle.

The concentric multielement detector is probably not practical because of the difficulty of manufacturing it. Because the concentric detectors contribute nothing to the modulation process, they can be replaced by a single detector. Such a single detector would have to be as large as the reticle. Because it is desirable to keep the detector small in order to minimize its noise, it would be necessary to use an optical condenser along

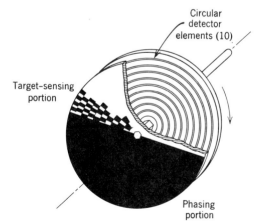

Figure 6.10 A ten-frequency reticle and multielement detector combination (adapted from Shapiro [32]).

with the smaller detector. Other reticles generating multiple frequencies are described in refs. 34–36.

Stationary Reticles

Thus far the discussion has been limited to *rotating reticle systems*. It is just as feasible to rotate the image optically with respect to a fixed reticle. Such *stationary reticle systems* offer additional flexibility in the types of modulation that can be produced and they offer the important feature that there is no loss of carrier for zero pointing error.

Rotation of the image is called *nutation*;[1] it can be accomplished in several ways [6, 13, 15, 37–40, 47–50]. A common method is shown in Figure 6.11. The lens is mounted so that it can be rotated about an axis normal to and passing through the center of the reticle. The lens, however, is displaced laterally a distance d so that the optical and rotational axes are parallel but not coincident. When the lens is rotated, the image follows a circular path called the *nutation circle*. When the pointing error is zero, the nutation circle is concentric with the rotational axis and with the center of the reticle. The linear diameter of the nutation circle is just twice the distance d by which the optics are displaced. The angular subtense of the nutation circle is

$$\delta_n = 2000 \frac{d}{f}, \qquad (6\text{-}3)$$

where δ_n is expressed in milliradians, d is the displacement of the lens, and f is the equivalent focal length measured in the same units as d. In a typical system using a lens having a focal length of 4 in. (10 cm), a nutation circle subtending 35 mrad (2 deg) would require that the lens be

---

[1]A term borrowed from astronomy, where it is used to describe the oscillation of the earth's polar axis about its mean position.

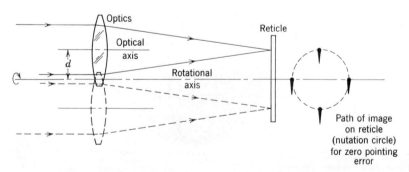

Figure 6.11 Rotation of decentered optics to produce nutation.

OPTICAL MODULATION 251

displaced by 0.070 in. (0.175 cm). The designer may put the required offset either in the metal mounting or in the lens. In the first case, the edge of the lens is ground concentric with its optical axis and the metal mount is machined eccentrically to provide the offset. Alternatively, the edge of the lens is ground so as to be concentric with the rotational axis, thus eliminating the need for an eccentric mount. Since the rotational speed of the lens, the *nutation frequency*, may be quite high, it is usually necessary to dynamically balance the mounted optical assembly. Although Figure 6.11 shows a lens, the principle is equally applicable to reflective optics [47].

Nutating systems are often referred to as *rotating field systems*. In Figure 6.11 the image is shown at several points around the nutation circle; we see that its spatial orientation does not change. Instead it undergoes a simple translation around the center of the nutation circle. Thus the designation of a rotating field is a misnomer if this term is taken literally.

When a pointing error exists, the nutation circle is no longer concentric with the center of the reticle. Turck [6] has used this effect in the tracking arrangement shown in Figure 6.12. A simple reticle with fan-blade segments is used for illustration; the reader will recognize that many of the patterns previously discussed can also be used. At (*a*) there is no pointing error. The image travels around a nutation circle that is concentric with

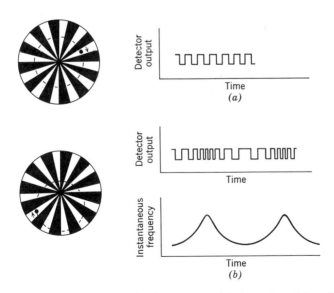

Figure 6.12 Nutating system generating frequency modulation (adapted from Turck [6]).

the center of the reticle. It encounters the chopping segments at a constant rate, and the detector output, shown at the right, is a train of pulses having the frequency given by (6-1). With a pointing error, as at (*b*), the nutation circle is no longer concentric with the center of the reticle. Since the width of the chopping segments decreases toward the center of the reticle, the image is chopped more rapidly near the center than near the edge. As a result, the detector output is a series of pulses of varying frequency. The lower part of the figure shows the instantaneous frequency of the pulses as a function of time. The fundamental chopping frequency of the reticle, as given by (6-1), has been frequency modulated; the phase of the modulation is proportional to the angular location of the image, and the magnitude of the modulation is proportional to the radial distance of the image from the axis of rotation. An external phase reference, such as the magnetic pip generator previously described, must be provided. Note that there is no loss of carrier for any position of the image on the reticle. This is virtually a necessity with high-precision tracking systems, which cannot tolerate the loss of error signals near zero pointing error.

Turck[6] has pointed out that the chopping segments in the reticle need not be bounded by straight lines. They can take almost any shape that will result in the desired frequency variation. He suggests the three types shown in Figure 6.13.

A hybrid system has been described by Gedance[41] that combines many of the desirable features of both the rotating and stationary reticle systems. These advantages are realized without moving the optics or the detector. The center of the reticle is displaced laterally from the optical axis, and the reticle is then rotated about the optical axis. The result is essentially equivalent to nutating the optics; the image path around the reticle and the output of the detector are similar to those shown in Figure 6.12. By simultaneously rotating the reticle about its own axis, the center frequency of the frequency-modulated signal can be increased. Gedance has pointed out that the frequency-modulated pulses may be considered equally well as a form of pulse-position modulation. Thus the designer has additional freedom in choosing the signal processing circuitry.

Figure 6.13 Some reticle patterns useful with nutating systems (adapted from Turck[6]).

OPTICAL MODULATION 253

Gedance[41] shows how a radial variation in the transmittance of a reticle can be used to generate an amplitude-modulated signal. Such a reticle is shown in Figure 6.14 along with a graph of its transmittance as a function of position on the reticle. The transmittance is high near the center of the reticle but decreases radially to almost zero at the edge. In the absence of pointing error, as at A, the output of the detector is constant. With a pointing error, the detector output is amplitude modulated at the nutation frequency. The manner in which the modulation is produced is easily understood with the aid of the reticle transmittance curve in Figure 6.14. The dotted lines show the equivalent path of the image with respect to the transmission gradient. In the absence of pointing error, the image travels along a path of constant transmittance; hence the detector output remains constant. With a pointing error, the image follows a path of constantly varying transmittance and the detector output is amplitude modulated. The amplitude of the modulation is a function of the radial displacement of the image from the optical axis and the phase of the modulation represents the angular coordinate of the image (measured about the center of the reticle). Such a gradient can be added to most of the recticles shown in this chapter. The gradient need not be linear and

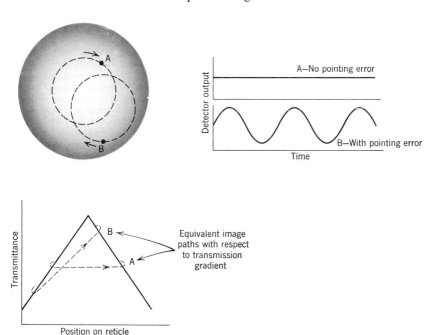

Figure 6.14 Use of a radial transmission gradient to generate amplitude modulation (adapted from Gedance[41]).

can assume any form dictated by the particular control system. Details of other stationary reticle systems can be found in [37-40].

Two-Color Reticles

Whenever the spectral distribution of the target and its background are quite different—for example, a turbojet against a background of reflected sunlight—two-color reticles are often proposed as a means of rejecting the background. A very simple example illustrates the concepts involved. Assuming that the sun is a 5900°K blackbody, a radiation slide rule, or Table 2.5, shows that the median wavelength of its spectral distribution is 0.69 μ. Thus exactly one-half of its radiant flux lies on either side of this wavelength. Assume that two filters are available, one a low-pass type transmitting all wavelengths shorter than 0.69 μ, the other a high-pass type transmitting all wavelengths longer than 0.69 μ. A reticle could be made by using the first filter for the normally clear portions of the pattern and the second filter for the normally opaque portions. When this reticle is used in a simple rotating reticle system with a detector that responds equally to all wavelengths, alternate elements of the reticle pattern transmit equal portions of the solar flux and no chopping occurs for anything reflecting solar radiation uniformly with wavelength. The reticle behaves simply as a neutral filter. For a turbojet, approximated by an 800°K blackbody, less than 10^{-6} per cent of the flux lies on the short wavelength side of 0.69 μ. Hence, for the turbojet, the reticle is effectively made up of clear and opaque segments, and the usual chopping action takes place. The result, at least in theory, is a complete rejection of solar radiation.

In practice such complete rejection is rarely achieved. Usually two narrow spectral bands are used rather than the entire spectrum. These bands may be in the same atmospheric window or in adjacent windows. For each spectral band, we calculate the integral of the product of the spectral radiance of the source to be rejected, the spectral transmittance of the intervening path through the atmosphere, the optics and the filter limiting the spectral bandpass, and the spectral response of the detector. In order to obtain the desired rejection, these integrals must be made equal to one another. The filter is essentially the only element that can be varied in order to equalize the integrals and thus balance the system. Since the spectral distribution of the source to be rejected and the transmittance of the atmosphere may vary in an unpredictable manner with time and temperature, it is virtually impossible to maintain the desired system balance for very long. Despite this problem, the difficulty encountered in constructing the reticle is probably the principal reason

OPTICAL MODULATION   255

why two-color systems have rarely received serious consideration. The segments in a reticle pattern may be only a few thousandths of an inch wide, and it is extremely difficult and expensive to make large numbers of such small filters, either individually or all at once. With systems using relatively long focal length optics, and consequently a larger diameter reticle, it may be practical to consider making such a two-color reticle.

A simplified two-color reticle has been made from two half-circular portions cut from different filters[42]. After cementing the two portions to form a disk, a conventional checkerboard pattern was deposited on it. Claims have been made for its usefulness in the detection of aircraft against sunlit backgrounds. Other two-color systems are described in [43–45].

6.4 TRACKING SYSTEMS WITHOUT RETICLES

By using multiple-element detectors it is possible to build tracking systems that do not use reticles. The basic principles involved have been known since the 1920's, or earlier[46]. One such system, using a four-element detector[47–49], is shown in Figure 6.15. Rotating optics are used with a detector consisting of four rectangular elements arranged in a cross-shaped pattern. In the absence of a pointing error, the center of the nutation circle coincides with that of the detector array. The image crosses the four elements at equal time intervals, and the detector output is a series of pulses occurring at a constant rate. With a pointing error, the nutation circle is not concentric with the center of the array and the time intervals between crossings of successive detector elements are no longer constant. By comparing these time intervals with reference signals derived from the rotating optics, the retangular components of the target

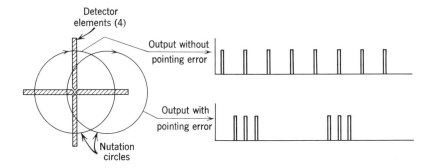

Figure 6.15 A pulse-modulation system without a reticle (adapted from Buntenbach[47]).

coordinates can be determined. Thus the combination produces a pulse position modulation in which variations in pulse position indicate the direction of the target. Similar performance can be obtained from a two-element array in which the elements are arranged in an L-shaped pattern (equivalent to removing one vertical and one horizontal element in Figure 6.15)[47, 50–52].

The background rejection characteristics of such pulse systems are excellent. Since the motion of the target image is across the short dimension of the detector elements, this dimension should meet the same criterion used in determining the width of reticle openings; that is, the width of the element should be from one to three times the diameter of the blur circle of the optics. Each pair of detector elements (those lying in the same straight line) can be oppositely biased so as to cancel signals from large straight edges, such as the horizon. Unfortunately, good rejection requires that the edges be accurately parallel to the detector elements, a condition difficult to maintain.

Tiny mirrors mounted on the tines of electrically driven tuning forks have been used to provide chopping or image motion over a reticle[53].

6.5 COMMENTS ON RETICLE DESIGN

At no other point in the design of an infrared system is there likely to be as large a gulf between the analytical and the hardware people as there is in the area of reticle design and comparative performance. At first glance the subject appears to be ideally suited to the mathematical approach. As one proceeds, more complicated mathematical techniques are required, and more than one promising design has become lost in an elegant set of equations having no solution.

If one is interested only in target detection and not in its direction, the analytical approaches described by Aroyan[7] are excellent. Supplementing these with probability distributions of background radiances and their spatial distributions can lead quite directly to very efficient space filtering. In a tracking system the designer is vitally interested in the *error response* of the system, that is, the relationship between pointing error and the error signals available at the output of the signal processor. The particular circuitry chosen for the signal processor and the tracking servo places strong demands on the choice of a reticle and the modulation it generates. It is extremely unlikely that a reticle designed by purely mathematical methods will meet all of these diverse requirements. Such designs do, however, offer a good starting point from which the design can be empirically modified to one having an error response curve of the desired shape and slope while still retaining good background rejection, good resolution against multiple targets, and the desired field of view. The exact procedures for refining a reticle design are usually regarded as proprietary by the organizations involved, and therefore each designer must develop his own solution to the problem. Graphical analysis, simulation techniques[54], and tests of the proposed reticle in the final system have proved to be useful approaches.

As shown in Section 6.3, typical tracking systems may employ a rotating reticle, a stationary reticle and nutating optics, or a combination of these to generate amplitude, frequency, or pulse modulation. Unfortunately there is very little information available in the open literature on the comparative performance of these combinations[23, 24, 55, 56]. For moderate signal levels, an FM system usually has a higher signal-to-noise ratio than an AM system. When the signal level drops so low that the limiter no longer functions, the performance of an FM system becomes inferior to that of an AM system. Therefore, if obtaining the maximum possible tracking range is the principal system requirement, it can probably best be met by the use of AM.

Some of the other factors to be considered in choosing a tracking system are shown in Table 6.1. The potentially poor performance of rotating-reticle systems in the presence of a high-radiance background has not been mentioned previously. Even though the optical condenser focuses the image of the entrance aperture on the detector, there is still an out-of-focus image of the reticle on the detector. Since it is likely that the sensitivity of the detector will vary across its surface, the out-of-focus image of the rotating reticle may generate spurious signals when the system views a high-radiance background. This difficulty is not encountered with a stationary reticle system since there is no relative motion between reticle and detector.

If several targets are in the field of view, most reticle systems will track the target that has the highest radiant intensity. If the targets have about the same radiant intensity, most reticle systems will track the effective radiation centroid. This can lead to serious difficulties if, for instance, the system is to be used to guide a missile against multi-engine aircraft. Proper design of the reticle pattern or of the shape of a superimposed radial transmission gradient can force the system to select one target rather than continuing to track the centroid[57]. An infrared guided missile fired toward the forward aspect of a turbojet target may consistently miss because its seeker does not see the hot tailpipe but, instead, tracks the centroid of the exhaust plume. Pulse modulation systems offer a possible solution because they can readily track the edge of a radiating source.

258   THE ELEMENTS OF THE INFRARED SYSTEM

TABLE 6.1   A COMPARISON OF SEVERAL TYPES OF RETICLES FOR TRACKING SYSTEMS

Type	Type of Modulation	Advantages	Disadvantages
Rotating reticle, fixed optics	AM	Simple mechanical construction	Loss of carrier in absence of pointing error
		Adequate discrimination against low-radiance backgrounds (when combined with optical filtering)	Relatively poor discrimination against high-radiance backgrounds
	FM	Same as for AM	Same as for AM
			Wide electrical bandwidth
Stationary reticle, nutating optics	AM	Excellent discrimination against high-radiance backgrounds	Moderately wide electrical bandwidth
		No loss of carrier with zero pointing error	Shift in apparent position of target for a large image size
		Relatively simple mechanical construction	
	FM or FM/AM	Same as for AM	Reversal of FM signal when image crosses center of reticle
			Abrupt shift in apparent position of target for a large image size
			Wide electrical bandwidth
Rotating reticle, nutating optics	FM	Good discrimination against high-radiance backgrounds	Highly complex mechanical construction
		No loss of carrier with zero pointing error	Wide electrical bandwidth

6.6 FABRICATION OF RETICLES

In many systems the reticle, or a mask placed immediately in front of it, acts as a field stop. The diameter of the field stop is

$$d = 0.0174 f\beta, \qquad (6\text{-}4)$$

where d and f are measured in the same units and β is the instantaneous field of view measured in degrees. This expression gives an error of less than 1 per cent provided that β is less than 8 deg. For larger values of β, use $d = 2f(\tan \beta/2)$, which is valid for all values of β. A convenient rule of thumb, derived from (6-4), is that the diameter of a reticle when used as a field stop is 0.017 in. per inch of focal length per degree of field of view. The reticles of typical infrared systems will rarely exceed a diameter of 0.5 in., and the width of the openings in the patterns may be only a few thousandths of an inch. Radial transmission gradients and semi-transparent phasing sectors may be formed from lines and spaces of varying widths combined to give the desired transmittance. The width of some of these shading lines may be no more than 0.0001 in.

Reticles are almost always produced by photographic processes. The processes are quite different, however, from those familiar to the ordinary industrial photographer or photoengraver. Stevens[58] gives an excellent discussion of the specialized techniques required. The first step is to make a master pattern. It is usually drawn by hand and is 10 to 100 times larger than the final reticle. Depending on the particular photographic process used, this master can be either a negative or a positive of the desired pattern. When finished, it is placed in an illuminated copy stand and photographed to give the required reduction in size. It is essential that the illumination be constant at all points on the master; variations greater than 2 parts in 500 are usually unacceptable. If the spectral bandpass of the system does not extend beyond 2.7 μ, the pattern can be photographed on commercially available high-resolution plates. After processing, the plates are cut and the reticles are edged to final diameter by conventional optical shop techniques. Tolerances on the concentricity of the pattern with respect to the ground edge may be as small as ±0.0005 in. It is relatively easy to damage the pattern during the edging process; consequently the designer should carefully define the allowable surface defects in the procurement specification.

Photographic plates absorb strongly beyond 2.7 μ. Figure 6.16 shows the transmittance of a Kodak high-resolution plate and of the glass substrate without emulsion. (The sample was prepared by clearing an unexposed plate in hypo, then washing and drying it in the normal manner). The water retained by the emulsion is responsible for most of the absorption around 3 μ. As a result of this absorption, the transparent spaces in the reticle pattern are no longer transparent at wavelengths beyond 2.7 μ.

Photoetching leaves no emulsion in the clear spaces of the pattern and can be used at any wavelengths for which transparent substrates are available. The first step in the photoetching process is to make a master negative by photographing the master pattern. This negative must be exactly the size required for the final reticle, and it should be completely free of any defects. Reticle blanks of suitably transparent materials should

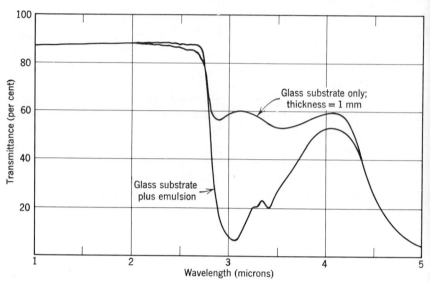

Figure 6.16 Transmittance of Kodak high-resolution photographic plates.

be well polished, their surfaces should be flat, parallel, and free of scratches and pits, and they should be edged to the desired reticle diameter. An evaporated metallic coating, usually of aluminum, is applied to one surface of the blank. Next a thin layer of photoresist is applied over the metallic coating. The prepared blank is placed in contact with the master negative and exposed through it to the ultraviolet from a mercury arc. Extreme care is required to ensure that the blank is held concentric with the pattern on the master negative in order to meet the tolerance limits on concentricity. Photoresist is a photosensitive material that is polymerized upon exposure to ultraviolet so that is is not affected by the acids used for etching. After the exposure the blank is washed to remove the photoresist from those portions of the pattern that were protected from the ultraviolet by the master negative. Immersion in an etchant removes the metal film from these areas. The final result is a reticle pattern etched in a metal film with no material in the clear spaces other than the substrate. With careful attention to detail in every step of the process, pattern openings as narrow as 0.00008 in. ($2\,\mu$) can be reproduced. The edges of such openings show no irregularities, even when viewed under a 900-power microscope.

It is not practical to apply an antireflection coating to the substrate exposed in open spaces of the reticle pattern, but the second surface of the substrate can be coated. Reflection losses can be entirely eliminated by etching the reticle pattern into a thin metal foil and then cementing it to a filter for support. The problem of ensuring concentricity of the reticle pattern about a reference axis may, however, be very difficult to solve when this technique is used.

REFERENCES

[1] A. H. Pfund, "Resonance Radiometry," *Science*, **69**, 71 (January 18, 1929).
[2] H. L. Clark, "Sun Follower for V-2 Rockets," *Electronics*, **23**, 71 (October 1950).
[3] H. A. Zahl, "Art of Locating Objects by Heat Radiation," U.S. Patent No. 2,392, 873, January 15, 1946, (filed April 3, 1934).
[4] H. G. Eldering, "The Theory of Optimum Spectral Filtering," *Infrared Phys.*, **4**, 231 (1964).
[5] N. Ginsburg, W. R. Fredrickson, and R. Paulson, "Measurements with a Spectral Radiometer" *J. Opt. Soc. Am.*, **50**, 1176 (1960).
[6] J. Turck, "Goniometer with Image Analysis by Frequency Modulation", U.S. Patent No. 2,967,247, January 3, 1961.
[7] G. F. Aroyan, "The Technique of Spatial Filtering," *Proc. Inst. Radio Engrs.*, **47**, 1561 (1959).
[8] L. M. Biberman, *Reticles in Electro-Optical Devices*, New York: Pergamon, 1966.
[9] T. R. Whitney, "Scanning Discs for Radiant Energy Responsive Tracking Mechanisms," U.S. Patent No. 2,972,276, February 21, 1961.
[10] J. Rabinow, "Noise Discriminating, High Gain Automatic Headlight Dimmer," U.S. Patent No. 2,892,124, June 23, 1959.
[11] D. W. Fisher, R. F. Leftwich, and H. W. Yates, "Survey of Infrared Trackers," *Appl. Opt.*, **5**, 507 (1966).
[12] H. S. Black, *Modulation Theory*, Princeton, N.J.: Van Nostrand, 1953.
[13] L. B. Scott, "Radiant Energy Tracking Apparatus," U.S. Patent No. 2,513,367, July 4, 1950 (filed May 26, 1948).
[14] T. H. Jackson, Jr., "Light Chopper," U.S. Patent No. 2,972,812, February 28, 1961 (filed March 14, 1949).
[15] V. E. Carbonara et al., "Star Tracking System," U.S. Patent No. 2,947,872, August 2, 1960.
[16] A. Robert and J. Deslaudes, "Tracking Devices," U.S. Patent No. 2,975,289, March 14, 1961.
[17] A. K. Chitayet, "Light Modulation System," U.S. Patent No. 3,024,699, March 13, 1962.
[18] L. Biberman and R. Estey, "Multislit Scanner," U.S. Patent No. 3,034,405, May 15, 1962 (filed October 13, 1953).
[19] E. G. Swann, "Electrical Gate Phase Discriminator," U.S. Patent No. 2,963,241, December 6, 1960 (filed August 11, 1953).
[20] F. H. Davis, "Scanner Disc," U.S. Patent No. 2,996,945, August 22, 1961 (filed September 12, 1955).
[21] G. F. Aroyan and S. H. Cushner, "Reticle Structure for Infrared Detecting System," U.S. Patent No. 3,144,555, August 11, 1964.
[22] D. J. Lovell, "Electro-Optical Position Indicator System," U.S. Patent No. 2,997,699, August 22, 1961.
[23] T. B. Buttweiler, "Optimum Modulation Characteristics for Amplitude-Modulated and Frequency-Modulated Infrared Systems," *J. Opt. Soc. Am.*, **51**, 1011 (1961).

[24] R. O'B. Carpenter, "Comparison of AM and FM Reticle Systems" *Appl. Opt.*, **2**, 229 (1963).
[25] T. G. B. Boydell, "Optical Sensing System," U.S. Patent No. 3,263,084, July 26, 1966.
[26] E. K. Sandeman, "FM Star-Lock System Using Mask with Linear Sectors," *IRE Trans. Aerospace and Navigational Electronics*, **9**, 24 (1962).
[27] E. K. Sandeman, "FM Star-Lock System Using Mask with Spiral Sectors," *IRE Trans. Aerospace and Navigational Electronics*, **9**, 35 (1962).
[28] H. Gabloffsky, "Photosensitive Means for Detecting the Position of Radiating or Reflecting Bodies," U.S. Patent No. 3,239,672, March 8, 1966.
[29] R. S. Estey, "Tracking the Partially Illuminated Moon," *IEEE Trans. Aerospace and Navigational Electronics*, **10**, 271 (1963).
[30] L. A. Iddings, "Scanning Devices for Optical Search," U.S. Patent No. 3,000,255, September 19, 1961 (filed May 31, 1955).
[31] M. Merlen, "Infrared Tracker," U.S. Patent No. 3,007,053, October 31, 1961.
[32] S. Shapiro, "Infrared Search and Tracking System Comprising a Plurality of Detectors," U.S. Patent No. 3,106,642, October 8, 1963.
[33] H. Dubner, J. Schwartz, and S. Shapiro, "Detecting Low-Level Infrared Energy," *Electronics*, **32**, 39 (June 26, 1959).
[34] S. Hansen, "Astrometrical Means and Method," U.S. Patent No. 2,941,080, June 14, 1960, (filed November 6, 1948).
[35] P. M. Cruse, "Apparatus for Processing Optically Received Electromagnetic Radiation," U.S. Patent No. 3,083,299, March 26, 1963.
[36] S. Jones, and L. Manns, "Signal Detector for Use with Radiation Sensor," U.S. Patent No. 3,134,022, May 19, 1964 (filed November 26, 1952).
[37] S. Hansen, "Star Tracking System," U.S. Patent No. 2,981,843, April 25, 1961 (filed September 2, 1947).
[38] R. A. Watkins, "Reticle System for Optical Guidance Systems," U.S. Patent No. 3,002,098, September 26, 1961.
[39] G. Jankowitz, "High Resolution Tracker," U.S. Patent No. 3,061,730, October 30, 1962.
[40] "Unique Reticle Improves Star Trackers," *Electronics*, **37**, 60 (February 14, 1964).
[41] A. R. Gedance, "Radiant Energy Angular Tracking Apparatus," U.S. Patent No. 2,942,118, June 21, 1960.
[42] L. N. McClusky, W. B. McKnight, and N. J. Mangus, Jr., "Rotary Radiation Discriminator," U.S. Patent No. 3,023,661, March 6, 1962.
[43] T. R. Whitney, "Radiant Energy Detection System for Suppressing the Effects of Ambient Background Radiation," U.S. Patent No. 3,144,554, August 11, 1964.
[44] R. K. Orthuber, F. F. Hall, and H. Emus, "Radiation Source Search System using an Oscillating Filter," U.S. Patent No. 3,144,562, August 11, 1964.
[45] C. B. Coleman, "Two Color Background Elimination Detector," U.S. Patent No. 3,107,302, October 15, 1963.
[46] H. A. Droitcour, "Apparatus for Automatically Training Guns, etc., on Moving Objects," U.S. Patent No. 1,747,664, February 18, 1930.
[47] R. W. Buntenbach, "Radiant Energy Translation System," U.S. Patent No. 3,069,546, December 18, 1962 (filed June 4, 1948).
[48] D. D. Wilcox, Jr., "Target Tracking System," U.S. Patent No. 2,994,780, August 1, 1961 (filed March 10, 1954).
[49] D. D. Wilcox, Jr., "Target Tracking System," U.S. Patent No. 2,997,588, August 22, 1961 (filed March 10, 1954).
[50] H. E. Haynes, "Optical Tracking System," U.S. Patent No. 3,117,231, January 7, 1964.

OPTICAL MODULATION 263

[51] T. W. Chew, "Homing System,," U.S. Patent No. 2,421,012, May 27, 1947.
[52] T. W. Chew, "Device for Automatic Homing of Movable Objects," U.S. Patent No. 3,112,399, November 26, 1963 (filed August 11, 1945).
[53] H. G. Lipson and J. R. Littler, "Tuning Fork Choppers for Infrared Spectrometers," *Appl. Opt.*, **5**, 472 (1966).
[54] H. F. Meissinger, "Simulation of Infrared Systems," *Proc. Inst. Radio Engrs.*, **47**, 1586 (1959).
[55] P. E. Mengers and K. B. O'Brien, "Analysis of Error Response of Amplitude Modulated Reticles," *J. Opt. Soc. Am.*, **54**, 668 (1964).
[56] A. R. Gedance, "Comparison of Infrared Tracking Systems," *J. Opt. Soc. Am.*, **51**, 1127 (1961).
[57] A. F. Nicholon, "Error Signals and Discrimination in Optical Trackers that see Several Sources," *Proc. IEEE*, **53**, 56 (1965).
[58] G. W. W. Stevens, *Microphotography*, New York: Wiley, 1957.

7

Introduction to Detectors

The technical literature abounds with a wide variety of connotations for the word *detector*. It is used to refer to a demodulator, to an indicator of a null or balance condition in a bridge circuit, to a mixer in a superheterodyne receiver, or to a device indicating the presence of almost any physical entity, such as radiant energy. For our purposes an infrared detector is simply a transducer of radiant energy. It converts radiant energy into some other measurable form; this can be an electrical current, a change in some physical property of the detector, or the blackening of a photographic plate.

Some of the detectors often used in the 0.2 to 50 μ region are shown in Figure 7.1. The arrows indicate the wavelength interval over which the response of each detector is at least 20 per cent of its maximum value (an arbitrary but useful criterion). The atmospheric transmission curve is included as a reminder of the location of the atmospheric windows. It is evident that most of the detectors shown in Figure 7.1 cover a limited wavelength interval and, in addition, many must be cooled to low temperatures in order to be useful.

Two groups of detectors are shown in Figure 7.1; those in the upper part of the figure are *imaging detectors* and those in the lower part are *point* or *elemental detectors*. This grouping is a convenient one when the objective is a system yielding a picture-like rendition of a scene. An imaging detector, such as photographic film, yields the picture directly. With a point detector, however, it is necessary to build up the picture by sequentially scanning the scene. The point[1] detector, when placed at an image plane, responds to the average irradiance at that particular point in the image. The imaging detector can be considered as a myriad of point detectors, each of which responds to the irradiance at a discrete point in the image. Hence observation time is a key difference between these two

---

[1]The term "point" is not used here in the strict mathematical sense. Instead it connotes a detector whose area is small with respect to the area of the image that is to be scanned.

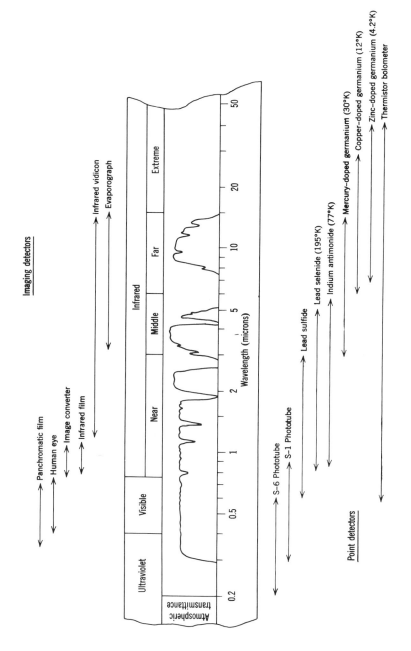

Figure 7.1 Representative detectors for the 0.2 to 50 μ region. (Detector operating temperature is 300°K unless otherwise noted.)

groups of detectors; the imaging type responds continuously to the entire image, but the point type must examine it sequentially. Note that it is possible for a given detector to change from one group to the other; that is, a group of point detectors can be assembled into a mosaic to form an imaging detector.

In many instances the system is designed to search for a target or to track it as it moves, rather than to provide an image of a scene. Under these conditions it is convenient to group detectors into two classes that differ by the physical mechanism involved in the detection process. In one group, called *thermal detectors*, the heating effect of the incident radiation causes a change in some electrical property of the detector. In the other group, called *photon* or *quantum detectors*, there is a direct interaction between the incident photons and the electrons of the detector material. Therefore the response of a thermal detector is proportional to the energy absorbed, whereas that of a photon detector is proportional to the number of photons absorbed. Thermal detectors have been used since the days of Herschel, but nearly all of the infrared-sensitive photon detectors now available have been developed since the 1940's.

The next four chapters are devoted to infrared detectors. This chapter covers the means used to describe the performance of a detector and the characteristics of specific types of detectors. Since electrical noise imposes the ultimate limit on the performance of a detector, the next chapter will be devoted to the characteristics of such noise. Any organization that expects to remain in the infrared system business will find it necessary to maintain a facility for measuring the characteristics of detectors, a subject to be covered in Chapter 9. Chapter 10 is devoted to a discussion of some of the engineering problems that are encountered in selecting and applying a detector to meet a specific system requirement.

7.1 HOW THE PERFORMANCE OF A DETECTOR IS DESCRIBED

The reader may by now have noticed a certain restraint in the use of the word "sensitivity." In fact, he may wonder why we don't follow the lead of many others, that is, drop our pretenses and admit that what we really wish to discuss is the sensitivity of detectors. Close scrutiny of the examples that are readily found in the literature reveals that the term "sensitivity" is often used ambiguously. It sometimes refers to the signal from a detector, and at other times it refers to the signal-to-noise ratio. Terms that eliminate this ambiguity will be introduced in this section.

Until recently there was little agreement concerning the terminology

INTRODUCTION TO DETECTORS 267

to be used in describing the performance of a detector. Jones[1] recalls that a detector was once described to him as having "a peak-signal-to-rms-noise-ratio of 23 db when there was 45 volts across the cell and 45 volts across the load resistor; when the source was at a temperature of 500°K, was 2.5 cm in diameter, and was 40 cm from the cell; the cell had a sensitive area of 9 mm^2; the radiation was square wave chopped at 450 cps, and the noise had a bandwidth of 9 cps." Such a mouthful might be even funnier if it wasn't so painfully true! Obviously, a standard set of test conditions and definitions of the quantities of interest was sorely needed. Although many workers have been active in this field, the contributions of R. Clark Jones are particularly noteworthy. For a summary of his work and an excellent bibliography of earlier contributions to this field, see refs. 2 and 3.

One of the simplest descriptions of detector performance is its *responsivity*, the detector ouput per unit input. Since most infrared detectors are used with a chopper, both the input and output are alternating quantities. The most useful measure of such alternating quantities is their root-mean-square (rms) amplitude. Since the signal ouput of the detector may contain higher-order harmonics generated by the chopping process, it is necessary to eliminate them by measuring the fundamental component only. For consistency, the input radiation is also measured in terms of the rms value of its fundamental component. The responsivity is

$$\mathscr{R} = \frac{V_s}{HA_d}, \qquad (7\text{-}1)$$

where the units of $\mathscr{R}$ are VW^{-1}, V_s is the rms value of the fundamental component of the signal voltage, H is the rms value of the fundamental component of the irradiance on the detector in W cm^{-2}, and A_d is the sensitive area of the detector in cm^2. The reader should note that it may be necessary to specify other conditions of measurement.[1]

The response time of a detector is characterized by its *responsive time constant*, the time that it takes for the detector output to reach 63 per cent[2] of its final value after a sudden change in the irradiance. For many detectors the response to a change in irradiance follows a simple

[1]Chief among these is a statement of how the operating point or bias of the detector is established and whether the signal voltage is the open circuit value measured across the detector or a value measured across the load circuit. As will be seen in Chapter 9, it is customary to select the value of the bias so as to maximize the detectivity and to assume that the signal voltage is an open circuit value.

[2]This definition is identical to that used to describe the charge or discharge of an RL or RC circuit; the figure of 63 per cent is the value of the quantity $(1 - 1/e)$.

exponential law, and the responsivity, chopping frequency, and time constant are related by

$$\mathscr{R}_f = \frac{\mathscr{R}_0}{(1+4\pi^2 f^2 \tau^2)^{1/2}}, \quad (7\text{-}2)$$

where $\mathscr{R}_f$ is the responsivity at the chopping frequency f, $\mathscr{R}_0$ is the responsivity at zero (or very low) frequency, and τ is the time constant of the detector. At low chopping frequencies, where $f \ll 1/2\,\pi\tau$, responsivity is independent of frequency, a fact that is often used to eliminate chopping frequency as a variable in measurements of responsivity.

Although responsivity is a convenient parameter, it gives no indication of the minimum radiant flux that can be detected. The missing information is, of course, the amount of noise in the output of the detector that will ultimately obscure the signal. A convenient means of introducing the noise is to use the concept of *noise equivalent power* (NEP), the radiant flux necessary to give an output signal equal to the detector noise. Since it is difficult to measure a signal when the signal-to-noise ratio is unity, it is customary to make the measurement at a high signal level and calculate the NEP from

$$\text{NEP} = \frac{HA_d}{V_s/V_n} = \frac{HA_d V_n}{V_s}, \quad (7\text{-}3)$$

where NEP is in watts and V_n is the rms value of the noise voltage at the output of the detector. This procedure assumes that the signal output of the detector is a linear function of the input, an assumption that is usually valid for signal-to-noise ratios of 10^3 or less. The electrical bandwidth of the circuit used to measure the noise must also be stated.

A similar measure, which is often used to describe the performance of an entire system, is *noise equivalent irradiance* (NEI), the radiant flux density (irradiance) necessary to give an output signal equal to the detector noise:

$$\text{NEI} = \frac{H}{V_s/V_n} = \frac{HV_n}{V_s}, \quad (7\text{-}4)$$

where the units of NEI are W cm^{-2}. When used to describe the performance of an entire system, NEI expresses the irradiance at the entrance aperture required for a signal-to-noise ratio of unity at the output of the system electronics.

When several detectors are compared, the one having the highest output for a given radiation input is said to have the best responsivity. However, when the detectors are compared in terms of their detecting ability, that is, in terms of the minimum detectable radiant flux, the

INTRODUCTION TO DETECTORS 269

best detector is the one with the lowest NEP. Many people inherently dislike a situation in which an improvement in the quantity of interest gives a lower value for the figure of merit. Jones[2] suggested that if the reciprocal of the NEP were used, which he proposed to call *detectivity*, this difficulty would be avoided since the better detector would have the higher detectivity. Hence the detectivity is

$$D = \frac{1}{\text{NEP}}, \quad (7\text{-}5)$$

where the units of D are W^{-1}.

How can the conditions of measurement be standardized so as to simplify the comparison of detectivities measured by different laboratories? To begin, we note that detectivity is a function of the following:

1. Wavelength of the incident radiation.
2. Temperature of the detector.
3. Chopping frequency.
4. Bias current applied to the detector.
5. Area of the detector.
6. Bandwidth of the circuit used to measure detector noise.

There is no apparent theoretical relationship between detectivity and wavelength or between detectivity and temperature of the detector. Hence both wavelength and detector temperature must be reported along with any measurement of detectivity. It is customary to make measurements at easily achievable and readily reproducible temperatures such as those of normal ambient conditions (300°K), solidified carbon dioxide (195°K), liquid nitrogen (77°K), liquid hydrogen (20°K), and liquid helium (4.2°K).

The simplest way to handle the variation of detectivity with chopping frequency is to select a frequency low enough to prevent any limitation by the time constant of the detector. This condition is met when $f \ll 1/2\pi\tau$. When a detector requires a bias current for its operation, it is customary to select the bias current that maximizes detectivity. Fortunately the maximum is single-valued and quite easy to find experimentally.

Extensive theoretical and experimental studies have shown that it is reasonable to assume that detectivity varies inversely as the square root of the area of the detector,[1] or

$$DA_d^{1/2} = \text{constant.} \quad (7\text{-}6)$$

---

[1] The reader is cautioned that this relationship may be in error for detectors that are long and narrow, that is, those in which the length is more than five times the width.

Since the noise in the output of the detector contains many frequencies, it is obvious that the noise voltage is a function of the electrical bandwidth of the circuitry used for its measurement. On the assumption that the noise voltage per cycle of bandwidth is independent of frequency (an assumption that will be examined in greater detail in Chapter 8), detectivity varies inversely with the square root of the electrical bandwidth Δf.[1]

Using these assumptions, Jones[3] introduced the quantity D^*, the detectivity referred to an electrical bandwidth of 1 Hz and a detector area of 1 cm^2:

$$D^* = D(A_d \Delta f)^{1/2} = \frac{(A_d \Delta f)^{1/2}}{\text{NEP}}. \qquad (7\text{-}7)$$

The symbol D^* is pronounced "dee-star" and its units are cm(Hz)$^{1/2}$ W^{-1} when A_d is measured in cm^2 and Δf is in Hz. A convenient way to remember the significance of D^* is to recall that it is the signal-to-noise ratio when one watt is incident on a detector having a sensitive area of 1 cm^2 and the noise is measured with an electrical bandwidth of 1 Hz. Thus D^* is a normalized detectivity that is particularly convenient for comparing the performance of detectors with different areas when used in circuits having different bandwidths.

The units of D^* are somewhat cumbersome. It has been suggested, and this author heartily concurs, that a cm(Hz)$^{1/2}$ W^{-1} should be defined as a jones; a fitting tribute to one who has done so much to bring order out of chaos in the field of infrared detectors.

Thus far the discussion has implied that in measuring D^*, NEP, or NEI, a blackbody is used as the source. Such measurements are known as blackbody D^*, and so on. To define the measurement conditions further, it is customary to write D^* followed by two numbers in parentheses. The first gives the temperature of the blackbody and the second gives the chopping frequency. Thus $D^*(500°\text{K}, 840)$ indicates a value of D^* measured with a 500°K blackbody at a chopping frequency of 840 Hz. Since the response of many detectors varies with wavelength, it is often desirable to report spectral values, that is, D^* at a specific wavelength. In this case the first number in the parentheses gives the wavelength at which the measurement was made, as $D^*(3.7\,\mu, 840)$. When the wavelength is that at which maximum detectivity occurs, it is shown as $D^*(\text{peak}, 840)$. Some workers include a third number in the parentheses, $D^*(500°\text{K}, 840, 1)$, to indicate that the bandwidth is 1 Hz.

[1] The bandwidth Δf is properly called the equivalent noise bandwidth. A discussion of means of measuring it will be deferred to Chapter 8, since it is not essential for an understanding of the present chapter.

INTRODUCTION TO DETECTORS 271

Such a notation is redundant because the unit bandwidth is implicit in the definition of D^*.

7.2 THERMAL DETECTORS

Because thermal detectors have been used since Herschel's discovery of the infrared portion of the spectrum, it is appropriate to consider them first. They are distinguished as a class by the observation that the heating effect of the incident radiation causes a change in some physical property of the detector. Since most thermal detectors do not require cooling, they have found almost universal acceptance in certain field and spaceborne applications in which it is impractical to provide such cooling. Because (theoretically) they respond equally to all wavelengths, thermal detectors are often used in radiometers. However, the practical limitations of available blackening materials often force one to modify the simple assumption that the detectivity of thermal detectors is independent of wavelength. The time constant of a thermal detector is usually a few milliseconds or longer, so that they are rarely used in search systems or in any other application in which high data rates are required. An excellent discussion of the physics of thermal detectors can be found in ref. 4; their detective properties are summarized in ref. 5.

The Thermocouple

Conceptually, the simplest thermal detector is a blackened receiver equipped with a device for sensing the increase in temperature resulting from the absorption of radiant energy. One of the most practical means of sensing this increase is the *thermocouple*, a junction between two metals that have a large difference in their thermoelectric power. Popular combinations are bismuth-silver, copper-constantan, and bismuth-bismuth tin alloy. Fine wires made of the two different materials are joined at one end to form a thermoelectric junction. The blackened receiver, which defines the sensitive area of the detector, is fastened directly to this junction. In a typical thermocouple the wires forming the couple are 3 or 4 mm long and about 25 μ in diameter. The free ends of the wires are fastened to relatively massive metallic supports to provide a reference junction at a constant temperature. The receiver is usually made from blackened gold foil about 0.5 μ thick. Fine quartz fibers are used to provide additional mechanical support for the assembly. Since they are extremely fragile, thermocouples are rarely used outside of the laboratory.

The electrical resistance of most thermocouples is very low; 1 to 10 ohms is the typical value. It is difficult to couple such a low resistance to a vacuum tube amplifier without using an impedance-matching transformer.

Because of the low signal levels involved, such transformers must be extremely well shielded and are, as a result, quite expensive. Since they have a low input impedance, transistorized amplifiers are well suited for use with thermocouples. Because thermocouple time constants range from a few milliseconds to several seconds, chopping frequencies are limited to 10 Hz or less. The thermocouple is one of the few infrared detectors that one can make experimentally without a great deal of expensive equipment. An excellent summary of constructional techniques is given by Strong[6].

The voltage developed at the thermoelectric junction is proportional to the increase in the temperature of the receiver. In a good thermocouple the minimum detectable signal corresponds to a temperature increase of about 10^{-6} °C.

Sanderson[7] has shown that the time constant of a thermocouple can be expressed as

$$\tau \approx \frac{C}{\Lambda}, \qquad (7\text{-}8)$$

where C is the thermal capacity of the junction and receiver assembly and Λ is the rate at which the assembly loses energy. Thus a short time constant requires a low thermal capacity and an efficient coupling of the detector to a heat sink. In a similar manner it can be shown that

$$\mathscr{R} \approx \frac{\tau}{C}. \qquad (7\text{-}9)$$

From this equation it is evident that we can always trade time of response for responsivity. Although they were derived for a thermocouple, these equations apply equally well to other types of thermal detectors; high responsivity requires a long time constant and a low heat capacity in the junction and receiver assembly.

The Thermopile

Several thermocouples can be connected in series to form a *thermopile*. The advantage of such a construction is that the voltages developed at each junction add so as to increase the responsivity. Similarly, the series connection increases the resistance of the detector and makes it easier to match it to an amplifier. Since the time constant of most thermopiles is several seconds, it is impractical to use a chopper with them.

It has been known for many years that thermocouples can be formed by evaporating overlapping films of antimony and bismuth[8]. Because such a construction is much more rugged than that of the traditional thermopile, it is not surprising that it should be resurrected and improved

INTRODUCTION TO DETECTORS 273

to make thermopiles for space applications [9, 10]. The overlapping areas that make up the junctions are formed on a thermally insulating layer set in the middle of one face of an aluminum block. The reference junctions are formed where the evaporated films contact the aluminum block. Because the junctions have a very low heat capacity, the time constant of an evaporated thermopile can be as low as 10 msec. In addition, the evaporation technique permits an extremely flexible control of the shape, size, and arrangement of junctions.

The Bolometer

Thermal detectors that change their electrical resistance when heated by the incident radiation are called *bolometers*. In their earliest form they consisted of a thin blackened strip of platinum foil connected in one arm of a Wheatstone bridge. Modern versions use other metals that have temperature coefficients of resistance of about 0.5 per cent per degree Centigrade. The blackened strip defines the sensitive area of the detector, and any change in its temperature unbalances the bridge. A typical bolometer consists of a pair of metal elements that are as nearly identical as possible. One, called the *active element*, is exposed to the incident radiation, and the other, called the *compensating element*, is carefully shielded from the incident radiation. By using these elements as two arms of a bridge, slow changes in the temperature of the detector will not affect the bridge balance.

In more recent bolometers the elements are made of semiconductor materials having temperature coefficients of resistance as high as 4.2 per cent per degree Centigrade. In the *thermistor bolometer* the elements are thin flakes formed by sintering a mixture of metallic oxides. These flakes are mounted on an electrically insulating substrate that is, in turn, mounted on a metallic heat sink. By using substrates having different thermal characteristics, it is possible to change the time constant of the detector from about 1 to 50 msec. Thermistor bolometers are relatively rugged, they require no cooling, and their high resistance makes it easy to match them to an amplifier. Consequently it is no surprise that the thermistor bolometer has been an almost unanimous choice for use in the first generation of infrared equipment for space vehicles. The properties of thermistor bolometers are described in refs. 11 and 12. Information on thermistor materials and how they are made into detectors is given in refs. 13-15. Since the thermistor material is not in itself a good absorber, it must be blackened during manufacture. Barnes and Wormser[16] have described the application of selectively absorbing coatings to thermistor bolometers to achieve a combination comparable to a blackened detector and a narrow bandpass filter.

A *superconducting bolometer* utilizes the tremendous change in resistance that occurs in the transition of certain metals and semiconductors from their normal to the superconducting state. Andrews et al.[17] used a ribbon of niobium nitride, which becomes superconducting at about 15°K. In the transition range, which is only a fraction of a degree wide, the temperature coefficient of resistance is about 5000 per cent per degree Centigrade. The control required to keep the temperature in the transition range makes it very unlikely that this detector will ever be useful outside the laboratory.

The *carbon bolometer*[18] has been used for spectroscopic investigations in the extreme infrared. The sensitive element is a slab cut from a carbon resistor and cooled to 2.1°K. Its D^* is at least an order of magnitude greater than that of a thermistor bolometer.

The *germanium bolometer*[19] is a single crystal of gallium-doped germanium cooled to 2.1°K. Its D^* is nearly two orders of magnitude greater than that of a thermistor bolometer, and since its spectral response extends beyond 1000 μ, it is equally suitable for detecting either infrared or microwaves.

The Pneumatic or Golay Detector

The *pneumatic detector* is based on the old principle of the gas thermometer. It consists of a radiation absorber placed in a gas chamber [20, 21]. The absorber is, in essence, a broad-banded radio antenna designed to match the impedance of free space[22]. It is heated by the incident radiation, which in turn heats the gas in the chamber. The resulting increase in pressure is observed optically by the deflection of a small flexible mirror. These detectors come within a half order of magnitude of the theoretically ultimate detectivity. They are, however, extremely fragile and are essentially useless for field applications.

The Calorimetric Detector

The *black radiation detector*, developed by Eisenman et al. [23, 24] is used as a blackness standard for determining the spectral response of other detectors. It is basically a fast-responding miniature calorimeter built in the form of a conical cavity. Calculations based on the theory of Gouffé (described in Chapter 3) show that the emissivity and hence the absorptance of this detector is probably greater than 0.995 from the visible to 40 μ.

Problems of Blackening Thermal Detectors

The constructional materials for thermal detectors are not in themselves good absorbers and must be blackened by applying an absorbing coating. An ideal black coating would have (1) uniformly high absorptance at all wavelengths, (2) negligible thermal capacity, (3) high thermal conductivity, and (4) no adverse effect on the electrical properties of the detector element. In addition, it must be possible to apply this coating without exposing the element to an undue danger of breakage. Obviously there is no such material available; hence whatever coating is used must necessarily be a compromise.

One of the simplest means of blackening is to apply a thin layer of soot from a candle flame or burning camphor. Such a film appears extremely black to the eye; indeed, its absorptance is about 0.99 in the visible. However, the absorptance decreases at longer wavelengths and rarely exceeds 0.5 at $10\,\mu$ [25]. Metallic blacks are formed by evaporating thin films of certain metals in a "poor" vacuum, that is, at a pressure of about 1 mm Hg. The best of these, gold black, has been reported to have an absorptance of 0.99 from 1 to $39\,\mu$ [26]. Unfortunately, gold black has a relatively low electrical conductivity and can be applied only to low-resistance detectors, such as thermocouples and thermopiles. For high-resistance detectors the most satisfactory blackening is done with black paints and lacquers.

One of the assumed advantages of a thermal detector is that its detectivity is independent of wavelength. However, recent measurements show that this ideal condition is rarely achieved. With the advent of photon detectors, which have higher detectivity (over limited wavelength intervals) and much shorter time constant, users began to demand that manufacturers increase the detectivity and shorten the time constants of thermal detectors. The reason for the manufacturer's dilemma is evident in (7-9); he can trade time constant for detectivity, but he cannot simultaneously improve both. Since blackening increases the thermal capacity of the detector, the manufacturer must tread the narrow path between a thin coating that reduces both time constant and detectivity and an overly thick coating that increases both quantities. Papers that discuss this problem [23, 24, 27] show measurements of detectors on which the blackening (by paint) is obviously too thin; at some wavelengths the response is no more than 30 per cent of the maximum value.

7.3 PHOTON OR QUANTUM DETECTORS

Most photon detectors have a detectivity that is one or two orders of magnitude greater than that of thermal detectors. This higher detectivity does not come for free, however, since many photon detectors will not function unless they are cooled to cryogenic temperatures. Because of the direct interaction between the incident photons and the electrons of the detector material, the response time of photon detectors is very short;

most have time constants of a few microseconds rather than the few milliseconds typical of thermal detectors. Finally, the spectral response of photon detectors, unlike that of thermal detectors, varies with wavelength.

The result of an interaction between photons and matter is called a *photoeffect*. Many such photoeffects have been described in the literature. We shall concern ourselves only with those that have proved most useful in the infrared portion of the spectrum. The reader who wishes to delve more deeply into the physics of the detection process will find excellent discussions in refs. 4 and 28.

If the incident photon transfers its energy to an electron in the detector material, this electron may have sufficient energy to escape from the surface. This is called the *photoelectric* or *photoemissive* effect. Since it is observable beyond the confines of the detector material, it can be classed as an external photoeffect. For wavelengths longer than about 1.2 μ, photons are not sufficiently energetic to free an electron from the surface. There are, however, a number of internal photoeffects in which the energy transferred from the photon raises an electron from a nonconducting to a conducting state and, in so doing, produces a charge carrier. The type of charge carrier depends on the characteristics of the detector material, which is nearly always a semiconductor. If the material is an intrinsic, or pure, semiconductor, the photon produces an electron-hole pair containing both a positively and a negatively charged carrier. If the material is an extrinsic, or impurity, semiconductor, the photons produce charge carriers of a single sign, that is, either positive or negative, but not both. If an electric field is applied by biasing the detector, changes in the number of charge carriers will vary the current flowing through the detector. This is called the *photoconductive* effect.

If the photon produces an electron-hole pair in the vicinity of a *p-n* junction, the electric field across the junction will separate the two carriers to give a photovoltage. This is termed the *photovoltaic* effect. No external bias supply is required for a photovoltaic detector since it is, in effect, furnished by the *p-n* junction.

When electron-hole pairs are formed near the surface of a semiconductor, they tend to diffuse deeper into the material in order to reestablish electrical neutrality. During this process the charge carriers can be separated by a strong magnetic field so as to give a photovoltage. This is known as the *photoelectromagnetic* effect.

The Photoelectric Detector

The photoelectric effect was first observed in 1887 by Hertz, but it was not until 1902 that Einstein was able to explain this process. Following a suggestion made two years earlier by Planck, he proposed that

INTRODUCTION TO DETECTORS

radiant energy exists in small packets, which we now call photons or quanta. The energy of a photon is

$$E = h\nu = \frac{hc}{\lambda}, \quad (7\text{-}10)$$

where h is Planck's constant, ν is the frequency, λ is the wavelength, and c is the velocity of light. When a photon collides with an electron in a metal, it may transfer its energy to the electron. If it does, the electron acquires all of this energy and the photon ceases to exist. This acquired energy may be sufficient to enable the electron to penetrate the potential barrier at the surface and to escape from the metal. Penetration of this barrier requires an amount of energy ϕ that is a characteristic of the material and is called its work function. Therefore the kinetic energy of the photoelectron as it leaves the surface is the difference between the energy gained from the photon and that used to overcome the work function:

$$E' = \frac{mv^2}{2} = h\nu - \phi = \frac{hc}{\lambda} - \phi, \quad (7\text{-}11)$$

where m is the mass of the electron and v is its velocity. This is Einstein's photoelectric equation. Since the energy of the photon varies with frequency, there is a low-frequency or long-wavelength limit beyond which this energy is less than that required to overcome the work function. The wavelength at which this occurs, called the *cutoff wavelength*, is

$$\lambda_c = \frac{1.24}{\phi}, \quad (7\text{-}12)$$

where ϕ is in electron volts. Of the elements that are photoemissive, the lowest work functions are found among the alkali metals. Cesium, with a work function of 1.9 ev, is the lowest and has a cutoff wavelength of 0.65 μ. Surfaces compounded of more than one material may have still lower work functions. The lowest value, 0.98 ev, is observed with a silver-oxygen-cesium surface. This is commonly called an S-1 surface and its cutoff wavelength is 1.25 μ. Hence the response of photoelectric detectors extends only a short way into the near infrared, and they are rarely used in the infrared systems described in this book.

In its simplest form a photoelectric detector consists of a photoemissive cathode and a plate, both housed in an evacuated glass envelope. An externally applied voltage maintains this plate about 100 V positive with respect to the cathode, so that essentially all of the photoelectrons are collected by the plate. Photoemissive surfaces with very low work functions also emit electrons by thermal excitation. Since these electrons are indistinguishable from photoelectrons, they ultimately limit the minimum detectable signal. The number of thermally emitted electrons can be reduced by cooling the photoemissive surface.

The Photoconductive Detector

The photon detectors that have proved to be most useful in the infrared make use of various internal photoeffects in semiconducting materials. To gain a better understanding of these effects, we shall briefly digress to discuss the band theory of solids, a convenient means of describing the way in which energy is distributed among the electrons in a solid. In the simple Bohr model of an atom the electrons surrounding the nucleus are restricted to discrete energy levels that are conveniently visualized in terms of discrete orbital diameters. Only the lower energy levels are occupied, unless the atom has been excited. When atoms are brought into close proximity, as they are in a solid, the discrete energy levels of the individual atoms broaden into nearly continuous bands that are separated from one another by energies that the electrons are forbidden to acquire because of quantum mechanical considerations. Although these energy bands appear to be continuous, detailed examination shows that they consist of discrete energy levels separated by as little as 10^{-14}ev. The highest energy band that is completely filled is called the *valence band*. The next higher allowed band, whether occupied or not, is called the *conduction band*. Only electrons in the conduction band can contribute to the electrical conductivity of the material.

A portion of the energy band structure for an electrical conductor is shown in Figure 7.2a.[1] The partially filled conduction band is the obvious indication of a conductor. In an electrical insulator, as shown in Figure 7.2b, there are just enough electrons to fill all of the energy levels in the valence band and the conduction band is empty. The forbidden energy band is wide, and it is unlikely that a valence electron can gain enough energy to raise it to the conduction band.

From an electrical standpoint, a semiconductor behaves as if its conductivity were intermediate between that of an insulator and that of a metallic conductor. In a pure, or intrinsic, semiconductor, shown in Figure 7.2c, the forbidden energy band is relatively narrow; it may be only a fraction of an electron volt wide, as compared to 3 or more ev for an insulator. As a result, even at room temperature some of the valence

---

[1]The reader will recognize that the energy band structure shown here is, in essence, merely a narrow cross section of the outermost orbits described by the Bohr model.

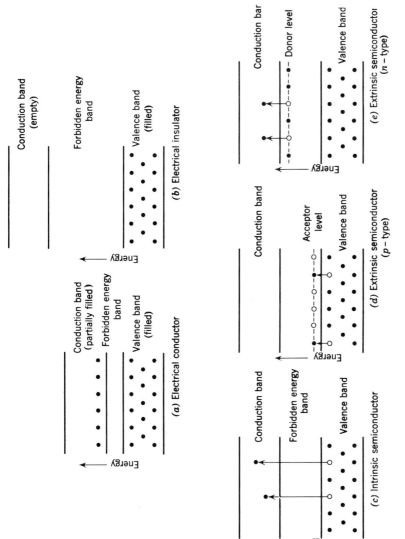

Figure 7.2 Energy bands in solids.

electrons can acquire sufficient energy to jump across the forbidden band to the conduction band. The sites formerly occupied by these electrons are positively charged and are called *holes*. In the presence of an electric or magnetic field, holes can flow through the material just as electrons do, although the two types flow in opposite directions. Hence, in a pure semiconductor the excitation of an electron into the conduction band creates an *electron-hole pair* of charge carriers, and both contribute to the conductivity. This is called *intrinsic conductivity*, and the material is termed an *intrinsic semiconductor*. Examples of such materials are pure crystals of silicon or germanium and compounds containing two components in stoichiometric proportions.

Valence electrons can acquire sufficient energy from incident photons for the production of electron-hole pairs. The result is a photon-induced variation in conductivity. A device based on this effect is called a photoconductive detector. As with a photoelectric detector, there is a long-wavelength limit beyond which a photon has insufficient energy to create an electron-hole pair. This cutoff wavelength is

$$\lambda_c = \frac{1.24}{E_g}, \qquad (7\text{-}13)$$

which is identical in form to (7-12) except that it uses E_g, the forbidden energy gap (or band) expressed in electron volts. All currently known intrinsic detectors have forbidden energy gaps in excess of 0.18 ev at room temperature. As a result, they are not usable beyond 7 μ. In general, the width of the forbidden gap is reduced by cooling, so that the cutoff wavelength is increased when the detector is cooled. Intrinsic photoconductive detectors include silicon, germanium, lead sulfide, lead selenide, indium arsenide, and indium antimonide [29–33].

Detectors with longer cutoff wavelengths require materials that have smaller forbidden energy gaps than those found in intrinsic semiconductors. The usual means of making a smaller forbidden gap is to add a small quantity of some impurity to an otherwise pure semiconductor, a process called *doping*.[1] The resultant materials are called *extrinsic* or *impurity-activated semiconductors*. In extrinsic materials, essentially only one type of charge carrier contributes to the conductivity; electrons are the

---

[1] Perhaps the following analogy will help the reader to better comprehand the very small quantities of impurities that are used. Germanium contains about 4×10^{22} atom cm^{-3}. Copper-doped germanium, a currently popular detector, is made by adding about 10^{16} atom cm^{-3} of copper to germanium, or about one impurity atom for every 4×10^6 atoms of germanium. Thus the task of finding the impurity atom is roughly comparable to the task of finding a particular individual in a city the size of Los Angeles.

INTRODUCTION TO DETECTORS 281

charge carriers in *n*-type material and holes are the carriers in *p*-type material.

Many infrared detectors made from extrinsic materials use germanium as the host material. The germanium atom has four valence electrons that form covalent bonds with four neighboring germanium atoms. If an impurity atom having, for instance, three valence electrons is added to the germanium, there will be only three covalent bonds formed. The required fourth bond must be provided by a nearby germanium atom. The result is an excess hole provided by the impurity. On an energy level diagram, such as that shown in Figure 7.2*d*, the energy level of the impurity lies just slightly above the top of the host's valence band. Consequently very little energy is required for an electron to jump from the valence band into the hole provided by the impurity. The holes left in the valence band become the charge carriers and the material is *p*-type. Such electron-deficient impurities are called *acceptors*, since they accept electrons from the host material. In a similar manner, impurities having five or more electrons may be used. Such impurities act as *donors* of electrons and the resulting materials are *n*-type(Figure 7.2*e*). While either *n*- or *p*-type materials can, in principle, be used, only *p*-type material is currently used for infrared detectors. Typical extrinsic detectors include germanium doped with gold, mercury, cadmium, copper. or zinc[34–38].[1] Many other dopants can, in theory, be used with germanium; the longest cutoff wavelength yet reported, 140 μ, occurs with boron [35].

Indium antimonide is the only known material in which still lower impurity levels can be introduced. When used in conjunction with a magnetic field, such detectors respond out to 8000 μ (8 mm), thus bridging the gap between the optical and the microwave regions of the spectrum[39–41].

It may, at times, be desirable to tailor the cutoff wavelength of a detector in order to meet a particular system requirement. This can, of course, be done by placing a short-wavelength-pass filter in front of the detector. It can also be done by doping an alloy of silicon and germanium. The cutoff wavelength for a given impurity can be shifted by varying the amount of silicon in the alloy[35, 42, 43]. In one published example, this technique was used to develop a photoconductive detector responsive to the 8 to 14 μ region while requiring a minimum of cooling[42].

A simple analysis shows how various material parameters affect the responsivity of a photoconductive detector. Assuming an extrinsic

[1]It is customary to designate such doped detectors as follows: Ge:Au, Ge: Hg, and so on, where the first symbol indicates the host material and the second symbol indicates the dopant.

material so that there is only one type of carrier, the conductivity in the absence of photon excitation is given by

$$\sigma = ne\mu, \quad (7\text{-}14)$$

where n is the number of free carriers per unit volume generated by thermal excitation, e is the electronic charge, and μ is the mobility of the free carrier. The steady-state increase in the number of free carriers produced by incident photons is

$$\Delta n = f\tau, \quad (7\text{-}15)$$

where f is the number of excitations per second per unit volume and τ is the lifetime of these carriers[45]. If the detector has an area of 1 cm² and a thickness of x cm, then

$$f = \frac{\eta Q}{x}, \quad (7\text{-}16)$$

when η is the quantum efficiency, that is, the number of charge carriers produced per photon, and Q is the incident photon flux. The responsivity will be proportional to the fractional change in conductivity per incident photon, or

$$\mathcal{R} = \frac{\Delta\sigma}{\sigma} = \frac{\eta\tau\mu e}{x\sigma}. \quad (7\text{-}17)$$

Thus high responsivity requires a high quantum efficiency, a long lifetime and a high mobility for the free carrier, a low conductivity in the absence of photon excitation, and a minimum thickness of detector.

Since the lifetime of the free carrier can, to a certain extent, be controlled by the method of preparation and the impurity content of the material, it is an extremely important parameter. In addition to its effect on the responsivity, the lifetime of the free carrier also determines the time constant of the detector[35]. The lifetime of the carrier is normally terminated by one of several recombination processes. These are extremely complex and will barely be outlined here; the reader interested in more detail can consult refs. 45 and 46. Conceptually, the simplest means of recombination would be for the excited electron to return to the hole it recently vacated. Since both the electron and the hole are moving, the probability of such a recombination is slight except for very high carrier densities. For low carrier densities the recombination is generally accomplished in a two-step process involving either recombination centers or trapping centers. Recombination centers lie deep within the forbidden band and have the ability to capture free carriers and hold them until an opposite number appears to complete

INTRODUCTION TO DETECTORS 283

the recombination. Trapping centers, on the other hand, lie relatively close to the edge of the forbidden band. They have a strong preference for one type of carrier. Since the probability of capturing the opposite type of carrier is low, it is quite likely that the trapped carrier will be reexcited thermally to the conduction band rather than suffer a recombination. During the time it is in the trap the carrier cannot, of course, contribute to the conductivity.

From (7-17) it is evident that high responsivity requires that a detector have a low conductivity in the absence of photon excitation. In essence, this means that one should minimize the number of free carriers generated by any means other than by the photons comprising the desired signal. Response at long wavelengths requires detector materials that have small forbidden energy gaps. Unfortunately, such small gaps permit large numbers of thermally induced carriers. Their number can, however, be reduced by cooling the detector. As a matter of fact, most of the photoconductive detectors responding to wavelengths beyond $3\,\mu$ will not work at all without cooling. This is because at room temperature all of the available charge carriers are thermally excited and there are none left for photon excitation. As a rough rule of thumb, no cooling is required for photon detectors that do not respond beyond $3\,\mu$; those responding from 3 to $8\,\mu$ require moderate cooling (77°K); those responding beyond 8μ require progressively lower temperatures that approach within a few degrees of absolute zero. As will be shown in Chapter 10, photons from the background may produce unwanted charge carriers. These can be reduced by shielding the detector.

Equation 7-17 also shows that a high responsivity requires a high quantum efficiency. Such a conclusion should be obvious; we wish to use all of the available photons as efficiently as possible. The semiconductor materials used to make photon detectors are relatively transparent, so that a fraction of the incident photons pass through the detector without any interaction. If these photons could be directed back through the detector, they would, in effect, be given a second chance. This can be accomplished by placing a reflector behind the detector or by using an integrating chamber to send the stray photons back through the detector. Similarly, an antireflection coating can be applied to the detector in order to prevent the loss of photons by reflection at its front surface [47]. Such simple expedients, in effect, improve the quantum efficiency and are a valuable supplement to the more esoteric efforts of the solid state physicist.

When considered as a circuit element, a photoconductive detector behaves much like a variable resistor. The detector is connected in series with a load resistor and a bias battery. Photon-induced changes in the conductivity of the detector modulate the current flowing through the detector and load resistor. The signal is taken across the load resistor and capacitively coupled into a preamplifier. Since the bias voltage across the load resistor can be millions of times greater than the signal voltage, it is imperative that the bias supply be as free as possible of noise and other spurious effects.

The Photovoltaic or p-n Junction Detector

A photovoltaic detector consists of a p-n junction formed in an intrinsic semiconductor. Incident photons produce electron-hole pairs that are, in turn, separated by the electric field at the junction so as to generate a photovoltage.

A p-n junction is formed at the boundary between a p- and an n-type region (when both are formed in the same host material). Such a junction can be formed during the growth of a crystal or by the diffusion of an impurity into the surface of a wafer cut from a crystal. A semiconductor crystal is grown by dipping a seed crystal into an appropriately doped melt and then withdrawing it at a carefully controlled rate. If the type of dopant is changed midway in the process, so that the melt is changed from an n- to a p-type, for instance, the finished crystal will contain a grown junction. The crystal can then be cut into wafers, each of which includes a portion of the junction. The junction in the wafer is long and narrow. Since the junction defines the shape of the sensitive area, detectors made from grown junctions are, by necessity, long and narrow. Such a shape is useful for matching the image of a spectrometer slit, but it is extremely awkward to use otherwise. This difficulty can be avoided by the use of a diffused junction, which is made by gaseous diffusion of the desired dopant directly into the surface of a wafer of semiconductor material. There is virtually no limitation on the size or shape of a diffused junction. Indeed, the gaseous diffusion process plays a key role in the current large-scale production of monolithic integrated circuits and planar transistors.

The typical nonohmic voltage-current characteristic of a photovoltaic detector is shown in Figure 7.3. When the detector is shielded from incident radiation, the V-I characteristic is that shown by curve 1. Incident radiation shifts the curve downward to that shown at 2. It is evident from these curves that the user of such a detector has available a broad range of possible operating points. If the detector is to be operated into a high impedance, corresponding to an operating point at A, it is the open-circuit voltage across the detector that is observed. Operation into a low impedance yields an operating point at B and it is the short-circuit current that is observed. Finally, the detector may be back biased so as to

INTRODUCTION TO DETECTORS 285

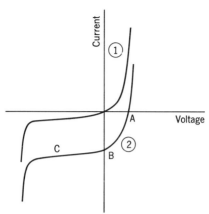

Figure 7.3 The current-voltage characteristic of a junction-type detector. Curve 1, no incident radiation. Curve 2, with incident radiation.

place the operating point at C. Here the detector functions as a high-resistance device and the photovoltage is developed across a load resistor placed in series with the detector. In general, photovoltaic detectors exhibit the highest detectivity when they are operated in the short-circuit or zero-bias mode at the point B.

The photovoltaic detector has the important advantage that it is a self-generating device, that is, it does not require a bias supply. This can sometimes result in a considerable reduction in circuit complexity. As will be shown in Chapter 10, the theoretical maximum-realizable detectivity is 40 per cent greater for a photovoltaic detector than it is for a photoconductive type. Since the device requires that the incident photons produce electron-hole pairs, only intrinsic semiconductor materials can be used for photovoltaic detectors. As previously discussed, intrinsic materials show no response at wavelengths longer than about 7 μ. Among the more popular photovoltaic detectors are silicon, indium arsenide, and indium antimonide [46-48].

The literature pertaining to photovoltaic detectors exhibits a very confusing set of terms. The term "photodiode" is used as a synonym for "photovoltaic detector." Such a usage is logical since one can readily perceive that photodiode combines the connotations of a photoeffect and the presence of a junction or a diode. The terms become confused when some workers try to say that a detector operated at the point A in Figure 7.3 is operating in the photovoltaic mode and operation at point B is in the photodiode mode. The situation becomes inexcusable when these same workers define operation at point C as being in the photoconductive mode.

To avoid this confusion in terminology, let us compare the photoconductive and the photovoltaic processes in intrinsic semiconductors. In both types the initial step is the production of electron-hole pairs by the incident photons. In the photoconductive detector the presence of the electron-hole pairs results in a change in the conductivity that can be observed by noting the variations in the current flowing through the detector from an external bias source. In the photovoltaic detector the electron-hole pairs are separated by the junction to form a photovoltage. Thus a photoconductive detector is distinguished by the presence of an external bias supply whereas the photovoltaic detector is distinguished by the presence of a p-n junction. The photovoltaic detector, because of the presence of the p-n junction, acts quite analogously to a photoconductive detector containing a built-in bias supply. We see now that the p-n junction detector operates as a photovoltaic device over the full range of operating points shown in Figure 7.3. Calling operation at the point C a photoconductive mode is particularly confusing since indium antimonide is made into both p-n junction photovoltaic detectors and intrinsic photoconductive detectors (containing no junction).

The Photoelectromagnetic Detector

A photoelectromagnetic detector consists of a wafer cut from a slab of intrinsic semiconductor material and a magnet. Incident photons produce electron-hole pairs, and these are separated by the externally applied magnetic field. Such detectors show response to $7\,\mu$ without requiring cooling and have very short time constants. They have enjoyed little popularity and are rarely used, principally because their detectivity is relatively low compared to that of photoconductive and photovoltaic types.

Spectral Response of Photon Detectors

The conventional way of describing the spectral response of a device, whether it be a detector or an entire system, is to plot its relative response as a function of wavelength *for a constant radiant flux per unit wavelength interval*. The response of a thermal detector is proportional to the energy absorbed; the spectral response plot is therefore simply a straight horizontal line (if it is assumed, as most manufacturers do assume, that the blackening process is somehow "magically" independent of wavelength and the manner of preparation). With photon detectors the situation is quite different. In the first place, (7-13) shows that there is a cutoff wavelength beyond which the detector does not respond. In addition, photon detectors are not energy sensitive; they respond, instead, to the total number of photons absorbed. Since the energy

INTRODUCTION TO DETECTORS 287

per photon is inversely proportional to wavelength, more photons are required at longer wavelengths in order to maintain a constant energy per unit wavelength interval. As a result, when the spectral response of a photon detector is plotted in the conventional manner, it appears to increase with wavelength until the cutoff is reached. The spectral responses of thermal and photon detectors are compared in Figure 7.4a. The response of most photon detectors approximates the idealized curve shown in this figure. The sharp break at the cutoff wavelength is usually more gradual than that shown because of the presence of slight impurities and defects in the detector material.

Another way to plot spectral response is to assume that the number of photons per unit wavelength interval remains constant. When plotted on this basis, the spectral responses of the two types of detectors are as shown in Figure 7.4b. Such a presentation, although interesting, is rarely used.

Fabrication of Photon Detectors

There is little likelihood that an experimenter can successfully build his own photon detectors. The equipment required to process semiconductor materials is expensive and of the type not normally found in most laboratories. The reader interested in learning more about how detectors are made will find the patent literature a valuable source of information; refs. 29, 31, 32, 37, 38, 43, 49, and 50 provide a good sampling of the available information.

7.4 THE COMPARISON OF DETECTORS

An infrared system designer must have reliable data on the characteristics of available detectors. In this section we shall take a preliminary

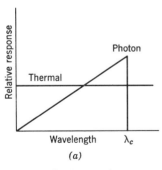 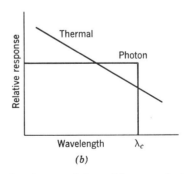

Figure 7.4 Idealized spectral response curves for photon and thermal detectors. In (a) the flux per unit wavelength interval is constant. In (b) the number of photons per unit wavelength interval is constant.

look at typical data. In Chapter 10 more detailed data are given for selected detectors.

Where does a designer find the necessary information on detector characteristics? The various manufacturers, of course, provide data on their own products, and there are occasional books and published papers reviewing the field [3–5, 28, 35, 36, 42, 51–59]. These sources rarely give the detailed information required for design purposes. In practice the designer makes a preliminary selection from the available data and then measures sample detectors in his own laboratory in order to get complete design information applicable to his particular circuit. It is for this reason that Chapter 9 contains a detailed discussion of methods of measuring the characteristics of detectors.

A designer must be particularly inquisitive about whether the data presented pertain to a detector that he can buy. The manufacture of detectors is still largely an art, and detector characteristics vary accordingly. There is, of course, a strong temptation for manufacturers to report their best efforts and to leave it to the potential customer to find out that "we made a detector like that once, but we have never been able to duplicate it." To provide a uniformly reliable source of detector data for the Department of Defense, the "Joint Services Infrared Sensitive Element Testing Program" has been conducted since 1952 by the U.S. Naval Ordnance Laboratory at Corona, California. This agency publishes test reports that are available to qualified personnel. The methods of measurement and of presenting data followed by this laboratory have been carefully standardized and are reported in the literature [60]. Despite the unquestioned excellence of their measurements, we must assume that manufacturers submit only their best detectors for evaluation.

Some of the more important characteristics of various detectors are given in Table 7.1. The data it contains were obtained from many sources. Since it is not intended to present data peculiar to the product of a particular manufacturer, many of the entries show a range of values; as a result, the reader should have a better indication of what he can expect to find in practice. Not all of the detectors shown are commercially available at present; some are listed only because of their historical interest. Most of the column headings are self-explanatory. The third column indicates the useful wavelength range for each detector. The wavelengths shown are those between which D^* exceeds 20 per cent of its peak value; this is an arbitrary, but still useful, criterion. Values of D^* are given for a 500°K blackbody and at the wavelength of peak response. The chopping frequency is indicated in parentheses after each value of D^*. The thermal detectors are shown separately from the

INTRODUCTION TO DETECTORS 289

photon detectors. Detectors in each group are further classified in accordance with their operating temperature. For an ideal thermal detector, that is, one that responds equally to all wavelengths, the blackbody and the peak D^* values are identical. The difference between the two values given for a particular detector is a measure of its deviation from a truly black detector.

Curves of D^* versus wavelength are shown in Figure 7.5 for many of the detectors listed in Table 7.1. Also shown is the theoretical limit of D^* for photoconductive and photovoltaic detectors viewing a hemispherical surround at a temperature of 300°K. The significance of this limit will be discussed more fully in Chapter 10. For the moment, however, note how many detectors are already quite close to the theoretical limit; several are within a factor of 2 and almost all are within a factor of 10 of this limit. The reader should note this fact carefully, because from time to time widespread publicity (usually generated in sales departments) is given to supersensitive detectors allegedly thousands of times more sensitive than those previously available[61].

7.5 OPTICALLY IMMERSED DETECTORS

A rather simple technique for increasing the D^* of a detector is to place it in optical contact with one surface of a lens. Those readers familiar with the oil-immersion objectives used by microscopists will recognize the similarity of the two techniques. The specific application to infrared detectors occurred to several workers in the early 1950's, but it appears that priority belongs to Cary[62, 63].

The principle is most easily explained by reference to Figure 7.6. Shown in Figure 7.6a are two rays focused on a detector, one passing through its center and the other passing through its edge. In Figure 7.6b a hemispherical lens made of a material having an index of refraction n_s has been placed in front of the detector. An air gap of no more than a few thousandths of an inch separates the detector from the plane surface of the lens. Since the rays from the edge of the lens approach the detector at a steep angle, they may, as the dashed ray shows, suffer total internal reflection at the plane surface of the lens and never reach the detector. This reflection can be prevented by placing the detector in optical contact with the plane surface of the lens. Such a detector is said to be *optically immersed;* this condition can be achieved by chemically depositing the detector on the plane surface of the lens, or it can be approximated by cementing the detector to the lens with a very thin layer of cement. In Figure 7.6c the detector is in optical contact with the lens, and even highly oblique rays will reach it since there is no air space at which in-

TABLE 7.1 SOME IMPORTANT CHARACTERISTICS OF INFRARED DETECTORS (DATA ADAPTED FROM PUBLICATIONS OF VARIOUS DETECTOR MANUFACTURERS)

Detector Material	Operating Mode[a]	Useful Wavelength Range[b] (μ)	Wavelength of Peak Response (μ)	Resistance (Ohm)[c]	Time Constant (μ sec)	D^* (500°K Blackbody) cm (Hz)$^{1/2}$ W^{-1} (at frequency indicated)[d]	D^* (At Peak Response) cm (Hz)$^{1/2}$ W^{-1} (at frequency indicated)[d]
THERMAL DETECTORS							
Room-temperature operation							
Thermocouple	thermoelectric	1–40	—	1–10	25,000	$3\text{–}12 \times 10^8 (5)$	$6\text{–}15 \times 10^8 (5)$
Evaporated thermopile	thermoelectric	1–40	—	100	5,000	$1 \times 10^8 (20)$	$2 \times 10^8 (20)$
Thermistor bolometer	bolometer	0.2–40	—	$1\text{–}5 \times 10^6$	2,000	$0.3\text{–}1.2 \times 10^8 (15)$	$1\text{–}3 \times 10^8 (15)$
Ferroelectric bolometer	bolometer	1–12	—	—	—	$1.1 \times 10^8 (100)$	—
Golay cell	gas expansion	5–1000	—	—	20,000	$1\text{–}5 \times 10^8 (10)$	$5\text{–}10 \times 10^8 (10)$
Low-temperature operation							
Niobium nitride bolometer (16°K)	superconducting	—	—	0.2	550	$5 \times 10^9 (360)$	$5 \times 10^9 (360)$
Carbon bolometer (2.1°K)	bolometer	40–100	—	0.12×10^6	10,000	$4 \times 10^{10} (13)$	$4 \times 10^{10} (13)$
Germanium bolometer (2.1°K)	bolometer	5–2000	—	12×10^3	400	$8 \times 10^{11} (200)$	$8 \times 10^{11} (200)$

PHOTON DETECTORS

Detector	Type	λ range (μm)	λ peak (μm)	Area (μm²)	Impedance	D*(λ_p)	D*(peak)
Room-temperature operation							
Silicon	PV	0.5–1.05	0.84	$0.1–1 \times 10^6$	100	$10^{10}–10^{11}(90)$	$1–5 \times 10^{12}(90)$
Lead sulfide (PbS)	PC	0.6–3.0	2.3–2.7	$0.5–10 \times 10^6$	50–500	$1–7 \times 10^8(800)$	$50–100 \times 10^9(800)$
Indium arsenide (InAs)	PV	1–3.7	3.2	20	~1	$1–3 \times 10^8(900)$	$3–7 \times 10^9(900)$
Indium arsenide (InAs)	PEM	1.4–3.8	3.4	—	~1	—	$6 \times 10^9(1000)$
Lead selenide (PbSe)	PC	0.9–4.6	3.8	$1–10 \times 10^6$	2	$0.7–2 \times 10^8(800)$	$1–4 \times 10^9(800)$
Indium antimonide (InSb)	PEM	0.5–7.5	6.2	20	~0.1	$0.8 \times 10^8(1000)$	$0.3 \times 10^9(1000)$
Operation at 195°K							
Lead sulfide (pbS)	PC	0.5–3.3	2.6	$0.5–5 \times 10^6$	800–4000	$0.7–7 \times 10^9(800)$	$20–70 \times 10^{10}(800)$
Indium arsenide (InAs)	PV	0.5–3.5	3.2	—	~1	$1–5 \times 10^9(1800)$	$3–25 \times 10^{10}(1800)$
Indium arsenide (InAs)	PEM	1.3–3.6	3.2	—	~1	$3 \times 10^9(1000)$	$20 \times 10^{10}(1000)$
Lead selenide (PbSe)	PC	0.8–5.1	4.2	10×10^6	30	$2–4 \times 10^9(800)$	$1–4 \times 10^{10}(800)$
Indium antimonide (InSb)	PC	0.5–6.5	5.1	20	~1	$1 \times 10^9(800)$	$0.5–0.9 \times 10^{10}(800)$

[a] PV = photovoltaic, PC = photoconductive, PEM = photoelectromagnetic.
[b] Wavelengths between which D^* exceeds 0.2 of its peak value.
[c] Value shown is for a square element. For PV detectors the value is the dynamic impedance dV/dI.
[d] All D^* values given for a detector viewing a hemispherical surround at a temperature of 300°K.

TABLE 7.1 Continued.

Detector Material	Operating Mode[a]	Useful Wavelength Range[b] (μ)	Wavelength of Peak Response (μ)	Resistance (Ohm)[c]	Time Constant (μ sec)	D^* (500°K Blackbody) cm (Hz)$^{1/2}$ W^{-1} (at frequency indicated)[d]	D^* (At Peak Response) cm (Hz)$^{1/2}$ W^{-1} (at frequency indicated)[d]
Operation at 77°K							
Lead sulfide (PbS)	PC	0.7–3.8	2.9	$1-10 \times 10^6$	500–3000	$3-8 \times 10^9 (800)$	$8-20 \times 10^{10} (800)$
Indium arsenide (InAs)	PV	0.6–3.2	2.9	10×10^6	~2	$3-8 \times 10^9 (1800)$	$20-70 \times 10^{10} (1000)$
Tellurium (Te)	PC	0.7–4	3.6	1×10^3	60	$3 \times 10^9 (3000)$	$5 \times 10^{10} (3000)$
Lead telluride (PbTe)	PC	1–5.4	5.0	$50-500 \times 10^6$	~5	$1 \times 10^9 (90)$	$0.8 \times 10^{10} (90)$
Lead selenide (PbSe)	PC	0.8–6.6	5.1	$5-10 \times 10^6$	40	$2-6 \times 10^9 (800)$	$1-3 \times 10^{10} (800)$
Indium antimonide (InSb)	PV	0.6–5.6	5.1	$1-50 \times 10^3$	~1	$3-20 \times 10^9 (900)$	$3-8 \times 10^{10} (900)$
Indium antimonide (InSb)	PC	0.7–5.9	5.3	$2-10 \times 10^3$	1–10	$3-10 \times 10^9 (900)$	$2-6 \times 10^{10} (900)$
Gold-doped germanium (p-type)	PC	1–9	5.4	$0.1-10 \times 10^6$	~1	$1-3 \times 10^9 (800)$	$0.3-1 \times 10^{10} (800)$
Gold-doped germanium (n-type)	PC	1–5.5	1.5	–	50	$0.5-2 \times 10^9 (90)$	$1 \times 10^{10} (90)$
Mercury-cadmium-telluride	PV	6–15	10.6	5–50	0.01	$10^9-10^{10} (900)$	–

Operation below 50°K					
Zinc-doped germanium – Silicon alloy(Ge–Se: Zn)/(48°K)	PC	2–15	10.5	0.1×10^6 ~ 1	5×10^9 (100)	1×10^{10} (100)
Mercury-doped germanium (Ge: Hg) (30°K)	PC	3–14	11	$2\text{–}100 \times 10^3$ ~ 1	$3\text{–}9 \times 10^9$ (900)	$1\text{–}1.5 \times 10^{10}$ (900)
Cadmium-doped germanium (Ge: Cd) (28°K)	PC	6–24	17	20×10^3 ~ 1	4×10^9 (400)	1.2×10^{10} (400)
Copper-doped germanium (Ge: Cu)(4.2°K)	PC	6–29	23	$0.5\text{–}1 \times 10^6$ ~ 1	$5\text{–}10 \times 10^9$ (1800)	$1.5\text{–}3 \times 10^{10}$ (1800)
Zinc-doped germanium (Ge: Zn)/(4.2°K)	PC	7–40	37	0.3×10^6 ~ 0.01	$3\text{–}4 \times 10^9$ (800)	$0.9\text{–}1.2 \times 10^{10}$ (800)

[a] PV = photovoltaic, PC = photoconductive, PEM = photoelectromagnetic.
[b] Wavelengths between which D^* exceeds 0.2 of its peak value.
[c] Value shown is for a square element. For PV detectors the value is the dynamic impedance dV/dI.
[d] All D^* values given for a detector viewing a hemispherical surround at a temperature of 300°K.

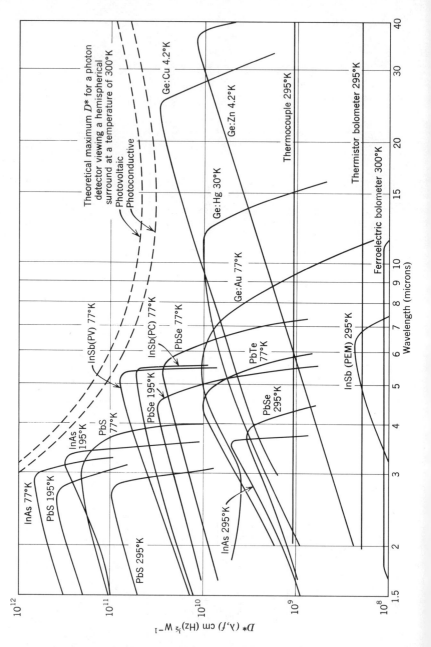

Figure 7.5 Comparison of the D^* of various infrared detectors when operated at the indicated temperature. Chopping frequency is 1800 Hz for all detectors except InSb(PEM) 1000 Hz; ferroelectric bolometer 100 Hz; thermocouple 10 Hz; thermistor bolometer 10 Hz. Each detector is assumed to view a hemispherical surround at a temperature of 300°K.

INTRODUCTION TO DETECTORS 295

ternal reflection can occur. The ray directed toward the center of the detector strikes it as before; since the surface of the lens is essentially concentric about the center of the detector, this ray is not refracted. The ray originally directed toward the edge of the detector is refracted by the lens and is focused considerably closer to the center of the detector. As a result, the original size of the detector can be reduced, and yet it will still cover the same instantaneous field of view. This reduction in size is important because in most detectors the internally generated noise is proportional to the square root of the detector area. When a concentric lens is used as an immersion element, both the linear dimensions of the detector and its noise are reduced by a factor of n_s. Since the same total flux falls on the detector, the signal is unchanged but both the signal-to-noise ratio and D^* are increased by a factor of n_s.

Figure 7.6d shows a hyperhemispherical immersion element designed to satisfy the optical aplanatic condition [64]. Here the linear dimensions of the detector are reduced by n_s^2. As a result, the signal-to-noise ratio and D^* are increased by n_s^2. This construction cannot be used with high speed optics since the aplanatic condition limits the lowest usable $f/$-number to a value of $n_s/2$. Therefore, with a hyperhemispherical element made of germanium, one cannot use optics that are faster than $f/2$.

Jones[65] discusses the theory of immersed detectors and points out that the conclusions given here are true only if the limiting noise in the output of the detector is internally generated. Fortunately this is the condition under which most detectors operate[66, 67]. If radiation or photon

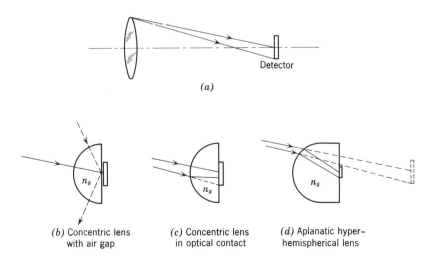

Figure 7.6 Optically immersed detectors.

noise sets the limit, there is, in general, no gain in detectivity caused by immersion.

The principal problem in fabricating an immersed detector is to achieve and maintain a true optical contact between the lens and detector. Most manufacturers of such devices seem to have this problem reasonably well under control. There are a number of requirements that the optical material of the lens must meet. It should have a high index of refraction, it should be electrically nonconductive so that it does not short out the detector, it should be chemically inert, and its coefficient of expansion should match that of the detector. Among the optical materials that have proved most successful are strontium titanate, arsenic trisulfide, and germanium.

7.6 IMAGING DETECTORS

With the exception of the image converter tube that was used in the Sniperscope during World War II, imaging detectors have seen very little application to infrared systems. At the present stage of their development, the performance of the available imaging detectors is simply not good enough to be competitive with devices using combinations of point detectors and mechanical or optical scanning. For an interesting critique of imaging detectors, see the paper by Weihe[68].

Infrared Film

Certain sensitizing dyes can be used to extend the response of photographic materials to 1.3 μ. Such sensitizings are available only in spectroscopic plates; the infrared film furnished for general photographic use does not respond beyond 0.9 μ. Unless elements of the scene are very hot, these films respond to variations in reflected illumination (as do ordinary films) rather than to thermally emitted radiation. Since many materials reflect quite differently in the near-infrared region than they do in the visible, such films have proved useful for detecting camouflage, forgeries, and alterations to works of art. The classic reference for photography by infrared is the book by Clark[69], which unfortunately has been out of print for some time. The current U.S. practice is well covered in [70] while [71] gives a valuable listing of the infrared-sensitive emulsions that are available throughout the world.

The Image Converter

An image converter tube is a photoemissive device that converts an infrared image into a visible image[72-74]. An infrared telescope using an electrostatic image converter tube[73] is shown in Figure 7.7. Either

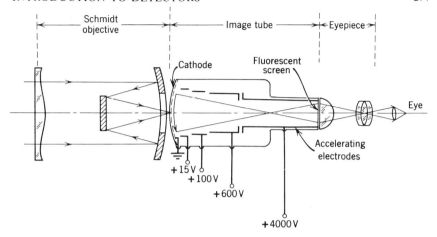

Figure 7.7 Infrared telescope using an electrostatic image tube (adapted from Morton and Flory [73]).

refractive or reflective objectives can be used; the Schmidt type shown here provides a very fast system. An image of the scene is formed on the cathode, which is a semitransparent silver-cesium oxide-cesium film with a maximum response at $0.85\,\mu$ and a cutoff wavelength of about $1.3\,\mu$. Photoelectrons leaving the cathode form an electron image of the scene that is electrostatically reimaged onto a fluorescent screen. When struck by an electron, this screen emits visible light. In this way the original infrared image is converted into a visible image. A magnifying eyepiece, which includes the glass hemisphere on the end of the tube, increases the apparent size of the image without appreciable loss in its apparent brightness.

Infrared telescopes of this sort were used during World War II for tank fire control systems and battlefield surveillance. Because of the relatively short wavelength response, most of these applications required that the scene be illuminated by an auxiliary source. The best of these were tungsten lamps fitted with filters that rejected all radiation below $0.8\,\mu$. Thus such sources could not be detected by the unaided eye. Interestingly enough, the wartime development of tungsten lamps sufficiently rugged for such illuminators gave, as a byproduct, the sealed-beam headlamp now used on automobiles.

The Vidicon

The vidicon [75–78] is a small television-type camera tube in which an electron beam scans a photoconductive target. It consists of an evacuated tube closed on one end by a flat, optically transparent faceplate. On the inner surface of the faceplate is a transparent electrically conducting film that serves as the signal electrode and as a support for a thin layer of photoconductive material. At the opposite end of the tube is an electron gun. The electron beam it produces is scanned over the photoconductive layer (usually called the target) by deflecting coils external to the tube. The signal electrode is kept at a potential that is about 20 V higher than that of the cathode. In the absence of any radiation the back of the target is kept at cathode potential by the scanning of the electron beam. Photons incident at a point on the target liberate carriers and cause local changes in the conductivity, thus reducing the potential difference between the front and back of the target. When the electron beam next scans across this point, enough electrons are deposited on the back surface of the target to restore the previous equilibrium condition. The electrical circuit is completed through the signal electrode, and the pulse generated by the charging electrons becomes the signal.

Various photoconductive materials can be used for the target. Spectral response and cooling requirements are similar to those already discussed in section 7.3 for photoconductive detectors. Redington and Van Heerden[79] report targets with response to 13 μ and give a detailed discussion of the difficulties in making targets for this region of the spectrum.

An important advantage of the vidicon is that it is a storage, or integrating, device. The potential difference between the front and back of the target at a given point is a function of the number of photons that strike the point between successive scans of the electron beam. Since there is very little lateral spreading of the resulting localized change in conductivity, only a small area, called the resolution element, is affected at any instant by the sharply focused electron beam. As a result, each resolution element integrates the number of incident photons in a complete scan period (1/30 sec in standard TV practice). A typical vidicon target has, in effect, about 10^5 resolution elements. To achieve the same theoretical efficiency with a mosaic would require 10^5 elemental detectors, each sampled at an interval of 1/30 sec. A single elemental detector could, in theory, be used with a scanning mechanism to cover the same angular field in the same scan period as the vidicon. However, a single detector would derive no benefit from integration. Since the signal-to-noise ratio is proportional to the square root of the integration time, the signal-to-noise ratio of the vidicon should be about 300 times that of the single detector.

Other well-known TV camera tubes, such as the image orthicon and the image dissector, are photoemissive devices; hence they are limited to a cutoff wavelength of 1.3 μ or less.

INTRODUCTION TO DETECTORS   299

The Photothermionic Image Converter

This device, known commercially as the Thermicon, is based on the thermal variation of photoemission. The target, called a retina in this tube, is a multilayer film a fraction of a micron thick. When the scene is imaged on the retina, a corresponding thermal image is produced. A beam from a flying-spot scanner moves across the retina and triggers the emission of photoelectrons. The number of photoelectrons emitted from any point is proportional to the temperature at this point. Garbuny et al.[80, 81] have described the Thermicon in detail. With one of their experimental tubes the detection of a large area object against a room-temperature background required a temperature differential of 10 to 12°C. The spectral response of the Thermicon extends to at least 12 μ.

The Evaporograph

The basic principal of the Evaporograph was developed in 1840 by Sir John Herschel. Since that time it has undergone a continuing series of improvements [82–84]. The Evaporograph makes use of the differential evaporation of an oil film when exposed to a heat pattern. The heart of the device is an extremely thin film blackened on one side with a material having a high infrared absorptance. The other side of the film is coated with a thin film of an oil that evaporates at a rate proportional to its temperature. An image formed on the absorbing film selectively heats the three-layer arrangement and causes a corresponding variation in the rate at which the oil film evaporates. The thermal image, which has been converted into a variable thickness oil film, can be seen by reflected light where the interference effects in the film make the image visible to the eye.

Temperature differentials as small as 1°C (at room temperature) should be detectable with the Evaporograph, although it may take a minute or more for a picture to be formed. The Evaporograph is not suited for field use because it requires a vacuum pump to keep the pressure in the oil chamber at the saturated vapor pressure of the oil.

The Infrared-Sensitive Phosphor

Phosphors are a relatively simple means of converting infrared radiation into visible light[85]. In use, a phosphor must first be excited by short wavelength radiation. After the visible afterglow has died out, the excited phosphor will emit visible light when stimulated by infrared radiation.

Phosphors underwent extensive development during World War II [86]. One of the better types used samarium-cerium activated strontium-sulfide, which gave a green glow when stimulated[87]. It had the convenient property that it could be excited by the alpha particles emitted by a small source containing a fraction of a milligram of radium. Other phosphors, using zinc sulfide, responded to 1.3 μ and beyond[88]. By cooling to liquid nitrogen temperature, the response of a selenide phosphor was extended to 2 μ [86]. Because of their simplicity and low cost, devices using phosphors are attractive to the military for the detection of infrared sources used with image converter tubes.

REFERENCES

[1] R. C. Jones, "A Method of Describing the Detectivity of Photoconductive Cells," *Rev. Sci. Instr.*, **24**, 1035 (1953).
[2] R. C. Jones, "Performance of Detectors for Visible and Infrared Radiation," in L. Marton, *Advances in Electronics*, New York: Academic, 1953, Vol. 5, p.1.
[3] R. C. Jones, "Phenomenological Description of the Response and Detecting Ability of Radiation Detectors," *Proc. Inst. Radio Engrs.*, **47**, 1495 (1959).
[4] R. A. Smith, F. E. Jones, and R. P. Chasmar, *The Detection and Measurement of Infra-Red Radiation*, London: Oxford University Press, 1957.
[5] R. DeWaard and E. M. Wormser, "Description and Properties of Various Thermal Detectors," *Proc. Inst. Radio Engrs.*, **47**, 1508 (1959).
[6] J. Strong, *Procedures in Experimental Physics*, Englewood Cliffs, N.J.: Prentice-Hall 1945, Chapter 8.
[7] J. A. Sanderson, "Emission, Transmission, and Detection of the Infrared," in A. S. Locke, ed., *Guidance*, Princeton, N.J.: Van Nostrand, 1955, Chapter 5.
[8] L. Harris and E. A. Johnson, "The Technique of Sputtering Sensitive Thermocouples," *Rev. Sci. Instr.*, **5**, 153 (1934).
[9] R. W. Astheimer and S. Weiner, "Solid-Backed Evaporated Thermopile Radiation Detectors," *Appl. Opt.*, **3**, 493 (1964).
[10] N. B. Stevens and D. D. Errett, "Thermopiles with Isothermal Collectors," *J. Opt. Soc. Am.*, **57**, 575 (1967).
[11] E. M. Wormser, "Properties of Thermistor Infrared Detectors," *J. Opt. Soc. Am.*, **43**, 15 (1953).
[12] I. M. Melman and I. M. Meltzer, "Status Report on Infrared Thermistor Detectors," *Proc. Natl. Electronics Conf.*, October 1962, p.556.
[13] J. A. Becker, "Bolometric Thermistor," U.S. Patent No. 2,414,792, January 28, 1947.
[14] E. M. Wormser and R. D. DeWaard, "Construction for Thermistor Bolometers," U.S. Patent No. 2,963,674, December 6, 1960.
[15] E. M. Wormser, "Bolometer," U.S. Patent No. 2,983,888, May 9, 1961.
[16] R. B. Barnes and E. M. Wormser, "Selective IR Detectors," U.S. Patent No. 2,891,913, April 25, 1961.
[17] D. H. Andrews, R. M. Milton, and W. DeSorbo, "A Fast Superconducting Bolometer," *J. Opt. Soc. Am.*, **36**, 354 (1946).
[18] W. S. Boyle and K. F. Rodgers, "Performance Characteristics of a New Low-Temperature Bolometer," *J. Opt. Soc. Am.*, **49**, 66 (1959).
[19] F. J. Low, "Low Temperature Germanium Bolometer," *J. Opt. Soc. Am.*, **51**, 1300 (1961).
[20] M. J. E. Golay, "Theoretical Considerations in Heat and Infrared Detection, with Particular Reference to the Pneumatic Detector," *Rev. Sci. Instr.*, **18**, 347 (1947).

INTRODUCTION TO DETECTORS

[21] N. A. Pankratov, "Nonselective Optico-Acoustic Radiation Detectors with Electrodynamic Microphone," *Optics and Spectroscopy*, **11**, 681 (1962).
[22] M. J. E. Golay, "Bridges Across the Infrared-Radio Gap," *Proc. Inst. Radio Engrs.*, **40**, 1161 (1952).
[23] W. L. Eisenman, R. L. Bates, and J. D. Merriam "Black Radiation Detector," *J. Opt. Soc. Am.*, **53**, 729 (1963).
[24] W. L. Eisenman and R. L. Bates, "Improved Black Radiation Detector," *J. Opt. Soc. Am.*, **54**, 1280 (1964).
[25] H. J. Keegan and V. R. Weidner, "Infrared Spectral Reflectance of Black Materials," *J. Opt. Soc. Am.*, **56**, 1453 (1966).
[26] L. Harris, R. T. McGinnies, and B. M. Siegel, "Preparation and Optical Properties of Gold Blacks," *J. Opt. Soc. Am.*, **38**, 582 (1948).
[27] R. Stair et al., "Some Factors Affecting the Sensitivity and Spectral Response of Thermoelectric (Radiometric) Detectors," *Appl. Opt.*, **4**, 703 (1965).
[28] P. W. Kruse, L. D. McGlauchlin, and R. B. McQuistan, *Elements of Infrared Technology*, New York: Wiley, 1962.
[29] R. Cooperstein, "Method of Forming a Photosensitive Layer of Lead Sulfide Crystals on a Glass Plate," U.S. Patent No. 3,018,236, January 8, 1962.
[30] T.H. Johnson, H. T. Cozine, and B. N. McLean, "Lead Selenide Detectors for Ambient Temperature Operation," *Appl. Opt.*, **4**, 693 (1965).
[31] B. N. McLean, "Method of Production of Lead Selenide Photodetector Cells," U.S. Patent No. 2,997,409, August 22, 1961.
[32] H. E. Spencer, "Chemically Deposited Lead Selenide Photoconductive Cells," U.S. Patent No. 3,121,022, February 22, 1964. (See also, No. 3,121,023, same title and date.)
[33] F. D. Morton and R. E. J. King, "Photoconductive Indium Antimonide Detectors," *Appl. Opt.*, **44**, 659 (1965).
[34] R. H. Bube, *Photoconductivity of Solids*, New York: Wiley, 1960, pp. 135, 136.
[35] H. Levinstein, "Extrinsic Detectors," *Appl. Opt.* **4**, 639 (1965).
[36] H. Levinstein, "Impurity Photoconductivity in Germanium," *Proc. Inst. Radio Engrs.*, **47**, 1478 (1959).
[37] E. Burstein, "Infrared Detector," U.S. Patent No. 2,671,154, March 2, 1954 (doped silicon or germanium, giving responses to 135 μ).
[38] E. Burstein, "Germanium Far Infrared Detector," U.S. Patent No. 2,816,232, December 10, 1957.
[39] E. H. Putley, "The Detection of Sub-mm Radiation," *Proc. IEEE*, **51**, 1412 (1963).
[40] E. H. Putley, "Indium Antimonide Submillimeter Photoconductive Detectors," *Appl. Opt.*, **4**, 649 (1965).
[41] K. A. Aganbekyan et al., "Receiver with an In Sb Detector for Investigating Absorption Spectra in the Submillimeter Wavelength Range," *Radio Eng. and Electron Phys.*, **7**, 1093 (1966).
[42] G. A. Morton, M. L. Schultz, and W. E. Harty, "Infrared Photoconductive Detectors Using Impurity-Activated Germanium-Silicon Alloys," *RCA Rev.*, **20**, 599 (1959).
[43] M. L. Schultz and W. E. Harty, "Ge, Si Alloy Semiconductor Detector for Infrared Radiation," U.S. Patent No. 3,105,906, October 1, 1963.
[44] P. W. Krúse, "Photon Effects in $Hg_{1-x}Cd_xTe$," *Appl. Opt.*, **4**, 687 (1965).
[45] A. Rose, "Performance of Photoconductors," *Proc. Inst. Radio Engrs.*, **43**, 1850 (1955).
[46] Bube, *op. cit.* (ref. 34), Chapter 10.
[47] J. R. Jenness, Jr., "Dual Photoconductive Infrared Detector," U.S. Patent No. 2,742,550, April 17, 1956.

[48] G. R. Pruett and R. L. Petritz, "Detectivity and Preamplifier Considerations for Indium Antimonide Photovoltaic Detectors," *Proc. Inst. Radio Engrs.*, **47**, 1524 (1959).
[49] H. J. Beaupre and C. M. Mesecke, "Infrared Detector and Method of Making Same," U.S. Patent No. 3,128,253, April 7, 1964. (diffused-junction indium antimonide).
[50] C. M. Mesecke, "Infrared Detector with PN Junction in Indium Antimonide," U.S. Patent No. 3,139,599, June 30, 1964.
[51] W. Beyen et al., "Cooled Photoconductive Infrared Detectors," *J. Opt. Soc. Am.*, **49**, 686 (1959).
[52] P. Bratt et al., "A Status Report on Infrared Detectors," *Infrared Phys.*, **1**, 27 (1961).
[53] N. C. Anderson, "Comparative Performance of Cooled IR Photoconductors," *Infrared Phys.*, **1**, 163 (1961).
[54] B. Kovit, "Infrared Detectors," *Space/Aeronautics*, **36**, 104 (November 1961).
[55] R. F. Potter and W. L. Eisenman, "Infrared Photodetectors: A Review of Operational Detectors," *Appl. Opt.*, **1**, 567 (1962).
[56] G. A. Morton, "Infrared Detectors," *RCA Rev.*, **26**, 3 (1965).
[57] R. A. Smith, "Detectors for Ultraviolet, Visible, and Infrared Radiation," *Appl. Opt.*, **4**, 631 (1965).
[58] E. H. Putley, "Solid State Devices for Infrared Detection," *J. Sci. Instr.*, **43**, 857 (1966).
[59] G. Chol et al., *Les Detecteurs de Rayonnement Infra-Rouge*, Paris: Dunod, 1960.
[60] R. F. Potter, J. M. Pernett, and A. B. Naugle, "The Measurement and Interpretation of Photodetector Parameters," *Proc. Inst. Radio Engrs.*, **47**, 1503 (1959).
[61] "Super-sensitive IR Detector," *Aviation Week*, **76**, 95 (June 11, 1962).
[62] D. S. Cary, "Optical System for Infrared Target Tracking Apparatus," U.S. Patent No. 3,002,092, September 26, 1961 (filed September 30, 1954).
[63] D. S. Cary, "Optically Immersed Photoconductive Cells," U.S. Patent No. 2,964,636, December 13, 1960 (filed June 5, 1956).
[64] A. C. Hardy and F. H. Perrin, *Principles of Optics*, New York: McGraw-Hill, 1932, p. 96.
[65] R. C. Jones, "Immersed Radiation Detectors," *Appl. Opt.*, **1**, 607 (1962).
[66] M. G. Dreyfuss, "Wedge-Immersed Thermistor Bolometer," *Appl. Opt.*, **1**, 615 (1962).
[67] S. S. Ballard and W. L. Wolfe, "Recent Developments in Infrared Technology," *Appl. Opt.*, **1**, 547 (1962).
[68] W. K. Weihe, "Classification and Analysis of Image Forming Systems," *Proc. Inst. Radio Engrs.*, **47**, 1593 (1959).
[69] W. Clark, *Photography By Infrared*, 2nd ed., New York: Wiley, 1946.
[70] *Infrared and Ultraviolet Photography*, Publication No. M-3, Eastman Kodak Co., Rochester, N.Y., 1961.
[71] S. M. Solovev, "Infrared Light-Sensitive Materials on the Market," *Zhurnal Navchnoi i Prikladnoi Fotografii i Kinematografii*, **4**, 385 (1959) (SLA: 60-16533).
[72] V. K. Zworykin and E. G. Ramberg, *Photoelectricity and its Application*, New York: Wiley, 1949, Chapters 9, 18.
[73] G. A. Morton and L. E. Flory, "An Infrared Image Tube and its Military Applications," *RCA Rev.*, **7**, 385 (1946).
[74] M. W. Klein, "Image Converters and Image Intensifiers for Military and Scientific Use," *Proc. Inst. Radio Engrs.*, **47**, 904 (1959).
[75] P. K. Weimer, S. V. Forgue, and R. R. Goodrich, "The Vidicon Photoconductive Camera Tube," *Electronics*, **23**, 70 (1950).

[76] F. A. Rosell, "Imaging-Tube Detectors," *Space/Aeronautics*, **36**, 113 (November 1961).
[77] N. L. Artem'yev, A. M. Gerasimova, and N. P. Stepchenkova, "Infrared Vidicon," *Tekhnika Kino i Televideniya*, **5**, 15 (1961) (OTS: 61-27701).
[78] N. L. Artem'yev and B. V. Kornilov, "A New Transmitting Television Tube – the Infracon," *Radio Eng. Electronic Phys.*, **10**, 1633 (1965).
[79] R. W. Redington and P. J. Van Heerden, "Doped Silicon and Germanium Photoconductors as Targets for Infrared Television Camera Tubes," *J. Opt. Soc. Am.*, **49**, 997 (1959).
[80] M. Garbuny, T. P. Vogl, and J. R. Hanson, "Image Converter for Thermal Radiation," *J. Opt. Soc. Am.*, **51**, 261 (1961).
[81] M. Garbuny, "Radiation Detector," U.S. Patent No. 3,034,010, May 8, 1962.
[82] M. Czerny, "Uber Photographie in Ultraroten," *Z. Physik*, **53**, 1 (1929).
[83] J. Lecomte, "The Evaporographic Method," *J. Phys. Radium*, **10**, 27 (1949).
[84] G. W. McDaniel and D. Z. Robinson, "Thermal Imaging by Means of the Evaporograph," *Appl. Opt.*, **1**, 311 (1962).
[85] B. Groth, "On Infrared Detectors Based on Phosphors," *Zeitschrift fur Naturforschung*, **16a**, 169 (1961) (SLA: 61-20826).
[86] B. O'Brien, "Development of Infrared Sensitive Phosphors," *J. Opt. Soc. Am.*, **36**, 369 (1946).
[87] F. Uhrbach, D. Pearlman, and H. Hemmendinger, "On Infrared Sensitive Phosphors," *J. Opt. Soc. Am.*, **36**, 372 (1946).
[88] G. R. Fonda, "Preparation and Characteristics of Zinc Sulfide Phosphors Sensitive to Infrared," *J. Opt. Soc. Am.*, **36**, 382 (1946).
[89] C. Verie and J. Ayas, "$Cd_xHg_{1-x}Te$ Infrared Photovoltaic Detectors," *Appl. Phys. Letters*, **10**, 241 (May 1, 1967).

8

Noise

In an infrared system, as in any information-transmitting system, spontaneous fluctuations that are called *noise* impose the ultimate limit on the transmission of information. Hence an understanding of noise is essential to an understanding of the limitations imposed on the performance of an infrared system. In its broadest sense the term "noise" refers to any spurious or unwanted signals in a system. Here, however, we are interested only in the random electrical fluctuations generated in circuit elements and not in externally generated effects such as static, power supply hash, and ignition noise.

Most physical quantities are noncontinuous, or granular in nature; for example, an electric current consists of a flow of electrons, each of which carries a discrete electrical charge. Variations in the rate at which electrons pass a point in a circuit are observed as variations in current or voltage and are referred to as noise. The system designer must ensure that all other types of noise are reduced to such an extent that the noise generated in the detector alone determines the ultimate limit to system performance.

Electrical noise is a random variable, the result of stochastic or random processes. At any particular instant, the amplitude of random noise is completely unrelated to its amplitude at any other time, either earlier or later. Consequently one cannot predict what the amplitude will be at a particular time. Instead one must use statistical methods to state the probability of observing specific amplitudes.

A record of the output from a random noise generator might look like that shown in Figure 8.1, which is a graph of the instantaneous voltage as a function of time. Because of the random nature of the noise, the voltage fluctuates about an average value V_{av}. How can one describe these variations? A simple average is meaningless since the mean of the fluctuations about V_{av} is zero. A measure that does have significance is the mean value of the square of the fluctuations about V_{av}, with the average taken over a time interval $\mathcal{T}$ that is much longer than the period of the

NOISE

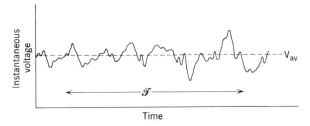

Figure 8.1 A record of random noise.

fluctuations. Mathematically, this is

$$\overline{v^2} = \overline{(v-V_{av})^2} = \frac{1}{\mathcal{T}}\int_0^{\mathcal{T}} (v-V_{av})^2\, dt, \qquad (8\text{-}1)$$

where v is the value of the voltage at time t. The quantity $\overline{v^2}$ is called the *mean-square* voltage fluctuation, and its square root $(\overline{v^2})^{1/2}$ is called the *root-mean-square* or rms voltage fluctuation.[1] The squaring operation in (8-1) removes the effect of negative values of $(v-V_{av})$ so that the mean-square and the rms are always positive quantities.

If two or more independent noise sources are present in a circuit, their net effect is found by adding their mean-square values. Since mean-square values are proportional to power, this is equivalent to saying that noise powers are additive but not noise voltages or currents. Hence, if two independent noise sources have rms noise voltages of 6 and $10\,\mu\text{V}$, respectively, their sum has an rms value of $(6^2+10^2)^{1/2}$ or $11.7\,\mu\text{V}$ rather than the $16\,\mu\text{V}$ given by simple addition. This result is true only if the noise voltages are independently generated, since phase information is lost in the mean-square calculation of (8-1). If the two noise sources are not independent, they are said to be correlated and the relative phase between them must be included in any calculations involving their sum.

There are many excellent books and papers on noise theory; a sampling of these is provided by [1–5].

8.1 TYPES OF NOISE

A complete list of all the types of noise would occupy several pages. The list given here is limited to those types that are most likely to be found in an infrared system and its component parts.

---

[1]The averaging bar is to indicate clearly that the value of the quantity is a function of the averaging time $\mathcal{T}$. Usually the value of $\overline{v^2}$ does not change for larger values of $\mathcal{T}$, but it may for shorter values.

Johnson or Thermal Noise

In 1928 Johnson showed experimentally that a resistor acts as a generator of noise having a mean-square voltage of

$$\overline{v^2} = 4kTR\Delta f, \qquad (8\text{-}2)$$

where k is Boltzmann's constant, T the temperature of the resistor in °K, R its resistance, and Δf the electrical bandwidth of the associated circuit. If the resistor is at room temperature ($T = 290°K$), then

$$\overline{v^2} = 1.60 \times 10^{-20} R\Delta f. \qquad (8\text{-}3)$$

This shows that the noise voltage is independent of frequency and, for a given temperature, depends only on the circuit resistance. If the noise source is an impedance, the noise voltage depends on the resistive part of the impedance and is independent of the capacitive or inductive part.

Either of two equivalent circuits can be used for analyzing the effect of such a noise generator. Figure 8.2a shows a voltage generator in series with a noiseless resistor. The application of Norton's theorem gives the equivalent circuit shown in Figure 8.2b, which is a current source in shunt with a noiseless conductance. By noting that $v^2 = i^2 R$ and $G = 1/R$, the value of the equivalent mean-square current source is

$$\overline{i^2} = 4kTG\Delta f, \qquad (8\text{-}4)$$

which, after substitution for $4kT$, becomes

$$\overline{i^2} = 1.60 \times 10^{-20} G\Delta f. \qquad (8\text{-}5)$$

For an idea of the magnitude of the quantities involved, assume a 1-megohm resistor (at room temperature) and a bandwidth of 5kHz. The rms values of noise voltage and current are

$$(\overline{v^2})^{1/2} = 9 \times 10^{-6} \text{ v}$$
$$(\overline{i^2})^{1/2} = 9 \times 10^{-12} \text{ amp.}$$

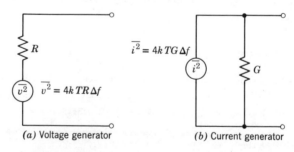

(a) Voltage generator   (b) Current generator

Figure 8.2 Equivalent circuits for noise generators.

NOISE

In order to draw the maximum power from a source, its resistance must be matched by that of the load.[1] Under matched conditions, the maximum available noise power is

$$w = \frac{1}{2}\frac{\overline{v^2}}{(2R)} = kT\Delta f. \tag{8-6}$$

Hence the maximum available noise power from a resistor is independent of its resistance and of the frequency and is a function of the bandwidth only. To generalize (8-6), write $w(f)$ for w, assume a unit bandwidth, and a resistor at room temperature; then

$$w(f) = 4 \times 10^{-21}. \tag{8-7}$$

Therefore the maximum available power from a resistor is 4×10^{-21} watts per Hz of bandwidth.

The reader may have noticed that (8-6) implies an infinite available power when the bandwidth is infinite, an obvious impossibility. The difficulty lies with the inadequacies of classical mechanics in explaining phenomena at very high frequencies. A derivation based on quantum mechanics shows that (8-6) should be written as

$$w(f) = \frac{hf}{e^{hf/kt} - 1}. \tag{8-8}$$

When hf/kT is small compared with unity, the exponential can be replaced by $1 + hf/kT$ and (8-8) reduces to (8-6). At room temperature the more exact expression need be used only at frequencies above 10^{12} Hz.

A noise source is said to be *white* if the mean-square voltage or current per unit bandwidth is independent of frequency. This is simply another way of stating that the *power spectrum*, the power per unit bandwidth, is *flat*, that is, independent of frequency. Thus it is evident from (8-2) that Johnson noise is white noise and from (8-6) that its power spectrum is flat.

Johnson noise occurs in all conducting materials. It is a consequence of the random motion of electrons through a conductor. The electrons are in constant motion, but they collide frequently with the molecules of the substance. Each free flight of an electron constitutes a minute current. The sum of all of these currents taken over a long period of time must, of course, be equal to zero. Over a short interval, the sum of all these currents is Johnson noise. Since the mean-square velocity of the electrons is proportional to absolute temperature, it is clear that Johnson noise

---

[1] More generally, the maximum-power transfer theorem states that the load impedance shall be the complex conjugate of the source impedance; that is, a source having an impedance of $R + jX$ transfers maximum power to a load of $R - jX$.

should vary with temperature; it is, in fact, often called thermal noise. Johnson noise is unaffected by current, from an external source, flowing through the conductor.

Shot Noise

The stream of electrons in a vacuum tube exhibits a noise caused by random fluctuations in the rate of arrival of the electrons at the collecting electrode. Schottky, who pointed out this effect in 1918, likened it to the noise of a hail of shot striking a target; hence the name *shot noise*.

The simplest form of shot noise is that observed in the temperature-limited diode. The circuit and the current-voltage characteristic of such a diode are shown in Figure 8.3. When the plate-to-cathode voltage is high enough, the plate current is a function of the emission, and hence of the temperature of the cathode. Under such temperature-limited conditions the mean-square current fluctuations are

$$\overline{i^2} = 2eI_{dc}\Delta f, \qquad (8\text{-}9)$$

where e is the charge on the electron (1.6×10^{-19} coulomb) and I_{dc} is the dc current flowing through the diode. For an idea of the magnitudes involved, assume that I_{dc} equals 2×10^{-3} amp and Δf is equal to 10 kHz; then $(\overline{i^2})^{1/2} = 2.5 \times 10^{-9}$ amp. If this current flows through a 10-kilohm resistor, the rms noise voltage is 25 μv. Since the Johnson noise in the resistor is 1.26 μv, its contribution to the total circuit noise is negligible.

It is evident from (8-9) that the power spectrum of shot noise is flat. This conclusion is true up to frequencies at which it is no longer possible to ignore the transit time of the electrons through the tube; this is a limitation of no particular interest to us.

Most vacuum tubes are used in a space-charge-limited rather than in a temperature-limited condition, and the shot noise is less than that given

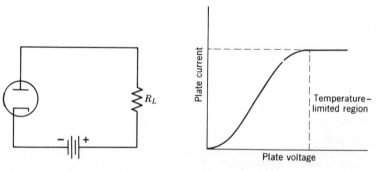

Figure 8.3 The temperature-limited diode.

NOISE 309

by (8-9) (a topic that will be discussed further in Chapter 12, where it will be related to the design of low-noise preamplifiers). The temperature-limited diode is, however, widely used for measurement of the noise characteristics of amplifiers.

Partition Noise

Partition noise is found in multigrid vacuum tubes and is caused by random fluctuations in the division of current between the screen grid and the plate. It is the reason that the noise in the output of a pentode is always higher than that of a triode. The power spectrum of partition noise is flat.

1/f Noise

At low frequencies there are several types of noise for which the power spectrum varies inversely with frequency. Since this $1/f$ noise is greater than, or in excess of, shot noise, it is sometimes called *excess noise*. A variety of names are used to designate the $1/f$ noise associated with specific devices; it is called *modulation noise* in semiconductors, such as transistors and photon detectors; *contact noise* in carbon resistors and their electrical contacts; and *flicker noise* in vacuum-tube cathodes. In most devices the $1/f$ noise becomes negligible with respect to other types of noise at frequencies above a few hundred Hz.

There is no simple expression for the mean-square voltage or current due to $1/f$ noise. However, the power spectrum can be described by an expression having the general form

$$w(f) = \frac{cI_{\text{dc}}^{\alpha}}{f^{\beta}}, \qquad (8\text{-}10)$$

where c is a proportionality constant, I_{dc} is the dc current flowing through the device, f is the frequency, and α and β are characteristic of the particular device. In most cases the value of α is very nearly equal to 2 whereas that of β can range from about 0.8 to 1.5.

Obviously, the $1/f$ characteristic cannot hold down to zero frequency, because (8-10) would require an infinite noise power. It is rather remarkable, however, that measurements on some semiconductors have shown that the $1/f$ characteristic extends to frequencies as low as 10^{-4} Hz. For further details on the theory of $1/f$ noise, the reader can consult [1] and [5].

Generation-Recombination Noise

In semiconductors (including most photon detectors), the major source of noise at intermediate frequencies, that is, those above which $1/f$

noise is negligible, is generation-recombination (gr) noise. It is due to fluctuations in the rate at which charge carriers are generated and recombined; it is therefore the semiconductor counterpart of the shot noise found in vacuum tubes. The power spectrum of gr noise is flat from low frequencies up to a frequency that is approximately equal to the inverse of the free carrier lifetime. Above this frequency the power spectrum falls off at approximately 6 dB per octave.

Radiation or Photon Noise

Even if all of the previously discussed sources of noise could be eliminated, there would still be noise in the output of a detector because of fluctuations in the incident flux. At the present state of the detector art, other types of noise predominate and radiation or photon noise is not observed. The power spectrum of radiation noise is flat. It appears now that the first application in which the performance of an infrared detector will be limited by radiation noise may be a space application using a radiation-shielded cooled detector and cooled optics.

Temperature Noise

Temperature noise is observed only in thermal detectors; it is caused by fluctuations in the temperature of the detector that are due, in turn, to fluctuations in the rate at which heat is transferred from the detector to its surroundings. The power spectrum is flat.

Summary — Noise in Detectors

The various types of noise found in detectors can be summarized by the generalized noise spectrum in Figure 8.4 (the absolute values of the

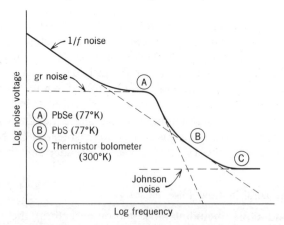

Figure 8.4 Generalized detector noise spectrum.

NOISE 311

frequency and noise voltage scales will be different for each type of detector). At low frequencies, $1/f$ noise is dominant. At intermediate frequencies, generation-recombination noise predominates, but it eventually gives way to $1/f$ noise at still higher frequencies. At very high frequencies, the noise spectrum flattens out and Johnson noise predominates [6]. For a given type of detector, only a portion of this generalized curve is observed in normal usage. For example, the noise at point A is typical of that observed with a lead selenide detector cooled to 77°K, and the lower $1/f$ portion is rarely seen. With lead sulfide the shoulder of the gr noise occurs at a very low frequency and the noise is typical of that found at the lower portion of the $1/f$ curve, at point B. A thermistor bolometer operates very close to Johnson noise, at the point C.

8.2 EQUIVALENT NOISE BANDWIDTH

In this and the previous chapters, there have been several references to circuit bandwidth but no precise definition has been given. The inference has been that bandwidth is synonymous with frequency interval. For calculations involving the transmission of noise power through a network or an amplifier, it is essential to have a more precise definition of bandwidth. If the power gain of an amplifier is constant between some upper and lower frequency and zero elsewhere, the definition of its bandwidth is simple. Since most amplifiers do not have such an ideal rectangular power gain characteristic, it is necessary to define an *equivalent noise bandwidth*

$$\Delta f = \frac{1}{\mathscr{G}(f_0)} \int_0^\infty \mathscr{G}(f)\, df, \qquad (8\text{-}11)$$

where $\mathscr{G}(f)$ is the power gain at frequency f and $\mathscr{G}(f_0)$ is the maximum value of the power gain. Whenever the symbol Δf appears in this book, it is used in the sense of an equivalent noise bandwidth. When signal bandwidths are discussed, they will be designated by some other symbol. Equation 8-11, in effect, replaces the bandpass of the actual circuit with an equivalent rectangular bandpass having a constant power gain $\mathscr{G}(f_0)$. The noise power transmitted by the equivalent bandpass is equal to that transmitted by the actual circuit; see Figure 8.5. The reader is cautioned to note again that it is the *power* gain and not the voltage gain that is used in defining equivalent noise bandwidth.[1]

---
[1]Since one is more likely to have data on the voltage gain of an amplifier or network than on the power gain, the equivalent noise bandwidth can be found by plotting the square of the voltage gain versus frequency (on linear scales), finding the area under the resulting curve, and constructing a rectangle having the same height and area.

THE ELEMENTS OF THE INFRARED SYSTEM

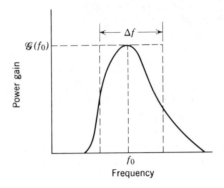

Figure 8.5 Equivalent noise bandwidth Δf.

Since the determination of equivalent noise bandwidth can be time consuming, it is of interest to see how well it is approximated by other measures. The most common definition of bandwidth is the frequency interval within which the power gain exceeds one-half of its maximum value. Since halving the power represents a change of 3 dB, it is often called the *3 dB bandwidth*. In this book, the 3 dB bandwidth will be designated by the symbol B. The difference between Δf and B is a function of the shape of the response curve of the particular circuitry. As the response curve becomes more rectangular, the difference between Δf and B tends to disappear. Table 8.1 shows a comparison of Δf and B for amplifiers with different interstage coupling networks [7].

Equivalent noise bandwidth, as defined by (8-11), assumes white noise; that is, the power spectrum of the noise is flat. Most photon detectors, however, exhibit $1/f$ noise that has a power spectrum that rises at low frequencies. If the system bandpass includes some of this $1/f$ noise, the equivalent noise bandwidth can be calculated from

$$\Delta f = \frac{1}{\mathscr{G}(f_0)\,\overline{v_0^2}} \int_0^\infty \overline{v^2}(f)\,\mathscr{G}(f)\,df, \qquad (8\text{-}12)$$

where $\overline{v^2}(f)$ is the mean-square noise voltage per unit bandwidth at the frequency f and $\overline{v_0^2}$ is the mean-square noise voltage per unit bandwidth measured at a frequency sufficiently high that the power spectrum of the noise is flat. Equation 8-12 can be used with noise having any specified departure from a flat spectrum. In general, the inclusion of $1/f$ noise within the bandpass tends to increase the equivalent noise bandwidth over that calculated for white noise.

NOISE 313

TABLE 8.1 COMPARISON OF EQUIVALENT NOISE AND
3 dB BANDWIDTHS FOR AMPLIFIERS WITH DIFFERENT
INTERSTAGE COUPLING NETWORKS[a]

Type of Coupling	Number of Stages	Equivalent Noise Bandwidth (Δf)	3 dB Band width (B)	Ratio ($\Delta f/B$)
Singly tuned	1	3.14	2.000[b]	1.57
	2	1.57	1.286	1.22
	3	1.18	1.02	1.16
	4	0.985	0.868	1.13
Doubly tuned	1	2.221	2.000[b]	1.11
	2	1.67	1.604	1.04
Triply tuned	1	2.096	2.000[b]	1.05
Quadruply tuned	1	2.038	2.000[b]	1.02
Quintuply tuned	1	2.020	2.000[b]	1.01

[a]Adapted from Lawson and Uhlenbeck [7].
[b]By definition.

8.3 THE STATISTICAL DESCRIPTION OF NOISE

Thus far the description of random noise has been in terms of mean-square values determined over periods that are long with respect to that of the individual fluctuations. For some applications it is necessary to know something about the probability distribution of peak amplitudes in random noise. Such information is essential, for example, in the interpretation of the output indications of a sensitive search system: is a given indication on the display really a target or is it merely a random noise peak?

Experiments have shown that the distribution of amplitudes in wideband random noise is accurately described by a normal, or Gaussian, distribution. The *probability density function is*

$$P(x) = \frac{1}{\sigma(2\pi)^{1/2}} \exp\frac{-(x-\bar{x})^2}{2\sigma^2}, \quad (8\text{-}13)$$

where $\bar{x}$ is the mean value of x and σ (often called the standard deviation) is the rms value of x, both determined over a period that is long with respect to that of the individual fluctuations. The probability density function gives the probability of observing a particular deviation from the mean. For convenience, the deviation is often expressed in units of σ by letting

$$t = \frac{x - \bar{x}}{\sigma} \quad (8\text{-}14)$$

and normalizing so that the area under the $p(x)$ curve is equal to unity:

$$p(t) = \frac{1}{(2\pi)^{1/2}} \exp \frac{-t^2}{2}. \qquad (8\text{-}15)$$

A curve of $p(t)$ as a function of t is shown in Figure 8.6a. The area under the curve between any two values of t represents the probability that an observation taken at random will lie between the two values chosen for t. Integrating $p(t)$ gives the *probability distribution function*

$$P(t) = \frac{1}{(2\pi)^{1/2}} \int_{-\infty}^{t} \exp \frac{-t^2}{2} dt, \qquad (8\text{-}16)$$

which is shown in Figure 8.6b. The symbol $P(t)$ represents the fraction of time that the event under observation has a value less than or equal to t. Tables of $p(t)$ and $P(t)$ are given in most engineering handbooks and in books on statistics.

How can these probability functions be used to estimate the peak values observed in random noise? First of all, the probability density curve extends infinitely in either direction along the t axis. As a consequence, *it is meaningless to describe random noise by its peak value*, because, if we are willing to wait long enough, there is theoretically no limit to the amplitude we may observe. It is convenient to consider t as a *peak factor* since it is the ratio of the instantaneous noise voltage to the (long-term) rms value of the same voltage. If $P(t)$ is evaluated between the limits of $\pm t$, a peak factor of unity, the result shows that 68 per cent of the noise peaks observed will not exceed the rms value of the noise (since by (8-14), t equals 1 when $(x-\bar{x})$ equals σ). Or, stated

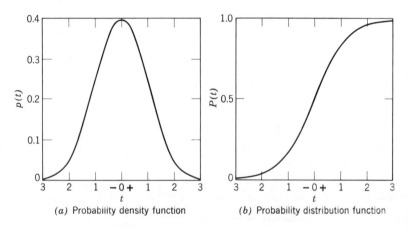

(a) Probability density function (b) Probability distribution function

Figure 8.6 Probability functions for Gaussian noise.

NOISE 315

another way, the noise peaks will exceed the rms value only 32 per cent of the time. Table 8.2 shows what per cent of the time a given peak value will be exceeded. If, for example, 1 million noise peaks are observed,

TABLE 8.2 PER CENT OF TIME A GIVEN PEAK VALUE IS EXCEEDED BY GAUSSIAN AND BY RAYLEIGH NOISE

| Peak Value | Per cent of Time Peak is Exceeded | |
(peak/rms)	Gaussian Noise	Rayleigh Noise
0.6745	50.0	—
1	31.7	60.6
1.18	—	50.0
2	4.54	13.6
3	0.26	1.11
4	0.0064	0.034

about 64 of them will have amplitudes that are greater than 4 times the rms value.

If random noise is passed through a narrow-bandpass filter, that is, one in which the bandwidth is small with respect to the center frequency, the noise at the output of the filter appears like a high-frequency carrier modulated by a low-frequency signal having a frequency that is approximately equal to the reciprocal of the 3 dB bandwidth of the filter. If this narrow-band noise is passed through an envelope detector, the output is a smooth curve through the positive peaks of the input waveform. If the input level is high enough so that the envelope detector operates over the linear portion of its characteristic, the noise peaks in the output are described by a Rayleigh distribution (for low-level inputs, the detector operates as a square-law device and the noise peaks retain their Gaussian distribution).

For a Rayleigh distribution the probability functions are

$$p(t) = \frac{t}{\sigma} \exp \frac{-t^2}{2} \tag{8-17}$$

$$P(t) = 1 - \exp \frac{-t^2}{2}. \tag{8-18}$$

These are shown in Figure 8.7. Due to the detection process, only positive values of t can occur. Peak factors for Rayleigh noise are given in Table 8.2, so that they can be compared with those for Gaussian noise.

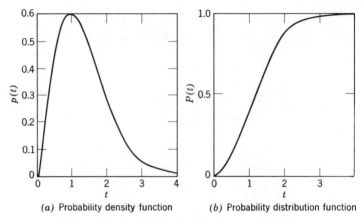

(a) Probability density function (b) Probability distribution function

Figure 8.7 Probability functions for Rayleigh noise.

8.4 METERS FOR THE MEASUREMENT OF NOISE

It is important to remember that different types of meters will, in general, give quite different readings when used to measure the rms value of random noise. There are basically three types of ac vacuum-tube voltmeters (VTVM): those responding to the peak, to the rms, or to the average value of the input waveform. Since the most common waveform is the sinusoid, these meters usually have scales calibrated to read the rms value of a sine wave. Because relatively few true-rms-responding[1] meters are found in the average laboratory, it is necessary to understand the limitations imposed by the other types [8–10].

Theoretically a meter used for measuring noise should have an infinite dynamic range, because any clipping of noise peaks will give an error in the meter reading. Unfortunately, a wide dynamic range makes it difficult to protect the meter from burnout by an inadvertent overload, and the manufacturer is forced to a compromise in the design of his meter. The extent of this compromise is given by the manufacturer's specification of a peak factor for the meter. This defines the maximum peak factor of the input waveform for which the meter reading will show no error because of its not responding to high-amplitude noise peaks. From Table 8.2 it is evident that a meter with a peak factor of 3 will ignore noise peaks that occur about 0.25 per cent of the time. The resulting error in the indicated voltage will be about 1.5 per cent (for a true-rms meter). Likewise a meter with a peak factor of 4 will ignore noise peaks that

---

[1] The word "true" is a deliberate redundancy to make certain that the reader understands that the reference is to the relatively rare type of meter that does, in fact, respond to the rms value of the input waveform.

NOISE
317

occur about 0.0064 per cent of the time, and the error in the indicated voltage will be about 0.5 per cent.

The mean-square and rms concepts introduced earlier in this chapter involve averages taken over a long peroid of time. The reader may logically ask, "How long?" A practical, but not very definitive answer is, "Long enough that the result is not appreciably different if the measurement is repeated for a longer period of time." This condition is achieved if the averaging time is at least ten times the reciprocal of the equivalent noise bandwidth of the circuit under test. The equivalent noise bandwidth of the meter should be greater than that of the circuit under test, a condition that is easy to achieve with most infrared systems. If it is not achieved, a correction is required [9].

Peak-Responding Meter

Here the advice is very simple: if at all possible, *never use a peak-responding meter to measure random noise*. Such a measurement is, at best, only a qualitative one since, in theory, there is no limitation on the observable peak value if one is willing to wait long enough. Most manufacturers of this type of meter can supply approximate correction factors if the meter must be used to measure noise.

Rms-Responding Meter

A true-rms meter is the most desirable type for making noise voltage measurements, but it is also the most expensive. Complex circuitry is required to, in effect, square the point-by-point voltage along one cycle of the waveform and then extract the square root of the average value of the squared quantity. However, commercially available meters of this type usually have a rated accuracy of ±1 per cent of full scale. Of the non-VTVM types, thermocouple, iron-vane, electrodynamometer, and electrostatic meters are all true-rms-responding. Unfortunately the sensitivity and frequency response of these types are rather limited.

Average-Responding Meter

The average-responding meter is by far the most common type of VTVM. The average value of an entire cycle of a sine wave is zero. However, the time average of the instantaneous value of the *absolute* voltage is not zero; it can be found by measuring the average value of the rectified waveform. In order to calibrate such a meter it is necessary to know the *form factor*, the ratio of the rms to the average value of the calibrating waveform. For a sinusoid the rms value is 0.707 of the peak amplitude, the average value of one-half cycle is 0.636 of the peak value and the form factor is 1.111. If the meter is calibrated to read the rms value of a sine wave, as most meters of this type are, the meter scale indicates values that are 1.111 times the average value to which the meter actually responds. For Gaussian noise, the form factor is 1.253 and the meter indication must be multiplied by 1.253/1.111 or 1.128 in order to obtain the rms value of the Gaussian noise. For Rayleigh noise the form factor is 1.128 and the meter indication must be multiplied by 1.128/1.111 or 1.015.

8.5 NOISE FIGURE

In discussing the problem of detecting a signal in the presence of noise, it is particularly convenient to speak of the ratio of the signal to the noise. Since the noise is random, it is impractical to speak of the instantaneous ratio of signal to noise. Instead we must use mean-square values that are obtained by averaging over a relatively long period of time. If both the signal and the noise are measured at the same point in the circuit, so that they appear across the same impedance, the ratio of the mean-square voltages is a power ratio that is called the *signal-to-noise (power) ratio*

$$\frac{S}{N} = \frac{\overline{v_s^2}}{\overline{v_n^2}}, \tag{8-19}$$

where $\overline{v_s^2}$ and $\overline{v_n^2}$ are the mean-square voltages due to the signal and the noise, respectively.

An ideal amplifier contributes no noise to signals passing through it. With such an amplifier, the signal-to-noise ratio measured at the output is identical to that measured at the input. Unfortunately all physically realizable amplifiers generate some internal noise. As a result the signal-to-noise ratio at the output is less than that at the input. Friis[11] proposed that the ratio of these two be called the *noise factor*:

$$F = \frac{S_i/N_i}{S_o/N_o}, \tag{8-20}$$

where the subscripts i and o indicate input and output, respectively. Thus noise factor is a convenient measure of the internal noise contributed by an amplifier. Since it is a ratio of powers, it is often convenient to express noise factor in decibels (10 log F); it is then called the *noise figure*. For an ideal noiseless amplifier, the noise factor is unity and the noise figure is zero.

If two devices are connected in series, such as two stages of an amplifier, it is not difficult to show that the over-all noise factor is

$$F = F_1 + \frac{(F_2 - 1)}{\mathcal{G}_1}, \tag{8-21}$$

where F_1 and F_2 are the noise factors of the first and second stages, respectively, and $\mathscr{G}_1$ is the power gain of the first stage. Hence, by making the power gain of the first stage high, the noise from the second stage contributes little to the over-all noise factor. Since most designers prefer to speak of noise figure rather than noise factor, (8-21) is a bit tedious to use since decibels must be converted to ratios that are then substituted in the equation and finally converted back to decibels. A nomogram published by Hudson[12] eliminates the need for these conversions and makes the calculation very simple.

In our discussion thus far, no particular limitations have been placed on bandwidth. As defined, the noise factor is, in effect, an average noise factor, where the average is taken over the entire equivalent noise bandwidth of the device. Alternatively one can define a *spot noise factor*, which is the noise factor measured in a unit bandwidth centered about a specific frequency.

There are some very subtle points involved in the derivation of the expression for the noise factor. The reader interested in delving further into this subject is referred to the excellent review paper by Goldberg[13], the paper by Cohn[14] that emphasizes some of the pitfalls in applying the theory to nonlinear devices, and the delightfully tongue-in-cheek paper by Greene[15] entitled "Noisemanship—The Art of Measuring Noise Figures Nearly Independent of Device Performance."

REFERENCES

[1] A. van der Ziel, *Noise*, Englewood Cliffs, N.J.: Prentice-Hall, 1956.
[2] J. R. Pierce, "Physical Sources of Noise," *Proc. Inst. Radio Engrs.*, **44**, 601 (1956).
[3] W. R. Bennett, "Methods of Solving Noise Problems," *Proc. Inst. Radio Engrs.*, **44**, 609 (1956).
[4] M. Schwartz, *Information Transmission, Modulation, and Noise*, New York: McGraw-Hill, 1959.
[5] W. R. Bennett, *Electrical Noise*, New York: McGraw-Hill, 1960.
[6] R. C. Jones, "Noise in Radiation Detectors," *Proc. Inst. Radio Engrs.*, **47**, 1481 (1959).
[7] J. L. Lawson and G. E. Uhlenbeck, *Threshold Signals*, New York: McGraw-Hill, 1950, p. 177.
[8] J. J. Davidson, "Average vs. RMS Meters for Measuring Noise," *IRE Trans. Audio*, AU-9, 108 (1961).
[9] E. Uiga, "How to Measure Noise with a VTVM," *Electronic Equipment Eng.*, **10**, 87 (October 1962).
[10] H. L. Roberts, "Which AC Voltmeter," *Electronic Design News*, **8**, 20 (November 1963).
[11] H. T. Friis, "Noise Figures of Radio Receivers," *Proc. Inst. Radio Engrs.*, **32**, 419 (1944).
[12] A. C. Hudson, "Nomogram for the Noise Figure of Two Cascaded Stages," *Microwave J.*, **7**, 84 (October 1964).

[13] H. Goldberg, "Some Notes on Noise Figures," *Proc. Inst. Radio Engrs.*, **36**, 1205 (1948).
[14] S. B. Cohn, "The Noise Figure Muddle," *Microwave J.*, **2**, 7 (March 1959).
[15] J. C. Greene, "Noisemanship—The Art of Measuring Noise Figures Nearly Independent of Device Performance," *Proc. Inst. Radio Engrs.*, **49**, 1223 (1961).

9

The Measurement of Detector Characteristics

Any organization that hopes to be successful in the infrared system business will soon find it necessary to establish its own facility for the measurement of detector characteristics. For a preliminary design, data published by manufacturers or those given in Section 10.4 are adequate. For a final design, one must know the characteristics of the detector as measured under the conditions peculiar to his application. Accordingly, this chapter describes the types of measurements to be made, the equipment needed, and the procedures for using it.

There will, however, be a departure from normal, in that some items of test equipment will be identified by manufacturer and model number. Identification of a particular device is not intended as an endorsement but simply as a convenience to the reader. Those familiar with the instrumentation field will recognize the instruments mentioned and will realize that, in most cases, there are instruments made by other manufacturers that will give comparable performance. For the newcomer, such identification may save many hours of searching for the supplier of some of the special items required.

9.1 QUANTITIES TO BE MEASURED

The various figures of merit used to describe the performance of a detector were discussed in Section 7.1. For convenience, they are summarized in Table 9.1. Inspection of this table shows that only five quantities need be known in order to calculate any of these figures of merit: the signal and the noise voltages, the equivalent noise bandwidth of the measuring circuit, the area of the detector, and the irradiance.

In practice the experimenter measures additional quantities in order to determine other characteristics of the detector. He also should record certain details describing the physical nature of the detector and the

THE ELEMENTS OF THE INFRARED SYSTEM

TABLE 9.1 FIGURES OF MERIT USED TO DESCRIBE THE PERFORMANCE OF A DETECTOR

Figure of Merit	Equation	Units
Responsivity	$R = \dfrac{V_s}{HA_d}$	VW^{-1}
Noise equivalent power	$NEP = HA_d \dfrac{V_n}{V_s}$	W
Noise equivalent irradiance	$NEI = \dfrac{NEP}{A_d}$	$W\,cm^{-2}$
Detectivity	$D = \dfrac{1}{NEP}$	W^{-1}
D-star	$D^* = \dfrac{(A_d \Delta f)^{1/2}}{NEP}$	$cm(Hz)^{1/2}W^{-1}$

A_d = Detector area, cm^2
H = Irradiance, $W\,cm^{-2}$
  (rms value of the fundamental component)
V_n = rms noise voltage
V_s = rms signal voltage (fundamental component)
Δ_f = Equivalent noise bandwidth, Hz.

conditions of measurement. Most laboratories find it worthwhile to prepare a data sheet for recording all quantities. As a guide to the contents of such a data sheet, the reader can consult Table 9.2 or refs. 1 and 2.

9.2 THE BASIC DETECTOR TEST SET

A block diagram of a basic test set for infrared detectors is shown in Figure 9.1. With it we can measure signal and noise voltages, time constant, frequency response, and detector resistance. From these we can calculate blackbody responsivity and D^*.

If one has the money to buy at one time all of the units shown in Figure 9.1, he should seriously consider the purchase of one of the commercially available test sets[3]. It is more likely that he will wish to start with a minimum amount of equipment and add units as the need arises. For this reason, the function of each block in Figure 9.1 will be examined in order that the reader may better decide which units are essential for his needs.

Blackbody Source and Blackbody Temperature Controller. The design, construction, and calibration of these units were covered in Section 3.1. Most measurements will be made at a blackbody temperature of 500°K, although there may be an occasional need for temperatures as high as

TABLE 9.2 QUANTITIES THAT MIGHT BE REPORTED ON A DETECTOR DATA SHEET (ADAPTED FROM [1] AND [2])

A. Description of Detector
Type of material
Manufacturer
Date of manufacture
Serial number
Window material
Characteristics of any integral optical filter
Size and nominal area of responsive element, cm, cm^2
Distance from window to plane of responsive element, cm
Field of view (if other than 180°)
Date of this test

B. Description of Test Conditions
Temperature of responsive element, °K
Bias current, μamp
Detector load resistor or type of coupling transformer
Temperature of blackbody source, °K
Chopping frequency, Hz
Irradiance at plane of responsive element, W cm^{-2}(rms)
Ambient radiation on detector, °K
Equivalent noise bandwidth of measuring circuitry, Hz
Ambient temperature, °C
Ambient relative humidity, per cent

C. Test Results
Dark resistance, ohms
Dynamic resistance, ohms
Signal voltage, rms volts
Noise voltage, rms volts
Wavelength of peak responsivity,
Chopping frequency for maximum D^*, Hz
Time constant, sec

D. Calculated Quantities
Responsivity, V W^{-1}
D^* (blackbody), cm(Hz)$^{1/2}$ W^{-1}
D^* (peak), cm(Hz)$^{1/2}$ W^{-1}
Ratio of D^* (peak) to D^* (blackbody)

900°K. Suitable commercial units include the Barnes Model 11-110, the Electro Optical Industries Model 141 blackbody and Model 202 controller, and the Infrared Industries Model 403 blackbody and Model 101 controller.

Limiting Apertures. These are used to provide an accurately known radiating area for the blackbody. Problems in their fabrication and maintenance have already been discussed in Section 3.1. Several apertures should be available, ranging from 0.25 to 5 mm in diameter. The commercial blockbodies listed above include such an assortment of apertures, arranged on a wheel so that they can be changed easily. If detectors having a response beyond 6 μ are to be measured, it will probably be necessary to cool the structure supporting the apertures.

Variable-Speed Chopper, Motor, and Speed Controller. At first glance this appears to be a relatively simple device; unfortunately, appearances are deceiving. Problems encountered with these units include difficulties in achieving the desired waveform of the chopped pulse, in maintaining constant chopping frequency, and in eliminating the transient noise

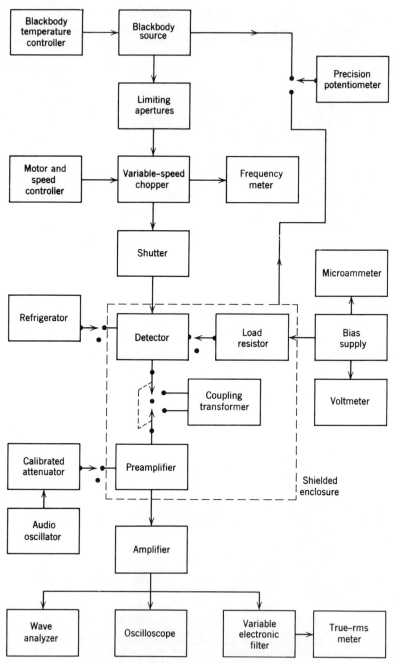

Figure 9.1 Basic test set for infrared detectors.

THE MEASUREMENT OF DETECTOR CHARACTERISTICS   325

spikes generated by motor-speed controllers. In testing photon detectors, it is desirable to have available chopping frequencies from 100 to 10,000 Hz. If thermal detectors are to be measured also, it is necessary that this range extend down to 5 Hz.

Knowledge of the waveform generated by the chopper is of vital importance. One of the assumptions made in Chapter 7 was that the irradiance on the detector would be specified in terms of the rms value of its fundamental component. If the chopping waveform is sinusoidal, this requirement is simple to meet. Unfortunately, choppers generating accurate sinusoidal waveforms are not easy to make [4–6]. It is relatively easy, however, to generate a square waveform. A Fourier analysis of a square wave shows that it consists of a fundamental and of odd-order harmonics. The rms value of the fundamental component is 0.45 times the peak-to-peak amplitude of the square wave. Hence the rms value of the fundamental component of the irradiance, when using a square-wave chopper, is

$$H = 0.45 \frac{\sigma T^4 A_s}{\pi d^2}, \qquad (9\text{-}1)$$

where T is the temperature of the blackbody source (°K), A_s is the area of the limiting aperture in front of the source (cm²), d is the distance from the limiting aperture to the plane containing the responsive element (cm), and σ is the Stefan-Boltzmann constant.

The limiting apertures at the source are nearly always circular. Chopping a circular aperture with a series of straight-sided slots cut in the rim of a disk gives an excellent approximation to a square wave, provided that certain precautions are observed. Figure 9.2 shows the geometry

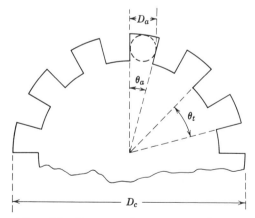

Figure 9.2   Geometry of a square-wave chopper.

involved. The quantity χ is defined as the ratio of the angles subtended by the aperture and by a tooth-slot pair,

$$\chi = \frac{\theta_a}{\theta_t}. \tag{9-2}$$

When $\theta_a \ll \theta_t$, the value of χ approaches zero and a square waveform is generated. Similarly, when θ_a is equal to $\theta_t/2$, the value of χ is 0.5 and a triangular waveform is generated. The rms value of the fundamental component of a triangular wave is 0.28 times its peak-to-peak amplitude. Thus, depending on the design of the chopper the rms value of the fundamental component, which will be called the *rms conversion factor*, may vary from 0.28 to 0.45 times the peak-to-peak value. The value of this rms conversion factor is given in Table 9.3 for various values of χ. For design purposes, it is convenient to rewrite (9-2) as

$$\chi = \frac{nD_a}{\pi(D_c - D_a)}, \tag{9-3}$$

where n is the number of teeth (or slots) around the disk, D_a is the diameter of the limiting aperture at the blackbody, and D_c is the diameter of the chopper disk.

In calculating the irradiance at the detector, most workers assume a square waveform and design their choppers so as to introduce no more than a stated error. Note, for instance, from Table 9.3, that at χ equal to 0.08, the rms conversion factor differs from that for a true square wave by 1 per cent. Substituting this value of χ in (9-2) gives an expression

TABLE 9.3 VALUES OF THE RMS CONVERSION FACTOR FOR VARIOUS CHOPPER GEOMETRIES[a]

χ	rms Conversion Factor
0 (square wave)	0.450
0.05	0.448
0.08	0.445
0.10	0.442
0.15	0.433
0.20	0.421
0.25	0.405
0.30	0.386
0.40	0.340
0.50 (triangular wave)	0.286

[a]Courtesy of Santa Barbara Research Center.

for the maximum number of teeth in a chopper that will have an rms conversion factor differing by less than 1 per cent from that of a true square-wave chopper,

$$n = 0.251 \frac{(D_c - D_a)}{D_a}. \qquad (9\text{-}4)$$

Assume, for example, a chopper with a diameter of 15 cm (6 in.) and an aperture with a diameter of 0.1 cm (0.040 in.); from (9-3) it is evident that this chopper must be limited to 37 teeth, or less, in order that the error in the assumption of a square wave may be less than 1 per cent. When used with a motor having a maximum speed of 166 rps (10,000 rpm) this chopper can generate a maximum chopping frequency of 6142 Hz.

Breakage of a chopper blade during operation can be a serious hazard to nearby personnel. Good design practice demands that a sturdy guard be used to enclose the blade completely. During the design of a chopper it is well to calculate the stresses it will experience at the maximum rotational rate and to include a generous safety factor. Circular saw blades are an excellent source of disks for choppers. For safety the teeth should be ground off; this must be done carefully so that the dynamic balance of the disk is not spoiled. Several manufacturers will sell the disks without teeth, thus simplifying their conversion to a chopper.

Maintaining an adequately constant chopping frequency has long been a vexing problem. Most workers make their measurements with a wave analyzer, which is essentially a voltmeter with a very narrow bandwidth (5 to 10 Hz) tunable to any desired frequency. Typical variable-speed motors have a speed stability of about ±1 per cent. At a chopping frequency of 1000 Hz, this results in a variation of ±10 Hz. Trying to keep a wave analyzer having a 5 Hz bandwidth tuned to such a variable chopping frequency has caused more than one head of hair to turn prematurely gray. One way to get a series of stable chopping frequencies is to use a synchronous motor with a gear or belt drive to the chopper. A far better solution is to use one of the newer wave analyzers that incorporate automatic frequency control.

Many motor speed controllers use silicon-controlled rectifiers that, depending on the circuit design, may generate all sorts of transients that will inevitably find their way into other parts of the test-set circuitry. Careful design, shielding, and filtering can minimize these transients, but not all manufacturers do this job as well as they might. Many workers, blessed with such noisy controllers, find themselves forced to momentarily turn off the motor and controller each time they make a measurement. It is well to note that some blackbody temperature controllers also generate transients that are equally difficult to eliminate.

The Electro Optical Industries Model 311 variable-speed chopper provides chopping frequencies ranging from 3 to 5000 Hz. The Infrared Industries Model 801 chopper has five blades that, in conjunction with a variable-speed motor, provide chopping frequencies ranging from 2 to 30,000 Hz.

Frequency Meter. Used to measure and to monitor the chopping frequency. Sometimes a frequency meter is included in the motor-speed controller. If not, the Hewlett-Packard Model 506A optical tachometer and either the Model 500C frequency meter or a digital counter can be used.

Shutter. An opaque shutter is used to block the flux from the blackbody in order that the noise from the detector can be measured in the absence of an input signal.

Detector. The principal requirement here is a means of holding the detector during test. Such items are usually custom made to fit the specific needs of the experimenter. It is particularly convenient if the detector mount can be moved either toward or away from the blackbody source. In this way detectors can be so positioned that the distance from the limiting aperture to the plane of the responsive element remains constant and one calculation of irradiance is valid for all measurements.

Load Resistor, Bias Supply, and Coupling Transformer. These blocks are intended to represent a variety of solutions to the problems of establishing the proper operating point for a detector and the means of coupling it to a preamplifier. These problems are discussed in detail in Section 9.3.

Preamplifier. The preamplifier is one of the most critical components in the entire test set. Unfortunately no single preamplifier can provide optimum performance with all of the available detectors. The primary requirement is for a noise figure that is so low that the noise from the detector is the limiting source of noise in the system. A low output impedance is desirable in order to permit the use of low-impedance connecting cables. Most designers prefer to build their own preamplifiers, although suitable ones can be purchased from various manufacturers of detectors. Additional information on preamplifiers will be found in Section 12.2.

Shielded Enclosure. At the very minimum, this enclosure should be a light-tight box painted black on the inside so as to shield the detector from the ambient illumination. In some test sets the blackbody source and chopper are also placed within this box. Some workers believe that an RF-tight shielded enclosure should be used, such as a metal box with metallic gasketing around all openings and with appropriate filtering for all leads that enter and leave the enclosure. Individual workers will have to find experimentally just how much shielding is required in their particular location.

Refrigerator. Many types of detectors must be cooled; see Chapter 11 for further details.

Calibrated Attenuator and Audio Oscillator. These provide a calibrating voltage of known amplitude and frequency for determining the voltage gain between the detector and the output-indicating device. A General Radio Type 546-C Audio Frequency Microvolter is a calibrated attenuator that can be used with an audio oscillator, such as the Hewlett-Packard Model 200CD, to provide accurately known audio-frequency voltages.

Precision Potentiometer. This, in conjunction with a thermocouple or a platinum resistance thermometer, is used to measure the temperature of the blackbody cavity and of the detector compartment. Either the Honeywell Model 2745 or the Leeds and Northrup Model 8686 is a suitable choice.

Amplifier. The level of the signal from the preamplifier is usually too low for direct application to the final readout device. Since most chopping frequencies lie in the audio range, any good audio amplifier can be used. Particularly recommended are amplifiers such as the Keithley Type 103R or the Tektronix Type 122 that permit a choice of upper- and lower-frequency cutoff characteristics. In the event that the Tektronix Type 122 is used, the reader may wish to make the low noise modifications that have been described in the literature [7, 8].

Oscilloscope. This is used to ensure that no overloading occurs at any point in the test set. Any general-purpose oscilloscope is suitable.

Wave Analyzer. This is a narrow-bandwidth voltmeter tunable from about 20 Hz to 50 kHz. The equivalent noise bandwidth varies with manufacturer but it is usually about 6 Hz. If at all possible, choose a wave analyzer that has automatic frequency control (see the remarks on this subject in the discussion of the variable-speed chopper). Since these instruments are not true-rms responding, the corrections discussed in Section 8.4 must be applied when noise voltages are read. Suitable wave analyzers are the General Radio Type 1900-A and the Hewlett-Packard Model 302A.

Variable Electronic Filter. It is often convenient to be able to select a particular bandwidth for the measuring circuit so that the noise can be measured in a bandwidth approximating that of the final system. Such control of the bandwidth can be had with the amplifiers mentioned above or with a variable electronic filter, such as the Spencer-Kennedy Laboratories Model 320.

True-rms Meter. This meter is used to measure both signal and noise. For the reasons set forth in Section 8.4, it should be of the true-rms-responding type. Suitable examples are the Ballantine Model 320, the Fluke Model 910A, the Hewlett-Packard Model 3400A, and the Keithley Model 121.

9.3 USE OF THE BASIC DETECTOR TEST SET

Let us suppose, for purposes of illustration, that we have been asked to measure the characteristics of a lead selenide detector when it is cooled to an operating temperature of 77°K.

Measurement of Detector Area

Since most detector elements are in a sealed package, their dimensions must be determined optically, if at all. A toolmaker's microscope is the best instrument for this job. After the linear dimensions of the detector have been determined, the same instrument, provided it has a calibrated vertical motion (that is, one that is parallel to the optical axis of the microscope), can be used to measure the distance from the protective window to the detector element. One focuses the microscope on the surface of the window (a few dust particles will make it easy to find), notes the scale reading, then shifts the focus to the detector element and again notes the scale reading. The difference between the two readings is the distance from the window to the element. For highest accuracy, it is necessary to make a small correction for the optical thickness of the window. Other alternatives for measuring the dimensions of a detector include an optical comparator or a microscope equipped with a micrometer eyepiece. If the window on the detector package is not visually transparent, there is little that one can do other than accept the values of area and distance that are given by the manufacturer.

Determining the Operating Point of a Detector

For a discussion of the means of establishing the operating point it is convenient to divide detectors into two classes, those that require an external bias supply and those that are self generating.

THE MEASUREMENT OF DETECTOR CHARACTERISTICS 331

The detectors that require an external bias supply include photoconductors, photoemitters, and bolometers. (Although the back-biased junction diode requires a bias supply, it is a photovoltaic device and is discussed with the self-generating detectors.) These detectors are connected in series with a bias battery and a load resistor, as shown in Figure 9.3. Changes in the incident flux vary the conductivity of the detector and,

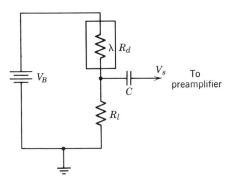

Figure 9.3 Method of applying an external bias to a photoconductive detector.

as a result, the current flowing through the circuit. The signal is taken from across the load resistor through a coupling capacitor so as not to disturb the dc conditions. The dc voltage across the load resistor is

$$V_{R_l} = V_B\left(\frac{R_l}{R_d+R_l}\right), \tag{9-5}$$

where V_B is the bias voltage, R_l is the resistance of the load resistor, and R_d is the resistance of the detector. Thus, when the resistance of the load resistor is made equal to that of the detector, one-half of the bias voltage appears across each.

When flux is incident on the biased detector, there is a change in its conductivity. The resulting change in signal voltage across the load resistor is found by differentiating (9-5), from which

$$V_s = \frac{V_B R_d R_l}{(R_d+R_l)^2}\frac{\Delta R_d}{R_d}. \tag{9-6}$$

This equation is an excellent means of analyzing the factors that lead one to a choice of bias voltage and load resistor (assume for the discussion that the value of $\Delta R_d/R_d$ remains constant). The initial criterion is to achieve a high signal level by choosing a high bias voltage. Having chosen the bias voltage, the maximum signal voltage occurs when the load resistor and detector are matched. There are at least three considerations that may force modification of these initial choices; these are (1) the maximum power that the detector can dissipate, (2) the establishment of an impedance match to the preamplifier that will result in a minimum noise figure, and (3) high frequency considerations.

As the bias voltage is raised, the detector must dissipate an increasing amount of power. In the absence of specific information from the manufacturer, a good rule of thumb is that no detector should ever exceed a power dissipation of 0.1 W cm^{-2}. Thus the maximum bias voltage that can be applied is

$$V_{B_{\max}} = \left[0.1 A_d \frac{(R_d + R_l)^2}{R_d}\right]^{1/2} \tag{9-7}$$

As an example of the magnitude of the quantities involved, a lead selenide detector when cooled to 77°K has a resistance of about 1 megohm. If this detector has an area of 1 cm^2 and is used with a matching load resistor, the maximum allowable bias is 630 V. Even with conservative design, it is unlikely that such a high value would be approached. Note, however, that one rarely finds a detector having an area as large as 1 cm^2; it is far more likely that the area will be 0.01 cm^2 or less, in which case the maximum bias voltage is 60 V.

The noise figure of a preamplifier is a function of the impedance connected across its input terminals. As is shown in Chapter 12, the optimum source impedance varies with the type and design of preamplifier. Hence the choice of preamplifier may dictate the choice of a load resistor for use with a particular detector.

Where very high chopping frequencies are to be used, it may be necessary to use a low value of load resistor. This is because most photon detectors have a distributed capacitance of 1 to 10 pF. The use of a high-value load resistor can result in an RC time constant in the input circuitry that is longer than the time constant of the detector. The only way to utilize the short time constant of the detector is to reduce the value of the load resistor. Such a mismatch reduces both the signal and the noise voltages from the detector so that the signal-to-noise ratio is not affected. However, if the mismatch is carried too far the signal may be lost in the noise of the preamplifier. The reduction in signal (or noise) caused by mismatching is less than one might guess. If the value of the load resistor is reduced to one-tenth that of the detector ($R_l = 0.1 R_d$) and the bias voltage is kept fixed, (9-6) shows that the signal voltage is reduced to one-third of the value it had for the matched condition.

Thus far the discussion has assumed that the bias voltage is held constant. It is, however, equally acceptable to assume that the bias current is fixed. With this assumption, the signal voltage is proportional to bias current, and again one must ensure that the maximum allowable

THE MEASUREMENT OF DETECTOR CHARACTERISTICS 333

power dissipation of the detector is not exceeded. Unlike the constant voltage case, the signal voltage is increased by using a value of load resistor that is greater than that of the detector. For example, if the relative signal voltage is 1 for a matched load resistor, and the bias current is kept constant, reducing the load resistor so that R_l is equal to 0.1 R_d will reduce the relative signal to 0.18. If the value of the load resistor is made infinite, the relative signal will increase to a value of 2. Obviously this condition is unattainable since the maintenance of a constant current with an infinite load resistance would require an infinite bias voltage.

Self-generating detectors include thermocouples, thermopiles, and photovoltaic and photoelectromagnetic types. Since the p-n junction or photovoltaic type is the only one of this group that is used extensively in infrared systems, it will be the subject of most of the discussion that follows. The various possible operating points for a photovoltaic detector were shown in Figure 7.3 and discussed briefly in the accompanying text. Such a detector can be operated as an open-circuit voltage generator by connecting it, through a coupling capacitor, to the terminals of a preamplifier that has a high-input impedance. Such operation is usually limited to detectors that have an area of 10^{-3} cm² or less and for which the short-circuit current does not exceed 1 μamp. As shown in Figure 9.4a, a photovoltaic detector can be transformer-coupled to an amplifier. The primary should have a low dc resistance so as to establish the operating point of the detector near to the short-circuit or zero-bias condition shown at point B in Figure 7.3. The ac impedance of the secondary is chosen to provide the optimum source impedance for the preamplifier. Suitable transformers include the Triad Geoformers (in particular, Model G-4) and certain of the shielded transformers made by UTC (such as the Model A-27). The principal disadvantage with transformer coupling is simply that the bulk of the transformer cannot be accommodated in some applications.

A very convenient means of operating a photovoltaic detector is shown in Figure 9.4b. The detector is back biased so as to place the operating point at any chosen point along the current-voltage characteristic. In general, the maximum value of D^* is observed when a photovoltaic detector is operated at the short-circuit or zero-bias condition. A variation of this arrangement is shown in Figure 9.4c, where the detector is back biased by the series supply and is operated in parallel with a load resistor that has a value much greater than the resistance of the detector. This arrangement gives considerable flexibility in providing the optimum source resistance for a preamplifier.

Finally, a photovoltaic detector can be used with a current-mode preamplifier that will automatically maintain the zero-bias operating point

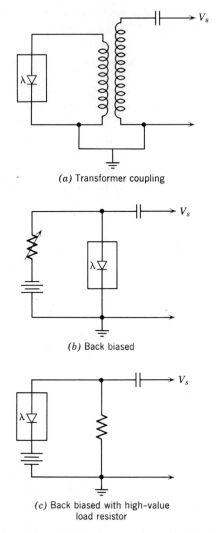

Figure 9.4 Methods of establishing the operating point of a photovoltaic detector.

despite changes in the level of the flux falling on the detector. Suitable current-mode preamplifiers are made by Perry Associates.

Determining the Operating Point for a Detector that Requires Bias

The test set should include an assortment of precision (1 per cent) low-noise load resistors. They may be wire wound or metal film types, but never ordinary carbon resistors. These resistors are usually mounted in shielded banana plugs so that they can be placed close to the terminals of the detector.

The bias supply should furnish a pure dc voltage that can be varied from about 1 to 500 V. Since the bias voltage is in series with the detector and must not introduce any spurious signals, it is almost impossible to obtain bias from a rectified and filtered ac source. Fortunately, a good solution is to use batteries. A simple switching circuit will permit the selection of any one of a wide variety of discrete voltages. It is wise to provide capacitive filtering for the batteries as further insurance against unwanted noise. Some batteries are noisier than others; it has been this author's experience that the quietest batteries are those manufactured by Burgess. It is well to pay a little extra in order to get precision meters (1 per cent) for the bias supply since they will be used to determine the resistance of detectors.

With a detector mounted in the test set, the first step is to determine the proper load resistor. For most measurements one will probably wish to use a matched load resistor. The approximate value of detector resistance may be known from past experience with detectors of the same type, it may be measured with an ohmmeter, or one can plug in any one of the available load resistors and calculate the resistance of the detector from the known value of resistance and the values of current and voltage indicated by the meters on the bias supply. Once the resistance of the detector is known, the load resistor that most nearly matches it is plugged into the test set and left in place for the remainder of the tests (unless it must be reduced in value to eliminate RC time-constant limiting by the input circuitry; see Section 9.5 for further advice on this point).

The next step is to determine the optimum bias, by measuring the signal and noise voltages as a function of the bias. *Optimum bias* is defined as that value of bias that maximizes the signal-to-noise ratio and hence the detectivity. For this measurement it is necessary to choose a chopping frequency, but the exact value chosen is of little concern since the optimum bias is not a function of the chopping frequency. The procedure is to measure the signal voltage for a series of successively increasing values of bias current (or voltage). Then close the shutter, so that the detector sees no chopped flux, and measure the noise voltage over the same range of bias current.

A typical plot of signal and noise voltages and of the calculated signal-to-noise ratio for various values of bias current is given in Figure 9.5. As shown there, the signal voltage increases linearly with bias current until the power dissipation limit is approached. At low values of bias current the noise voltage increases less rapidly than the signal voltage does, but at high bias currents the noise increases much more rapidly than the signal. The result is a rather broad maximum in the curve of signal-

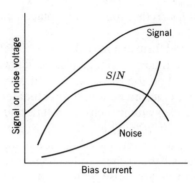

Figure 9.5 Determination of optimum bias current for a lead selenide detector cooled to 77°K.

to-noise ratio as a function of bias current. In an electrically noisy environment, such as in an aircraft, the limiting system noise may not be from the detector. Under these conditions, a higher-than-optimum bias current will increase not only the signal but the signal-to-noise ratio of the entire system and, if carried far enough, it may restore the detector as the limiting source of noise.

Determining the Operating Point for a Self-Generating Detector

Since a photovoltaic detector is a junction device, it is meaningless to speak of its dc resistance. Instead the quantity of interest is the *dynamic impedance*; the slope, dV/dI, of the current-voltage curve at the operating point. The dynamic impedance can be measured with an impedance bridge, provided that the proper operating point can be established. It is, however, quite easy to measure dynamic impedance with the test set. For convenience, transformer coupling, as shown in Figure 9.4a, can be used to establish the desired zero-bias condition. Select a suitable chopping frequency and measure the signal at the output of the test set. Connect a potentiometer across the detector and vary its setting until the signal from the detector is reduced to one-half of its original value. At this point the resistance of the potentiometer is equal to the dynamic impedance of the detector.

The method of determining the optimum operating point for a photovoltaic detector should be obvious from the discussion of Figure 9.4. The optimum operating point for any photovoltaic detector is almost always found at, or very near, the zero-bias condition. With the use of back bias, the operating point can be shifted and the optimum operating point found in much the same manner that was used for the photoconductive detector shown in Figure 9.5.

Calibrating the Amplification of the Test Set

Most of the figures of merit used to describe detectors require only the ratio of voltages and not their absolute values. However, to ensure that the measurements will be of the greatest utility, it should be possible to calibrate the test set so that the measured values can be converted to the actual values existing at the output terminals of the detector. The audio oscillator and the calibrated attenuator, shown in Figure 9.1, are included to permit the measurement of the amplification between the detector and the output indicator. A typical calibrated attenuator is the General Radio Type 546-C Microvolter. Figure 9.6 shows how the calibrating voltages can be introduced into a circuit containing a photoconductive detector and bias supply. The 100 kilohm resistor provides the proper load for the microvolter. The calibrating voltages are developed across the 100 ohm resistor, which is so small with respect to the value of R_l and R_d that it does not disturb the established bias condition. This general arrangement can be used to introduce calibrating signals into any detector circuit. For low-resistance detectors it is necessary to reduce the value of the 100 ohm resistor to 1 ohm or less so as to not disturb the established operating point. Note, however that the 100 kilohm and 100 ohm resistors provide an additional attenuator across the microvolter. The voltage amplification of the measuring circuitry is, of course, equal to the ratio of the voltage read on the output indicator to the calibrating voltage introduced at the detector.

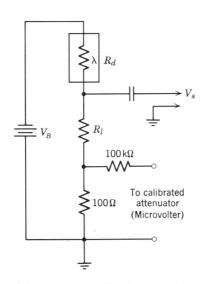

Figure 9.6 Method of introducing a calibrating voltage into a detector circuit.

Before proceeding with other measurements, one should check to see whether the noise voltage read at the output of the test set is from the detector or whether a significant portion is from the preamplifier. Many authors suggest doing this by measuring the output noise with the detector connected to the preamplifier and comparing it to the noise measured with the input terminals of the preamplifier shorted together. The implication is that the second measurement shows the noise originating in the preamplifier whereas the first measurement shows both the detector and the preamplifier noise. The trouble with this method is that preamplifier noise depends, to some extent, on the impedance of the source connected to its input. As a result, the noise contributed by the preamplifier is not the same with a shorted input as it is with a detector across the input.

A more desirable way to evaluate the relative contributions of the detector and the preamplifier to the noise is to measure the noise at the output of the preamplifier, first with the detector connected in the normal manner and then with the detector replaced by a low-noise resistor having the same resistance as that of the detector. Although the equations that describe the results of these two measurements are most easily written in terms of noise power and power gain (because noise powers, rather than noise voltages, are additive), the experimenter will almost certainly measure the rms value of the output noise voltages rather than the noise powers. The rms value of the output noise voltage, with the detector connected to the input, is

$$V_{0_d} = (V_a^2 + V_d^2 A^2)^{1/2}, \qquad (9\text{-}8)$$

where V_a is the noise voltage contributed by the preamplifier, V_d is the noise voltage from the detector, and A is the voltage amplification of the preamplifier. Similarly, when the resistor is connected in place of the detector, the rms value of the output noise voltage is

$$V_{0_r} = (V_a^2 + V_r^2 A^2)^{1/2}, \qquad (9\text{-}9)$$

where V_r is the rms value of the Johnson noise from the resistor; from (8-2) its value is $4kTR\Delta f$. Substituting this value for V_r and combining (9-8) and (9-9) gives

$$V_d = \left(\frac{V_{0_d}^2 - V_{0_r}^2}{A^2} + 4kTR\Delta f\right)^{1/2} \qquad (9\text{-}10a)$$

$$V_d = \left(\frac{V_{0_d}^2 - V_{0_r}^2}{A^2} + 1.6 \times 10^{-20} R\Delta f\right)^{1/2}. \qquad (9\text{-}10b)$$

The preamplifier noise *referred to its input*, so as to be directly comparable with the detector noise, is

$$V_{a_{\text{in}}} = \left(\frac{V_a^2}{A^2}\right)^{1/2} = \left(\frac{V_{0_d}^2}{A^2} - V_d^2\right)^{1/2}. \quad (9\text{-}11)$$

The experimenter will have to judge for himself, on the basis of these measurements, whether it will be necessary to correct the values of detector noise read on the test set in order to eliminate the effect of the preamplifier noise.

Measurement of Frequency Response

This test determines the variation of signal voltage as a function of chopping frequency. Select a low chopping frequency (50 Hz for a photon detector, for example). Tune the wave analyzer to this frequency and record the signal voltage. Repeat this process at successively higher frequencies. All measurements are normalized to the initial low-frequency measurement and plotted as shown in Figure 9.7. This curve can be used to determine the time constant of the detector provided that there is no limiting by the RC time constant of the input circuitry.

Measurement of the Detector Noise Spectrum

In Section 8.1 the power spectrum of random noise was defined as the mean-square noise voltage per unit bandwidth. Since most meters respond to, or are calibrated to read, root-mean-square voltages, it is more convenient to measure the root-power spectrum. The wave analyzer is ideal for this measurement because of its very small equivalent noise bandwidth, typically about 6 Hz in most commercial instruments. The exact value can be obtained from the manufacturer's instruction manual, by measuring the actual response curve and applying (8-11), or by measuring the noise voltage from a known resistance placed across the input terminals of the preamplifier and then solving (8-3) for the band-

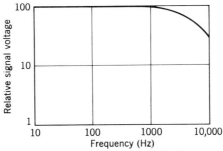

Figure 9.7 Frequency response of a lead selenide detector cooled to 77°K.

width. This value need be determined only once, since it is a constant for a given wave analyzer and is independent of the frequency to which the analyzer is tuned.

With the shutter closed so that the detector receives no chopped input signal, and with bias applied to the detector, the wave analyzer is successively tuned to the same frequencies as those that were used in measuring the frequency response, and the noise voltages are recorded.[1] Since the noise voltage is proportional to the square root of the bandwidth used to measure it, the observed values must be divided by the square root of the equivalent noise bandwidth of the analyzer in order to obtain the root-power spectrum. A plot of this spectrum for a cooled lead selenide detector is shown in Figure 9.8. At low frequencies $1/f$ noise predominates, but above 100 Hz the noise is of the generation-recombination type.

Calculation of the Various Figures of Merit

The measurements described up to this point provide all of the information needed to calculate the various figures of merit shown in Table 9.1. Figure 9.9 shows the result of such a calculation; it is a plot of the variation of D^* with chopping frequency for a cooled lead selenide detector. It is evident that the optimum chopping frequency is 2000 Hz or higher.

9.4 THE MEASUREMENT OF SPECTRAL RESPONSE

The basic detector test set measures total-energy or blackbody characteristics of detectors. In addition, one usually needs to have information about the spectral response of a detector, that is, the manner in which

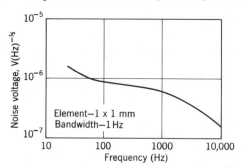

Figure 9.8 Typical root-power spectrum of the noise from a lead selenide detector cooled to 77°K.

[1]Because of the very narrow bandwidth of the wave analyzer, this procedure gives results that are essentially equally accurate for flat or nonflat power spectra.

THE MEASUREMENT OF DETECTOR CHARACTERISTICS 341

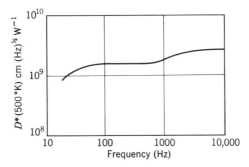

Figure 9.9 Variation of D^* with frequency for a lead selenide detector cooled to 77°K.

its responsivity and D^* vary with wavelength. The spectral-response test set shown in Figure 9.10 is used to obtain such data. In essence, it provides radiant flux in a very narrow spectral band centered about any desired wavelength and compares the response of the detector under test with that of a reference detector that is assumed to respond equally to all wavelengths.

The monochromator and reference detector are the heart of the spectral-response test set; they can be procured in several different

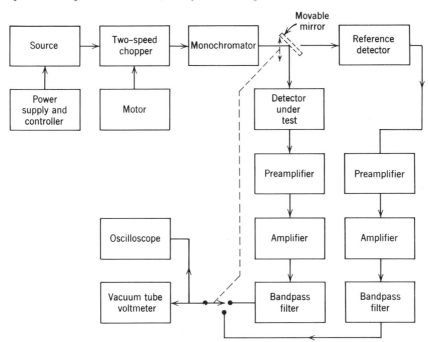

Figure 9.10 Spectral-response test set.

ways. The simplest is to buy a complete instrument, such as the Perkin–Elmer Model 98. This includes a Globar source and power supply, a 13 Hz chopper, a monochromator covering the wavelength range from 2 to 15 μ, a thermocouple detector, a preamplifier, and a narrow-bandpass amplifier tuned to 13 Hz. An alternative approach that involves less expense is to purchase a Leiss double monochromator and a thermocouple detector, and then either build or purchase a suitable source and power supply, a chopper assembly, and an amplifier for the thermocouple. Both monochromators mentioned above have a notably low content of scattered flux in their exit beams, a most important requirement for this application.

A thermocouple is a practical choice for the reference detector. The basic assumption is, of course, that this detector is truly black so that its response is independent of wavelength. The thermocouple meets this requirement as well as any commercially available detector. The blackening is apparently quite good out to about 8 μ, but it becomes progressively less efficient at longer wavelengths. Eisenman et al.[9,10] have devoted considerable effort to developing a better black detector, and their papers give excellent measurements of a variety of nominally black detectors.

Using a thermocouple as the reference detector dictates a chopping frequency of about 13 Hz. Unfortunately this frequency is too low to be practical with most photon detectors. A convenient solution is to provide two choppers in series, one at 13 Hz and the other at 200 to 300 Hz. They can be mounted on coaxial shafts each belt-driven by a separate motor. No unfavorable cross modulation results since both detector channels include bandpass filters centered about their respective chopping frequencies.

The beam from the monochromator normally falls on the reference detector. However, a plane mirror can be inserted to deflect this beam to the detector under test. A single voltmeter is switched so as to read the output of either detector. It is particularly convenient if the mirror-shifting mechanism is mechanically ganged with the switch that connects the meter to the proper detector. Since there is no requirement to measure noise with this meter, almost any general-purpose VTVM such as the Hewlett-Packard Model 400D can be used. To simplify Figure 9.10, it is assumed that such items as bias supplies, load resistors, coupling transformers, and detector cooling will be supplied as required.

There are three functional controls on the monochromator. One control selects the center wavelength of the exit beam. The width of the exit slit determines the width of the spectral interval about this center wavelength. Finally, the width of the entrance slit controls the amount of flux passing through the monochromator. The reference detector monitors the flux in the exit beam. Ideally a constant flux per unit wavelength interval should be maintained for any selected center wavelength. This can be done as follows: (1) Set the monochromator for some convenient wavelength. (2) From data supplied by the manufacturer, set the width of the exit slit to pass a predetermined wavelength interval (0.05 μ, for example). (3) Adjust the width of the entrance slit to give a convenient signal from the reference detector.

After adjusting the monochromator as described above, the movable mirror is inserted into the beam and the signal from the detector under test is recorded. This procedure is repeated as often as is necessary to find the limits of the detector response. The recorded data are examined to find the wavelength at which the detector has its maximum response, and all other values are normalized to this value. The wavelength at which the maximum occurs is called the peak wavelength and is the wavelength at which responsivity and D^* reach their maximum values. A plot of the relative spectral response of a cooled lead selenide detector is shown in Figure 9.11.

Spectral measurements are simple to make because they are comparison measurements and involve no absolute calibration. However, it is necessary to find a way to relate these spectral measurements to the absolute measurements made earlier with a blackbody source. Specifically, we want to take the relative response curve of Figure 9.11 and replace the relative response scale with an absolute scale of D^*. Fortunately such a step is not difficult.

Let us examine just what it is that we have measured with the basic detector test set. The irradiance at the detector is

$$H = 0.45 \frac{A_s W}{\pi d^2}, \tag{9-12}$$

where W is the radiant emittance of the blackbody source (most workers have agreed on the use of a 500°K source for detector measurements). Replace W by its equivalent integral form and multiply the expression by the area of the detector so as to give the total number of incident watts:

$$P = \frac{0.45 A_s A_d}{\pi d^2} \int_0^\infty W_\lambda d\lambda. \tag{9-13}$$

If the detector under test had been truly black, all of these watts would have been equally effective in contributing to the signal. Since the response of the detector varies with wavelength, as shown in Figure

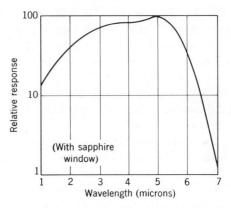

Figure 9.11 Relative spectral response of a lead selenide detector cooled to 77°K.

9.11, a watt at 5μ, for instance, contributes much more to the signal than does a watt at 7μ. Hence the number of watts that actually contributed to the signal is

$$P = \frac{0.45 A_s A_d}{\pi d^2} \int_0^\infty s(\lambda) W_\lambda d\lambda, \tag{9-14}$$

where $s(\lambda)$ is the relative spectral response of the detector. A perfectly black detector will have a response that is proportional to (9-13) and a nonblack, or selective detector, will have a response that is proportional to (9-14). It is convenient to define the ratio of (9-14) to (9-13) as an *effectiveness factor*:

$$\gamma = \frac{\int_0^\infty s(\lambda) W_\lambda d\lambda}{\int_0^\infty W_\lambda d\lambda}, \tag{9-15}$$

Thus γ is a measure of the effectiveness with which a detector uses the incident flux. A little thought shows that $1/\gamma$ is the ratio of D^* measured at the peak of the response curve to that measured with a blackbody. For a truly black detector, γ is equal to 1 and peak D^* is equal to blackbody D^*. For a lead selenide detector (77°K) having the relative response shown in Figure 9.11, the value of γ is about 0.25, and the quantities measured at the peak ($5\,\mu$) are four times those measured with a 500°K blackbody. From Figure 9.9, the value of the blackbody D^* is about 3×10^9 cm(Hz)$^{1/2}$ W^{-1}. Multiplying this by $1/\gamma$ gives 1.2×10^{10} cm(Hz)$^{1/2}$ W^{-1} for the value of the detectivity at the peak of the response curve.

THE MEASUREMENT OF DETECTOR CHARACTERISTICS 345

If the reader recalls the practical definition of D^*, it will help him to understand the distinction between peak and blackbody quantities. The symbol D^* is the signal-to-noise ratio at the output of the detector resulting from an incident flux of 1 W. Hence, if the lead selenide detector mentioned above received 1 W of flux from a 500°K blackbody, the signal-to-noise ratio at its output would be 3×10^9. If, somehow, this 1 W could be transformed so that it occupied an infinitesimally narrow wavelength interval centered around 5μ, the signal-to-noise ratio would be 1.2×10^{10}.

In order to simplify the calculation of γ, write

$$W_\lambda d\lambda = \frac{W_\lambda W_{\lambda_m} d\lambda}{W_{\lambda_m}}, \qquad (9\text{-}16)$$

where W_{λ_m} is the spectral radiant emittance at the peak of the blackbody curve. Substituting this in (9-15) and replacing the integral in the denominator by W, the radiant emittance of a blackbody, gives

$$\gamma = \frac{W_{\lambda_m}}{W} \int_0^\infty \frac{s(\lambda) W_\lambda d\lambda}{W_{\lambda_m}}. \qquad (9\text{-}17)$$

Using the values of W and W_{λ_m} from (2-9) and (2-11), and noting that 500°K is the standard blackbody temperature for testing detectors, gives

$$\gamma = 0.1135 \int_0^\infty \frac{s(\lambda) W_\lambda d\lambda}{W_{\lambda_m}}. \qquad (9\text{-}18)$$

This is a convenient form to use for calculations since the values of W_λ/W_{λ_m} can be read from the GE or Thornton radiation slide rules, or from most of the tables mentioned in Section 2.7. The integration is carried out numerically, usually at intervals of 0.1 μ.

9.5 THE MEASUREMENT OF TIME CONSTANT

The time constant provides a convenient number to describe the speed with which a detector responds to a change in incident flux. For many detectors the variation of response with frequency can be described by

$$\mathscr{R}_f = \frac{\mathscr{R}_0}{(1 + 4\pi^2 f^2 \tau^2)^{1/2}}, \qquad (9\text{-}19)$$

where τ is the time constant. Taking f_1 as the frequency at which the responsivity is 0.707 times its low-frequency value, the time constant is given by

$$\tau = \frac{1}{2\pi f_1}. \qquad (9\text{-}20)$$

From the frequency response curve for the cooled lead selenide detector, shown in Figure 9.7, the value of f_1 is about 3500 Hz; the calculated value of the time constant is 45 μsec.

Time constants determined in this way are best thought of as *effective* time constants, values that can be achieved in practical circuits. One must remember that all detectors have an unavoidable shunt capacitance. Combining a large load resistor with this shunt capacitance forms an RC circuit that may limit the high-frequency response of the detector. When this occurs, the observed time constant bears little relationship to the true value. Therefore, in measuring the frequency response of a detector with a short time constant, it is advisable to use a small load resistor, one having no more than one-tenth the resistance of the detector. Alternatively, one can start with a matched load resistor and keep reducing its value until further reductions yield no change in the high-frequency response of the detector.

Table 7.1 shows that some detectors have time constants that are significantly less than 1 μsec. Few laboratories are equipped to measure such short time constants properly because the usual mechanical choppers become impractical at the high frequencies required. Since an injection diode can be modulated quite easily at frequencies well into the megahertz region, it appears to offer an excellent solution to the problem of obtaining the necessary short pulses. An InAs diode driven by a square-wave generator, such as the Tektronix Model 106, has been used with success. One word of caution, however: the time constant of some detectors can vary with wavelength. In doped-germanium detectors, for instance, the time constant measured at a wavelength of 1.8 μ is typical of the host material and is considerably longer than that measured at longer wavelengths, which is due to the dopants.

9.6 THE MEASUREMENT OF DETECTOR RESPONSE CONTOURS

Many detectors show a nonuniform response from point to point across their sensitive surface. Such effects are of little concern if the image always covers the entire detector. However, if the image is smaller than the detector the nonuniform response can introduce serious problems in the signal processing.

To examine a detector for such effects, it is mounted on a support that can be moved in either of two orthogonal directions. A reflecting-microscope objective is used to form a small image of a source on the detector. The detector is moved in some orderly fashion, and the response is noted at various points over its sensitive area. These measurements can be plotted on a large-scale drawing of the detector, and points of similar response can be joined by lines representing contours of equal response[1, 11].

REFERENCES

[1] R. F. Potter, J. M. Pernett, and A. B. Naugle, "The Measurement and Interpretation of Photodetector Parameters," *Proc. Inst. Radio Engrs.* **47**, 1503 (1959).
[2] R. C. Jones, D. Goodwin, and G. Pullan, "Standard Procedure for Testing Infrared Detectors and for Describing Their Performance," Office of Director of Defense Research and Engineering, Washington D.C., September 12, 1960.
[3] P. R. Bradshaw, "Improved Checkout for IR Detectors," *Electronic Industries*, **22**, 82 (October 1963).
[4] W. Wallin, "Sinusoidal Light Chopper," U.S. Patent No. 2,813,460, November 19, 1957.
[5] R. B. McQuistan, "On an Approximation to Sinusoidal Modulation," *J. Opt. Soc. Am.*, **48**, 63 (1958).
[6] R. B. McQuistan, "On Radiation Modulation," *J. Opt. Soc. Am.*, **49**, 70 (1959).
[7] J. J. Brophy, "Low Noise Modifications of the Tektronix Type 122 Preamplifier," *Rev. Sci. Instr.*, **26**, 1076 (1955).
[8] W. L. Eisenman, "Tektronix Type 122 Preamplifier Modification," *Appl. Opt.*, **4**, 512 (1965).
[9] W. L. Eisenman, R. L. Bates, and J. D. Merriam, "Black Radiation Detector," *J. Opt. Soc. Am.*, **53**, 729 (1963).
[10] W. L. Eisenman and R. L. Bates, "Improved Black Radiation Detector," *J. Opt. Soc. Am.*, **54**, 1280 (1964).
[11] M. N. Markov and E. P. Kruglyakov, "Zonal Sensitivity of PbS Photoconductors," *Optics and Spectroscopy*, **9**, 538 (1960).

10

Modern Detectors and the Ultimate Limits on their Performance

The three previous chapters were devoted to a discussion of the various types of infrared detectors, their associated noises, and the methods used to measure their characteristics. This chapter contains engineering data on some of the more popular detectors and a discussion of the problems involved in choosing a detector for a specific application.

In order to make the maximum use of these data, it is necessary that the designer first understand the factors that will ultimately limit the D^* of a detector. From the discussion in Chapter 8 it is quite obvious that such a limit must exist, since detectors are not noise-free devices. An ideal photon detector would count all incident photons that have a wavelength shorter than its cutoff wavelength. Similarly, the ideal thermal detector would absorb all of the incident flux. These ideal detectors, since they generate no noise of their own, would be limited by noise associated with the photons that the detector receives from its surroundings, which are usually called the *background*. When the D^* of a detector is limited by the noise associated with photons from the background, the detector is said to be *background limited*. For a photon detector, the term *blip detector* (background-limited photodetector) is often used [1, 2]. Manufacturers have worked long and hard to develop background-limited photon detectors. The first indications of their success began to appear in the late 1950's. Today several types of detectors capable of background-limited operation are available in production quantities.

10.1 BACKGROUND-LIMITED PHOTON DETECTORS

In the early stages of their development most photon detectors are likely to show excessive noise, that is, more noise than theoretical considerations predict. Most of this excess noise is usually traced to

DETECTORS AND THEIR PERFORMANCE LIMITS 349

peculiarities in the production process, and much engineering effort may be required to eliminate it.

Having successfully eliminated the sources of excess noise, the remaining noise from the detector will consist of $1/f$ and generation-recombination noise. The $1/f$ noise varies inversely with frequency and in a good quality modern detector becomes appreciable only below frequencies of 500 to 1000 Hz. By using chopping frequencies and amplifier passbands above this region, the effect of $1/f$ noise is minimized and the only remaining source of noise is from the fluctuations in the generation and recombination of charge carriers. In the absence of a signal input, charge carriers are produced by photons from the background and by thermal excitation arising from vibrations of the crystalline lattice. By cooling the detector, the number of charge carriers produced by lattice vibrations can be made negligible. Under these conditions, the number of charge carriers is a function only of the photons arriving from the background, and the detector is background limited. Fluctuations in the rate at which the background photons arrive give rise to fluctuations in the rate at which charge carriers are generated. In photoconductive detectors there are also fluctuations in the rate at which the charge carriers recombine. Since photovoltaic detectors do not exhibit these fluctuations in the recombination process, their limiting noise will be about 40 per cent less than that of photoconductive detectors.

For a background-limited *photoconductive* detector the maximum theoretical value of D^* at a particular wavelength is

$$D_\lambda^* = \frac{\lambda}{2hc}\left(\frac{\eta}{Q_b}\right)^{1/2}, \tag{10-1}$$

where η is the quantum efficiency and Q_b is the photon flux from the background. Substituting for h (Planck's constant) and c (velocity of light), and assuming that λ is expressed in microns, gives

$$D_\lambda^* = 2.52 \times 10^{18} \lambda \left(\frac{\eta}{Q_b}\right)^{1/2}. \tag{10-2}$$

For a *photovoltaic* detector this expression is increased by the square root of 2 because of the absence of recombination noise, so that

$$D_\lambda^* = 3.56 \times 10^{18} \lambda \left(\frac{\eta}{Q_b}\right)^{1/2}. \tag{10-3}$$

The simplest application of (10-2) and (10-3) is to use them to calculate the value of D^* at the spectral peak. For a photon detector, the spectral peak coincides with the cutoff wavelength. The term Q_b (given in photon $cm^{-2} sec^{-1}$) includes all of the photons from zero wavelength up to the

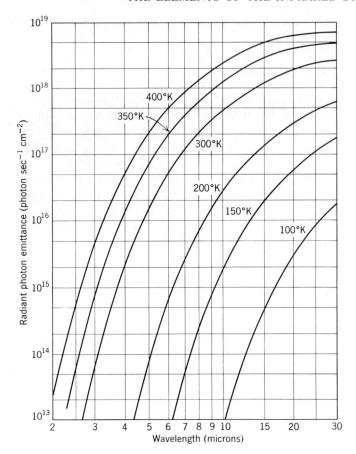

Figure 10.1 Radiant photon emittance from wavelength zero to the indicated wavelength as a function of temperature.

cutoff wavelength. The value is given in Figure 10.1 for a detector viewing a hemispherical blackbody surround at the temperature shown. Alternatively, the value of Q_b can be read from the Thornton radiation slide rule or found in [3 and 4]. It is generally assumed that the quantum efficiency η has a constant value for all wavelengths up to the cutoff wavelength. For an ideal detector, η would equal unity; for the high-quality photon detectors that are available today, values of η range from 0.1 to 0.4. Figure 10.2 shows the theoretical value of D^* at the spectral peak for ideal background-limited photon detectors viewing a hemispherical surround at a temperature of 300°K. These curves appear in Figure 7.5 so that they can be compared with the curves of spectral

DETECTORS AND THEIR PERFORMANCE LIMITS 351

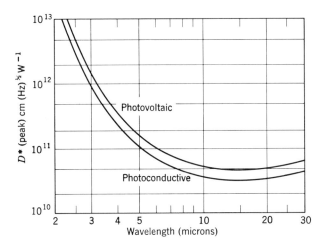

Figure 10.2 Theoretical value of D^* at the spectral peak for background-limited photon detectors viewing a hemispherical surround at a temperature of 300°K.

D^* for currently available detectors. It is remarkable that so many of the detectors approach the theoretical limit so closely.

As shown in Figure 10.2, the value of D^* at the spectral peak has a minimum at about 14 μ for a background temperature of 300°K (which is essentially the temperature of terrestrial backgrounds). As the cutoff wavelength is decreased below 14 μ, the number of background photons seen by the detector decreases and the value of peak D^* increases rapidly. When the cutoff wavelength is extended beyond 14 μ, the number of background photons seen by the detector increases, but much more slowly than does the number of signal photons, and as a result, the value of peak D^* increases toward longer wavelengths as well.

At first glance it is surprising to note from (10-2) and (10-3) that the D^* of a background-limited detector is proportional to wavelength. The explanation, which has already been discussed in Section 7.3, is that it is a consequence of defining D^* in terms of signal-to-noise ratio per watt rather than per unit photon flux.

The D^* of a background-limited detector can be increased by radiation shielding that reduces the photon flux from the background. In principle, the two methods that are used are quite simple: a cooled shield to limit the angular response of the detector and a cooled filter to limit its spectral response. In the first method the detector is placed within a cooled shield that has an aperture just large enough to admit the conical bundle of flux from the optics. This significantly reduces the background photon flux because the shielded detector can receive photons only from the solid angle subtended by the optics, rather than from an entire hemisphere (assuming, and most reasonably so, that the number of photons emitted from the cooled shield is negligible). Thus, under background-limited conditions D^* is a function of the angular field of view of the detector. In order to remove this dependence, Jones[5] suggested a new figure of merit D^{**} (pronounced "dee-double-star"). D^* is the measured detectivity referred to unit bandwidth and unit area; D^{**} includes these reference conditions and also a reference to a field of view having an (effective weighted) solid angle of π steradians (which is approximately equivalent to a hemispherical surround). D^{**} is defined as

$$D^{**} = \left(\frac{\Omega}{\pi}\right)^{1/2} D^*, \qquad (10\text{-}4)$$

where Ω is the effective weighted solid angle that the detector sees through the shield. In an arbitrary case the determination of Ω requires a quadruple integration. The result is a value averaged over the total solid angle seen by the detector, averaged over the entire area of the detector, and weighted by the projected area of the detector as seen from every point within the aperture. If the detector is Lambertian, that is, a perfectly diffusing surface (a condition approximated by most detectors), if it has circular symmetry, and if the solid angle subtended by the aperture at the detector can be represented as a cone with a half-angle θ, the relationship between Ω and θ is

$$\Omega = \pi \sin^2 \theta. \qquad (10\text{-}5)$$

Substituting this value in (10-4) gives

$$D^{**} = D^* \sin \theta. \qquad (10\text{-}6)$$

Even though the assumptions upon which (10-5) is based are rarely satisfied in practice, the approximation to the value of Ω is quite good, and one would probably never carry out the quadruple integration to get the true value. For a detector without a shield, θ equals 90°, Ω equals π, and D^{**} is equal to D^*. Figure 10.3 shows D^* and D^{**} for a cooled lead selenide detector and values of θ ranging from 4.5° to 90°. It is evident that D^* varies with θ as predicted by (10-6), and that D^{**} is essentially independent of θ (the decrease at very small values of θ is probably due to an excess noise source that is not apparent until the generation-recombination noise has been markedly reduced by shielding). When a detector is equipped with a cooled shield, the value of Q_b in (10-1) through (10-3) is that read from Figure 10.1 multiplied by $\sin^2 \theta$. The scale across the top of Figure 10.3 shows the equivalent f/number for the various apertures [$f/\text{no} = 1/(2 \sin \theta)$].

DETECTORS AND THEIR PERFORMANCE LIMITS

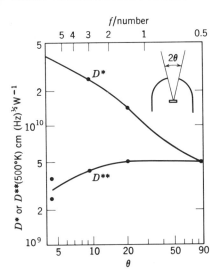

Figure 10.3 D^* and D^{**} of a lead selenide detector cooled to 77°K (data courtesy of Santa Barbara Research Center).

The second technique for reducing the background photon flux is to use a cooled optical filter in front of the detector. The spectral bandpass of this filter is chosen so as to reject as much of the background photon flux as possible while rejecting a minimum of the target flux. If the system is to be used within the earth's atmosphere, the filter bandpass will lie within one of the atmospheric windows in order to minimize the flux contributed by the intervening atmosphere. This is the reason for our insistence in Chapter 4 that the system designer needs information on the transmission characteristics of the atmospheric windows rather than on the absorption characteristics of the regions between. The value of the background photon flux for any filter bandpass can be found from Figure 10.1 by taking the difference between the values of Q_B at the upper and lower cutoff wavelengths. As an example, for a detector with a cutoff wavelength of 7 μ viewing a 300°K background, the value of Q_b is 1.1×10^{17} photon cm^{-2} sec^{-1}. If a filter with a bandpass extending from 4 to 5 μ is placed in front of the detector, the value of Q_b is reduced to 1.1×10^{16} photon cm^{-2} sec^{-1}. Of course the filter must be cooled so that it does not contribute to the background photon flux.

The means for incorporating radiation shielding into a coolable detector package [6] are shown in Figure 10.4. The detector element, shield, and filter are all in good thermal contact with the coolant chamber. Thus all are at essentially the same temperature.

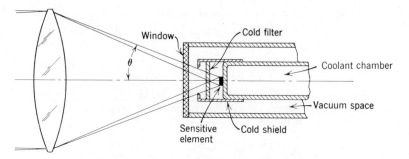

Figure 10.4 Radiation-shielded detector package.

The effectiveness of radiation shielding in increasing the D^* of a cooled lead selenide detector is shown in Figure 10.5. Curve A is for an unshielded detector. Curves B, C, and D are for the same detector but with a shield having successively smaller apertures. Curve E is the same shield that was used for curve D, but a cooled quartz filter has been added to limit the cutoff wavelength to $5\,\mu$[7]. As shown, the radiation shielding has increased the peak value of D^* by an order of magnitude. Translated into terms of system performance, such a gain would approximately treble the maximum detection range of a system.

One might suppose from (10.1) that there is no limit on D^*. Apparently, as long as the background flux can be reduced, the value of D^* should increase. Such cannot be the case, however, since even if background flux is entirely eliminated, there will still be photon noise associated with

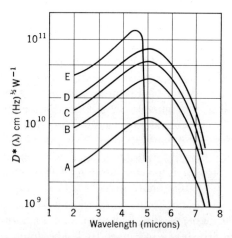

Figure 10.5 Effect of radiation shielding on a cooled lead selenide detector. The same detector was used in each example; only the shielding was changed. (Adapted from Bode [7]).

the signal. It appears unlikely that systems operating within the earth's atmosphere can reach such a limit, because even when extensive radiation shielding is used, the background flux on the detector is still many times the flux from the target. On the other hand, for systems operating in space, where the background has an effective temperature close to absolute zero and the optics can be cooled to 100°K or less, the photon noise associated with the signal can, indeed, limit system detectivity. Published measurements showing the increase in D^* as the background flux is reduced do not extend beyond a decrease of three orders of magnitude in the flux. Thus the ultimate limit to the process is still not evident. The limitations that have been noted seem to be due to excess noise sources that do not become evident until the generation-recombination noise is reduced by shielding. Such noise sources can probably be reduced once their mechanism is understood, but this will require a substantial effort on the part of detector manufacturers. An example of the progress that has been made with cooled lead selenide is shown in Figure 10.6. On the left are data reported in 1959 showing that reduction of the background photon flux below 5×10^{15} photon cm^{-2} sec^{-1} causes no further increase in D^*. On the right are data reported in 1963[7] showing that D^* for more recent detectors continues to increase for background fluxes as low as 5×10^{13} photon cm^{-2} sec^{-1}. For an unshielded detector ($\theta = 90°$), such a low flux requires a background temperature of 150°K. If the background temperature is 200°K, a shield is required with an aperture for which $\theta = 2.8°$ (an angle this small would limit the f/number of the optical system to about $f/10$). Other combinations of background temperature and shield aperture needed to obtain the corresponding background photon flux are shown across the bottom of this figure. The 1959 curve appears to follow the $(Q_b)^{-1/2}$ relation predicted by the theory, at least over a portion of its extent. The 1963 data are best fitted by a curve having a slope proportional to $(Q_b)^{-0.37}$. This departure from theory is probably peculiar to lead salt detectors since their mechanism of photoconductivity is more complicated, and consequently less well understood, than that of extrinsic photoconductors.

The circuit designer should not forget that radiation shielding increases the resistance of a detector in proportion to $(Q_b)^{-1}$ or to $(\sin^2 \theta)^{-1}$. This increase must be allowed for in the design of the matching network between the detector and the preamplifier.

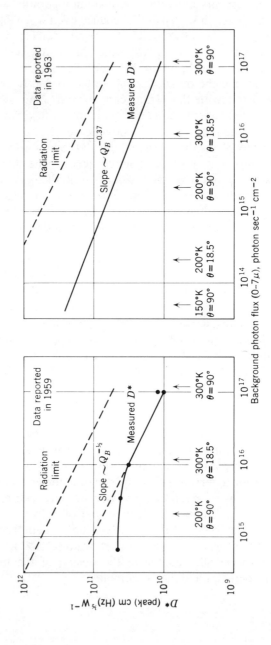

Figure 10.6 Peak D^* for cooled lead selenide detectors as a function of the background photon flux. The arrows show the combination of background temperature and shield aperture needed for various values of photon flux. (Data adapted from Bode[7] and courtesy of Santa Barbara Research Center).

DETECTORS AND THEIR PERFORMANCE LIMITS 357

10.2 LIMITATIONS ON THE PERFORMANCE OF THERMAL DETECTORS

Despite the fact that they are historically much older than photon detectors, thermal detector development probably has not received its proportionate share of research funds. As a result, thermal detectors, are not as highly developed as they might be. Havens [8–10] has estimated the minimum noise equivalent power attainable with bolometers and thermocouples at room temperature, viewing a 300°K background. His estimate, usually known as *Haven's limit*, is not intended to show a fundamental limit, but is instead an optimistic engineering estimate extrapolated from a study of available detectors. It has proved to be a remarkably good estimate, since no bolometers or thermocouples operating at room temperature have yet exceeded it[11].

The original conclusion by Havens was that the minimum detectable energy is about the same for all types of thermal detectors and that it is

$$(\Delta P)(\tau) = 3 \times 10^{-12} \text{ joule}, \tag{10-7}$$

where ΔP is the minimum detectable power in watts, τ is the duration of the radiation pulse in seconds, and the area of the detector is assumed to be 1 mm². Assuming that ΔP is equal to the noise equivalent power, that τ is the responsive time constant of the detector, and converting to an area of 1 cm², gives

$$\text{NEP} = \frac{3 \times 10^{-11}}{\tau}. \tag{10-8}$$

D^* was defined in (7-7) as

$$D^* = \frac{(A_d \Delta f)^{1/2}}{\text{NEP}}.$$

Since (10-8) is already written for an area of 1 cm², the term A_d can be dropped. The relationship between the time constant and the bandwidth of the detector is[1]

$$\Delta f = \frac{1}{4\tau} \tag{10-9}$$

Making the indicated substitutions, the value of Haven's limit (in terms of blackbody D^*) is

$$D^* = 1.67 \times 10^{10} \tau^{1/2}. \tag{10-10}$$

[1]There are several subtle considerations involved in this relationship; the reader interested in them can consult [9–11].

A high-quality thermistor bolometer will have a D^* of 1×10^8 cm(Hz)$^{1/2}$ W^{-1} and a time constant of 2 msec. For such a time constant, the value of Haven's limit is 7.5×10^8 cm(Hz)$^{1/2}$ W^{-1}. Thus the thermistor bolometer is within a factor of 7 of Haven's limit.

Jones[9–11] has derived a thermodynamic limit for thermocouples and bolometers that is given by

$$\text{NEP} = \frac{2.76 \times 10^{-11}}{\tau^{1/2}} \qquad (10\text{-}11)$$

for a detector having an area of 1 cm^2. Proceeding as before, and assuming that Johnson noise is the limiting type of noise, the maximum value of D^* is

$$D^* = 1.81 \times 10^{10}. \qquad (10\text{-}12)$$

From this expression it is seen that the thermistor bolometer is 2 orders of magnitude away from the thermodynamic limit. It is evident from a comparison of (10-10) and (10-12) that Haven's limit approaches the thermodynamic limit when the time constant is 1 sec.

10.3 CONSIDERATIONS IN THE SELECTION OF A DETECTOR

The problems the system engineer faces in selecting a detector are, of course, only one part of the system integration process. The first efforts of the system engineer are usually to define the mission of the system and the characteristics of the threat. From this stage it is a logical step to a preliminary estimate of the most desirable operating wavelength. With this information, the system engineer can look at general collections of data on detectors, such as those shown in Table 7.1 and Figure 7.5. Almost immediately a new problem arises: can a cooled detector be used? The answer is not always an easy one. Modern cooling devices are no longer the mystery they once were, but they still pose problems of size, weight, power drain, maintenance, reliability, and cost. At this point, Figure 10.7 is likely to be quite useful. The vertical columns show the three principal atmospheric windows, and the horizontal rows show the four most common operating temperatures for detectors. Each box contains a plot of D^* versus wavelength for the detectors that respond in that window and at that operating temperature. This figure shows, for instance, that if the decision has been made to use the 8 to 13 μ window without cooling the detector, the only suitable detector is a thermistor bolometer. If the situation is reevaluated and it is decided to supply cooling, then the figure shows that supplying a means for cooling to 195°K is useless since no detector is available for operation at 195°K in this window. Further cooling to 77°K makes it possible to use gold-

DETECTORS AND THEIR PERFORMANCE LIMITS 359

doped germanium, while the optimum detectors for this window must be cooled to temperatures below 35°K. Having narrowed the choice of detector to just a few types, one can turn to the detailed data sheets in Figures 10.9–10.16 and to papers, such as those in [12–20], for additional information. Eventually one type emerges as the leading candidate and it is used in preliminary design studies to see how well system design objectives are met. When a final decision has been reached as to detector type the circuit designer will need to have considerably more data on its characteristics. The most satisfactory solution is to order sample quantities of the detector from various manufacturers and make measurements under conditions that simulate those expected in the final system.

The perceptive reader may have already noticed that *the figures of merit used to describe detector performance fail to give any indication of the absolute value of either the signal or the noise levels.* Such measures as D^* and NEP are a delight to the solid-state physicist, but are far from satisfactory for the circuit designer, who needs more information than that conveyed by a simple ratio of signal to noise. The anticipated signal levels are essential to the designer because he must provide sufficient dynamic range in his signal processing circuitry to ensure that no saturation occurs. Information on both the level and the power spectrum of the noise is needed in order to design the optimum preamplifier. It is a curious fact that so many writers on detectors fail to recognize the basic inadequacy of detector data that do not go beyond NEP and D^*.

No satisfactory data have appeared in the literature on the range over which the response of a detector is linearly proportional to the irradiance. It appears that most photon detectors are linear over a range that extends about four orders of magnitude above the NEI for the detector. Apparently thermal detectors are far superior in this respect. The circuit designer will probably have to make the necessary measurements himself, since most detector manufacturers do not have such information available.

In order to help substantiate various spaceborne equipment designs, it would be nice to have data available on the effects of ionizing radiation on infrared detectors. Unfortunately few such data are available in the published literature. Two papers[21, 22] have investigated the effect of exposure to the Van Allen belts on the life of lead sulfide detectors. They conclude that lifetimes of two years should be possible. Another paper [23] provides cursory data on the behavior of lead sulfide, indium antimonide, and thermistor bolometers when exposed to gamma and neutron radiation.

It appears that arrays of detectors will play an increasingly important role in new infrared system designs. An excellent survey of the state of

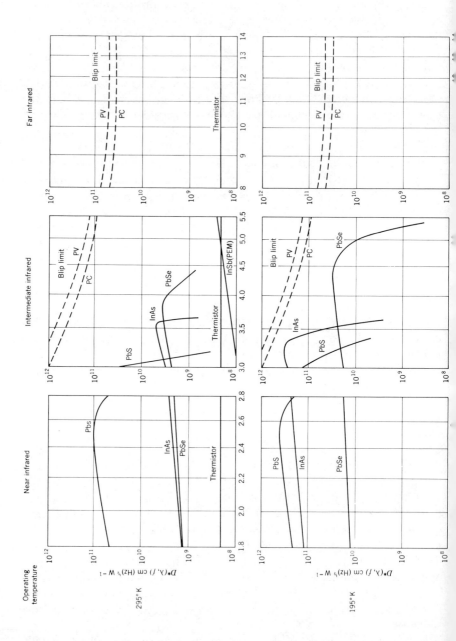

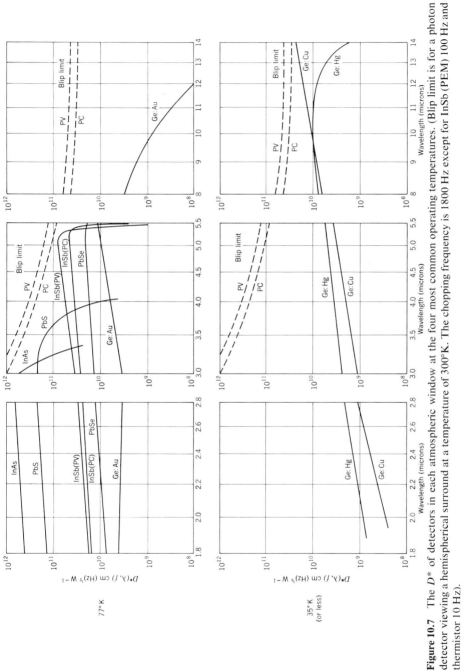

Figure 10.7 The D^* of detectors in each atmospheric window at the four most common operating temperatures. (Blip limit is for a photon detector viewing a hemispherical surround at a temperature of 300°K. The chopping frequency is 1800 Hz except for InSb (PEM) 100 Hz and thermistor 10 Hz).

the art for arrays of indium antimonide, lead sulfide, lead selenide, copper-doped germanium, and mercury-doped germanium is given in [24].

The choice of a package for a detector is an important consideration and one that should be examined early in the design process. All too often the detector is shoehorned into whatever space is left over after completion of the mechanical design. The result may be an unduly expensive package or, in some cases, one that is impossible to make. While each manufacturer has what he calls standard packages, there is no industry-wide standard. For this reason most packages turn out to be custom designs. Some indication of the wide variety of packages that a single manufacturer has produced is given in Figure 10.8.

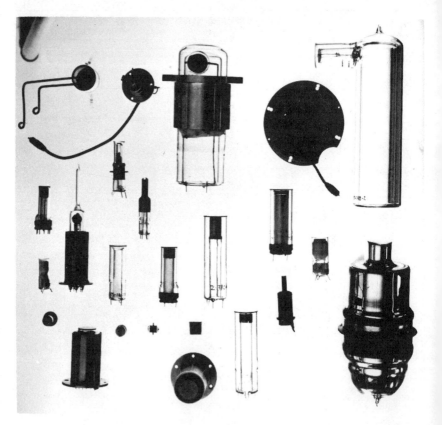

Figure 10.8 Typical coolable detector packages (courtesy of Santa Barbara Research Center).

DETECTORS AND THEIR PERFORMANCE LIMITS

10.4 ENGINEERING DATA ON SELECTED DETECTORS

Data on the characteristics of eight of the most useful currently available detectors are given in Figures 10.9–10.16. This information is not intended to be representative of any particular manufacturer's product. Instead, it refers to good state-of-the-art detectors that can be produced in reasonable quantities; it can be used with confidence for preliminary design purposes.

The detector resistance is given for a square element. This is particularly convenient since the resistance of a square element is independent of its linear dimensions. The resistance of rectangular elements is proportional to the ratio of length to width (where length is the distance between the signal electrodes). Thus if a particular data sheet shows a detector with a resistance of 1 megohm, one would expect that a similar detector having a length that is 4 times its width should have a resistance of 4 megohms. If necessary, the manufacturer can interleave the electrodes to reduce the resistance of the detector. By convention, the values of resistance that are reported are those measured with the detector in the dark so that it receives no flux except that from the room-temperature background.

The values of responsivity are 500°K blackbody values for a 1 mm^2 element under optimum bias conditions. They should be considered as only an indication of the order of magnitude of the values that can be obtained.

Indium Arsenide (InAs)

Type:	Photovoltaic.
Element sizes available:	Standard elements are circular, ranging from 0.5 to 3 mm in diameter.
Time constant:	Less than 2 μsec at all temperatures.
Dynamic impedance:	5–10 × 10^3 ohm at 195°K. 3–6 × 10^4 ohm at 77°K.
Responsivity:	3 × 10^{-1} VW^{-1} at 295°K. 5 × 10^2 VW^{-1} at 195°K. 5 × 10^3 VW^{-1} at 77°K.
$\dfrac{D^*\text{ (peak)}}{D^*\text{ (500°K)}}$:	65 at 77°K, 40 at 195°K.

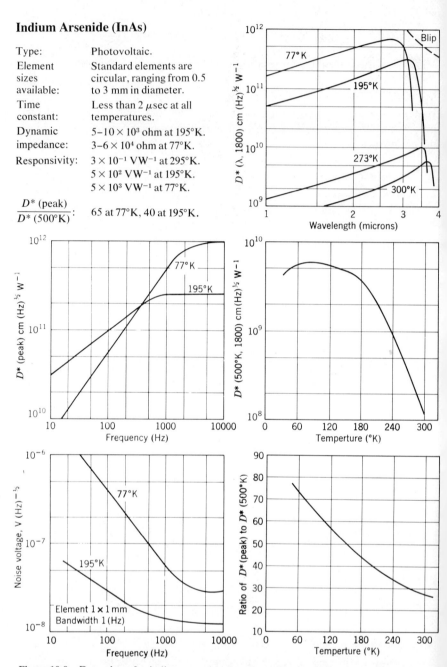

Figure 10.9 Data sheet for indium arsenide detectors. (Data adapted from publications of the Santa Barbara Research Center and Texas Instruments, Inc.).

Lead Sulfide (PbS)

Type: Photoconductive.
Element sizes available: From 0.01 × 0.01 to 25 × 25 mm. Rectangular elements from 0.01 mm wide. For arrays, minimum element spacing is 0.01 mm.

Temp. (°K)	Time constant (μsec)	Resistance (megohms)	$\mathscr{R}$ (VW^{-1})	D^* (peak) $\overline{D^*(500°K)}$
295	50–500	0.3–0.6	4 × 10^3	90
195	800–4000	3–6	3 × 10^5	25
77	500–3000	6–12	2 × 10^5	60

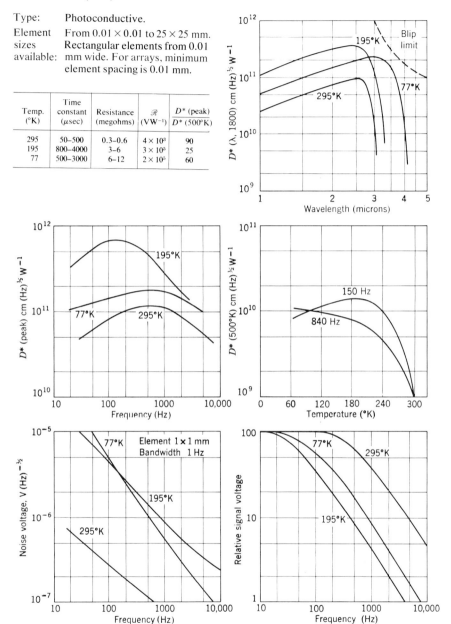

Figure 10.10 Data sheet for lead sulfide detectors. (Data adapted from publications of Infrared Industries and the Santa Barbara Research Center).

Lead Selenide (PbSe)

Type: Photoconductive. Material is optimized during manufacture for operation at ambient (ATO), Intermediate (ITO) and low (LTO) temperatures.

Element sizes available: From 0.25 × 0.25 to 10 × 10 mm. Rectangular elements from 0.075 mm wide. For arrays, minimum element spacing is 0.025 mm.

Type	Time constant (μsec)	Resistance (megohms)	$\mathscr{R}$ (VW^{-1})	$\dfrac{D^*\text{ (peak)}}{D^*\text{ (500°K)}}$
ATO	2	2–5	2×10^3	17
ITO	30	6–12	5×10^4	5.5
LTO	40	0.5–2	2×10^5	3.5–4

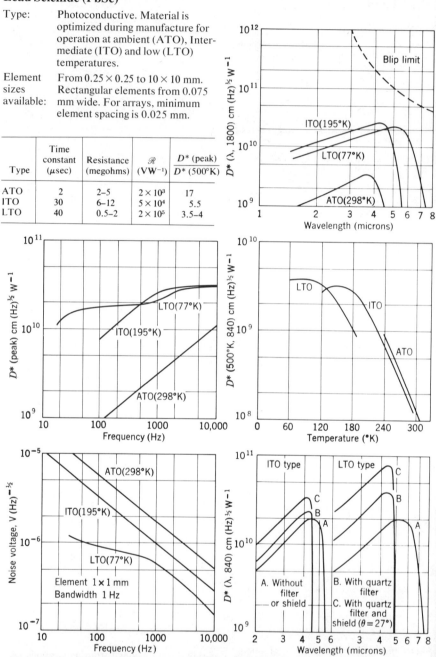

Figure 10.11 Data sheet for lead selenide detectors. For additional data, see Figs. 10.3, 10.5 and 10.6. (Data adapted from publications of the Santa Barbara Research Center).

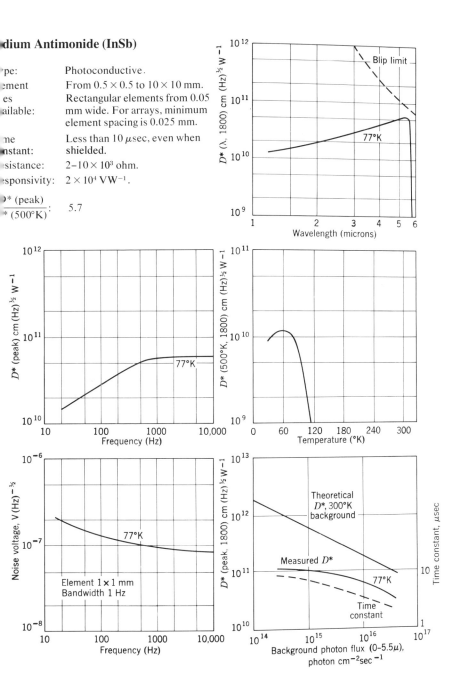

Figure 10.12 Data sheet for photoconductive indium antimonide detectors. (Data adapted from publications of the Santa Barbara Research Center and Texas Instruments Inc.).

Indium Antimonide (InSb)

Type: Photovoltaic.
Element sizes available: From 0.1×0.1 to 10×10 mm. Rectangular elements from 0.1 mm wide. Circular elements with diameters from 0.1 to 10 mm.
Time constant: Less than 1 μsec.
Dynamic impedance: 2–5×10^4 ohm.
Responsivity: 2×10^4 VW^{-1}.
$\dfrac{D^*\text{(peak)}}{D^*\text{(500°K)}}$: 6.2

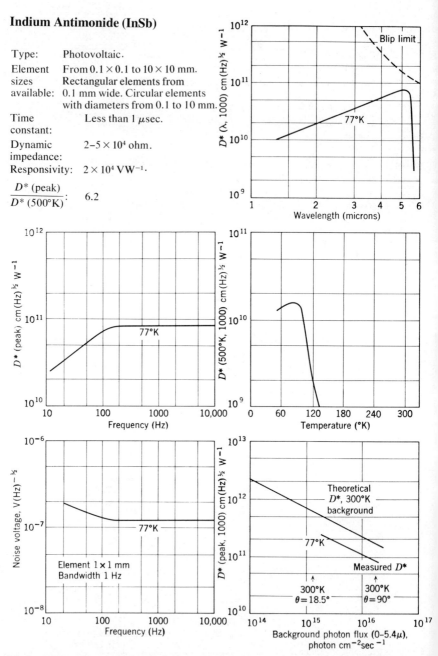

Figure 10.13 Data sheet for photovoltaic indium antimonide detectors. (Data adapted from publications of Philco Corp. and Texas Instruments, Inc.).

Mercury-Doped Germanium (Ge:Hg)

Type:	Photoconductive.
Element sizes available:	Standard elements are circular ranging from 0.3 to 3 mm in diameter.
Time constant:	Less than 0.1 μsec for all temperatures below 25°K.
Dynamic impedance:	Varies with bias, shielding, etc. Typically 0.5 megohm for $\theta = 30°$.
Responsivity:	1.5×10^5 VW^{-1} ($\theta = 30°$).
$\dfrac{D^*\text{ (peak)}}{D^*\text{ (500°K)}}$:	1.8 (IRTRAN-2 window).

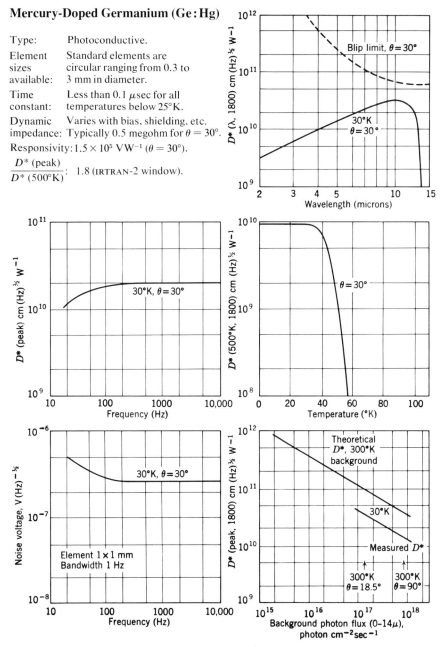

Figure 10.14 Data sheet for mercury-doped germanium detectors. (Data adapted from publications of the Santa Barbara Research Center and Texas Instruments Inc.).

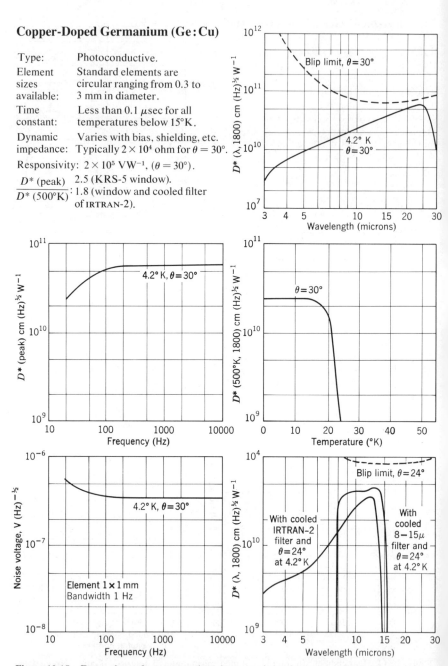

Figure 10.15 Data sheet for copper-doped germanium detectors. (Data adapted from publications of the Santa Barbara Research Center and Texas Instruments Inc.).

Thermistor

Type: Bolometer.

Element sizes available: From 0.1×0.1 to 2.5×2.5 mm. Rectangular elements from 0.1 mm wide. Germanium-immersed elements from 0.1×0.1 to 1×1 mm.

Time constant: 1–5 msec.

Resistance: 0.1 mm wide. Germanium-immersed elements, resistances of 0.27, 0.5, and 2.7 megohms are available.

Responsivity: 50 VW^{-1}.
10^3 VW^{-1} (Germanium-immersed element).

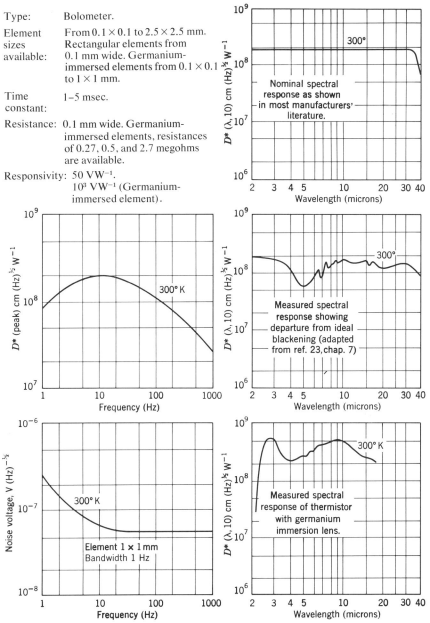

Figure 10.16 Data sheet for thermistor bolometers. (Data adapted from publications of Barnes Engineering Co. and Servo Corporation of America).

REFERENCES

[1] E. Burstein, G. S. Picus, and N. Sclar, "Optical and Photoconductive Properties of Silicon and Germanium," in *Photoconductivity Conference*, New York: Wiley, 1956, p. 353.
[2] R. L. Petritz, "Fundamentals of Infrared Detectors," *Proc. Inst. Radio Engrs.*, **47**, 1458 (1959).
[3] A. N. Lowan and G. Blanch, "Tables of Planck's Radiation and Photon Functions," *J. Opt. Soc. Am.*, **30**, 70 (1940).
[4] I. H. Swift, "Performance of Background-Limited Systems for Space Use," *Infrared Phys.*, **2**, 19 (1962).
[5] R. C. Jones, "Proposal of the Detectivity D^{**} for Detectors Limited by Radiation Noise," *J. Opt. Soc. Am.*, **50**, 1058 (1960).
[6] T. H. Johnson and G. D. Hayball, "Radiation Shielding for Infrared Detectors," U.S. Patent No. 3, 103, 585, September 10, 1963.
[7] D. E. Bode, "Lead Selenide Infrared Detectors," *Proc. Natl. Electronics Conf.*, 1963, p. 630.
[8] R. R. Havens, "Theoretical Comparison of Heat Detectors," *J. Opt. Soc. Am.*, **36**, 355 (1946).
[9] R. C. Jones, "The Ultimate Sensitivity of Radiation Detectors," *J. Opt. Soc. Am.*, **37**, 879 (1947).
[10] R. C. Jones, "A New Classification System for Radiation Detectors," *J. Opt. Soc. Am.*, **39**, 327 (1949).
[11] R. C. Jones, "Factors of Merit for Radiation Detectors," *J. Opt. Soc. Am.*, **39**, 344 (1949).
[12] N. C. Anderson, "Comparative Performance of Cooled Infrared Photoconductors," *Infrared Phys.*, **1**, 163 (1961).
[13] J. N. Humphrey, "Optimum Utilization of Lead Sulfide Infrared Detectors under Diverse Operating Conditions," *Appl. Opt.*, **4**, 665 (1965).
[14] T. H. Johnson, H. T. Cozine, and B. N. McLean, "Lead Selenide Detectors for Ambient Temperature Operation," *Appl. Opt.*, **4**, 693 (1965).
[15] D. E. Bode, T. H. Johnson, and B. N. McLean, "Lead Selenide Detectors for Intermediate Temperature Operation," *Appl. Opt.*, **4**, 327 (1965).
[16] F. D. Morton and R. E. J. King, "Photoconductive Indium Antimonide Detectors," *Appl. Opt.*, **4**, 659 (1965).
[17] D. E. Bode and H. A. Graham, "A Comparison of the Performance of Copper-Doped Germanium and Mercury-Doped Germanium Detectors," *Infrared Phys.*, **3**, 129 (1963).
[18] P. W. Kruse, "Photon Effects in $Hg_{1-x}Cd_xTe$," *Appl. Opt.*, **4**, 687 (1965).
[19] G. R. Pruett and R. L. Petritz, "Detectivity and Preamplifier Considerations for Indium Antimonide Photovoltaic Detectors," *Proc. Inst. Radio Engrs.*, **47**, 1524 (1959).
[20] E. M. Wormser, "Properties of Thermistor Infrared Detectors," *J. Opt. Soc. Am.*, **43**, 15 (1953).
[21] R. R. Billups and W. L. Gardner, "Radiation Damage Experiments on PbS Infrared Detectors," *Infrared Phys.*, **1**, 199 (1961).
[22] R. P. Day, R. A. Wallner, and E. A. Lodi, "An Experimental Study of Energetic Proton Radiation Effects on IRI Lead Sulfide Infrared Detectors," *Infrared Phys.*, **1**, 212 (1961).
[23] E. W. Kutzscher, "The Behavior of Infrared Detection Elements Under Nuclear Radiation Exposure," *Proc. Inst. Radio Engrs*, **47**, 1520 (1959).
[24] T. H. Johnson, "Detector Array Technology," *Infrared Phys.*, **5**, 1 (1965).

11

Techniques for Cooling Detectors

The rapid development of cooled photon detectors that started in the mid-1950's has, in turn, fostered the development of suitable cooling devices. The basic principles of such devices have been known for many years, but the requirements peculiar to infrared systems placed a new premium on extreme miniaturization, minimum power consumption, simplicity of maintenance, and high reliability. This chapter will show to what extent these objectives have been met.

The temperatures needed for cooling detectors are, for the most part, lower than those produced by conventional refrigeration machinery. The term *cryogenic*[1] has become identified with this low temperature regime. Most authors consider that the cryogenic region starts at about 125°K and extends to absolute zero [1].

11.1 PACKAGING COOLED DETECTORS

When an infrared detector is to be cooled, it is mounted in an insulated package or dewar flask (commonly referred to simply as a dewar). Two examples of such packages are shown in Figure 11.1. As illustrated there, a dewar is a double-walled container in which the space between the walls is evacuated. The reader will recognize that the common "thermos bottle" is a dewar. Detector-packaging technology has advanced to the point where glass dewars with a diameter of 0.5 in. and a length of 1.25 in. can be mass produced.

The packages shown in Figure 11.1 can be used for cooling to temperatures as low as 50°K. The inner stem, which forms the coolant chamber, is usually made of kovar sealing glass; the outer shell is made of glass, pyrex, or metal. If the package is to be exposed to temperatures that are well above the normal ambient, the thermal insulating capabilities of

---

[1]From the Greek word *cryo* meaning extreme cold and the suffix *gen* meaning producer of. "Cryogen" is a noun; "cryogenic" is an adjective.

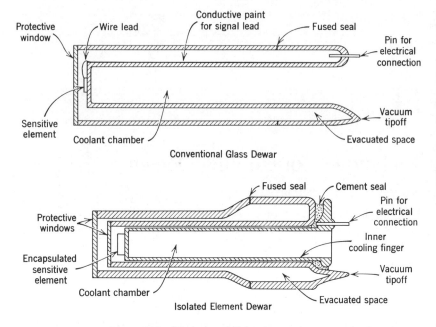

Figure 11.1 Typical cooled detector packages.

the package can be improved by aluminizing or silvering the inner surfaces (those that define the evacuated volume). For cooling to temperatures below 50°K, more complex designs are required; such dewars typically have a diameter of about 4 in. and a length of about 10 in. The variety of coolable packages that are available is shown in Figure 10.8.

The sensitive element is a small slice of semiconductor material or a chemically deposited film on a thin substrate. To ensure an efficient thermal contact the sensitive element is cemented to the end of the coolant chamber. Wires with a diameter of 0.001 in. are soldered or welded to contacts on the sensitive element and then cemented to the inner stem. In order to minimize *microphonics*, that is, spurious signals generated during vibration, it is essential to keep the unsupported leads as short as possible. A strip of conductive paint is used to connect the pin forming the external electrical connection to the leads from the detector. This paint is not subject to microphonics and has a low thermal conductivity that minimizes the heat leak to the detector. Figure 10.4 shows a cooled detector package with radiation shielding.

The protective window over the sensitive element should have high transmission over the spectral bandpass of the system; the use of antireflection coatings is standard practice. Some window materials can be fused directly or through intermediate fused seals to the outer shell of the dewar. Soldered or cemented joints are sometimes used, but they are usually not as reliable as a fused seal. A designer may be tempted to replace the protective window with a filter or reticle, but such designs are usually ill-advised. Filters and reticles are relatively fragile; damage to either can render useless the entire detector package rather than just the filter or reticle.

Since some types of sensitive elements can be damaged by excessive heating, considerable ingenuity is required in the final assembly of a dewar. The window and most of the outer shell form one subassembly, and the inner stem, pins, and a short length of the outer shell form the other. These two subassemblies are fused together on the outer shell at a point about one-fourth of the distance from the pins to the window. During evacuation the entire package is heated in order to drive off adsorbed gases. Sometimes this bakeout is not as thorough as desired because the sensitive element cannot withstand the temperatures involved. The manufacturer may then include a "getter" as an aid to achieving and maintaining a satisfactory vacuum. For additional details of the packaging process, see refs. 2 and 3.

The long-term stability of lead selenide is significantly increased if it is packaged in the isolated-element package shown in Figure 11.1, in which the detector element is isolated from the vacuum in the dewar. The sensitive element is sealed in a small, unevacuated capsule that is cemented to an inner cooling finger. The final step in the assembly process is to cement this cooling finger into the dewar. Since the dewar is fabricated without the sensitive element in place, an optimum bakeout can be used. Such isolated-element dewars have proved to be extremely reliable.

Although the detector package is remarkably rugged, it is still necessary to use care in handling it. The area around the pins is the most susceptible to damage. When leads are being soldered to these pins, the heat applied must be kept to a minimum to prevent stresses in the seals where the pins pass through the dewar wall. Since this operation is so critical, serious consideration should be given to having the manufacturer supply the detector package complete with leads. Excessive mechanical pressure applied to the pins can also cause cracks and loss of vacuum. Liquid coolants should never contact the outside of the package. The cooling element must be inserted into the inner stem with care to avoid cracking this stem or knocking the bottom out of the coolant chamber.

Sometimes the area around the pins may become so cold that moisture from the surrounding air condenses and causes a partial short circuit across the pins. A thin film of silicone grease applied to this area will prevent such condensation. If the vaporized coolant is not reused, it is often possible to arrange for it to flow over the pins and act as a barrier between the cold surface and the surrounding air. In normal operation the window is not cooled to any significant extent and there is no problem with moisture condensing on it. If condensation is observed on the window, it is a sure sign that the dewar has lost its vacuum. When this happens, the detector package must be replaced, since without a vacuum the heat load is so large that the sensitive element cannot be cooled to the desired operating temperature.

11.2 LOW-TEMPERATURE COOLANTS

In the earlier discussion of cooled detectors, it was noted that the operating temperatures were dictated by the availability of specific coolants. Physical properties of some of the more useful low-temperature coolants are given in Table 11.1. Except for ice and solid carbon dioxide, all of

TABLE 11.1 PHYSICAL PROPERTIES OF LOW-TEMPERATURE COOLANTS

Coolant	Boiling Temperature[a] (°K)	Refrigeration Capacity (W hr liter^{-1})	Weight-Density (lb liter^{-1})
Ice	273.2[b]	—	—
Solid carbon dioxide	194.6[c]	—	—
Liquid oxygen	90.2	67.6	2.51
Liquid argon	87.3	63.5	3.07
Liquid nitrogen	77.3	44.4	1.78
Liquid neon	27.1	28.9	2.67
Liquid hydrogen	20.4	8.79	0.156
Liquid helium	4.2	0.71	0.275

[a] At a pressure of 760 mm Hg.
[b] Melting point.
[c] Sublimation point.

the coolants listed are liquefied gases. Of these, the most commonly used are liquid nitrogen and liquid helium. The others are used less often because they are not readily available, they are expensive, or they constitute a safety hazard.

In the handling of liquefied gases, great care must be taken to prevent them from contacting the skin, since they can cause severe burns.

TECHNIQUES FOR COOLING DETECTORS 377

The moisture normally present in the pores of the skin is almost instantly frozen; this freezing is accompanied by a disruption of the cellular structure and a loss of blood circulation. When the area thaws, there may follow an abnormal accumulation of blood, clots may form, and if the condition is allowed to go unattended, gangrene may result. After an accidental contact, the affected part should be thawed by flushing it with cold water; do not rub or massage it. Report immediately to a medical facility for further treatment. If the liquid contacts the eyes, do not flush them with water or rub them. Get medical attention immediately. A face mask or goggles should always be worn when one is handling liquefied gases [4, 5].

Since the dewar is not a perfect thermal insulator, it provides a heat load for the cooling system and work must be expended in order to maintain the detector at the desired operating temperature. For dewars of the type shown in Figure 11.1, which are designed for use with liquid nitrogen, the heat load will range from 0.1 to 0.5 W. With the larger, more efficient dewars designed for use with liquid helium, the heat load will range from 0.01 to 0.1 W. The actual heat leak to the detector in these larger dewars is much greater than this, but most of it is removed by the exhaust gas.

The third column in Table 11.1 shows the refrigeration capacity of each of the liquefied gases. It is the heat input necessary to convert a liter of the liquid to its gaseous phase. The units, W hr liter^{-1}, may seem strange, but they are very convenient for calculating the length of time that a quantity of liquid will keep a detector cool. A comparison of the refrigeration capacities of the various liquids shows why the dewars used at the very lowest temperatures must present the smallest heat loads. A liter of liquid nitrogen, for instance, has about 60 times the refrigeration capacity of a liter of liquid helium.

11.3 OPEN-CYCLE REFRIGERATORS

The simplest method of cooling a detector is to pour liquefied gas into its coolant chamber. Such a method is often employed in the laboratory, but it is obviously impractical for equipment that is to be used in the field. The next step is to use an open-cycle refrigerator, that is, one in which no attempt is made to collect and reuse the refrigerant after it has absorbed heat from the load.

Liquid-Transfer Refrigerators

A liquid-transfer refrigerator consists of a dewar for storing liquefied gas and a feedline through which the liquid is supplied to the detector.

Figure 11.2 shows a unit of this type that is designed to store and deliver liquid nitrogen for airborne infrared systems.

The storage vessel is usually a double-walled dewar in which the space between the walls is evacuated. Thermal insulation is provided by the vacuum or by one of several insulations designed for use in a vacuum [6, 7]. When it is desired to transfer liquid, the vent valve is closed and the heater is turned on. A small quantity of liquid is vaporized, builds up pressure in the dewar, and causes liquid to flow through the transfer line to the detector. The pressure required is very low; it need be only 1 to 3 psi greater than that outside the dewar.

The transfer line operates in a completely unexpected manner. The optimum transfer line is an uninsulated line made of rubber, metal, or synthetic elastomer and not the heavily insulated type that one would expect intuitively. Miller[8] has described the resulting *Leidenfrost* or *two-phase flow* that results and has emphasized the importance of the supply boss, a short metallic fitting that is in good thermal contact with the outer wall of the storage vessel. This boss acts as a heat exchanger; it draws heat from the surroundings and transfers it to the stream of liquid nitrogen. Some of this liquid is thereby vaporized and introduces a two-phase mixture of liquid and gaseous nitrogen into the transfer line. Careful observations of the flow, using a transparent transfer line, show that it consists of a suspension of liquid droplets in a moving gas stream. The droplets do not touch the interior wall of the transfer line but are separated from it by a thin film of gas. For efficient transfer, it is essential

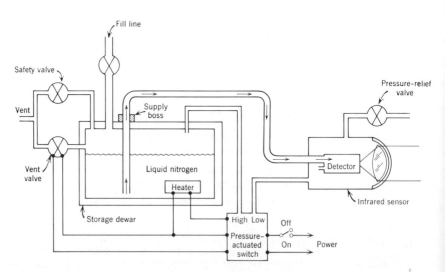

Figure 11.2 Liquid-transfer refrigerator.

TECHNIQUES FOR COOLING DETECTORS

to maintain this gas film. Therefore the interior wall of the line should be smooth, the line should contain no abrupt bends, and any couplings in the line should not break up the gas film. In general, the transfer efficiency of such a line is an increasing function of the velocity of the gas stream and a decreasing function of the length of the line, and there is an optimum diameter for the line. Goodenough and Swift[9] report that the optimum inner diameter of an aluminum transfer line for liquid nitrogen is between 0.05 and 0.10 in. If the inner diameter of the line is smaller than 0.05 in., it begins to approach the droplet diameter, and as a result the droplets touch the inner wall of the line. Figure 11.3 shows the transfer efficiency of a 0.075 in.-diameter aluminum transfer line as a function of length, for two different pressures in the storage dewar. With a line 5 ft long and a pressure differential of 1.5 psi, 1 liter of liquid nitrogen was enough to cool a detector for 5 hours (in a package similar to those shown in Figure 11.1).

The pressure-control components shown in Figure 11.2 are dictated by the airborne application of the unit. The key requirement is to keep the pressure in the storage dewar higher than the outside pressure at all times. If, for instance, the unit is designed to maintain a constant pressure differential, the pressure in the dewar will fall as the aircraft climbs. When the aircraft descends, the pressure must increase. Since the heater may not be able to vaporize enough gas to keep pace with the necessary increase in pressure, the unit must suck in outside air. The moisture

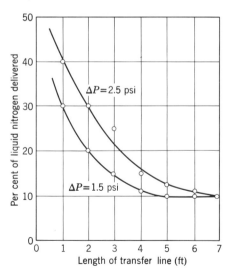

Figure 11.3 Percent of liquid nitrogen transferred through an aluminum tube with an inner diameter of 0.075 in. (adapted from data in Goodenough and Swift[9]).

in this air will freeze out and eventually stop the operation of the unit. This difficulty is avoided by including an absolute pressure reference so that the system always maintains a pressure that is 1 to 3 psi greater than that at sea level.

Since the transfer line is small and flexible, it is easy to cool a detector mounted on a moving structure. Because this line will not transmit vibrations to the detector, microphonics are not a problem. The obvious disadvantage of this type of unit is that the storage dewar must be refilled periodically[10,11]. One might think that the logistics of providing liquid nitrogen would be difficult. They certainly are in some situations, but they are often quite simple around military aircraft since many such aircraft carry liquid oxygen and a simple converter for the pilot's breathing supply. The base, therefore, has facilities for producing liquid oxygen and personnel skilled in handling it. Liquid-oxygen machines also produce liquid nitrogen, which is normally used to precool the incoming oxygen, but it can also be drawn off and used to cool infrared detectors. Walker [12] has designed a liquid-transfer cooler that uses liquid oxygen from the pilot's supply and returns the spent gas for the pilot to breathe.

Joule–Thomson Refrigerators

Joule–Thomson Refrigerators use a Joule–Thomson *cryostat*, a miniature gas liquefier that can be placed directly in the coolant chamber of a detector package.[1] Details of the cryostat and its operating principles are shown in Figure 11.4. High-pressure gas, cooled by expansion at the throttle valve, flows back through the counter-current heat exchanger and precools the incoming gas until the gas is liquefied as it leaves the throttle valve.

The Joule–Thomson process has been used to liquefy gases for many years. The development of a miniaturized cryostat for use with infrared detectors was carried out in England during the early 1950's by Parkinson of RRE. In its most common form, a cryostat consists of a cylindrical mandrel carrying a helically wound coil of finned metal tubing [13–15]. In a typical commercially available cryostat, the mandrel is 1.5 in. long and the diameter of the mandrel plus finned tubing is about 0.2 in. The finned tubing serves as the counter-current heat exchanger. The simplest throttle valve is a nozzle of 0.001 to 0.003 in. diameter formed at the

[1]Unfortunately the word "cryostat" means different things to different people. The infrared community has for many years called the Joule-Thomson liquefier a cryostat. Cryogenic specialists use the word to refer to a regulator for low temperatures. They consider that a dewar is a cryostat, since it is a means of stabilizing the temperature of an object at a given value. Because workers in the infrared field, as well as manufacturers of the device, are unanimous in calling the miniature liquefier a cryostat, this usage will be adopted here.

TECHNIQUES FOR COOLING DETECTORS

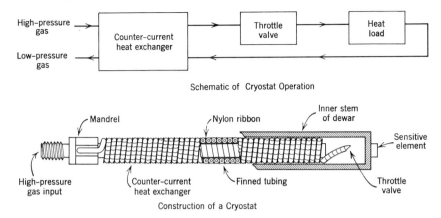

Figure 11.4 The Joule–Thomson cryostat.

end of the tubing. Some more recent designs omit the nozzle and throttle throughout the length of the tubing. The sacrifice in efficiency is quite small, but the possibility of dirt plugging the nozzle is greatly reduced. The cryostat slides snugly into the inner stem of the dewar, and the stream of liquefied gas is directed toward the back of the surface carrying the sensitive element. After expansion, the cooled gas flows along the heat exchanger and extracts heat from the incoming gas. In a well-designed heat exchanger, the gas leaving the cryostat is almost at room temperature. Figure 11.5 shows the stream of liquid nitrogen leaving a cryostat. The cryostat shown there throttles throughout the length of the tubing and therefore has no nozzle. The tubing has an outer diameter of 0.020 in. and an inner diameter of 0.010 in. An alternative type of construction for a cryostat has been described by Pastuhov[16].

In order to liquefy a gas, heat must be withdrawn from it. If this is done by an unrestrained expansion or *throttling* of the gas, as in the cryostat, the Joule–Thomson coefficient of the gas becomes an important design parameter. It is given by

$$\mu = \left(\frac{\partial T}{\partial P}\right)_h, \qquad (11\text{-}1)$$

where T is the temperature, P is the pressure, and h the enthalpy (a measure of the total heat content of the gas), which remains constant during the process. The value of μ is a function of both temperature and pressure, and it must have a positive value if cooling is to take place. Typically, the value of μ is positive at low temperatures, negative at high temperatures, and zero at the inversion temperature. Hence, to be useful in a cryostat, a gas must be cooled below its inversion tempera-

Figure 11.5 A Joule–Thomson cryostat in operation (courtesy of Santa Barbara Research Center, Goleta, California).

ture before it is throttled. Table 11.2 gives the inversion temperature of several commonly used gases. Note that for neon, hydrogen, and helium the inversion temperature is below room temperature and these gases cannot be used in a cryostat without precooling them. A dual cryostat can be designed in which the first stage liquefies nitrogen that is then used to precool hydrogen or neon for use in the second stage[16].

To produce temperatures in the 80°K region, the only practical gases for a cryostat are nitrogen (77°K) and argon (87°K). As the data in Chapter 10 show, the D^* of most suitable detectors will be the same at either temperature (the one exception to this statement is photoconductive indium antimonide; see Figure 10.12). However, the *cooldown*

TABLE 11.2 APPROXIMATE INVERSION TEMPERATURES OF COMMONLY USED GASES

Gas	Inversion Temperature (°K)
Oxygen	890
Argon	720
Nitrogen	620
Neon	225
Hydrogen	205
Helium	50

time, the length of time it takes before liquefaction begins, is quite different for the two gases. The Joule–Thomson coefficient for argon is about double that for nitrogen. Hence the cooldown time for argon is about half that for nitrogen. Evers[14] describes a nitrogen cryostat having a cooldown time of 2 min; with argon it should be about 1 min. Cryostats can be designed to have cooldown times as short as 5 sec. Cooldown time will, of course, vary with the temperature of the input gas; the warmer the gas is, the longer it will be before liquefaction takes place.

The gas supplied to the cryostat must be at a high pressure; most cryostats will cease functioning if the gas pressure drops below 1000 psi. The maximum usable pressure is about 6000 psi for nitrogen and 7500 psi for argon. Table 11.3 shows what the rate of gas flow must be for nitrogen and argon cryostats to produce 1 W of refrigeration.

Since the diameter of the throttle valve is only 0.001 to 0.003 in., it is imperative that the gas and all supply lines be kept scrupulously clean. A sintered-metal filter at the input to the cryostat will prevent particles larger than $5\,\mu$ in diameter from reaching the valve. The gas will almost certainly contain water vapor that will become frozen and plug the valve. The most effective means of removing the water vapor is to use a filter containing Molecular Sieve (a product of Linde Company, a division of Union Carbide)[17]. Many people erroneously believe that cryostats cannot be operated for more than a few minutes without becoming plugged. This is not true; if one exercises reasonable care and follows the precautions mentioned here, a cryostat will operate satisfactorily for tens of hours.

Although most cryostats are designed for use with nitrogen or argon, they can also be used with the various members of the Freon ® family

TABLE 11.3 RATE OF GAS FLOW TO PRODUCE 1 W OF
REFRIGERATION WITH A JOULE-THOMSON CRYOSTAT

Gas	Temperature of Gas Supplied to Cryostat (°K)	Flow Rate,[a] Liter min^{-1}		
		$P = 1500$ psi	$P = 3000$ psi	$P = 4500$ psi
Nitrogen	350	3.6	2.4	1.9
Argon	350	2.8	1.5	1.2
Nitrogen	295	2.4	1.5	1.2
Argon	295	1.9	1.1	0.8
Nitrogen	273	2.0	1.2	1.0
Argon	273	1.6	0.9	0.7

[a] Reduced to standard temperature and pressure.

to produce a wide range of temperatures. These include Freon 12 (243°K), Freon 22 (233°K), Freon 13B1 (216°K) Freon 13 (192°K), and Freon 14 (145°K). Liquid carbon dioxide can be expanded through a nozzle to form carbon dioxide snow at a temperature of 195°K; no heat exchanger is needed[18, 19].

Solid-Refrigerant Coolers

The need to cool infrared detectors on space vehicles has led to interest in a cooler that can operate for several months, or even years, in the space environment[20]. Such a cooler must operate under zero-gravity conditions, consume minimum power, be highly reliable, and weigh as little as possible. One solution to this problem consists of an insulated container for storing a solid refrigerant and a thermally-conducting rod to transfer heat from the detector to the refrigerant. The temperature of the refrigerant is a function of the pressure maintained in the storage container; in space a relief valve can be used to maintain the desired pressure.

At least six acceptable solid refrigerants are available. Table 11.4 shows the range of temperatures that each provides and the pressure needed to achieve these temperatures. The solid refrigerant is formed by first filling the storage tank with the desired refrigerant in liquid form. Next a vacuum pump reduces the pressure in the tank so that the liquid boils. Since the storage container is an excellent thermal insulator, the energy necessary to support the boiling must come from the liquid; the temperature of the liquid, therefore, drops until a solid is finally formed.

Gross and Weinstein[20] have calculated the weight of solid refrigerant

TABLE 11.4 ATTAINABLE TEMPERATURES AND REQUIRED STORAGE PRESSURES FOR SOLID REFRIGERANTS[a]

Refrigerant	Temperature Range (°K)	Range of Storage Pressure (mmHg)
Methane	90–67	80–1
Argon	83–55	400–1
Carbon monoxide	68–51	10–1
Nitrogen	62–47	73–1
Neon	24–16	260–1
Hydrogen	14–10	56–1.7

[a] Adapted from data in Gross and Weinstein[20].

and storage container necessary to cool a 0.1 W heat load for one year; see Table 11.5.

TABLE 11.5 WEIGHT OF SOLID REFRIGERANT AND CYLINDRICAL STORAGE CONTAINER NEEDED TO COOL A 0.1 W HEAT LOAD FOR ONE YEAR. LENGTH OF CONTAINER IS EQUAL TO ITS DIAMETER AND THE TEMPERATURE OF ITS OUTER SHELL IS 300°K[a]

Refrigerant	Temperature (°K)	Weight of Solid Refrigerant and Storage Container (lb)
Methane	88	26
Argon	84	66
Carbon monoxide	68	59
Nitrogen	61	62
Neon	24	119
Hydrogen	13	66

[a] Adapted from data in Gross and Weinstein[20].

Radiative-Transfer Coolers

Another means of cooling detectors in space vehicles is to transfer the heat from the detector and radiate it to outer space. In principle, the detector is fastened to one end of a copper bar that extends through the side of the vehicle. The outer end of the bar is formed into a hemispherical bulb that is blackened to give it a high emissivity, and is so oriented that it sees only outer space. Highly efficient radiation shields are needed to prevent the bar from seeing the vehicle on which it is mounted. The requirement that the hemispherical bulb see only outer space is not difficult to meet if the vehicle is space stabilized.

A detailed analysis of the factors involved in the design indicates that with a radiative-transfer cooler it should be possible to cool a detector to a temperature as low as 100°K. Several satellite-borne infrared sensors have used such a cooler to maintain detectors at 195°K [21–24].

Comparison of Typical Open-Cycle Refrigerators

The characteristics of typical open-cycle refrigerators are shown in Table 11.6.

TABLE 11.6 CHARACTERISTICS OF TYPICAL OPEN-CYCLE REFRIGERATORS[a]

	Liquid Transfer	Joule–Thomson		Solid Refrigerant		Radiative Transfer
Operating temperature (°K)	77	77	87	13	88	195
Heat load (W)	0.25	0.25		0.1		0.1
Standby time	24 hr	5 yr		N.A.		No limit
Cooldown time (min)	5	2	1	N.A.		600
Operating time	4 hr	3 hr		1 yr		No limit
Weight, including refrigerant (lb)	8	15		66	26	5
Power consumption	5 W intermittently, to vaporize liquid	5 W momentarily, to open valve at start		—		—

[a]Data adapted from Gross and Weinstein[20], publications of Hughes Aircraft Co., and of Santa Barbara Research Center.

11.4 CLOSED-CYCLE REFRIGERATORS

With the refrigerators discussed thus far there is no attempt made to collect and reuse the spent refrigerant. In a closed-cycle refrigerator the spent refrigerant is collected and fed back through the refrigeration system. Hence the operation of a closed-cycle refrigerator is analogous to that of the ordinary home refrigerator. A great many refrigeration cycles are known[25], but few are practical when the refrigerator must be scaled down to the small size required by many infrared system applications. Those that are practical include the Joule–Thomson, Claude, Stirling, and derivatives of the Stirling cycle. A considerable developmental effort is being expended on refrigerators of these types. Since

TECHNIQUES FOR COOLING DETECTORS 387

there has not yet been a large-scale procurement of any such refrigerator, there is no existing body of performance data against which a system engineer can judge the conflicting claims of competing manufacturers.

A figure of merit that is commonly used to express the effectiveness of a refrigeration system is the *coefficient of performance* (COP). It is defined as the ratio of refrigeration produced to the net power supplied to the refrigerator from all external sources. Although it is a dimensionless quantity, it is convenient to think of COP in terms of watts of cooling per watt of input power. In principle COP should be a foolproof means of comparing competing coolers; in practice it rarely is, unless one can find exactly what has or, more likely, what has not been included in summing the net input power. For instance, since these refrigeration cycles include a compression step, the heat of compression must be removed by using a pump to circulate a coolant or a fan to circulate cooling air. In either case the power consumed by the pump (typically 50 W) or by the fan (typically 100 W) should be included in the calculation of the power input. The reader is warned that not all manufacturers include such items in their advertised values of COP.

Joule-Thomson (Closed-Cycle) Refrigerators

The operation of the Joule-Thomson cryostat was described in Section 11.3. It can be converted to closed-cycle operation by adding a compressor to repressurize the gas leaving the cryostat. The compressor is the critical element in such a refrigerator since it must produce high pressures while introducing virtually no contaminants. An oil-lubricated compressor is desirable, but its operating time is limited by the lifetime of the filters necessary to prevent contamination by the oil. Dry, or unlubricated, compressors eliminate the oil problem, but have a relatively short life because of the high wear rates involved. Diaphragm pumps would provide an ideal choice if a suitable long-lived diaphragm material could be found.

Since the line connecting the compressor to the liquefier carries only high-pressure gas and needs no thermal insulation, it can be of almost any length. This means that in aircraft or ship installations the compressor can be located at some remote point where space is not at a premium. Since the volume of low-pressure gas leaving the liquefier is several hundred times that of the high-pressure gas at the input, the return line to the compressor is relatively large and rigid. As a result, this type of refrigerator is difficult to use with scanning systems in which the detector is mounted on a moving structure. A method that minimizes this problem is to insert a cold box close to the detector. The high-pressure gas is liquefied in this cold box, and the liquid is transferred to the detector through flexible lines. This arrangement represents, in effect, an open-cycle Joule–Thomson unit combined with a liquid-transfer cooler. It also offers the possibility of being a direct replacement for previously installed liquid-transfer coolers.

The primary advantage of a Joule–Thomson closed-cycle refrigerator is the simplicity of construction in its cold portions, because it contains no moving parts that are at a low temperature. Since the throttling is thermodynamically irreversible, the Joule–Thomson process is, even under ideal conditions, two to three times less efficient than the Claude or Stirling cycles. The COP for a Joule–Thomson refrigerator is about 1/100 at a refrigerant temperature of 77°K. This type of refrigerator is best suited for use with systems that are severely space limited in the region of the detector, and for refrigerant temperatures of 77°K and higher.

Claude Refrigerators

The Claude refrigerator is similar to the Joule–Thomson in that both use a counter-current heat exchanger to bring the refrigeration potential to the desired load temperature. In the Claude cycle, however, the irreversible Joule–Thomson expansion is replaced by a more efficient thermodynamically reversible engine expansion. During expansion the gas does work against an expansion engine, which transfers this work to an area at ambient temperature and thereby creates a refrigeration potential in the load area. The operation of the expansion engine is analogous to that of a steam engine, which accepts steam at a high pressure and temperature, extracts and delivers work, and finally exhausts the steam at a lower temperature and pressure. In the Claude refrigerator the gas from the expansion engine is used to cool the load and is then injected into the heat exchanger to cool the incoming gas. No liquid is produced during this process. A single-stage Claude refrigerator that uses helium as the working gas can cool to about 35°K. Three stages in cascade can reach a temperature of 10°K; the helium can then be passed through a Joule–Thomson cryostat to reach a temperature of 4.2°K. The COP is about 1/50 at 40°K, 1/100 at 20°K, and 1/200 at 4.2°K.

Because it is a reversible process, the expansion in the Claude cycle can produce the same refrigeration potential as the Joule–Thomson cycle does but at a much lower pressure ratio. Therefore the compressor in the Claude cycle must produce pressures of only 10 to 15 atmospheres and compressor wear is correspondingly less. The type of gas used is not critical, provided that it does not liquefy. The lighter gases, such as helium, are generally preferred since they are more efficient in a heat exchanger because of their lower density and greater thermal conductivity. Since it contains no extremely small orifices that can become clogged, the Claude refrigerator can tolerate gas containing about 10 times more vaporized contaminants than can be tolerated by a Joule–Thomson refrigerator.

Unfortunately, in the Claude cycle many of the moving parts must operate at low temperatures. Obtaining good mechanical action and long life with miniaturized components in a cryogenic environment is an extremely difficult engineering problem. The expansion engine can be either a reciprocating or a turbine type. Both types pose design problems because of the small sizes required. Since the detector is mounted adjacent to the expansion engine, vibration-induced microphonics are likely to be a serious problem. For additional information on refrigerators of this type, see refs. 26 and 27.

Stirling Refrigerators

The Stirling refrigerator is thermodynamically more complex than either of those previously considered[28–30]. It consists of two sets of pistons and cylinders and a thermal regenerator through which the gas passes back and forth between the cylinders. One piston operates as a compressor and the other as an expander. The thermal regenerator serves the same purpose as the heat exchanger used in the Claude and Joule–Thomson cycles; that is, it brings the refrigeration potential to load temperature. It is the most critical element in the Stirling refrigerator and ultimately limits the lowest temperature attainable to about 15°K. For maximum efficiency the regenerator should have a high heat capacity and a large area for efficient heat transfer. It is usually made in the form of a chamber filled with a finely divided mass, such as lead-plated copper gauze or lead shot. Lead is a favorite material since it is one of the few that maintains a high specific heat at temperatures as low as 15°K.

Because of the regenerator the Stirling refrigerator is remarkably tolerant of contaminants in the working gas. An analysis of its action shows that vaporized contaminants condense on the regenerator packing when the gas flows to the expander and that they are flushed away when the direction of flow reverses.

A very low pressure ratio is used in the Stirling refrigerator; it is approximately 3 to 1 with a high pressure that is typically 15 atmospheres. The combination of low pressure and low pressure ratio ensures a long operating life for moving parts. This advantage, combined with the absence of valves and the freedom from the effects of gas contamination, indicates that the Stirling refrigerator should have a long and reliable operating life. The COP is about 1/30 for a refrigerant temperature of 77°K.

One disadvantage of the Stirling refrigerator is the possibility of microphonics, since the detector is mounted on the expander. Fresh et al.[31] have reported measurements of the noise spectra generated by a Stirling refrigerator when used to cool a mercury-doped germanium detector. Their measurements show that the noise due to the refrigerator may be as much as double that from the detector alone, at frequencies below 400 Hz and above 2000 Hz.

As in the Claude refrigerator, the working gas, which is usually helium, does not liquefy. A single stage will produce temperatures as low as 20°K. The temperature at the expander is the equilibrium temperature that results from the refrigeration capacity of the cooler and the magnitude of the heat load. Hence, by including a small electrical heater as part of the heat load, any desired temperature to as low as 20°K can be maintained. For still lower temperatures, additional stages can be cascaded. In one such three-stage Stirling refrigerator[32,33] capable of producing 1 W of refrigeration at 4.2°K, the temperature at the compressor is 300°K, whereas it is 100°K at the first expander, 40°K at the second, and 15°K at the third, where the gas feeds a Joule–Thomson cryostat to produce a temperature of 4.2°K.

Cowans and Walsh[34] analyzed the requirements for airborne infrared systems and devised an interesting variant of the Stirling refrigerator for use when the detector is mounted on a moving structure. A simple heat exchanger added to the sealed refrigerator liquifies air. Conventional liquid-transfer techniques are then employed to carry the liquid to the detector. Such a system enjoys the inherently high efficiency of the Stirling cycle, entails no ground-support problems, and does not impair the performance of a scanning infrared system.

Refrigerators Using Other Refrigeration Cycles

The Solvay-modified Stirling refrigerator consists of two units: a compressor, that can be located at any convenient point, and a cryoengine that contains the expander and detector. The cryoengine fits in a cylindrical package that has a diameter of 2 in., a length of 10 to 15 in., and a weight of 2 to 4 lb. It is small enough that it can often be mounted directly on a moving structure. The lines connecting the cryoengine to the compressor can be any desired length. The COP is about 1/300 at 30°K and 1/1250 at 12°K.

Like the Solvay, the Gifford–McMahon refrigerator employs a remotely located compressor[35]. Development efforts to date have been devoted to large refrigerators, weighing about 900 lb, for use in fixed terrestrial locations. The COP is about 1/500 at 14°K. Nadol'nikov et al.[36] have built a miniature version of this refrigerator that reaches a temperature of 50°K. The unit is incomplete, since it does not include a compressor;

TECHNIQUES FOR COOLING DETECTORS 391

high-pressure gas is obtained from commercially supplied storage cylinders. The paper gives an indication of the size of the unit, but not of its weight.

Comparison of Typical Closed-Cycle Refrigerators

The characteristics of some commercially available closed-cycle refrigerators are shown in Table 11.7.

11.5 SOLID-STATE REFRIGERATORS

Solid-state refrigerators use various thermoelectric and thermomagnetic effects that have been observed for many years. The relatively recent interest in semiconductors has sparked a renewed effort to utilize these effects. A few years ago solid-state refrigerators were billed as the answer to a cooling engineer's prayers. It is now clear that what they have to offer is of limited utility for cooling detectors, and there seems to be little reason to think that this conclusion will be modified very much in the future.

Thermoelectric Refrigerators

In 1821 Seebeck observed the current caused by heating the junction of two dissimilar conductors, which is the basis for the operation of the thermocouple. In 1834 Peltier found that when current flows in a circuit consisting of two dissimilar metals, heat is absorbed at one junction and released at the other. Since one junction is heated while the other is cooled, the Peltier effect can be used for either heating or cooling by simply reversing the current's direction. In 1838 Lenz succeeded in freezing a drop of water at a Peltier junction. Because the efficiency of these early thermoelectric coolers was so very low, they were, for nearly a century, regarded as mere curiosities, and it is only the recent availability of appropriate semiconductor materials that has rescued them from this category. The reader interested in learning more of their development and of some of their uses should see [37–39].

The suitability of materials for thermoelectric devices can be judged by their *thermoelectric figure of merit*:

$$z = \frac{\alpha^2}{\rho k}, \qquad (11\text{-}2)$$

where α is the Seebeck or thermoelectric coefficient in $V(°K)^{-1}$, ρ is the resistivity in ohm cm, and k is the thermal conductivity in $W(°K)^{-1}$ cm^{-1}. Semiconductors provide the highest values of z [40]. For bismuth telluride, the most commonly used material, z is equal to 3×10^{-3} $(°K)^{-1}$. It may be doped to form either a p- or an n-type material. Small bars of p- and

TABLE 11.7 CHARACTERISTICS OF COMMERCIALLY AVAILABLE CLOSED-CYCLE REFRIGERATORS[a]

	Joule–Thomson	Stirling	Stirling with External Air Liquefier and Liquid Transfer	Solvay	Stirling, Two-Stage	Solvay, Two-Stage	Solvay, Three-Stage plus Joule–Thomson Stage
Operating temperature, °K	81	77	77	28	25	12	4.2
Refrigeration capacity (W)	2	15	10	2	2.5	0.4	2 @ 4.2°K 9 @ 15°K 15 @ 50°K 50 @ 150°K
Cooldown time (min)[b]	3	3	3	20	15	35	90
Power consumption (W)	350	500	450[c]	575[c]	450[c]	575[c]	6600
Weight (lb)	16	15	13	30	10	30	300
Method of removing heat of compression	Air, with self-contained blower	Air, with self-contained blower	Liquid, 0.5 GPM from external supply	Liquid, 2.5 GPM from external supply	Liquid, 0.5 GPM from external supply	Liquid, 2.5 GPM from external supply	Air, with self-contained blower

[a] Data adapted from publications of the Hughes Aircraft Co. and of the Santa Barbara Research Center.
[b] Depends on detector package.
[c] Does not include power required to remove heat of compression.

TECHNIQUES FOR COOLING DETECTORS 393

n-type material are soldered to copper straps to form a thermoelectric couple. The maximum temperature differential than can be achieved with such a couple occurs for a zero heat load, and is

$$\Delta T_{\max} = 0.5 z T_c^2, \qquad (11\text{-}3)$$

where T_c is the temperature of the cold junction in °K. Using the value of z stated above, the maximum temperature differential across such a couple is about 75°C when the hot junction is at 300°K. In practice it is found that temperature differentials of 60 to 65°C can be achieved with an infrared detector as a heat load.

In order to reach still lower temperatures, several couples can be cascaded, that is, the hot junction of one couple placed in good thermal contact with the cold junction of the next. A three-stage device is shown in Figure 11.6. The total temperature differential across the three stages is 140°C; when the temperature of the hot junction is 335°K, for example, that of the cold junction is 195°K, the sublimation temperature of carbon dioxide. This device is 1 in. square and 0.75 in. high (without thermal insulation) and requires a current of 4 amp from a 6 V source. Because of its size it will not fit in most detector dewars. Instead, various foamed insulations are used to form the package containing the detector and thermoelectric cooler[41]. Variables in the design of such a cooler include

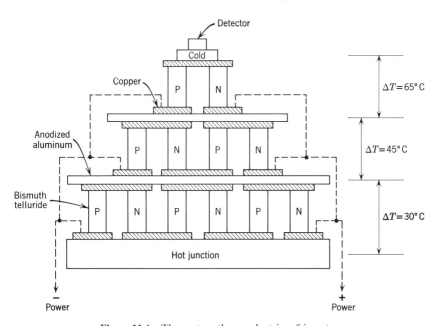

Figure 11.6 Three-stage thermoelectric refrigerator.

the current, the ratio of the length to the area of each arm in the couples, the heat load, the variation in z with temperature, and whether maximum cooling or maximum efficiency is required[42].

One cannot continue the cascading process indefinitely in order to obtain still lower temperatures; three or four stages are about the practical limit, beyond which efficiency becomes hopelessly low. Thermoelectric refrigerators therefore do not produce temperatures that are sufficiently low to be of interest in most detector applications. A point that has been overlooked by many advocates of this type of device is that a thermoelectric refrigerator is actually a heat pump. Heat on the order of tens of watts must be removed from the hot junction. If the detector and cooler are mounted on a moving structure, it is virtually impossible to remove such quantities of heat. In other locations it may not be impossible to remove the heat, but it may still be inconvenient. Since the bismuth telluride used for the couples is very difficult to fabricate in the shape needed for low-current operation, many of the thermoelectric devices on the market require extremely high currents and low voltages, 100 amp at 0.5 V, for instance — a very difficult way to have to supply power. The refrigerator shown in Figure 11.6, which requires 4 amp at 6 V, is an exception because its manufacturer, the Santa Barbara Research Center, has developed proprietary techniques for fabricating the couples.

Thermomagnetic Refrigerators

It is not at all clear that these refrigerators have any potential as a means for cooling infrared detectors. They use semiconductor materials placed in a strong magnetic field to produce cooling by virtue of the Ettingshausen effect[43,44].

11.6 INTEGRATING THE DETECTOR AND REFRIGERATOR

The rapid development of practical refrigerators for infrared detectors has brought with it a host of new problems concerned with the way in which a detector and refrigerator can best be integrated into a working unit. Unless the interface problems are identified and their solutions specified early in the development cycle, the system engineer may find, all too late, that the detector procured from manufacturer A will not perform acceptably with the cooler from manufacturer B. Until the cryogenic engineer and the infrared system engineer understand each other's problems far better than they do today, it is probably a good precaution to procure both the detector and the refrigerator from a single supplier, and to hold him responsible for achieving optimum performance from the combination.

The type of information that the system engineer should provide includes the type of detector, the desired operating temperature and an absolute tolerance on this temperature, the oscillations in detector temperature that can be tolerated, the heat load to be refrigerated (this includes the radiation load falling on the cold spot, the bias power, and the heat conducted down the leads to the detector), the cooldown time, the power and space available for the refrigerator, the minimum distance between the detector and the refrigerator, and the acceptable excess noise at the output of the detector caused by operation of the refrigerator.

Given this information, the cryogenic engineer can start his design work. His problems include the choice of a refrigerator type that is most likely to meet the system requirements, the means of mounting the detector to the refrigerator (so as to achieve the best thermal coupling while introducing a minimum of excess noise in the detector circuit), the means of best reducing and controlling the various heat loads imposed on the refrigerator, the best means of bringing the leads from the detector to the outside world, and methods for maintaining any vacuum integrity required by the refrigerator.

Relatively little discussion of these problems has appeared in the literature. For what is available, see refs. 28, 34, and 45.

REFERENCES

[1] R. W. Vance and W. M. Duke, *Applied Cryogenic Engineering*, New York: Wiley, 1962, p. xi.
[2] G. D. Hayball, E. W. Peterson, and T. H. Johnson, "Bolometer," U.S. Patent No. 3,052,861, September 4, 1962. (A most misleading title since the invention pertains to improved detector packages.)
[3] M. F. Amsterdam, "Cooled Infrared Radiation Detector," U.S. Patent No. 3,114,041, December 10, 1963.
[4] J. T. Harvell, "Working With Cryogenics," *Microwaves* 3, 22 (April 1964).
[5] M. G. Zabetakis, "Hazards in the Handling of Cryogenic Fluids," in K. D. Timmerhaus, ed., *Advances in Cryogenic Engineering*, New York: Plenum, 1963, Vol. 8, p. 236.
[6] J. Rousseau, "Cryogenic Storage Vessels," *Space/Aeronautics*, 37, 61 (March 1962).
[7] R. H. Kropschodt, "Low-Temperature Insulation," in R. W. Vance and W. M. Duke, *Applied Cryogenic Engineering*, New York: Wiley, 1962, Chapter 6.
[8] L. E. Miller, "Method and Apparatus for Transfer of Cryogenic Liquid," U.S. Patent No. 3,126,711, March 31, 1964.
[9] J. G. Goodenough and I. H. Swift, "Low Temperature Liquid Transfer Apparatus," U.S. Patent No. 2,966,893, August 22, 1962.
[10] G. C. Haettinger, R. P. Skinner, and R. A. Trentham, "Design Considerations for Cryogenic Liquid Refill Systems for Cooling Infrared Detection Cells," in K. D. Timmerhaus, ed., *Advances in Cryogenic Engineering*, New York: Plenum, 1961, Vol. 6, p. 354.

[11] J. S. Tyler and J. A. Potter, "Infrared Detector Refrigerators," in K. D. Timmerhaus, ed., *Advances in Cryogenic Engineering,* New York: Plenum, 1961, Vol. 6, p. 363.
[12] H. R. Walker, "Liquid Oxygen Cooler for Airborne Infrared Cells," U.S. Patent No. 3,016,716, January 16, 1962.
[13] S. Gross, "Infrared Sensor Cooling by the Joule-Thomson Effect," *Infrared Phys.,* **6**, 47 (1966).
[14] D. D. Evers, "Cryogenic Refrigerating Means," U.S. Patent No. 3,018,643, January 30, 1962.
[15] H. P. Wurtz, "Complete Liquefaction Cryostat," *Infrared Phys.,* **1**, 197 (1961).
[16] A. Pastuhov and F. J. Zimmerman, "Miniature Refrigeration Device," U.S. Patent No. 2,909,908, October 27, 1959.
[17] D. W. Breck and J. V. Smith, "Molecular Sieves," *Sci. Am.,* **200**, 85 (1959).
[18] D. H. Dennis and L. V. Hebenstreit, "Cooling Apparatus," U.S. Patent No. 3,067,589, December 11, 1962.
[19] F. L. Hutchens, "Cooling Device for Infrared Detector," U.S. Patent No. 3,079,504, February 26, 1963.
[20] U. E. Gross, and A. I. Weinstein, "A Cryogenic-Solid Cooling System," *Infrared Phys.,* **4**, 161 (1964).
[21] "Nimbus IR Sensor to Map Cloud Cover," *Electronic Design,* **12**, 25 (June 22, 1964).
[22] W. Nordberg, "Research With Tiros Radiation Measurements," *Astronautics and Aerospace Eng.,* **1**, 76 (April 1963).
[23] "Spectrophotometer to Investigate Existence of Vegetation on Mars," *Aviation Week,* **74**, 139 (May 1, 1961).
[24] "IR Device Built to Detect Any Vegetation on Mars," *Electronic Design,* **9**, 6 (April 12, 1961).
[25] S. C. Collins and R. L. Canaday, *Expansion Machines for Low Temperature Processes,* London: Oxford University Press, 1958.
[26] H. R. Senf, "Masers for Radar Systems Applications," *IRE Trans. Mil. Electronics,* **5**, 58 (1961).
[27] "Cryogenic Refrigerator Weighs Fifty Pounds," *Space/Aeronautics,* **39**, 107 (April 1963).
[28] J. W. L. Kohler, "The Stirling Refrigeration Cycle," *Sci. Am.,* **212**, 119 (April 1965).
[29] S. F. Malaker and J. G. Daunt, "Miniature Cryogenic Engine," U.S. Patent No. 3,074,244, January 22, 1963.
[30] F. K. Dupre and A. Daniels, "Miniature Refrigerator Opens New Possibilities for Cryo-Electronics," *Signal,* **20**, 10 (September 1965).
[31] D. L. Fresh, F. K. Dupre, and J. P. Walker, Jr., "Long Infrared Wavelength Detection with Mercury Doped Germanium and Stirling Cycle Cooling," *Proc. Natl. Electronics Conf.,* 1963 p. 635.
[32] S. F. Malaker and J. G. Daunt, "Closed Cycle Cryogenic System," U.S. Patent No. 3,128,605, April 14, 1964.
[33] S. F. Malaker and J. G. Daunt, "Multi-Stage Cryogenic Engine," U.S. Patent No. 3,147,600, September 8, 1964.
[34] K. W. Cowans and P. J. Walsh, "Continuous Cryogenic Refrigeration for 3-to-5-Micron Infrared Systems," in K. D. Timmerhaus, ed., *Advances in Cryogenic Engineering,* New York: Plenum, 1965, Vol. 10, p. 468.
[35] W. E. Gifford and T. E. Hoffman, "A New Refrigeration System for 4.2°K," in K. D. Timmerhaus, ed., *Advances in Cryogenic Engineering,* New York: Plenum, 1961, Vol. 6, p. 82.

[36] A. G. Nadol'nikov, V. G. Fastovskii, and Y. V. Petrovskii, "A Miniature Refrigerating Machine," *Cryogenics*, **6**, 342 (1965). (This paper originally appeared in *Pribory i Tekhnika Eksperimenta*, **6**, 188 (1963).)
[37] A. F. Joffe, "The Revival of Thermoelectricity," *Sci. Am.*, **199**, 31 (1958).
[38] A. F. Joffe, *Semiconductors and Their Use*, 3d ed., Moscow: Foreign Languages Publishing House.
[39] A. G. Heaton, "Thermoelectric Cooling: Material Characteristics and Applications," *Proc. Inst. Elec. Engrs.*, **110**, 1277 (1963).
[40] R. W. Ure, Jr., "Theory of Materials for Thermoelectric and Thermomagnetic Devices," *Proc. IEEE*, **51**, 699 (1963).
[41] R. W. Ure, Jr. and E. V. Somers, "Self-Cooled Infrared Detection Cell, "U.S. Patent No. 3,103,587, September 10, 1963.
[42] M. B. Grier, "Lumped Parameter Behavior of the Single-Stage Thermoelectric Microrefrigerator," *Proc. Inst. Radio Engrs.*, **47**, 1515 (1959).
[43] R. Wolfe, "Magnetothermoelectricity," *Sci. Am.*, **210**, 70 (1964).
[44] S. R. Hawkins, "Low Temperature Ettingshausen Coolers," in K. D. Timmerhaus, ed., *Advances in Cryogenic Engineering*, New York: Plenum, 1964, Vol. 9, p. 367.
[45] J. N. Crouch, "Cryogenic Cooling for Infrared," *Electro-Technology*, **75**, 96 (May 1965).

12

Signal Processing and Displays

The signal processor receives the low-level signal from the detector, amplifies it, limits the bandwidth, extracts the information the signal contains, and delivers this information to the final control device or display. For the most part, the techniques employed are quite similar to those used with radar, sonar, and television. The purpose of this chapter is to review these techniques and to see how they are applied.

In the late 1950's some of the more resourceful circuit designers began to introduce transistors, printed circuits, and other miniaturized components into their infrared systems. There were, of course, some growing pains in the process, but the reward was a spectacular decrease in system size and weight and, sometimes, an increase in reliability. The early 1960's saw the development of microelectronics and its inevitable impact on system design[1, 2]. It is clear that the military will increase the use of microelectronics in infrared systems[3-7] and their application to nonmilitary projects will not be far behind. In addition to further reductions in system size and weight, microelectronics appears to offer a genuine potential for increased reliability along with decreased cost.

It is not yet clear how a designer can best use microelectronics in infrared systems. The U.S. Department of Defense has issued a policy that will ensure that all of its new developmental programs give serious consideration to the use of microelectronics[8]. The goals of this policy on the use of microelectronics are to maximize reliability, minimize cost, provide a measure of logistic self-support, and encourage the use of throwaway packages. The use of standard microelectronic circuits is not favored, because it may place undesirable constraints on new system designs. The use of hybrid circuits is encouraged because of the wider variety of circuit components that are available.

Since it seems certain that digital signal processing techniques will be used much more widely with microelectronic circuits, most older circuit designers will have to supplement their experience in analog circuits with new knowledge concerning digital circuitry.

SIGNAL PROCESSING AND DISPLAYS 399

Some system engineers will have to face the problem of when and how to introduce microelectronics into an existing design. The original Redeye infrared-guided missile used transistorized electronics. Less than a year after this version went into production it was superseded by a model that used microelectronics[9]. The transition was so rapid because the original circuits were not redesigned to take the fullest advantage of the advantages offered by microelectronics. Instead, hybrid replicas of the existing discrete component circuits were built. The final result was a total of 26 monolithic or hybrid circuits for the new production model.

12.1 GENERAL CONSIDERATIONS

The circuit designer must first estimate the signal levels and the gains required at the various stages of the signal processor. Unless system sensitivity is to be limited, there must be sufficient gain to bring signals that are comparable to the level of the noise from the detector up to such a level that they can be viewed on a display or used to actuate control circuitry. From the data sheets in Chapter 10 it is evident that detector noise levels of a few tens of microvolts can be expected. Typical displays may require 100 V or more. Hence a net voltage gain of about 10^7 is indicated. Since the detector is usually located in a confined area and may be on a moving structure as well, little of the signal processing can be done at the detector. A small preamplifier, placed close to the detector, provides sufficient amplification to allow the signal to be transmitted through low-impedance shielded cables to a point where there is room for the rest of the signal processor.

Having determined the conditions for low signal levels, the designer must next look at what happens when high signal levels are encountered. With some systems that detect a target at a great distance and eventually approach quite close to it, the signal level may change by a factor of 10^4 to 10^6. To provide this much dynamic range in the signal processor is often an exceedingly difficult task. If the signal ultimately goes to a display, there are also problems in matching the large dynamic range of the signal with the limited dynamic range of the display, without losing much of the information the signal contains. If the required dynamic range is not too great, conventional automatic gain control (AGC) techniques can be applied to one or more stages of the signal processor. Additional dynamic range can be obtained by applying AGC to the detector bias in order to reduce the responsivity of the detector. When still greater dynamic range is needed, logarithmic amplifiers are sometimes used.

Depending on the location of the system, that is, whether it is in an aircraft or in a laboratory, it may be necessary to pay close attention to minimizing the introduction of extraneous noise into the circuitry. Such noise can come from microphonics, electrostatic or electromagnetic pickup, ground loops, poorly filtered power supplies, and stray rectification of RF energy from nearby transmitters. Low-impedance coaxial cables are available that have very low susceptibility to microphonics [10]. Circuit grounds should be located so as to prevent the formation of ground loops[11]. Considerable experimentation during the first installation may be required to establish the desired conditions. Shielded interconnecting cables can be especially troublesome; it may be necessary to leave the shields floating at one end or the other. If RF pickup is a problem, the preamplifier and other low-level circuitry must be shielded. Each power lead passing through a shield should have an efficient low-pass filter[12]. Disk-ceramic feed-through capacitors and ferrite-bead feed-through combinations can be used. The pulse repetition frequency of nearby radars may be within the electrical bandpass of the infrared system. If shielding, filtering, and other suppression techniques do not eliminate the pickup, it is probable that the infrared detector is demodulating the RF and thereby introducing the interference directly at the input of the signal processor. In such cases, the only solution is to relocate either the infrared sensor or the radar.

12.2 PREAMPLIFIERS

The preamplifier is probably the most critical element in the entire signal processor. Its noise figure must be low so that detector noise is the limiting noise in the system. In addition, the designer must make certain that the low noise figure is preserved when the preamplifier is in its final operational environment.

The desired voltage gain of the preamplifier depends on the system application. It is affected by the distance from the preamplifier to the rest of the signal processor, the mechanical and electrical environment, and the maximum undistorted output swing of the preamplifier. To maintain a low noise figure for the system, a minimum voltage gain of 10 in the preamplifier will ensure that the noise contribution of succeeding stages is negligible. If other noise sources are anticipated, such as microphonics or cable pickup, the gain should be increased. In most system applications the designer will find that voltage gains of 30 to 100 are most desirable. AGC should not be applied to a preamplifier if one wishes to maintain the lowest possible noise figure. Most transistorized preamplifiers begin to saturate when the output exceeds a few volts; a conservative designer will assume a maximum output swing of 1 V, and about 100 V for vacuum tubes. The dynamic range of the preamplifier is therefore limited by the detector noise level at its input and the maximum allowable output swing.

It is desirable that the output impedance of the preamplifier be quite low, 100 to 1000 ohms, in order to minimize the pickup in low-noise shielded cables running to the rest of the signal processor. In systems with a high information rate, a low output impedance ensures that the capacitance of the cable will not attenuate the high-frequency components of the signal.

Preamplifiers Using Vacuum Tubes

The noise at the output of a preamplifier comes from two sources, Johnson noise developed in the resistance of any elements connected to its input, and shot noise developed within the tube. The mean-square value of the Johnson noise voltage from a resistor (see Section 8.1) is

$$\overline{v^2} = 4kTR\Delta f, \qquad (12\text{-}1)$$

where k is Boltzmann's constant, T is the temperature of the resistor and R is its resistance, and Δf is the electrical bandwidth. In an ideal noiseless amplifier, such as that shown in Figure 12.1 there are two sources of Johnson noise in the input circuit; R_g the grid resistor and R_s the resistance of the signal source connected to the input. Since the amplifier is assumed to be noiseless, the noise factor is determined by the input circuit. The mean-square noise voltage at the input terminals of the amplifier is

$$4kT\Delta f \frac{R_g R_s}{R_g + R_s}, \qquad (12\text{-}2)$$

and the mean-square signal voltage is

$$V_s^2 \frac{R_g^2}{(R_g + R_s)^2}. \qquad (12\text{-}3)$$

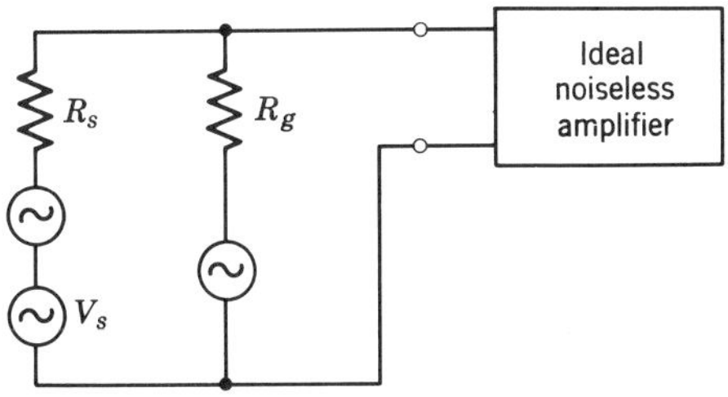

Figure 12.1 Equivalent circuit for finding the noise factor of a noiseless amplifier.

An expression for noise factor was given in (8-20); substituting (12-2) and (12-3) into that expression gives

$$F = 1 + \frac{R_s}{R_g}. \tag{12-4}$$

This expression is important because it relates noise factor to the ratio of the source resistance and the input resistance of the amplifier. Most circuit designers, when asked to match a detector and a preamplifier for minimum noise factor, will (probably influenced by the maximum power transfer theorem) assume that R_s should equal R_g. Equation 12-4 shows that for such a matched condition, the noise factor of a perfectly noiseless amplifier is 2 (noise figure is 3 dB). The noise factor can be reduced, however, by making $R_g \gg R_s$. Hence even a perfectly noiseless amplifier has a noise factor that is greater than unity, unless R_s is negligible with respect to R_g. The specific conclusion from this is that for minimum noise factor, the input resistance of a preamplifier should be very much higher than the resistance of the detector connected to its input.

Equation 12-4 was derived for an ideal noiseless amplifier; it can be extended to cover the more practical case of a noisy amplifier. The noisy amplifier exhibits shot noise at its output. This noise can be represented by an equivalent noise resistance placed in series with the grid of the tube. The value of the equivalent noise resistance is such that the Johnson noise it generates, when multiplied by the voltage amplification of the tube, is equal to the noise voltage observed at the output of the tube. The equivalent circuit for finding the noise factor of a noisy amplifier is shown in Figure 12.2. The noisy amplifier has been represented by a noiseless amplifier with an additional Johnson noise source in its grid circuit. Proceeding as before, the mean-square noise voltage at the input terminals of the amplifier is

$$4kT\Delta f \left(\frac{R_g R_s}{R_g + R_s} + R_{eq} \right), \tag{12-5}$$

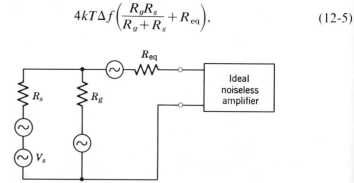

Figure 12.2 Equivalent circuit for finding the noise factor of a noisy amplifier.

SIGNAL PROCESSING AND DISPLAYS 403

and the mean-square signal voltage is

$$V_s^2 \frac{R_g^2}{(R_g+R_s)^2}. \tag{12-6}$$

The noise factor is

$$F = 1 + \frac{R_s}{R_g} + \frac{R_{eq}}{R_s}\left(1 + \frac{R_s}{R_g}\right)^2. \tag{12-7}$$

Minimum noise factor requires a low value of R_{eq} and a high value of R_g. With a matched input ($R_s = R_g$) the noise factor is greater than 2. It is possible, however, to find an optimum value of R_s, for which the noise factor will have a minimum value, by differentiating (12-7):

$$R_{s_{opt}} = R_g\left(\frac{R_{eq}}{R_{eq}+R_g}\right)^{1/2}. \tag{12-8}$$

Substituting this value into (12-7) gives the minimum noise factor attainable by mismatching:

$$F(\min) = 1 + \frac{2R_{eq}}{R_g} + 2\left[\frac{R_{eq}}{R_g}\left(1 + \frac{R_{eq}}{R_g}\right)\right]^{1/2}. \tag{12-9}$$

Before making use of this expression, it is necessary to find appropriate values for R_{eq}, the equivalent noise resistance. The required expressions were derived many years ago in a remarkable series of papers by Thompson, North, and Harris[13]. They showed that for frequencies above a kilohertz, or so, so that flicker noise is not a factor, the equivalent noise resistance of a triode is

$$R_{eq_t} = \frac{2.5}{g_m}, \tag{12-10}$$

where g_m is the grid-plate transconductance. In pentodes an additional source of noise, called partition noise, is caused by fluctuations in the division of tube current between the screen grid and the plate. Therefore a pentode is always noisier than a triode. The equivalent noise resistance of a pentode is

$$R_{eq_p} = R_{eq_t}\left(1 + \frac{8I_c}{g_{m_p}}\right), \tag{12-11}$$

where the sub-subscripts p and t indicate corresponding quantities for the tube connected as a pentode or a triode, respectively, and I_c is the screen grid current.

Since a low noise factor requires a low value of equivalent noise resistance (see (12-9)), the designer should choose a triode tube having a high transconductance. Triodes are available that have an equivalent noise resistance as low as 200 ohms. Using this value and assuming a value of 10^6 for R_g, the minimum noise factor is 1.028 (0.1 dB) for the optimum source resistance. From (12-8), the optimum source resistance is 1.4×10^4 ohms. The minimum value of the equivalent noise resistance for pentodes is about 700 ohms.

It is now possible to relate the effect of noise factor, the resistance match between the detector and preamplifier, and the selection of an operating point for the detector (as discussed in Section 9.3). If a detector and its load resistor are substituted for the signal source in Figure 12.2, the resulting equivalent circuit is shown in Figure 12.3. Note that R_s has been replaced by the parallel combination of R_l and R_d. Suppose that the detector has a resistance of 10^6 ohms. In the previous calculation involving a triode, the optimum source resistance was 1.4×10^4 ohms. This value can be achieved by selecting a value of 1.4×10^4 ohms for R_l, since the resistance of the parallel combination consisting of the detector and load resistor will be essentially 1.4×10^4 ohms. Having selected the load resistor so as to minimize the noise factor of the preamplifier, the designer can vary the bias current through the detector (by changing the bias voltage) so as to obtain the maximum signal-to-noise ratio, as described in Section 9.3. Further study of Figure 12.3 shows that R_d, R_l, and R_g are in parallel across the input of the amplifier (this fact is, of course, used in the derivation of (12-5) through (12-9)). The mean-square noise voltage from the input circuitry is then equal to $4kT\Delta f R_p$, where R_p is the resistance of the parallel combination of R_d, R_l, and R_g. The mean-square tube noise is $4kT\Delta f R_{eq}$. Hence it is evident that as long as $R_p \gg R_{eq}$, the noise in the output of the amplifier will be that due to the input circuitry and not that due to the internal noise of the amplifier. As a final refinement, it should be noted that most detectors show excess noise. As a result, the equivalent noise generator representation for them should have a resistance that is larger than R_d.

Thus far it has been assumed that there is no flicker noise (a form of $1/f$ noise) present in the output of the amplifier. With vacuum tubes, flicker noise becomes evident at frequencies below 500 to 1000 Hz and its effect is to raise the noise factor of the amplifier. Flicker noise varies with the total current through the tube and for most tubes it is a minimum for a current of about 1 ma[14,15]. A designer is forced to make his own measurements since the tube manuals rarely give data on flicker noise. Among the tubes that have been found to have low flicker noise are the 6AG5, 6AQ8, 6BK7A/B, 6CB6, 12AX7, EF37, and EF38.

Depending on the location of the preamplifier, microphonic noise caused by vibration of the tube elements may be troublesome. Subminiature

SIGNAL PROCESSING AND DISPLAYS 405

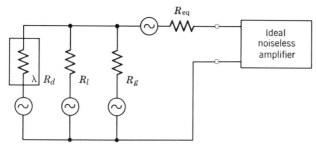

Figure 12.3 Equivalent circuit for finding the noise factor of an amplifier and photoconductive detector.

tubes, such as the 6533 and the 6247, are less sensitive to microphonics, but their noise factors are not quite as low as those of the group mentioned above. Because of these conflicting requirements regarding noise factor, flicker noise, and microphonics, a circuit designer may have to make several compromises to achieve adequately low preamplifier noise under actual operating conditions.

Preamplifiers Using Transistors

A plot of noise factor versus frequency for a transistor shows three well-defined regions, much as does a similar plot for a vacuum tube. At low frequencies the noise factor increases because of $1/f$ noise. A high-gain planar transistor is out of the $1/f$ region above 500 Hz; for other types, $1/f$ noise extends to a few kilohertz. Above this frequency the noise factor is nearly constant up to a frequency of about $0.1 f_\alpha$, where f_α is the frequency at which α, the short-circuit common-base current gain is 3 dB less than its low-frequency value. Beyond this frequency the noise factor increases rapidly, but since infrared systems rarely work at such high frequencies, this increase can be ignored.

For vacuum tubes it is possible to derive a simple circuit model in which the various sources of noise within a tube are represented by a single source of noise in series with the grid. Such a model can be applied to transistors, but it gives a very poor physical picture since in transistors there are more sources of noise and there are various degrees of correlation between each. A more meaningful model uses the transistor T-equivalent circuit with the various noise sources in series or shunt with the three elements of the transistor[16]. In the model used by Van der Ziel[15], shot and $1/f$ noise-voltage generators are in series with the emitter, a thermal noise-voltage generator is in series with the base, and shot and $1/f$ noise-current generators are in parallel with the collector. This model has been simplified for use at frequencies above the $1/f$ region[17, 18]. Details of these treatments will not be given here, but the conclusions will be summarized. For a low noise factor, a transistor should have a small base-spreading resistance, an α close to unity, and a high f_α, and it should be operated with a low emitter current (some types of transistors, which should be avoided, have a low current gain at low emitter currents).

As was the case with vacuum tubes, there is an optimum value of source resistance to give the minimum noise factor with transistors. This is shown in Figure 12.4. For transistors the optimum source resistance is in the range from 10^2 to 10^4 ohms, for vacuum tubes and field-effect transistors it ranges from 10^4 ohms and up. The optimum source resistance is about equal to the source resistance required for maximum gain in the common-emitter configuration. As a result, most low-noise transistorized preamplifiers use the common-emitter configuration.

The simple equivalent circuit shown in Figure 12.5 can be used to give a reasonably accurate calculation of noise figure and matching conditions for a transistor. Comparison of this figure with Figure 12.3 shows that the two are essentially the same, so that (12-7) to (12-9) can be applied to the case of matching a detector to a transistor amplifier. The only changes required are to note thar R_{eq} is now the equivalent noise resistance of the transistor and R_i is the input resistance of the transistor. For transistors, the value of R_i is several orders of magnitude lower than is the corresponding quantity R_g for vacuum tubes. As before, the optimum source resistance is given by (12-8) and the minimum noise factor is given by (12-9). The noise factor for matched input ($R_s = R_i$) is found from (12-7) to be

$$F = 2 + 4\frac{R_{eq}}{R_s'}, \qquad (12\text{-}12)$$

where R_s' is the effective parallel resistance of the detector and its load resistor. Hence, for the matched condition the lowest noise factor is, as before, equal to 2 and is obtained only when R_{eq} is negligible with

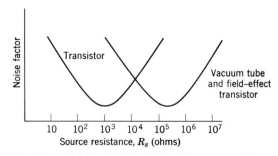

Figure 12.4 Variation of noise factor with source resistance.

SIGNAL PROCESSING AND DISPLAYS 407

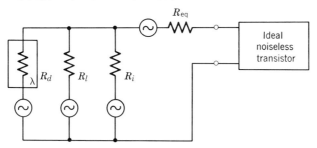

Figure 12.5 Equivalent circuit for finding the noise factor of a transistorized amplifier and a photoconductive detector.

respect to R'_s. Therefore the mismatch resulting from the use of the optimum source resistance, given by (12-8), can provide no more than a halving of the noise factor. Curves showing the effect of mis-matching on noise factor are given in ref. 19.

Although relatively little information has appeared in the unclassified literature about the performance of transistorized preamplifiers in infrared systems, it is clear that they can be used with any of the available detectors and will yield noise figures comparable to those obtained with the best vacuum tubes. An example of their actual performance is shown in Figure 12.6, from which it is evident that by a proper choice of transistor, preamplifiers having noise figures of 1 dB can be realized over a wide frequency range. The optimum source resistance is shown for each transistor. Further design details may be found in [16, 19–23]. Field-effect transistors have proved useful in preamplifiers because of their low noise factor and high input impedance; for additional information, see [20, 24 and 25]. Design principles for logarithmic amplifiers are given in [26–28]. Techniques for measuring the noise factor of transistorized preamplifiers are discussed in [29–31].

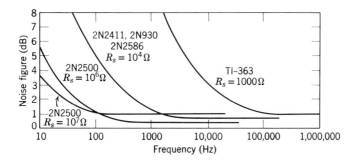

Figure 12.6 Noise figures of typical transistorized preamplifiers using various types of transistors (data adapted from publications of Texas Instruments, Inc.).

Preamplifiers Using Microelectronics

From the few details that have appeared in the unclassified literature, it is evident that microelectronics can be used with excellent results in preamplifiers for infrared systems[3-5,32]. One such low-noise preamplifier[3] contains five transistors on a silicon chip measuring 0.11 by 0.095 in. Both thin-film and monolithic techniques were used. The preamplifier described in refs. 4 and 5 was developed for a system using a 100-element detector. A noise figure of less than 2 dB was obtained with a voltage gain of 93 dB. Each packaged preamplifier occupies less than 0.03 in.[3] and weighs less than 1.5 g.

12.3 ADDITIONAL CONSIDERATIONS IN SIGNAL PROCESSING

Signal processing following the preamplifier usually consists of additional amplification, bandwidth limiting, detection, and control initiation or display of the final output signal. Detailed examples of the signal processing in specific infrared systems can be found in the patents referenced in Chapter 6.

Determination of the optimum bandwidth is an important decision and one that the system engineer must examine carefully. The optimum bandwidth depends upon the spectral characteristics of both the signal and the noise. The spectral characteristics of signals are described in Chapter 6, and the characteristics of the noise found in infrared systems are discussed in Chapters 8 and 10.

In most pulse systems there is rarely a need to retain the exact shape of a pulse as it passes through the signal processor. An exception would be a high-precision tracker designed to track the edge of a pulse rather than its centroid. In most other applications the centroid is adequate for tracking signal generation. If the peak amplitude of the pulse is preserved through the signal processor, it may be used for radiometric purposes.

The response of an amplifier or filter to a rectangular pulse is related inversely to its bandwidth. If a pulse is passed through an amplifier that has a very narrow bandwidth—that is, the response time is much longer than the duration of the pulse—the peak output signal power is proportional to the square of the bandwidth. As the bandwidth is increased, so that the response time is decreased, the peak output signal power soon reaches a constant value independent of the bandwidth. Since the output noise power increases linearly with bandwidth, there is an optimum bandwidth for which the ratio of peak pulse power to noise power is a maximum. For a rectangular pulse this maximum occurs when

$$B\tau_d \approx 0.5, \qquad (12\text{-}13)$$

where B is the 3 dB bandwidth of the circuit and τ_d is the time duration of the rectangular pulse.

The exact value of the constant in (12-13) depends on the shapes of the pulse and of the bandpass of the amplifier. The maximum signal-to-noise ratio is obtained for a matched filter, one whose transfer function is the complex conjugate of the Fourier transform of the input pulse. The ideal matched filter is not physically realizable. For the networks that are realizable, the signal-to-noise ratio is 1 or 2 dB less than it would be with the matched filter[33]. There are several excellent references that describe the determination of optimum bandwidth[33–35].[1] Since the curves of signal-to-noise ratio for various shapes of pulse and bandpass have fairly broad maxima, it is possible to say that the optimum bandwidth for any pulse detection problem will lie within the limits of $B\tau_d$ equal to 0.25 to 0.75. Figure 12.7 shows the effect of bandwidth on the shape and amplitude of a rectangular pulse after passage through a network. For values of $B\tau_d < 0.5$, the peak amplitude of the pulse decreases and its length increases. For values of $B\tau_d > 0.5$, the peak amplitude is essentially constant but the pulse shape becomes more nearly rectangular. Hence accurate pulse reproduction will require a bandwidth given by $B\tau_d \approx 4$.

In a system that uses a reticle to generate a modulated carrier, the modulation forms sidebands about the carrier frequency. The center frequency of the passband for such a system is made equal to the carrier

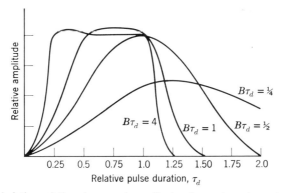

Figure 12.7 Variation of the shape and amplitude of a rectangular pulse after passing through a network having the bandwidth shown (adapted from Lawson and Uhlenbeck [35]).

---

[1] In applying the information contained in these references, the reader must be aware that the bandwidth used in ref. 35 is that of a radar IF, while the bandwidth used in refs. 33 and 34 is a video bandwidth. The video bandwidth is one half of the IF bandwidth.

frequency, and the bandwidth is made wide enough to accept the desired number of sidebands. Since the sidebands in an AM system occur at multiples of f_r, the rotational frequency of the reticle, the minimum bandwidth will be equal to $2f_r$.

Some spectrometers and radiometers use synchronous rectification [36, 37] in which a reference pulse is used to gate, or turn on, a diode rectifier in synchronism with the signal. At low values of signal-to-noise ratio, the synchronous rectifier is approximately 3 dB better than other types of rectifiers.

12.4 MULTIPLE-CHANNEL SYSTEMS

When a multielement detector array is used, there are, of course, as many signal channels as there are elements in the array. Retention of the separate channels throughout the signal processor is undesirable from the standpoint of cost, size, and weight. One possible solution is to use *time multiplexing,* that is, a periodic sampling of each channel. In effect, the single channel leaving the multiplexer is time-shared by all of the channels from the detector. The principles of time multiplexing have been highly developed by workers in the field of telemetry. Present techniques require signal levels of about 0.1 V or more before the sampling can take place. Attempts to switch directly at the output of an infrared detector without intervening amplification are usually unsuccessful because of the transients that are introduced during the switching process and the false signals, called offset voltages, that are generated. These transients and offset voltages are usually at the millivolt level, and are therefore many times greater than the noise level at the detector. It is possible to use a clean-up gate to eliminate most of these transients, but some signal is lost also and the signal-to-noise ratio suffers accordingly.

In the absence of suitable low-noise switches, most multiple-channel infrared systems use a preamplifier for each detector element in order to raise the signal level to at least 0.1 V so that it can be sampled without causing an appreciable decrease in the signal-to-noise ratio. Such an arrangement is clearly a deterrent to the development of systems to exploit the advantages of multielement detectors. Besides the obvious cost of the preamplifiers, there are severe problems in packaging them. The preamplifier assembly eventually becomes so large that it can no longer be placed near the detector. The resulting profusion of connecting cables brings with it new problems of reliability and weight, and may also limit the high-frequency response. The application of microelectronic techniques will simplify, but not eliminate, this problem. The real solution will be a means of multiplexing directly at the detector. There are recent

SIGNAL PROCESSING AND DISPLAYS   411

indications that new solid-state switching techniques may make this possible. The catalog of at least one detector manufacturer[1] contains a cautiously optimistic offering of low-level switching systems for photoconductive detectors; switching takes place at a 1 megahertz rate, and the decrease in signal-to-noise ratio due to switching transients is said to be negligible.

12.5 DISPLAYS

After the completion of signal processing, it may be necessary to display the result for a human observer who is responsible for the final decision concerning the presence or absence of a signal. Since a human observer is involved, any analysis of the detection problem is incomplete unless it considers the psychophysical characteristics of the human eye. Under these added constraints an already difficult problem becomes almost insoluble, and it is not surprising that much of the work in this field has been experimental rather than analytical.

The relatively few types of display that are employed with infrared systems are all well known because of their use in other fields[38–40]. Among these are cathode-ray tubes, photographic materials, banks of neon lamps, indicating meters, and pen-type recorders. A cathode-ray tube[41–43] is used in many systems. Its major drawbacks include a very limited dynamic range or gray scale and low screen brightness. The cathode-ray storage tube offers a far brighter screen, high resolution, the ability to store indefinitely any portion of the information displayed, and a selective erasure capability[41, 42, 44]. When a greater dynamic range is required, photographic techniques are available similar to those used for facsimile picture reproduction[45, 46]. The signal modulates the brightness of a glow tube, which is then imaged as a spot on the film. This spot is moved across the film by a mirror synchronized with the system scanning motion. Motion at right angles to that of the spot is provided by moving the film. By using Polaroid film, the picture is available almost as soon as scanning has been completed. When the system uses a multielement detector, a simple yet effective display is a bank of neon lamps [47]. For each element in the detector there is a corresponding lamp in the display. The recent introduction of transistorized lamp drivers may make such displays much more popular.

What are some of the problems affecting an observer's ability to distinguish a signal from the noise on a display? The cathode-ray tube, which has been studied in great detail[48], will provide a basis for the

---

[1]The Santa Barbara Research Center, Goleta, California.

discussion; however, it will be seen that many of the conclusions have a far broader applicability. First, how does one define a minimum detectable signal? Since the problem is a statistical one, the definition is meaningful only when we are forced to limit the number of observations, or the time available for the observations. Thus, in effect, it is not the noise that limits the minimum detectable signal; it is instead the imposition of a limited time of observation. An observer trying to detect a signal is forced to make a guess or a bet as to its presence. For a particular set of conditions there is a specifiable probability of success, that is, of the guess being correct. As shown in Figure 12.8, this probability increases rapidly as the signal-to-noise ratio increases. From such a curve the minimum detectable signal can be defined in terms of the signal-to-noise ratio required for a given probability of success. The values associated with such a curve will vary with the observer and with the viewing conditions, but its shape will remain relatively unchanged. An analysis of many such curves indicates that when an observer views a cathode-ray display, a ratio of peak signal voltage to rms noise voltage of 4 to 6 is required for a 90 per cent probability of detection.

The output of the signal processor, as seen on the display, consists of random noise interspersed with occasional signals. An observer viewing this display is faced with the problem of selecting a point at which the display amplitude is so large that he feels justified in betting that this is due to a combination of signal plus noise rather than to noise alone. Since he can never be certain that the observed peak is not due to noise alone, it is impossible to define an absolute minimum value for the signal-to-noise ratio required for detection. The observer's judgement can, of course, be influenced by prior knowledge about the probability of the presence of a signal. If, for instance, he knows that a signal will appear at a given point about half of the time, he will probably guess that it is present each time the amplitude at that point is just a little greater than the average. On the other hand, if he knows that the signal will be present only about 1 per cent of the time, he will probably set his standards of acceptance much higher.

The frequency of the fluctuations of the noise seen on the display is

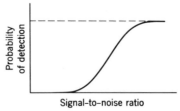

Figure 12.8 Probability of detection as a function of signal-to-noise ratio.

very nearly equal to the bandwidth (in Hz) of the circuit. In other words, noise amplitudes observed at time intervals of $1/B$ seconds are statistically independent, and the noise statistics given in Section 8.3 can be applied to find the probability of observing a noise peak of any desired amplitude. For a pulse-type system, consisting of a detector, amplifier, filter, and display, the noise at the display will have a Gaussian distribution. For a 2 kHz bandwidth, there will be approximately 2000 noise pulses per sec. The probability that a noise peak will exceed t times the rms value of the noise is given by (8-15) (or values may be found in Table 8.2). The probability of a peak exceeding a value of t equal to 3 is 0.0026. With a 2 kHz bandwidth this value corresponds to 5 such peaks per sec. If t is equal to 4, there will be about 8 such peaks per min. If t is equal to 5, there will be about 11 such peaks per hr. Finally, if t is equal to 6, the interval between such peaks will be about 23 hr. From these examples it is easy to see the usefulness of a *threshold detector*, a detector (rectifier) that is biased so that it provides an output only when the signal exceeds a specific value. In the preceeding example, for instance, if it was desired that the system give no response until a peak exceeded 6 times the rms noise level ($t = 6$), there would be 1 response approximately every 23 hr to a random noise peak. Such a response is called a *false alarm*, and the average time between false alarms is called the *false alarm time*.[1]

Thus far the discussion has assumed that the observer must make his decision on the basis of a single observation. However, observers viewing a display rarely make their decisions on a single observation. Instead they wait to see whether the apparent signal appears on several successive traces. Provided that there is sufficient amplification in the signal processor, noise fluctuations will be observed on each trace, but their position will vary at random from trace to trace. If two successive traces are stored before display and integrated point by point, the displayed result will still show fluctuations, but they will be much smaller than those in either of the original traces. If a signal is present on successive traces, the integration process will increase its amplitude, provided that the signal occurs at the same point on each trace. Studies indicate that within reasonable limits the signal-to-noise ratio increases with the square root of the number of scans that are integrated. The cost and complexity of the circuitry needed to perform the integration usually limits the number of situations in which this technique can be applied. An observer can accomplish much the same result by watching the

---

[1]Statistically speaking, false alarm time is the time interval within which there is a 50 per cent probability of a noise peak exceeding a given threshold value. This is essentially the same as the paraphrased definition given in the text.

deflection at a particular point on several successive traces and performing his own integration. Unfortunately, the human observer is a rather imperfect integrator, and the potential gain from integration is usually observed for only the first few scans.

A threshold detector is also useful in automatic tracking systems in order to reduce the possibility of tracking a false target. The threshold detector is often a Schmitt trigger[49] with its triggering level set at the desired threshold. Signals exceeding this level open a signal gate and allow a pulse to pass. If these pulses are registered in a logic network, the logic can be designed to require some set number of pulses in successive frames, from the same point in the field of view, before indicating the presence of a target. In effect, the logic network performs the integration process by replacing it with a coincidence requirement. Note, however, that the threshold setting is a very delicate one; changing the threshold setting t from 3 to 6 changes the false alarm time from a fraction of a second to nearly a day.

REFERENCES

[1] "Special Issue on Integrated Electronics," *Proc. IEEE*, **52** (December 1964).
[2] W. C. Hittinger and M. Sparks, "Microelectronics," *Sci. Am.*, **213**, 56 (November 1965).
[3] "Silicon Integrated Circuits Steal The Show," *Electronic Design*, **12**, 12 (April 13, 1964).
[4] C. P. Hoffman and R. F. Higby, "Application of Molecular Engineering Techniques to Infrared Systems," *Proc. IEEE*, **52**, 1731 (1964).
[5] C. P. Hoffman and M. N. Giuliano, "Micropower Molecular Amplifiers for Infrared Search-Track System Application," *Solid State Design*, **6**, 24 (March 1965).
[6] "Microcircuits for Infrared Recon," *Aviation Week*, **84**, 109 (June 6, 1966).
[7] W. E. Montgomery et al., "Application of Integrated Electronics to Military Communications and Radar Systems," *Proc. IEEE*, **52**, 1721 (1964).
[8] "New Designs Must Consider Microelectronics," *Electronic Engr.*, **26**, 24 (June 1967).
[9] "Fast Transition to Integrated Circuits," *Electronic Equipment Eng.*, **14**, 20 (October 1966).
[10] Z. Neumark, "Tribo-Electricity and Microphonic Noise in Infrared Systems," *Infrared Phys.* **4**, 67 (1964).
[11] EID Staff, "Grounding Low-Level Instrumentation Systems," *Electronic Instrument Digest*, January 1966, p. 16.
[12] A. L. Albin and E. I. Busch, "Miniaturized Components for EMI Suppression in Photo-Optical Control Systems," *Soc. Phot. Instr. Engrs. J.*, **4**, 228 (1966).
[13] B. J. Thompson, D. O. North, and W. A. Harris, "Fluctuations in Space-Charge-Limited Currents at Moderately High Frequencies," *RCA Rev.*, **4**, 441 (1940); **5**, 106, 244, 505 (1941).
[14] J. G. Van Wijngaarden, K. M. Van Vliet, and C. J. Van Leeuwen, "Low Frequency Noise in Electron Tubes," *Physica*, **18**, 689 (1952).
[15] A. Van der Ziel, "Noise Aspects of Low-Frequency Solid-State Circuits," *Solid State Design*, **3**, 20 (March 1962).

[16] J. F. Pierce, *Transistor Circuit Theory and Design*, Columbus, Ohio: Charles E. Merrill Books, 1963, Chapter 10.
[17] E. G. Nielsen, "Behavior of Noise Figure in Junction Transistors," *Proc. Inst. Radio Engrs.*, **45**, 957 (1957).
[18] A. E. Sanderson and R. G. Fulks, "A Simplified Noise Theory and Its Application to the Design of Low-Noise Amplifiers," *IEEE Trans. Audio*, **AU-9**, 106 (1961).
[19] W. A. Rheinfelder, *Design of Low-Noise Transistor Input Circuits*, New York: Hayden, 1964.
[20] J. J. Rado, "Designing Input Circuits with Lowest Possible Noise," *Electronics*, **36**, 46 (August 2, 1963).
[21] G. R. Pruett and R. L. Petritz, "Detectivity and Preamplifier Considerations for Indium Antimonide Photovoltaic Detectors," *Proc. Inst. Radio Engrs.*, **47**, 1524 (1959).
[22] F. Schwarz and A. Ziolkowski, "2-Channel Infrared Radiometer for Mariner 2," *Infrared Phys.*, **4**, 113 (1964).
[23] C. M. Verhagen, "Minimum Noise-Setting of Transistors," *Proc. IEEE*, **54**, 83 (1966).
[24] E. G. Fleenor, "Low-Noise Preamplifier," *Electronics*, **36**, 67 (April 12, 1963).
[25] "Low Noise, 60 dB Amplifier Uses FET's," *Electronic Design News*, **10**, 41 (March 1965).
[26] S. J. Solms, "Logarithmic Amplifier Design," *IRE Trans. Instrumentation*, **I-8**, 91 (1959).
[27] D. S. Grey, P. Mark, and S. Haskell, "New Class of Wide-Range Logarithmic Circuit for a Light-Intensity Meter," *J. Opt. Soc. Am.*, **50**, 40 (1960).
[28] R. Hirsch, "Four Pitfalls in Using Log Amplifiers," *Electronic Equipment Eng.*, **14**, 122 (August 1966).
[29] P. J. Beneteau, "Narrow-Band and Wide-Band Noise Figures," *Proc. IEEE*, **49**, 1954 (1953).
[30] K. Schjonneberg and U. S. Davidson, "Better Ways to Measure Noise in Transistors," *Electronic Design*, **11**, 46 (March 15, 1963).
[31] W. B. Mitchell, "How to Specify and Measure Transistor Noise," *Electronic Design*, **10**, 50 (July 19, 1962).
[32] J. R. Maticich, "Design Considerations for an Integrated Low-Noise Preamplifier," *Proc. IEEE*, **53**, 605 (1965).
[33] M. Schwartz, *Information Transmission, Modulation, and Noise*, New York: McGraw-Hill, 1959, section 6.2.
[34] S. Goldman, *Frequency Analysis, Modulation, and Noise*, New York: McGraw-Hill, 1948, section 4.11.
[35] J. L. Lawson and G. E. Uhlenbeck, *Threshold Signals*, New York: McGraw-Hill, 1950, section 8.6.
[36] M. D. Liston, "Modulated Heat Ray Detector," U.S. Patent No. 2,442,298, May 25, 1948.
[37] V. J. Coates and L. B. Scott, "Radiation Comparison Systems," U.S. Patent No. 3,039,353, June 19, 1962.
[38] R. K. McDonald, "Infrared Data Presentation," *Proc. Inst. Radio Engrs.*, **47**, 1572 (1959).
[39] I. Stambler, "Information Display" *Space/Aeronautics*, **40**, 90 (October 1963).
[40] F. A. Muckler and R. W. Obermayer, "Information Display," *Intl. Sci. Tech.*, August 1965, p. 34.
[41] A. S. Kramer, "Cathode-Ray Storage Tubes for Direct Viewing," *Electronics*, **32**, 40 (January 23, 1959).

[42] M. D. Harsh, "Display and Storage Tubes," *Electronic Industries,* **25**, 54 (April 1966).
[43] J. J. Josephs, "A Review of Panel-Type Display Devices," *Proc. Inst. Radio Engrs.,* **48**, 1380 (1960).
[44] N. H. Lehrer, "Selective Erasure and Nonstorage Writing in Direct-View Halftone Storage Tubes," *Proc. Inst. Radio Engrs.,* **49**, 567 (1961).
[45] R. Astheimer and E. M. Wormser, "Instrument for Thermal Photography," *J. Opt. Soc. Am.,* **49**, 184 (1959).
[46] R. W. Astheimer and E. M. Wormser, "Image Transducer," U.S. Patent No. 2,895,049, July 14, 1959.
[47] "Fifty Winking Lights," *Sci. Am.,* **205**, 71 (December 1961).
[48] Lawson and Uhlenbeck, *op. cit.* (ref. 35), chapters 8, 9.
[49] T. J. Galvin, R. A. Greiner, and W. B. Swift, "Switching Levels in Transistor Schmitt Circuits," *IRE Trans. Instrumentation,* **I-9**, 309 (1960).

# 13

The Analysis of Infrared Systems

The 12 preceding chapters have examined in detail the various elements that make up the infrared system. Both this chapter and the one that follows are concerned with the functional relationships between these elements when they are integrated into a complete system. An understanding of these relationships is essential to the system engineer, for he will need them to predict the performance of his system and to evaluate the effect of design tradeoffs [1].

The exact equations expressing the performance of infrared equipment are complex, require numerous integrations, and are extremely tedious to evaluate without a computer. In this chapter emphasis is placed on the use of engineering approximations, which considerably simplify the equations but still retain sufficient accuracy for system analysis and performance prediction.

13.1 THE GENERALIZED RANGE EQUATION

Of interest in many applications is the maximum range at which an infrared system can detect or track a target. A generalized range equation will therefore be derived and used to show the design tradeoffs available to the system engineer. The only prior assumption is that the limiting source of system noise is that from the detector. If this condition is not met because of poor design or improper installation procedures, the detection range will be less than that predicted by these equations.

The spectral irradiance from an unresolved target, that is, one that does not fill the instantaneous field of view, is

$$H_\lambda = \frac{J_\lambda \tau_a(\lambda)}{R^2}, \quad (13\text{-}1)$$

where J_λ is the spectral radiant intensity of the target, $\tau_a(\lambda)$ is the spectral transmittance of the path between the sensor and the target, and R is the range (distance) to the target. The spectral radiant power incident on the detector is

$$P_\lambda = H_\lambda A_0 \tau_0(\lambda), \tag{13-2}$$

where A_0 is the area of the entrance aperture of the optics and $\tau_0(\lambda)$ is the spectral transmittance of the sensor (including protective windows, optics, filters, reticle substrates, and blockage by secondary mirrors). The signal voltage from the detector is

$$V_s = P_\lambda \mathcal{R}(\lambda), \tag{13-3}$$

where $\mathcal{R}(\lambda)$ is the spectral responsivity of the detector.

Thus far the derivation applies only to an infinitesimal spectral interval centered about the wavelength λ. For any other spectral interval the signal voltage can be found by integrating over the interval

$$V_s = \frac{A_0}{R^2} \int_{\lambda_1}^{\lambda_2} J_\lambda \tau_a(\lambda) \tau_0(\lambda) \mathcal{R}(\lambda) d\lambda. \tag{13-4}$$

The signal-to-noise ratio is found by introducing the rms value of the noise from the detector:

$$\frac{V_s}{V_n} = \frac{A_0}{V_n R^2} \int_{\lambda_1}^{\lambda_2} J_\lambda \tau_a(\lambda) \tau_0(\lambda) \mathcal{R}(\lambda) d\lambda. \tag{13-5}$$

Unfortunately this equation cannot be solved implicitly since the atmospheric transmittance term $\tau_a(\lambda)$ is a function of both wavelength and range. The other wavelength-dependent terms complicate the evaluation of the expression because of the integrations involved. These difficulties can be avoided by passing from the integral form to one in which the various wavelength-dependent terms are replaced by their averaged or integrated values taken over the spectral bandpass of the sensor. This procedure assumes a rectangular spectral bandpass having a transmittance of τ_0 between λ_1 and λ_2 and a transmittance of zero outside of this interval. This approximation is similar to that used in Section 8.2 for the equivalent noise bandwidth. The term $J_\lambda d\lambda$ is replaced by J, the radiant intensity between λ_1 and λ_2. On the further assumption that the target is a blackbody, the value of J can be calculated quite easily with a radiation slide rule. The term $\tau_a(\lambda)$ is replaced by τ_a, the average value of the atmospheric transmittance between λ_1 and λ_2 for some assumed distance. Appropriate values of τ_a can be calculated by the methods given in Section 4.7 or estimated directly from Figure 4.15. Similarly, $\mathcal{R}(\lambda)$ is replaced by $\mathcal{R}$, the average value of the responsivity between λ_1 and λ_2. Unless one of the functions changes rapidly within the spectral bandpass,

THE ANALYSIS OF INFRARED SYSTEMS   419

these approximations introduce very little error. Introducing these changes, and solving for the range, gives

$$R = \left[\frac{A_0 J \tau_a \tau_0 \mathscr{R}}{V_n(V_s/V_n)}\right]^{1/2}. \quad (13\text{-}6)$$

The performance of the detector is more appropriately described by D^* than by its responsivity. From Table 9.1,

$$\mathscr{R} = \frac{V_s}{H A_d} \quad (13\text{-}7)$$

$$D^* = \frac{V_s(A_d \Delta f)^{1/2}}{V_n H A_d}, \quad (13\text{-}8)$$

from which

$$\mathscr{R} = \frac{V_n D^*}{(A_d \Delta f)^{1/2}}, \quad (13\text{-}9)$$

where A_d is the area of the detector and Δf is the equivalent noise bandwidth. In conformity with the previous assumptions, D^* is considered to be the average value between λ_1 and λ_2. Its value may be estimated from a curve of D^* versus wavelength or from the peak value of D^* and the relative-response curve. If the instantaneous field of view of the sensor is $\omega(\mathrm{sr})$, the area of the detector is, from (5-20),

$$A_d = \omega f^2, \quad (13\text{-}10)$$

where f is the equivalent focal length of the optics. It is convenient to characterize the optics by their numerical aperture (see (5-14) and (5-16)):

$$(\mathrm{NA}) = \frac{D_0}{2f}, \quad (13\text{-}11)$$

where D_0 is the diameter of the entrance aperture of the optics. By substituting these values and replacing A_0 by $\pi D_0^2/4$, the range equation becomes

$$R = \left[\frac{\pi D_0(\mathrm{NA}) D^* J \tau_a \tau_0}{2(\omega \Delta f)^{1/2}(V_s/V_n)}\right]^{1/2}. \quad (13\text{-}12)$$

When the range equation is used to find the maximum detection or tracking range, the term (V_s/V_n) represents the minimum signal-to-noise ratio required for the system to function properly. In systems that use a reticle to provide a modulated carrier, both V_s and V_n are considered to be rms quantities. In pulse systems, V_s is usually considered to be a peak value and V_n an rms value. The appropriate corrections will be introduced in the equations for each type of system. For assessing the effect of various parameter changes, it is convenient to define an *idealized* range as that range for which V_s/V_n is equal to unity:

$$R_0 = \left[\frac{\pi D_0(\text{NA})D^*J\tau_a\tau_0}{2(\omega\Delta f)^{1/2}}\right]^{1/2}, \qquad (13\text{-}13)$$

where R_0 is given in cm when D_0 is in cm, D^* is in cm(Hz)$^{1/2}$W^{-1}, J is in W sr^{-1}, ω is in sr, Δf is in Hz, and (NA), τ_a, and τ_0 are dimensionless.

If a target is at the idealized range, the irradiance at the entrance aperture is called the noise equivalent irradiance (NEI); from (13-1) and 13-13) it is

$$\text{NEI} = H_0 = \frac{2(\omega\Delta f)^{1/2}}{\pi D_0(\text{NA})D^*\tau_0}. \qquad (13\text{-}14)$$

Although the term NEI was originally introduced in Section 7.1 as a means of characterizing a detector, it has a far broader usage in characterizing any sensor in terms of the input required to give an output equal to the system noise.

Tradeoff Analysis

In order to see more clearly how the various factors affect the maximum detection range, it is convenient to regroup the terms as follows:

$$R = [J\tau_a]^{1/2}\left[\frac{\pi}{2}D_0(\text{NA})\tau_0\right]^{1/2}[D^*]^{1/2}\left[\frac{1}{(\omega\Delta f)^{1/2}(V_s/V_n)}\right]^{1/2}. \qquad (13\text{-}15)$$

| target and atmospheric transmittance | optics | detector | system characteristics and signal processing |

The first term pertains to the radiant intensity of the target and the transmittance along the line of sight. The system engineer has virtually no control over either quantity, although it can be argued that he does in effect, have, some control by virtue of his choice of a spectral bandpass for the sensor. When specific data on target characteristics are not available, the procedures given in Section 3.4 can be used to provide an estimated value of the radiant intensity.

The second term contains various factors that characterize the optics. The numerical aperture has a theoretical maximum value of unity. In practice it rarely exceeds 0·5. Reasoning intuitively, we are likely to conclude that the maximum detection range should vary directly with the diameter of the entrance aperture rather than with its square root, as shown here. The usual argument is that the signal-to-noise ratio must be proportional to the square of the diameter of the entrance aperture

THE ANALYSIS OF INFRARED SYSTEMS   421

because the power incident on the detector is proportional to the area of the aperture. This reasoning overlooks the fact that when scaling up an optical design it is usually necessary to maintain a constant value for the numerical aperture. Hence scaling up the diameter of the optics requires a corresponding increase in the focal length. Increasing the focal length requires, in turn, that the linear dimensions of the detector be similarly increased in order to maintain a constant field of view. Finally, the noise from the detector increases as the square root of its area. As a result, the signal-to-noise ratio increases in proportion to the scaling factor, and the maximum detection range increases in proportion to the square root of the scaling factor. Since the weight of the sensor is roughly proportional to the cube of the scaling factor, the maximum detection range is approximately proportional to the sixth root of the sensor weight.

The third term pertains to the characteristics of the detector. Since (as Figure 7.5 shows), many detectors are already quite close to the theoretical limit of D^*, there is little prospect of large increases in the maximum detection range by future improvements in detectors. The exception is, of course, the gain resulting from radiation shielding of the detector.

The fourth term contains factors describing system and signal processing characteristics. It shows that decreasing either the field of view or the bandwidth increases the maximum detection range, but not very rapidly because of the fourth root dependence. The reason for this gain is, of course, the trading of information rate for detection range.

13.2 THE GENERALIZED RANGE EQUATION FOR A BACKGROUND-LIMITED DETECTOR

In Chapter 10 it was shown that some photon detectors can be operated in a background-limited condition, that is, one in which the detector noise is due to fluctuations in the rate at which carriers are generated by photons from the background and subsequently recombined. The paper by Swift[2] is an excellent reference to specific design techniques for such systems.

The D^* of a photoconductive Blip detector is

$$D_\lambda^* = \frac{\lambda}{2hc}\left(\frac{\eta}{Q_b}\right)^{1/2}, \qquad (13\text{-}16)$$

where h is Planck's constant, c the velocity of light, η the quantum efficiency, and Q_b the background photon flux. Assuming that D_λ^* is the value at the spectral peak, it is necessary to multiply it by s, the value of the relative spectral response averaged over the spectral bandpass of the sensor (assuming that the relative spectral response is normalized to the spectral peak). The resulting value of D^* is then compatible with the similar approximations that have been made in deriving the generalized range equation. When a cooled shield is used to limit the field of view of the detector, the total number of photons falling on the detector is proportional to $\sin^2\theta$, where θ is the half-angle of the cone subtended at the detector by the shield aperture, and

$$D^* = \frac{s\lambda}{2hc}\left(\frac{\eta}{\sin^2\theta Q_b}\right)^{1/2}. \qquad (13\text{-}17)$$

From (5-15), for a system operating in air, the numerical aperture is

$$\mathrm{NA} = \sin u', \qquad (13\text{-}18)$$

where u' is the half-angle of the cone of rays converging at the focal point. If the shield aperture just accepts the entire cone of rays from the optics, u' is equal to θ and the idealized range with a background-limited detector is

$$R_0(\mathrm{Blip}) = \left[\frac{\pi D_0 J \tau_a \tau_0 s\lambda}{4hc}\left(\frac{\eta}{Q_b \omega \Delta f}\right)^{1/2}\right]^{1/2}. \qquad (13\text{-}19)$$

The important point here is that the numerical aperture has dropped out of the range equation. Thus, with a Blip detector the maximum range is influenced by the diameter of the optics but not by their speed. The term Q_b represents the photon flux received from a hemispherical surround; its value can be obtained from Figure 10.1 or from the Thornton radiation slide rule.

After the shield aperture is matched to the optics, there are several other design steps that can be taken to further reduce the background photon flux:

1. Use spectral filtering to limit the spectral bandpass to a region most favorable to the target flux. The bandpass should exclude those wavelengths at which the atmosphere absorbs strongly, since they contribute to the background noise without enhancing the signal. Detailed conditions for the selection of an optimum filter are given by Swift[2] and Kleinhans[3].

2. Minimize the emissivity of all optical elements.

3. Reduce the size of any optical support members that can be seen by the detector. As a general rule, any such members should be highly reflective (to reduce their emissivity) and be formed from a section of a sphere with its center of curvature at the detector.

4. Cool all optical elements and their support members seen by the detector.

Assuming that the number of photons from the cooled shield is negligible, the background flux on the detector $Q_{b\theta}$ is that received through the aperture in the shield. It comes from two sources: (1) the background against which the target is viewed and any flux emitted or scattered toward the system by the intervening atmosphere, and (2) the optics and their supports seen by the detector. If the target background and the optical elements are considered to be graybodies with a known temperature and emissivity, the value of $Q_{b\theta}$ can be found from

$$Q_{b\theta} = \sin^2\theta \int_{\lambda_1}^{\lambda_2} (Q_{tb}(\lambda)\epsilon_{tb} + Q_o(\lambda)\epsilon_o) d\lambda, \qquad (13\text{-}20)$$

where the subscripts tb and o refer to the target background and to the optics, respectively, and ϵ is their emissivity. It is important to note that $\sin^2\theta$ has already been included in (13-19) (through (13-17)); therefore, when substituting for Q_b in (13-19) its proper value is given by $Q_{b\theta}/\sin^2\theta$ (or simply the value of the integral in (13-20) without the multiplication by the $\sin^2\theta$ term). As before, the values of Q_{tb} and Q_o can be found from Figure 10.1 or the Thornton radiation slide rule.

It is well to emphasize again that (13-19) is valid only if (1) the detector receives flux solely through the optics by virtue of a cooled shield, (2) the detector is operated in the Blip condition, and (3) the limiting system noise is from the detector. When these conditions are met, the maximum detection range is

$$R_0(\text{Blip}) \approx \frac{J^{1/2}}{Q_b^{1/4}}. \qquad (13\text{-}21)$$

Thus for a background-limited system the maximum detection range is proportional to the square root of the radiant intensity of the target and inversely proportional to the fourth root of the background flux on the detector. At the present time, there are probably no systems that achieve true background-limited performance, but it seems certain that some systems now being planned for use in space will achieve it.

13.3 THE RANGE EQUATION FOR SPECIFIC TYPES OF SYSTEMS

The generalized range equation developed in Section 13.1 can, in principle, be applied to any type of infrared system. In this section, the proper modifications will be introduced to develop range equations for a search system and for several types of tracking systems. Succeeding sections will develop appropriate equations for line-scan thermal mappers and for radiometers.

Search Systems

Most search systems use either a single detector or a linear array of detectors that are scanned optically or mechanically so as to cover the search field. Such systems are pulse systems since they depend on the scan motion to generate a pulse whenever the instantaneous field of view passes over a target. The performance of search systems has been treated in detail by Genoud[4]; McFee[5] has described system design considerations.

With a pulse system the most meaningful interpretation of the signal-to-noise ratio (V_s/V_n) in (13-12) is that it should represent the ratio of peak signal to rms noise; therefore V_s should be replaced by V_p. Unfortunately V_s is equal to V_p only when the pulse is sinusoidal or when the bandwidth of the signal processor is infinite. In most pulse systems the pulse is nearly rectangular and, therefore, contains a broad spectrum of frequencies, some of which are attenuated by the limited bandwidth of the signal processor. As a result, V_p is less than V_s, and V_p/V_s is a convenient measure of how much signal is lost as the pulse passes through the signal processor. To account for this loss, it is necessary to include V_p/V_s in the expression for the idealized range:

$$R_0(\text{search}) = \left[\frac{\pi D_0(\text{NA}) D^* J \tau_a \tau_0 V_p}{2(\omega \Delta f)^{1/2} V_s}\right]^{1/2}. \tag{13-22}$$

The number of resolution elements in the search field is

$$r = \frac{\Omega}{\omega C}, \tag{13-23}$$

where Ω is the size of the search field (sr) and C is the number of independent detector elements (each covering an instantaneous field of view ω sr). The time rate of search is

$$\dot{\Omega} = \frac{\Omega}{\mathscr{T}}, \tag{13-24}$$

where $\mathscr{T}$ is the *frame time*, the time required to scan the entire search field. The duration of the target pulse is equal to the time required for the image of the target to pass across the detector. It is called the *dwell time* and is given by

$$\tau_d = \frac{\mathscr{T}}{r} = \frac{\omega C}{\dot{\Omega}}, \tag{13-25}$$

from which

$$\omega = \frac{\tau_d \dot{\Omega}}{C}, \tag{13-26}$$

and the idealized range is

$$R_0(\text{search}) = \left[\frac{\pi D_0(\text{NA})D^* J\tau_a\tau_0 V_p}{2(\dot{\Omega}\tau_d\Delta f/C)^{1/2}V_s}\right]^{1/2}. \quad (13\text{-}27)$$

Four of these terms — V_s, V_p, Δf, and τ_d — are all, in effect, related to the characteristics of the signal processor. Following Genoud[4], these terms can be replaced by the pulse visibility factor,

$$v = \left(\frac{V_p}{V_{ss}}\right)^2 \frac{1}{\tau_d \Delta f}, \quad (13\text{-}28)$$

which is used to describe the effectiveness of a signal processor in separating signals from noise. The term V_p is the peak amplitude of the pulse at the output of the signal processor, and V_{ss} is the peak amplitude that would be observed if the pulse suffered no loss in the signal processor. Therefore V_p/V_{ss} is the same as the term V_p/V_s that appears in (13-22) and (13-27), and the idealized range becomes

$$R_0(\text{search}) = \left[\frac{\pi}{2}D_0(\text{NA})D^* J\tau_a\tau_0\right]^{1/2}\left[\frac{vC}{\dot{\Omega}}\right]^{1/4}. \quad (13\text{-}29)$$

The value of v depends on the spectral characteristics of the noise and on the bandwidth of the filtering used in the signal processor. For linear processing, the value of v cannot exceed 2^1; in practice its value is likely to be in the range of 0.25 to 0.75. The value of v can be calculated from the transfer function of the system electronics.

It is evident from (13-29) that the idealized range for a search set is proportional to the fourth root of the number of elements in the detector. Arrays containing from 50 to 100 elements are now within the capability of most detector manufacturers. Hence designers of search sets will probably turn increasingly toward the use of arrays, since they may well be the least expensive means of obtaining greater range.

By applying the statistical ideas discussed in sections 8.3 and 12.5, it is possible to estimate the signal-to-noise ratio required for specified values of the probability of detection and the false-alarm time. If the time available for an observation is equal to the frame time, and if there is a single target within the search field, there will be just one signal pulse during the observation interval. The probability of this signal being detected is called the *single-look probability of detection*. Curves showing the single-look probability of detection as a function of the peak-signal-to-rms-noise ratio V_p/V_n and the ratio of actual to idealized range R/R_0

[1]This can be demonstrated as follows: for optimum signal processing, $V_p/V_{ss} \approx 1$. If the shape of the amplifier pass band is such that the equivalent noise bandwidth Δf is equal to the 3 dB bandwidth B, then, from (12-13) $\tau_d\Delta f \approx 0.5$ and $v \approx 2$.

are given in Figure 13.1. The parameter n' on the curves is the total number of noise pulses occuring during an interval equal to the false-alarm time. On the assumption that the number of noise pulses per second is approximately equal to the system bandwidth,

$$n' \approx \tau_{fa} \Delta f. \qquad (13\text{-}30)$$

If, for instance, the system bandwidth is 2000 Hz and the false-alarm time is 50 sec, the value of n' is 10^5 and the curves show that a 90 per cent probability of detection requires a signal-to-noise ratio of 5.7 or a range to the target that is equal to $0.42 R_0$. It is evident from the curves that these values do not change very rapidly with a change in the false-alarm time, $cf.$, increasing the false-alarm time to 50×10^3 sec, so that n' is 10^8, only increases the required signal-to-noise ratio to 6.9.

Stirling[6] has extended the statistical treatment to include the effect of changes in target range between successive scans, target scintillation, and target range at initiation of scan.

Tracking Systems that use Reticles

The function of a tracking system is to track or follow a particular target and to generate information about the spatial location of the target that can be used for homing or navigational purposes. Such trackers were discussed in Chapter 6. In essence, they include an optical assembly mounted on gimbals, a signal processor, and a servo control. If, for instance, the tracker is for homing, that is, for the guidance of a missile to a target, the tracker must continually measure the coordinates of the

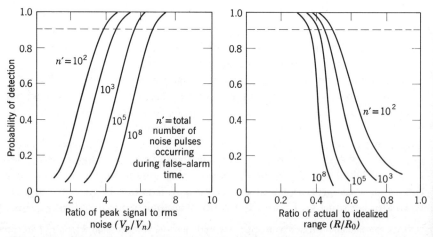

Figure 13.1 Single-look probability of detection for a pulse-type search system (adapted from Genoud[4]).

target with respect to the missile. In practice, a reticle in the optical assembly generates error signals that are proportional to the angle between the line of sight and the axis of the tracker. These error signals are fed to a servo control that then causes the tracker to follow the target. For obvious reasons, very little information on such homing devices has appeared in the literature. The one notable exception is the excellent book by Kriksunov and Usol'tsev[7].

A typical tracker has a circular instantaneous field of view (sr) that is defined by the reticle acting as a field stop. The reticle has an effective transmittance τ_r that is equal to 0.5 for an FM type and 0.25 for an AM type.[1] In the simplest case the detector is circular in shape, it is placed directly behind the field stop, and the idealized range is

$$R_0(\text{track}) = \left[\frac{\pi D_0(\text{NA}) D^* J \tau_a \tau_0 \tau_r V_p}{2(\omega \Delta f)^{1/2} V_s} \right]^{1/2}. \qquad (13\text{-}31)$$

Not all detectors can be made in a circular shape. If the detector is square and circumscribes the circular area of the field stop, it will contribute more noise and the range will, as a consequence, be about 10 per cent less than that given by (13-31).

Most tracking systems that use a reticle also use a field lens to form a reduced image of the field stop and permit the use of a smaller detector, as described in Section 5.5. From (5-29) it is evident that the diameter of the detector (and hence its noise) is reduced by the ratio of the numerical aperture of the primary optics to that of the field lens; therefore

$$R_0(\text{track}) = \left[\frac{\pi D_0(\text{NA})_f D^* J \tau_a \tau_0 \tau_r V_p}{2(\omega \Delta f)^{1/2} V_s} \right]^{1/2}, \qquad (13\text{-}32)$$

where $(\text{NA})_f$ is the numerical aperture of the field lens. Note that with a field lens the range depends on the diameter, but not the speed, of the primary optics.

The term V_p/V_s has been left in (13-31) and (13-32) to emphasize again the effect of the signal processing. An exact value, applicable to all situations, cannot be given for this term; the reader will have to evaluate it for his specific system[8]. In tracking systems the signal-to-noise ratio is usually understood to be the ratio of rms signal to rms noise. Therefore the signal processing factor V_p/V_s will have to include a factor that converts the irradiance from the peak value used in these equations to the rms value of its fundamental component. For square-wave chopping, this

---

[1]The average transmittance for one traverse around an FM reticle, such as that shown in Figure 6.8, is 0.5. For an AM reticle, such as that described in the text accompanying Figure 6.14, an additional transmission loss of 0.5 is required in order to produce the modulation.

conversion factor is equal to 0.45 ($2^{1/2}/\pi$), as given in Section 9.2. It is also necessary to specify at what point the signal-to-noise ratio is measured; that is, is it a predetection or a postdetection value? Satisfactory tracking typically requires a predetection signal-to-noise ratio of about 5 for an AM system and somewhat higher for an FM system. However, because of the very narrow bandwidth involved, an AM system will track satisfactorily with a postdetection signal-to-noise ratio of 0.5.

Tracking Systems that use Pulse Position Modulation

Tracking systems of this type do not use a reticle but, instead, use a cross-array detector to generate the pulse position modulation. They are described in Section 6.4. Opposite pairs of detector elements are connected to form an information channel. Each such pair covers an instantaneous field α by β (rad). The idealized range for one channel is

$$R_0(\text{track}) = \left[\frac{\pi D_0(\text{NA})D^* J \tau_a \tau_0 V_p}{2(\alpha\beta\Delta f)^{1/2} V_s}\right]^{1/2}. \qquad (13\text{-}33)$$

Since the value of $\alpha\beta$ is usually much less than ω and there is no loss term τ_r caused by the reticle, one is likely to conclude that the pulse position tracker will track to a greater range than will a tracker using a reticle. In practice this is rarely true because the bandwidth of the pulse position type is much greater than that of the reticle type, and because the reticle type can benefit from the use of a field lens whereas the pulse position type cannot.

As an example of the bandwidths involved, assume that both the pulse and the reticle-type trackers use nutating optics and that the nutation frequency f_n is 100 Hz for each. If the reticle tracker uses AM, the modulation frequency is 100 Hz. From the information in Section 12.3, the minimum bandwidth (predetection) required is 200 Hz. For the pulse tracker, if the length of the detector (β) is equal to the diameter of the nutation circle, the dwell time is equal to $\alpha/\pi\beta f_n$. A typical value for α is 1 mrad and for β is 35 mrad. With these values, the dwell time is equal 9.3×10^{-5} sec. From (12-13) the optimum bandwidth for such a pulse is 5500 Hz. Hence the bandwidth of the pulse-type tracker is about 27 times that of the reticle-type tracker.

13.4 LINE-SCAN THERMAL MAPPING SYSTEMS

Line scanners were originally developed to provide thermal maps for reconnaissance. The military potential of such devices is obvious. In the future there may be considerable demand for similar equipment for use in medical examinations, nondestructive testing, and remote sensing of the environment. In essence, a thermal mapper is much like the scanner shown in Figure 5.17. A movable mirror placed in front of the sensor directs the line of sight into any desired scan pattern. In a typical airborne installation the optical axis of the sensor is parallel to the direction of flight. A mirror placed at 45 deg to the vertical directs the line of sight to the ground below. As the mirror rotates about an axis parallel to the optical axis of the sensor, it causes the line of sight to scan across the ground at right angles to the direction of flight. The motion of the aircraft provides scanning in the direction of flight. With a properly designed system a series of contiguous line scans are thereby obtained, and they can be displayed to form a thermal map of the terrain[9].

Such a map can be likened to a photograph. The ease with which an object can be identified is a function of the contrast between it and the background (the many psychophysical factors involved are not properly a part of this book and will therefore be ignored). In a photograph the contrast between objects is a function of their reflection characteristics and the illumination of the scene. In a thermal map the contrast depends on differences in the radiant flux emitted by the objects. Unlike the previous systems, the target fills the field of view at all times. Under these conditions the radiometric quantity of interest is the radiance

$$N = \frac{W}{\pi}, \qquad (13\text{-}34)$$

where W is the radiant emittance. The irradiance at the sensor is independent of the distance to the source (if atmospheric attenuation is ignored) and is

$$H = N\omega. \qquad (13\text{-}35)$$

When two areas differ in radiance, this difference may be due to a difference in temperature, in emissivity, or in both, so that

$$\Delta N = \Delta N_T + \Delta N_\epsilon. \qquad (13\text{-}36)$$

When evaluating the performance of a thermal mapper, one customarily assumes that the source is a blackbody; in effect, this forces the interpretation that variations in radiance are due only to temperature differences. On the basis of this assumption, the system performance can be characterized by its *noise equivalent differential temperature* (NEΔT), which is the temperature differential between two adjacent elements in the scene that will give a signal equal to the system noise. Since in any terrestrial scene the differences in emissivity are likely to be as important as those in temperature, the concept of NEΔT is, at best, rather artificial. However, it is widely used and is a convenient means of comparing the performance of two or more thermal mappers.

If the instantaneous field of view in the direction of flight is α (rad) for each element of the detector, the length of the image of the detector array projected on the ground (parallel to the direction of flight) is equal to αhC, where h is the altitude of the aircraft and C is the number of elements in the array. This length is the distance the aircraft must move between successive scans in order to ensure that there are no gaps or overlaps in the coverage. The aircraft velocity for such contiguous scanning must be

$$v = n\alpha hC, \qquad (13\text{-}37)$$

where n is the number of scans per second. The angular scanning rate is

$$\dot{\theta} = 2\pi n = \frac{2\pi}{\alpha C} \frac{v}{h}. \qquad (13\text{-}38)$$

The quantity v/h is an important parameter in reconnaissance work since it describes the angular rate (rad sec^{-1}) at which a point on the ground appears to pass beneath the aircraft. To retain its identity and to prevent it from being confused with other usages for the individual symbols, let $\mathscr{V}$ be equal to v/h. For a detector element subtending an angle β(rad) in the direction of scan, the detector dwell time is

$$\tau_d = \frac{\beta}{\dot{\theta}} = \frac{\alpha\beta C}{2\pi\mathscr{V}} = \frac{\omega C}{2\pi\mathscr{V}}, \qquad (13\text{-}39)$$

since the solid angle ω is equal to $\alpha\beta$. For reasonable pulse fidelity the bandwidth should be

$$\Delta f = \frac{1}{2\tau_d} = \frac{\pi\mathscr{V}}{\omega C}. \qquad (13\text{-}40)$$

The noise equivalent irradiance for an infrared sensor is given by (13-14). For the reasons already stated in the discussion of search systems, the factor V_p/V_s must be added in order to account for the effect of the signal processor, and the NEI is

$$H_0 = \frac{2(\omega\Delta f)^{1/2}}{\pi D_0(\text{NA})D^*\tau_0(V_p/V_s)}. \qquad (13\text{-}41)$$

Substituting the value of ω, from (13-39), and v, the visibility factor as defined in (13-28), gives

$$H_0 = \frac{2}{D_0(\text{NA})D^*\tau_0}\left(\frac{2\mathscr{V}}{\pi c v}\right)^{1/2}. \qquad (13\text{-}42)$$

The relationship between the irradiance at the sensor and the radiance of the scene within the instantaneous field of view is given by (13-35), and

THE ANALYSIS OF INFRARED SYSTEMS 431

the noise equivalent differential temperature is

$$\text{NE}\Delta T = \frac{H_0}{\omega \tau_a N} \tag{13-43}$$

or

$$\text{NE}\Delta T = \frac{2}{D_0(\text{NA})D^* \omega \tau_a \tau_0 \Delta N} \left(\frac{2\mathscr{V}}{\pi C v}\right)^{1/2}. \tag{13-44}$$

The differential radiance due to a unit temperature change ΔN must be evaluated before (13-44) is used. Differential radiance (discussed in Section 2.8) is the partial derivative of the Planck function taken with respect to temperature. For a particular set of conditions its value can be calculated by differentiating the Planck function, weighting it by the relative spectral response of the detector and the spectral transmittance of the atmospheric path and the optics, and integrating the resulting function over the spectral bandpass of the system:

$$\Delta N = \int_{\lambda_1}^{\lambda_2} \left(\frac{\partial N_\lambda}{\partial T}\right) \tau_a(\lambda) \tau_0(\lambda) s(\lambda) d\lambda. \tag{13-45}$$

Evaluating this expression is time-consuming without a computer. Fortunately the methods described in Section 2.8 give the value of the derivative with negligible error. The usual engineering approximation assumes that average values of τ_a, τ_0, and s taken over the spectral bandpass of the system can be used as before, so that

$$\Delta N = \tau_a \tau_0 s \int_{\lambda_1}^{\lambda_2} \left(\frac{\partial N_\lambda}{\partial T}\right) d\lambda. \tag{13-46}$$

For convenience, the value of the integral is given in Table 13.1 for the three atmospheric windows and for a temperature of 300°K.

TABLE 13.1 VALUES OF DIFFERENTIAL RADIANCE ΔN FOR THE THREE ATMOSPHERIC WINDOWS AND FOR A TEMPERATURE OF 300°K

Spectral Interval (μ)	Differential Radiance (W cm^{-2}sr^{-1} °K^{-1})
2.0–2.5	6.0×10^{-9}
3.2–4.8	5.2×10^{-6}
8–13	7.4×10^{-5}

It is unfortunate that an improper procedure for evaluating the derivative has become firmly established in the literature. The reader is warned to beware, since this procedure can result in large errors. It starts with the Stefan-Boltzmann relationship, written to express the radiance

$$N = \frac{\sigma T^4}{\pi}, \tag{13-47}$$

and assumes that for small temperature differences, the change in radiance is

$$\frac{dN}{dT} = \frac{4\sigma T^3}{\pi}. \tag{13-48}$$

This expression is correct, *but only for the total change in radiance.* The error in the procedure occurs when (13-48) is applied to describe the condition within a limited spectral interval, such as an atmospheric window. A radiation slide rule is used to calculate the fraction of the total flux within the window; it is then incorrectly assumed that this fraction can be applied to (13-48) to find the change in radiance within the window. For the 8 to 13 μ window the error from this assumption is about 20 per cent (for $T = 300°K$), whereas it is about 250 per cent for the 3 to 5 μ window.

13.5 RADIOMETRY

The systems discussed up to this point are intended to detect or track targets. There are, however, many applications in which it is necessary to measure the radiometric properties of a target or its temperature. Any radiometric device responds to the irradiance at its entrance aperture, which is related to instrument and signal-processing parameters and the signal-to-noise ratio by

$$H = \frac{2(\omega \Delta f)^{1/2}}{\pi D_o (NA) D^* \tau_0} \left(\frac{V_s}{V_n}\right) \left(\frac{V_s}{V_p}\right) \tag{13-49}$$

(which comes by way of (13-41) with a few minor changes). If the target is not resolved, that is, if its image is smaller than the detector, then

$$H = \frac{J\tau_a}{R^2} \tag{13-50}$$

and

$$J = \frac{2R^2(\omega \Delta f)^{1/2}}{\pi D_0(NA) D^* \tau_a \tau_0} \left(\frac{V_s}{V_n}\right) \left(\frac{V_s}{V_p}\right). \tag{13-51}$$

In many measurement situations the value of τ_a is unknown. It is then customary to consider that τ_a is equal to unity and to call the resulting value of J the *apparent radiant intensity* of the target.

THE ANALYSIS OF INFRARED SYSTEMS

For an unresolved target

$$J = \frac{\sigma T^4 A_s}{\pi}, \qquad (13\text{-}52)$$

where A_s is the area of the source. Substituting,

$$T = \left[\frac{2R^2(\omega \Delta f)^{1/2}}{\sigma A_s D_0(\text{NA})D^*\tau_a \tau_0}\left(\frac{V_s}{V_n}\right)\left(\frac{V_s}{V_p}\right)\right]^{1/4}. \qquad (13\text{-}53)$$

The temperature calculated from this expression agrees with the physical temperature of the target only if the value of τ_a is accurately known and if the target is a blackbody (if it is a graybody, its emissivity must be included in the denominator). Otherwise the temperature given by (13-53) is called the *apparent radiation temperature*. Unfortunately a calculation of the temperature requires that both the distance to the target and its radiating area be known.

If the target is an extended source, that is, if it more than fills the field of view so that its image is larger than the detector, H is equal to $N\omega\tau_a$ (from (13-35)) and

$$N = \frac{2}{\pi D_0(\text{NA})D^*\tau_a \tau_0}\left(\frac{V_s}{V_n}\right)\left(\frac{V_s}{V_p}\right)\left(\frac{\Delta f}{\omega}\right)^{1/2}. \qquad (13\text{-}54)$$

As before, N can represent either the radiance or the apparent radiance, depending on the value assumed for τ_a. Since N is equal to $\sigma T^4/\pi$, the temperature of the target is

$$T = \left[\frac{2}{\sigma D_0(\text{NA})D^*\tau_a \tau_0}\left(\frac{V_s}{V_n}\right)\left(\frac{V_s}{V_p}\right)\left(\frac{\Delta f}{\omega}\right)^{1/2}\right]^{1/4}. \qquad (13\text{-}55)$$

It is important to note that when the target fills the field of view, it is not necessary to know its distance or its area in order to measure its temperature radiometrically (as long as τ_a is known or can be assumed to be equal to unity). This is a particularly valuable characteristic of a radiometer and one that has been exploited for industrial process control, nondestructive testing, and the determination of the temperatures of astronomical bodies.

The calibration curve of a radiometer can be calculated from the preceding equations. Such a calibration is checked by using a known source, typically one of the blackbodies described in Section 3.1. For calibrations at temperatures lower than room temperature, it may be necessary to build one's own source. The various types of cavities described in Section 3.1 can be cooled by an appropriate liquid coolant. A thin-walled metal can painted with a black granular-pigment paint also makes a fine source when it is filled with several gallons of coolant. Since the emissivity of these paints usually exceeds 0·9 and is reasonably independent of wavelength, these sources are good approximations to a blackbody. If the temperature of the coolant is below the dew point of the surrounding air, frost may form on the outside of the can. Since the emissivity of frost is higher than that of paint, the formation of frost improves the approximation to a blackbody. When one is using a very low temperature source, it is important to ensure that it does not reflect radiation from a warmer source and invalidate the calibration. This precaution is particularly important when one is trying to use the surface of a liquefied gas as the calibration source.

The radiometric calibration procedure consists of measuring the signal voltage[1] at the output of the radiometer as a function of some radiometric input, such as power, irradiance, or apparent radiance. Responsivity is the proportionality constant between the input and output quantities. The three types of responsivity are defined as

$$\mathcal{R}_P = V/P = \text{power responsivity, VW}^{-1}$$

$$\mathcal{R}_H = V/H = \text{irradiance responsivity, VW}^{-1}\text{cm}^{-2}$$

$$\mathcal{R}_N = V/N = \text{radiance responsivity, VW}^{-1}\text{ cm}^{-2}\text{sr}^{-1}.$$

The power responsivity was used in Section 7.1 to describe the performance of a detector. It is rarely used for any other purpose.

The irradiance responsivity is used in measuring the apparent radiant intensity of targets that are too small to fill the field of view of the radiometer. If a target at a distance R produces a voltage V at the output of the radiometer, the apparent radiant intensity is

$$J = R^2 V \mathcal{R}_H. \tag{13-56}$$

The radiance responsivity is used in measuring the apparent radiance of targets or backgrounds that fill the field of view of the radiometer. It is customary to define background radiation as the radiation from any source other than the desired target. Defined in this way, it includes radiation from localized sources as well as that arising along the line of sight from atmospheric emission and scattering. Because of this nonlocalized characteristic, one rarely associates a distance with a background.

With a radiometer having an entrance aperture of area A_0 and a field of view ω, the three responsivities are related by

$$\mathcal{R}_N = \omega \mathcal{R}_H = \omega A_0 \mathcal{R}_P. \tag{13-57}$$

---

[1] It is actually the signal-to-noise ratio that should be measured. However, since many radiometric measurements are made at a high signal-to-noise ratio, it is permissible to measure only the signal voltage.

THE ANALYSIS OF INFRARED SYSTEMS   435

A variety of calibration techniques are described in the literature [10-15]. Among the most popular are the use of a point source either at a great distance or with a collimator (Section 5.11), the use of a large-area source at such a distance that it fills the field of view, and the use of various techniques in which a source having a small area is placed close to the entrance aperture while the radiometer is focused at infinity.

For a comprehensive treatment of radiometric theory and such applications as filter, temporal or ac, and interferometric radiometry, see [16].

13.6 THE SPECIFICATION OF SYSTEM PERFORMANCE

As described in Chapter 1, performance specifications are determined early in the system engineering process. They provide the system engineer with quantitative measures for system inputs and outputs. They are used during the design phase to assess the results of the tradeoff studies and are used to provide a final evaluation of the system operating in its normal working environment. Since there are few standards for performance specifications, each system engineer must fend for himself. In the limited space available here, we can only mention a few of the problems and their solutions and point out the lone paper that exists on the subject [17].

Conceptually, the problem of describing the performance of infrared equipment is more difficult than describing that of RF equipment, which is usually described in terms of its response to a single-frequency signal. Infrared equipment, on the other hand, responds to a signal consisting of a broad band of wavelengths, which makes it difficult to describe the performance by a single number. Ideally it is desirable to specify a continuous function that describes the characteristics of the equipment over its spectral bandpass. In practice one can make judicious approximations in order to allow the use of single-number descriptions having sufficient accuracy for system analysis and performance estimates. Such single-number descriptions as the noise equivalent irradiance and the noise equivalent differential temperature have been used in preceding sections. Perhaps the greatest difficulty with such single-number descriptions is that no uniform set of approximations is used to derive them, and many workers fail to adequately define the approximations they use.

The most fundamental description of an infrared device is a curve showing the noise equivalent irradiance as a function of wavelength. It is important to recognize that such a curve is not a distribution curve, such as a curve of radiant emittance versus wavelength. The NEI curve simply defines the irradiance at each wavelength necessary to give a signal output equal to the system noise. It can be summarized in several ways; one such way is to state that the NEI has an average value of 10^{-10} W cm^{-2} over a spectral interval extending from 3 to 5 μ. It might also be stated that the peak value of the NEI is 10^{-11} W cm^{-2} μ^{-1}. When using NEI as a description of system performance, one must, as has already been emphasized, define the input and output relationships. Thus, for a pulse-type tracking system it is customary to relate the ratio of peak signal to rms noise at the system output to the irradiance at the entrance aperture. For a tracking system the specification should relate the irradiance to the angular tracking rate. Such characteristics can be accurately measured, and they can be explicitly defined to minimize the chances of misinterpretation by those responsible for testing.

There are two very simple measurements that one can make on any infrared sensor. The first is to measure its response to a known value of irradiance from a blackbody having a specified temperature. The second is to measure the relative response of the sensor as a function of wavelength. With these two measurements it is possible to convert the relative response curve to an absolute response curve by the procedure described in Section 9.4.

In production testing it is rarely practical to measure the relative spectral response of each sensor, and the ability to characterize the performance by a single number becomes essential. The solution is to specify the minimum acceptable output for a particular irradiance at the entrance aperture. Since most sensors respond over a relatively narrow spectral interval, not all of the incident flux is effective in producing a signal output. For this reason, many workers describe the irradiance in terms of its effective value

$$H_{\text{eff}} = \int_{\lambda_1}^{\lambda_2} H_\lambda s(\lambda) \tau_0(\lambda) d\lambda. \tag{13-58}$$

Since $s(\lambda)$ and $\tau_0(\lambda)$ can vary from one sensor to the next, one must recalculate the value of the integral for each sensor or else ignore the problem by adopting some average value for $s(\lambda)$ and $\tau_0(\lambda)$. In order to avoid such an ambiguity, it is preferable to specify the irradiance by its total, rather than by its effective, value. To minimize the possibility of misinterpreting the value of the irradiance, it is suggested that the test specification take the following form:

The ratio of peak signal to rms noise, as measured at the output of the signal processor, shall be greater than 50 when the entrance aperture of the sensor is irradiated by collimated radiant flux from a blackbody at a temperature of $800 \pm 5°K$ and the total irradiance is 5×10^{-8} W cm^{-2}.

Tests of systems in the field run the gamut from laboratory-type tests to a simple check of response to a lighted cigarette. From the viewpoint of the user, particularly a military user, it is usually far more important to know whether a system is or is not working than to know that its response has dropped 10 per cent because of some extraneous noise pickup. From this viewpoint the lighted cigarette is a devastatingly simple go/no-go test. Calibration packages that consist of a collimator, a source, and several apertures can be tailored to the requirements of a particular system and can provide in-service calibrations to an accuracy of ±20 per cent. For some applications it is practical to build a simple source (usually a miniature incandescent lamp) into the sensor in order that it may be used for a go/no-go test at any time.

REFERENCES

[1] R. D. Hudson, Jr., "Passive Optical Intercept Techniques," *Proc. Natl. Aerospace Electronics Conf.*, 1963, p. 54.
[2] I. H. Swift, "Performance of Background-Limited Systems for Space Use," *Infrared Phys.*, 2, 19 (1962).
[3] W. A. Kleinhans, "Optimum Spectral Filtering for Background-Limited Infrared Systems," *J. Opt. Soc. Am.*, 55, 104 (1965).
[4] R. H. Genoud, "Infrared Search-System Range Performance," *Proc. Inst. Radio Engrs.*, 47, 1581 (1959).
[5] R. H. McFee, "Infrared Search System Design Considerations," *Proc. Inst. Radio Engrs.*, 47, 1550 (1959).
[6] N. C. Stirling, "Detection Range Prediction for Infrared Detection Systems," *Proc. IEEE*, 51, 1327 (1963).
[7] L. Z. Kriksunov and I. F. Usol'tsev, *Infrared Equipment for Missile Homing*, Moscow: Voenizdat, 1963 (OTS: 64-31416).
[8] R. O'B Carpenter, "Comparison of AM and FM Reticle Systems," *Appl. Opt.*, 2, 229 (1963).
[9] M. R. Holter and W. L. Wolfe, "Optical-Mechanical Scanning Techniques," *Proc. Inst. Radio Engrs*, 47, 1546 (1959).
[10] F. E. Nicodemus, "Radiometric Calibration for Extended-Source Measurements," *J. Opt. Soc. Am.*, 43, 547 (1953).
[11] E. E. Bell et al., "Spectral Radiance of Sky and Terrain at Wavelengths between 1 and 20 microns. I. Instrumentation," *J. Opt. Soc. Am.*, 50, 1308 (1960); "II. Sky Measurements," 50, 1313 (1960); "III. Terrain Measurements," 52, 201 (1962).
[12] N. Ginsburg, W. R. Fredrickson, and R. Paulson, "Measurements with a Spectral Radiometer," *J. Opt. Soc. Am.*, 50, 1176 (1960).
[13] C. Ravitsky, C. Cumming, and T. S. Moss, "Standard Procedure for Target and Background Infrared Measurements," Office of the Director of Defence Research and Engineering, Washington, D.C., April 1961.
[14] G. Kelton et al., "Infrared Target and Background Radiometric Measurements— Concepts, Units, and Techniques," *Infrared Phys.*, 3, 139 (1963).
[15] H. F. Gilmore, "The Determination of Image Irradiance in Optical Systems," *Appl. Opt.*, 5, 1812 (1966).
[16] F. E. Nicodemus, "Radiometry," in R. Kingslake, ed., *Applied Optics and Optical Engineering*, New York: Academic, 1967, Vol. 4, Chapter 8.
[17] R. W. Powell, "Criteria for the Performance of Infrared Systems," *J. Opt. Soc. Am.*, 50, 660 (1960).

14

The Design of an Infrared Search System

Let us assume that after a lengthy sales campaign a major airline has decided to equip each of its long-range jet transports with an infrared search system. The moment of truth has now arrived! We must build and deliver the equipment that is so glowingly described in our sales literature. Many man-hours of engineering time have been consumed in order to reach this point, and many more hours will be required before the last system is delivered and finally installed. We will examine the entire process by, in effect, peering over the shoulder of the system engineer as he goes about his tasks.

14.1 PRELIMINARY STUDIES

Our fictitious company had already established itself as a major supplier of infrared systems for the military. In casting about for ways to increase its business, marketing surveys indicated that long-range jet transports represented a good potential market for infrared search systems. The opportunity to diversify out of the military market provided an added incentive to achieve a strong position in this new field.

The identification studies showed that pilots were experiencing certain problems during their operations near terminals. A detailed analysis of these problems led to the conviction that an infrared search set could supplement the ground-based air traffic control system by providing information on other traffic in the vicinity in a form that could be easily interpreted by a pilot. Hence it would be most useful during climb-out, letdown from altitude, and the time spent in holding patterns. While the aircraft is cruising at assigned altitude, the search system might provide early warning of the appearance of other aircraft in the vicinity and thereby reduce the possibility of collision. Finally, the search system might

THE DESIGN OF AN INFRARED SEARCH SYSTEM

provide a crude ground mapping capability for emergency navigation if the regular navigational equipment becomes inoperative.[1]

In general, the proposed search system should be highly reliable and easily maintainable, should require a minimum of special ground support equipment and no specially trained personnel for its operation. It should be possible to perform a simple go/no-go test of system operation while the system is on the ground or in the air. Installation procedures should require an absolute minimum of structural changes to the aircraft in order to accommodate the system. The signal processor and power supplies should be packaged in small modules in order that they may be scattered about the aircraft to take advantage of existing bits of space without the need to relocate previously installed equipment.

Since one of the major functions of the search system is to warn of the presence of other aircraft in the vicinity, it is essential that it have a wide search field. It was decided to use a gimbaled telescope assembly (optics and detector) that could be driven to scan any desired pattern. With a hemispherical irdome, such a sensor can scan over a total angle of 140 deg, which is an ample azimuthal coverage for this application. An elevation coverage of 10 deg was considered to be adequate provided that the pilot could, at his option, position the 10 deg by 140 deg search field from 30 deg below the horizontal to 30 deg above it. An initial estimate of the angular resolution requirements showed that an instantaneous field of view of 0.25 deg by 0.25 deg was desirable. A gimbaled telescope assembly can be driven in a linear scan motion as fast as 250 deg sec^{-1}. In the interest of long operating life, it was decided that the scan rate for the preliminary design should not exceed 100 deg sec^{-1}. There is, of course, some *dead time* at the end of each scan line, while the scan motion is reversed. In a well-designed system the dead time is usually about 5 per cent of the total scan time. Since the range equation depends on the scanning rate, the simplest way to account for the dead time is to assume that the search field is a bit wider than it actually is. Accordingly, the scanning rates for the preliminary design were calculated on the basis of an azimuthal coverage of 150 deg, although only 140 deg was to be displayed. With a single detector, it would require 60 sec to scan the entire search field. With a ten-element linear array of detectors, the frame time could be reduced to 6 sec, a time commensurate with the intended application. Because these jet transports already have a cathode-ray tube in

---

[1]The writer asks that the reader grant him poetic license, since he, in fact, sees little justification for adding such a search set to a jet transport. An exception would be a search set that could also function as a detector of clear air turbulence, a subject that is discussed in Section 17.3.2. The analysis given in this paragraph merely serves as an introduction to the more important discussion of the system design process that is the heart of this chapter.

the cockpit for the weather radar, it can be time-shared to display the output of the infrared search set, since the two systems are not required to operate simultaneously.

The sensor is the only part of the system that must extend through the skin of the aircraft. A study of the Boeing 707 (Figure 3.8) shows that there are relatively few possible locations for the sensor. The best location is on the nose, but this spot is already taken by the weather radar. One just aft of the radome, either on the top or on the bottom of the nose, appears feasible, although the radome may protrude into the search field at extreme elevation or depression angles. Locations further aft are unsuitable because of obscuration by the fuselage. A possible exception is a location just above the top of the pilot's windshield; from Figure 3.8 it appears that obscuration by the nose might be tolerable. An aerodynamically clean installation could be made by mounting the sensor in the leading edge of the wing, within a few feet of the point at which the wing joins the fuselage. If this location is chosen there must be two sensors, one on either side of the fuselage, in order that each may scan approximately half of the search field. A wing-tip installation is subject to considerable obscuration by the fuselage and is likely to experience high acceleration forces because of the relative flexibility of the wings, and it is unlikely that there will be sufficient room near the sensor for the installation of a refrigerator for the detector. A final, but even less desirable, alternative is to mount the sensor on top of the vertical stabilizer. On the Boeing 707 such a location might interfere with the pitot tube, which is already located there. On other aircraft the stabilizer might require structural modification to support the extra load imposed by the sensor. There is also considerable obscuration by the fuselage and wings, and the possibility that the sensor might scan across the engines and temporarily saturate the system. It was therefore strongly recommended that a location be found near the nose. A final approval of the location selected would have to await a detailed study, by the customer, of the equipment already located in this area, the structural problems of accommodating the sensor, and any undesirable aerodynamic effects resulting from its placement in this location.

The data in Section 3.4 indicate that the peak of the spectral distribution curve for the radiant energy from jet engines lies in the 3 to 5 μ region, which coincides nicely with the atmospheric window. Figure 10.7 shows that high-performance detectors for this window must be cooled to approximately 77°K. For use in commercial transports, any type of cooler other than a closed-cycle refrigerator would probably place an intolerable burden on ground service facilities. As shown in Table 11.7, a Stirling refrigerator is available commercially, but it adds 15 lb to the

THE DESIGN OF AN INFRARED SEARCH SYSTEM 441

system weight and requires an additional 500 W of power. Figure 10.7 shows that there are other detectors that will operate at higher temperatures and still cover the 3.2 to 4.8 μ window, but their detectivity is so low that they are impractical for this application.

What other alternatives are available? Is it possible to avoid the use of a cooled detector? Referring again to Figure 10.7, we see that high-performance detectors for the 8 to 13 μ window must be cooled to even lower temperatures. Also, both their detectivities and the effective radiant intensity of the target are lower than they are in the 3.2 to 4.8 μ window. The situation looks a bit more promising for an uncooled detector and the 2 to 2.5 μ window, since the detectivity of uncooled lead sulfide is higher than that of any of the cooled detectors used in the 3.2 to 4.8 μ window. Unfortunately the radiant intensity of the target is considerably lower in the 2 to 2.5 μ window, and thus the choice of uncooled lead sulfide imposes a penalty on the maximum detection range. Taken alone, such a penalty in detection range might be an acceptable price to pay in order to eliminate the need for cooling the detector. Unfortunately, operation in the 2 to 2.5 μ window is likely to be limited during the daytime by sunlight reflected from the background. As Figure 3.15 shows, the spectral radiance of typical backgrounds is a minimum in the 3.2 to 4.8 μ window and it increases rapidly below 3 μ because of reflected sunlight.

To summarize: The most favorable choice of window and detector is the 3.2 to 4.8 μ window, with a detector cooled to 77°K. The second choice is the 2 to 2.5 μ window, with an uncooled lead sulfide detector. Adoption of the second choice implies a willingness on the part of the designer to pay the price of shorter detection ranges and higher false-alarm rates during daytime operation in order to eliminate the cooling requirements.

14.2 SYSTEM SYNTHESIS AND ANALYSIS

On the basis of the above decisions it is possible to take a first cut at the system design and to make a preliminary estimate of its performance.

The first step is to define the shape and size of the sensor. The preliminary estimate, which was based on past experience with similar units and the analysis of the available locations, was that the sensor package would be a cylinder 6.5 in. in diameter with a hemispherical irdome mounted on one end. The total length of this package, including the irdome, is 14 in.

Next a decision must be made on the specific type of detector to be used. Since the preliminary studies clearly indicated the advisability of using the 3.2 to 4.8 μ window and a detector cooled to 77°K, its use was assumed in the preliminary design. From Figure 10.7 the possible choices for a detector include photovoltaic or photoconductive indium antimonide and lead selenide. The detailed data sheets for these detectors (Figures 10.11–10.13) show the relative advantages and disadvantages of each. When these three are compared on the basis of D^*, photovoltaic indium antimonide is clearly the best detector. A possible difficulty with this conclusion arises when the responsivities are compared and it is noted that value for lead selenide is 10 times higher than it is for the other two types of detector. In an ideal operating environment, that is, one in which there are no extraneous sources of noise, the designer is interested in the signal-to-noise ratio at the detector, as described by D^*, and not the signal level, as described by responsivity. An aircraft, on the other hand, is a relatively hostile environment for sensitive, low-noise equipment. In such a noisy environment there are advantages to choosing the detector that has the highest signal level (responsivity) because its use will make it easier to achieve the goal of a detector-noise-limited system. The severity of the noise pickup can be determined only after the equipment has been installed in its final operating environment. On this basis, and since it represents conservative design, lead selenide was selected as the detector for the preliminary design and for the early prototypes with the expectation that it would be replaced by photovoltaic indium antimonide if the installation proved to be sufficiently free of extraneous noise pickup. Fortunately these two types of detector can be interchanged without making any other modifications in the system except for changing the preamplifier.

A hemispherical irdome is needed to accommodate the wide azimuthal scanning angle. Since it mounts on the end of a cylindrical package that has a diameter of 6.5 in., the diameter of the irdome is, likewise, 6.5 in. and the radius of its outer surface is 3.25 in. Because it is not mounted inside the aircraft, the irdome must be capable of withstanding the effects of various aerodynamic forces, abrasion by dust, dirt, and insects during takeoff and landing, and erosion by rain. Of the preferred optical materials shown in Figure 5.21, silicon has proved to be an excellent material for such an application. When given an antireflection coating, its transmission is excellent in the 3 to 5 μ region. A hemispherical dome of 6.5 in.-diameter is well within the capabilities of material suppliers and the optical shops that will do the grinding and polishing. The thickness of the irdome must be sufficient to withstand the anticipated aerodynamic forces. A calculation by the usual methods of stress analysis indicates that for silicon a thickness of 0.120 in. is adequate for subsonic flight. The inner and outer surfaces should, for optical reasons, be accurately

THE DESIGN OF AN INFRARED SEARCH SYSTEM

concentric. A discussion with personnel from an optical shop indicates that a tight tolerance on concentricity will be much easier to hold if the thickness is increased to 0.150 in. Therefore the outer surface of the irdome is given a radius of 3.25 in. and the inner surface a radius of 3.10 in.

The optics can be either reflective or refractive. A wide-field mechanical scanning system is of the simplest form if the sensitive element of the detector is located at the gimbal center (the point at which the gimbal axes intersect) and if the irdome is concentric about this center. Because refractive optics lead to a somewhat simpler mounting, they were chosen for use in the search system. Since the front surface of the lens need be no more than 0·1 in. from the inner surface of the irdome and it is possible to design the lens so that the second principal point coincides with its front surface, the focal length of the lens is set equal to 3·0 in. Silicon is an appropriate material for the lens. Because each detector element subtends an angle of 0·25 deg (4·4 mrad), the f/number of the lens must be selected so as to achieve a blur circle no larger than this. From Figure 5.8 and (5–27), an f/2.3 silicon lens has a blur circle that is 2.5 mrad in diameter. The optical designer attached to the project team suggested that an f/2 silicon lens could achieve this size blur circle if the second surface of the lens was made aspheric. Hence the lens was defined as an f/2 silicon element with a focal length of 3.0 in. and a diameter of 1.5 in.

From the data sheet for lead selenide (Figure 10.11) it is estimated that with a cooled filter and a shield matching the f/2 lens, a detectivity of 5×10^{10} cm(Hz)$^{1/2}$ W^{-1} can be achieved in the 3.2 to 4.8 μ region. Since the detector elements are to subtend an angle of 0.25 deg by 0.25 deg when used with a lens of 7.6 cm (3 in.) focal length, each element is $(7.6 \times 0.01745 \times 0.25)$ or 0.033 cm on a side. Figure 10.11 also shows that elements of this size are well within current manufacturing capabilities.

The sensor, as it looks at this stage of the design, is shown in Figure 14.1. As seen there, it consists of the irdome, lens, detector, gimbal structure, azimuth and elevation drive motors and associated drive circuitry, a 10-channel preamplifier, and a flexible line to deliver the coolant to the detector. Not shown are resolvers for determining the instantaneous position of the telescope assembly and a vertical gyro to provide a reference for stabilization purposes. Since ample driving power is available for the telescope assembly, it is feasible to use a liquid-transfer system to deliver the coolant to the detector. This implies that the refrigerator can be either the closed-cycle Joule-Thomson type or the Stirling type with an external liquefier (both types are described in Section 11.4). It is estimated that the sensor package weighs about 15 lb.

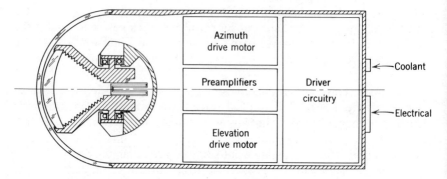

Figure 14.1 Tentative sensor design.

The sensor is now sufficiently well defined that its maximum detection range can be estimated by using the methods developed in Chapter 13. Before doing this, the values of several additional parameters must be calculated. The transmittance of the optics τ_0 depends on the transmittances of the irdome and the lens.[1] From the information in Section 5.8, it is estimated that a single-layer antireflection coating applied to each surface of the irdome and lens will result in a transmittance of 0·9 per element in the 3.2 to 4.8 μ spectral bandpass. Thus the value of τ_0 is (0.9×0.9) or 0.81.

The normal cruising altitude for a commercial jet transport is about 30,000 ft. In order to make a preliminary estimate of the atmospheric transmittance τ_a, assume a detection range of 30 (statute) miles. From Figure 4.4 the precipitable water content at this altitude is approximately 0·035 mm kft^{-1} (in temperate zones during the summer). Hence there is approximately 5.5 mm of precipitable water in the 30 mile path. Figure 4.15 shows that the transmittance at sea level along a path having this water content is about 0.7. Since the altitude of the desired path is 30,000 ft, the transmittance will be greater than 0.7. However, there is no reason to determine a more accurate value at this time, since it is based on a purely arbitrary choice of a 30 mile path.

The effective search field is to be 10 by 140 deg, with a frame time of 6 sec. Using 150 deg for the width of the search field (to account for the scan dead time, the time rate of search is, from (13-24),

$$\dot{\Omega} = \frac{\Omega}{\mathscr{T}} = \frac{(10 \times 150)(3.045 \times 10^{-4})}{6}$$
$$= 7.61 \times 10^{-2} \text{ sr sec}^{-1},$$

[1] The transmittance of the protective window on the detector package and of the cooled filter could be added to this list except that, in practice, these are normally included in the measured value of D^* for the detector.

THE DESIGN OF AN INFRARED SEARCH SYSTEM

where the factor 3.045×10^{-4} is used to convert square degrees to steradians. From (13-25), the detector dwell time is

$$\tau_d = \frac{\omega C}{\dot{\Omega}} = \frac{(0.25 \times 0.25 \times 3.045 \times 10^{-4})(10)}{7.61 \times 10^{-2}}$$

$$= 2.50 \times 10^{-3} \text{ sec}.$$

From (12-13), the optimum 3 dB bandwidth is

$$B = \frac{1}{2\tau_d} = \frac{1}{2 \times 2.5 \times 10^{-3}} = 200 \text{ Hz}.$$

Assuming, as before, that the 3 dB bandwidth is equal to the noise equivalent bandwidth, the value of Δf is 200 Hz.

The exact value of the pulse visibility factor v can only be calculated after the transfer function of the system electronics has been determined. Prior experience with similar systems shows that the value is likely to lie between 0.25 and 0.75. Assume as a starting point that the value of v is 0.5.

The characteristics of the preliminary design are summarized in Table 14.1.

TABLE 14.1 CHARACTERISTICS OF THE PRELIMINARY DESIGN FOR AN INFRARED SEARCH SYSTEM

Quantity	Symbol	Value	Units
Diameter of entrance aperture	D_o	3.8	cm
Numerical aperture	NA	0.25	—
Detectivity	D^*	5×10^{10}	$cm(Hz)^{1/2} W^{-1}$
Transmittance of the optics	τ_o	0.81	—
Pulse visibility factor	v	0.5	—
Number of detector elements	C	10	—
Instantaneous field of view per detector element	ω	1.90×10^{-5} (0.25×0.25)	sr deg
Search field (effective)	Ω	0.42 (10×140)	sr deg
Frame time	$\mathcal{T}$	6	sec
Search rate	$\dot{\Omega}$	7.61×10^{-2}	sr sec^{-1}
Electrical bandwidth	Δf	200	Hz
Spectral bandpass	—	3.2–4.8	μ

Since the Boeing 707 commercial jet transports are used the world over and we have already made an extensive study of their radiating characteristics, it is appropriate to use them as a means of describing the performance of the search system. Table 3.2 contains data on the effective radiant intensity of the Boeing 707-320 turbojet at maximum cruise thrust. From equation 13-29 the idealized range is

$$R_0 = \left[\frac{\pi}{2} \times 3.8 \times 0.25 \times 5 \times 10^{10} \times 2100 \times 0.7 \times 0.81\right]^{1/2} \left[\frac{0.5 \times 10}{7.61 \times 10^{-2}}\right]^{1/4}$$

$R_0 = 2.68 \times 10^7$ cm $= 149$ (statute) miles.

For a false-alarm time of 500 sec, the number of noise pulses is given by (13-30) as

$$n' = \tau_{fa}\Delta f = 500 \times 200 = 10^5.$$

From Figure 13.1, for a probability of detection of 0.9, the ratio of the actual to the idealized range is 0.41; therefore

$$R_{0.9} = 0.41\ R_0 = 68\ \text{(statute) miles}.$$

Figure 13.1 also shows that, for this false-alarm time, a signal-to-noise ratio of 5.6 would be required in order to observe a detection range equal to R_0.

Because the alternative of using an uncooled lead sulfide detector and the 2 to 2.5 μ window has not yet been ruled out, it is of interest to see what the maximum detection range is with this combination. From Table 3.2 the total radiant intensity of the Boeing 707-320 is 7900 W sr^{-1}. A simple calculation with a radiation slide rule (see Example 2.7.1) indicates that the effective radiant intensity in the 2 to 2.5 μ window is 300 W sr^{-1}. Figure 10.10 shows that a detectivity of 9×10^{10} cm (Hz)$^{1/2}$ W^{-1} is feasible. From Figure 4.15 the transmittance of the atmosphere should be about 0.9. For this spectral region a fused silica irdome is a good choice. As shown in Figure 5.22, its transmittance is 0.95, and there is little to be gained from giving it antireflection coating. The silicon lens used in the other design could be retained. Since the spectral interval is narrower than before, it should be possible to achieve a transmittance (after antiflection coating) of 0.95. Similarly an interference filter to limit the spectral bandpass to 2 to 2.5 μ would have an average transmittance of 0.85. The transmittance of the optics τ_0 is, therefore, (0.95 $\times$ 0.95 $\times$ 0.85) or 0.77. Making these substitutions in (13-29) and assuming, as before, a 90 per cent probability of detection and a false-alarm time of 500 sec, the maximum detection range for the Boeing 707-320 is 38 (statute) miles. Thus a system using an uncooled lead sulfide detector in the 2 to 2.5 μ window has hardly more than half the detection range of a similar system using a cooled lead selenide detector in the 3.2 to 4.8 μ window. One way of regaining the loss in detection-range capability would be to increase the diameter of the optics in the lead

THE DESIGN OF AN INFRARED SEARCH SYSTEM 447

sulfide system by $(68/38)^2$ or 3.2 times. Such an increase would require that the irdome have a diameter of nearly 21 in., and the entire sensor would be ridiculously large for the intended application.

There is one serious defect in the design that has been overlooked so far: the 200 Hz bandwidth is far too narrow. Since this is a pulse and not a carrier system, the bandwidth Δf is a video bandwidth and therefore it extends from dc to 200 Hz. At such low frequencies, $1/f$ noise from the detector will be excessive and the detection range will be considerably less than that calculated. There are several ways to eliminate this design defect. These include making the detector elements narrower or increasing the scan rate so as to make the dwell time shorter. Either action will widen the bandwidth so that the low frequencies, where the $1/f$ noise is excessive, can be eliminated with a high-pass filter. Another approach, but one that would require a major redesign, is to add a reticle to the sensor and convert it to a modulated-carrier system. The narrow bandwidth could then be retained, but it would be centered around the carrier frequency, which would be high enough to be clear of the $1/f$-noise region.

14.3 TRADEOFF STUDIES AND FINAL SYSTEM DESIGN

Despite the evident defect due to the excessive $1/f$ noise, the preliminary design is still very encouraging. The next step is to begin a series of engineering tradeoff studies in which the design is modified and then reanalyzed to determine the effect of the modification on the performance. Ultimately this should lead to a design that meets the system objectives and one that can be successfully implemented within the cost and delivery constraints imposed by the customer.

After several meetings with engineering representatives of the airline and of the aircraft manufacturer, it was agreed that the sensor could be mounted on top of the nose just aft of the radome. It was decided, however, that in this location the diameter of the sensor package could not exceed 5.0 in. At the same time a careful study was made of the possible locations for a closed-cycle refrigerator. Adequate space for such a refrigerator was found in one of the wheel wells. Since there was room for a liquefier immediately behind the sensor package, it was decided to use a closed-cycle Stirling refrigerator with an external liquefier. This decision cleared the way for the use of a cooled detector and the 3.2 to 4.8 μ atmospheric window.

A more careful analysis of the angular resolution requirements indicated that a greater resolution was highly desirable. Airline pilots suggested that the ability to resolve the individual engines on other aircraft in a holding pattern would help them identify the type of aircraft and thus plan their own flight actions. In the preliminary design the azimuthal resolution is 0.25 deg. As shown in Table 3.1, the distance between the engines on the same wing of a Boeing 707 is 38 ft. To resolve the individual engines requires that this aircraft be no more than 8700 ft from the sensor. Pilots felt that this distance should be increased to at least 25,000 ft. This can be done by reducing the width of each detector. It was decided to retain 10 elements in the detector array but to make each subtend only 0.05 deg in azimuth. With this change the individual engines on a Boeing 707 can be resolved at a distance of at least 40,000 ft. This reduction in the width of the detector also helps eliminate the narrow-bandwidth problem.

The decision to limit the diameter of the sensor package to 5.0 in. limits the radius of the outer surface of the irdome to 2.5 in. Despite the smaller size, it was decided that the thickness of the irdome should not be changed from its previous value of 0.150 in. Thus the radius of the inner surface is 2.35 in.

The single silicon lens used in the preliminary design can not be used with the narrower detector elements because its blur circle is too large. To meet this requirement it was necessary to use a two-element lens, with one element made of arsenic trisulfide and the other of IRTRAN-1. With two elements at his disposal, the optical designer was able to increase the speed of the lens to $f/1.5$ (NA = 0.33) and still achieve a blur circle with a diameter of 0.05 deg. It was decided that the clearance between the front surface of the lens and the inner surface of the irdome could be reduced to 0.075 in. Since the lens was designed to have its second principal point coincident with its first surface, the focal length is 2.275 in. (5.8 cm). The transmittance of the optics τ_0 was estimated to be 0.77 by assuming that the coated silicon irdome and the uncoated[1] IRTRAN-1 element each have a transmittance of 0.9 and that the transmittance of the coated arsenic trisulfide element is 0.95.

As a result of reducing the size of the irdome and the lens, the telescope assembly, gimbal structure, and scan drive motors also become smaller. The new sensor package has a diameter of 5 in., a length of 11 in. (including the irdome), and a weight of 11 lb.

Because the size of the telescope assembly and gimbal structure were reduced, it is possible to increase the angular scanning rate to 185 deg sec^{-1} without unduly shortening the operating life of the sensor. As a result of this faster scanning rate, the frame time decreased to 3.25 sec. Following the changes in scanning rate and detector size, the optimum

---

[1] Since the index of refraction of IRTRAN-1 is so low, the slight gain in transmittance from an antireflection coating does not justify the expense involved.

THE DESIGN OF AN INFRARED SEARCH SYSTEM

system bandwidth became 1850 Hz (the calculation of this value will be presented later). This made it possible to use a high-pass filter to reject the lower frequencies and their accompanying $1/f$ noise. With an electrical bandpass extending from 650 to 1850 Hz, laboratory measurements of the noise from various samples of lead selenide detectors showed that the previously excessive $1/f$ noise content had been satisfactorily eliminated. An order was placed for lead selenide detector arrays to support the flight test evaluation of the system. At the same time an order was placed for a similar number of photovoltaic indium antimonide detector arrays. This order represented a calculated gamble that the indium antimonide could eventually replace the lead selenide.

The use of a high-pass filter to eliminate the $1/f$ noise removes some of the signal frequencies as well, and as a result, the signal processing is somewhat less efficient than it was before. The pulse visibility factor, which is a measure of the efficiency of the signal processing, has a value of 0.35 for the final choice of electrical bandpass.

When the maximum detection range was calculated for the preliminary design, the estimated value of the atmospheric transmittance was based on a range of 30 miles. Since the calculated value for the detection range was about 68 miles, we probably should have adopted a more appropriate value for the atmospheric transmittance and recalculated the detection range. Accordingly, assume a range of 75 miles. Proceeding as before, the new estimated value of τ_a is 0.6 (for the reasons already stated, this is probably a very conservative estimate).

Upon completion of the mechanical and electrical design, three engineering models were built. One model was installed in an aircraft for flight tests. All signal leads were carefully shielded, and the power leads were equipped with low-pass filters. Considerable noise pickup was noted during the initial flights; it was eventually traced to ground loops in the cables connecting the sensor to the signal processor. Many hours of patient experimenting were required to eliminate these ground loops. Eventually the day arrived when a series of measurements, made while the system was airborne, confirmed that the limiting noise was that from the detector. It was therefore decided to replace the lead selenide detector with a photovoltaic indium antimonide detector. When this detector is equipped with a cooled filter and shield, its detectivity is 8×10^{10} cm $(Hz)^{1/2}$ W^{-1} for the 3.2 to 4.8 μ spectral bandpass of the system.

The characteristics of the final design are summarized in Table 14.2. Repeating the original calculation for these new system characteristics gives

$\dot{\Omega} = 0.14$ sr sec^{-1}
$\tau_d = 2.71 \times 10^{-4}$ sec
$\Delta f = 1846$ Hz
$R_0 = 2.96 \times 10^7$ cm $= 184$ (statute) miles
$\tau_{fa} = 500$ sec
$n' = 9.2 \times 10^5$
$R_{90} = 0.40\, R_0 = 74$ (statute) miles.

The maximum detection range for a Boeing 707-320B (turbofan) is (see the data in Table 3.2)

$$R_{90} = 74 \left(\frac{1520}{2100}\right)^{1/2} = 63 \text{ (statute) miles.}$$

These ranges pertain, of course, to tail aspect only. Since there are no data available to show the variation in radiant intensity with aspect angle, it is not possible here to calculate the maximum detection range at other aspects. A rough rule of thumb is that the maximum detection range at nose aspect for a subsonic jet is about one-tenth that at tail aspect. Therefore, detection ranges of about 6 miles should be attainable at nose aspect. At beam aspect the detection range should be about 30 miles.

The elements comprising the final design of the search system and their

TABLE 14.2 CHARACTERISTICS OF THE FINAL DESIGN FOR AN INFRARED SEARCH SYSTEM

Quantity	Symbol	Value	Units
Diameter of entrance aperture	D_0	3.8	cm
Numerical aperture	NA	0.33	—
Detectivity	D^*	8×10^{10}	cm(Hz)$^{1/2}$W^{-1}
Transmittance of the optics	τ_0	0.77	—
Pulse visibility factor	v	0.35	—
Number of detector elements	C	10	—
Instantaneous field of view per detector element	ω	3.8×10^{-6} (0.05×0.25)	sr deg
Search field (effective)	Ω	0.42 (10×140)	sr deg
Frame time	$\mathcal{T}$	3.25	sec
Search rate	$\dot{\Omega}$	0.14	sr sec^{-1}
Electrical bandwidth	Δf	1846	Hz
Spectral bandpass	—	3.2–4.8	μ

THE DESIGN OF AN INFRARED SEARCH SYSTEM

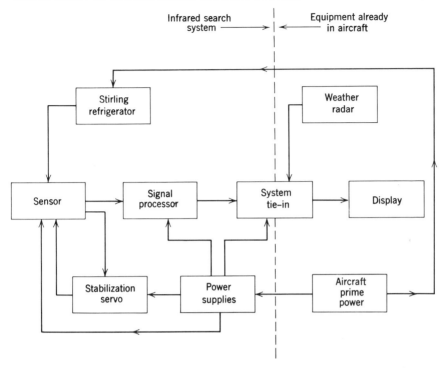

Figure 14.2   The infrared search system and its interconnections with the aircraft.

tie-in with the aircraft are shown in Figure 14.2. Preamplifiers in the sensor package raise the signal level so that it can be transmitted over shielded cables to the signal processor. After further amplification, the output of each detector channel is sampled and applied to the Z-axis of the display to provide intensity modulation. To ensure that a pulse is sampled at its peak value, each detector channel is sampled every 50 μsec. (The pulse length, which is equal to the dwell time, is 271 μsec.) A logarithmic amplifier following the multiplexer compresses the dynamic range of the pulses to about 20 dB in order that they may be compatible with the dynamic range of the display. A threshold detector, using a Schmidt trigger with a fixed triggering level, opens a signal gate and allows a pulse to pass to the display whenever the signal-to-noise ratio exceeds a value of about 5.5.

A stabilization servo minimizes the effects of aircraft motions on the search pattern. No details will be given here, because they are influenced by the particular means used to integrate this equipment with the elements already in the aircraft system.

The block labeled "system tie-in" is the electrical interface between

TABLE 14.3 ESTIMATED WEIGHT, VOLUME, AND POWER CONSUMPTION OF THE INFRARED SEARCH SYSTEM (FINAL DESIGN)

Unit	Weight (lb)	Volume (ft^3)	Power Consumption (W)
Sensor	11	0.15	55
Signal processor, stabilization servo, and power supplies.	16	0.40	80
Tie-in	3	0.10	15
Stirling refrigerator	15	0.15	550
Total	45	0.80	700

the infrared search system and the cathode-ray-tube display already in the aircraft. It includes a switch by which the pilot can select either infrared or radar input for the display. For an infrared input, the display is a modified C type that shows the position of the target in azimuth and elevation. The display is intensity modulated and can also be deflection modulated in order to increase the visibility of a target on the display.

The estimated weight, volume, and power consumption of the system are shown in Table 14.3.

2
The Applications of Infrared Techniques

15

An Introduction to the Applications of Infrared Techniques

The first part of this book has described the elements of the infrared system, their interactions with one another, and the skills that are needed to design and build such systems. The rest of the book is intended to stimulate the reader to apply this information by showing how others have used infrared techniques for the solution of various problems that arise in the military, industrial, medical, and scientific fields.

To one who has studied the literature, it is quite apparent that many of the applications for infrared techniques have been known for half a century or more. The changes that we observe in a survey of the literature are often nothing more than changes in the elements of the solution to a particular problem. It is not surprising that after a short lecture on the characteristics of the infrared portion of the spectrum, almost anyone can think of many applications for a radiation that is emitted by virtually all objects in the universe and that cannot be detected by human sensory organs. As early as 1910, many workers were intrigued by the possibility of heat seekers and were proposing a wide variety of infrared search devices. This interest was fostered by the availability of the selenium detector. Many of the basic techniques used today for generating tracking signals and suppressing the effects of unwanted backgrounds were conceived in the 1910 era. This era also gave birth to the myth that persists even today that infrared radiation has a magical ability to penetrate fog, rain, and clouds. The nighttime iceberg detector that excited so much interest after the sinking of the Titanic in 1912 was an impractical device because of the limitations of the components available at that time. With the components available today, such an iceberg detector is small enough to be held in one hand. Unfortunately, infrared radiation does not penetrate present-day fogs any better than it did in 1912, and the business of iceberg detection is better left to other sensors that are not limited by bad weather.

In the early 1920's the availability of the thallous sulfide detector encouraged a new generation of workers to reinvent infrared search devices. Similar cycles of reinvention occurred in the early 1940's as a result of the development of the image converter tube, in the late 1940's when the lead sulfide detector appeared, in the mid-1950's when photon detectors sensitive in the 3 to 5 μ window were developed, and in the early 1960's when small and reliable cooling devices became available for these 3 to 5 μ detectors. Such cycles of reinvention are, of course, observed in any rapidly developing technology. For this reason it is well to remember when searching for a new way to solve a problem that one of the most fruitful means of attack may be a careful study of the prior literature in the field.

The reader who delves into the remainder of this book will not find the usual stereotyped collection of how-to-do-it ideas interlarded with photographs and descriptions taken from the manufacturers' literature. He will find, instead, the most comprehensive set of annotated references available anywhere that deal with the *published* literature of the infrared field. The references are carefully categorized so that it is easy to find those that pertain to the problem at hand. In many instances, virtually all of the open literature is listed that is relevant to a particular application. This provides a rarely observed historical perspective and allows the reader to trace the efforts of those who have preceded him and to examine in detail how various techniques have been applied and with what success. Because of the wealth of ideas contained in the references and the unique way in which they are presented, the reader should find it stimulating to do a little random browsing through the references in fields outside his own immediate interests. Since the references are to the published literature, any reader with a little persistence can obtain them. He will experience none of the frustrations that arise when one turns up a promising new reference and then finds that it cites an internal report of the XYZ Company or, worse yet, an unpublished communication between inconsiderate authors.

Chapter 16, on the military applications of infrared techniques, is, of course, limited by the dictates of military security. Some of the articles cited in that chapter are very short, and they may contain only a sentence or two on a particular application. However short they may be, these citations make it possible to mention many of the military applications, while ensuring that there is no violation of security restrictions.

Many of the references given here are to literature published outside

INTRODUCTION   457

of the United States. This merely underscores the fact that no nation has a monopoly on the applications of infrared techniques. Among the countries represented are Australia, Czechoslovakia, England, France, Germany, Holland, Hungary, Japan, Poland, Rumania, and the Soviet Union. Translations are available for most of the non-English-language references that are cited here.[1]

The Soviet literature is of particular interest because it offers a possible insight into the infrared capabilities of one of the major contemporary powers. The Soviets are not signatories to the International Copyright Convention, and they do not hesitate to lift figures and large blocks of text from the world's literature without the usual courtesy of acknowledging the source—a practice that most readers will loathe. It is quite evident, however, that the Soviets are remarkably efficient at sifting through the non-Soviet literature and making their findings readily available to those who need this information. In the United States we have nothing even remotely comparable to this Soviet capability. In general, the Soviet literature contains excellent treatments of theoretical matters, but their treatment of the applications of this theory is almost invariably taken straight from the pages of such well-known Western publications as *Aviation Week*. As a result, the student of the Soviet literature sees a peculiarly incomplete picture. On one side he sees an obviously strong capability in infrared technology, and on the other side he sees a determined Soviet effort to prevent the rest of the world from knowing very much about what they are doing with this capability. There are, however, enough cracks in this wall of secrecy to show that the Soviets have successfully applied their knowledge of infrared techniques to the guidance of missiles, to fire control, night driving, communications, process control, meteorological observations (both earth bound and satellite-borne), and to medical diagnostics.

The process of annotating such a large collection of references must inevitably reveal the interests and biases of the annotator. This author has been an enthusiastic missionary for infrared techniques since the days of World War II, when he had a chance to learn the combat effectiveness of the infrared sniperscope. His background is strongly system-oriented, and he is happiest when up to his elbows in hardware. If the reader sees such an image reflected in the annotations, this is the explanation of its origin.

---

[1]These are indicated by the inclusion of a CFSTI, OTS, or SLA number at the end of the reference. The translations are available from the Clearinghouse for Federal Scientific and Technical Information. For further details concerning this service, refer to Appendix 4.

15.1 THE APPLICATIONS OF INFRARED TECHNIQUES

The infrared applications matrix shown in Figure 15.1 is a convenient framework within which to examine the various applications of infrared techniques. Across the top of this matrix are listed the potential users of such techniques. These are grouped into four categories that include the military, industrial, medical, and scientific users. The reader may find it convenient to consider them not as users but as potential customers. At the side of the matrix are indicated the kinds of functions for which infrared equipment is peculiarly well adapted. The result is a logical classification scheme for the entire infrared field in the form of a four-by six-element matrix.[1]

Because of space limitations, the applications shown in each element of the matrix are merely a sample of what has been done. In addition, the matrix serves as a guide to the organization of the annotated references that compromise the bulk of the remaining chapters. Military applications are discussed in Chapter 16, industrial applications in Chapter 17, medical applications in Chapter 18, and scientific applications in Chapter 19. Each chapter is divided into six sections that correspond to the functions shown in the matrix. When appropriate, sections are subdivided in order to more clearly separate the references that pertain to specific applications.

Since the applications matrix is such a convenient guide to the chapters that follow, it is appropriate to examine more closely the rationale for the choice of the functions in the matrix. If it is assumed that the target radiates at least some of its flux in the spectral region of interest, infrared equipment has the potential capability of detecting the presence of the source, determining its direction with respect to some reference, and determining its distance. Hence the first function in the matrix is called *search*, *track*, and *range*. Infrared equipment can also measure the flux from the target in a broad spectral interval, an operation designated as *radiometry* in the matrix, or it can measure the flux in a narrow spectral interval, which is called *spectroradiometry*. If the target is an extended source, the infrared equipment can also be used to determine the geometrical or spatial distribution of the flux, which is called *thermal imaging* in the matrix.

If the target is a poor radiator in the spectral region of interest, perhaps because of its low temperature or low emissivity, the infrared equipment can respond to the flux that the target reflects from an auxiliary source of illumination. A photographic system, for instance, responds to the

---

[1] I am indebted to W. E. Mathews of the Hughes Aircraft Company for his assistance in the development of this matrix.

INTRODUCTION 459

sunlight reflected from a scene. In the matrix this function is called *reflected flux*. As a final alternative, one can place a source at the target, at the observer, or at both locations and arrange to have the infrared equipment sense such things as the motion or the modulation of one or more of the sources. Examples of such a mode of operation include tracking systems in which a beacon is placed on the object to be tracked, and communication systems. In the matrix this function is designated as a *cooperative source*.

With any classification scheme, and this one is no exception, it is necessary to make a few more or less arbitrary assignments. Among those made here is the decision that both emission and absorption spectroscopy belong in the spectroradiometry category. Similarly, reflection spectroscopy is placed in the reflected-flux category. The field of communications has been divided into terrestrial communications that are assigned to the military user and space communications that are assigned to the scientific user.

A complete coverage of the applications of infrared spectroscopy to the analysis of organic compounds would require a volume by itself. Since it is rarely employed by the infrared system designer, this subject will not be discussed here; the interested reader can consult one of the standard reference works on the subject (see [43–45] for Chapter 2).

15.2 MISCELLANEOUS REFERENCES

The references in this chapter are, in general, too broad to be placed in any of the following chapters. Ballard and Wolfe [1, 2] provide excellent state-of-the-art reports on infrared techniques and their applications up to 1962. References 3–17 (arranged chronologically) contain interesting surveys of the past, present, and future applications of infrared techniques. References 18 and 19 are rather specialized reviews of the Soviet literature on infrared sensors. Dulles [20] gives a fascinating account of the period in the early 1950's when those shady figures who have intelligence for sale decided that information on infrared was an important and lucrative commodity.

Figure 15.1 INFRARED APPLICATIONS MATRIX.

	Military — Chapter 16	Industrial — Chapter 17
1. Search, track, and range	Intrusion detection Bomber defense Missile guidance Navigation and flight control Proximity fuses Ship, aircraft, ICBM, and mine detection Fire control Aircraft collision warning	Forest fire detection Guidance for fire-fighting missiles Fuel ignition monitor Locating hidden law violators Monitoring parking meters Detect fires in aircraft fuel tanks
2. Radiometry	Target signatures	Detection of hot boxes on railroad cars Noncontact dimensional determination Process control Measurement of the temperature of brake linings, power lines, cutting tools, welding and soldering operations, and ingots
3. Spectro-radiometry	Terrain analysis Poison gas detection Target and background signatures Fuel vapor detection Detection of contaminants in liquid oxygen piping	Detection of clear-air turbulence Analysis of organic chemicals Gas analysis Determination of alcohol in the breath Discovery of leaks in pipelines Detection of oil in water Control of oxygen content in germanium and silicon
4. Thermal imaging	Reconnaissance and surveillance Thermal mapping Submarine detection Detection of underground missile silos, personnel, vehicles, weapons, cooking fires, and encampments Damage assessment	Nondestructive testing Inspection Locating piping hidden in walls and floors Inspection of infrared optical materials Detect and display microwave field patterns Study efficiency of thermal insulators
5. Reflected flux	Night driving Carbine firing Intrusion detection Area surveillance Camouflage detection Station keeping Docking and landing	Industrial surveillance and crime prevention Examination of photographic film during manufacture Detection of diseased trees and crops Travelling matte photography Automatic focusing of projectors
6. Cooperative source	Terrestrial communications Command guidance for weapons Countermeasures for infrared systems Range finding Drone command link Intrusion detection	Intrusion detection Automobile collision prevention Traffic counting Radiant heating and drying Data link Intervehicle speed sensing Aircraft landing aid Cable bonding

Medical — Chapter 18	Scientific — Chapter 19	
Obstacle detection for the blind	Satellite detection Space vehicle navigation and flight control Horizon sensors Sun followers for instrument orientation Studies of the optical structure of the horizon	1. Search, track, and range
Measurement of skin temperature Early detection of cancer Monitor healing of wounds and onset of infection, without removing bandages Remote biosensors Studies of skin heating and temperature sensation	Measurement of lunar, planetary, and stellar temperatures Remote sensing of weather conditions Study of heat transfer in plants Measurement of the earth's heat balance	2. Radiometry
Detection and monitoring of air pollution Determination of carbon dioxide in the blood and in expelled air	Determination of the constituents of earth and planetary atmospheres Detection of vegetation or life on other planets Terrain analysis Monitor spacecraft atmospheres Zero-G liquid level gauge Measurement of magnetic fields	3. Spectro-radiometry
Early detection and identification of cancer Determination of the optimum site for an amputation Localization of the placental site Studies of the efficiency of Arctic clothing Early diagnosis of incipient stroke	Earth resource surveys Locate and map the gulf stream Detect forest fires from satellites Study volcanoes Detect and study water pollution Locate crevasses Sea-ice reconnaissance Petroleum exploration	4. Thermal imaging
Measurement of pupillary diameter Location of blockage in a vein Monitoring eye movements Study the nocturnal habits of animals Examination of the eye through corneal opacities Monitoring healing processes	Detection of forgeries Determine thickness of epitaxial films Determination of the surface constituents of the moon and the planets Gem identification Analysis of water quality Detection of diseased crops	5. Reflected flux
Ranging and obstacle detection for the blind Heat therapy	Space communications Understand the mechanism of animal communication Peripheral input for computers Study the nocturnal habits of animals Terrain illumination for night photography	6. Cooperative source

REFERENCES

[1] S. S. Ballard, ed., "Special Issue on Infrared Physics and Technology", *Proc. Inst. Radio Engrs.*, Vol.47, September 1959. Details the state of the art as of 1959.
[2] S. S. Ballard and W. L. Wolfe, ed., "Special Issue on Infrared," *Appl. Opt.*, Vol. 1, September 1962. Supplements reference 1 and updates it to 1962.
[3] M. Deribere, *Les Applications Pratiques des Rayons Infrarouges (The Applications of Infrared Rays)*, Paris: Dunod, 1954.
[4] G. B. B. M. Sutherland, "Infrared Radiation", *J. Inst. Elec. Engrs. (London)*, **105B**, 306 (1958). A very perceptive article by one of the major contributors to the field.
[5] M. Borchert and K. Jubitz, *Infrarot Technik (Infrared Technology)*, Berlin: VEB Verlag Technik, 1958. Describes the state of the art in Germany as of 1958.
[6] W. N. Arnquist, "Survey of Early Infrared Developments" *Proc. Inst. Radio Engrs.*, **47**, 1420 (1959). An excellent reference for the period from about 1915 to 1955.
[7] M. A. Bramson and A. Ye Kalikeyev, *Infrared Technology of Capitalist States*, Moscow: Sovietskoye Radio, 1960. The Soviet view of the art in the Western world.
[8] C. M. Cade, "Infrared Radar, Surveillance and Communications," *British Communications and Electronics*, **7**, 414–418, 510–517 (1960). Contains a good survey of the British infrared industry. Covers radiometry, process control, communication, and military applications.
[9] "Infrared – The New Horizon," *Electronic Equipment Engineering*, **9**, 44 (May 1961). A what-lies-ahead article.
[10] "Seeing Red," *Time*, March 31, 1961, p. 70. A news magazine's assessment of the wonders of infrared and some of the companies that are translating these wonders into reality.
[11] R. D. Hudson, Jr., and E. I. Staff, "The Optical Electronic Spectrum," *Electronic Industries*, **21**, 112 (April 1962). Features several quick-reference charts showing the portion of the spectrum that is generally used for various military, industrial, medical, and scientific applications of infrared techniques.
[12] T. Burakowski, J. Gizinski, and A. Salas, *Podczerwien i jej Zastosowania (Infrared and Its Applications)*, Warsaw: Ministry of National Defense, 1963. The state of the art as seen from Poland.
[13] D. A. Findlay, "Infrared," *Electronic Design*, **11**, 32 (October 25, 1963). A general-coverage, what-lies-ahead article.
[14] J. King, "Infrared," *Intl. Sci. Tech.*, April 1963, p. 26. An excellent article that contains an extremely broad coverage of the applications of infrared techniques.
[15] V. Kotesovec, "All Union Conference on the Technique of Infrared Radiation and its Utilization in Science and Industry," *Jemna Mechanika a Optika*, Vol. 8, No. 2, 1963 (OTS: 64-211). A report of a meeting sponsored by the Zapotocky Military Academy and the Czech Academy of Science in Novermber 1962. Several review papers were given, and then the attendees split into working groups and discussed infrared physics, sources of infrared, infrared equipment, scientific and medical applications, and heating by infrared.
[16] B. Kovit, "Infrared in Space," *Space/Aeronautics*, **39**, 56 (January 1963). How big is the market? What are the applications? What are the problems? A good survey article.
[17] L. Eisner, "Infrared: What's Ahead," *Electronic Products*, **7**, F2, February 1965.
[18] "Soviet Infrared Sensors: 1. Thermal Sensors," Washington, D.C.: Library of Congress, 1965 (CFSTI: 65-63277). A review of Soviet articles on infrared sensors published since the mid-1950's. Only thermal sensors are covered.

[19] C. Shishkevish, "Soviet Infrared Sensors: 2. Photon Detectors, Comprehensive Report," Washington, D.C.: Library of Congress, 1965 (CFSTI: 65-64120). Covers all pertinent articles up to February 1965.
[20] A. Dulles, *The Craft of Intelligence*, New York: Harper and Row, 1963, p. 215. The author describes a period in the early 1950's when nearly every intelligence source seemed to be trying to sell the latest inside information on infrared.

16

Military Applications of Infrared Techniques

It is easy to understand the appeal that infrared techniques have for the military system planner. Since infrared radiation cannot be detected by the human eye, it offers the possibility of seeing in the dark, of detecting targets by their self-emission, and of communicating by secure means.

The military organizations of the United States, Great Britain, and Germany began experimenting with infrared equipment soon after 1900. During World War I, experimental blinker signaling, voice communication, and search equipment were developed. Although neither side put any infrared equipment into production before the end of the war, the experience gained with the experimental equipment clearly demonstrated the military potential of infrared technology.

The development of the infrared image converter tube on the eve of World War II made it possible for Germany to be the first country to deploy infrared equipment in the field. Early in the war the Germans concluded that the Allies were using infrared equipment to detect Nazi U-boats and aircraft. Because of this erroneous conclusion, the Germans concentrated much of their research and development on infrared equipment and on means of countermeasuring it. By contrast, the Allies concentrated their efforts on the development of radar. As a result, the Germans lost the war but they clearly won the battle of the infrared. Details of the German efforts can be found in [16.0] [1]–[5].

Relatively little information has appeared concerning the British use of infrared during World War II. Reference 16.0 [6] summarizes the prewar developments and also notes that the first air-to-air detection of another aircraft occured in 1937.

In the United States the World War II research and development effort was managed by the National Defense Research Committee (NDRC) of the Office of Scientific Research and Development (OSRD). As originally organized in 1940, NDRC had a section called Infrared

MILITARY APPLICATIONS OF INFRARED TECHNIQUES 465

Devices. In 1942, NDRC was reorganized and Section 16.4 (Infrared) of Division 16 (Optics and Camouflage) was given responsibility for the application of infrared techniques to the solution of pertinent military problems. The summary technical reports of Division 16 were declassified in 1958 (16.0 [7] and [8]) and they are the prime source of information on the development of infrared equipment in the United States during World War II. The work of Division 16 is also summarized in refs. 16.0 [9] and [10]. The development of infrared equipment by the U.S. Army Signal Corps is described briefly in ref. 16.0 [11].

The convincing World War II demonstration of the effectiveness of infrared techniques ensured continuing military support for development in this field. Virtually all of the world's major military organizations now use infrared equipment in one form or another (16.0[12]–[15]). These applications are summarized in Figure 15.1 and are described in greater detail in the references that make up the rest of this chapter.

Of particular interest in the post–World War II period is the emergence of the Soviet Union as a major proponent of infrared equipment. Apparently the Soviets had no infrared capability until they acquired a large part of the German scientific know-how at the end of the European phase of World War II. A sampling of their publications (16.0 [16]–[25]) shows that the Soviets now have a genuine familiarity with infrared techniques. Do not, however, expect to find very much in these publications about what the Soviets are doing with infrared; the applications that they describe are all taken from Western literature.

While browsing through these references and pondering the military applications of infrared, do not overlook the fundamental changes that are occurring in the way wars are fought. It seems quite clear that we are now in an era of limited wars. It is not easy to define what is meant by limited war, but Deitchman (16.0 [26]) points out that the essence of most definitions refers to any military action that does not threaten the *immediate* destruction of one or more of the world's major powers. Most of the conventional wars in recent history have been fought on the plains of Western Europe, and our strategy and tactics are largely derived from this experience. The number of limited wars that have occurred depends, of course, on the exact definition one chooses. Deitchman (16.0 [26]) has carefully documented a total of 32 limited wars in the period from 1945 to 1962, for an average frequency of two such conflicts per year. It seems quite certain that future conflicts of this type will continue to occur in the rain forests and mountainous terrain of underdeveloped countries rather than on the more familiar plains of Western Europe. Much of the fighting will be done by local troops supported by the major powers. The heavy involvement of local troops has already forced the weapon-system designer to consider less sophisticated systems that the local troops can understand, operate, and maintain. If the reader still has doubts about the impact of limited war on weapon-system design, a few minutes' study of some of the communist literature on the subject (16.0 [27]) should be enough to convince him of the impact of limited war on the strategy and tactics that were assumed in the design of today's weapon systems.

16.0 GENERAL

[1] E. A. Underhill, "German Applications of Infrared in World War II," *Proc. Natl. Electronics Conf.*, **3**, 284 (1947). Describes developments in lead sulfide detectors and image converter tubes and their application to night viewing, communication, tank fire control, ship and aircraft detectors, proximity fuses, and missile guidance. An improved type of snooperscope is described that has a range of 300 m. A viewer with an image converter tube and a 25 cm $f/1$ lens could detect objects having a temperature as low as 200°C and could detect the exhaust stubs on aircraft at a distance of 20 miles. Methods of packaging and cooling lead sulfide detectors are described.

[2] V. Krizek and V. Vand, "The Development of Infra-Red Technique in Germany," *Electronic Eng.*, **18**, 316 (1946). Developments include thallium sulfide, lead sulfide, lead selenide, and lead telluride detectors, image converter tubes, and infrared iconoscopes. A dewar flask for cooling lead sulfide detectors to 195°K is described and illustrated.

[3] F. E. Jones, "Infrared—Its Problems and Possibilities," *British Communications and Electronics*, **4**, 36 (1957). Contains a general discussion of the possibilities of infrared equipment for radar-type and military applications. Reports the German development in 1942 of a bomber detection system that used a lead sulfide detector and a 150 cm mirror. In 1943 the Germans developed KID, a scanning search system for airborne fire control. It used a 5 in. mirror and had a range of 5 miles (target unspecified). Late in 1943 the Germans developed Madrid, a lead-sulphide-equipped missile seeker.

[4] R. Lusar, *Waffen und Geheim-Waffen des 2 Weltkrieges und Ihre Weiterentwicklung*, Munich: J. F. Lechmanns, 1959. A translated version is available; *German Secret Weapons of the Second World War*, New York: Philosophical Library, 1963. Among other things, this book mentions the Hs 293, a guided bomb that could use infrared guidance; the BV246, a glider bomb that used infrared guidance; Madrid, an infrared seeker for the guidance of small air-to-air missiles; Pistole, an infrared proximity fuse; infrared image converters for seeing in the dark.

[5] "Infrared Equipment for Military Purposes," *Engineering*, **163**, 258 (April 4, 1947). Describes German World War II developments in infrared communication and night viewing equipment. This equipment was used in the Mediterranean and African theaters as early as 1941 and on the Eastern front against the Russians in 1942–1944, but it was not used on the Western front.

[6] R. V. Jones, "Infrared Detection in British Air Defense, 1935–38," *Infrared Phys.*, **1**, 153 (1961). Summarizes early British experience. For specific data on search systems see ref. 16.1.1 [24].

[7] "Summary Technical Report of Division 16, NDRC; Non-Image Forming Infrared," Vol. 3, Office of Scientific Research and Development, Washington, D.C., 1946. This and the following reference are the prime source of detailed information on the World War II development of infrared equipment in the United States. This volume covers

16.0 GENERAL

near-infrared sources, near-infrared transmitting filters, near-infrared detecting devices, near-infrared recognition and code communication systems, near-infrared systems for determining range and direction, glider position indicators, cloud attenuation meters, far-infrared detectors, far-infrared receiving systems, and a bibliography of NDRC reports.

[8] "Summary Technical Report of Division 16, NDRC; Image Forming Infrared," Vol. 4, Office of Scientific Research and Development, Washington, D.C., 1946. This and the preceding reference are the prime source of detailed information on the World War II development of infrared equipment in the United States. This volume covers infrared image tubes and electron telescopes, Metascopes, infrared-sensitive phosphors, infrared sources, ultraviolet sources and filters, autocollimators, antiglare devices, miscellaneous optical developments, and a bibliography of NDRC reports.

[9] C. G. Suits, G. R. Harrison, and L. Jordan, eds., *Science in World War II—Applied Physics*, Boston: Little, Brown, 1948. Tells the story of NDRC Div. 16 and the development of infrared equipment in the United States during World War II. Among the developments described are infrared sources, narrowband filters, imaging and nonimaging detectors, infrared-sensitive phosphors, Metascopes, infrared electron telescopes, and infrared gas analyzers. See also the two preceding references.

[10] O. S. Duffendack, "Wartime Developments in the Detection and Measurement of Thermal Radiation," *Engineers Dig.*, **3**, 483 (1946). Describes U.S. development during World War II of bolometers for the 8 to 13 μ window. The author was Chief of Section 16.4 (Infrared) of NDRC.

[11] D. Terrett, *The United States Army in World War II: The Signal Corps*, Office of the Chief of Military History, Department of the Army, Washington, D.C., 1956. The U.S. Army Signal Corps was investigating infrared equipment for the detection of ships and planes in the mid-1920's. In 1926 an aircraft was detected by an active infrared system. In 1932 a blimp was tracked to a distance of 6300 ft by an active system. Because of the lack of high-power sources, further work with these systems was dropped. In 1935 a passive system using a bolometer was able to detect the SS Mauretania at a distance of 13 miles, the SS Normandie at a distance of 17 miles, and the SS Aquitania at a distance of 10 miles in a fog. The equipment could distinguish the Mauretania's dummy smokestack from the three real ones.

[12] P. J. Klass, "Exclusive Report on Infrared: 1. Infrared Challenges Radar's Monopoly. 2. IR System Designer Faces Many Hurdles. 3. Funds Lag for Basic Research on Infrared," *Aviation Week*, **66**, 50 (March 4, 1957); **66**, 78 (March 11, 1957); **66**, 89 (March 18, 1957). Covers military applications up to 1957 and also makes some surprisingly accurate predictions for the future.

[13] L. W. Nichols et al., "Military Applications of Infrared Techniques," *Proc. Inst. Radio Engrs.*, **47**, 1611 (1959). A survey article covering missile seekers, fire control, bomber defense, airborne early warning, ballistic-missile detection, near- and far-infrared viewers, reconnaissance, communication, range instrumentation, and foreign military applications.

[14] L. H. Dulberger, "Advanced Strategic Bombers," *Space/Aeronautics*, **45**, 62 (June 1966). These aircraft may carry infrared-guided missiles for defensive armament, infrared terminal homing for offensive missiles, offset infrared line scanners for bombing control, and infrared sensors to warn of the approach of hostile aircraft or missiles.

[15] "SAT Supplies Infrared Systems," *Aviation Week*, **86**, 320 (May 29, 1967). A detailed discussion of France's sole supplier of infrared systems for missile guidance and aerial reconnaissance. For information on specific items, see refs. 16.1.4[52], 16.4.1[31], and 16.6.4[13].

[16] I. A. Margolin and N. P. Rumyantsev, *Fundamentals of Infrared Technology*, Moscow: Military Publishing House, 1955 (SLA:59-18105). Discusses the physics of thermal radiation, detectors, optical materials, optical systems, and atmospheric transmission, and includes a bibliography of Soviet references. A second edition is available; see ref. 16.0 [18].

[17] V. P. Moskovskii and P. T. Astashenkov, eds., *Modern War Technology*, Moscow: Military Publishing House, 1956 (OTS:MCL-110/1). Discusses infrared command guidance by the use of coded flares on missiles for tracking, active infrared systems, and passive search systems for ship and aircraft detection. Reports that active infrared systems were used during World War II for tank fire control, air-to-air IFF, marking safe paths through minefields, marking landing areas and assembly points for troops, and for ship-to-ship and ship-to-shore communications. A passive ship detection system is described that used a bolometer and a 60 cm mirror and could detect an average size ship at a distance of 20 km. Angular accuracy was 1 mrad. The authors claim that lead sulfide detectors were used as early as 1941 for the detection of aircraft. Such a system is described as using a lead sulfide detector cooled to a temperature of 195°K, a reticle to generate an 800 Hz carrier with a 15 Hz modulation, and a 1.5 m mirror. The display used neon lamps to show the coordinates of the target. The detection range (at night) was 12 km against a bomber. A ship search unit using a lead sulfide detector is claimed to have a range of 35 km against a large ship. The authors speculate on the possibility of using lead sulfide detectors in missile-guidance systems.

[18] I. A. Margolin and N. P. Rumyantsev, *Fundamentals of Infrared Technology*, 2d ed., Moscow: Military Publishing House, 1957. (Chapter 13 has been added: "Infrared Instruments for Military Purposes." A translation is available, SLA: 60-13034.) For the contents of the first edition, see ref. 16.0[16]. The new chapter discusses the military advantages of infrared devices, infrared communication systems, search systems for aircraft and ships, infrared photography, reconnaissance photography from great distances with infrared film, infrared-sensitive phosphors, infrared guidance units and proximity fuses for missiles, infrared rangefinders (active type, using a pulsed source), image converters, and the Metascope. A coastal search system, called Donau 60, has a 60 cm mirror, a bolometer detector, a 12 Hz chopper, and a galvanometer combined with a fluorescent screen for a display. Large ships can be detected by Donau 60 at a distance of 30 km. The authors report that U.S. infrared-guided missiles have a range of 18 km, which they consider to be inadequate.

[19] I. A. Margolin and N. P. Rumjancew, *Podstawy Techniki Podczerwieni* (Technique of Infrared Rays), Warsaw: Ministry of National Defense, 1957. Translated from the first Soviety edition; see ref. 16.0[16].

[20] V. Ye. Kichka, *Infrared Rays in Warfare*, Moscow: Military Publishing House, 1958. (Chapter 7, entitled "Instruments for Recording Infrared Rays" is available in translation, SLA: 61-19370.) Discusses infrared devices, such as the Metascope, that use phosphors for detection; communication systems; search systems for aircraft and iceberg detection; automatic tracking systems; proximity fuses; and image converters.

[21] K. M. Listov and K. N. Trofimov, *Radio and Radar Technology and its Applications*, Moscow: Military Publishing House, 1960 (CFSTI: 65-61976). Contains chapters on infrared technology, radar and infrared equipment for the Navy, and radar and infrared equipment for land forces.

[22] Yu. A. Ivanov and B. W. Tyapkin, *Infrared Technology in Military Science*, Moscow: Sovietskoye Radio, 1963. Discusses the physics of radiation, sources of infrared, atmospheric transmission, infrared optical materials, infrared detectors, image con-

16.1.1 SEARCH SYSTEMS

verters, surveillance equipment, search systems, infrared seekers for missiles, performance of infrared equipment, and infrared countermeasures. One chapter on the detection of ballistic missiles is available in translation; see ref. 16.1.1 [34].

[23] Y. P. Safronov, Y. G. Andrianov, and D. S. Iyevlev, *Infrakrasnaya Tekhnika v Kosmose (Infrared Technology in Space)*, Moscow: Voyenizdat, 1963. Among the military uses of infrared technology in space are the early detection of ballistic missiles, missile guidance, communication, reconnaissance, and surveillance. The book is intended for military staff officers.

[24] T. Burakowski and A. Sala, *Noktowizja (Night Vision)*, Warsaw: Ministry of National Defense, 1965. One of the best books available on the problems of seeing at night. Covers the physics of infrared radiation, sources, optical elements for night vision devices, the characteristics of the human eye, display techniques, image converters, thermal imaging devices, and television-type night viewers. See ref. 16.5.1 [31] for specific details on image converter devices.

[25] K. Lajos, *Infravoros Felderites es Alcazas (Infrared Discovery of Camouflage)*, Budapest: Zrinyi Military Publishing House, 1966. One of the better books available on night vision devices. The author is captain in the Hungarian Army Engineer Corps, hence the emphasis on camouflage detection. The book discusses the physics of radiation, emissivity, image converters, infrared photography, near-infrared reflection spectra (in great detail), and passive detection systems. It describes the German World War II Kiel IV infrared search set that was able to detect a bomber at a range of 32 km.

[26] S. J. Deitchman, *Limited War and American Defense Policy*, Cambridge, Mass.: M.I.T. Press, 1964, Chapter 1. An excellent dissertation on the theory, consequences, and conduct of limited warfare.

[27] M. Tse-tung, *On Guerilla Warfare*, New York: Praeger, 1961. There is no comparable single source that outlines official Soviet views on irregular warfare in the way that Tse-tung has given the Chinese communist policy. This book, however, serves as the theoretical basis for Soviety policy. It was originally published in 1937.

16.1 SEARCH, TRACK, AND RANGING APPLICATIONS

Passive systems for the detection and tracking of ships, aircraft, missiles, surface vehicles, and personnel have been of great importance to the military system planner. Such equipment began to appear before World War I and has been steadily improved since that time. The characteristics of the atmospheric windows and the advantages of limiting the instrument's spectral bandpass were not well understood until after World War II. As a result, much of the early equipment operated without optical filtering of any sort and was so susceptible to saturation by reflected sunlight that it could be used only at night. Most present-day equipment will operate equally well during the day or at night.

16.1.1 Search Systems

[1] R. D. Parker, "Thermic Balance or Radiometer," U.S. Patent No. 1,099,199, June 9, 1914. Small coils forming the arms of a Wheatstone Bridge are placed at the focus of a parabolic mirror to act as a detector. Two coils are shielded and two are exposed. The

resulting bridge unbalance is observed with a galvanometer. Suggested application is as an iceberg detector.
[2] L. Bellingham, "Means for Detecting the Presence at a Distance of Icebergs, Steamships, and other Cold or Hot Objects," U.S. Patent No. 1,158,967, November 2, 1915. A manually scanned mirror and thermopile combination.
[3] S. O. Hoffman, "The Detection of Invisible Objects by Heat Radiation," *Phys. Rev.*, **14**, 163 (1919). Describes equipment that uses a cross-array thermopile detector and a galvanometer for readout. It could detect a man at a distance of 600 ft and an airplane at a distance of 1 mile. This equipment is also described in the next reference.
[4] S. O. Hoffman, "Method of and Apparatus for Detecting and Observing Objects in the Dark," U.S. Patent No. 1,343,393, June 15, 1920. For performance data, see the previous reference.
[5] J. H. Hammond, Jr., "Optical Instrument," U.S. Patent No. 1,542,937, June 23, 1925 (filed May 8, 1920). A system to detect ships or surfaced submarines. One of the earliest proposed applications for the Case thallous sulfide detector, the first photon detector for the infrared. System output is a variable-intensity audio tone.
[6] J. V. Bayer, "Fog Penetrating Televisor," U.S. Patent No. 1,876,272, September 6, 1932.
[7] A. S. Fitzgerald, "Photo Electric System," U.S. Patent No. 2,016,036, October 1, 1935. Designed to detect motion within a field of view, such as by an intruder. Optically superimposes a pair of complementary checkerboard reticles to balance the system. Any subsequent motion within the field of view causes an imbalance and a signal output.
[8] H. A. Zahl, "Art of Locating Objects by Heat Radiation," U.S. Patent No. 2,392,873, January 15, 1946 (filed April 3, 1934). An early patent of fundamental importance. Discusses detection of ships, aircraft, and dirigibles. Describes problems with backgrounds and methods, such as a differential connection of detectors, to minimize their effects.
[9] "Night Vision with Electronic Infrared Equipment," *Electronics*, **19**, 192 (June 1946). Reports that the Germans used aircraft and ship search systems during World War II. The detector in these systens was lead sulfide (Elac) cooled to 195°K by carbon dioxide snow. These detectors were being produced at the rate of 1000 per month at the end of the war.
[10] R. D. Kell and W. A. Tolson, "Heat Responsive Indicator," U.S. Patent No. 2,422,971, June 24, 1947. Uses a rotating, tilted secondary mirror for image nutation. Target position is indicated on a cathode ray tube. The Wolff thermal detector is described, in which the flux enters a cell bounded by a pair of thin diaphragms and is absorbed by powdered carbon. Heating the carbon causes it to give off a gas, increasing the pressure in the cell and deforming the diaphragms. These diaphragms are part of a capacitor; hence the deformation is sensed as a change in capacitance.
[11] L. Hammond, "Radiant Energy Detecting Apparatus," U.S. Patent No. 2,423,885, July 15, 1947 (filed February 3, 1941). Describes a device that can be used in an airplane for finding ships or for aerial torpedo guidance. Uses a multiple-element detector and the motion of the aircraft to generate a "push-broom" scan. A differential connection of the detector elements is used to minimize background effects. Both positive and negative target contrast are discussed.
[12] W. A. Tolson, "Heat Detecting System," U.S. Patent No. 2,431,625, November 25, 1947. A pair of sensors placed, for example, on the wing tips of an aircraft provide a search function and a means of triangulation ranging. Target coordinates are displayed on a cathode-ray tube. This system merely combines two of the sensors described in ref. 16.1.1[10].

16.1.1 SEARCH SYSTEMS 471

[13] E. J. Martin and G. G. Scott, "Sealed Heat Ray Detector," U.S. Patent No. 2,491,192, December 13, 1949 (filed November 11, 1944). A submarine-mounted infrared device for detecting ships. Unit is hermetically sealed so that it is not damaged when the submarine is submerged. Entrance window is made of sodium chloride (rock salt) protected by a thin sheet of chlorinated rubber. The detector is a thermopile.

[14] C. Orlando, "Moving Object Indicator," U.S. Patent No. 2,774,961, December 18, 1956. A device to detect the motion of an object within a selected field of view. Images of the field of view are formed on a pair of wedges having a linear variation in density with position on the wedge. On one wedge the density increases from top to bottom, and on the other the density increases from bottom to top. The system is initially balanced, and then any motion within the field of view results in opposite effects at the two wedges and unbalances the system.

[15] L. E. Ravich, "Radiation Detection Methods and Devices," U.S. Patent No. 2,834,891, May 13, 1958. Simple devices using phosphors for the detection of the infrared sources associated with active viewing system, such as sniperscopes.

[16] "Infrared Detection from Recon Satellites," *Space/Aeronautics*, **31**, 151 (April 1959). An elementary introduction to the subject.

[17] H. Dubner, J. Schwartz, and S. Shapiro, "Detecting Low-Level Infrared Energy," *Electronics*, **32**, 38 (June 26, 1959). Describes several multiple-detector arrangements, including those in refs. 16.1.1 [26], and the possibility of using a mosaic infrared tube.

[18] E. Kosyrev, "Heat Direction Finding—Means of Reconnaissance," *Voyennyye Znaniya*, **35**, 19 (1959) (OTS: 61-28337). Describes the military possibilities of infrared search sets ("direction finders" in the author's terminology). These search sets use photon detectors, choppers (frequency of 800 Hz), transistorized electronics, a neon lamp display, and manual tracking. Range against an unspecified target is 20 km and angular accuracy is close to 0.1 deg. The author claims that infrared gunsights and search sets are used extensively in the Soviet Air Force. A typical installation in an all-weather interceptor has a detection range of 8 to 10 km. Infrared seekers for guided missiles have ranges of a few kilometers. Some highly sensitive search sets are said to be useful in detecting submerged submarines (see ref. 16.4.1 [2]). The author is a major in the Soviet Engineering Corps.

[19] R. Lusar, *Waffen und Geheim-Waffen des 2 Weltkrieges und Ihre Weiterentwicklung*, Munich: J. F. Lechmanns, 1959. A translated version is available: *German Secret Weapons of the Second World War*, New York: Philosophical Library, 1963. Describes German World War II infrared search systems for aircraft detection, regulation of crossroads traffic during blackouts, and the discovery of infiltrating patrols. One such system, the Donau 60, had 60 cm optics and a chopping frequency of 12 Hz. It had a detection range of 7 km against tanks and 20 km against ships.

[20] "Industry Observer," *Aviation Week*, **75**, 23 (July 3, 1961). A report that U-2 aircraft equipped with infrared equipment are flying picket missions off the coast of the Soviet Union. The equipment is said to provide an infrared detection capability similar to the warning function to be provided by the Midas satellite.

[21] L. A. Iddings, "Scanning Devices for Optical Search," U.S. Patent No. 3,000,255, September 19, 1961 (filed May 31, 1955). A point-source sensor for fire control purposes. Device uses a coaxial pair of scanning reticles. One reticle has a single radial slot and the other has a one-turn spiral slot. One reticle rotates slightly faster than the other, and the effect is to generate a spiral scan of the object space. The target coordinates are displayed on a cathode-ray tube.

[22] "Midas Detects Titan," *Aviation Week*, **75**, 33 (October 30, 1961). States that the infrared sensors aboard the Midas satellite picked up a Titan minutes after it was launched.
[23] "Fifty Winking Lights," *Sci. Am.*, **205**, 71 (December 1961). Describes and pictures an infrared scanning head having a 10 in. aperture $f/0.76$ Schmidt objective and a 50-element linear array of detectors. Display is a set of 50 gas-discharge lamps each of which glows when its associated detector element is irradiated.
[24] R. V. Jones, "Infrared Detection in British Air Defense, 1935–38," *Infrared Phys.*, **1**, 153 (1961). Developed a search system with an 11 cm aperture and a bolometer detector. It was ground tested in 1936 and detected an aircraft at a distance of 1 mile during the day and 2 miles at night. An improved system was flight tested in 1937 and showed detection ranges of about one-third mile against an aircraft. This may be the first time that one aircraft was detected in flight from another by the use of infrared equipment.
[25] H. F. Hicks, Jr., "Hemispherical Scanning System," U.S. Patent No. 3,023,662, March 6, 1962. Shows how two reflecting elements can be combined to scan the entire hemisphere above a point.
[26] H. A. Dubner, "Multi-Cell IR Search System," U.S. Patent No. 3,041,460, June 26, 1962. Describes a search system that uses a linear array of detectors and an alternate configuration that uses a mosaic of detectors. The bias is commutated so that only one element of the detector is operative at any instant of time.
[27] P. J. Klass, "Lack of Infrared Data Hampers Midas," *Aviation Week*, **77**, 54 (September 24, 1962). A comprehensive discussion of the problems encountered by the Midas system. They are attributed to a lack of data on the radiating characteristics of targets and terrestrial backgrounds when viewed through the earth's atmosphere and to a lack of data on the transmission characteristics of the atmosphere.
[28] P. J. Klass, "Cold Clouds Troubling Horizon Sensors," *Aviation Week*, **77**, 50 (October 1, 1962). A continuation of the discussion contained in the previous reference. This article also discusses problems with infrared horizon sensors.
[29] V. D. Sokolovskii, ed., *Soviet Military Strategy*, Moscow: Voenizdat, 1962. A translation is available by H. S. Dinerstein, L. Goure, and T. W. Wolfe, Englewood Cliffs, N.J.: Prentice-Hall, 1963. It is quite evident from the text that the authors are aware of and fear the capabilities of the infrared detection system carried aboard the Midas satellite.
[30] "Midas Reduced to Experimental Status," *Aviation Week*, **78**, 40 (February 18, 1963). Originally scheduled to become operational in 1964, the Midas early warning satellite has been reduced to an experimental program. Program schedule did not allow for adequate measurements of the radiation from the terrestrial background against which the ICBM rocket plume had to be detected.
[31] W. E. Osborne, "Infrared Mine Detector a Reality," *Electronics*, **36**, 54 (August 2, 1963). Proposes that a sensitive radiometer can be used to detect explosive mines buried in the ground. Claims that a simulated mine buried at a depth of 18 in. was readily detected for at least a week after burial. Radiometer used a lead telluride detector cooled to 195°K.
[32] W. E. Osborne, "Infrared System Detects Intruders," *Electronics*, **36**, 26 (August 16, 1963). A passive intrusion detector using the motion of the intruder to provide a chopped signal.
[33] R. D. Hudson, Jr., "Passive Optical Intercept Techniques," *Proc. Natl. Aerospace Electronics Conf.*, 1963, p. 54. An introduction to the principles of the design of infrared search systems. Describes a search device with an input aperture diameter of

16.1.1 SEARCH SYSTEMS

3 in., a single lead selenide detector, and a spectral bandpass of 3.2 to 4.8 μ. Maximum detection range against a single-jet target is calculated to be 50 miles with a probability of detection equal to 0.9 and a false alarm time of 20 sec. The sensor weighs 25 lb and occupies less than 0.5 ft^3. The signal processor, servo drive, power supplies, and detector cooling system weigh a total of 35 lb and occupy less than 1.5 ft^3.

[34] Yu. A. Ivanov and B. V. Tyapkin, *Infrared Technology in Military Science*, Moscow: Sovietskoye Radio 1963. (Chapter 9, section 4, "On the Possibility of Detection of Ballistic and Controlled Rockets," is available in translation, OTS: TT-64-71400.) Concludes that infrared systems can detect ballistic missiles during boost and reentry. Data quoted is from U.S. publications that pertain to measurements made during Operation Gaslight. Quotes an erroneous U.S. report about excess radiation from a satellite. For the other contents of this book, see ref. 16.0[22].

[35] D. E. Fink, "New Missile Warning Satellite Succeeds," *Aviation Week*, **80**, 33 (February 3, 1964). The ballistic missile alarm satellite system, a follow-on to the Midas program, carried infrared sensors that made a number of detections of liquid-fueled and solidfueled ICBM launches.

[36] R. K. Orthuber, F. F. Hall, and H. Emus, "Radiation Source Search System Using an Oscillating Filter," U. S. Patent No. 3,144,562, August 11, 1964. Describes the conventional two-color reticle systems and their disadvantages. Proposes the addition of a third filter to provide a simple means of changing the balance between the other two filters.

[37] M. M. Postan, D. Hay, and J. D. Scott, *Design and Development of Weapons*, London: Her Majesty's Stationery Office and Longmans, Green and Co., 1964, p. 444. Describes British experiments in 1941 with a ship detector using thermocouple detectors.

[38] "Airborne Ground-Fire Detector," *Aviation Week*, **82**, 60 (March 8, 1965). Describes a proposed program to assess the feasibility of an airborne system to detect and locate sources of ground fire by their infrared radiation.

[39] "Industry Observer," *Aviation Week*, **82**, 15 (March 29, 1965). Describes the Air Force's 461 program, another attempt toward fulfilling the Midas early warning mission by the use of unmanned satellites carrying infrared sensors.

[40] B. Miller, "Tactical Radar Homing Programs Pushed," *Aviation Week*, **84**, 78 (March 21, 1966). Describes an electro-optical countermeasures receiving set that detects the infrared radiation from burning fuels in aircraft and missile engines. An associated countermeasures dispenser ejects flares to spoof incoming infrared-guided missiles. A picture of the system appeared in the same magazine; **80**, 24 (May 25, 1964).

[41] "Satellite Competition," *Aviation Week*, **84**, 29 (April 25, 1966). Describes the integrated satellite system, or Program 266, an outgrowth of the defunct Midas infrared missile early warning satellite.

[42] B. Miller, "Wide Use of F-111A Mk. 2 Avionics Seen," *Aviation Week*, **84**, 93 (June 6, 1966). The weapon system includes a countermeasures receiving set, located on the rear of the vertical stabilizer, that uses a cryogenically cooled detector. The receiver detects the infrared radiation from the engines of hostile aircraft or missiles. The output of the receiver initiates countermeasure dispensers. For other details of the weapon system, see refs. 16.1.3 [18] and 16.6.3 [7].

[43] P. Alelyunas, "Synergetic Satellites," *Space/Aeronautics*, **46**, 52 (October 1966). Discusses Program 461, the successor to Midas, and speculates not only that improved infrared sensors can detect ICBM launches but that they can track the launch to determine whether it is a threatening ICBM or a nonthreatening vehicle in an orbital trajectory.

[44] "Industry Observer," *Aviation Week*, **85**, 23, (November 21, 1966). The integrated reconnaissance/earlywarning satellite (Project 266) will carry infrared missile-launch and ground-explosive-damage sensors.

[45] "New USAF Satellite to Include Infrared, Photo Scanning Gear," *Aviation Week*, **85**, 23 (December 26, 1966). The Integrated Satellite (formerly Project 266) is to improve the early warning capability of the defunct Midas satellite. Infrared scanners will provide an early warning of hostile ICBM launches, and they could also be used for postattack intelligence. It is estimated that the optics for the scanner will have a diameter of about 3 ft.

[46] "USAF Reconnaissance Satellites," *Aviation Week*, **86**, 116 (March 6, 1967). Describes 3 satellites launched during 1966 that appear to be carrying experimental infrared sensors for detecting hostile missiles.

[47] F. Leary, "Finding the Enemy," *Space/Aeronautics*, **47**, 92 (April 1967). Discusses passive infrared intrusion detectors. One unit, described in detail, uses a thermistor bolometer and has sufficient sensitivity to detect a 1°C temperature differential at a distance of 750 ft. (The article infers that a 1°C temperature differential is equivalent to a human being.) The field of view is about 3 by 9 mrad. A sketch shows how several of these units might be used for perimeter security. See also ref. 16.5.1 [35].

[48] "Industry Observer," *Aviation Week*, **86**, 13 (May 1, 1967). A note that the U.S. Army is considering the development of an airborne infrared system for the detection of small-arms fire from the ground.

[49] "USAF, Industry Studying ABM Concepts," *Aviation Week*, **86**, 84 (May 15, 1967). Describes new studies of the use of orbiting vehicles for intercepting ballistic missiles during their boost phase. Infrared sensors would provide both search and target-tracking functions. Detection range might be as much as 1000 miles.

[50] "Stability Augmentation Planned for B-52s," *Aviation Week* **86**, 37 (June 12, 1967). Among the major items proposed for a modification program is the addition of the AN/ALR-13 passive infrared warning system to alert the crew to the presence of enemy aircraft or missiles. The ALR-13 was deve oped for the F-111A, and further details are given in ref. 16.1.1 [42].

[51] "Industry Observer," *Aviation Week*, **87**, 13 (October 2, 1967). Reports that the U.S. Army is studying an infrared gun-flash detection system for base security in Vietnam. The system would give the angular coordinates and the distance to hostile guns after they had fired their initial salvos. The equipment would be effective against rockets, artillery, and recoilless rifles, but not against mortars.

[52] "Filter Center," *Aviation Week*, **87**, 97, December 11, 1967. Reports that the USAF will choose an infrared countermeasures system for its B-52 bombers after a flight evaluation. Competing systems are the ALR-23, already in production for the F-111A, and the ALR-21. Both systems warn pilots of missiles launched against them.

[53] "Electronic Warfare Gains Key Viet Role," *Aviation Week*, **88**, 48 (January 1, 1968). An exceptionally detailed coverage of electronic countermeasure systems. Data tables show plans to add either the ALR-21 or the ALR-23 infrared missile launch detection systems to the B-52.

[54] "Military Intrusion Detector Becomes Commercial Spin-Off," *Aerospace Tech*, **21**, 30 (March 11, 1968). Describes a passive intrusion detector that can detect personnel at ranges up to 1000 ft and vehicles at over 2500 ft under reasonable weather conditions. The unit can be used for monitoring trails and harbor traffic, protecting unattended missile silos, and triggering booby traps. It is expected that this unit will be available commercially for a price of $6,000. The detector head weighs 3 lbs, uses thermistor bolometers in the 8 to 14 μ region to cover a 2.5 by 7.5 mrad field of view,

and has a 2 week life time on its self contained batteries. Target discrimination is aided by limiting the electrical bandpass of the signal processor to 0.2 to 2 Hz. This removes the effect of slow changes in the background.
[55] B. Miller, "Advanced F-111 System Nears Production," *Aviation Week*, **89**, 68 (September 9, 1968). Reports that the ALR-23 infrared countermeasures receiver is mounted atop the F-111's vertical stabilizer. For a photo showing the installation, see the same magazine; **89**, 23 (July 22, 1968).

16.1.2 Track Systems

[1] H. A. Droitcour, "Apparatus for Automatically Training Guns, etc., on Moving Objects," U.S. Patent No. 1,747,664, February 18, 1930. Of historical interest because of its use of a four-element cross-array detector.

[2] M. J. E. Golay and H. A. Zahl, "System for Detecting Sources of Radiant Energy," U.S. Patent No. 2,424,976, August 5, 1947 (filed June 12, 1939). A device for tracking ships or aircraft. Uses a multielement array of pneumatic detectors.

[3] D. Dederer, "Infrared Eye," *Control Engrg*. **5**. 24 (November 1958). A device used to track missiles at low altitudes when tracking radars are ineffective because of ground clutter. Tracker uses four sets of optics and a single reticle to generate both coarse and fine azimuth and elevation error sensing. Tracking accuracy is ± 0.01 mrad, field of view is 4 deg, and the time constant of the error signal circuitry is $400\,\mu$sec. The optical unit, weighing 22 lb, is mounted on the antenna of a tracking radar. The detector is lead sulfide.

[4] A. R. Gedance, "Radiant Energy Angular Tracking Apparatus," U.S. Patent No. 2,942,118, June 21, 1960. Uses a dual rotation of a reticle to give the effect of a rotating-reticle nutating-optics system. The reticle rotates about its center and, simultaneously, about an axis that is displaced from the optical axis of the system.

[5] A. Robert and J. Deslaudes, "Tracking devices," U.S. Patent No. 2,975,289, March 14, 1961. Uses a pair of rotating choppers. One chopper, consisting of radial sectors, is located at the focal plane. The second chopper is an opaque disk with a single hole whose circumference passes through the center of rotation of the disk. The second chopper is located approximately halfway between the optics and their focal plane.

[6] D. D. Wilcox, Jr., "Target Tracking System," U.S. Patent No. 2,994,780, August 1, 1961 (filed March 10, 1954). A four-element cross-array detector and image nutation is used to generate error signals. The next reference appears to be material that was divided out of the application for this patent.

[7] D. D. Wilcox, Jr., "Target Tracking System," U.S. Patent No. 2,997,588, August 22, 1961 (filed March 10, 1964). See comments in the previous reference.

[8] M. M. Merlen, "Infrared Tracker," U.S. Patent No. 3,007,053, October 31, 1961. A device for tracking low-flying aircraft, rockets, or missiles when tracking radars are adversely affected by ground clutter. Uses a two-frequency reticle to generate error signals.

[9] H. S. Snyder and R. F. Hummer, "Infrared Tracker," U.S. Patent No. 3,012,148, December 5, 1961 (filed July 2, 1951). Uses a pair of overlapping chopper disks to generate error signals.

[10] "Twin Tracker Follows Missiles," *Electronics*, **35**, 32 (November 30, 1962). A device to track missiles during boost and reentry. Tracking accuracy is ± 0.2 mrad. Uses 12 in. optics and lead sulfide and thermistor detectors.

[11] "Infrared Tracking," *Aviation Week*, **77**, 55 (December 17, 1962). Describes an infrared tracker designed to continuously position a pedestal with an error of less than 0.2 mrad. See also ref. 16.3.1 [8].
[12] R. W. Buntenbach, "Radiant-Energy Translation System," U.S. Patent No. 3,069,546, December 18, 1962 (filed June 4, 1948). An excellent description of tracking systems that use 4-element cross-shaped or 2-element L-shaped arrays of detectors. Claims that these systems using thermistor detectors can detect the radiation from a man's hand at a distance of 500 ft and the radiation from a large area heat source at a distance of 10 miles or more.
[13] "IR Radar Tracks Rocket from Pad," *Electronic Design*, **11**, 18 (March 29, 1963). Describes an infrared tracker that will track ballistic missiles during the early, high-acceleration portion of their flights when the angular rates are too high for tracking radars. An infrared source placed near the base of the missile permits lock-on prior to launch. The title appears to be a misnomer, because the tracker operates passively.
[14] T. P. Dixon, "A High Resolution Infrared Tracking System," *Infrared Phys.*, **3**, 27 (1963). A device to provide reliable tracking data during the early portion of a missile launch when tracking radars experience difficulty because of multipath ground and sea return. Reticle generates an FM signal carrying target coordinate information. Detector is of uncooled lead sulfide. Optical unit, mounted on the antenna of a tracking radar, has an entrance aperture of 4.5 in. and weighs 17 lb. The electronic unit weighs 50 lb. The effective NEI (2.1 to 2.6 μ) of the system is 10^{-10} W cm^{-2}. Tracking accuracy is ± 0.25 mrad. The same system is described in the next reference.
[15] T. P. Dixon, "New Infrared System Tracks Missiles Against Bright Florida Sky," *Electronics*, **36**, 39 (April 19, 1963). See the previous reference.
[16] R. W. Astheimer, "Photosensitive Image Motion Detector," U.S. Patent No. 3,090,869, May 21, 1963. A device to track a short-range wire-trailing rocket so as to generate error signals that can be used for guidance of the rocket.
[17] H. E. Haynes, "Optical Tracking System," U.S. Patent No. 3,117,231, January 7, 1964. A tracking device that uses an L-shaped 2-element detector to generate error signals.
[18] "Infrared Aims Laser, Locks Radar on Evading Aircraft," *Machine Des.*, **36**, 12 (June 18, 1964).
[19] T. R. Whitney, "Radiant Energy Detection System for Suppressing the Effects of Ambient Background Radiation," U.S. Patent No. 3,144,554, August 11, 1964. A device that operates in two spectral bands in order to achieve good discrimination against backgrounds. Two detectors are required.
[20] "Mobile Radar Fire Control System," *Aviation Week*, **82**, 117 (June 14, 1965). Pictures the Super Fledermaus mobile radar fire control system modified to provide guidance for newly developed antiaircraft missiles. Modifications include the addition of an infrared tracker.
[21] "Infrared-Aimed Laser Radar Prototyped for Army," *Electronic Design*, **14**, 50 (January 4, 1966). The infrared tracker uses a silicon photovoltaic detector and operates in the 0.85 to 1.15 μ region (and is probably seriously hampered by reflected sunlight during daytime operation). Tracking accuracy is said to be better than ± 10 arc sec. For a description of the ranger, see ref. 16.6.2 [5].
[22] D. Marquis, "Optical Tracking; A Brief Survey of the Field," *Appl. Opt.*, **5**, 481 (1966). Introduction to a special issue on trackers. Makes a brief mention of infrared tracking systems.
[23] D. W. Fisher, R. F. Leftwich, and H. W. Yates, "Survey of Infrared Trackers," *Appl. Opt.*, **5**, 507 (1966). An excellent survey article with an extensive tabulation of the characteristics of many infrared trackers.

16.1.3 Search and Track Systems

Infrared search and track equipment is becoming increasingly important as a component of the fire control system of tactical aircraft. In general, the infrared equipment has been used to complement existing radars rather than to replace them. The following are some of the reasons for including infrared search and track equipment in new aircraft fire control systems or for adding it to systems that are already operational.

1. Passive detection capability makes it possible to complete an attack without the warning often given by radar. Sometimes the target can be deceived by deliberately leaving the radar in a search mode and using the infrared equipment to press home the attack.

2. There is the possibility of greater detection range than that provided by the accompanying radar. This situation is especially likely in low-altitude attacks when the radar is limited by ground clutter and in attacks directly on the tail of the target.

3. The excellent angular resolution of infrared equipment is valuable in evaluating the size of an incoming raid and in assigning individual targets to interceptor aircraft.

4. The infrared equipment is not affected by the electronic countermeasures that the target may employ against radar. If the enemy knows that both radar and infrared equipments are in use, he may be forced to carry countermeasures for both types of equipment.

5. Since infrared equipment is much less complex than radar equipment, it is potentially more reliable.

Remember, however, that since infrared equipment does not have an all-weather capability, it is unlikely that such equipment will displace radar from airborne fire control systems.

[1] "Gyroscopic IR Seeker-Tracker for ICBM and Satellite Defense," *Space/Aeronautics*, **34**, 195 (November 1960). Entire device, consisting of optics, cooled detectors, preamplifier, and signal processor, is mounted in a 3-degree-of-freedom gyro and rotated at 600 rpm. Tracker has a 12 in. input aperture and uses a 24-element array of cooled zinc-doped germanium detectors. Designers claim a range of more than 50 miles against a small, cool satellite, a range of thousands of miles against an ICBM during its boost phase, and a range of 125 miles against an ICBM nose cone (using a 24 in. input aperture).

[2] "Bomber IR System Being Designed to Track Enemy While Scanning," *Electronic Design*, **9**, 18 (April 26, 1961). Describes an infrared track-while-scan system to warn bombers of interceptors and missiles. Sensor uses four rotating lenses to scan a field of view that is 60 deg in azimuth and 40 deg in elevation. The detector is liquid-nitrogen-cooled indium antimonide. Detection range is said to be up to 20 miles. For photographs and drawings, see ref. 16.1.3[6].

[3] "Infrared—The New Horozon," *Electronic Equipment Eng.*, **9**, 44 (May 1961). Describes and shows several diagrams of a rotating-eyeball search and track system that uses three rotating lenses and a liquid-nitrogen-cooled detector.

[4] H. W. Berry, "Infrared Receiver," U.S. Patent No. 2,987,622, June 6, 1961 (filed February 23, 1954). Describes a search and track device of the rotating-eyeball type. This patent is especially notable for the detailed discussion of the signal processing circuitry.

[5] "Interim Interceptor Modification...," *Aviation Week*, **75**, 23 (August 21, 1961). Reports that F-101, F-102, and F-106 aircraft are being modified to give them infrared capability for acquisition, lock-on, and completion of approach. An item in the same magazine at **85**, 77 (October 24, 1966) reports upgrading of the F-106 by the addition of a closed-cycle cooler for the infrared detection system.

[6] B. Kovit, "Track-While-Scan IR System," *Space/Aeronautics*, **36**, 123 (August 1961). Contains three pages of photographs and cross-sectional drawings of the track-while-scan system that is further described in ref. 16.1.3[2].

[7] R. D. Barbera, "Infrared Automatic Acquisition and Tracking System," *IRE Trans. Mil. Electronics*, **MIL-5**, 312 (1961). Device uses a liquid-nitrogen-cooled lead selenide detector, an FM-generating reticle, and an external scan mirror. Tracking range against a jet aircraft is said to be 10 miles. Equipment is also used for measurements of missile plumes and reentry objects.

[8] "Infrared Systems," *Electronic Design*, **10**, 30 (July 5, 1962). A brief item concerning an infrared search and track system for the Phantom II jet fighter. The indium antimonide detector contains two elements. A photograph of the system is shown in ref. 16.1.3[12].

[9] P. M. Cruse, "Apparatus for Processing Optically Received Electromagnetic Radiation," U.S. Patent No. 3,083,299, March 26, 1963. Uses a multifrequency reticle to permit tracking a selected target while continuing to search the field of view. Requires only a single detector and no mechanical or optical scanning motion.

[10] S. Shapiro, "Infrared Search and Tracking System Comprising a Plurality of Detectors," U.S. Patent No. 3,106,642, October 8, 1963. Uses a multifrequency reticle and multiple detectors to permit simultaneous search and track.

[11] A. F. Fairbanks, "Infrared Searching and Tracking Apparatus Utilizing a Spiral Scan," U.S. Patent No. 3,116,418, December 31, 1963 (filed October 5, 1954). A device having a large field of view for search purposes and a means of reducing the field of view when tracking a selected target.

[12] "Nose of a Hunter," *Aviation Week*, **80**, 86 (April 27, 1964). A photograph of an infrared detection system mounted in a chin mount, that is, beneath the radome, of an F4-B. Further information is contained in ref. 16. 1. 3[8].

[13] P. J. Klass, "Microcircuit Technology shows Versatility," *Aviation Week*, **80**, 59 (June 1, 1964). Describes an infrared search and track system that makes extensive use of microelectronic circuitry. One hundred detectors are used and are packaged in a 10 by 10 element matrix. See also refs. 16. 1. 3[14] and [16].

[14] A. Corneretto, "Microelectronics Reinforces Bid for Avionics," *Electronic Design*, **12**, 6 (June 8, 1964). For details, see refs. 16. 1. 3[13] and [16].

[15] G. F. Aroyan and S. H. Cushner, "Reticle Structure for Infrared Detecting System," U.S. Patent No. 3,144,555, August 11, 1964. Describes a track-while-search device using multiple combinations of collecting optics and detectors with a single reticle. Contains a detailed discussion of the properties of reticle patterns formed from various combinations of involute curves.

[16] C. P. Hoffman and R. F. Higby, "Application of Molecular Engineering Techniques to Infrared Systems," *Proc. IEEE*, **52**, 1731 (1964). Describes a search and track system

developed to exploit the potential of microelectronics. The system has a 100-element detector, a field of view of 0.4 deg per detector element, a tracking accuracy of 1 mrad, and it consumes 6.2 W. Peak values of NEP are 5×10^{-13} W cm^{-2} at 2.5 μ with uncooled lead sulfide, 5×10^{-14} W cm^{-2} at 2.5 μ with cooled lead sulfide, and 1×10^{-12} W cm^{-2} at 4.3 μ with cooled lead selenide. See also refs. 16.1.3[13] and [14].

[17] "Camera Rigged Beneath Nose of a USAF/McDonnell F-4C," *Aviation Week*, **84**, 45, January 31, 1966. Photograph shows a close-up of an infrared detector bay beneath the nose of an F-4C.

[18] B. Miller, "Wide Use of F-111A Mk. 2 Avionics Seen," *Aviation Week*, **84**, 93 (June 6, 1966). Weapon system includes an infrared search and detection subsystem to complement the attack radar. In use, one sensor could search a wide field of view and the other a narrow field of view. The infrared sensor could take over if the radar was jammed. For other details of the weapon system, see ref. 16.1.1[42].

[19] "IR Search Unit Mounted on Draken," *Aviation Week*, **84**, 93 (June 27, 1966). Shows a photograph of an infrared search unit mounted on the underside of the nose of a Swedish Air Force J35F Draken jet interceptor. Output of the search unit is displayed for the interceptor pilot as an aiming dot for lining up the aircraft in a manner similar to a radio-controlled intercept. Additional information is given in the same magazine; **84**, 23 (May 16, 1966). Unit is referred to as a tracking device having a range of 12.5 to 18.7 miles.

[20] B. Miller, "Crucial Testing Phase Nears for Phoenix," *Aviation Week*, **87**, 59 (September 11, 1967). Describes the infrared search and track set that augments the fire control radar in the Phoenix system. The infrared equipment is contained in a chin-mounted pod just aft of the radome. It provides substantially better angular resolution than does the radar. A self-contained refrigerator is part of the system. Under some circumstances, targets can be located and tracked and missiles launched without the aid of the fire control radar.

[21] "Filter Center," *Aviation Week*, **88**, 71 (February 26, 1968). Reports that two firms are competing in the development of advanced search and track sensors operating in the far infrared region. They are to be used on space vehicles for tracking hostile vehicles in space.

[22] B. Miller, Avionics Extends Military Helicopter Role," *Aviation Week*, **88**, 186 (June 24, 1968). Shows a photograph of a forward-looking infrared system (FLIR) mounted on a UH-1 armed helicopter. The FLIR is designed to provide nighttime target acquisition and fire control.

16.1.4 Weapon Guidance

The infrared-guided missile is probably the best known military application of infrared. Although the U.S. Sidewinder missile seems to receive most of the publicity, at least seven countries (see Table 16.1) have missiles that are thought to employ infrared guidance. The information contained in Table 16.1 is not of the type that is readily available in the daily newspaper. It has been culled from many publications, and some of the entries may be speculative.

[1] H. Centervall, "Aerial Torpedo," U.S. Patent No. 1,388,932, August 30, 1921 (filed July 27, 1916). Uses a four-element or quadrant detector to provide right-left and up-down steering information.

[2] J. H. Hammond, Jr., "Torpedo," U.S. Patent No. 2,164,916, July 4, 1939 (filed January 5, 1934). Describes an influence fuse that uses a side-looking photocell. When the

torpedo passes under a ship, the reduced illumination on the photocell causes it to trigger the explosive.

[3] M. A. McLennan, "Radiant Energy Responsive Directional Control," U.S. Patent No. 2,403,387, July 2, 1946. A bomb-guidance system using a four-element detector and a simple on-off control system.

[4] T. W. Chew, "Homing System," U.S. Patent No. 2,421,012, May 27, 1947. A dual-frequency chopper generates separate carrier frequencies for right-left and up-down channels. Uses a graded-density filter over each detector to generate an amplitude modulation that is proportional to the magnitude of the pointing error.

[5] H. F. Rost et al., "Self-Steering Device," U.S. Patent No. 2,242,193, July 15, 1947. Contains several sets of simultaneously rotating optical assemblies and associated detectors. An RF source and appropriate receivers are provided for times when bad weather restricts use of the optical devices.

TABLE 16.1 MISSILES THAT ARE THOUGHT TO EMPLOY INFRARED GUIDANCE[a]

Country	Missile	Mission	Status	Range, Statute Miles
France	Matra 511	Air to air	Obsolete	—
	Matra 530	Air to air	Production	13
Great Britain	Firestreak	Air to air	Production	5
	Red Top	Air to air	Production	7
	Taildog	Air to air	Development	0.9
Italy	C-7 (Rigel?)	Air to air	Development	6.2
Japan	XAAM-A-3	Air to air	Development	3+
Sweden	Rb 28	Air to air	Development	5
United States	Falcon (AIM-4C)	Air to air	Production	5.7
	Falcon (AIM-4D)	Air to air	Production	>5.7
	Falcon (AIM-4F)	Air to air	Production	8
	Sidewinder (AIM-9B)	Air to air	Production	2.3
	Sidewinder (AIM-9D)	Air to air	Production	11.5
	Redeye (FIM-43A)	Surface to air	Production	2
	Chaparral (MIM-72A)	Surface to air	Production	—
USSR	Anab	Air to air	Production	—
	Ash	Air to air	Production	—
	Atoll	Air to air	Production	—
	Awl	Air to air	Production	—
	Golem-3	Surface to air	Production	7.5
	M100A	Air to air	Production	5
	SA-3	Surface to air	—	—
	Samovar	Surface to air	Production	3
	T-8	Surface to air	Production	15
	Type 2	Air to air	Development	5.5

[a]Data adapted from refs. 16.1.4 [9, 19, 33, 41, 44, 46, 47, 49, 52, 53, 55, and 67].

16.1.4 WEAPON GUIDANCE

[6] L. Hammond, "Steering Control Apparatus," U.S. Patent No. 2,439,294, April 6, 1948 (filed November 27, 1943). Image nutation is produced by a mirror whose reflecting surface is not normal to its axis of rotation.

[7] R. B. Dow, *Fundamentals of Advanced Missiles*, New York: Wiley, 1958. Covers the basic principles of the design of guided missiles. Chapter 6 is devoted to infrared techniques.

[8] T. J. Lauroesch, "Rotary Scanning Device," U.S. Patent No. 2,873,381, February 10, 1959. Uses a rotating scanner, an optical derotation device, and a pair of fixed detectors.

[9] "Soviets Mass Produce Infrared Guided Missile," *Prod. Engrg.*, **30**, 21 (March 2, 1959). Describes the Samovar, an infrared-guided surface-to-air missile designed for use against low-altitude supersonic aircraft. Useful range is 3 miles. The article claims the detector is a zinc sulfite (sic) type (which may be a good indication of the accuracy of the rest of the article).

[10] R. W. Ketchledge, "Radiation Sensitive Scanning System," U.S. Patent No. 2,897,369, July 28, 1959 (filed December 28, 1944). Uses separate pulse-type scanners for azimuth and elevation indication.

[11] W. J. Haywood, Jr., "Infrared Missile Homing," *Space/Aeronautics*, **32**, 131 (August 1959). Discusses the elementary design principles of passive guidance systems.

[12] A. Nyman and F. E. Null, "Heat Seeker with Proportional Control," U.S. Patent No. 2,903,204, September 8, 1959 (filed November 8, 1946). The detectors are biased from an ac rather than a dc source. Sixteen detectors cover the field of view, and each detector is biased at a different frequency. Hence any signals are coded, through the bias frequency, with directional information. The description implies that the device has a tracking range of 5 miles.

[13] F. E. Null, and L. J. Pleshek, "Heat Sensor," U.S. Patent No. 2,911,167, November 3, 1959. Combines a radial scanning slot and a rotary scanning drum to determine target coordinates. The two detectors are connected so as to null out spurious signals from the aerodynamically heated irdome.

[14] V. I. Marisov and I. K. Kucherov, *Guided Missiles*, Moscow: Ministry of Defense Press, 1959 (CFSTI: 64-21461). An exhaustive treatment of the principles of guided missile design and the characteristics of the world's guided missiles. Notably lacking is any mention of Soviet developments. There is a good discussion of infrared guidance systems and methods of calculating their performance. Typical detectors described are lead sulfide and lead telluride, both cooled to 90°K. The book is intended primarily for the military reader.

[15] H. Dubner, "Optical Design for Infrared Missile Seekers," *Proc. Inst. Radio Engrs.*, **47**, 1537 (1959). Describes a number of optical systems that can be used for infrared-missile guidance. The optical resolution is shown for each type.

[16] C. Dawson, "Firestreak," *Space/Aeronautics*, **33**, 169 (June 1960). Describes the guidance unit and the control electronics of the British Firestreak, an air-to-air infrared-guided missile. The guidance unit is unique in that it is the only known example of the use of a nonspherical irdome to reduce the aerodynamic heating of the dome. Eight flat plates are bonded together to form a segmented conical dome. The reticle is an FM type. The missile also uses an infrared proximity or influence fuse. For further details, see ref. 16.1.4[47].

[17] "Practice Runs on Aircraft Halted After Sidewinder Destroys B-52," *Aviation Week*, **74**, 30 (April 17, 1961). An infrared-guided Sidewinder missile, inadvertently fired from an F-100, hit the exhaust section of an engine on the B-52, causing it to explode and destroy the aircraft.

[18] "Photos of Soviet Aircraft Show New Missiles," *Aviation Week*, **75**, 26 (July 24, 1961). Early photos of the Soviet MiG-21 aircraft equipped with Atoll infrared-guided air-to-air missiles. For additional photos and descriptions see the same magazine; **75**, 25 (July 10, 1961); **84**, 27 (February 14, 1966); **84**, 79 (February 21, 1966); **84**, 25 (May 2, 1966). These missiles have been used in Vietnam; **84**, 37 (May 16, 1966).

[19] "1961 Missile Handbook — Russia, Intelligence Report," *Aircraft and Missiles*, July 1961, p. 51.

[20] S. M. MacNeille, "Target-Seeking Head for Guided Missile," U.S. Patent No. 2,997,594, August 22, 1961. Uses a separate set of detectors for sensing up-down and right-left pointing errors. The chopper is a drum with slots cut in its periphery. Pointing error results in the generation of an FM signal

[21] D. S. Cary, R. E. Kessel, and S. M. MacNeille, "Targer-Seeking Head for Guided Missile," U.S. Patent No. 2,997,595, August 22, 1961. Describes a gyro-stabilized seeker using separate detectors to sense up-down and right-left pointing errors.

[22] R. A. Watkins, "Reticle System for Optical Guidance Systems," U.S. Patent No. 3,002,098, September 26, 1961. A rotating prism is used to nutate the target image on a rotating reticle. The phase and magnitude of the resulting FM are proportional to target coordinates.

[23] P. J. Klass, "Competition Slashes Sidewinder 1-A Price," *Aviation Week*, **76**, 89 (April 9, 1962). Shows a photograph of the guidance and control unit and describes production problems.

[24] L. M. Biberman and R. S. Estey, "Multislit Scanner," U.S. Patent No. 3,034,405, May 15, 1962 (filed October 13, 1953). A crucially important patent because it discloses the concept of the balanced reticle, that is, one in which both halves of the reticle have the same average transmittance. Spectra are shown for signals from sky and cloud backgrounds and from point targets. Describes a gyro-stabilized telescope assembly for an infrared guidance unit.

[25] A. V. Thompson, "Passive Radiation Proximity Detector," U.S. Patent No. 3,036,219, May 22, 1962. Several detectors and their associated optics are arranged to view successive segments of the surrounding space. A target passing through this space generates a series of pulses. The circuitry can be set to respond to a particular pulse repetition rate for detonating a warhead or the pulses can be used to determine the speed of the target.

[26] S. C. Argyle, "Optical Scanning Device," U.S. Patent No. 3,068,740, December 18, 1962. Describes an image nutator consisting of a rotating tilted plate and a rotating tilted lens to compensate for the aberrations of the plate.

[27] T. G. Jones, Jr. and F. E. Null, "Photosensitive Ground Target Seeker," U.S. Patent No. 3,088,034, April 30, 1963 (filed January 8, 1957). Describes a target seeker for missiles or a fire control sensor for fighter aircraft in which an optically generated cross-course scan is combined with the motion of the aircraft to scan a two-dimensional area. The system is specifically designed for use against tanks.

[28] "Mirage 3 Shows Varied Weapons, Auxiliary Rocket," *Aviation Week*, **79**, 62 (July 1, 1963). The Mirage 3 is shown carrying a Matra 530 air-to-air infrared-guided missile. An article in the same magazine [**82**, 147 (June 14, 1965)] identifies the French manufacturer of the infrared guidance head.

[29] "Japan Begins Missile Development Tests," *Aviation Week*, **79**, 34 (July 15, 1963). Contains a short description of the XAAM-A-3, an infrared-guided missile of the Sidewinder type. A picture of this missile is shown in the same magazine; **79**, 46 (July 29, 1963).

16.1.4 WEAPON GUIDANCE

[30] T. W. Chew, "Device for Automatic Homing of Movable Objects," U.S. Patent No. 3,112,399, November 26, 1963 (filed August 11, 1945). Nongimbaled seeker for an aerial bomb. Uses a four-element detector, dual frequency carrier, and a wind-driven chopper.

[31] S. P. Willits, "Automatic Optical Guiding System," U.S. Patent No. 3,139,246, June 30, 1964. Describes a system for detecting the relative angular motion between a vehicle and a preselected target and a means for utilizing this information to guide the vehicle toward the target. The signal processing is described in detail.

[32] "Adaptive Long Range IR Missile Tracker," *Aviation Week*, **81**, 71 (August 31, 1964). Indicates that research and development are under way on an adaptive missile tracker.

[33] H. D. Watkins, "Redeye Nearing Tactical Prototype Stage," *Aviation Week*, **81**, 20(October 5, 1964). Describes results of the first public firings of Redeye, a shoulder-fired, anti-aircraft, infrared-guided missile. Targets were an OH-13 helicopter, an infrared-augmented drone, and an F-9F aircraft. Average firing ranges were 5000 ft. Missile is 4 ft long and has a diameter of 3 in. Quantity-production cost of missile and launcher estimated to be $3000. Additional photos showing Redeye firings and target intercepts can be found in the same magazine; **78**, 33 (March 18, 1963); **80**, 28 (April 20, 1964); and **84**, 80 (April 4, 1966). Pictures of the helicopter intercept appeared in the Los Angeles *Times* on September 25, 1964.

[34] C. E. Dunning, "Optical System for Identifying and Tracking Source of Infrared Radiation Emission," U.S. Patent No. 3,160,751, December 8, 1964. Describes an FM missile-guidance system that incorporates means of identifying a target by the use of a simple spectrum analyzer. Ring-shaped bandpass filters, concentric with one another and with the optical axis, are located at the input aperture. A field lens images the filters onto a concentric array of detectors, so that each detector receives only the flux passing through its corresponding filter. The signal from each detector is thus proportional to the flux within the bandpass of its filter.

[35] L. Z. Kriksunov and I. F. Usol'tsev, *Infrared Equipment for Missile Homing*, Moscow: Military Publishing House, 1964 (OTS:64-31416). The best reference available on the design of infrared seekers for guided missiles. Unfortunately there is no date on this book. Judging from the references, it was probably completed in late 1963 or early 1964.

The book includes chapters on homing systems for guided missiles, the principles of optical (infrared) target seekers, the design of target seekers, detectors, optical systems, target radiation, and the testing of infrared devices. The appendix contains tables of blackbody functions, atmospheric transmission data, a table showing the characteristics of detectors, and an example of the calculation of the maximum detection range for an infrared seeker.

The authors state that this book "...systematizes *foreign experience* in the development of infrared homing devices of guided missiles according to *materials published in the foreign periodical press*." For this reason, there are places in the book where the coverage is very uneven. Notably lacking is material on detector cooling techniques that are suitable for missile use and information on background-limited detectors. It is curious that there is no mention of silicon as an infrared-transparent optical material, despite the fact that germanium is mentioned. With the exception of the book by Biberman ([8], Chapter 6) there is nothing in the open literature that can match the present book's chapter on reticles.

[36] W. C. Wetmore, "Sweden Maintains A37 Program Funding," *Aviation Week*, **82**, 94 (April 5, 1965). Armament is the Rb. 28, an infrared-guided contact-fused missile.

The Rb. 27, a radar-guided version of the same missile, is said to be much more expensive. Additional information is given in the same magazine; **79**, 57 (October 21, 1963).
[37] P. Alelyunas, "Air-to-Ground Missiles," *Space/Aeronautics*, **44**, 60 (November 1965). State-of-the-art article.
[38] P. Alelyunas, "Air-to-Air Missiles," *Space/Aeronautics*, **44**, 68 (November 1965). State-of-the-art article.
[39] B. Miller, "Studies Seek Improved ASM Guidance," *Aviation Week*, **83**, 77 (December 6, 1965). Describes the possibility of using infrared guidance to home on the heat generated by SAM radars. Notes potential difficulties in background discrimination and vulnerability to decoying by heat sources. Also describes development of a non-scanning seeker that uses a mosaic of detectors operating in the 8 to 14 μ region.
[40] "First Launch Photos of Chaparral Missile," *Aviation Week*, **84**, 63 (January 17, 1966). A mobile system using the infrared-guided Sidewinder missile for the protection of front line areas against low-altitude aircraft. An additional photo and description appeared in the same magazine: **83**, 19 (July 26, 1965).
[41] C. M. Plattner, "SAMs Spur Changes in Combat Tactics, New Equipment," *Aviation Week*, **84**, 26 (January 24, 1966). Describes the Russian SA-3, an advanced surface-to-air missile equipped with infrared homing.
[42] "First Sidewinder Fired Downs MiG-21," *Aviation Week*, **84**, 25 (May 2, 1966). An infrared-guided Sidewinder missile entered the tailpipe of a MiG-21 over Vietnam. Similar notes appear in the same magazine: **83**, 25 (April 12, 1965); **83**, 27 (June 21, 1965); **85**, 29 (July 18, 1966); and **85**, 23 (October 17, 1966). Similar items appeared in the Los Angeles *Times* on July 12, 1965, and April 27, 28, May 1, 13, 27, June 24, and September 17, 1966.
[43] T. G. B. Boydell, "Optical Sensing System," U.S. Patent No. 3,263,084, July 26, 1966. Describes an FM reticle and gives other details on an infrared guidance unit for an air-to-air missile.
[44] F. Leary, "Tactical Air Defense," *Space/Aeronautics*, **46**, 78 (September 1966). A broad survey of tactical air defense with some discussion of infrared-homing Sidewinder, Chaparral, and Redeye missiles. Claims that Redeye has a range of well over 2 miles against a jet target.
[45] "Fast Transition to Integrated Circuits," *Electronic Equipment Eng.*, **14**, 20 (October 1966). Describes the successful efforts to convert the Redeye infrared-guided missile from a model that used transistors and discrete circuitry to a model that uses integrated circuits. Redeye is designed for use by foot soldiers against low-flying aircraft. It is shoulder fired and weights 29 lb.
[46] L. H. Dulberger, "Advanced Interceptor Aircraft," *Space/Aeronautics*, **46**, 54 (November 1966). A good survey article with considerable emphasis on infrared-guided missiles and search sets for interceptors. Contains an interesting discussion of the limitations of infrared-guided missiles. These include performance degradation in fog and rain, the limitation to tail attack, the existence of a minimum firing range, and the possibility of the seeker locking onto interfering backgrounds. Article also describes Russian interceptors and their infrared-guided missiles. These include the Atoll missile on the MiG-21 and the newer Awl missile on the MiG-23.
[47] J. W. R. Taylor, ed., *Janes All the World's Aircraft*, New York: McGraw-Hill, 1966–1967, p. 431. Gives an excellent photograph and description of the infrared-guided air-to-air Firestreak missile. The photo is of particular interest, because it shows the unique conical irdome that is made from 8 flat plates bonded together. This is the only known attempt to achieve the optimum aerodynamic shape for an irdome on an infrared-guided missile. An ever better picture of the irdome appears on p. 450 of

16.1.4 WEAPON GUIDANCE 485

the 1964–1965 edition of Janes. For further details of the Firestreak's infrared fuse, see ref. 16.1.6 [7].

[48] C. Brownlow, "USAF Boosts North Viet ECM Jamming," *Aviation Week*, **86**, 22 (February, 6, 1967). Contains an analysis of combat experience over North Vietnam with Sidewinder infrared-guided missiles. The effectiveness of this missile suffers when the launch aircraft is pulling a high-g load (in a tight turn). An improved version will be capable of launch during a 6-g turn.

[49] "Specifications," *Aviation Week*, **86**, 171 (March 6, 1967).

[50] "Industry Observer," *Aviation Week*, **86**, 23 (April 10, 1967). States that the Russian Atoll A-2 air-to-air infrared-guided missile is 105 in. long and has a 6 in. diameter.

[51] "USAF F-4C Carries Varied Ordnance Loads," *Aviation Week*, **86**, 75 (April 24, 1967). Shows an excellent full-color photograph of a Sidewinder air-to-air infrared-guided missile.

[52] "SAT Supplies Infrared Systems," *Aviation Week*, **86**, 320 (May 29, 1967). Article has a photograph and describes the infrared guidance unit for the Matra 530 air-to-air missile. This unit operates in the 4 to 5 μ region with a detector cooled by liquid nitrogen. With this guidance unit the missile can be fired day or night and at almost any target aspect angle except for a direct nose-on attack. An earlier infrared guidance unit for the Matra 511 air-to-air missile could be used only at night.

[53] C. Brownlow, "Soviet Air Force Unveils Advanced Designs for Expanded Limited War Capability," *Aviation Week*, **87**, 32 (July 17, 1967). Describes the Yak-28P all-weather interceptor and notes that standard armament includes both Atoll and Anab infrared-guided air-to-air missiles. A photograph accompanying the article shows that the underwing mountings are quite different for the two missiles. The article also notes that the Fiddler, a long-range interceptor, carries infrared-guided Ash air-to-air missiles. Photographs in a later issue of the same magazine, **87**, 19 (July 24, 1967), show these two aircraft in flight, carrying both Ash and Anab missiles. Apparently both of these missiles are available in either an infrared- or a semiactive radar-guided version.

[54] W. C. Wetmore, "Israelis Display Soviet-Built Atoll Missile," *Aviation Week*, **87**, 65 (July 24, 1967). Shows an excellent photograph of two Atoll missiles and contains considerable information about the missile and its operational use. Unfortunately, the infrared guidance units were covered during the display. The missiles bear a remarkable resemblance to the U.S. Sidewinder. The article states that an Iraqi MiG-21 fired an Atoll missile that homed in and detonated near the tailpipe of an Israeli Mirage 3CJ. This apparently indicates that the Atoll missile includes some sort of influence fuse, a conclusion that appears to be borne out by a study of the photograph accompanying this article.

[55] "Eleventh Annual World Aerospace Encyclopedia," *Technology Week*, **20** (July 31, 1967).

[56] W. C. Wetmore, "Israelis Class MiG-21C as Efficient, High-Altitude Fighter," *Aviation Week*, **87**, 50 (July 31, 1967). Shows a static display of a MiG-21C (previously delivered by a defecting Iraqi pilot) and armament captured from the Egyptians by the Israelis. The Atoll infrared-guided missiles shown in one of the pictures appear to be the same missiles described in ref. 16.1.4[54]. The left-side gun fairing on the aircraft has been removed to make room for the avionics for the Atoll missiles.

[57] "A-4Gs, S-3Es Bought by Australian Navy," *Aviation Week*, **87**, 21 (July 31, 1967). New aircraft delivered to the Australian Navy carry Sidewinder infrared-guided missiles. A photograph accompanying the article shows that four Sidewinders can be carried on each aircraft.

[58] "First Pictures Show AIM-4D on USAF F-4," *Aviation Week*, **87**, 25 (August 7, 1967). Reports that the AIM-4D was adapted from earlier models of the Falcon to make it suitable for use against high-speed, maneuvering targets, such as fighters. Flight tests of the missile included launches during tight turns and during approaches from the front, rear, and side of targets.

[59] "Filter Center," *Aviation Week*, **87**, 85 (August 7, 1967). Reports the flight testing of a long-wavelength infrared-homing guidance system. This would be used to guide an antisatellite payload against a hostile satellite.

[60] E. Regelson, "Impact of IR Devices on Aircraft Design Trends," *Astronautics and Aeronautics*, **5**, 60 (August 1967). Discusses the effect that infrared-guided missiles have had on the design of aircraft. The subject is obviously fraught with security restrictions, and the author has been unable to go into any significant detail on what can be done to reduce the emission from aircraft power plants.

[61] "Washington Roundup," *Aviation Week*, **87**, 25 (October 9, 1967). Reports that Sweden has placed an $8 million order for more than 1000 Redeye missiles. The Redeyes will be designated the Robot 69 and will be assigned to infantry brigades to replace some of the 20 mm automatic guns now used for antiaircraft protection.

[62] "Filter Center," *Aviation Week*, **87**, 97 (November 6, 1967). Reports a study of advanced air-to-air missile guidance concepts. One such concept is a dual-mode seeker that would permit the missile to home on either reflected radar energy or infrared energy radiated by the target.

[63] "Tactical Arms Shown in Soviet Exercises," *Aviation Week*, **88**, 44 (February 19, 1968). Shows a photograph of Atoll infrared-guided missiles mounted on a Soviet MiG-21 fighter.

[64] M. P. London, "Tactical Air Superiority," *Space/Aeronautics*, **49**, 62 (March 1968). Contains one of the best discussions available on the guns or missiles debate arising from the Vietnam conflict. Reports that the infrared-guided Sidewinder must be fired from the tail of the target, that its maximum effective range is about 8000 ft, and that this range is less if the launch aircraft is undergoing lateral acceleration. There is a minimum range inside which Sidewinder cannot be fired. The article states that this minimum range is beyond the maximum range of aircraft guns. Sidewinder is credited with at least 3 kills in Vietnam.

[65] "Industry Observer," *Aviation Week*, **89**, 11 (July 29, 1968). Reports developmental work on an all-aspect infrared seeker for the U.S. Navy short-range air-to-air missile. The missile would be useful at ranges of 1000 to 10,000 ft.

[66] R. D. Archer, "The Soviet Fighters," *Space/Aeronautics*, **50**, 64 (July 1968). The Soviets are known to have at least three infrared-guided, air-to-air missiles, Atoll, Anab, and Ash. A table in this article shows that the Atoll is carried on the MiG-19, 21F and 21PF. The Anab is carried on the MiG-23, Faithless, Flogger, SU-9 and 11, and the YAK-28. The Ash is carried on the Fiddler.

[67] W. C. Wetmore, "Britain Seeks Strong Missile Export Role," *Aviation Week*, **89**, 21 (September 23, 1968). Describes Taildog, a short-range air-to-air, infrared-guided, dogfight missile. It is designed to fill the gap between guns and unguided rockets and longer range guided missiles. It is believed that the infrared guidance unit is derived from that of the Firestreak missile. Maximum range is 0.9 mile and the minimum range is not given.

16.1.5 Navigation and Flight Control Systems

Despite the fact that the star trackers used in most navigation systems operate in the visible portion of the spectrum, the details of their con-

16.1.5 NAVIGATION AND FLIGHT CONTROL SYSTEMS

struction, reticle design, and signal processing are readily adaptable to infrared equipment. Hence, many of the references that follow deal with star trackers.

[1] H. Konet, "Photoelectric Drift Indicator," U.S. Patent No. 2,425,541, August 12, 1947. Uses a reticle consisting of parallel straight lines to determine the true ground track relative to the heading of an aircraft.

[2] R. E. Jasperson, "Automatic Navigational Director," U.S. Patent No. 2,444,933, July 13, 1948. Principal emphasis is on the implementation of a navigational system that uses optically observed celestial references and continuously computes the position of a ship or aircraft.

[3] R. H. Varian, "Radiant Energy Directional Apparatus," U.S. Patent No. 2,462,925, March 1, 1949. A star tracker that uses a cross-array detector and a tilted rotating mirror to produce nutation of the image.

[4] R. J. Herbold, "Anticollision Apparatus," U.S. Patent No. 2,489,223, November 22, 1949. Describes a terrain-avoidance device for low-flying aircraft that uses a two-color system to differentiate between skylight and radiation from the terrain. Assumes (incorrectly) that infrared is not attenuated by clouds or fogs.

[5] L. B. Scott, "Radiant Energy Tracking Apparatus," U.S. Patent No. 2,513,367, July 4, 1950. A star tracker that uses a simple two-sector reticle. Also shows a means of eliminating the reticle by using a rotating dove prism and a beam splitter.

[6] J. M. Slater, "Stable Reference Apparatus," U.S. Patent No. 2,740,961, April 3, 1956 (filed July 9, 1947). Describes an infrared device that senses the radiation discontinuity at the earth's horizon in order to establish the local vertical. Contains a good description of the entire stabilization system.

[7] O. T. Schultz, L. B. Scott, and W. G. Wing, "Navigation System," U.S. Patent No. 2,762,123, September 11, 1956 (filed May 26, 1948). Gives an excellent discussion of the design philosophy for a complete navigation system. The star tracker uses a simple two-sector reticle or, alternatively, the dove prism and beam splitter described in ref. 16.1.5 [5].

[8] W. E. Osborne, "Airborne Infrared Warning System Measures Range," *Electronics*, **30**, 190 (July 1, 1957). Describes a proximity-warning system for preventing aircraft collisions. Since there is a sensor on each wingtip, there is the possibility of determining range by triangulation. There is no evidence that this device has ever been tested.

[9] "Airlines to Test Infrared Warning," *Electronics*, **31**, 8 (June 6, 1958). Describes an infrared device weighing 30 lb that is designed to give a minimum of 20 sec warning for aircraft collision avoidance.

[10] R. M. Burley, "Navigation Device," U.S. Patent No. 2,867,393, January 6, 1959 (filed January 11, 1950). A star tracker that uses a two-sector reticle. An alternate configuration uses a reflecting rather than a transmitting reticle.

[11] H. Blackstone, "Optical Scanner for Determining Velocity-Altitude," U.S. Patent No. 2,878,711, March 24, 1959. Describes a means of determining v/h by sampling the output from a rotating-eyeball scanner used for aerial reconnaissance. The signals from two detector elements spaced along the direction of flight are autocorrelated to yield the value of v/h.

[12] H. Blackstone and F. G. Willey, "Airborne Scanner for Determining v/h Rate," U.S. Patent No. 2,878,712, March 24, 1959. An extension of the previous reference. The separation between the two detector elements used to determine v/h can be varied manually.

[13] H. Blackstone, "Optical Scanner for Determining Velocity-Altitude," U.S. Patent No. 2,878,713, March 24, 1959. A further extension of the basic principles set forth in the two previous references.
[14] H. Blackstone, "Radiant Energy Ground-Clearance Meter," U.S. Patent No. 2,882,783, April 21, 1959. Describes a passive altimeter consisting of a velocity-responsive device coupled to a velocity-altitude-responsive device so as to yield the quotient of their outputs.
[15] B. Miller, "IR Velocity/Height Computer Readied," *Aviation Week*, 72, 81 (May 23, 1960). A v/h sensor to provide signals to control the film speed in image-motion-compensating systems for aerial photography. Two sensors placed at an angle to one another view the ground beneath an aircraft. The time delay between signals from a ground object as seen by the two sensors is inversely related to the v/h ratio. The unit can be operated at altitudes ranging from 500 to 100,000 ft and v/h ratios of 0.01 to 10 rad sec^{-1}.
[16] S. Hansen, "Astrometrical Means and Method," U.S. Patent No. 2,941,080, June 14, 1960 (filed November 6, 1948). A star tracker that uses a two-frequency reticle to generate phase modulations indicative of the azimuth and elevation pointing errors.
[17] W. B. Greenlee and V. A. Miller, "Star Sensing System," U.S. Patent No. 2,943,204, June 28, 1960 (filed July 11, 1950). A star tracker that uses a two-frequency reticle.
[18] C. C. Baum, "Scanner for Optical Systems," U.S. Patent No. 2,946,893, July 26, 1960 (filed December 13, 1950). Describes an electromagnetic drive for the reticle in a star tracker. It is claimed that this drive is free of any vibration and is strictly synchronous.
[19] V. E. Carbonara et al., "Star Tracking System," U.S. Patent No. 2,947,872, August 2, 1960. Uses an optical wedge for image nutation and a simple two-sector reticle.
[20] R. H. Ostergren, "Moving Field Scanner," U.S. Patent No. 2,967,246, January 3, 1961 (filed February 18, 1952). A star tracker that uses a rotating, tilted, flat plate and a rotating octagonal prism to scan the image of the field of view over a small aperture placed in front of the detector.
[21] L. Kaufold and C. H. Getz., "Fan Scanner Celestial Detector System," U.S. Patent No. 2,968,735, January 17, 1961 (filed August 17, 1953). This appears to be the same device that is described in ref. 16.1.5 [26] with the exception that the reticle has been changed to a balanced AM type. Considerable information is given on the signal processing.
[22] T. H. Jackson, Jr., "Light Chopper," U.S. Patent No. 2,972,812, February 28, 1961 (filed March 14, 1949). A star tracker that uses a reflective reticle to generate a pulse-width modulation.
[23] S. Hansen, "Star-Tracking System," U.S. Patent No. 2,981,843, April 25, 1961 (filed September 2, 1947). Uses an FM reticle.
[24] D. J. Lovell, "Electro-Optical Position Indicator System," U.S. Patent No. 2,997,699, August 22, 1961. A star tracker using a balanced FM reticle.
[25] A. Lovoff, "Method of and Means for Horizon Stabilization," U.S. Patent No. 2,999,161, September 5, 1961 (filed October 10, 1945). Describes an infrared device to locate the horizon and thus to establish the local vertical. Two detectors are used in a radiation-balance arrangement.
[26] P. H. Taylor, "Tracking Telescope," U.S. Patent No. 3,015,249, January 2, 1962 (filed March 14, 1949). A star tracker with a reticle that has a single radial slot. The drawings that accompany this patent are exceptionally well done.
[27] F. G. Willey and T. A. Westover, "Aircraft Velocity-Altitude Radio Meter," U.S. Patent No. 3,018,555, January 30, 1962 (filed December 12, 1956). A further extension of the basic principles set forth in refs. 16.1.5 [11]–[13]. A means of compensating for drift angle is included.

16.1.6 RANGING SYSTEMS 489

[28] A. K. Chitayet, "Light Modulation System," U.S. Patent No. 3,024,699, March 13, 1962 (filed June 15, 1956). A star tracker that uses two rotating reticles to generate a double modulation.

[29] G. Jankowitz, "High Resolution Tracker," U.S. Patent No. 3,061,730, October 30, 1962. A high-precision star tracker in which a rotating dove prism and a drum reticle produce an FM signal.

[30] O. A. Becklund et al., "Method and Apparatus for Determining Altitude," U.S. Patent No. 3,076,095, January 29, 1963 (filed September 5, 1956). A passive altimeter consisting of a sensor to track a source of radiation on the ground and a computer to determine the absolute altitude from the tracking rate and aircraft velocity. Uses a checkerboard reticle with superimposed radial density gradient.

[31] H. R. Hulett, "Electrooptical Light-Detecting Apparatus," U.S. Patent No. 3,080,484, March 5, 1963 (filed December 6, 1951). A star tracker that uses Maksutov optics and tilts the corrector plate to produce image nutation. Contains a detailed discussion of the signal processing problems.

[32] L. Kaufold, "Automatic Multiple Grid Scanning Tracker," U.S. Patent No. 3,088,033, April 30, 1963 (filed August 31, 1953). A star tracker that uses an electromagnetically driven plane scan mirror to move the image of a star over a reticle. The reticle divides the field of view into four sectors. Each sector contains a different number of straight-line elements so that a pointing error results in a pulse-length modulation. A lead sulfide detector is cooled to 203°K by frozen ethylene dichloride. Contains a particularly complete discussion of the signal processing.

[33] S. Jones and L. Manns, "Signal Detector for Use with Radiation Sensor," U.S. Patent No. 3,134,022, May 19, 1964 (filed November 26, 1952). A star tracker using a simple radially segmented reticle and a phase-sensitive rectifier. An alternate arrangement uses a two-frequency reticle to indicate the angular position of the target.

[34] E. O. Frye and D. E. Killham, "Aircraft Collision Avoidance Systems," *IEEE Spectrum*, 3, 72 (January 1966). No mention of infrared techniques, but the discussion of the system aspects is excellent.

16.1.6 Ranging Systems

From the military standpoint, one of the shortcomings of passive infrared equipment is that it does not readily provide information on the distance to a target, as radar does. Triangulation rangefinders that are similar in principle to those used in the visible portion of the spectrum are feasible, but they are rarely used because of the difficulty of ensuring that both halves of the rangefinder are looking at the same target. For this reason, one finds occasional attempts to build infrared rangefinders that derive range from measurements of angular rates, flux changes, or time-to-go determinations. Active rangefinders that operate as infrared radars are discussed in section 16.6.2

[1] W. A. Tolson, "Heat Detecting System," U.S. Patent No. 2,431,625, November 25, 1947. See ref. 16.1.1 [12].

[2] J. R. Esher, Jr., "Closure Time Computer," U.S. Patent No. 2,993,121, July 18, 1961 (filed December 18, 1953). A passive closure time computer for aircraft use. Consists of a simple sensor to develop a voltage proportional to the flux received from the target and a means of taking the derivative of the logarithm of this voltage, which is, as the patent shows, related to closure time.

[3] J. Strong, "Infrared Range Finder," U.S. Patent No. 3,005,913, October 24, 1961 (filed September 18, 1951). An infrared rangefinder operating on the same principle as the conventional triangulating optical rangefinder.

[4] F. L. Hutchens and J. W. Ingels, "Infrared Range Finder," U.S. Patent No. 3,014,131, December 19, 1961. A passive rangefinder that, in effect, rotates two laterally separated sensors at a constant rate and measures the time delay between pulses caused by each sensor scanning across the target.

[5] P. J. Ovrebo and R. C. Wood, "Passive Infrared Ranging Device Using Absorption Bands of Water Vapor or Carbon Dioxide," U.S. Patent No. 3,103,586, September 10, 1963. If the spectral distribution of the radiation from the target and the spectral absorptive properties of the intervening atmosphere are known, a measurement of the received signal at two wavelengths, chosen to be near an absorption band, can be used to calculate range. Suggests a multicolor chopper or a rotating interference filter.

[6] J. R. Jenness, Jr., and F. J. Shimukonis, "Apparatus for Passive Infrared Range Finding," U.S. Patent No. 3,117,228, January 7, 1964 (filed October 12, 1956). A passive infrared rangefinder for airborne use. Principle is to measure received signal in two narrow spectral bands and use the ratio of the two signals to indicate range. This particular system must be calibrated against each target type prior to use, making its use rather limited.

[7] J. W. R. Taylor, ed., *Janes All the World's Aircraft*, New York: McGraw-Hill, 1966–1967, p. 431. Gives an excellent photograph and description of the infrared-guided air-to-air Firestreak missile. The photograph shows the two ring-shaped windows that house the infrared fuze. As the missile nears the target, signals from the two fuze systems give a continuous indication of range and bearing of the target so that the warhead can be detonated at the desired range. A similar fuze system is evident in the photograph of the Red Top missile that appears on the same page. For further details on the Firestreak's unique irdome, see ref. 16.1.4[16].

[8] "Europe Pushing Low-Altitude Anti-Aircraft Capabilities," *Aviation Week*, **87**, 48 (July 3, 1967). Article states that the Swiss Indigo surface-to-air missile utilizes an infrared proximity fuse. This fuse is not evident in the photographs of the missile that are shown in this article.

16.2 RADIOMETRIC APPLICATIONS

There is relatively little evidence in the literature of a military interest in radiometry. This apparent disinterest may be attributable to the classification scheme used here, since it is not always easy to separate radiometric, spectroradiometric, and thermal imaging applications.

16.2.1 Measurement of Flux

[1] "Floating Laboratory Gathers Re-Entry Data," *Elec. Eng.*, **80**, 39 (August 1961). A description of the infrared radiometric equipment used to record missile reentry data.

[2] W. N. Arnquist and D. D. Woodbridge, "Re-Entry Radiation from an IRBM," *IRE Trans. Mil. Electronics*, **Mil-5**, 19 (1961). Describes the infrared radiometric instrumentation used in the Gaslight program to measure the radiation from Jupiter, Thor, and Polaris missiles during reentry. The radiant intensities (lead sulfide region, 1.8 to 2.6 μ) of three Jupiter reentry bodies were 14×10^3 W sr^{-1}, 7.37×10^5 W sr^{-1}, and 4.29×10^5 W sr^{-1}.

16.3.1 TARGET AND BACKGROUND SIGNATURES

[3] "Mobile Tracking Lab," *Aviation Week*, 77, 103 (December 10, 1962). Pictures and briefly describes a mobile infrared tracking and measurement laboratory designed to obtain infrared signatures of aircraft or reentry vehicles. Equipment includes an automatic infrared tracker for target acquisition and several infrared radiometers.

[4] R. V. Meyer, "A Spectral Radiometer for Reentry Measurements," *Appl. Opt.*, 5, 159 (1966). Describes the design of radiometers that are to measure the total power radiated from an object that has a long thin geometry, such as a reentry event. The spectral bandpass is relatively wide, and the device is perhaps best described as a filter radiometer.

[5] P. E. Schumacher, "Narrow Field-of-View, High Radiant-Heat-Flux Sensor," *J. Opt. Soc. Am.*, 57, 1411 (1967). Describes a specialized radiometer for the measurement of the flux from rocket engine exhaust plumes. The radiometer uses reflective-refractive optics and a thin-foil thermocouple detector. The spectral response extends from 0.2 to 6 μ and the field of view is 6 deg.

16.3 SPECTRORADIOMETRIC APPLICATIONS

In this category one finds extensive use of spectroradiometers in the continuing search for useful target signatures, that is some unique characteristic of the radiation from a target that can be used as a positive identification of this target in the future. In general, such signatures include the spectral, spatial, and temporal characteristics of the target radiation.

16.3.1 Target and Background Signatures

[1] C. D. LaFond, "IR Identifies Missiles by Plumes," *Missiles and Rockets*, 6, 22 (May 23, 1960). A device using a scanning monochromator to cover the range from 0.3 to 3.6 μ. Missile performance data are being accumulated for use in identifying missiles by their exhaust-plume signature. Another description of this device appears in the next reference.

[2] "Infrared System Studies Missile Plumes," *Instr. Soc. Am. J.*, 7, 61 (July 1960). Describes a rapid-scan spectrometer that scans from 0.3 to 3.6 μ at a rate of 2.5 to 180 scans per second.

[3] L. G. Mundie et al., "An Airborne Spectroradiometer," *J. Opt. Soc. Am.*, 50, 1187 (1960). Describes a multiple-channel tracking spectroradiometer that covers the 0.25 to 15 μ region. The primary collecting optics have a diameter of 20 in. The target-tracking accuracy is 0.1 mrad. By the use of reflective choppers, dichroic filters, and multiple exit slits, five types of data are acquired simultaneously; these include spectral data in the 0.25 to 0.6 μ, 0.6 to 5 μ, and 5 to 15 μ regions, and radiometric data in the 0.25 to 0.6 μ and 0.6 to 15 μ regions. Instrument is to be used to acquire spectroradiometric data from man-made objects in space and from astronomical bodies.

[4] B. Miller, "Rocket Study to Assist Missile Detection," *Aviation Week*, 75, 77 (July 17, 1961). Describes a program to determine the optimum infrared and/or ultraviolet wavelengths for detecting and identifying hostile ballistic missile launches. Measurements will be made of the radiation from terrestrial and sky backgrounds and various missiles during their launch phase.

[5] M. E. Seymour, "Observing Missile Plumes with Image Tubes," *Electronics* 34, 70 (October 20, 1961). Examines the possibility of detecting missile plumes at low

altitude in the visible or the infrared. The plume model is a blackbody at a temperature of 2000°K and a total radiant intensity of 10^6 W sr^{-1}. Author claims that actual missiles emit within about an order of magnitude above or below this value.

[6] R. C. Barbera, "Infrared Automatic Acquisition and Tracking System," *IRE Trans. Mil. Electronics*, **Mil-5**, 312 (1961). Describes a device that is used for the study of radiation signatures from missile plumes and reentry objects. For further details, see ref. 16.1.3 [7].

[7] B. Miller, "USAF Explores Missile Plume Radiation," *Aviation Week*, **76**, 63 (April 2, 1962). A description of Project TRUMP (target radiation measurement program). Most of the effort is devoted to measurements of plume radiation made from above the earth's atmosphere. A limited number of background measurements are planned. Data should be of interest to the Midas program.

[8] "Infrared Tracking," *Aviation Week*, **77**, 55 (December 17, 1962). Pictures and briefly describes infrared tracking and spectroradiometric equipment installed on Ascension Island. Spectral range is 0.24 to 14 μ. See also ref. 16.1.2 [11].

[9] N. W. Rosenberg, W. M. Hamilton, and D. J. Lovell, "Rocket Exhaust Radiation Measurements in the Upper Atmosphere," *Appl. Opt.* **1**, 115 (1962). Describes a grating spectrometer that uses an image intensifier orthicon and covers the 0.4 to 0.65 μ region. Emission spectra are given for a kerosene-lox missile, and they show a fairly continuous emission similar to that of a blackbody at a temperature of 2000°K.

[10] "Detecting Missiles by Infrared Emission," *Aviation Week*, **78**, 93 (March 4, 1963). Reports that military agencies are increasing their support for studies of the infrared emission characteristics of targets and backgrounds. Programs mentioned are investigating infrared radiation from clouds, spectral radiance of exhaust plumes from large rocket engines, high-altitude rocket plume phenomena, and target recognition techniques.

[11] "Modified KC-135 Used in RAMP," *Aviation Week*, **78**, 113 (April 8, 1963). Shows an over-all photograph of the aircraft and a close-up of the radiometers, spectrometers, and so on mounted thereon. Aircraft windows for various sensors are of quartz. RAMP stands for radiation airborne measurement program.

[12] "Spectrum Analysis During Re-Entry Done Automatically," *Electronic Design*, **11**, 24 (May 24, 1963). System performs a 40-channel spectral analysis 10 times per second. Spectral coverage extends from 0.3 to 4 μ. System utilizes a 48 in. diameter telescope.

[13] "Infrared Spectrophotometer Carried in Airborne Pod," *Aviation Week*, **79**, 107 (October 14, 1963). Pictures an airborne infrared spectrophotometer housed in an 8 in. diameter pod for use on high-altitude high-speed jet aircraft. Spectral region covered is 2 to 6 μ.

[14] "U-2s Gathering Data on Missile Exhausts," *Aviation Week*, **79**, 53 (November 18, 1963). Describes a program in which spectrometers and radiometers are carried aboard U-2 aircraft to gather data on the infrared characteristics of the exhausts from ballistic missiles. The spectrometer and the radiometer weigh approximately 400 lb each.

[15] "Project Glow to Aid ABRES Program," *Missiles and Rockets*, **14**, 27 (March 2, 1964). A program to gather ultraviolet, visible, and infrared spectral signatures of reentry vehicles.

[16] W. H. Gregory, "DOD, NASA Agree on Gemini Experiments," *Aviation Week*, **80**, 38 (June 1, 1964). Experiments D-4 and D-7 deal with radiometric measurements of space backgrounds and objects in space. The spectrometers cover the ultraviolet, visible, and infrared portion of the spectrum. For a picture of one of the spectrometers, a neon-immersed cryogenic infrared interferometric spectrometer, see the same magazine; **80**, 77 (June 8, 1964).

16.3.2 MISCELLANEOUS   493

[17] J. M. Satterfield, "The NASA-LRC Telespectrograph Test and Operation Results," *Soc. Phot. Instr. Engrs. J.*, **3**, 163 (June–July 1965). Describes a tracking telespectrograph used to track the reentry flight of a space vehicle and make a continuous measurement of the spectral distribution of the flux received in the 0.22 to 5.5 μ region. Contains a good description of the design, but no results of the measurements are included. Collecting optics have a diameter of 36 in.

[18] "Gemini 5 Sustains Accelerated U.S. Pace," *Aviation Week*, **83**, 24 (August 30, 1965). Describes infrared measurements made during the Gemini 5 flight. An interferometric spectrometer was used to make measurements of the radiation signatures of a Minuteman 1 missile during launch, earth and sky backgrounds, a rocket motor launch on a sled test track, and an accompanying rendezvous pod. Additional discussion appears in the same magazine; **83**, 26 (September 27, 1965).

[19] E. J. Bulban, "Shirtsleeve Garb Eases Tasks in Gemini 7," *Aviation Week*, **83**, 30 (December 13, 1965). Reports measurements of the infrared radiation from a Polaris A-3 made during the Gemini 7 flight. A further report in the same magazine [**83**, 53 (December 27, 1965)] indicates that additional measurements were made on a reentry body from a Minuteman 1 missile, a rocket sled, the launch of the Gemini 6 vehicle, and the Gemini 6 during rendezvous.

[20] "Two Missiles Fired in Tests at Vandenburg," *Los Angeles Times*, March 26, 1966. Reports the launch of a Titan 2 ICBM followed a few seconds later by a Nike-Javelin high-altitude probe. The Nike-Javelin trailed the Titan so as to measure the radiation from its exhaust plume.

[21] R. J. Condon, "Three-Channel IR-UV Rocket Radiometer," *Infrared Phys.*, **6**, 167 (1966). Describes a three-channel radiometer having narrow-band responses peaked at 0.275, 2.7 and 4.3 μ. The respective detectors are a solar-blind photomultiplier, lead sulfide, and lead selenide. Small lamps associated with each channel allow in-flight calibration. This radiometer was designed to be ejected from an Atlas booster rocket at an altitude above 300,000 ft. A spin imparted to the package caused the radiometer to make successive scans of the plume from the rocket motor.

[22] "Industry Observer," *Aviation Week*, **87**, 13 (July 3, 1967). Reports that the U.S. Air Force is considering timing its launches of test and training missiles so that they cannot be observed during their powered phase by Soviet satellites.

[23] P. J. Klass, "Soviet Payloads Overfly Nike-X Test Site," *Aviation Week*, **87**, 81 (December 11, 1967). Nine single-orbit objects launched by the Soviet Union and tentatively identified as part of the Fractional Orbit Bombardment System (FOBS) passed over the U.S. test site at Kwajalein. The article speculates that this may represent a quick-reaction reconnaissance capability to monitor Nike-X tests. One possibility is that the FOBS vehicle might carry equipment for making infrared signature measurements of the U.S. reentry vehicles.

[24] "Industry Observer," *Aviation Week*, **83**, 13 (February 26, 1968). Reports that the USAF has dropped plans for satellite measurement of the earth's long-wavelength infrared radiance. Such data are said to be available from several meteorological satellite programs.

[25] "Appollo/Saturn 502 Space Vehicle," *Aviation Week*, **88**, 29 (April 15, 1968). A striking photograph that shows how big the exhaust plume can be from a large rocket. When photographed the rocket was at an altitude of nearly 40 miles. The diameter of the plume appears to be at least 10 times the diameter of the rocket.

16.3.2 Miscellaneous

[1] L. B. Carpenter, S. Hormats, and H. Tannenbaum, "Long Path Infrared Detection of Atmospheric Contaminants," U.S. Patent No. 2,930,893, March 29, 1960. A

device to detect minute quantities of atmospheric contaminants, particularly highly toxic chemical warfare agents. For nerve gases such as sarin, the device can detect a quantity as small as 10^{-8} g per liter of air. Detection is accomplished by monitoring the transmittance in three narrow spectral intervals. For the detection of sarin, one of these intervals is centered in an absorption band at 9.8 μ and the other two are in absorption-free regions at 9.25 and 10.4 μ. Absorption path lengths of several hundred yards are used. Generally, a chopper and collimating optics are used to project a beam to a remotely located mirror from which the beam is reflected to the absorption sensor.

[2] S. Wallack, "Infrared Detector for Missile Fuel Vapors," *Electronic Equip. Engr.*, **9**, 49 (May 1961). A device that compares the transmission at two selected wavelengths and correlates this with the amount of absorber present. Device will detect concentrations as low as 4 ppm.

[3] G. Alexander, "Atlas Accuracy Improves as Test Program is Completed," *Aviation Week*, **78**, 54 (February 25, 1963). Describes the use of an infrared spectrophotometer for the detection of hydrocarbon contaminants in liquid oxygen systems.

[4] W. S. Beller, "Forthright CBR Policy Urged," *Missiles and Rockets*, **16**, 27 (April 19, 1965).

[5] F. J. Granzeier, "Toxic Weapons," *Indust. Research*, August 1965, p. 69. Both this and the preceeding reference contain excellent summaries of the properties and characteristics of chemical warfare agents. Neither reference mentions the use of infrared as a means of detection.

[6] H. R. Carlon, "The Apparent Dependence of Terrestrial Scintillation Intensity upon Atmospheric Humidity," *Appl. Opt.*, **4**, 1089 (1965). Describes LOPAIR (long path infrared), a device for the detection of toxic gases through their characteristic absorption in the 10 μ wavelength region. Device has a 30 cm input aperture and uses a thermistor bolometer.

16.4 THERMAL IMAGING APPLICATIONS

Infrared equipment seems to have acquitted itself quite well in providing a method by which the military can find an enemy at night and keep track of his activities. The techniques described here are all passive, in contrast to the active techniques that are discussed in section 16.5.

16.4.1 Reconnaissance

[1] "Hottest Front in the Cold War," *Electronics*, **32**, 22 (July 10, 1959). Describes the shooting down of a U.S. Navy P4M reconnaissance patrol plane over international waters by two MiG jet fighters. Infrared scanning equipment was carried in place of the front gun turret on the P4M. This equipment was used to look for hot areas that might be caused by stack gases from surface ships or snorkel gases from submarines. Over land, the infrared equipment can pick up underground installations or detect various kinds of camouflaged activity.

[2] E. Kosyrev, "Heat Direction Finding—Means of Reconnaissance," *Voyennyye Znaniya*, **35**, 19 (1959) (OTS: 61-28337). The author claims that modern airborne infrared search sets can detect submarines submerged as deep as 40 m by the temperature difference of 0.05 to 0.5°C that exists between the surface wake and the surrounding water.

16.4.1 RECONNAISSANCE

[3] J. Holahan, "Undersea Weapon Systems, State of the Art; Detection and Communications," *Space/Aeronautics,* **33**, 58 (January 1960). Infrared sensors detect temperature differences in the surface water directly over or in the wake of a submarine. Such sensors can detect a temperature difference of 0.01°C. The problems in using such sensors appear to be due to high seas, rain, and fog.

[4] "HRB-Singer Infrared Recon Photographs May Have Anti-Sub Warfare Application," *Aviation Week,* **72**, 76 (February 22, 1960). Shows a thermal image of Manhattan Island that was made in 1958. Electrical power plants, ships, and the Times Square area are clearly visible. The equipment can detect the slight difference in temperature between the water in a ship's wake and that near the ship, so that the direction of the ship's motion can be determined. Examples of thermal imagery from the early 1950's are shown.

[5] H. A. Nye, "The Problem of Combat Surveillance," *IRE Trans. Mil. Electronics,* **MiL-4**, 551 (1960). A short tutorial treatment of the characteristics of modern warfare that contribute to the difficulty of combat surveillance.

[6] C. M. Cade, "Infrared Radar, Surveillance, and Communications," *British Communications and Electronics,* **7**, 414–418, 510–517 (1960). Speculates on the possibility of submarine wake detection by infrared reconnaissance equipment.

[7] "Satellite Needs Push Reconnaissance Gains," *Aviation Week,* **74**, 26 (May 22, 1961). An estimate of the improvements in reconnaissance techniques that are expected within the next decade. States that present infrared reconnaissance equipment has given 100 ft resolution at an altitude of 40,000 ft. Predicts that by 1970 infrared equipment should be able to provide 90 ft resolution from an altitude of 300 mi.

[8] C. M. Cade, "Infrared Scanners for Airborne Reconnaissance," *British Communications and Electronics,* **8**, 94 (1961). A good summary of the design principles of thermal mappers.

[9] L. H. Dulberger, "Gains in ASW Surveillance Threaten Subs Invulnerability," *Electronics,* **36**, 10 (September 6, 1963). Describes the use of infrared scanners to detect the heat in a submarine's wake. Author estimates that the temperature differential between the wake and the surrounding water is 0.5 to 1°C.

[10] "Navy Pressed to Enlarge Anti-Submarine Capabilities," *Aviation Week,* **80**, 84 (March 16, 1964). States that a submerged submarine leaves a surface wake as the water it heats rises to the surface. The temperature difference between the wake and surrounding water is only a fraction of a degree.

[11] W. Wright, "USAF Seeking Improved Reconnaissance Aids to Conduct Demanding New Mission," *Aviation Week,* **80**, 88 (April 6, 1964). Shows a thermal image of Dallas' Love Field. The hot exhaust plumes trailing a taxiing jet transport are clearly visible, as are runways, parked aircraft, terminal buildings, and a buried pipeline.

[12] J. Fusca, "Space Surveillance", *Space/Aeronautics,* **41**, 92 (June 1964). A survey of the possibilities of satellite-borne reconnaissance and surveillance systems. Infrared systems are expected to play a major role and this article attempts to explain why.

[13] "Industry Observer," *Aviation Week,* **81**, 13 (November 2, 1964). Announces a program to study multiple-band infrared reconnaissance over the 1 to 20 μ region for the detection of underground missile silos, infantry weapons, and other targets of significance.

[14] C. Brownlow, "Vietnam Spurs Tactical, Hardware Shifts," *Aviation Week,* **82**, 18 (May 3, 1965). Describes the South Vietnam rain forests with trees that are 120 ft high and densely foliaged. Because traditional methods of reconnaissance are often ineffective, infrared sensors are being studied as a possible solution.

[15] F. Leary, "Search for Subs," *Space/Aeronautics*, **44**, 58 (September 1965). An excellent tutorial treatment of the problems of detecting, locating, and classifying submarines. Principal emphasis is on sonar methods. Among the nonacoustic methods that are mentioned is the possibility of detecting the infrared emission from submarine wakes. It is claimed that the passage of a submarine through the water raises the local water temperature by about 1°.

[16] C. M. Plattner, "Tactical Raids by B-52's Stun Viet Cong," *Aviation Week*, **83**, 17 (November 29, 1965). Mentions the use of infrared sensors for damage assessment and indicates that the Viet Cong are beginning to develop appropriate countermeasures.

[17] L. Dulberger, "Targeting for Air Attack," *Space/Aeronautics*, **44**, 84 (November 1965). A state-of-the-art article. Infrared techniques are emphasized. Article includes an example of the imagery from an infrared thermal mapper.

[18] C. M. Plattner, "Limited-War Concepts Weighed in Battle," *Aviation Week*, **84**, 42 (January 31, 1966). Discusses the problems of nighttime target acquisition by forward air controllers. Indicates that currently operational infrared systems have no real-time readout capability.

[19] C. M. Plattner, "Mohawk Helps Confirm Army Air Concept," *Aviation Week*, **84**, 70 (February 28, 1966). A detailed study of the use of the OV-1 Mohawk aircraft in Vietnam for armed reconnaissance and infrared and radar surveillance. Infrared surveillance has been one of the primary tools in providing target information for B-52 raids.

[20] B. Kovit, "New Anti-Sub Aircraft," *Space/Aeronautics*, **45**, 58 (February 1966). One of the means currently used for the detection of submarines is an infrared sensor to locate the difference in surface temperature that is due to the upwelling of water heated by the submarine's passage. Such equipment is reported to have the capability of detecting a temperature difference of 0.005°. Future development of this equipment is uncertain because high seas, rain, fog, and haze affect its operation.

[21] C. Brownlow, "Limited War Problems Challenge Industry," *Aviation Week*, **84**, 26 (March 14, 1966). Describes the problems of detecting enemy operations under a jungle canopy. Infrared systems have been tried for this task in Vietnam, but their performance leaves something to be desired. There is hope that infrared sensors may be capable of detecting the heat from the cooking fires of the Viet Cong.

[22] "F-4K Version for RAF will Carry British-Made Reconnaissance Pod," *Aviation Week*, **84**, 82 (June 27, 1966). Describes a line-scanning device suitable for operation in the visible or in the infrared. Self-emitted radiation can be utilized for passive operation, or a narrow beam of light can be used to illuminate the ground in synchronism with a suitable detector. An example of the imagery from this system is given.

[23] "Real-Time Infrared Reconnaissance," *Aviation Week*, **85**, 81 (July 11, 1966). A description of the Army's future real-time, ground-based infrared station that will extend to ground observers a real-time tactical infrared capability. Input is via data link from the Mohawk OV-1C.

[24] "Modified OV-1C's Set for Autumn Viet Duty," *Aviation Week*, **85**, 31 (August 15, 1966). The infrared sensors on the OV-1C are capable of detecting a hibachi cooking fire, a recoilless rifle, or a truck that has been parked for as long as 16 hr. An item in the same magazine, at **85**, 25 (August 1, 1966), notes that modified OV-1C's will carry a closed-cycle cooler for the infrared system.

[25] "Filter Center," *Aviation Week*, **85**, 69 (December 5, 1966). Reports that forward-looking infrared reconnaissance sets are being produced for the USAF F/RF-4 aircraft. A note in the same magazine, at **85**, 83 (December 12, 1966), indicates that there are severe vibration problems with this equipment.

16.5 APPLICATIONS INVOLVING REFLECTED FLUX   497

[26] G. Alexander, "Sensors Developed in Simulated War Area," *Aviation Week*, **85**, 34 (Mid-December 1966). Describes a special test area used to support both development and operational testing of a wide variety of airborne image-forming sensors. Among the targets are villages, trucks, sampans, simulated surface-to-air missile sites, troops, underground bunkers, cooking fires, and other typical Vietnamese ground targets.

[27] "Tactical Strike," *Space/Aeronautics*, **47**, 105 (January 1967). A summary of the U.S. tactical strike experience in Vietnam. States that considerable R & D effort is going into the development of more sensitive infrared equipment to assist in the problems of seeing at night. Such equipment, using film readout, is said to have revealed Viet Cong cooking fires and truck engines.

[28] C. Brownlow, "Needs Outpace Strong Viet Recon Gains," *Aviation Week*, **86**, 19 (March 13, 1967). Describes the AAS-18 thermal mapper and its use in Vietnam for the detection of boats on waterways, cooking fires beneath the jungle canopy, warm engines of supply trucks, and the determination of whether above-ground petroleum storage tanks contain fuel. The sensor scans a 120 deg angle.

[29] F. Leary, "Finding the Enemy," *Space/Aeronautics*, **47**, 92 (April 1967). A survey article that discusses a wide variety of sensors (including infrared) for battlefield reconnaissance and intrusion detection. See also ref. 16.5.1 [35].

[30] J. F. Mason, "Mohawk Proves a Good Scout," *Electronics*, **40**, 147 (May 15, 1967). Contains information about the AN/AAS-14 infrared detecting set and its tactical applications in Vietnam. The equipment uses two detectors and has separate cockpit displays for each. The system provides a permanent film record and also transmits a data-annotated picture to a ground station. The picture on the cockpit display is available for study within 2 min.

[31] "SAT Supplies Infrared Systems," *Aviation Week*, **86**, 320 (May 29, 1967). Indicates that the French are using infrared reconnaissance equipment in reconnaissance drones and in the Mirage 3R aircraft. The data can be transmitted to the ground by radio to give ground commanders a quick-look capability.

[32] "Filter Center," *Aviation Week*, **87**, 85 (August 7, 1967). Discusses the possibility of using a 15-MHz-bandwidth data link to relay, in near-real time, information gathered by airborne infrared surveillance equipment.

[33] B. Miller, "Joint Reconnaissance Data System Pushed," *Aviation Week*, **88**, 65 (February 26, 1968). Describes the Joint Services In-Flight Data Transmission System (Jifdats). Among the infrared sensors (thermal mappers) that will be tied into Jifdats are the AN/AAS-18, 21, and 24.

16.5   APPLICATIONS INVOLVING REFLECTED FLUX

This section covers the various night-viewing devices that use image converter tubes and a source of illumination. Most of the military organizations of the world use some type of image converter equipment. Table 16.2 summarizes the typical characteristics of such equipment.

The principal references on infrared photography and its application to reconnaissance are found in section 19.5.1. The few references given here are concerned with applications peculiar to the military, such as camouflage detection.

TABLE 16.2 TYPICAL CHARACTERISTICS OF NIGHT-VIEWING EQUIPMENT THAT USES INFRARED IMAGE CONVERTERS[a]

Application	Maximum Useful Range (ft)	Power Input to Illuminator (W)
Carbine aiming	300–450	10 to 30
Night driving	150–600	10 to 100
Battlefield surveillance	600–2500	200 to 1500
Tank fire control	2000–6000	1000 to —
Ship surveillance	5000–40000	—
Beacon detection	50,000	—

[a]Data adapted from refs. 16.5.1 [2, 4, 19, 27, 31, and 34].

16.5.1 Applications of Image Converter Tubes

[1] T. C. Lengnick, "Discharge Tube," U.S. Patent No. 1,936,514, November 21, 1933 (filed November 1, 1928). An early image converter tube consisting of a selenium cell and a gaseous discharge lamp. The tube is shown combined with an infrared searchlight to make a ship detector. The description of the use of this tube is a fine example of the misconception held by early workers that infrared can readily penetrate fog, clouds, artificial smoke screens, and so on.

[2] "Night Vision with Electronic Infrared Equipment," *Electronics*, **19**, 192 (June 1946). Describes the German Bildwandler (infrared image converter tube). During World War II this tube was used in three types of infrared equipment: *Nachtfahrgerat* (night driving equipment for tanks), *Zielgerat* (rifle-sighting equipment), and *Muecka* (an experimental system to identify friendly aircraft). The *Nachtfahrgerat* permitted driving under blackout conditions at normal daytime speeds. The roadway was clearly illuminated for 300 ft and most obstacles could be seen at 600 ft. Illumination was provided by 100 or 200 W headlamps fitted with removable infrared-transmitting filters. A simple vibrator power supply furnished 6 to 8 kV for the tube. The *Zielgerat* was used on the Russian front. It is said to have an effective range of 300 ft.

[3] G. A. Morton and L. E. Flory, "Infrared Image Tube," *Electronics*, **19**, 112 (September 1946). Describes the type IP25 image converter tube. This tube was used in practically all of the U.S. World War II infrared viewing equipment. Applications include night driving and flying, signalling, marker light identification, reconnaissance, and gun aiming.

[4] E. Stolzenberger, "Army Removes Night Sight Devices from Secret List," *Broadcast Engrs. J.*, **13**, 9 (September 1946). The sniperscope, when mounted on a 0.30 cal. carbine, allowed accurate firing at distances of 150 to 200 ft.

[5] "Radioactive Infrared Detector," *Electronics*, **19**, 142 (October 1946). Describes the Metascope, a simple device using a phosphor as a detector. The Metascope can be used by paratroops to observe infrared sources marking landing areas, by infantry patrols for identification at night, and for the detection of active sources required for the operation of image converter systems.

16.5.1 APPLICATIONS OF IMAGE CONVERTER TUBES

[6] G. E. Brown, "Military Application of Infrared Viewers," *Proc. Natl. Electronics Conf.*, **2**, 181 (1946). Describes several types of infrared equipment developed during World War II by the Engineer Board of the U.S. Army Corps of Engineers. The Metascope is a simple hand-held viewer that uses an infrared-sensitive phosphor. It was used to detect enemy active infrared systems and to locate infrared beacons placed on landing beaches. The beacon is a tungsten lamp fitted with an infrared-transmitting filter. A coding motor, adjustable for any letter of the alphabet, turned the beacon on and off for identification purposes. The snooperscope and sniperscope are described. An infrared transit was developed that consisted of an infrared viewer mounted on an alidade. With appropriate auxiliary equipment, such as range poles, beacons, and tapes, surveying could be carried out at night without breaching visual security.

[7] G. A. Morton and L. E. Flory, "An Infrared Image Tube and its Military Applications," *RCA Rev.* **7**, 385 (1946). Describes the work that lead to the development of the 1P25 infrared image converter tube. Applications for this tube include telescopes for observing infrared signaling, marker beacons, and battlefield reconnaissance, the snooperscope (Molly) for area surveillance, the Sniperscope (Milly) for carbine firing, and night driving systems. The design of high-voltage power supplies for image-converter tubes is also discussed.

[8] E. R. Blout et al., "Near Infrared Transmitting Filters," *J. Opt. Soc. Am.*, **36**, 460 (1946). Dyed-plastic filters for the sources used with active viewing devices.

[9] J. H. Shenk et al., "Plastic Filters for the Visible and Near Infrared Regions," *J. Opt. Soc. Am.*, **36**, 569 (1946). See the previous references.

[10] E. K. Kaprelian, "Recent and Unusual German Lens Designs," *J. Opt. Soc. Am.*, **37**, 466 (1947). German wartime developments in lenses for use with image converter tubes.

[11] T. H. Pratt, "An Infrared Image Converter Tube," *J. Sci. Instr.*, **24**, 312 (1947).

[12] T. H. Pratt, "The Infra-Red Image Converter Tube," *Electronic Engineering*, **20**, 274, 314 (1948). Begins with an excellent tutorial treatment of image converter tubes. The second part of the article gives a detailed description of the design of an image-converting viewer that was used by the British Navy in the Mediterranean area as early as 1941. It was used to detect homing beacons on parent vessels, to guide the return of attack craft, and to assist preinvasion reconnaissance parties returning to their pickup ships. The RAF used similar viewers, starting in 1942, for the recognition of friendly aircraft. Night fighters were equipped with viewers, and bombers were equipped with a tail-mounted, coded, infrared beacon. In 1944 viewers were fitted to the rear turrets of night bombers to enable rear gunners to differentiate between bombers with forward-shining beacons and German night fighters. The British Army used similar viewers for night driving, thus allowing convoys to move in complete darkness at near-daylight speeds, but without the danger from aerial observation. Medical applications of image converters are discussed in ref. 18.5.1[2].

[13] V. K. Zworykin and E. G. Ramberg, *Photoelectricity*, New York: Wiley, 1949, Chapter 18. Discusses the development and use of infrared communication systems and image converter tubes.

[14] F. Eckart, *Elektronenoptische Bildwandler und Rontgenbildverstarker (Electron-Optical Image Converters and X-Ray Image Intensifiers)*, Leipzig: Johann Barth, 1956.

[15] T. Fujii and H. Kojima, "R-F Power Supply for Infrared Viewers," *Electronics* **30**, 163 (November 1, 1957).

[16] M. W. Klein, "Image Converters and Image Intensifiers for Military and Scientific Use," *Proc. Inst. Radio Engrs.*, **47**, 904 (1959). Describes the improvements that have occurred in image converter tubes since World War II.

[17] C. M. Cade, "Infrared Navigation Aids," *British Communications and Electronics*, **6**, 592 (1959). Describes the use of image converters and beacons for ship docking and aircraft landing systems.

[18] A. Mikhaylov, "Noctovision Devices for Land Armies," *Voyennyye Znaniya*, **35**, 17 (1959) (OTS: 62-32728). Contains a detailed description of night vision devices that use infrared image converter tubes. Applications such as night driving, rifle firing, and fire control are described.

[19] R. Lusar, *Waffen und Geheim-Waffen des 2 Weltkrieges und Ihre Weiterentwicklung*, Munich: J. F. Lechmanns, 1959. A translated version is available: *German Secret Weapons of the Second World War*, New York: Philosophical Library, 1963. Describes German usage of infrared image converters during World War II. When the Allies gained air superiority, image converters allowed the nighttime movement of V-2 weapons across Germany and Holland to their launching ramps. German U-boats were equipped with image converters for night operations. These units had effective ranges up to 3000 m. In 1944 German tanks equipped with image converters were able to score hits against other tanks at ranges up to 3000 m, on approximately two-thirds of their tries.

[20] E. Nanas, "5-Year NATO Demand Seen for IR Units," *Electronic News*, March 27, 1961. Author concludes that NATO will be a major customer for infrared weapon systems over the next five years. Tentative agreement has been reached within NATO on standardization for infrared image converters. U.S. Army has placed a multimillion dollar contract for approximately 5000 transistorized infrared weapon sight systems. The rifle-mounted sighting unit weighs one-third less than earlier sniperscopes. A transistorized power supply provides 16 kV at 1 μamp. A single 1.5V C-cell furnishes all power and has an operating lifetime of 40 hr. The range of the sighting unit is believed to be in excess of 300 ft.

[21] J. F. Menke, "Combination of an Optical and Electronic-Optical System of a Resolving Power Tuned to Each Other with a Subminiature High Voltage Supply," U.S. Patent No. 2,982,861, May 2, 1961. An image converter system that uses Maksutov optics. The $f/0.7$ optics have a focal length of 35 cm and a blur circle that is about 35 μ in diameter. The image converter tube has a resolution of 30 lines per mm. Small, transistorized, high-voltage power supplies are mounted behind the mirror. The entire device, including power supplies, has a length of 85 cm, a diameter of 65 cm, and a weight of 80 kg.

[22] "Viewing Unit Uses Infrared," *Electronic Design*, **9**, 26 (June 21, 1961). Pictures and describes a compact infrared viewing device, called the Metascope, that consists of optics, an image converter tube, battery, solid-state power supply and infrared illuminator. Calling this device a Metascope is an error, since the World War II Metascope used an infrared-sensitive phosphor. For example, see ref. 16.5.1 [6].

[23] "Infrared Binoculars Pass Their Tests," *Machine Des.*, **33**, 12 (July 20, 1961). Shows several pictures of helmet-mounted binoculars fitted with image converter tubes for night driving.

[24] "Troops Use IR Device for Night Operations," *Electronic Industries*, **20**, 211 (September 1961). The device described here is the Metascope; see ref. 16.5.1 [6].

[25] K. Junge et al., *Unsichbares Licht-Nachtschen (Invisible Light-Night Vision)*, Berlin: Deutscher Militarverlag, 1961. Published in East Germany and written for military personnel. The authors do an excellent job in covering the basic physics necessary for an understanding of image converter systems. There is a detailed discussion of how these systems are used in the following tactical missions: reconnaissance, fire control, battlefield communications, amphibious operations, movement of troops and vehicles, and air drops. The authors even have some friendly advice on how the foot soldier can

16.5.1 APPLICATIONS OF IMAGE CONVERTER TUBES

best camouflage himself against detection by such systems. It is claimed that in East Germany there are only 35 to 45 days per year in which fogs reduce visibility below 1000 m. Hence image converter systems for night driving and small arms fire can be used virtually on a year-round basis. The reader is told that there is a longer-wavelength portion of the infrared and that it is used for passive detection systems, but no examples are given.

[26] R. E. Young, "Aircraft Landing Techniques and the Future," *Electronic Eng.*, **33**, 496 (1961). Suggests the use of an image converter mounted in the aircraft and one or more infrared beacons mounted on the ground.

[27] C. M. Cade and C. J. Hart, "Infrared Applications in Navigation," *British Inst. Radio Engrs. J.*, **23**, 477 (1962). Describes a viewer equipped with an image converter tube. With a 1 kW searchlight, small vessels could be seen at a distance of about 1.4 miles. Also describes the use of such viewers as docking aids.

[28] "Battlefield Lit Up by Tank's IR Beam," *Electronic Design*, **12**, 27 (April 27, 1964). Describes a 70-million-candlepower searchlight that is part of a night vision fire control package for Army tanks. A series of filters, controlled from within the tank, makes it possible to use the device as an infrared or as a visible-light illuminator. In the infrared mode the tank commander uses an image converter viewer.

[29] "Soviet Missile Parade Includes Scaled-up Sandal, New Transporter," *Aviation Week*, **81**, 68 (November 30, 1964). One of many similar examples showing the use of blackout headlights, that is, filtered tungsten lamps as illuminators for image converter devices, on Soviet military vehicles.

[30] "Airborne Searchlight," *Aviation Week*, **83**, 20 (November 29, 1965). Describes a dual-mode xenon searchlight originally developed for Army tank use and recently proposed for airborne use. The searchlight can be operated in either an infrared or a visible mode. Usual procedure is to operate in the infrared mode for target search and acquisition and then to switch to the visible mode in order to illuminate the target for further action by ground- or air-support units.

[31] T. Burakowski and A. Sala, *Noktowizja (Night Vision)*, Warsaw: Ministry of National Defense, 1965. One of the best books available on the problems of seeing at night. For a detailed listing of the contents, see ref. 16.0 [24]. The treatment of image converters is excellent. Table 2 shows the characteristics of American night viewers for the observation of terrain at night. Such viewers include the snooperscope with a range of 100 m and a power input to the illuminator of 100 W and a large surveillance viewer with a range of 1000 m and a power input of 1500 W to the illuminator. Table 3 shows German, French, and U.S. viewers for infantry use in firing weapons. (Tables 2 and 3 appear on pages 138 and 144 of the reference.)

[32] O. Penkovskiy, *The Penkovskiy Papers*, Garden City, N.Y.: Doubleday, 1965, p. 142. While discussing the means that Soviet intelligence officers use to signal each other prior to a meeting, Penkovskiy mentions the use of invisible infrared signals produced by a pocket flashlight equipped with a special infrared filter. Such signals can be received with Soviet binoculars, Model B-I-8, that incorporate a special phosphor for converting infrared to visible light.

[33] F. Leary, "Tactical Command and Control," *Space/Aeronautics*, **45**, 94 (April 1966). Among the new U.S. Army combat surveillance devices is a 2 lb hand-held Metascope that can detect infrared beacons at a distance of 2 miles and a helmet-mounted infrared binocular with a detection range of 165 ft.

[34] K. Lajos, *Infravoros Felderites es Alcazas*, (*Infrared Discovery of Camouflage*), Budapest: Zrinyi Military Publishing House, 1966. One of the better books available on night vision devices. The author is a captain in the Hungarian Army Engineer Corps, hence the emphasis on camouflage detection. For the complete contents, see

ref. 16.0 [25]. The book contains an unusually complete listing of infrared image converter devices and data on the reflectance characteristics of paints and clothing materials in the 0.4 to 1.6 μ region.
[35] F. Leary, "Finding the Enemy," *Space/Aeronautics*, **47**, 92 (April 1967). Describes the infrared weaponsight for the M-14 rifle. The sight has a range in excess of 200 yd, a field of view of 8 deg, an angular resolution of 0.2 mrad, and it weighs 5.5 lb. A small power supply furnishes 16 kV from a 1.5 V battery for the image converter tube. The illuminator weighs an additional 5.5 lb. The unit costs more than $2000.
[36] "Compact Searchlights Designed by Xerox for Viet Battle Needs," *Aviation Week*, **86**, 23 (June 26, 1967). Describes the development of a family of compact searchlights for tactical warfare use. They can be operated in a visible mode or in an infrared mode, presumably for use with image converters. The electroformed reflectors can be deformed so as to vary the shape of the beam. Several of these units are pictured in the same magazine at **87**, 71 (July 3, 1967).
[37] G. D. Friedlander, "World War II: Electronics and the U.S. Navy," *IEEE Spectrum*, **4**, 56 (November 1967). The subtitle is "Radar, Sonar, Loran, and Infrared Techniques," but only a small portion of the article pertains to infrared equipment. Among the infrared devices developed by the U.S. Navy during World War II are equipment for the transmission of voice, teletype, and Morse code during periods of enforced radio silence; homing and recognition kits provided underwater demolition teams; and devices to give bearing and range information for ship station keeping. Though not detailed here, the homing and recognition kits contained infrared beacons and Metascopes for finding the beacons.

16.5.2 Infrared Photography

[1] L. E. Bowles, "Infrared Camouflage," *Aero Dig.* **42**, 159 (February 1943). A good discussion of the problems of camouflage against infrared film.
[2] E. I. Stearns, "Infrared Reflectance in Textiles," *Am. Dyestuff Reptr.*, **33**, 131 (March 13, 1944). Discusses problems that arise in the use of textiles for camouflage against infrared film.
[3] E. K. Kaprelian, "Recent and Unusual German Lens Designs," *J. Opt. Soc. Am.*, **37**, 466 (1947). Describes a lens having a focal length of 3 m that was used with infrared film and appropriate filters to photograph English coastal installations from German positions on the French coast.
[4] R. Thoren, "Photo-Interpretation in Military Intelligence," *Photogrammetric Eng.*, **23**, 428 (1952). This is perhaps the best unclassified, English-language discussion of military photointerpretation. Both infrared and camouflage detection films are discussed.
[5] R. J. Mayer, "The Near-Infrared Fluorescence of Green Leaves," *Infrared Phys.* **5**, 7 (1965). Author suggests that this effect could be a factor in the photographic detection of camouflage, because the fluorescence would be indistinguishable from reflected sunlight. See refs. 19.5.1 [15] and [16].
[6] "Soviets May Unveil Advances in ICBM's," *Aviation Week*, **86**, 108 (March 6, 1967). Reports that the Soviets are using infrared photography from reconnaissance satellites to obtain data on U.S. installations.

16.6 APPLICATIONS INVOLVING A COOPERATIVE SOURCE

This section includes terrestrial communications, (active) ranging devices, infrared countermeasures, and command guidance for tactical missiles.

16.6.1 Terrestrial Communications

The use of the near-infrared portion of the spectrum for communications dates back to World War I; such equipment reached a high state of development during World War II. It was, however, limited to rather low information-transmission rates because the sources (usually xenon arcs) could not be modulated at a rate in excess of a few kilohertz. Since it is easy to modulate an injection diode at very high frequencies, there is a renewed interest in near-infrared communication and data transmission systems.

[1] T. W. Case, "Infrared Telegraphy and Telephony," *J. Opt. Soc. Am.*, **6**, 398 (1922). Describes a device capable of sending code signals over a distance of 18 miles.

[2] R. Mechau, "Receiver for Heat Signs," U.S. Patent No. 1,639,411, August 16, 1927 (filed August 13, 1921). Describes a sevice that can be used for detecting ships or infrared signaling units. Discloses the idea of using identical detectors, placed adjacent to each other, so that one detector views the target and a small portion of the background while the other views a small portion of the background adjacent to the target. The detectors are connected so that their outputs oppose each other. Hence background signals are canceled, but there is no effect on the target signal.

[3] H. V. Hayes, "Radiant Energy Signalling," U.S. Patent No. 1,954,204, April 10, 1934. Describes a complete system for signaling over an infrared beam. It consists of a source and chopper, collecting optics, and the Hayes pneumatic detector. Usable range is said to be several miles.

[4] G. Krawinkel and W. Kronjager, "A Transmitting System Using Modulated Infrared Waves," German Patent No. K 148,518, November 11, 1937.

[5] W. M. Hall, "Radiant Energy Receiver," U.S. Patent No. 2,115,578, April 26, 1938. A pneumatic detector for an infrared signaling system. An improvement on the Hayes patent, ref. 16.6.1 [3].

[6] D. G. Hull, "The German Army Speech-on-Light Signalling Apparatus," *Electronic Eng.*, **16**, 185 (1943). Describes a communication system, the *Lichtsprecher*, that was adopted by the German Army in 1935. Its existence was a closely guarded secret until October 1942, when the British captured one of the systems in the battle of Alamein. The maximum effective range was about 5 miles. Atmospheric scintillation and low-frequency flutter caused by air currents near the ground were limiting factors at long ranges. Their effects were minimized by setting the low-frequency cutoff of the receiving amplifier at about 800 Hz. The detector was a Thalofide cell. Switch-selected filters permitted system operation with white, red, or near infrared light. Modulation is accomplished by mechanically varying the contact between two prisms. The upper-frequency limit of the modulator is about 2.2 kHz.

[7] P. M. G. Toulon, "Orientation System," U.S. Patent No. 2,369,622, February 13, 1945. A device to keep the transceivers of an infrared communication system pointed at each other independent of any motions of the ship or aircraft upon which they are mounted. Two beams are used, one for orientation and one for communication. The beams are chopped at different frequencies in order to provide differentiation between the two.

[8] J. M. Fluke and N. E. Porter, "Some Developments in Infrared Communications Components," *Proc. Inst. Radio Engrs.*, **34**, 876 (1946). The components described are intended for use in the 0.8 to 1.2 μ region. Coverage includes tungsten and cesium-vapor lamps, infrared-transmitting filters, lamp power supplies, image-converter tubes, thallous sulfide photocells, and photocell preamplifiers.

[9] W. S. Huxford and J. R. Platt, "Survey of Near Infrared Communication Systems," *J. Opt. Soc. Am.*, **38**, 253 (1948). Limited to voice and code systems. Discusses methods of producing modulated beams, filters to provide military security, and suitable detectors. Contains an excellent bibliography.

[10] H. S. Snyder and J. R. Platt, "Principles of Optical Communication Systems," *J. Opt. Soc. Am.*, **38**, 269 (1948). Discusses the factors that determine the range of an optical communication system.

[11] H. B. Briggs, J. R. Haynes, and W. Shockley, "Infrared Energy Source," U.S. Patent No. 2,683,794, July 13, 1954. Describes an injection diode, that is, one in which infrared radiation is produced by the recombination of electron-hole pairs in semiconductors. Contains an excellent description of the physical principles involved.

[12] R. L. Wallace, Jr., "Valve for Infrared Energy," U.S. Patent No. 2, 692, 950, October 24, 1954. This device utilized controlled-carrier injection in germanium or silicon to vary the transmittance and thus to allow their use as a modulator in an infrared communication system.

[13] P. Aigrain and C. B. a la Guillaume, "Infra-Red Emitting Device," U.S. Patent No. 2,861,165, November 18, 1958. Describes the construction of an injection diode in which the semiconducting material is shaped so that flux losses caused by total internal reflections are minimized.

[14] C. M. Cade, "Infrared Radar, Surveillance, and Communications," *British Communications and Electronics*, **7**, 414, 510 (1960). The author notes that communication systems should not operate in the ultraviolet because the human eyeball, while not sensitive in the ultraviolet, does fluoresce so that an ultraviolet system may not be visually secure.

[15] "Design of Infrared Telephone Achieves a $20 Retail Price," *Electronic Design*, **9**, 17 (January 4, 1961). Device uses a flashlight bulb, an aluminized mylar diaphragm, a 2 in. parabolic mirror, a lead sulfide detector, and a three-stage transistorized amplifier. The mylar diaphragm is used as a reflecting element in the transmitter. The diaphragm is deformed by the speaker's voice and gives about 40 per cent modulation. Ranges of 300 to 500 yd are claimed.

[16] "Honeywell Gun Transmits Sound on Infrared Beam," *Wall Street J.*, February 8, 1961. The same device that is described in ref. 16.6.1[18]. A pair of the units costs about $15,000 but will probably cost less than $1000 when the company gets into mass production.

[17] P. W. Kruse, and L. D. McGlauchlin, "Solid-State Modulators for Infrared Communications," *Electronics*, **34**, 117 (March 10, 1961). The modulator described here is probably the one that is used in the device described in the preceding and the following references.

[18] "Infrared Communications System," *Electronic Equipment Engineering*, **9**, 40 (March 1961). Describes a transmitter/receiver unit for an infrared communication system. Transmitter contains a tungsten lamp, germanium modulator, and optics. Receiver consists of collecting optics and a lead sulfide detector. Some portions of the electronic circuitry are common to both the transmit and receive functions. Units are shaped and aimed like a gun. Hand-held units have a useful range of three miles, and larger units can be used up to 20 miles.

[19] N. C. Beese, "Light Sources for Optical Communication," *Infrared Phys.*, **1**, 5 (1961). Describes several types of modulated light sources that have been used in infrared communication systems. Detailed information is given on the characteristics of cesium-vapor and high-intensity, short-arc xenon lamps.

[20] "Infrared for Aircraft Communication," *Aviation Week*, **77**, 59 (July 23, 1962).

16.6.1 TERRESTRIAL COMMUNICATIONS

A jam-resistant communication system with a range of 100 miles. The system features automatic search, acquisition, and lock-on. The beam width is about 0.1 deg.

[21] N. H. Koch, "Designing Infrared Communication Systems," *Electronic Design*, **10**, 60 (August 2, 1962). The author derives an equation that gives the maximum range of an infrared communication system.

[22] "New Jam-Resistant Communication Technique," *Electronic Industries*, **21**, 205 (September 1962). The system already described in ref. 16.6.1 [20].

[23] C. M. Cade and C. J. Hart, "Infrared Applications in Navigation," *British Inst. Radio Engrs. J.*, **23**, 477 (1962). Mentions a narrow-beam infrared communications device that could be used at distances up to 1 mile when hand-held and up to 5 miles when mounted on a tripod.

[24] T. S. Moss, "Methods of Modulating Infrared Beams," *Infrared Phys.*, **2**, 129 (1962). Various methods of modulating infrared radiation are discussed, with emphasis on systems that can operate at very high frequencies, that is, up to 10^{10} Hz.

[25] R. J. Keyes et al., "Modulated Infrared Diode Spans 30 Miles," *Electronics*, **36**, 38 (April 5, 1963). System utilizes a modulated gallium arsenide injection diode that radiates at 0.84 μ. The $f/1.25$ transmitting optics have a diameter of 5 in. and produce a beam width of 2 mrad.

[26] "IR Communication Between 2 Planes to Undergo Tests," *Electronic Design*, **11**, 26 (April 26, 1963). Indicates that the system described in refs. 16.6.1 [20] and [22] is about to undergo flight testing.

[27] E. J. Chatterton, "Optical Communications Employing Infrared Emitting Diodes and FM Techniques," *Proc. IEEE*, **51**, 612 (1963).

[28] J. K. Buckley, "Xenon Arc Lamps for Modulation," *Illum. Engr.*, **58**, 365, (1963). Discusses and illustrates the parameters that dictate the design of an electronically modulated xenon arc lamp.

[29] "Corner Reflector Allows Light-Beam Voice Messages," *Electronic Design*, **12**, 22 (June 8, 1964). A unique one-way communication system that uses a light source and collimating optics, optics for collecting the reflected beam, a special corner reflector for modulating and returning the beam, a detector, and a transistorized audio amplifier. A corner reflector has three mutually perpendicular reflecting surfaces and has the characteristic that an incident beam is returned in a direction exactly opposite to that from which it arrived. In the corner reflector used here, one reflecting surface is a thin aluminized membrane that is displaced by the voice and thus modulates the beam. While the system described here works in the visible, the principle is equally applicable to the infrared.

[30] J. P. Gordon, "Optical Communication," *Intl. Sci. Tech.*, August 1965, p. 60. A good tutorial treatment.

[31] M. King, "Some Parameters of a Laser-Type Beyond the Horizon Communication Link," *Proc. IEEE*, **53**, 137 (1965). Expressions for the information capacity of a scatter-propagation link are derived from consideration of the scattering properties of clouds and hazes and from other system parameters.

[32] S. Kainer, "Laser Beam Security," *Proc. IEEE*, **53**, 1752 (1965). Most designers assume that an optical communication link is secure when the diameter of the beam is approximately the same as the diameter of the collecting mirrors. The author points out that this assumption is true only for propagation in a vacuum. Within the atmosphere, scattering processes make the beam visible for long distances out of the direct line of sight. Author's calculations show that for typical atmospheric conditions the beam can be intercepted within a cone of directions whose half-angle of 10 deg. in the forward and backward direction.

[33] E. J. Chatterton, "Semiconductor Laser Communications Through Multiple-Scatter Paths," *Proc. IEEE*, **53**, 2114 (1965).

[34] A. T. Davies, "A Technique for the Transmission of Digital Information over Short Distances Using Infrared Radiation," *Radio and Electronic Engr.*, **29**, 369 (1965). Uses a modulated injection diode as a source.

[35] "Russians Show Laser Telephone," *Aviation Week*, **84**, 77 (January 3, 1966). A brief item mentioning a laser communication system in operation in Moscow over a path several kilometers long.

[36] D. Buhl and L. Spinazze, "A Practical Infrared TV System," *Electronic Industries*, **25**, 48 (1966). A misleading title, since the article describes an infrared telemetry link utilizing a modulated injection diode.

[37] B. Cooper, "Optical Communications in the Earth's Atmosphere," *IEEE Spectrum*, **3**, 83 (July 1966). A good tutorial treatment.

[38] T. E. Walsh, "Gallium-Arsenide Electro-Optic Modulators," *RCA Rev.*, **27**, 323 (1966) Describes practical electro-optical modulators for use from 0.9 to 16 μ.

[39] I. P. Kaminow and E. H. Turner, "Electrooptic Light Modulators," *Appl. Opt.*, **5**, 1612 (1966); *Proc. IEEE*, **54**, 1374 (1966). A survey of electrooptic light modulation by means of the Pockels and Kerr effects in crystals. All available data on electrooptic materials are tabulated. Design considerations and operating principles are outlined.

[40] D. W. Peters, "Infrared Modulator Utilizing Field-Induced Free Carrier Absorption," *Appl. Opt.*, **6**, 1033 (1967). Contains a theoretical analysis and experimental measurements of a modulator that uses field-induced free carrier absorption in germanium.

16.6.2 Ranging

[1] D. D. Withem, "Distance Measuring Device," U.S. Patent No. 2,216,716, October 1, 1940. A device for measuring the altitude of an aircraft. It consists of a source mounted near the tail of the aircraft and pointed at the ground and an angle-measuring sensor mounted near the nose of the aircraft.

[2] E. G. H. Mobsby, "Detection of Objects by Electromagnetic Rays," U.S. Patent No. 2,237,193, April 1, 1941. Describes a rangefinding device that can be used as an aircraft altimeter. A variable-speed chopper is common to both the transmitter and the receiver. The speed of the chopper is adjusted so that, during the time required for a chopped pulse to reach a target and be reflected back to the receiver, the chopper rotates just far enough to move a tooth into the position formerly occupied by a slot. Range can be calculated from chopper speed and the known system parameters. An infrared-filtered tungsten source is used to provide visual security.

[3] W. L. Hyde, "Optical Height Finders," U.S. Patent No. 3,000,256, September 19, 1961. An aircraft altimeter for use at altitudes of 200 ft or less. Consists of a light source at the tail of an aircraft, pointed toward the ground, and an angle-measuring sensor near the nose of the aircraft.

[4] M. R. Klop, "Laser Ranging," *Space/Aeronautics*, **39**, 89 (April 1963). A good tutorial article.

[5] "Infrared-Aimed Laser Radar Prototyped for Army," *Electronic Design*, **14**, 50 (January 4, 1966). Describes and pictures a laser ranging device that is aimed by an infrared tracker. The laser is a continuous-wave helium-neon type emitting at 0.63 μ. Demonstrated range resolution is less than 1.5 ft at a distance of 7 miles. For a description of the tracker, see ref. 16.1.2[21].

[6] "Multi-Function Diode Laser Radar Concept," *Aviation Week*, **84**, 86 (May 2, 1966). Describes a feasibility study of an injection diode ranging system that can also perform altimetry, communications, and target illumination functions.

16.6.3 Infrared Countermeasures

[1] "Tactical Laser Weapons," *Aviation Week*, **79** 95 (November 18, 1963). States that laser beams can damage or destroy infrared sensors and suggests their use as a countermeasure against infrared-guided missiles.

[2] "New Penetration Aid Being Tested," *Missiles and Rockets*, **15**, 9 (October 19, 1964). Describes the use of infrared flares ejected from a reentering nose cone so as to provide one or more additional objects having approximately the same radiation signature as the warhead.

[3] O. W. Fix, "Infrared Countermeasure System Using Radial Shutter Array for Light Modulation," U.S. Patent No. 3, 169, 165, February 9, 1965. Describes a device that permits 100 per cent modulation of an infrared source. The modulated source is carried in an aircraft and is intended to be used as a countermeasure against infrared-guided missiles.

[4] B. Miller, "Varied Tactical Uses Developing for Laser," *Aviation Week*, **82**, 39 (May 31, 1965). Describes a program to determine laser damage effects in infrared detectors.

[5] B. Miller, "Tactical Radar Homing Programs Pushed," *Aviation Week*, **84**, 78 (March 21, 1966). Describes an airborne countermeasure system consisting of a receiver that detects the infrared radiation from burning fuels in aircraft and missile engines and a dispenser that drops flares so as to spoof infrared-guided missiles. See also ref. 16.1.1 [40].

[6] L. H. Dulberger, "Advanced Fighter-Attack Aircraft," *Space/Aeronautics*, **45**, 80 (April 1966). One of the problems for these aircaft is the avoidance of infrared-guided missiles. Among the countermeasure techniques under consideration is that of shifting the wavelength of the radiation from turbojets to longer wavelengths where present-day infrared seekers do not respond. Apparently this shift in wavelength would be accomplished by cooling and shielding the exhaust and by redesign of the exhaust nozzle.

[7] B. Miller, "Wide Use of F-111A Mk2 Avionics Seen," *Aviation Week*, **84**, 93 (June 6, 1966). In this weapon system, a countermeasures receiving set initiates countermeasures at the approach of hostile aircraft or missiles. Countermeasures include flares to spoof infrared missiles.

[8] L. H. Dulberger, "Advanced Strategic Bombers," *Space/Aeronautics*, **45**, 62 (June 1966). Discusses techniques for countering infrared sensors. These include shielding of turbojet exhausts, designing engine nozzles to shift heat wavelengths to longer frequencies, and deploying pyrotechnic flares.

[9] F. Leary, "Tactical Aircraft Survivability," *Space/Aeronautics*, **47**, 68 (June 1967). Describes the possible use of wing-tip-mounted ramjets fired at random so as to confuse incoming infrared-guided missiles. Also describes efforts to suppress some of the infrared radiation from aircraft engines. A modification kit for the UH-1 helicopter adds 43 lb and causes a power degradation of 4.3 per cent. It consists of a means for bleeding excess compressor air and mixing it with the hot exhaust gases from the turbine. For thrusting engines, a telescoping shield has been suggested as a means of limiting the radiation from the tailpipe to a very narrow angle.

[10] "Filter Center," *Aviation Week*, **88**, 65 (March 4, 1968). Mentions a modified version of the ALE-29, a dispenser which ejects chaff and infrared flares for aircraft protection.

[11] B. Miller, "Major Role in Electronic Air War Earned by Radar Chaff," *Aviation Week*, **88**, 55 (March 11, 1968). Shows a sketch of the AN/ALE-32 dispenser that can eject either infrared flares or chaff.

16.6.4 Command Guidance

Many tactical missiles intended for antitank or antiaircraft missions use infrared as an element of their command guidance. Most of these missile systems consist of a missile, one or more tail-mounted flares, and a dual optical sight. When the missile has been fired, the infrared tracker automatically tracks the flares. Differences between the line of sight of the optics and that of the infrared tracker are converted into steering-correction signals that are transmitted over the command link to the missile. The command link may consist of trailing wires, radio, or a modulated infrared beam. Some of the characteristics of missiles of this type are shown in Table 16.3.

[1] V. P. Moskovskii and P. T. Astashenkov, eds., *Modern War Technology*, Moscow: Military Publishing House of the USSR, Ministry of Defense, 1956 (OTS:MCL-110/1), Chapter 5. Describes the use of tail-mounted flares or beacons on missiles in order to facilitate tracking of the missile by optical or infrared equipment. See ref. 16.0[17] for a more complete listing of the contents of this book.

[2] "Infrared—The New Horizon," *Electronic Equipment Eng.*, **9**, 44 (May 1961). In air-to-air refueling operations the pilot of the refueling plane must guide the tip of his probe into the drogue unit that is trailed behind the tanker plane. This article describes a system for controlling airfoils on the drogue so as to direct the drogue toward an infrared light source mounted on the probe. The heart of the device is a four-quadrant mosaic detector that generates the steering signals for the drogue. The sensor, electronics, and servo control are mounted on the drogue and are powered by an alternator and an air-driven turbine.

[3] H. A. Wagner, "Optical Sighting Device," U.S. Patent No. 2,994,245, August 1, 1961 (filed June 28, 1955). Describes a tracking device for a missile command-guidance

TABLE 16.3 TACTICAL MISSILES THAT ARE THOUGHT TO USE INFRARED AS AN ELEMENT OF THEIR COMMAND GUIDANCE[a]

Country	Missile	Mission	Status	Range, statute miles	
				Min.	Max.
France	AS 20	Air to surface	Production	—	4.3
	AS 30	Air to surface	Production	—	7.5
	SS 11 B-1	Surface to surface	Production	0.3	
	SS 11 TCA(Harpon)	Surface to surface	Production	0.3	
	SS 12	Surface to surface	Production	—	3.7
Great Britain	Rapier	Surface to air	—	—	1.9
International	HOT	Surface to surface	Development	0.05	2.4
	Milan	Surface to surface	Development	0.05	1.2
	Roland	Surface to air	Development	0.3	3.7
United States	AGM-22A(SS 11)	Surface to surface	Production	0.3	1.9
	Shillelagh	Battlefield support	Production	—	10
	Teton	Air to surface	Development	—	—
	Dragon	Surface to surface	Development	—	—

[a]Data adapted from refs. 16.1.4[47, 49, 55] and 16.6.4[14, 15, 16].

16.6.4 COMMAND GUIDANCE

system. The key element in the system is a monochromatic flare on the missile to provide a spectrally unique target for tracking. The tracker includes a plane mirror on an electromagnetically actuated torsion mount.

[4] "Infrared Tracker," *Aviation Week*, **77**, 102 (December 3, 1962). Describes an infrared tracker for guiding a short-range missile to its target. Missile carries a flashing (modulated) infrared source, and the tracker is designed to respond only to that modulation frequency. A production version of the tracker would weigh 15 lb, have a 20 deg acquisition field and a 1 deg tracking field, a tracking accuracy of 0.1 mrad, and would require 4 W.

[5] R. W. Astheimer, "Photosensitive Image Motion Detector," U.S. Patent No. 3,090,869, May 21, 1963. A device that will automatically track a short-range wire-trailing weapon and generate error signals that can be used to control the weapon.

[6] J. A. Houston, "Electro-Optics, Modern War Game Munitions," *Electronics*, **37**, 27 (March 6, 1964). An invisible-light version of the typical shooting gallery. Rifles are equipped with a near-infrared source (filtered tungsten lamp) to produce a collimated beam. Targets carry a near-infrared detector for scoring "hits."

[7] M. Getler, "Europeans Show New Tactical Missiles," *Missiles and Rockets*, **17**, 16 (June 21, 1965). Photos and limited description of the HOT and Milan antitank missiles and the Roland surface-to-air missile. Each uses infrared for guidance or tracking. Development is a joint French-German effort. For an additional description of HOT and Milan, see *Aviation Week*, **81**, 19 (December 7, 1964), and the following reference.

[8] W. C. Wetmore, "Europeans Pushing Joint Missile Efforts," *Aviation Week*, **83**, 69 (July 5, 1965). Describes Roland, a ground-to-air antiaircraft missile equipped with infrared tracking flares, Milan, an antitank missile that uses an infrared tracker in its control system, and HOT, an antitank missile with optical-infrared guidance.

[9] "Semi-Active Infrared Helicopter Missile Guidance," *Aviation Week*, **84**, 77 (January 3, 1966). Describes the forthcoming evaluation of a semiactive infrared guidance system that lets helicopter-launched missiles follow an infrared beam pointed at the target.

[10] "Industry Observer," *Aviation Week*, **84**, 23 (February 14, 1966). A note concerning a new infrared beam-riding, air-to-surface missile called the Teton.

[11] "Shillelagh Guidance Units in Production," *Aviation Week*, **84**, 106 (June 13, 1966). See also the error-correcting letter at **85**, 134 (July 18, 1966). Shows several pictures of various portions of the guidance system for the Shillelagh guided missile. The missle uses an infrared command-guidance link.

[12] J. W. Sarnow, "Infrared Emitters," in F. B. Pollard and J. H. Arnold, Jr., eds., *Aerospace Ordnance Hardware*, Englewood Cliffs, N.J.: Prentice-Hall, 1966, Chapter 9. A survey of infrared-emitting flares and pyrotechnic devices. Applications include infrared augmentation devices for ground and air targets and tracking flares for missiles, drones, and sounding rockets. A table of commercially available flares shows that such flares have effective radiant intensities (1.8 to 2.7 μ) ranging from 150 to 1000 W sr^{-1}.

[13] F. Leary, "The Army's Surface Missiles," *Space/Aeronautics*, **47**, 70 (March 1967). Brief descriptions of the various missiles that use some form of infrared tracking.

[14] "SAT Supplies Infrared Systems," *Aviation Week*, **86**, 320 (May 29, 1967). States that the French company SAT is developing the infrared-optical tracker for the Roland air-defense missile and the Milan and HOT antitank missiles. The article also indicates that an infrared link has been used instead of a radio link for the remote control of drones and missiles.

[15] "Europe Pushing Low-Altitude Anti-Aircraft Capabilities," *Aviation Week*, **87**, 48 (July 3, 1967). A surprisingly detailed discussion of missiles that use an automatic

infrared tracker to generate steering commands for transmission to the missile. Among the missiles that use such a guidance system are the SS. 11 B-1, Harpon (SS. 11 TCA), Milan, HOT, Roland, and Rapier, Several of the photographs in the article show details of the missile-mounted flares, but the infrared tracker is not visible.

[16] "Army Studies Funding Request for Dragon Missile Production," *Aviation Week*, **87**, 81 (August 21, 1967). Currently in the development stage, Dragon is an antitank medium assault weapon. The guidance system consists of a combined telescopic sight and infrared tracker, an infrared source on the missile, and a trailing wire for transmitting guidance commands to the missile.

[17] "Army to use Lasers in Simulated Combat," *Aviation Week*, **88**, 83 (January 29, 1968). The U.S. Army is planning to use gallium-arsenide injection-diode lasers mounted on small arms so that maneuvering troops can fire weapons at each other under simulated combat conditions. The laser beam is coded to identify the party firing the weapon. Participants will wear omnidirectional receivers which will verify, decode, and store all hits. This information will eventually be transmitted to a computer for scoring purposes. An item in the same magazine indicates that a contract has been let for the development of this system: **89**, 59 (September 2, 1968).

[18] "Nord Missiles Arm Libyan Patrol Boat," *Aviation Week*, **89**, 17 (July 1, 1968). Shows a test firing of an SS11 or an SS12 wire-guided missile from a Libyan Patrol Boat. This marks a new use for this type of missile. The infrared controller is not visible in this photograph.

17
Industrial Applications of Infrared Techniques

It has been demonstrated time and again that infrared techniques offer useful solutions to a variety of industrial problems. Although the potential value of this market is high (from $25 to $50 million in 1968), few companies have made more than a minimal market penetration. One observes two kinds of companies either established or trying to become established in this market.

On the one hand are the old-line process-control instrument companies that have offered simple infrared radiometers for several decades. These companies have developed a sensitivity to industry's needs and serve them with a knowledgeable sales force, an efficient distribution network, and nationwide repair facilities. Unfortunately these companies have not been able to acquire much capability in the newer infrared techniques.

On the other hand, one finds the companies that have prospered by serving the military market and have, as a result, acquired a great deal of proficiency in the latest infrared techniques. Some of these companies have tried to diversify by moving into the industrial market with products that are, in essence, merely warmed-over military designs. These companies have found that it is virtually impossible to penetrate the industrial market with an engineering and sales force that is geared to the military market, a limited distribution network, and inadequate repair facilities. It appears that several of these companies have profited from this experience and are ready for another try with a product line designed specifically for the industrial market and supported by adequate sales, distribution, and repair functions.

17.1 SEARCH, TRACK, AND RANGING APPLICATIONS

Infrared techniques appear to be attractive for the early detection of forest fires. A simple scanner placed on a high point in the local terrain could continuously scan many square miles of forest. The time required for one complete azimuthal scan need only be of the order of tens of minutes. The use of integrated circuitry would permit operation from batteries that are recharged by solar cells. A companion transistorized transmitter would alert a central monitoring facility if a fire was discovered. Since infrared-guided missiles find a fire to be a very attractive target, it is not surprising that some thought has been given to having such missiles carry a fire-suppressing payload.

Infrared equipment is used for the early detection and rapid extinguishing of fires in aircraft fuel tanks. Ruggedized lead sulfide detectors mounted inside each tank detect any flame and activate an extinguishing system. The detonation time of fuel-air mixtures is from 3 to 8 msec, and typical infrared-actuated extinguishing systems respond in 1 to 3 msec.

17.1.1 Search Systems

[1] G. A. Barker, "Apparatus for Detecting Forest Fires," U.S. Patent No. 1,959,702, May 22, 1934. The preferred form of the infrared search set is a scanner, but a multidetector nonscanning device is also described.

[2] B. N. Watts and J. R. Howells, "Apparatus for the Detection of Infrared Radiation," U.S. Patent No. 2,794,926, June 4, 1957. A fire-detection device that uses two lead sulfide detectors. The detectors are oppositely biased so that changes in the ambient temperature do not generate an output signal. Only one detector is exposed to the scene, so that a fire within its field of view generates an output signal.

[3] "Infrared Missiles May Fight Forest Fires," *Electronics*, **32**, 69 (March 30, 1959). Proposes a fire-fighting system consisting of an infrared search set that would automatically trigger an infrared-guided missile loaded with fire-suppressing chemicals. The search set would be located on a hilltop, have a detection range of 20 miles, and would transmit an alarm over a solar-powered transistorized transmitter. The article states that tests show that a Sidewinder missile will unerringly seek out a fire.

[4] W. E. Osborne, "Farewell to Free Time on City Parking Meters," *Electronics*, **37**, 72 (December 28, 1964). Suggests a simple infrared sensor to detect the hot exhaust pipe of a car as it is being parked and to automatically start the timing mechanism of a parking meter.

[5] "Explosion Suppression System Planned for TWA's Jet Aircraft," *Aviation Week*, **83**, 39 (August 9, 1965). Describes a system to prevent the ignition of fuel vapor at surge-tank vents by lightning strikes or static electricity. This system depends on the fact that the ignition of volatile fuel mixtures requires from 3 to 8 msec but detectors and their circuitry can respond much more rapidly.

[6] R. A. Wilson, "The Remote Surveillance of Forest Fires," *Appl. Opt.*, **4**, 899 (1965). Contains very little applications information. The author of the following reference claims that the calculation of NEΔT is in error.

[7] E. W. Bivans, "Calculation of the Noise Equivalent Temperature of a Radiometer," *Appl. Opt.*, **5**, 1857 (1966). Author claims that the equations for NEΔT in the previous reference are in error and they should be divided by 2.5.

[8] I. Burleigh, "Radar May Track Criminals, Officer Says," *Los Angeles Times*, March 24, 1967. Discusses the use of infrared equipment to aid police officers in locating hidden suspects.

[9] F. Leary, "Tactical Aircraft Survivability," *Space/Aeronautics*, **47**, 68 (June 1967). Describes several fast-response fire suppression systems for use in aircraft fuel tanks. A fuel-air explosion can be detected in 100 μsec and the tank flooded with a suppressant in less than 5 msec.
[10] W. E. Osborne, "Infrared Detection of Humans," *Electronic Communicator*, **2**, 18 (July/August 1967). Concludes that an infrared device with 2 in. diameter optics has a theoretical maximum detection range of 3000 ft for an average human being. This article omits so many system parameters that it is impossible to tell whether its claims are at all valid.
[11] S. A. Yefsky, ed., *Law Enforcement Science and Technology*, Washington, D.C.: Thompson Book Co., 1967. The Law Enforcement Assistance Act of 1965 provides federal funding for research and development of new techniques for law enforcement. Approximately $15 million was spent during fiscal years 1966 and 1967. This book appears to be a prime reference for anyone hoping to enter this market. A number of its articles describe the use of infrared systems for surveillance and detection of criminals and their activities.

17.2 RADIOMETRIC APPLICATIONS

From the standpoint of the industrial user, the most important attribute of infrared equipment is its ability to measure temperature from a distance without physical contact. Equation 13-55 shows that when the object of interest fills the field of view, it is not necessary to know its distance or its area in order to measure its temperature radiometrically. Temperatures determined in this way are called equivalent blackbody or apparent radiation temperatures; it is well to note again that they depend on the following assumptions: (1) The object radiates as a blackbody. (2) The object fills the field of view of the radiometer. (3) Atmospheric absorption along the line of sight is negligible. (4) The responsivity of the detector is independent of wavelength. If any of these conditions are not satisfied, the corrections indicated in section 13.5 must be applied.

17.2.1 Measurement of Temperature

A problem that has long plagued the railroads, and one that has been solved by infrared radiometry, is the detection of overheated journal bearings, commonly called hot boxes, on railroad cars. A hot box, if neglected, can cause a wheel to lock, shearing the axle and derailing the car. An overheated bearing will ignite the packing in the journal box and cause smoke. Visual observation of this smoke, the traditional method employed for hot-box detection, is obviously not very satisfactory. A radiometer mounted beside the track, so as to view each journal box as the train passes, can easily detect overheating and signal an alarm. Unfortunately the process is complicated by the fact that both roller and sleeve-type bearings are used on railroad cars and the roller bearings operate at a considerably higher temperature. Sleeve-type bearings normally operate at a temperature of less than 5°C above ambient, and their operation becomes critical when their temperature reaches 15 to 20°C above ambient. Roller bearings normally operate at a temperature that is about 20°C above ambient, and their operation becomes critical when their temperature reaches 60 to 80°C above ambient. Hence the mere indication of a temperature considerably above ambient is ambiguous. Since it can be assumed that the bearings on opposite ends of the same axle are of a similar type, one means of resolving this ambiguity is to mount radiometers on both sides of the track in order to view both bearings simultaneously. Simple logic circuitry is used to reject all observations in which both bearings are hot.

Scanning radiometers (as well as thermal imaging devices) are finding increasing application in the study of the thermal conditions in electronic circuitry. The introduction of integrated circuits has fostered the development of radiometric microscopes that can sense temperature differences of a fraction of a degree between areas that are separated by less than 0.001 in. Such equipment does not disturb the electrical and thermal conditions in the circuit and automatically accounts for the effects of heat sinks, heat sources, and unusual geometrical arrangements. There is increasing evidence that failure-prone components can be spotted by radiometric methods and replaced before their failure can cause damage.

In a lighter vein, perhaps, is the sex differentiator displayed a few summers ago at one of the major electronic shows. The subject stepped onto a platform of a properly impressive machine and after a brief interval a light appeared to signify either male or female. The secret was a simple radiometer, using a cooled lead telluride detector, mounted about a foot above the platform and arranged so as to view the subject's legs. With such an arrangement, a woman's bare legs give a larger signal than that from a man's legs that are covered by his trousers. Such a scheme appears to be beautiful in its simplicity, but the poor booth attendants lived in mortal fear that some high-ranking military officer would appear attired in summer shorts.

[1] J. T. Nichols, "Temperature Measuring," U.S. Patent No. 2,008,793, July 23, 1935. A radiometer for use in steel-rolling mills to indicate the uniformity with which the strip is heated. It can also be used to control the operation of a flying shear.

[2] W. H. Gille, "Temperature Measuring Apparatus," U.S. Patent No. 2,096,323, October 19, 1937. Uses a reflective chopper so that the flux from the object can be compared with the flux from an internal tungsten lamp.

[3] J. Strong, "A New Radiation Pyromter," *J. Opt. Soc. Am.*, **29**, 520 (1939). Uses a thermopile detector and a galvanometer readout. Response is limited to a narrow band around 8.8 μ by a restrahlen (selective reflection) effect.

17.2.1 MEASUREMENT OF TEMPERATURE

[4] H. W. Russell, "Optical Pyrometer," U.S. Patent No. 2,237,713, April 8, 1941. A two-color radiometer. Equations describing the theory of two-color devices are derived.

[5] T. R. Harrison, "Industrial Use of Radiation Pyrometers under Non-Blackbody Conditions," *J. Opt. Soc. Am.*, **55**, 708 (1945). Discusses methods of correcting for emissivity so that the temperature of gray bodies can be measured radiometrically.

[6] R. C. Parker and P. R. Marshall, "The Measurement of the Temperature of Sliding Surfaces with Particular Reference to Railway Brake Blocks," *Proc. Inst. Mech. Engrs.*, **158**, 209 (1948). Describes the use of a high-speed radiometer for analyzing transient temperatures in brake shoes. A lead sulfide detector is used.

[7] W. S. Gorrill, "Industrial High Speed Infrared Pyrometer," *Electronics*, **22**, 112 (March 1949). Describes the use of an infrared radiometer to measure the temperature of a soldered seam on a tin can. The cans move past the radiometer at rates as high as 6 cans per second. The radiometer uses a thermistor bolometer chopped at a frequency of 300 Hz. Temperatures from 205 to 370°C can be measured with an accuracy of ±3°C.

[8] J. R. Leslie and J. R. Wait, "Detection of Overheated Transmission Line Joints by Means of a Bolometer," *Trans. Am. Inst. Elec. Engrs.*, **68**, 64 (1949). Describes the use of an infrared radiometer to detect hot joints in an overhead power-transmission line. The heating is caused by a gradual increase in the resistance of the joint. Such heating can cause early failure of the joint and the consequent necessity of shutting down the line for repairs. For an interesting discussion of an even simpler detection technique, see ref. 17.2.1 [12].

[9] J. C. Mouzon and C. A. Dyer, "Low Temperature Radiation Pyrometry in Industry," *J. Opt. Soc. Am.*, **39**, 203 (1949). Describes a radiometer for the measurement of surface temperatures from 40 to 200°C and some of the problems in its use.

[10] A. F. Gibson, "A Two-Colour Infrared Radiation Pyrometer," *J. Sci. Inst.*, **28**, 153 (1951). Uses a lead sulfide detector and covers the range of 200 to 1250°C with an accuracy of ±3°C.

[11] R. A. Huggins, I. B. Roll, and H. Udin, "Measurement of Transient Surface Temperatures," *Rev. Sci. Instr.*, **23**, 467 (1952). Discusses the measurement of temperatures produced during spot welding. The radiometer that is described uses a lead sulfide detector, and it can measure temperatures as low as 100°C.

[12] P. J. Sandiford, "Simple Method for Detecting Hot Joints on Electrical Transmission lines," *Rev. Sci. Instr.*, **23**, 644 (1952). Discusses the earlier use of a radiometer (ref. 17.2.1 [8]) to detect hot joints and then describes a method that makes use of the irregular refraction of light passing through the heated air just above a defective joint. An observer uses a telescope to observe an object in the background through the heated air and thus to detect the shimmering or irregular refraction. With a 10x telescope mounted on a tripod and a suitable background object, a joint having a temperature 50°C above ambient is easily seen. Scratch one promising application for infrared radiometry!

[13] E. C. Pyatt, "A Brightness Temperature Pyrometer Using a Photoconductive Cell," *J. Sci. Instr.*, **29**, 125 (1952). Covers the range from 150 to 500°C. An improved version is described in the following reference.

[14] E. C. Pyatt, "Simplified Brightness Temperature Pyrometer using Photoconductive Cell," *J. Sci. Instr.*, **29**, 414 (1952). Describes a radiometer to cover the temperature range of 300 to 720°C with a ±5°C reproducibility. A lead sulfide detector is used to compare the flux from the unknown with that from a built-in tungsten lamp.

[15] E. M. Wormser and R. D. Hudson, Jr., "Infrared Radiometer for Radiation and Temperature Measurements of Remote Objects," Symposium on Molecular Structure

and Spectroscopy, Ohio State University, Columbus, Ohio, June 14–18, 1954, paper 05. Original announcement of what is still one of the most successful commercially available radiometers. It uses two thermistor bolometers, one covering 0.5 to 30 μ and the other being equipped with a filter slide to permit examining any selected portion of the spectrum. The field of view of the radiometer can be from 0.1 to 0.5 deg and is determined by choice of size of detector. Among the applications discussed is that of monitoring the temperature of paper as it passes through a gas-fired dryer on a high-speed printing press.

[16] R. B. Sims and J. A. Place, "A Surface Scanning Pyrometer," *J. Sci. Instr.*, **31**, 293 (1954). Uses a Nipkow disk to generate a spiral scan. With a photomultiplier tube the temperature range is 900 to 1150°C. Probable error in temperature is less than 0.5 per cent.

[17] R. A. Bracewell, "An Infrared Radiation Pyrometer for Measuring the Temperature of Brake Linings," *Electronic Engr.*, **27**, 238 (1955). A small radiometer is described that uses a lead sulfide detector. Response time of the radiometer is 40 μsec. The brake linings are viewed through a slot cut in the brake drum. Lining temperatures as high as 1000°C have been observed.

[18] J. D. Harmer and B. N. Watts, "An Infrared Radiation Pyrometer," *J. Sci. Inst.*, **32**, 167 (1955). A general-purpose radiometer using a lead sulfide detector to cover the range of 200 to 1000°C with an accuracy of ±5°C and a response time of less than 0.1 sec.

[19] P. C. Patin and R. L. Simonin, "Process and Apparatus for Detecting the Overheating of Railway-Vehicle Axle Boxes along the Line," French Patent No. 1,133,052, November 12, 1956.

[20] E. M. Wormser, "Total-Radiation Pyrometer," U.S. Patent No. 2,798,962, July 9, 1957 (filed June 30, 1951). Describes several versions of a commercially available radiometer. In one of the versions a reflective chopper permits comparison of the flux from the unknown object with that from a built-in fixed-temperature blackbody.

[21] A. A. Obermaier, "Radiation Pyrometers," U.S. Patent No. 2,800,023, July 23, 1957. Two-color radiometers.

[22] S. N. Howell, "Hot-Box Detector," U.S. Patent No. 2,829,267, April 1, 1958. The radiometer response is gated by an acoustic sensor that controls the radiometer's signal processing circuitry.

[23] R. K. Orthuber and C. V. Stanley, "Apparatus and Method for Detecting Overheated Journal Boxes," U.S. Patent No. 2,856,539, October 14, 1958. A trackside radiometer equipped with a shutter that is opened by a wheel passing over a switch on the track. The shutter is open for only 5 msec so that the radiometer sees only the journal housing and no other parts of the train.

[24] H. D. Warshaw, "Infrared Detection Apparatus," U.S. Patent No. 2,856,540, October 14, 1958. A simple hot-box detector. The bolometer detector is connected as one arm of a dc bridge circuit so as to eliminate slow drifts in output caused by changes in the ambient temperature.

[25] G. S. Reichenbach, "Experimental Measurement of Metal Cutting Temperature Distributions," *Trans. ASME*, **80**, 525 (1958). Author used a radiometer with a lead sulfide detector to measure the shear-plane temperature in a metal chip and the clearance-face temperature of a cutting tool.

[26] S. N. Howell, "Optical Pyrometer," U.S. Patent No. 2,869,369, January 20, 1959. Describes one of the well-known commercially available radiometers.

[27] "Electronic Man-Woman Detector," *Radio-Electronics*, **30**, 53 (February 1959). This is the sex differentiator described in the introductory material at the beginning of this section.

17.2.1 MEASUREMENT OF TEMPERATURE

[28] R. W. Astheimer and E. M. Wormser, "High-Speed Infrared Radiometers," *J. Opt. Soc. Am.*, **49**, 179 (1959). Describes radiometers having either 4 or 8 in. diameter Cassegrain optics and thermistor detectors. Their response is uniform from 0.7 to 35 μ. The NEP is approximately 10^{-10} Wcm^{-2}, which corresponds to a NEΔT of 0.01°C. Both the theory of operation and calibration techniques are discussed.

[29] H. L. Berman and G. F. Warnke, "Use of Infrared Techniques in Industrial Instrumentation," *IRE Trans. Industrial Electronics*, **PGIE-11**, 15 (1959). Discusses the use of radiometers for process and quality control, nondestructive testing, temperature measurement, and control of spot welders.

[30] H. L. Berman and G. F. Warnke, "Using Infrared Techniques in Industrial Processes," *Automatic Control*, **11**, 1 (1959). Coverage is similar to that of the previous reference.

[31] W. Derganc, "Pyrometer," U.S. Patent No. 2,920,485 January 12, 1960. Uses two simultaneously chopped detectors and a synchronous rectifier to give a dc output whose magnitude is proportional to the temperature of the object being viewed.

[32] M. Buchner and F. Loy, "Device for Detecting Abnormal Heating of Axle Boxes of Rolling Stock," French Patent No. 1,232,431, April 25, 1960.

[33] R. W. Astheimer, "Electrical Signal Offsetting Apparatus," U.S. Patent No. 2,963,910, December 13, 1960 (filed August 15, 1955). Describes the optical configuration and the signal processing of one of the more popular commercially available radiometers. By means of a reflective chopper the incoming flux is compared with the flux from a built-in blackbody reference. For the greatest accuracy the temperature of the reference should be very nearly equal to that of the unknown. Rather than change the temperature of the reference, this patent discloses a means of achieving the same result electrically.

[34] T. R. Harrison, *Radiation Pyrometry and its Underlying Principles of Radiant Heat Transfer*, New York: Wiley, 1960. Emphasizes the physical bases of radiometric measurements and their application to industrial process control.

[35] "Polyethylene Windows Pass Infrared Radiation," *Industrial Research*, February 1961, p. 61. It is often difficult to monitor the true temperature of electronic equipment when it is installed in its operating position since the enclosure is not transparent to infrared and removing a panel may upset the thermal conditions. Tests have shown that commercial-grade 4 mil polyethylene has a total transmittance of 93 per cent and it can be used as a temporary window.

[36] S. N. Howell, "Pyrometer Construction," U.S. Patent No. 2,976,730, March 28, 1961. Relates to improved sighting means for the radiometers described in refs. 17.2.1 [20] and [31]. A sighting telescope mounted externally on the radiometer has its line of sight shifted by small mirrors so as to coincide with the optical axis of the radiometer. The principle is similar to that used by the same author in ref. 17.2.1 [26].

[37] C. A. Gallagher and W. M. Pelino, "Hot-Box Detector," U.S. Patent No. 2,999,152, September 5, 1961. Describes a bidirectional device, that is, one that is independent of the direction of travel of the train.

[38] R. W. Astheimer and R. F. Leftwich, "Contrast Radiometer with Background Elimination," U.S. Patent No. 3,012,473, December 12, 1961. Pertains to an improved reticle and detector combination to permit cancellation of background signals received by a radiometer. Background rejection ratios of 10,000 to 1 are claimed. The reticle is divided into alternately reflecting and transmitting segments and is used with two detectors; one detector senses the reflected flux and the other senses the transmitted flux.

[39] B. T. Chao, H. L. Li, and K. J. Trigger, "An Experimental Investigation of Temperature Distribution at Tool Flank Surface," *Trans. ASME*, **B83**, 496 (1961). A radiometer with a lead sulfide detector was used to study the temperature of metal-cutting tools.

[40] D. K. Wilburn, "Survey of Infrared Inspection and Measuring Techniques," *Matls. Res. Stand.*, **1**, 528 (1961).

[41] A. C. Rudomanski and R. D. DeWaard, "System for Measurement of Air Temperature at Ground Level," U.S. Patent No. 3,058,346, October 16, 1962. Describes a system for measuring the air temperature along an airport runway. Blackbody radiators are placed along the runway and a radiometer in the control tower is used to scan them. Means are described to ensure that the blackbodies assume the same temperature as that of the surrounding air without errors caused by sunlight and other factors.

[42] S. M. Kozel, "About a Certain Method of Recording Weak Infrared Radiation," *Fiziko-Tekhnicheskii Instit.*, **8**, 73 (1962) (OTS: 63-23188). Describes a radiometer that uses a phosphor as a detector.

[43] W. M. Pelino, "Infrared Hot Box Detectors," *Proc. Natl. Electronics Conf.*, 1962, p. 236. A relatively short tutorial treatment.

[44] R. Ramanadham and A. V. S. Murty, "A Thermistor Thermometer for Field Studies," *J. Sci. Industrial Research*, **21D**, 429 (1962).

[45] D. M. Goodman, "An Electro-Optical Test System," *IRE Trans. Mil. Electronics*, **MIL-6**, 310 (1962). Proposes distributing a host of sensors throughout a system so that all parts of the system can be monitored continuously. In some cases, fiber optics are used to bring "test points" out to centrally located sensors.

[46] R. Stapelfeldt, "Hot Bearing Detector Circuits," U.S. Patent No. 3,076,090, January 29, 1963. This device views a portion of the wheel near the axle rather than the journal box. It is claimed that this eliminates uncertainties caused by thermal gradients across the journal box.

[47] "Infrared," *Control Eng.*, **10**, 20 (February 1963). Describes a number of commercially available infrared devices for industrial process control. These include radiometers, radiation micrometers, and traffic counters.

[48] F. Schwarz, "Radiation Detector Systems," U.S. Patent No. 3,081,399, March 12, 1963. Describes a hot-box detector that uses ac bias and thermistor bolometers. The discussion of the use of ac bias is particularly valuable.

[49] C. G. Kaehms, "Infrared Hotbox Detection System," U.S. Patent No. 3,086,108, April 16, 1963. For a description, see ref. 17.2.1 [60].

[50] F. W. Woltersdorf, "Method for Hot Box Detection," U.S. Patent No. 3,100,097, August 6, 1963. Concerned with differentiating between a normally hot roller bearing and an abnormally hot sleeve bearing. This is done by using two trackside sensors to view the journal boxes simultaneously on both ends of the same axle. If one of the boxes is much hotter than the other, it can be assumed that its bearing needs immediate attention.

[51] R. J. Hance, "Rotary Kiln Shell Temperature Scanning System," U.S. Patent No. 3,101,618, August 27, 1963. Describes the use of a scanning radiometer to detect hot spots on the outer surface of a rotary kiln.

[52] W. M. Pelino, "Hot-Box Detector Alarm Circuit," U.S. Patent No. 3,108,772, October 29, 1963. A pair of sensors are used to view the journal boxes simultaneously at opposite ends of the same axle.

[53] W. M. Pelino, "Hotbox Detector," U.S. Patent No. 3,108,773, October 29, 1963. This device uses a pair of differentially connected detectors so as to effectively limit the response of the radiometer to the hottest portion of the journal box.

[54] H. F. Miserocchi, "Application of Infrared in the Steel Industry," *Iron and Steel Engrg.*, **40**, 89 (1963).

[55] E. G. Menaker, "The Fundamentals of Infrared Hotbox Detection," *IEEE Trans. on Applications and Industry*, **17**, 178 (1963). A comprehensive, tutorial treatment.

17.2.1 MEASUREMENT OF TEMPERATURE

[56] R. S. Adhav and J. G. Kemp, "Infrared Radiometer," *J. Sci. Instr.*, **40**, 26 (1963). Describes a radiometer for use in the food-processing industry.

[57] D. C. Belsey and W. P. Gabriel, "A Direct Indicating Infrared Radiation Meter," *J. Sci. Instr.*, **40**, 526 (1963). A device for measuring the irradiance from industrial heaters.

[58] M. Walker, J. Roschen, and E. Schlegel, "An Infrared Scanning Technique for the Determination of Temperature Profiles in Microcircuits," *IEEE Trans. Electron Devices*, **ED-10**, 263 (1963). Describes a scanning radiometer that uses a cooled indium antimonide detector. The device has been used to examine the thermal conditions in microelectronic circuitry. The instantaneous field of view, measured in the plane of the circuit, is 1.4×10^{-3} in^2. The NEΔT is about 0.5°C at a temperature of 180°C.

[59] R. H. Kennedy, "Infrared Temperature Measurement and Its Application to Textile Finishing Operations," *Am. Dyestuff Rep.*, **53**, 56 (January 20, 1964).

[60] C. G. Kaehms, "Infrared Hotbox Detection by Measuring the Difference in Radiated Energy from Two Areas of the Journal," U.S. Patent No. 3,119,017, January 21, 1964. Uses two radiometers to view areas near the top and bottom of the journal box. The detectors are connected in opposition so that a signal is generated only when a temperature difference exists between the two areas. The entire system includes means of activating the sensor as a train approaches, means of gating the sensor in synchronism with the passage of each journal box through the field of view, means of marking the hot box, and means of alerting the train crew to the problem. This patent was divided out of the original application for ref. 17.2.1 [49]. The major difference between the two seems to be the inclusion of a cooled housing for the sensors in the later patent.

[61] L. G. Rubin, "Measuring Temperature," *Intl. Sci. Tech.*, January 1964, p. 74. A good tutorial article.

[62] W. N. Redstreake, "Hot Targets Flash Flaw Patterns," *Iron Age*, **193**, 65 (March 5, 1964). Describes a device used to monitor strip temperature in a steel-rolling mill. The device has proved of value since cool spots in the strip and the cool tail are harder to roll and often result in an overgage condition. The author notes one major difficulty: most instruments of this type are of a custom design and they lack versatility.

[63] P. J. Klass, "Use of Infrared Testing Technique Grows," *Aviation Week*, **80**, 82 (May 4, 1964). Describes the use of an infrared radiometer to measure the temperature profiles across the components on a circuit board. The article states that at least one company plans to use the technique for production-line testing of avionic equipment for the Saturn S 1-C booster and associated ground support equipment. The article contains some excellent illustrations of the results of typical measurements.

[64] "Infrared Looks Into Integrateds," *Electronic Design*, **12**, 70 (May 11, 1964). Describes the application of rapid-scan radiometers to the examination of the thermal conditions in microelectronic circuitry.

[65] GSE Staff, "Infrared: The New Technique for Electronic Testing," *Ground Support Equipment*, **6**, 9 (May/June 1964). Describes an advanced concept wherein all of the circuit modules in a complex system are linked to a scanning radiometer through individual fiber-optic bundles. By sequentially scanning each bundle, overheated conditions can be detected long before module failure occurs.

[66] L. E. Florant, "Testing Using Infrared," *Instrum. Soc. Am. J.*, **11**, 61 (July 1964). Describes a radiometer that uses a thermistor detector and has a NEΔT of 0.02°C. The use of the radiometer to detect voids, unbonds, and resin-free areas in glass-filament-wound rocket motor cases is described.

[67] "Hidden Flaws," *Prod. Engrg.*, **35**, 73 (October 26, 1964). Describes the use of a scanning radiometer to determine temperature profiles in resistors.

[68] N. Engborg, F. Schwarz, and M. G. Lowenstein, "Temperature Compensated Radiometer System," U.S. Patent No. 3,161,771, December 15, 1964. Describes means of temperature-compensating a feedback resistor in a preamplifier so as to stabilize the temperature-dependent output of a thermistor bolometer.

[69] "Thermal Plots Check Circuits," *Electronic Design*, **13**, 58 (January 18, 1965). Shows excellent examples of thermal plots obtained with new rapid-scan radiometers from two different manufacturers.

[70] "Coast Guard Needs Better Electronic Gear," *Electronic Design*, **13**, 21 (March 15, 1965). Discusses the efforts of the Coast Guard to find improved means for the detection of icebergs. Infrared radiometry easily distinguishes bergs from ships and field ice in clear weather. Since fog is present much of the time, infrared radiometry is judged to be of limited usefulness.

[71] R. F. Leftwich, "Infrared Radiometers with External Chopping and Elimination of Chopped Radiation from Instrument Walls and Components," U.S. Patent No. 3,175,092, March 23, 1965. A radiometer in which the chopper is placed ahead of the entrance aperture. As a result, stray radiation from the walls, optics, and so on, is not chopped and does not cause an output signal.

[72] "Infrared Finds the Hot Spots," *Iron Age*, **195**, 146 (April 8, 1965). This article stresses the usefulness of being able to measure the temperature of industrial equipment by radiometric means. Among the applications noted are the monitoring of flow through piping systems and the detection of any blockage, the detection of areas of inefficient operation in heat exchangers, the detection of local hot spots in motors, transformers, and cable splices, and the mapping of piping systems that are buried in walls or floors.

[73] P. W. Montgomery and R. L. Lowery, "Turbojet Temperature by IR Pyrometry," *Instrum. Soc. Am. J.*, **12**, 61 (April 1965). Describes an infrared radiometer mounted in the side of a turbojet engine so that the temperature of the turbine blades can be monitored.

[74] G. Revesz and B. G. Marks, "Testing Components by Thermal Plotter," *Electro-Technology*, **76**, 50 (August 1965). Describes the use of two new rapid-scanning radiometers to measure temperature profiles in microelectronic circuitry.

[75] "SST to Challenge Non-Destructive Testing," *Aviation Week*, **83**, 43 (December 27, 1965). Describes the use of an infrared radiometer to monitor resistance spot-welding and fusion-welding processes. The method shows promise in indicating weld quality, but much additional work must be done before the method is suitable for production.

[76] G. Bauer, *Measurement of Optical Radiations*, New York: Focal Press, 1965. Written for a German audience. Covers radiometry and spectroradiometry out to 15 μ.

[77] A. L. Pachynski, Jr., "Journal Temperature Data Transmission System using Pulse Duration Modulation," *Proc. Natl. Electronics Conf.*, 1965, p. 180. Describes a commercially available system for transmitting the data from an infrared hot box detector to a central location.

[78] B. Myers, "Heat Detector Uses Fiber Optics," *Electronic Design*, **14**, 66 (January 18, 1966). A simple radiometer, consisting of a sapphire fiber-optic light pipe and an infrared detector, is used to measure the temperature of a transistor header during die-bonding operations.

[79] C. W. Briggs, "Infrared Sensors and Glass Melting Control," *Ceramic Ind.*, **86**, 44 (March 1966).

[80] H. S. Rickert, "Modern Controls for Atmosphere Heat Treating: A User Compares Infrared and Carbon Resistance Units," *Metal Prog.*, **89**, 67 (May 1966).

[81] N. K. Koebel, "Modern Controls for Atmosphere Heat Treating: Using Dew Point and Infrared Systems," *Metal Prog.*, **89**, 70 (May 1966).

17.2.1 MEASUREMENT OF TEMPERATURE

[82] J. A. Rose, "Infrared Controls Solder Temperature," *Electrical Design News*, **11**, 19 (September 14, 1966). A simple radiometer, using a lead sulfide detector, monitors the temperature of a joint during bonding, brazing, or soldering. The radiometer output controls the energy supplied to the joint so that all joints are made at the same temperature despite variations in the size of leads or bonding pads.

[83] G. A. Hornbeck, "Optical Methods of Temperature Measurement," *Appl. Opt.*, **5**, 179 (1966). An excellent survey of the available techniques.

[84] E. L. Geery et al., "Nozzle Tube-Wall Temperature Measurement by Radiometric Techniques," *J. Spacecraft and Rockets*, **3**, 1144 (1966). Describes a radiometer to measure the nozzle tube-wall temperature without disturbing the flow of hot gas inside or the flow of coolant on the outside of the nozzle. The device uses a color-coding reticle and an uncooled lead selenide detector.

[85] W. O. Hamlin, "Industrial Non-Contact Temperature Control," *Electronic Communicator*, **2**, 13 (January/February 1967). Describes the use of infrared radiometers for process control. Applications include monitoring the temperature during the growth of epitaxial layers on silicon, during welding operations, and during the firing of the helix windings for traveling wave tubes.

[86] "Infrared Unit Spots Circuit Production Flaws," *Electronic Design*, **15**, 32 (February 15, 1967). An automatic scanning radiometer compares thermal profiles of each production unit with profiles from a unit known to be good. The unit takes 45 sec to scan a circuit board and 55 sec for the profile comparison. The radiometer uses a mercury-doped germanium detector. Linear resolution on the work piece is 0.020 in. The unit picks out flaws in about 3 per cent of circuit boards that have already passed all conventional electrical tests. The unit is particularly useful in spotting components having the wrong power rating and in locating poor mechanical connections to heat sinks.

[87] L. Hamiter, "Infrared Techniques for Reliability Enhancement of Microelectronics," *Semiconductor Prod.*, **10**, 41 (March 1967). A review of the application of infrared radiometric and thermal imaging equipment to the study of temperature distributions in integrated circuits.

[88] R. Vanzetti, "Infrared Exposes Hidden Circuit Flaws," *Electronics*, **40**, 100 (April 3, 1967). A tutorial article that concentrates on the analysis and application of the data from scanning radiometers rather than on the design of the radiometers.

[89] "IR Scanner Improves Production Testing," *Electronic Engr.*, **26**, 25 (April 1967). Describes an automatic infrared scanner used for production testing of circuit boards. Thermal profiles are compared with standard profiles stored in a tape memory. Deviation from the standard profiles causes rejection. The scanner uses a mercury-doped germanium detector and a closed-cycle cooler. The scanner takes 45 sec to scan 180 scan lines on a 5 by 7 in. circuit board with a linear resolution of 0.5 mm on the board. NEΔT is 0.1°C.

[90] R. Vanzetti, "A Revolutionary New Test Approach: The Compare System," *Electronic Communicator*, **2**, 5 (September/October 1967). Another description of the system described in the previous reference.

[91] H. F. John, "Silicon Power Device Material Problems," *Proc. IEEE*, **55**, 1249 (1967). Among the specialized techniques for studying defects in silicon are the observation of the 9 μ absorption band for the determination and control of the oxygen content of silicon, transmission microscopy using an image converter tube to reveal copper precipitates and crystallographic imperfections, and infrared radiometers for the determination of radiation profiles. There are several commercially available radiometers capable of a spatial resolution of several microns and temperature resolution of 0.5°C at 300°C. The radiometric techniques show great promise in the selection of high-

reliability devices and in the study of failure modes associated with bonding and interconnects for integrated circuits.

[92] R. J. Boncuk, "Determination of the Current Distribution in Power Transistors by Use of Infrared Techniques," *Proc. IEEE*, **55**, 1486 (1967). Describes the use of an infrared radiometric microscope to measure the temperature distribution under an emitter finger of an interdigitated power transistor. The measurements are in good agreement with calculations. The commercially available radiometric microscope that is used can measure the temperature of a 0.0003 in. spot within an accuracy of $\pm 0.5°C$. The author simplified the measurement by painting the device with black paint so as to ensure that the emissivity was essentially unity.

[93] T. P. Murray, "Polaradiometer — A New Instrument for Temperature Measurement," *Rev. Sci. Instr.*, **38**, 791 (1967). This paper reviews various means of making radiometric measurements of temperature that are independent of the surface characteristics of an object. The method proposed here is to examine the sum of the radiation emitted by a test surface and the radiation from a blackbody reflected from the test surface, not for intensity but for polarization. The instrument that is described has a spectral bandpass of 1.8 to 2.7 μ and is useful over a temperature range of 200 to 500°C. Results are presented of measurements of the temperature of various types of steel samples.

[94] "Temperature Measurement Without Contact," *Electronic Design News*, **12**, 54 (December 1967). Describes a hand-held radiometer that uses a thermistor bolometer. One unique feature is a projected spot of light to show the area being measured.

[95] W. O. Hamlin, "Infrared Temperature Measurements," *Electronics World*, **79**, 34 (February 1968). Gives a cursory discussion of infrared radiometers and then describes how they can be used to monitor the temperature of plastics prior to molding, to monitor the temperature of TNT while mixing, to control the temperature of steel strip in a bonding process, and to check the temperature of welds.

[96] P. J. Klass, "Infrared System Spots Production Flaws," *Aviation Week*, **88**, 101 (May 20, 1968). A status report on the system that is described in ref. 17.2.1 [86]. This article contains new information on the problem of obtaining standard infrared profiles of circuit boards. The examples of what defects this system has spotted are impressive. It is estimated that copies of this system could be sold commercially for $100,000 each.

[97] R. Compton, "Trends in Computer-Aided Testing," *Electronic Design News*, **14**, 22 (September 1968). Another description of the system that is discussed in ref. 17.2.1 [86] and in the preceeding reference.

17.2.2 Position Sensing

There have been a few attempts to use infrared equipment to make non-contact measurements of the width of steel strip in rolling mills and of the diameter of wires during drawing operations. These attempts appear to have been successful, but there seems to have been little effort to exploit them.

[1] F. Offner, "Method and Apparatus for Measuring the Dimensions of Objects by the Radiation Differential Between the Object to be Measured and a Comparison Object," U.S. Patent No. 2,488,430, November 15, 1949. A simple reticle consisting of an opaque disk with a single pinhole aperture scans the image of the object to be measured. If the signal pulses are rectified, the resulting output is proportional to the width of the object. Similarly, the peak amplitude of the pulses is proportional to the temperature of the object.

17.3.1 MEASUREMENT OF TEMPERATURE

[2] "Measuring Width in a Hot Strip Mill," *Metallurgia*, **57**, 307 (1958). Uses a pair of infrared sensors to monitor the two edges of the hot strip.

[3] F. J. Danks, "Infrared Gage Measures Hot Steel Strip Width," *Electronics*, **33**, 65 (October 21, 1960). Describes a gage that measures the width of a steel strip to an accuracy of ±1/8 in. The strip is at a temperature of 1150°C and is moving at speeds up to 3000 ft min^{-1}. A lead sulfide detector and a 12-sided mirror are used to scan each edge of the strip. The signal processing is well described, and complete circuit diagrams are given.

[4] "Infrared–The New Horizon," *Electronic Equipment Eng.*, **9**, 44 (May 1961). Describes an infrared gauging system for continuously monitoring the width of a hot strip in a steel-rolling mill. The strip is scanned by means of a rotating octagonal reflector. The signal from the scanner is differentiated so as to produce a pair of pulses that mark the two edges of the strip. These pulses trigger a bistable multivibrator to form pulses with lengths proportional to strip width. Any variation of pulse length from a preset value generates correction signals for the rolling mill. Accuracy of the device is said to be ±1/8 in.

[5] R. W. Astheimer and M. M. Merlen, "Infrared Comparison Micrometer," U.S. Patent No. 3,003,064, October 3, 1961. A single detector scans the moving material. The circuitry measures the width of the pulses caused by scanning across the two edges. Pulse clipping is used to eliminate the effects of temperature variations.

[6] C. D. Bryant, "Non-Contact Dimensional Measurements by Optical and Electronic Techniques," *IRE Trans. Industrial Electronics*, **1E-9**, 1, (1962). An excellent discussion of the basic techniques for such measurements.

[7] J. L. Murphy, "Extended Radiation Micrometer Gage," U.S. Patent No. 3,093,742, June 11, 1963. A modification of the device described in ref. 17.2.2[5] to permit the measurement of very wide moving strips. In essence, this modification splits the scan into two scans that cover a very narrow width at each edge of the moving strip.

[8] R. W. Astheimer and M. M. Merlen, "Infrared Comparison Micrometer," U.S. Patent No. 3,097,298, July 9, 1963. A device to monitor the width of a hot moving wire by scanning across the wire and comparing the resulting signal to that from a scan across a precision reference slit placed at right angles to the wire. By placing the reference slit over the hot wire, the amplitudes of the two scan signals are independent of variations in the temperature of the wire.

[9] W. Dougherty et al., "Noncontacting Digital Length/Width Gage for Hot Slabs," *Iron and Steel Eng.*, **44**, 96 (August 1967).

17.3 SPECTRORADIOMETRIC APPLICATIONS

The principal application in this category is infrared spectroscopy, a subject that is, in general, outside the scope of this book. Other applications include temperature measurement, process control, detection of environmental pollution, and the detection of clear air turbulence.

17.3.1 Measurement of Temperature

The simplest way to measure temperature (as shown by the large number of references in section 17.2.1) is to use a radiometer that responds to the total radiation. It is also feasible to measure temperature by making measurements of the flux in two or more narrow spectral bands. Such narrow-band methods are advantageous when the emissivity of the object is not known or when the object is transparent in some portions of the spectrum.

[1] R. Stair, "Filter Radiometry and Some of Its Applications," *J. Opt. Soc. Am.*, **43**, 971 (1953). Shows how spectroradiometry can be done with a radiometer and a narrow-band optical filter.
[2] F. J. Larsen, "Turbine Blade Temperature Control Apparatus," U.S. Patent No. 2,687,611, August 31, 1954. A dual-filter radiometer to monitor the temperature of turbine blades and control the fuel input to the turbine so as to not exceed the allowable operating temperature. The two narrow-band filters are equipped with variable-area apertures that can be adjusted to equalize the flux passing through the two aperture/filter combinations. Knowing the characteristics of the filters and the two areas, it is possible to calculate the temperature of the turbine blades.
[3] W. M. Flook, Jr., and D. D. Friel, "Radiation Temperature Measurement," U.S. Patent No. 2,909,924, October 27, 1959. Describes what is, in effect, a filter radiometer that can be used to measure the temperature of hot, semitransparent plastic films.
[4] A. C. Rudomanski, R. D. DeWaard, and E. M. Wormser, "Selective Radiation Detector and Free-Air Thermometer," U.S. Patent No. 3,091,693, May 28, 1963. Describes a device for measuring the free-air temperature from a supersonic aircraft. A highly selective detector measures the radiation from the $14\,\mu$ emission band of carbon dioxide. The detector is a thermistor bolometer with a talc filter and a mica window. The optical train contains a reflecting surface of polished magnesium oxide. As a result, the spectral response of the detector is limited to the region from 14 to 15 μ. The detector is located at the bottom of a reference blackbody cavity to simplify the comparison of the $14\,\mu$ atmospheric emission with that from a reference blackbody.

17.3.2 Miscellaneous

One of the more recent applications of infrared equipment is for the early detection of clear-air turbulence (CAT), that is, conditions in which an aircraft encounters severe turbulence despite the fact that visual and radar observations show no evidence of any convective cloud activity in the area. There are numerous reports of damaged aircraft, injured passengers and, in some instances, lost aircraft because of encounters with CAT. Although it can occur at any altitude, CAT is most often observed between 25,000 and 50,000 ft. CAT normally occurs in thin horizontal sheets that are about 2000 ft high, 10 to 20 miles wide, and 50 to 100 miles long. Since these CAT layers are about 2°C warmer than the surrounding nonturbulent air, it is tempting to try and detect them with infrared equipment. Results seem to indicate that infrared methods are among the best detection schemes yet tried.

[1] W. G. Fastie and A. H. Pfund, "Selective Infrared Gas Analyzers," *J. Opt. Soc. Am.*, **37**, 762 (1947). Discusses the principles of nondispersive gas analyzers.
[2] J. A. Sanderson, "The Diffuse Spectral Reflectance of Paints in the Near Infrared," *J. Opt. Soc. Am.*, **37**, 771 (1947). Describes the development of infrared-reflecting

17.3.2 MISCELLANEOUS

paints and equipment for measuring their reflectance. The paints were developed as a means of controlling the temperature of surfaces by either reducing their absorption of solar radiation or controlling their self-emitted radiation.

[3] R. C. Beitz, "Spectrophotometer," U.S. Patent No. 2,630,736, March 10, 1953 (filed September 16, 1947). Gives a detailed description of the operating principles of a commercially available rapid-scan spectrophotometer.

[4] W. Kaiser, P. H. Keck, and C. F. Lange, "Infrared Absorption and Oxygen Content in Si and Ge," *Phys. Rev.*, **101**, 1264 (1956). Describes the use of characteristic absorption bands to determine the oxygen content of silicon and germanium material for transistor manufacture.

[5] W. Kaiser and P. H. Keck, "Oxygen Content of Silicon Single Crystals," *J. Appl. Phys.*, **28**, 883 (1957). Describes the use of the 9 μ absorption band to control the oxygen content of silicon.

[6] A. E. Martin, A. M. Reid, and J. Smart, "Infra-red Gas Analyzers for Plant Control," *Research*, **11**, 258 (1958). Describes the basic principles of nondispersive gas analyzers and their application to industrial process control.

[7] G. W. Bethke, "Ultrarapid-Scan Infrared Spectrometer," *J. Opt. Soc. Am.*, **50**, 1054 (1960). Describes a low-resolution spectrometer that uses an indium antimonide detector and achieves scan rates up to 20,000 scans per second. The bibliography is excellent.

[8] G. Romans, "Methods of Detecting Pipe Line Leaks," U.S. Patent No. 3,032,655, May 1, 1962. Describes a means of detecting hydrocarbon-gas leakage from buried pipelines with an airborne infrared sensor. Hydrocarbons absorb strongly at 3.4 μ. If a simple search set having a narrow spectral bandpass centered at 3.4 μ is used to view the sunlit ground, the cloud of gas above a leaking pipeline can be detected by its absorption of the signal from the ground.

[9] J. A. Fusca, "Clear Air Turbulence," *Space/Aeronautics*, **42**, 60 (August 1964). An excellent tutorial treatment. Suggests the possibility of using an infrared device to detect the temperature gradients that seem to be associated with clear-air turbulence.

[10] A. S. Curry, "Science Against Crime," *Intl. Sci. Tech.*, November 1965, p. 39. Describes how infrared spectroscopy has become a useful tool in the fight against crime.

[11] P. J. Klass, "Airborne Turbulence Sensor Still Sought," *Aviation Week*, **84**, 97 (March 14, 1966). Describes and shows a picture of an experimental scanning radiometer said to be capable of detecting temperature differentials as small as 0.2°C at a distance of 25 miles. This radiometer weighs 15 lb. An operational version of the device would determine temperature by taking the ratio of the radiation at two different wavelengths. These wavelengths would be varied automatically with the altitude of the aircraft.

[12] R. T. H. Collis, "Clear Air Turbulence Detection," *IEEE Spectrum*, **3**, 56 (April 1966). A good discussion of the physical phenomena involved. Contains only a limited mention of the possibility of detection by infrared sensors.

[13] "Instrument Measures Oil Pollution of Water," *Frontier*, **27**, 23 (Winter 1966). Describes an oil-in-water detection instrument for monitoring pollution of harbors. Instrument is a two-channel filter radiometer that continuously compares the transmittance of a sample of clean water with a sample of the water under surveillance.

[14] S. M. Norman and N. H. Macoy, "Inference of Clear Air Turbulence by Means of an Airborne Infrared System," *J. Aircraft*, **3**, 289 (1966). Describes a device for detecting clear air turbulence at distances of up to 10 miles. It consists of a novel spectrometer that scans from 13.5 to 14.5 μ at the edge of the carbon dioxide absorption band. Since the atmospheric transmission varies markedly with wavelength in this spectral region,

the device sees a progressively shorter atmospheric path during the course of the scan. If a temperature discontinuity, such as is characteristic of some types of clear air turbulence, exists ahead of the aircraft, it can be detected by the change in output of the device when compared with the output under normal conditions. Horizontal temperature gradients of 3 to 5°C over distances of 3 to 20 miles were observed in flight tests.

[15] M. G. Beard, "Status Report on Latest Development in Clear Air Turbulence," *J. Aircraft*, **3**, 443, (1966). A status report on new developments in CAT forecasting and the methods of detecting it. Results to date seem to show that there is at least a 2°C temperature difference between the turbulent and nonturbulent air when CAT exists. The lack of a range indication is considered the major disadvantage of infrared methods. The author feels that infrared sensors are so well developed that they fall in the category of a solution looking for a problem.

[16] "Airline Observer," *Aviation Week*, **86**, 46 (January 30, 1967). Announcing an FAA-sponsored test program for the remote detection of clear air turbulence with an infrared scanning radiometer.

[17] R. D. Hibben, "IR Device Tested as Turbulence Detector," *Aviation Week*, **86**, 61 (February 20, 1967). Contains a lengthy description and detailed photos and drawings of an infrared interferometric spectrometer for the detection of clear-air turbulence. The temperature gradients thought to be typical of clear-air turbulence are detected by scanning the 12.7 to 13.7 μ region. The experimenters predict detection ranges of 25 miles but ranges of only about 3 miles have been observed in a current flight test program.

[18] "Pan Am Plans to Evaluate Turbulence Detector on 707," *Aviation Week*, **86**, 35 (April 10, 1967). System consists of an infrared radiometer, data processor, and display. Temperature differentials of 1.5°C will trigger the unit. The display shows both direction and magnitude of the turbulence. The system has received 145 hr of flight test over the continental United States. Up to 4 min of advance warning has been demonstrated.

[19] "Spanning the Infrared Spectrum," *Electronics*, **40**, 159 (May 15, 1967). Describes a spectroradiometer that uses a rotating circularly varying interference filter as the "dispersing" element. The center wavelength transmitted by the filter varies linearly with the angular rotation of the filter. Effective spectral bandwidth is 1 to 2 per cent of the center wavelength. The filters are available to cover the region from the visible to 27 μ, although several filters are needed to cover this total range.

[20] "Infrared Sending Expo 67," *Electronic Design News*, **12**, 18 (June 1, 1967). Two infrared data links were in daily use at Expo 67. The links used injection diodes to transmit digital data from the Montreal stock exchange and from a remotely located computer.

[21] "Breatha-laser," *Electronic Design News*, **12**, 29 (June 1, 1967). Describes the use of a helium neon laser at 3.391 μ to detect alcohol vapor in the breath at concentrations as low as 50 ppm.

[22] "Airline Observer," *Aviation Week*, **87**, 47 (July 10, 1967). Contains a report that recent flight tests of an infrared scanning radiometer have shown that the device is very effective in locating clear-air turbulence. Detections occurred at about 8.5 mi. An earlier report on the same device is in ref. 17.3.2[17].

[23] "Filter Center," *Aviation Week*, **87**, 93 (August 28, 1967). Reports Pan American flight tests showed that an experimental infrared clear-air turbulence detector consistently gave 3 to 4 min advance warning. The device did, however, give a number of false alarms because of radiation from the earth when the 707 aircraft was in a turn or in descent.

17.3.2 MISCELLANEOUS

[24] "NASA Abandons Laser Radar for CAT Detection, But Three Passive Infrared Systems Show Promise," *Laser Focus*, **3**, 18 (August 1967). Tests in Canada, using an infrared scanning radiometer, showed that clear-air turbulence could be detected at a distance of nearly 10 miles. A current British program is using a spectroradiometer covering the 1 to 25 μ region. A recently completed flight test in the United States demonstrated the detection of CAT at a distance of up to 48 miles. This equipment is expected to cost less than $25,000.

[25] C. O. Peterson, Jr., W. V. Dailey, and W. G. Amrhein, "Applying Non-Dispersive Infrared to Analyze Polluted Stack Gases," *Instrumentation Technology*, **14**, 45 (August 1967). Describes the operation of nondispersive infrared analyzers and how they are used to detect highly corrosive nitrogen dioxide in stack gas. The method will detect concentrations as low as a few parts per million.

[26] "Filter Center," *Aviation Week*, **87**, 64 (November 27, 1967). Reports that Pan American flight tests of an infrared clear-air turbulence detector have been resumed. The improved model is gyro-stabilized to prevent false alarms during turns. See also refs. 17.3.2[18] and [23].

[27] H. F. John, "Silicon Power Device Material Problems," *Proc. IEEE*, **55**, 1249 (1967). Describes the use of the 9 μ absorption band as a means of determining and controlling the oxygen content of silicon. The method is limited to a sensitivity of about 10^{16} oxygen atoms/cm^3.

[28] P. J. Klass, "Infrared Clear Air Turbulence Unit," *Aviation Week*, **88**, 80 (May 6, 1968). Contains a lengthy discussion of the flight testing of the device that has been referred to in refs. 17.3.2[18], [23], and [26]. During the most recent tests, on Pan American's regularly scheduled flights, 14 incidents of CAT were encountered, of which the system detected 13. But there were five times when the system predicted CAT and none was encountered. The system triggers an alarm if it detects a 4° temperature change in successive 16-mile segments ahead of the aircraft. The entire system weighs just under 50 lb and requires 20 W of 28 VDC and 75 VA of 115 VAC, 400 Hz power. The sensor is cylindrical with a diameter of 6 in. and it protrudes 5.5 in. above the fuselage. The thermistor detector is not cooled and the spectral response of the system extends from 13 to 14 μ. Keeping insect splatters off of the sensor window has been a problem that was finally solved by a small v-shaped screen atop the fuselage just ahead of the sensor.

[29] "Infrared CAT Tamer," *Industrial Research*, p. 21 (July 1968). A description of the clear air turbulence detector that is discussed in the previous reference.

[30] "Airlines to Test New Turbulence Device," *Aviation Week*, **89**, 33 (August 19, 1968). Three airlines plan to test a third generation infrared detector of clear air turbulence that was previously described in ref. 17.3.2[17]. The sensor has a maximum diameter of 7 in. and a length of 12.5 in. The portion that protrudes through the aircraft skin has a diameter of 3.25 in. and a length of 3.25 in. A possible design change would eliminate the internal reference blackbody and substitute a motor-driven filter wheel carrying a pair of filters to permit the measurement of nearby air at 15 μ and distant (40 miles) air at 14.1 μ.

17.4 THERMAL IMAGING APPLICATIONS

This category has undergone a rapid growth because of the successful application of thermal imaging devices to nondestructive testing and inspection. Several companies are supplying this equipment, and a rather healthy competition has developed. Since radiometers are also used for nondestructive testing and inspection, the reader with an interest in these subjects should examine the references in Section 17.2.1 as well as those given here.

17.4.1 Nondestructive Test and Inspection

[1] J. Evans, "System for Forming Images of Heat Radiating Objects," U.S. Patent No. 2,403,066, July 2, 1946. Describes a thermal imaging device consisting of a linear array of detectors, a Nipkow scanning disk, and an oscillating mirror. The image is displayed on a cathode-ray tube.

[2] W. A. Tolson, "Image-Forming Heat Detector," U.S. Patent No. 2,435,519, February 3, 1948. Employs a pneumatic detector as an image converter. A thin membrane formed into a spherical mirror has its surface deformed by a thermal image formed on it. The deformations can be seen visually.

[3] F. Urbach, N. R. Nail, and D. Perlman, "The Observation of Temperature Distributions and of Thermal Radiation by Means of Non-Linear Phosphors," *J. Opt. Soc. Am.*, **39**, 1011 (1949). This article shows that the temperature-dependence of the luminescent efficiency of certain phosphors can be used to observe, measure, and record temperatures and temperature distribution. Temperature differences as small as 4°C can be observed.

[4] H. Gobrecht and W. Weiss, "On Photographic Recording in Infrared by the Method of Czerny (Evaporography)," *Zeitschrift für Angewandte Physik*, **5**, 207 (1953) (SLA: 62-10425).

[5] D. Z. Robinson et al., "Method and Apparatus for Long Wave-Length Infra-Red Viewing," U.S. Patent No. 2,855,522, October 7, 1958 (filed April 30, 1953). Describes an improved version of the Evaporograph.

[6] R. W. Astheimer and E. M. Wormser, "Image Transducer," U.S. Patent No. 2,895,049, July 14, 1959. Gives an excellent description of the design details of the commercially available thermal imaging device that is referred to in the next reference. The recorder consists of a glow modulator tube whose brightness is modulated by the output of the radiometer, suitable optics, a small scan mirror that is attached to the back of the radiometer's scan mirror, and a device to hold Polaroid film. Thermal pictures are available 10 sec after the completion of scanning.

[7] R. W. Astheimer and E. M. Wormser, "Instrument for Thermal Photography," *J. Opt. Soc. Am.*, **49**, 184 (1959). Describes the combination of an 8 in. scanning radiometer and a recording device that can produce thermal pictures. A large plane mirror placed in front of the radiometer is driven so as to produce a raster scan. Typically, the unit can scan a field of view that is 10×10 deg with an angular resolution of 1 mrad in a time of 17 min. Under these conditions the NEΔT is 0.04°C if an immersed thermistor bolometer is used for a detector. With an unimmersed thermistor, the NEΔT is 0.15°C. The unit described is commercially available. The details of the recorder are covered more fully in the previous reference.

[8] "Infrared Camera Spots Malfunctions," *Electronic Design*, **9**, 12 (December 6, 1961). Describes the use of a thermal imaging device to detect overheated components on circuit boards.

[9] D. K. Wilburn, "Survey of Infrared Inspection and Measuring Techniques," *Matls. Research and Standards*, **1**, 528 (1961).

[10] "Circuit Inspections Speeded with Infrared," *Electronics*, **35**, 72 (July 6, 1962). A short but well-illustrated article.

17.4.1 NONDESTRUCTIVE TEST AND INSPECTION

[11] M. J. Cohen, "Measuring Systems Using Infra-Red Radiation," U.S. Patent No. 3,043,956, July 10, 1962. Describes means of testing materials by subjecting them to modulated infrared radiation. The method is based on the measurement of the time delay of a radiation pulse in passing through the material under test. The time delay is a function of the thermal impedance of the material which is, in turn, a function of the thickness and several other material parameters. The transmission of the pulse is observed with a thermal imaging device.

[12] P. J. Klass, "Component Failures Predicted by Infrared," *Aviation Week*, **77**, 85 (December 3, 1962). Describes the use of a thermal imaging device to spot potential component failures by means of their abnormal heat distributions. There is some evidence to suggest that a 5 min examination with the device can pinpoint transistors that will be the first to fail several thousand hours later.

[13] G. W. McDaniel and D. Z. Robinson, "Thermal Imaging by Means of the Evaporograph," *Appl. Opt.* **1**, 311 (1962). States that production versions of the Evaporograph can detect a temperature difference of 1°C against a 300°K background. Resolution is 10 lines per millimeter for a temperature difference of 10°C. The article describes the use of the Evaporograph for nondestructive testing, studies of heat exchangers and the temperature distributions in wind tunnel models, and the location of thin spots in furnace walls and faulty joints in power lines.

[14] G. W. McDaniel and A. P. DiMattia, "Inspecting Infrared Optical Materials and Systems by Means of the Evaporograph," *Appl. Opt.* **1**, 483 (1962). Describes the use of the Evaporograph for the quality control of visually opaque optical materials, for the inspection of antireflection coatings and optical filters, for the examination of thermal images produced by optical systems, and for the determination of optical birefringence.

[15] G. A. W. Boehm, "Reliability Engineering," *Fortune*, **67**, 124 (April 1963). Contains some excellent examples of thermal images of a typical circuit board.

[16] R. Vanzetti, "Infrared Techniques Enhance Electronic Reliability," *Solid State Design*, **4**, 29 (August 1963). An excellent tutorial article on the use of thermal imaging devices for nondestructive testing of electronic circuitry.

[17] G. P. Faerman, "Images Obtained by Exposing Emulsions to Radiant Energy of Far Infrared Region by Evaporography," *Zh. Nauch. i Prikl. Fotogr. i Kinematogr.*, **8**, 153 (1963) (OTS: TT-64-10893).

[18] W. McGonnagle and F. Park, "Nondestructive Testing," *Intl. Sci. Tech.*, July 1964, p. 14. A good survey of the field including the use of thermal imaging techniques.

[19] "Detecting Hot Spots (or Cool Spots) in Breadboards," *Electronic Equipment Engineering*, **12**, 24 (October 1964). Describes a commercially available infrared scanner that can detect abnormally hot or cold areas on a circuit board. Linear resolution on the circuit board is 1 mm. and the minimum detectable temperature is about 0.1°C.

[20] "Failure-Prone Resistors Spotted in Fast IR Tests," *Electronic Equipment Engineering*, **12**, 19 (November 1964). Describes the use of a thermal imaging device to scan resistors and detect abnormal heating patterns. Examples are given of certain failure mechanisms that have been discovered by this method. The article points out that the gold and silver bands used in color coding have a low emissivity and will give a false indication of a temperature discontinuity.

[21] W. S. Beller, "Navy Sees Promise in Infrared Thermography for Solid Case Checking," *Missiles and Rockets*, **16**, 22 (January 4, 1965). Describes the use of a thermal imaging device to detect delaminations, voids, undesired inclusions, and cracks in the motor casings of Polaris missiles. The motor case is placed in a warming room and allowed to stabilize at a temperature that is several degrees above the ambient temper-

17.4.1 NONDESTRUCTIVE TEST AND INSPECTION

ature in the examination room. When removed from the warming room, the case begins to cool. The cooling process is monitored with the thermal imaging device. Defects in the case show up as distortions in the thermal image. The price of the thermal imaging device (thermograph) is quoted as $26,000. A letter containing additional explanatory information appeared in the same magazine at **16**, 4 (February 8, 1965).

[22] "Program—1965 Spring National Convention of the Society for Nondestructive Testing," *Materials Evaluation*, **23**, 25 (January 1965). It is interesting to note that 40 per cent of the papers at this meeting were devoted to the uses of infrared techniques for nondestructive testing.

[23] R. B. Barnes, M. C. Banca, and N. E. Engborg, "Infrared Thermograph," U.S. Patent No. 3,169,189, 9 February 1965. Describes an improvement to the thermal imaging device of refs. 17.4.1 [6] and [7]. The improvement is a means of projecting a small spot of light onto the scene so as to show at all times the exact point in the scene that is being scanned.

[24] "IR Scanner Displays Thermal Map in Just 2.5 Seconds," *Electronic Design*, **13**, 24 (April 12, 1965). Describes an (apparently) experimental scanning radiometer that scans an 18×20 deg field of view in 2.5 sec with a 350 line resolution. The detector is cooled indium antimonide. The NEΔT is 0.5°C.

[25] R. Dobriner, "IR Testing of Microelectronics Surges," *Electronic Design*, **13**, 6 (April 26, 1965). A good survey of the continuing development of the field.

[26] "Detecting Hot Spots in Integrated Circuits," *Electronic Design News*, **10**, 102 (April 1965). Describes a rapid-scanning radiometer and its use in detecting hot spots in electronic circuits. The scanner uses a cooled indium antimonide detector and scans a 20×20 deg field of view in 5 sec with a resolution of 150 lines. The NEΔT at 300°K is 0.7°C. If the scan motion is disabled, a single point can be monitored and temperature transients of 10 μsec can be detected.

[27] "Thermography—Infrared Photography," *Electronic Design News*, **10**, 110 (May 1965). An article that shows examples of the use of thermograms for the examination of electronic chassis, individual circuit components, and microcircuitry.

[28] R. Vanzetti, "Infrared for Electronics Reliability," *Electronic Design News*, **10**, 42 (July 1965). An excellent tutorial article that fittingly complements the author's earlier paper, ref. 17.4.1 [16].

[29] D. W. Ballard, "Non-Destructive Testing," *Industrial Research*, **7**, 69 (October 1965). Describes the use of thermal imaging for the detection of hot spots in discrete and microelectronic components, the detection of flaws in bonded structures, and the quality control of weldments.

[30] L. M. White and R. W. Jones, "Infrared Fault-Finder Pinpoints Defects in Multilayer Boards," *Electronics*, **38**, 96 (November 29, 1965).

[31] "Filter Center," *Aviation Week*, **86**, 57 (January 9, 1967). A thermal imaging device will be used in the reliability testing of avionics subsystems for the USAF C-5A transport. Thermal profiles of the equipment made under normal operating conditions will be used as reference guides throughout the life of the equipment. The thermal scanner has an NEΔT of 0.6°C.

[32] "Infrared Cameras at Work," *Electronic Products*, **9**, 80 (May 1967). Shows examples of the thermal imagery that results from studies of diode heating, foam-insulated railroad cars, anti-icing panels, circuit boards, and RF dummy loads.

[33] F. Davis, "Liquid Crystals: A New Tool for NDT," *Research/Development*, **18**, 24 (June 1967). One of the best articles available on cholesteric liquid crystals. The color of these crystals will shift through the entire visible spectrum for temperature changes of 2 or 3°C. Specific temperatures can be measured to an accuracy of 0.1°C. The article includes some excellent color photographs of the crystals.

17.5.1 APPLICATIONS OF IMAGE CONVERTER TUBES

[34] "Liquid Crystals Plot the Hot Spots," *Electronic Design*, **15**, 71 (September 13, 1967). The authors have applied the liquid-crystal technique to the examination of a wide variety of microcircuitry. Their results show that this technique may be a serious competitor for infrared thermal imaging devices. The cholesteryl esters (liquid crystals) cost about \$25 for 50 mL, enough material to test 5000 cm^2 of surface.

[35] "Current Infrared Papers," Society for Nondestructive Testing, Inc., 914 Chicago Ave, Evanston, Ill., 1967. A collection of 24 papers from recent conventions of the society that cover the application of infrared techniques to nondestructuve testing.

[36] P. H. B. Koch and H. Oertel, Jr., "Microwave Thermography," *Proc. IEEE*, **55**, 416 (1967). Describes the use of a Czerny Evaporograph in studies of microwave optics. Use of the evaporograph permits immediate visual display of field patterns in the neighborhood of propagation singularities or obstacles, such as the focal plane of antennas, apertures, diffracting edges, and so on. Approximately 1 to 3 min is required to form an image.

[37] S. Schwartz, ed. *Integrated Circuit Technology*, New York: McGraw-Hill, 1967 Chapter 7. This chapter is entitled "Infrared Testing and Mask Alignment," a curious juxtaposition of subjects. Covers the use of thermal imaging devices for the examination of integrated circuits.

[38] E. Wormser, ed., "Nondestructive Testing Issue," *Appl. Opt.*, **7** (September 1968). Contains 23 papers on nondestructive testing with infrared thermal imaging devices. Individual papers have not been included in the references in this book and this note was added during proofing.

17.5 APPLICATIONS INVOLVING REFLECTED FLUX

Among the applications that fall in this category are the use of image converter tubes for industrial surveillance and crime prevention, examination of photographic film during manufacture, and observation of conditions at the junctions of injection diodes. Infrared photography appears to be finding an increasing application in the assessment of agricultural areas and the detection of diseased crops. The general references on infrared photography are in Section 19.5.1.

17.5.1 Applications of Image Converter Tubes

[1] A. Vasko, "Investigation of Near Infrared Radiations by Means of Image Converters," *Nature*, **158**, 235 (1946).

[2] W. C. Dash, "Copper Precipitation on Dislocations in Si," *J. Appl. Phys.*, **27**, 1193 (1956). Observation method is a microscope equipped with an image converter tube.

[3] H. Arnet and H. Baechthold, "The Use of Night Vision Devices in the Police Services," *Kriminalistik*, **10**, 434 (1956) (OTS: 59-13524). As one would expect, the German police are well advanced in the application of image converter devices to night surveillance.

[4] P. K. Oshchepkov, "Infrared Microscope," *Akad. Nauk SSSR, Vestnik*, **28**, 64 (1958) (SLA: 59-22635). Describes a microscope fitted with an image converter tube.

[5] L. Dun, G. Lushnikov, and A. Iakobson, "Introscopy" ("Flaw Detection"), *Znanie-Sila*, **11**, 48 (1960) (OTS: TT62-11082). Describes in an elementary manner Soviet work with acoustic, X-ray, and infrared devices for the detection of flaws in materials. The infrared device is a microscope fitted with an image converter tube. The only

application for the infrared microscope described here is for the detection of flaws in germanium and silicon crystals.

[6] A. Kojima, A. Kasahara, and S. Saruta, "Microscope System for Use with Both Infrared and Visible Rays," U.S. Patent No. 3,124,682, March 10, 1964. Describes the addition of an image converter tube to a microscope.

[7] A. A. Gutkin et al., "Detection of p-n Junctions in Gallium Arsenide by Means of Mik-1 Infrared Microscopes," *Pribory i Tekhnika Eksperimenta,* **9**, 184 (1964). Translation in *Instruments and Experimental Techniques,* April 1965, p. 1102. This method detects the recombination radiation from an injection diode and permits the location of the p-n junction to within $\pm 0.5\ \mu$.

[8] "Night Light," *Electronics,* **38**, 34 (March 22, 1965). Describes a commercially available image converter device for police use. The device weighs 2 lb and has a useful range of 100 yd. It does not include an illuminator. Price of the viewer is under $500.

[9] D. Krawlow, "Agent Tells of Training to Wiretap, Pick Locks," *Los Angeles Times,* July 20, 1965. This front-page article carries a large photograph showing a U.S. Senator peering through a sniperscope. Internal Revenue agents are accused of using the sniperscope for night surveillance of suspected tax evaders.

[10] "Fast Infrared Telescope," *Sky and Telescope,* **33**, 17 (January 1967). Describes an $f/1$, 20 in.-diameter, catadioptric optical system for use with an image converter tube.

[11] H. F. John, "Silicon Power Device Material Problems," *Proc. IEEE,* **55**, 1249 (1967). Describes the use of infrared transmission microscopy with an image converter to study copper precipitation on dislocations and general crystallographic defects in silicon.

17.5.2 Infrared Photography

[1] M. Jorg, "Über Weitere Anwendung der Ultrarotphotographie in Kriminalistik und Medizin," *Phot. Korr.,* **74**, 148 (1938). Demonstrates that obliterated tattoo patterns can be seen in infrared photographs.

[2] U. Ploger, "Determination of Surface Temperatures by Means of Infrared Photographs," *Arch. Eisenhuettenw.* **31**, 87 (1960) (SLA: 63-18426). Used infrared photography to determine the temperatures of furnace exteriors in a steel mill. The calibration technique is also described.

[3] *Photography in Law Enforcement,* Rochester, N. Y.: Eastman Kodak Co., 1960. Applications of infrared photography are described.

[4] W. Beyer, "Travelling-Matte Photography and the Blue-Screen System," *J. Soc. Motion Picture and Television Engrs.,* **74**, 217 (1965). A tutorial paper on the traveling-matte process, a motion picture technique to photographically place an actor in front of any desired background. Infrared film is used in one of the intermediate production steps.

17.5.3 Miscellaneous

[1] G. B. Harrison, R. J. Hercock, and R. G. Horner, "Testing of Photographic Films, Plates, and Papers," U. S. Patent No. 2,393,631, January 29, 1946. Describes a means of examining photographic film during manufacture in order to detect flaws without fogging the film. A small beam from a tungsten lamp, filtered to remove any visible light, is scanned across the film. The reflected flux is picked up by a suitable detector and the resulting signals are displayed on a cathode-ray tube.

[2] T. Motiu, "The Humidity Determination of Tobacco by Means of Infrared Radiation," *Industria Alimentara* (Rumania), **11**, 282 (1960) (CFSTI: K-H: 11816). Employs the same general principle of operation as that described in ref. 17.5.3 [4]

[3] W. W. Sanville, "IR Scanner Reads Freight Car Numbers," *Control Eng.*, **9**, 147 (April 1962). Describes a system that can automatically identify freight cars moving at speeds of 5 to 60 mph. An infrared source illuminates the passing freight car, which carries a special coded plate. Variations in the flux reflected from the plate are sensed by an infrared scanner. The coded plate contains a pattern of alternately high- and low-reflectance areas. If these areas are chosen properly, the output of the scanner is a series of pulses in standard teletype code. It is evident that the author firmly believes that infrared can penetrate fog or other bad weather. This system can also be used in conjunction with an infrared hot-box detector to furnish an immediate identification of the car with the hot box.

[4] R. C. Ehlert, "Infrared Reflection and Absorption System for Measuring the Quantity of a Substance that is Sorbed in a Base Material," U. S. Patent No. 3,150,264, September 22, 1964. A device to measure the amount of a substance in a base material. In the example discussed here, the device is used to determine the water content of a moving strip of wet paper during manufacture. The paper is illuminated by an incandescent lamp, and the reflected energy is viewed by two narrow-band sensors. One sensor responds at $1.94\,\mu$ in the middle of a water absorption band. The other sensor responds at $1.63\,\mu$ where water has little absorption. The ratio of the signals from the two sensors is proportional to the water content of the paper. Accuracy is said to be ± 0.1 per cent.

[5] "Industrial Electronic Developments," *Electronic Industries*, **23**, 8 (September 1964). Describes a commercially available version of the device disclosed in the previous reference.

[6] A. J. Buetler, "Infrared Backscatter Moisture Gage," *Tappi*, **48**, 490 (1965).

[7] "IR Gage Hooks up with Process Computer for Complete Closed Loop Feedback Control of Paper Drying Operation," *Instrumentation Technology,* **14**, 13 (February 1967). Heart of the system is an infrared moisture gage. The gage takes continuous measurements of moisture content with a 0.1 per cent precision. See also ref. 17.5.3 [4].

[8] "Automatic Focusing 35 mm Slide Projector," *Phot. Appl. in Sci. and Tech.*, **1**, 15 (Summer 1967). An infrared beam is reflected from the surface of the film onto silicon photodiodes. Any buckling or shift of the film results in a displacement of the beam, which can be detected by the photodiodes and converted to a correction signal for the focusing mechanism.

17.6 APPLICATIONS INVOLVING A COOPERATIVE SOURCE

This category includes such applications as intrusion detection, blind-landing and collision-avoidance systems, intervehicle speed sensing, and methods of bonding cables. Passive intrusion detectors are discussed in Section 16.1.1.

17.6.1 Intrusion Detection

[1] E. S. Carey, "Waterfront Protection System," U.S. Patent No. 2,212,200, August 20, 1940. Describes a device to detect the intrusion, from the water side, of wharves or waterfront property. The usual source and detector placed at opposite ends of the area to be protected will be ineffective at times because the height of the beam above the water will vary with the state of the tide. The solution adopted here is to place the source and detector on small floating platforms.

[2] P. M. Farmer, "Invisible Light for Protection Against Sabotage," *Aero Dig.*, **39**, 171 (August 1941). Describes various beam patterns that can be used to protect harbors and other large areas from unauthodized intrusion.

[3] S. Bagno and J. Fasal, "Intruder Alarm Uses Phase-Sensitive Detector," *Electronics*, **31**, 102 (February 14, 1958). An intruder alarm that uses a modulated tungsten lamp and a photodiode detector.

[4] J. G. W. Lee, "Narrow Beam Radiation Scanned Pattern Alarm System," U.S. Patent No. 3, 120, 654, February 4, 1964. Describes an optical means of providing a large number of beams from a single source and using these to completely surround the area to be protected. The sensor uses a lead sulfide detector operating in the 2.2 to 2.7 μ region.

[5] "Infrared Protective Device," *Engineer*, **218**, 739 (1964). Describes a commercially available (in France) intrusion alarm consisting of a modulated tungsten lamp and a photodiode detector. The lamp and detector can be placed as far apart as 250 m.

[6] B. Miller, "Vietnam War Needs Spur Laser Research," *Aviation Week*, **84**, 236 (March 7, 1966). Describes a battery-powered intrusion alarm that uses a gallium arsenide injection diode as a source.

[7] J. Vollmer, "Applied Lasers," *IEEE Spectrum*, **4**, 66 (June 1966). Pictures and describes an intrusion alarm that uses an injection diode. Peak power output is 8 W at a rate of 200 pulses per sec. The beam from the transmitter is 3 by 0.3 mrad. The receiver consists of a silicon photodiode, a low-noise preamplifier, and an alarm relay. The size of the transmitter is $2 \times 2 \times 1$ in. and the receiver appears to be about twice this size. The transmitter requires a 300 V battery.

17.6.2 Miscellaneous

[1] F. M. Pottenger, Jr., and J. R. Balsley, "Airplane Navigating Apparatus," U.S. Patent No. 2, 070, 178, February 9, 1937. Describes a blind-landing device for aircraft. Sensors on each wing tip track a modulated source at the runway threshold. Distance is determined automatically by a simple computer that uses the known distance between the sensors and the angle measured by each sensor. Unfortunately the utility of the device is based on the author's belief that infrared can penetrate fog without difficulty.

[2] J. D. Hall, *Industrial Applications of Infrared*, New York: McGraw-Hill, 1947. Describes the use of infrared heaters to achieve better and more economical surface finishes and other industrial heating applications.

[3] E. C. Malins, "Infrared Rays and Some of their Applications," *Technisch Wetenschappelijk Tijdschrift*, **16**, 284 (December 1947). The principal emphasis is on infrared lamps for paint drying.

[4] H. Manders, *Infrarood Stralers (Infrared Radiators)*, Amsterdam: Edition Diligentia, 1948. A study of heat sources for industrial drying.

[5] M. LaToison, *Chauffage et Séchage par Lampes à Rayonnement Infrarouge* (Heating and Drying by Means of Infrared Lamps), Paris: Editions Eyrolles, 1953. A translation of an updated version is available; see ref. 17.6.2 [12].

[6] "Infrared Eye Monitors Traffic," *Electronics*, **33**, 94 (September 16, 1960). Describes an infrared traffic detector that uses a chopped infrared beam and a sensor to detect the flux reflected from passing cars. By tuning the sensor's amplifier to the chopping frequency, the system is immune to false signals from sunlight, headlights and streetlights. The device can be used to count traffic or to actuate traffic signals.

[7] "IR Detector For Traffic Ok'd by State," *Electronic News*, March 22, 1961. Describes an infrared traffic detector that has been approved by the Bureau of Public Roads for use on federally sponsored roads. This is probably the device that is described in the previous reference.

17.6.2 MISCELLANEOUS

[8] P. J. Klass, "FAA Stresses Proximity Warning Systems," *Aviation Week*, **77**, 51 (July 30, 1962). This article describes experiments with infrared proximity warning systems. Early experiments indicated that there was not enough energy radiated by small aircraft to make a passive infrared system practical. Later experiments with a rotating beacon indicated that detection ranges of up to 6 miles in daytime and 9 miles at night could be achieved. A supplementary article appeared in the same magazine at **77**, 37, (October 22, 1962). Both articles recognize that an infrared system cannot be considered as an all-weather system.

[9] "IR Tracker Fixes on Modulated Beam," *Electronic Design*, **10**, 38 (November 22, 1962). Describes a ground-mounted infrared tracker and an aircraft-mounted source and their integration into an automatic aircraft landing system. Proponents of this idea apparently overlook the difficulties with infrared sensors in bad weather when such a landing system would be needed most.

[10] W. Summer, *Ultra-Violet and Infra-Red Engineering*, New York: Interscience, 1962. The title is misleading, since there is little mention of any application involving wavelengths longer than $1.2\,\mu$. The author, a chemical engineer, is interested in describing such applications as heating, drying, sterilizing, night driving, and signaling. There is an interesting discussion of possible health hazards from exposure to intense sources of infrared radiation.

[11] D. W. DeWerth, "A Study of Infrared Energy Generated by Radiant Gas Burners," Research Bulletin 92, American Gas Association Laboratories, Cleveland, Ohio, 1962. Tabulates the radiating characteristcs of gas-fired and elctric-powered sources suitable for industrial heating applications. A curve of the spectral distribution (1.4 to $16\,\mu$) is given for each source. A particularly valuable feature of this report is the data given on the absorption spectra of materials that might be cooked, dried, or heated by infrared sources.

[12] M. LaToison, *Infrared and its Thermal Applications*, Eindhoven, the Netherlands: N. V. Philips Gloeilampenfabrieken, 1964. Industrial heating and drying applications. See also ref. 17.6.2 [5].

[13] E. D. Mills, N. A. Sullivan, and J. W. Meyer, "An Experimental Infrared Radar," *Microwave J.*, **9**, 33 (February 1966). This device uses a pulsed gallium arsenide injection diode as a transmitter (at $0.84\,\mu$) and a photomultiplier tube and 5 ft diameter mirror as a receiver. Maximum range appears to be about 1600 ft.

[14] E. Herbert, "Traffic Safety," *Intl. Sci. Tech*, September 1966, p. 42. Describes an experimental infrared intervehicle sensing system that continuously measures the distance between two cars. Each car carries on its rear bumper an infrared source modulated at a frequency that is proportional to the speed of the car. Similarly, each car has on its front bumper an infrared sensor to detect the modulated signal from the car ahead. The signal processor interprets the received signal level as a measure of the distance to the other car and compares the speeds of the two cars to indicate whether the following car is overtaking. A radar system proved unreliable because of its response to signals reflected from the pavement.

[15] "Practical Injection Lasers Open Wide Range of Applications," *Laser Focus*, **3**, 13 (January 1967). Describes two new injection diodes, one with a 2 W output, the other with a 50 W output, and the potential application for automobile collision warning, burglar alarms, secure communications, and the myth that won't die—fog penetrators. In the collision warning system, the injection laser would be mounted on the front of the car and function as a simple pulse radar.

[16] "Infrared Beams May Stop You Next Time You Speed," *Microwaves*, **6**, 6 (April 1967). Proposes an infrared radar to detect speeders.

[17] S. B. Ruth, "IR Method Bonds Flexible Cables," *Electronic Engr.*, **26**, 21 (April 1967). Uses a tungsten lamp and ellipsoidal reflector to solder as many as 50 connections in 10 sec.
[18] "Focused Infrared Solders Flat Cable," *Electronic Equipment Eng.*, **15**, 37 (June 1967). Device uses a tungsten-halogen lamp and an elliptical reflector. Using flat cable, as many as 50 connections can be soldered in 10 sec.
[19] P. J. Klass, "Infrared Under Study for Pilot Warning," *Aviation Week*, **87**, 56 (December 25, 1967). Reports a symposium at which the FAA and various airspace users outlined their desires for a low-cost pilot-warning indicator. One infrared system that is described uses a high-intensity xenon discharge lamp, which has an intense emission line at $0.9\ \mu$, and a silicon detector. Detection ranges up to 8 miles have been demonstrated, both day and night.
[20] "Lasar Radar Detector," *Industrial Research*, December 1967, p. 34. The first project to be tested at the Department of Transportation's new center in Denver is a laser detector to warn railroad engineers of track obstacles as small as an ice cube. Pulsed gallium arsenide injection diodes are placed at 360 m intervals so that trains traveling 200 mph will be given up to 5 miles of warning.

18

Medical Applications of Infrared Techniques

Thanks to the efforts of a relatively small number of workers, it is becoming increasingly evident that infrared devices can provide practical solutions to numerous medical problems. Most of these successful applications capitalize on the ability of infrared devices to measure temperature without contact. Since many bodily ills produce temperature changes, equipment that can sense these changes provides useful diagnostic assistance. Because the skin is relatively transparent in the near infrared, photographs made on infrared film, or observations with image converter devices, yield valuable information about skin disorders, healing of wounds, and the condition of the superficial venous system.

Although a great deal of research is being devoted to chemotherapy as the most promising approach to the cure of cancer, means of detecting cancer in its early stages and of accurately delineating its extent are sorely needed. Infrared radiometers and thermal imaging devices have been applied to this task and appear to have established themselves as one of the more reliable methods available. Apparently the higher metabolic rate in a tumor causes an increase in the local temperature. As a result, the temperature of the skin overlying the tumor may be as much as several degrees warmer than that of the surrounding skin.[1] Early investigators used radiometers to scan the suspected area and to measure the observed temperature differences. References to this work are found in Section 18.2. More recently, thermal imaging devices that were originally developed for the military have found great favor for this application because the thermal pictures that they present are so very convenient to interpret. References to this work are found in Section 18.4.

Local metabolic activity ceases when the blood supply is stopped; the

---

[1]In practice, most clinicians compare the temperatures of contralaterally symmetrical areas of the skin.

tissue then dies. Nerve tissue dies in about 4 min, while other kinds of tissue may live for 12 to 15 min. The local temperature at any point in the body depends on several factors, the most important of which are the heat carried into and away from the area by the blood and the heat generated by the local metabolic activity. Hence, when the blood supply is interrupted and metabolic activity ceases, the temperature of the affected area falls. Here then is a potentially interesting application for radiometers and thermal imaging devices, the differentiation of living tissue from dead tissue. Such a device, provided that it has a real-time display capability, could be extremely useful in emergency wards, where the surgeon must cut away the dead tissue before sewing up a patient's wound.

The potential dollar value of the market for infrared medical devices makes it attractive for new specialty companies or as a means of diversification for companies already established in the military market. Unfortunately the successful penetration of this market will require overcoming many of the same problems that were noted in Chapter 17 in connection with penetrating the industrial market. Since most engineers have little understanding of medical instrumentation, it is evident that the successful development team must include capable medical personnel. Such equipment must have an unusually high degree of reliability, be simple enough for operation by noninstrumentation personnel, be noncontaminating, and be operable in explosive atmospheres. The thermal imaging devices that are currently in use for medical investigations are too expensive for purchase by the individual physician or surgeon and are, as a result, found only in hospitals or clinics. It is hoped that further development will change this situation.

18.1 SEARCH, TRACK, AND RANGING APPLICATIONS

There has been some effort devoted to making an infrared obstacle detector for the blind. One such passive device is described in the references given here; active devices to accomplish the same purpose are described in Section 18.6.1.

18.1.1 Obstacle Detection (Passive)

[1] H. E. Kallman, "Optical Automatic Range Determining Device," U.S. Patent No. 2,524,807, October 10, 1950. Describes an automatic ranging device that can be used as an obstacle detector for the blind. The novel feature is the use of a reticle that scans in depth in order to locate the plane of sharpest focus. The reticle blank forms a one-turn helix. The openings in the reticle are comparable in size to the blur circle of the optics. As the reticle scans in depth along the optical axis, pulses are generated in the output of the detector. The sharpest pulse occurs when the chopping takes place at the

plane of best focus. The pitch of the reticle helix can be selected in conjunction with the focal length of the optics so as to give a convenient relationship between reticle rotational angle and range.

[2] H. E. Kallman, "Optar, a Method of Optical Automatic Ranging, as Applied to a Guidance Device for the Blind," *Proc. Inst. Radio Engrs.*, **42**, 1438 (1954). See the previous reference for a description of this device.

18.2 RADIOMETRIC APPLICATIONS

In addition to their use for the early detection of cancer, radiometers are being used as noncontact biosensors for astronauts, for studies of the physiological mechanisms of temperature sensation, to monitor healing of wounds, and to detect the onset of infection in wounds.

18.2.1 Measurement of Temperature

[1] J. D. Hardy, "The Radiation of Heat from the Human Body. I. An Instrument for Measuring the Radiation and Surface Temperature of the Skin," *J. Clin. Invest.*, **13**, 593 (1934). Describes the development of the first Hardy dermal radiometer, an instrument that has served as a prototype for many later investigators. For further details, see ref. 18.2. 1[3].

[2] J. D. Hardy and C. Muschenheim, "The Radiation of Heat from the Human Body. IV. The Emission, Reflection, and Transmission of IR Radiation by the Human Skin," *J. Clin. Invest.*, **13**, 817 (1934).

[3] J. D. Hardy and G. F. Soderstrom, "An Improved Apparatus for Measuring Surface and Body Temperature," *Rev. Sci. Instr.*, **8**, 419 (1937). Describes the Hardy dermal radiometer, an instrument that can measure surface temperatures to an accuracy of $\pm 0.01°C$. The radiometer consists of a blackened conical cavity with a thermopile detector located at the apex of the cone. The open end of the cone is placed against the surface to be measured. The indicating device is a portable galvanometer. The radiometer is calibrated with a Leslie Cube, a massive cube with a conical cavity cut in one face. The temperature of the cube is monitored by a thermocouple. This radiometer has long been the principal instrument for studies of the temperature of the human skin.

[4] J. D. Hardy, "The Radiating Power of Human Skin in the Infra-Red," *Am. J. Physiol.*, **127**, 454 (1939).

[5] R. H. Richards, A. M. Stoll, and J. D. Hardy, "The Panradiometer: An Absolute Measuring Instrument for Environmental Radiation," *Rev. Sci. Instr.*, **22**, 925 (1951). An improvement on the basic Hardy dermal radiometer.

[6] A. M. Stoll and J. D. Hardy, "A Method of Measuring Radiant Temperature of the Environment," *J. Appl. Physiol.*, **5**, 117 (1952). Uses a modified Hardy dermal radiometer.

[7] A. M. Stoll, "A Wide-Range Thermistor Radiometer for the Measurement of Skin Temperature and Environmental Radiant Temperature," *Rev. Sci. Instr.*, **25**, 184 (1954). Describes a modified version of the Hardy dermal radiometer. The modifications include the use of a thermistor bolometer as a detector and electronic amplification.

[8] J. D. Hardy and A. M. Stoll, "Radiometric Methods for Measurement of Skin Temperature," in *Methods in Medical Research*, Chicago, Year Book Publishers, 1954, Vol. 6, p. 85. Gives an excellent summary of Hardy's work and of the state of the art as of 1954.

[9] R. N. Lawson, A. L. Saunders, and R. D. Cowen, "Breast Cancer and Heptaldehyde," *Can. Med. Assoc. J.*, **75**, 486 (1956). Contains the first published indication that breast tumors can be detected by radiometry. The measurements reported here indicate that the temperature rise associated with breast tumors averages about 1°C. Clinical experience suggests that there is some correlation between the temperature rise and the degree of malignancy.

[10] W. E. Osborne, "Infrared Detector Aids Medical Diagnosis," *Electronics*, **30**, 155 (October 1957). This author claims to have used a small radiometer (lead sulfide detector) to sense minute temperature changes in the head or body that are produced by emotion or deep thought. Complete circuit diagrams are given for the radiometer electronics.

[11] E. Hendler and J. D. Hardy, "Infrared and Microwave Effects on Skin Heating and Temperature Sensation," *IRE Trans. Medical Electronics*, **ME-7**, 143 (1960). Describes an investigation of the physiological mechanisms of temperature sensation. The foreheads of 7 subjects were heated by various means while the temperature of the skin was monitored by an infrared radiometer. The radiometer has an NEΔT of 10^{-3}°C. For another mention of the same work, see ref. 18.2.1 [15].

[12] K. F. L. Williams, F. L. Williams, and R. S. Handley, "Infra-red Radiation Thermometry in Clinical Practice," *Lancet*, **1**, 958 (1960).

[13] K. F. L. Williams, F. L. Williams, and R. S. Handley, "Infra-red Thermometry in the Diagnosis of Breast Disease," *Lancet*, **2**, 1378 (1961). Authors report that they were able to make a correct diagnosis of cancer in 54 of 57 cases based on the radiometric measurement of temperature rises in the affected area. A temperature increase of 1°C was considered to be a significant indication of malignancy.

[14] R. D. Elam, D. W. Goodwin, and K. F. L. Williams, "Optical Properties of the Human Epidermis," *Nature*, **198**, 1001 (1963).

[15] J. D. Hardy, ed., *Temperature—Its Measurement and Control in Science and Industry*, New York: Reinhold, 1963; see E. Hendler, J. D. Hardy, and D. Murgatroyd, "Skin Heating and Temperature Sensation Produced by Infrared and Microwave Irradiation," p. 211. Covers the same material as does ref. 18.2.1 [11]. In addition, this paper shows that the spectral reflectance of white skin and of Negro skin is essentially identical at wavelengths longer than 2 μ.

[16] "Where it's a Little Warmer," *Sci. Am.*, **210**, 59 (April 1964). Describes a simple radiometer consisting of an IRTRAN-4 lens, a thermistor bolometer in a bridge circuit, and a microammeter. This radiometer has been used in clinical practice to monitor the healing of wounds, to detect the onset of infection in a bandaged wound without having to remove the bandage, to measure the temperature of palpable masses found in breasts, and to detect temperature rises in ulcerated lesions.

[17] "Remote Biosensors Sought for Astronauts," *Aviation Week*, **82**, 36 (April 19, 1965). Reports the award of a contract for the development of a family of sensors that can monitor an astronaut's physical condition without contact. Among the sensors is an infrared radiometer to measure skin temperatures.

[18] E. E. Brueschke, J. D. Haberman-Brueschke, and J. Gershon-Cohen, "Infrared Thermoprofile Analysis in Clinical Medicine," *Am. J. Med. Electronics*, **4**, 65 (1965).

[19] R. F. Rushmer et al., "The Skin," *Science*, **154**, 343 (1966). An excellent tutorial paper that gives a detailed description of the structure of the human skin. See also ref. 18.4.1 [50].

18.3 SPECTRORADIOMETRIC APPLICATIONS

Most of the applications that fall in this category are applications of infrared spectroscopy, a subject that is not covered in this book. The references in this section describe gas analyzers for physiological studies and contaminant detectors for air pollution studies.

18.3.1 Miscellaneous

[1] L. E. Baker, "A Rapidly Responding Narrow-Band Infrared Gaseous CO_2 Analyzer for Physiological Studies," *IRE Trans. Bio-Medical Electronics*, **BME-8**, 16 (1961). The rapid determination of carbon dioxide in expired air is one of the more important measurements in respiratory physiology. This paper shows that an infrared gas analyzer provides an excellent means of making this measurement. For further information on gas analyzers, see refs. 17.3.2 [1] and [6].

[2] E. R. Stephens, "Long-Path Infrared Spectroscopy for Air Pollution Research," *Infrared Phys.*, **1**, 187 (1961). Contaminant levels as low as 0.1 ppm can be detected by this method.

[3] D. J. Hunt, "Laser Action in Human Breath and its Use for Monitoring the Carbon Dioxide Content of Air," *J. Sci. Instr.*, **44**, 408 (1967). Laser action at $10.6\,\mu$ has been obtained from the carbon dioxide in exhaled human breath. The author shows that the total radiant energy from this laser is proportional to the carbon dioxide content. Such a device could be used in medical research, in safety level monitoring, or to measure the carbon dioxide content in deep-sea diving equipment.

18.4 THERMAL IMAGING APPLICATIONS

It is well known by now that numerous bodily ills give rise to thermal anomalies that can be detected by thermal imaging or radiometric devices. Since the thermal imaging devices provide a near-real-time display (Polaroid film or cathode-ray tube), they have proved to be valuable diagnostic aids for the physician or surgeon. Among the more successful applications[1] of this equipment are early detection of breast cancers, determination of the extent of burns and frostbite, pregnancy determination, studies of rheumatoid arthritis, determination of the optimum site for an amputation, placental location, and the early diagnosis of incipient stroke. In general, workers in the medical field refer to thermal imaging devices as "thermographs" and to the resulting thermal images as "thermograms."

It is worth repeating the warning that infrared equipment measures temperatures that are characteristic of surface conditions only. Hence their use is limited to surface phenomena or to near-surface conditions whose effect can be observed at the surface. As a result, the infrared equipment that has been used for the early detection of cancer has been

---

[1] Dr. R. Bowling Barnes of the Barnes Engineering Co. has been most generous in bringing to my attention many of the references in this section.

most successful in the detection of the near-surface tumors that occur in cancer of the breast. Deep-seated thermal anomalies, such as a severely inflamed appendix, will occasionally show on thermal imagery, but usually not until long after the condition has been identified by more conventional tests.

18.4.1 Diagnostic Assistance

[1] F. Urbach, N. R. Nail, and D. Perlman, "The Observation of Temperature Distributions and of Thermal Radiations by Means of Non-Linear Phosphors," *J. Opt. Soc. Am.*, **59**, 1011 (1949). Describes methods of making thermal images by the use of phosphors applied directly to the skin. The minimum detectable temperature difference is about 2°C.

[2] R. N. Lawson, "Implications of Surface Temperature in the Diagnosis of Breast Cancer," *Can. Med. Assoc. J.*, **75**, 309 (1956). This paper (in conjunction with ref. 18.2.1 [9], contains the first report that carcinoma (cancer) of the breast raises the temperature of the overlying skin; thus this paper forms the foundation for all of the later work that has been done in this area. Lawson used an Evaporograph for some of his work. He suggests that it might be the means of making extensive surveys in the same manner that chest X-rays are used for the early detection of tuberculosis. The temperature measurements reported in this paper were made with a thermocouple placed in contact with the skin. In a series of 26 patients who were later proved to have breast cancer, the average temperature rise in the area of the tumor was 1.2°C and all of the observations fell between 0.7 and 2°C.

[3[R. N. Lawson, "Thermography—A New Tool in the Investigation of Breast Lesions," *Can. Services Med. J.*, **13**, 517 (1957). Describes early experiments with the Evaporograph and with hand-held thermocouples for the measurement of temperature differentials. More recently, extensive work has been done with the thermal imaging device described in ref. 17.4.1 [7].

[4] R. N. Lawson, "A New Infrared Imaging Device," *Can. Med. Assoc. J.*, **79**, 402 (1958).

[5] R. W. Astheimer and E. M. Wormser, "Instrument for Thermal Photography," *J. Opt. Soc. Am.*, **49**, 184 (1959). Describes a commercially available thermal imaging device that has been used extensively in cancer detection and other diagnostic applications. See ref. 17.4.1 [7] for a description of the device.

[6] R. N. Lawson, G. D. Wlodek, and D. R. Webster, "Thermographic Assessment of Burns and Frostbite," *Can. Med. Assoc. J.*, **84**, 1129 (1961). Reports that a thermal imaging device was 90 per cent accurate in predicting the extent of 55 skin burns and that it correctly predicted the onset of gangrene in all 27 frostbite cases that were studied.

[7] "Infrared Camera Measures Temperatures in Arctic Survival Tests," *J. Opt. Soc. Am.*, **51**, 1316 (1961). Describes the use of a thermal imaging device to measure the variation in body temperatures during Arctic survival studies. The device takes 7 min to scan the individual subject, and the minimum detectable temperature difference is 0.1°C.

[8] C. M. Cade, "Seeing by Heat Waves," *J. Sci. Industr. Research*, **20A**, 624 (1961). Describes a thermal imaging device that uses a cooled indium antimonide detector. The output is displayed on electrochemical paper with a resolution of 100 lines per inch. The NEΔT as measured on the display is about 1°C. The use of this device as diagnostic aid is described in ref. 18.4.1 [14].

18.4.1 DIAGNOSTIC ASSISTANCE

[9] G. C. Shrivastava, and J. H. Quastec, "Malignancy and Tissue Metabolism," *Nature*, **196**, 876 (December 1, 1962). Discusses the high glycolytic activity in a malignant tumor. This may be responsible for the temperature rise observed at the tumor or, alternatively, the temperature rise may result from a combination of metabolic factors and the increased supply of blood to the tumor.

[10] J. H. Veghte and G. Solli, "Determining Arctic Clothing by Means of Infrared Radiometry," *Military Medicine*, **127**, 242 (1962). Reports that thermal imagery clearly shows the excessive heat loss from the head and the extremities, and the high heat losses associated with wrinkles or areas in which clothing is compressed.

[11] N. F. Boas and W. Z. Lane, "Infrared Thermography of the Human Body in Rheumatoid Arthritis," *Arthritis and Rheumatism*, **6**, 763 (1963). Arthritis is an inflammatory process, and the resulting temperature increases can be detected with thermal imaging devices. Apparently the magnitude of the temperature increases can be correlated with the discomfort experienced by the patient.

[12] R. N. Lawson and M. S. Chughtai, "Breast Cancer and Body Temperature," *Can. Med. Assoc. J.*, **88**, 68 (1963). The authors believe that their work, extending over a period of six years, shows that the temperature increases observed at the site of a tumor are due to local exothermic (metabolic) processes and not to an increase in the blood flow. A table shows detailed temperature measurements of 54 female patients who had palpable masses in the breast. Pathology indicated that 19 of the masses were malignant tumors and that the remainder were benign. The average temperature over the malignancies was 36.9°C, and over the benign tumors it was 35.4°C. On the assumption that a temperature rise of more than 1°C indicates a malignancy, 34 of the 35 benign tumors would have been diagnosed correctly as would 14 of the 19 malignant tumors.

[13] R. B. Barnes and J. Gershon-Cohen, "Clinical Thermography," *J. Am. Med. Assoc.*, **185**, 949 (1963). A short survey article.

[14] K. F. L. Williams, C. M. Cade, and D. W. Goodwin, "The Electronic Heat-Camera in Medical Research," *J. Brit. Inst. Radio Engrs.*, **25**, 241 (1963). Describes the application of a thermal imaging device to the diagnosis of breast diseases, the determination of the optimum site for an amputation, the location of blockage in a vein, monitoring the progress in a skin graft, and studies of thyroid disorders. The scanning device is described in ref. 18.4.1[8].

[15] R. B. Barnes and J. Gershon-Cohen, "Thermomastography," *J. Albert Einstein Medical Center*, **11**, 107 (1963). The authors define thermomastography as the technique of determining the infrared radiation from the breasts. Work to date indicates that temperature differences greater than 1°C between contralaterally symmetrical areas of the skin are strongly indicative of an abnormal condition. The observation of such temperature differences appears to be a particularly promising method for diagnosis of diseases of the breast, such as cancer, and for the differentiation of benign from malignant tumors.

[16] R. P. Goldman, "New Medical Device Spots Hidden Diseases," *Sat. Eve. Post*, September 28, 1963, p. 44. An excellent discussion of the pioneering work of Barnes and Gershon-Cohen in applying thermal imaging devices to medical diagnosis.

[17] R. B. Barnes, "Thermography of the Human Body," *Science*, **140**, 870 (1963). An outstanding discussion of the use of thermal imaging devices for medical diagnosis. This paper is required reading for anyone with an interest in this field.

[18] J. Gershon-Cohen et al., "Thermography of the Breast," *Am. J. Roentgenology*, **91**, 919 (1964). A survey paper that contains a particularly good description of the physical process involved in the production of heat by the body.

[19] J. Gershon-Cohen and R. B. Barnes, eds., "Thermography and its Clinical Applications," *Ann. N.Y. Acad. Sci.*, **121**, 1 (1964). Contains 29 papers that were presented during a two-day symposium. This monograph is the most complete reference available on the subject. The various authors indicate that thermography is of value in such fields as breast diseases, cancer detection, dermatology, gynecology, infections, obstetrics, oncology, orthopedics, peripheral and cerebrovascular diseases, and veterinary medicine.

[20] R. J. Young, "Application of Thermography to the Problem of Placental Localization: Preliminary Communication," *Brit. Med. J.*, **2**, 978 (1964).

[21] J. D. Haberman-Brueschke and J. Gershon-Cohen, "Thermographic Localization of the Placental Site," *J. Albert Einstein Medical Center*, **12**, 248 (1964). Demonstrates that the placenta can be located thermographically without maternal or fetal hazard. In 48 of 54 cases the thermographic indication was correct.

[22] J. T. Crissey et al., "A New Technique for the Demonstration of Skin Temperature Patterns," *J. Invest. Dermatology*, **43**, 89 (1964). Discusses the application of thermography to the determination of the degree and depth of burns and frostbite injuries and to the determination of the most favorable time for skin grafting.

[23] K. F. L. Williams, "The Heat Camera in Medicine," *New Scientist*, **400**, 162 (1964).

[24] K. F. L. Williams, "Pictorial Heat Scanning," *Physics in Medicine and Biology*, **9**, 433 (1964). An excellent tutorial article on the clinical applications of thermal imaging devices.

[25] J. Gershon-Cohen and J. D. Haberman, "Thermography," *Radiology*, **82**, 280 (1964).

[26] J. L. Fergason, "Liquid Crystals," *Sci. Am.*, **211**, 76 (August 1964). Liquid crystals, such as cholesterol esters, exhibit changes in reflectance when heated or cooled through a transition region above their melting points. The crystals are transparent except for a narrow band about $0.02\,\mu$ wide that shifts from the ultraviolet to the near infrared as the temperature of the crystal falls through the transition temperature. Solutions of the crystals can be painted directly on an object so as to provide a qualitative picture of the thermal conditions. Since this technique reveals temperature differences as small as a fraction of a degree, it may provide stiff competition for the more conventional thermal imaging devices.

[27] "Medical Electronics," *Electronic Design News*, **10**, 99 (December 1965). A broad survey of the field with a short description of the use of thermal imaging devices in medicine.

[28] B. Bernard, "The Promise of Infrared," *Electronic Design News*, **10**, 150 (December 1965). Surveys the application of thermal imaging techniques to medical diagnosis.

[29] E. Brueschke, J. D. Haberman-Brueschke, and J. Gershon-Cohen, "Infrared Thermoprofile Analysis in Clinical Medicine," *Am. J. Medical Electronics*, **4**, 65 (1965). Describes automatic methods of obtaining isotemperature curves from ordinary thermograms.

[30] D. A. Kliot and S. J. Birnbaum, "Thermographic Studies of Wound Healing," *Am. J. Obstetrics and Gynecology*, **93**, 515 (1965).

[31] J. Gershon-Cohen and J. D. Haberman, "Medical Thermography," *Am. J. Roentgenology*, **94**, 735 (1965).

[32] A. Smessaert et al., "Use of Thermography for Evaluation of Sympathetic Blocks," *Am. J. Surgery*, **109**, 594 (1965).

[33] R. N. Lawson and L. L. Alt, "Skin Temperature Recording with Phosphors," *Can. Med. Assoc. J.*, **92**, 255 (1965). Describes the use of a zinc-cadmium sulfide phosphor. Under ultraviolet illumination, warm areas fluoresce less than cooler areas. The method has been applied to the detection of breast cancer.

[34] H. M. Wright and I. M. Korr, "Neural and Spinal Components of Disease: Progress in the Application of Thermography," *J. Am. Osteopathic Assoc.*, **64**, 918 (1965).

[35] D. D. Delahanty and J. R. Georgi, "Thermography in Equine Medicine," *J. Am. Veterinary Med. Assoc.*, **147**, 235 (1965).

[36] J. D. Haberman-Brueschke et al., "Effect of Smoking on Skin Temperature as Demonstrated by the Thermograph," *J. Albert Einstein Medical Center*, **13**, 205 (1965). The abrupt temperature changes due to vasoconstriction after smoking can be demonstrated by thermography. Temperature drops of 1.5°C are typical.

[37] J. Gershon-Cohen, J. D. Haberman-Brueschke, and E. E. Brueschke, "Obstetric and Gynecologic Thermography," *Obstetrics and Gynecology*, **26**, 842 (1965). Among the last 100 cases of breast cancer seen by the authors, 96 showed a hot spot by thermography, with a temperature rise greater than 1°C. The thermographic hot spot was the sole available clue to an underlying cancer in seven cases, despite painstaking physical and roentgenologic examinations. Since not all hot spots signify cancer, it is necessary to supplement thermography with other techniques.

[38] A. Gandy et al., "Les Cameras A Infra-Rouge Serviront-Elles au Diagnostic Des Tumeurs," *Presse Medicale*, **73**, 629 (1965).

[39] J. Gershon-Cohen, J. D. Haberman-Brueschke, and E. E. Brueschke, "Medical Thermography: A Summary of Current Status," *Radiologic Clinics of N. Am.*, **3**, 403 (1965). An excellent paper. Contains a survey of available equipment for thermography and its application to such tasks as placental localization, pregnancy determination, and the detection of breast diseases, cancer in other parts of the body, orthopedic abnormalities, dermatologic disturbances, urologic conditions, and acute and chronic thoracic conditions. The authors believe that a good thermogram should show a minimum temperature difference of 0.25°C, an angular resolution of 1 mrad, and a frame time (total scanning time) of no more than 2 or 3 min.

[40] J. D. Wallace et al., "Rapid Scanning Thermography in Suspect Breast Evaluation," *Radiology*, **84**, 132 (1965).

[41] E. H. Wood, "Thermography in the Diagnosis of Cerebrovascular Disease," *Radiology*, **85**, 270 (1965). Describes efforts to provide an early diagnosis of incipient stroke by the use of thermography. The normal values of facial temperatures were established by making thermograms of approximately 1000 healthy subjects. This report is based on the study of 117 patients with intra- or extracranial cerebrovascular abnormalities and 25 patients with brain tumors and miscellaneous neurological disorders. Of the 64 patients known to have abnormalities of the extracranial portion of the carotid artery, 56 showed abnormal thermograms. Only about 10 per cent of the patients with intracranial abnormalities showed abnormal thermograms. The author concludes that thermography provides a valid means of detecting approaching stroke. A preliminary report on this work appeared in the same journal at **83**, 540 (1964).

[42] A. G. Swearingen, "Thermography: Report of the Radiographic and Thermographic Examinations of the Breasts of 100 Patients," *Radiology*, **85**, 818 (1965). Thermograms gave no false-negative indications but there were a few false-positive indications. An addendum, reporting observations on a total of 250 patients, indicates that two false-negative indications were found. When thermograms were used in conjunction with mammograms, the true-positive rate increased significantly.

[43] C. M. Cade and A. E. Hanwell, "Thermal Patterns," *Sci. J.*, **2**, 62 (February 1966). An excellent survey article that gives the British viewpoint on the usefulness of thermography in medicine. Photographs are shown of commercially available scanners from Sweden, England, and the United States.

[44] P. M. Johnson, D. G. Bragg, and J. J. Sciarra, "Placental Localization: A Comparison

of Radiopharmaceutic and Thermographic Methods," *Am. J. Roentgenology*, **96**, 681 (1966).

[45] R. N. Lawson, "A Technique for Identifying Suitable Veins for Blood Sampling in Breast Cancer," *Can. Med. Assoc. J.*, **94**, 451 (1966).

[46] J. F. Connell et al., "Thermography in the Detection of Breast Cancer," *J. Am. Cancer Soc.*, **19**, 83 (1966). The authors' experience with more than 300 patients indicates that thermography is an extremely useful aid in cancer detection, in evaluation of benign conditions in the breast, and as a means of following the efficiency of tumor therapy. The authors believe the method may become the first line of attack in a widescale detection program. The method is simple and easily taught, and may be used repeatedly without injury to the patient. All suspicious indications can be referred for mammographic examination.

[47] R. N. Lawson and E. Pederson, "Imaging of Human Surface Temperatures," *J. Soc. Motion Picture and Television Engrs.*, **75**, 641 (1966). Describes a technique that is competitive to thermography: heat-quenched ultraviolet-sensitive phosphors and a variable-threshold television system to enhance image contrast. The system can detect a temperature difference of about 1°C.

[48] S. J. Birnbaum, "Breast Temperature as a Test for Pregnancy," *Obstetrics and Gynecology*, **27**, 378 (1966). Pregnancy can be detected within a few weeks of conception by the changes that occur in the surface temperature of the breasts. Tests on 50 pregnant and 50 nonpregnant patients indicate that in the nonpregnant state the breast tends to be slightly cooler (0.4°F) than the chest wall. During pregnancy the temperature of the breasts is approximately 1.5°F warmer than that of the chest wall. This temperature change appears within three to four weeks of conception and remains constant until delivery. Oral contraceptives may give a false indication of pregnancy when this method is used.

[49] C. E. Crandell and R. P. Hill, "Thermography in Dentistry. A Pilot Study," *Oral Surgery, Oral Medicine, and Oral Pathology*, **21**, 316 (1966). This study suggests that thermography is of limited value in dentistry. Thermograms showed slight indications of abscesses with the mouth closed. With the mouth open, oral air and salivation prevented any useful results.

[50] R. F. Rushmer, *et al.*, "The Skin," *Science*, **154**, 343 (1966). Contains two pictures in color that show the use of color changes in liquid crystals as a means of temperature indication. The crystals, cholesterol esters, exhibit a full spectrum of color as their temperature changes from 28 to 31°C. Above this temperature they are colorless. See also ref 18.2.1 [19].

[51] J. Gershon-Cohen, "Medical Thermography," *Sci. Am.*, **216**, 94 (February 1967). This article contains some very excellent illustrations of typical thermograms.

[52] R. B. Barnes, "Medical Thermography," *J. Opt. Soc. Am.*, **57**, 1410 (1967). Describes a commercially available thermal imaging device, the MI-A Thermograph, that scans a 10 by 20 deg field of view with a resolution of about 2 mrad. Temperatures can be determined to an accuracy of ±0.1°C. This paper describes the correlations that have been found between localized thermal abnormalities and pathological conditions.

[53] "Every Man His Own Furnace," *Life*, **64**, 83 (April 5, 1968). This article contains four thermograms in color, a notable advance in display technology. In these thermograms the temperature information is conveyed by color rather than by shades of gray. The warmest areas are shown as red, intermediate temperature areas are yellow, and cool areas are green. The full page thermogram of a seated male graphically demonstrates many of the comments that are made in Section 3.4 on personnel as infrared targets. In this example the exposed skin of the face and hands has the

18.5.1 APPLICATIONS OF IMAGE CONVERTER TUBES

highest temperature (28°C), followed closely by those areas where the clothing fits tightly, such as a man's shirt across his shoulders. The coolest areas are those in which the clothing fits loosely, such as a man's trouser legs (23°C in this thermogram).

[54] R. B. Barnes, "Diagnostic Thermography," *Appl. Opt.*, **7**, 1673 (1968). A superb paper that is certain to become a classic in its field. Discusses the basis for diagnosis by thermography, available equipment, examination techniques, and the history of medical thermography. Among the specific applications that are discussed are cancer, obstetrics and gynecology, peripheral vascular disease, cerebrovascular disease, trauma and wound healing, orthopedics, arthritis and dermatology. A number of excellent illustrations accompany this article.

18.5 APPLICATIONS INVOLVING REFLECTED FLUX

Human skin is relatively transparent at the shorter infrared wavelengths. The spectral transmittance curve of the human cheek, for instance, shows that the maximum transmittance, about 20 per cent, occurs at $1.15\,\mu$ and that there is a sharp cutoff beyond $1.4\,\mu$ because of water in the body. For a discussion of the importance of these facts in heat therapy, the reader is referred to Section 2.8. Because of the relative transparency of the skin at the shorter wavelengths, the superficial veins can be examined by means of image converter tubes or with the aid of photographs taken with infrared-sensitive film. The nonoxygenated blood flowing in the veins has a high absorption in the near-infrared region and thus photographs quite dark. Since oxygenated blood flowing in the arteries is quite transparent in this spectral region, it does not appear in infrared photographs or on image converter tubes. Several diseases, such as axillary thrombosis (a formation of a blood clot in the circulatory system) and cirrhosis (a degenerative disease, especially of the liver, marked by excessive formation of connective tissue), cause serious disturbances of the normal venous structure. The progress being made in treating a patient for varicose veins is often followed by infrared photography or with image converter tubes, since scabs are quite transparent in the near infrared.

18.5.1 Applications of Image Converter Tubes

[1] A. Vasko and M. Peleska, "Visual Diagnosis of Eye Diseases by Means of Infrared Radiation," *Brit. J. Ophthal.*, **31**, 419 (1947). The first reported use of an image converter tube for the examination of the human eye through corneal opacities and for the study of eye movements in complete darkness.

[2] T. H. Pratt, "The Infra-Red Image Converter Tube," *Electronic Engineering*, **20**, 274, 314 (1948). Begins with an excellent tutorial treatment of image converter tubes. The military applications cited are discussed in ref. 16.5.1 [12]. Viewers with image converter tubes have been used for observing rats in the dark during an investigation of the spread of typhus in the Far East, and for observing the behavior of malaria-carrying mosquitoes in the dark. Similar viewers have also proved useful for examination of the retina of the eye through certain types of corneal opacities.

[3] M. Koomen, R. Tousey, and H. A. Knoll, "An Infrared Pupillometer," *J. Opt. Soc. Am.*, **38**, 719 (1948). Describes the use of an image converter tube to measure the size of the pupil in the human eye without affecting its state of adaptation.
[4] H. Knoll, "An Infrared Skiascope," *Am. J. Optom.*, **30**, 346 (1953).
[5] M. Marshall, "Infrared Image Converter in Rod Scotometry," *Brit. J. Ophthal.*, **37**, 316 (1953).
[6] N. B. Chin and R. E. Horn, "Infrared Skiascopic Measurements of Refractive Changes in Dim Illumination and in Darkness," *J. Opt. Soc. Am.*, **46**, 60 (1956).
[7] A. Ogg, "Examination of the Eye with Infrared Radiation," *Brit. J. Ophthal.*, **42**, 306 (1958). Author adapted a British image converter tube (CV-147) to a standard slit-lamp microscope and used it to study eyes with visually opaque corneas. In one instance an unsuspected malignancy was discovered.
[8] J. Friedman, "Penetration of Corneal Opacities by Infrared Electronics," *IRE Trans. Med. Electronics*, **ME-7**, 182 (1960). Describes the use of an image converter device to examine the anterior chamber of the eye beneath relatively dense corneal opacities. The type 6032 image converter tube used here has a resolution of 18 lines per mm, which is about 50 per cent better than that usually observed with the more common type 1P25 tube. This article contains an excellent bibliography.
[9] S. Tasaka et al., "Infrared Microscope and its Application to Medical Use," paper presented at International Conference on Medical Electronics, New York, N.Y., July 16, 1961. Reviewed in *Electronics*, **34**, 41 (August 25, 1961). Describes a microscope fitted with an image converter tube that has been used to distinguish normal cells from cancer cells, examine the internal structure of bone tissue, and study details of the iris of the eye.
[10] J. M. Davies and A. Levine, "Infrared Image-Converter Method of Observing Eye Motion in Flash Blindness Experiments," *J. Opt. Soc. Am.*, **55**, 1670 (1965).
[11] D. G. Green and F. Maaseidvaag, "Closed-Circuit Television Pupillometer," *J. Opt. Soc. Am.*, **57**, 830 (1967). Describes a near-infrared pupillometer with a conventional television monitor as well as automatic digital readout of pupil diameter. The size of the pupil is found by processing the video signal so as to count the number of scan lines that cross the image of the pupil.

18.5.2 Infrared Photography

[1] C. H. Cartwright, "Infra-Red Transmission of the Flesh," *J. Opt. Soc. Am.*, **20**, 81 (1930). The author measured the transmittance of his own cheek by placing a source in his mouth and analyzing the transmitted energy with a spectrometer. He found that the maximum value of the (external) transmittance was 20 per cent and that it occurred at a wavelength of $1.15\,\mu$. When these measurements are corrected for reflection losses at this wavelength the (internal) transmittance is about 30 per cent.
[2] H. Haxthausen, "Infrarotes Photographieren in der Dermatologie," *Dermatol. Wochschr.*, **97**, 1289 (1933). Contains the first mention of the fact that the near-surface venous structure can be photographed on infrared film.
[3] H. M. Dekking, "Infrarot Photographie des Auges," *Arch. f. Ophthal.*, **130**, 373 (1933); **133**, 20 (1934). Probably the first report of the use of infrared photography to examine the human eye through a corneal opacity.
[4] I. Kugelberg, "Der Augenhintergrund in Infrarotem Licht," *Acta Ophthal.*, **12**, 179 (1934). A paper of fundamental importance for its conclusion that the near-infrared region, such as that covered by infrared film, is a poor choice for the diagnosis of fundus diseases. The fundus is highly reflective in this region of the spectrum, and it is extremely difficult to observe any of its details.

[5] W. A. Mann, "Infrared Photography of the Eye," *Arch. Ophthal.*, **13**, 985 (1935).
[6] J. B. Feldman, "A Review of Infra-Red Photography with Reference to its Value in Ophthalmology," *Arch. Ophthal.*, **15**, 435 (1936).
[7] L. C. Massopust, "Infrared Photographic Study of the Changing Pattern of the Superficial Veins in a Case of Human Pregnancy," *Surgery, Gynecology, and Obstetrics*, **63**, 86 (1936).
[8] M. Nagel and A. Klughardt, "On the Measurement of the Twilight Pupil of the Eye by Means of Infrared Photography," *Zeitschrift für Physik*, **101**, 373 (1936) (SLA: TT-61-14346).
[9] E. E. Wilson, "The Changes in Infra-Red Photographs Taken During the Treatment of Varicose Veins," *Am. J. Surg.*, **37**, 470 (1937).
[10] F. Ronchese, "Infra-Red Photography in the Diagnosis of Vascular Tumors," *Am. J. Surg.*, **37**, 475 (1937).
[11] W. A. Gorman and A. Hirsheimer, "A Study of the Superficial Venous Pattern in Pregnant and Non-Pregnant Women by Infra-Red Photography," *Surgery, Gynecology, and Obstetrics*, **68**, 54 (1939).
[12] H. L. Gibson, "Infrared Photography of Patients," *Radiography and Clinical Photography*, **21**, 72 (1945). One of the finest survey articles available on the medical applications of infrared photography.
[13] W. Clark, *Photography by Infrared*, 2d ed., New York: Wiley, 1946. Chapter 9, entitled "Medical Infrared Photography", is still one of the best basic references on the various medical phenomena that can be observed in the near infrared.
[14] L. C. Massopust, *Infrared Photography in Medicine*, Springfield, Ill.: C. C. Thomas, 1952.
[15] M. A. Rosenbloom, "Infrared Photography of the Female Breast," *Obstetrics and Gynecology*, **2**, 603 (1953).
[16] L. C. Massopust and W. D. Gardner, "The Infrared Phlebogram in the Diagnosis of Breast Complaints," *Surgery, Gynecology, and Obstetrics*, **97**, 619 (1953).
[17] J. Friedman, "Penetration of Corneal Opacities by Infrared Electronics," *IRE Trans. Med. Electronics*, **ME-7**, 182 (1960). The first part of this paper traces the application of infrared photography to medicine and ophthalmology and includes a good bibliography on the subject. The applications of image converter tubes are discussed in ref. 18.5.1[8].
[18] H. L. Gibson, *The Photography of Patients*, 2d ed., Springfield, Ill.: C. C. Thomas, 1960, p. 165.
[19] H. Nelson, "World War II Infrared Color Film Tried as Weapon Against Disease," *Los Angeles Times*, October 31, 1966. Describes the use of false-color film for the early detection of hemorrhaging in hemophiliacs.
[20] *Medical Infrared Photography*, Publication N-1, Rochester, N.Y.: Eastman Kodak Co., 1967. In addition to the use of conventional infrared films, this book stresses medical uses of false-color film and infrared luminescence.

18.5.3 Miscellaneous

[1] "Sensor Detects Eyeball Movement," *Electronic Design*, **14**, 77 (March 1, 1966). Describes a means for extracting useful signals from the self-controled movements of the human eye. Device consists of a small infrared source (tungsten lamp with filter) and a cadmium selenide detector, both mounted on eyeglass frames. The detector senses the location of the spot on the eyeball from which light from the source is reflected.

[2] J. Merchant, "Oculometer For Hands-Off Pointing and Tracking," *Space/Aeronautics*, **49**, 92 (February 1968). A more complete description of the item described in the previous reference. The objective is eye control of various pointing and tracking tasks, such as weapon aiming. It is claimed that when a human observes small target detail the axis of the eye is pointed with an accuracy of 3 mrad.

18.6 APPLICATIONS INVOLVING A COOPERATIVE SOURCE

The development of the injection diode has stimulated interest in its use for an (active) obstacle detector for the blind. For passive obstacle detectors, see Section 18.1.1.

18.6.1 Obstacle Detection (Active)

[1] J. M. Benjamin, Jr., and T. A. Benham, "Electronic Obstacle Detector for the Blind," *J. Franklin Inst.*, **271**, 526 (1961). Uses a near-infrared source, a chopper, and a silicon photodiode to pick up the reflected flux. The output of the signal processor causes vibration of a stimulator located in the handle of the unit. The user can learn to interpret the vibration rate in terms of the distance to the obstacle. The unit has been tested at distances of 2 to 8 ft.

[2] "Seeing Eye Cane Made With Lasers," *Electronic Design News*, **12**, 22 (August 1967). Pulses from a pair of injection diode lasers are aimed so as to be reflected from points on the ground that are 3 and 6 ft from the cane. Two detectors sense the reflected beams and activate two blunt pins in the handle of the cane, causing them to vibrate against the user's hand. If there is an obstacle in either beam, the beam is not reflected back to its detector and the corresponding pin stops vibrating. The cane weighs 5 lb, and the rechargeable batteries can provide 10 hr of operation. The diodes are pulsed at a rate of 20 Hz and a pulse length of 80 nsec. Peak power output is 5 W.

19

Scientific Applications of Infrared Techniques

The scientist has long been an enthusiastic user of infrared techniques. Since the mid-1800's the literature has been filled with accounts of scientists using infrared radiometers to measure lunar and planetary temperatures, and spectrometers to investigate molecular structure and to identify organic materials in locations stretching from the laboratory to the atmospheres of other planets. As we have already seen, it was not until the early 1900's that engineers began to appropriate some of these infrared techniques and adapt them to the solution of military, industrial, and medical problems. Today, of course, the distinction between scientist and engineer has become blurred and the two often work side-by-side, so that new scientific discoveries are converted to practical applications in an astonishingly short time.

Because infrared techniques are particularly well adapted to observation from a distance, it is not surprising to find them so widely utilized in the current space programs. Hence the majority of the references in this chapter are concerned with means of increasing our knowledge of space. The term "remote sensing" has recently become fashionable and its practitioners have tried hard to sell it as a fundamentally new technique. This is somewhat amusing, because ever since Lord Rosse made the first radiometric determination of the moon's temperature in 1869, astronomers and astrophysicists have attached radiometers and spectrometers to their telescopes and gone about their business of gathering information on the universe by remote sensing techniques.

In analyzing various astronautical missions it is evident that astrionics, the application of electronics to astronautical systems, performs three major roles: guidance and control, scientific instrumentation, and communications. The guidance and control system establishes the trajectory appropriate for the mission, monitors any deviation from this trajectory, generates the required corrective commands, and maintains the proper attitude for particular subsystems or for the entire space vehicle. The scientific instrumentation fulfills the primary reason for most astronautical missions—that of learning more about space. Practically all of our information about the space environment comes from electronic instruments that sense some quantity, measure it to a (hopefully) known accuracy, and provide an output indication suitable for transmission back to earth. In the communication system, astrionics provides the link between the vehicle and the earth. Its primary purpose is to return the data from the scientific instrumentation; additionally it may be used to transmit command signals, for space-vehicle-to-space-vehicle communications, and for transmission of data on vehicle performance.

The instrumented spacecraft is obviously a powerful research tool. Its sensors can look down on the earth, studying it and what lies between; they can sample their immediate environment, making *in situ* measurements along extended trajectories; they can look out into space and examine the sun, moon, planets, and stars without suffering degradation from the earth's atmosphere; they can provide close-up views of other celestial bodies. From the beginning, infrared sensors have been among the most important instrumentation carried on each mission. They have measured the temperature of astronomical bodies, determined the constituents of their atmospheres and the characteristics of their surface materials, measured atmospheric heat balance phenomena, spotted terrestrial storms in their formative stage, and in general opened entirely new vistas in meteorological research.

Perhaps the single most dominant factor underlying the entire astrionics market is the almost unheard-of demand placed on reliability and the high cost of achieving it. Successful launches and the attainment of accurate trajectories are little solace for a dead vehicle that cannot measure data and transmit it back to earth. Designers are continually asked for more accurate, lighter-weight equipment with lower power consumption, to operate reliably in the relatively unknown environment of space for periods of time extending from a few days to several years. Regardless of its other capabilities, the company that refuses to recognize the importance of reliability is doomed to early failure in the astrionics market.

19.1 SEARCH, TRACK, AND RANGING SYSTEMS

The applications in this section include such things as satellite detection and navigation, attitude control, rendezvous, and docking systems for space vehicles.

19.1.1 Search and Track Systems

Although the feasibility of detecting satellites with earth-based infrared equipment has been demonstrated, there is little indication that this method will ever be competitive with radar techniques. Once again, the lack of an all-weather capability penalizes the infrared system. For a short discussion of the factors that affect the temperature of a satellite, see the introductory material in Section 19.2.

[1] "Sputnik is Radiating Strong Infrared Signal," *Aviation Week*, **67**, 31 (November 4, 1957). Reports that the radiation from Sputnik persists at night and is greater than can be accounted for from known sources, leading to the suspicion that there is some exotic form of radiation being emitted. In retrospect, this obviously erroneous report shows that even the infrared community did not escape the Sputnik hysteria. The radiometer used an uncooled lead sulfide detector. The field of view was 2 by 26 deg, with the long side set at right angles to the predicted flight path.

[2] P. J. Ovrebo, R. Astheimer, and E. Wormser, "Astronautical Applications of Infrared Techniques," *Proc. Inst. Radio Engrs.*, **47**, 1625 (1959). Describes the equipment used to track an early Sputnik satellite. The tracker had a 6 in. aperture and used a lead sulfide detector cooled to 195°K. Detection was by reflected sunlight.

[3] I. H. Swift, "System Design Using Background-Limited Infrared Detectors," *Military Systems Design*, **4**, 27 (May–June 1960). Essentially the same paper as that described in ref. 19.1.1[6].

[4] "50 Winking Lights," *Sci. Am.*, **205**, 71 (December 1961). Describes an infrared scanner that uses a 50-element array of lead sulfide detectors. The optics are an $f/0.76$ Schmidt system with a 10 in. aperture. The display is a series of 50 neon lamps, each lamp corresponds to one element of the detector.

[5] F. F. Hall, Jr., and C. V. Stanley, "Infrared Satellite Radiometry," *Appl. Opt.*, **1**, 97 (1962). Describes the design of an infrared satellite-tracking radiometer and reports measurements of the radiation from several orbiting satellites. The radiometer has an aperture of 50.8 cm, a nutating involute reticle for tracking, and uses either cooled lead sulfide or cooled indium antimonide detectors. The instantaneous field of view is 1/4 by 1/4 deg. The peak NEI with the lead sulfide detector (195°K) was 9×10^{-15} W cm^{-2}. The NEI with the indium antimonide detector (77°K) was 1.8×10^{-12} W cm^{-2} (effective, 1.8 to 6 μ). The observed irradiances ranged from 2×10^{-12} to 1×10^{-14} W cm^{-2} (1 to 3 μ) and 3×10^{-12} W cm^{-2} (2 to 6 μ). At no time was there any evidence of any exotic or unexplained radiation, as had been reported earlier in ref. 19.1.1[1].

[6] I. H. Swift, "Performance of Background-Limited Systems for Space Use," *Infrared Phys.*, **2**, 19 (1962). This paper is probably the best available reference on system design with background-limited detectors. Describes a search device that has an aperture of 60 cm, uses a five-element copper-doped germanium detector, and scans a 10-deg-square field of view in 1 sec. The pulse visibility factor of the electronics is equal to 1.6. The detection range, outside of the earth's atmosphere, is 90 miles (signal-to-noise ratio of 5 for a 1 m^2 target at a temperature of 300°K). The author states that this is the equivalent of detecting a refrigerator ice cube at a distance of 3 miles.

[7] M. E. Seymour, "Infrared Detection of Satellites and Space Stations," *Military Systems Design*, **7**, 22 (June 1963). Discusses satellite temperature, radiant intensity, and possible detection ranges.

[8] Y. P. Safronov, Y. G. Andrianov, and D. S. Iyevlev, *Infrakrasnaya Tekhnika v Kosmose (Infrared Technology in Space)*, Moscow: Voyenizdat, 1963. Authors conclude

that infrared technology will be useful for observation, search, and communication in space for both scientific and military purposes.

19.1.2 Navigation and Flight Control Systems

A typical interplanetary mission involves three distinct phases: a launch phase with early flight in the vicinity of the takeoff planet, a midcourse phase well away from any celestial body, and an approach phase that may end with an actual landing, an orbiting, or merely a flyby of the planet. Each phase may require different navigational techniques, often resulting in multiple systems located in different stages of the vehicle.

Space navigation and terrestrial navigation are conceptually quite different. The terrestrial navigator attempts to determine his true position on the surface of the earth. The space navigator determines the deviations of his vehicle from a carefully precalculated trajectory. The lines of position and dead reckoning used on earth are not at all the same as the differential correction techniques used in space. Space navigation is based on residuals, on the differences between observations as they would have appeared from a vehicle traveling the precalculated orbit, and the observations as they actually appear from the navigator's vehicle.

Let us adopt three definitions:

1. *Space navigation* — the observation and reduction of data to give the deviation of a vehicle from its expected trajectory.

2. *Orientation* — the observation and reduction of data to give the deviation of the vehicle reference axes from their predetermined position.

3. *Guidance* — the process of redirecting the vehicle in response to the indications of the navigation and orientation sensors.

By using these definitions, navigation and orientation become processes of observation and computation and guidance covers the means for achieving the required changes in vehicle flight path and orientation. Since optical sensors excel at providing angular information, there are many applications for them in the navigation and orientation areas and probably none at all in the guidance area.

What applications are there for infrared sensors in each of the phases of interplanetary flight? There seem to be very few applications during the launch phase; an infrared horizon scanner could be used for orientation, but it is more likely that an on-board inertial system, or information derived from ground-based tracking systems, would be used. Similarly, the navigational aspects are handled by ground-based tracking or on-board inertial systems.

The midcourse phase will almost certainly use infrared and optical

19.1.2 NAVIGATION AND FLIGHT CONTROL SYSTEMS

sensors for navigation and orientation. Navigation will be done by star trackers, sophisticated descendants of the terrestrial navigator's sextant. Orientation will be done by sun, earth, or planet sensors, star trackers, or combinations of two or more of these. One might assume that inertial devices would be strong contenders for these functions. The operating lifetime of many current inertial components is simply not compatible with the relatively long flight times associated with the midcourse phase. Additionally, the errors caused by drift of inertial components increase directly with time unless some external reference (such as a star tracker) is used to reset them periodically. Inertial devices can provide accurate indications of linear or angular velocities and accelerations for short periods of time. Hence they are used during initial stabilization, for retaining attitude information when the optical sensors may be temporarily inoperative, and for autopilot reference during guidance maneuvers. The sun, being relatively easy to identify, provides a primary reference and is often used to align two of the space vehicle axes. For nearby trajectories the earth can provide the third reference. Other trajectories will require a celestial reference; several current systems use Canopus because of its favorable location (nearly 90 deg from the plane of the ecliptic). For trajectories away from the sun, the earth provides a poor reference. It has a low brightness, will often be seen as a partially illuminated crescent, and at times the sun (some 12 magnitudes brighter than the earth) will fall into the field of view of the sensor. Once again Canopus or other celestial references provide the solution.

During the approach phase the navigation will probably be handled by other sensors, with the possible exception of an infrared v/h sensor for detecting lateral motion at landing. A horizon scanner could be used to provide range information derived from measurements of planetary subtense. One way of doing this would be to add logic circuitry to the horizon scanner used to control vehicle orientation. If the purpose of the mission is to orbit the planet, rather than to land on it, the horizon scanner appears to be the only satisfactory means of determining the vehicle orientation.

One additional mode of operation that should be treated here is orbital rendezvous or intercept. Purposes range from refueling, reprovisioning, and crew exchange for friendly space vehicles, to inspection and destruction of strange or hostile vehicles. Rendezvous systems will probably combine radar and infrared sensors. Passive infrared equipment will be advantageous for long-range search and target acquisition, since it requires far less power than active systems. After acquisition, the infrared search and track device will continue to provide angular tracking information, but either a radar or an active infrared (laser) device will probably be required for range and range-rate information. In the terminal phase, the final few hundred feet, the radar blind range will be reached and infrared or optical methods of ranging down to a few inches will be required. Obviously, rendezvous with a friendly target will be far easier than with a noncooperative target.

If the purpose of the mission is to orbit a planet, the moon, and so on, most attitude control systems will require a determination of the local vertical. The infrared horizon scanner appears to be the natural and perhaps the only choice for this function.

Conceptually, an infrared horizon scanner is a very simple device. All it need do is sense the radiation discontinuity between the cold background of space and the warm edge of a planet or of the moon. If the direction to at least three points on the horizon can be observed, then it should be possible to determine the direction to the center of the body (the local vertical). Unfortunately the solution is not often this simple, particularly if the body being observed has an atmosphere. The difficulty lies in the fact that there is rarely as sharp a discontinuity at the horizon as the designer would like. The relative sharpness changes markedly with the wavelength of observation. A glance at the references in this section will show how much effort has been expended in trying to find the optimum spectral interval for observing the earth's horizon.

Design approaches for horizon scanners fall into two distinct categories: scanning systems that bisect the angle between successive scans of the horizon or which edge-track the planet, and systems that use radiometric-balancing techniques. The early horizon sensors were of the scanning type, but designers seem to be favoring the radiometric-balancing type for the newer designs; perhaps this is because of their lack of moving parts and consequent promise of longer and more reliable life. The basic technique is to arrange several detectors so that they are equally illuminated when the scanner is pointed directly at the center of the target. Any deviation from this condition results in an imbalance in the illumination on the detectors, which can be sensed and used to generate error signals.

First-generation horizon scanners designed for earth-orbital missions used the 8 to 13 μ atmospheric transmission window (or, in some cases, a spectral bandpass extending from about 2 to 15 μ). This was an apparently logical choice of spectral region because the scanner would see only the self-radiated energy from the earth and would not be troubled by changes in illumination in going from day to night. Over cloudless areas the scanner would see all the way down to the surface of the earth, whereas in other areas it would see down as far as the tops of clouds.

The problems these early scanners encountered, and their generally

19.1.2 NAVIGATION AND FLIGHT CONTROL SYSTEMS

unsatisfactory performance, provide an interesting object lesson in the perils of designing equipment without a complete understanding of the phenomena to be measured. These units used a conical scan in which the scan element passed from cold outer space across the warm earth and back to space. If the earth is considered as a sphere with a temperature of 280°K and no atmosphere, the signal level from the scanner should increase sharply as the scan element crosses the horizon and remain high until the scan element again crosses the horizon and passes to outer space. The resulting (ideal) pulse, shown in Figure 19.1a, determines the thermal boundaries of the earth. By using two sensors, one scanning about the roll axis and the other about the pitch axis, two sets of thermal boundaries are determined. These boundaries represent the angles subtended by the earth along the pitch and roll axes; bisecting these angles determines the direction to the earth's center, or the local vertical.

When one adds an atmosphere and clouds to this simple model, the shape of the scanner pulse becomes more complex and the process of bisecting it becomes more difficult. The thermal gradient at the horizon is diffused, and the steepest portion of the gradient occurs at the tropopause rather than at the true physical horizon. Since clouds are usually colder than the earth's surface (because they are at a higher altitude), the signal level should drop whenever the scan element crosses a cloud, giving the serrated pulse shown in Figure 19.1b. In order to minimize the errors caused by pulse serration, the width of the pulse is measured at some fraction of the maximum amplitude of the pulse. The particular level at which the pulse is sampled is called the *signal slice*. If the serration because of a cloud is deep enough to protrude into the signal slice, the pulse bisector no longer represents the local vertical (Figure 19.1c). The data available to the designers of this early horizon scanner indicated that it was unlikely that cloud temperatures would dip lower than 240°K. In

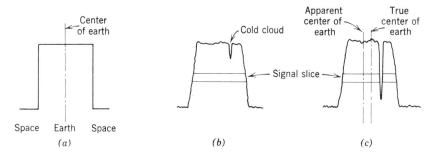

Figure 19.1 Signal pulses from a conical-scan horizon sensor. (*a*) Ideal pulse. (*b*) Actual pulse showing location of the signal slice and the effect of a cold cloud. (*c*) Error caused by an unusually cold cloud.

order to avoid false indications from cold high-altitude clouds, the signal slice was set far enough down on the pulse so that temperatures lower than 220°K were required to give a false indication.

Experience soon showed that there were serious errors in the indications of these horizon scanners. It was assumed that cold clouds were the problem, and the "quick fix" was obvious — lower the signal slice. This reduced but did not eliminate the problem. The original design placed the signal slice at about 50 per cent of the nominal peak value; it was eventually lowered to 25 per cent of nominal. Finally, instrumentation for making measurements of cloud-top temperatures was carried on an orbital flight, and it indicated that the signal level from some very high altitude clouds fell as low as 5 per cent of nominal, corresponding to a cloud temperature of 170°K.

[1] H. L. Clark, "Sun Follower for V-2 Rockets," *Electronics*, **23**, 71 (October 1950). Describes an automatic tracking device to keep a spectrograph in a spinning rocket pointed at the sun. Probably the first description in the U.S. literature of the use of a reticle in such a device.

[2] A. Goddard, Jr., et al., "Balloon-Borne System for Tracking The Sun," *Rev. Sci. Instr.*, **27**, 381 (1956). Describes a biaxial pointing control to maintain the image of the sun on the entrance slit of a spectrometer.

[3] V. P. Petrov, "Orientation in the Cosmos," *Nauka i Zhizn*, **25**, 7 (1958) (SLA: 59-19524). Discusses various methods of orienting a satellite in space. Infrared techniques are prominently mentioned.

[4] J. H. McLeod, "Axicons and Their Uses," *J. Opt. Soc. Am.*, **50**, 166 (1960). Describes the optical principles of the conical elements that form the heart of the scanner described in refs. 19.1.2[10] and [20].

[5] S. W. Spaulding, "Television and Lunar Exploration," *J. Soc. Motion Picture and Television Engrs.*, **69**, 39 (1960). Details an infrared scanner for vehicle orientation and determination of the range to the lunar surface. The scanner uses thermistor bolometers, Cassegrain optics with a 4.5 cm diameter and a 4.5 cm focal length, a field of view 2.5 deg square, and a scan rate of 3 Hz. The scanner weighs 10 lb, occupies 150 in^3 and consumes 20 W.

[6] W. Beller and R. Van Osten, "Mariner Carries Planet Fly-By Hopes," *Missiles and Rockets*, **8**, 14 (January 2, 1961). Describes the planet sensor that will be used for midcourse and terminal guidance on the Venus flyby.

[7] B. Kovit, "IR Horizon Sensor Guides Planetary Orbiting," *Space/Aeronautics*, **35**, 131 (February 1961). Contains a good description and many photographs of the first operational horizon scanner. A rotating germanium prism generates a conical scan that intersects the horizon twice during each scan. The signal processor compares the time interval between alternate intercepts with a fixed reference. Separate sensors are used to detect yaw and pitch errors. See ref. 19.1.2[11] and [13] for further details.

[8] "Infrared Radiation of Stars Being Charted as Navigation Aid," *Electronic Design*, **9**, 14 (June 21, 1961). Describes measurements made in the "middle-wavelength" infrared for the compilation of a star atlas for space navigation.

[9] E. M. Wormser, "IR Instrumentation — A Sure Bet For Spacecraft," *Space/Aeronautics*, **36**, 127 (August 1961). A survey of infrared instrumentation for space vehicle navigation.

19.1.2 NAVIGATION AND FLIGHT CONTROL SYSTEMS

[10] "Novel Infrared Horizon Indicator Designed Without Moving Parts," *Aviation Week*, **75**, 72 (September 4, 1961). See refs. 19.1.2[4] and [20].

[11] F. Schwarz and W. Chou, "Infrared Circuits in Tiros Weather Satellites," *Electronics*, **34**, 136 (September 29, 1961). Discusses the horizon scanners for Tiros satellites. These scanners provide two kinds of information, the spin rate of the satellite and the attitude of the satellite with respect to the local vertical. The scanners consist of a germanium lens with an area of 1.3 cm^2 and a germanium-immersed thermistor bolometer. The over-all effective f/no is f/0.21. The spectral bandpass extends from 1.8 to 20 μ. A circuit diagram of the signal processor is included. See also refs 19.1.2[7] and [13].

[12] J. S. Albus, "Digital Solar Aspect Sensors," *Astronautics*, **7**, 30 (January 1962). Uses an array of silicon photodiodes and produces a Gray-coded binary output.

[13] M. M. Merlen, "Horizon Sensor," U.S. Patent No. 3,020,407, February 6, 1962. The basic patent covering the horizon scanner that was used on the Mercury and Tiros programs. See refs. 19.1.2[7] and [11].

[14] "Communications Provide MA-6 Narrative," *Aviation Week*, **76**, 30 (February 26, 1962). A transcript that includes the pilot's remarks about his troubles with the infrared horizon scanners on the first U.S. manned orbital flight.

[15] "Nimbus Delay," *Aviation Week*, **76**, 32 (June 4, 1962). Indicates a six-month slippage in the Nimbus launch date because of inability of the infrared horizon sensors to distinguish between clouds and the true horizon. Original specifications called for the sensors and control system to align the vehicle within 1 deg of the local vertical. Experience indicates that the true figure is "several degrees."

[16] W. Gillespie, Jr., N. M. Hatcher, and R. J. Guillotte, "Infrared Scanner," U.S. Patent No. 3,038,077, June 5, 1962. A horizon tracker that scans simultaneously in two orthogonal planes.

[17] B. Miller, "Radar, Infrared Studies for Rendezvous," *Aviation Week*, **77**, 54 (July 16, 1962). Article discusses several types of sensors, including infrared, for space vehicle rendezvous.

[18] P. J. Klass, "Cold Clouds Troubling Horizon Sensors," *Aviation Week*, **77**, 50 (October 1, 1962). An excellent discussion of the problems encountered with horizon scanners on several of the early U.S. space programs. Unexpectedly cold, high-altitude clouds caused serious errors in the indications of the scanners.

[19] D. Goetze, "Accuracy and Range of Infrared Horizon Sensors as Limited by Detector Noise," *ARS J.*, **32**, 1039 (1962). Derives expressions for the alignment accuracy and the maximum useful working distance for infrared horizon scanners that are limited by detector noise.

[20] J. Killpatrick, "Inside-Out Horizon Scanner," *Appl. Opt.*, **1**, 147 (1962). An optical system focuses an image of the earth onto four detectors. Pitch and roll errors can be sensed from the unbalances in the signals from opposite pairs of detectors. The unique feature of this design is the use of a conical optical element to form an "inside out" image of the earth. For another report on this scanner, see ref. 19.1.2[10]. The optical principles of conical elements are discussed in ref. 19.1.2[4].

[21] R. A. McGee, "An Analytical Infrared Radiation Model of the Earth," *Appl. Opt.*, **1**, 649 (1962).

[22] R. A. Hanel, W. R. Bandeen, and B. J. Conrath, "The Infrared Horizon of the Planet Earth," *J. Atmos. Sci.*, **20**, 73 (March 1963). Contains calculations of the appearance of the earth's horizon as viewed from space in five spectral intervals. These intervals are the water vapor band from 6.33 to 6.55 μ, the ozone band from 8.9 to 10.1 μ, the atmospheric window from 10.75 to 11.75 μ, the carbon dioxide band from 14 to 16 μ, and the water band from 21 to 125 μ. The last two are judged to be the most suitable for horizon scanner use.

[23] "Cooper Flies Almost Perfect 22 Orbits," *Aviation Week*, **78**, 26 (May 20, 1963). Describes a series of horizon-definition experiments carried on aboard one of the Mercury test vehicles. In the series described here, infrared film was used to investigate the 0.7 to 1.2 μ region.

[24] T. H. Chin, "Spacecraft Stabilization and Attitude Control," *Space/Aeronautics*, **38**, 88 (June 1963). Various operational and conceptual systems are compared on the basis of weight, power consumption, and reliability. The author concludes that infrared horizon scanners have an accuracy of about ±1 deg and a relatively low reliability.

[25] S. Ramsey, "Cosmonauts Describe Flights, Experiments," *Aviation Week*, **79**, 28 (July 1, 1963). Describes flight experience with Vostok 5 and 6. Among the experiments were observations in the infrared of the earth and its horizon.

[26] B. Miller, "New Horizon Sensors Planned for Agena D," *Aviation Week*, **79**, 82 (September 30, 1963). Describes a series of horizon scanners developed for the Agena D, Gemini, and OGO Programs. Each scanner operates in the spectral region around 15 μ and has a claimed accuracy of ±0.1 deg.

[27] "Celestial Tracker Issue," *IEEE Trans. Aerospace and Navigational Electronics*, Vol. 10, September 1963. An excellent reference on the principles of star tracker design.

[28] C. V. Stanley and R. K. Orthuber, "Infrared Energy Tracking System Comprising Oscillatory Scanner," U.S. Patent No. 3,107,300, October 15, 1963. Describes a horizon scanner in which four small mirrors are driven simultaneously by a magnetic voice coil assembly. The four detectors, one associated with each mirror, view the horizon at 90 deg azimuthal intervals. As the mirrors oscillate, the image of the horizon is swept across each detector. The resulting pulses from each of the detectors will be time coincident only when the body or planet is centered on the optical axis of the system.

[29] W. R. Bandeen, B. J. Conrath, and R. A. Hanel, "Experimental Confirmation from the Tiros 7 Meteorological Satellite of the Theoretically Calculated Radiance of the Earth Within the 15-Micron Band of Carbon Dioxide," *J. Atmos. Sci.*, **20**, 609 (1963).

[30] J. W. Woestman, "Earth Radiation Model for Infrared Horizon Sensor Applications," *Infrared Phys.* **3**, 93 (1963). Develops an analytical model of the earth's horizon as seen from space. Shows that the gradient between the sky background and the earth's limb is sharpest when viewed in the long-wavelength absorption bands. Author does not attempt to determine the optimum wavelength interval.

[31] K. J. Stein, "Avionics Changes Included in Tiros Wheel," *Aviation Week*, **81**, 44 (August 3, 1964). Describes the use of a horizon scanner to trigger a vidicon-camera shutter at the instant that the camera is looking down at the earth.

[32] B. Miller, "New IR Space Sensors Add to Reliability," *Aviation Week*, **81**, 68 (August 31, 1964). Describes two second-generation horizon scanners. One is a long-lived, high-reliability device with a relatively coarse accuracy of ±3 deg in the location of the local vertical. The second unit, called the Reliable Earth Sensor, has a quoted accuracy of ±0.05 deg. A single lens images the earth onto a cross-shaped array of detectors. Opposite detectors are connected as two arms of two separate bridges. When the image of the earth is centered on the detector array, both bridges are balanced. Any displacement of the image unbalances one or both of the bridges.

[33] D. Q. Wark, J. Alishouse, and G. Yamamoto, "Variation of the Infrared Spectral Radiance Near the Limb of the Earth," *Appl. Opt.*, **3**, 221 (1964). Concludes that the most favorable spectral intervals for horizon scanner use are 30.8 to 33.3 μ and 14.8 to 15.4 μ.

[34] R. W. Astheimer and S. Weiner, "Solid-Backed Evaporated Thermopile Radiation

19.1.2 NAVIGATION AND FLIGHT CONTROL SYSTEMS

Detectors," *Appl. Opt.*, **3**, 493 (1964). Describes the design and the fabrication of thermopile detectors for a radiometric-balance horizon sensor.
[35] "Radial Thermopile IR Detector," *Appl. Opt.*, **3**, 1234 (1964). One of the applications described for this 40-element detector is a no-moving-parts horizon sensor.
[36] S. F. Singer, *Torques and Attitude Sensing in Earth Satellites*, New York: Academic Press, 1964. Discusses the theoretical considerations of spectral radiance and their application to the choice of an optimum spectral interval for infrared horizon scanners, the design of horizon scanners, and earth-scan signal relationships in the Tiros radiation experiment.
[37] C. D. La Fond, "Probe Designed to Give Better Horizon Definition," *Missiles and Rockets*, **16**, 31 (January 25, 1965). Describes Project Scanner, a high-altitude probe containing a dual radiometer that is designed to scan the horizon in two wavelength intervals as part of a study to determine the optimum spectral interval for horizon scanners. The range from 14 to 40 μ can be utilized. A schematic drawing of the radiometer is also included.
[38] "Scanner Spacecraft to Measure IR Radiation," *Aviation Week*, **82**, 53 (March 1, 1965). See also the same magazine at **85**, 23, October 17, 1966, and the previous reference.
[39] "Infrared Sensor Controls Spacecraft Orientation," *Electronic Design News*, **10**, 6 (May 1965). Describes a modified version of the horizon scanner in ref. 19.1.2[16]. Device uses a thermistor bolometer, four germanium lenses, and parabolic mirrors. The local vertical is determined within an accuracy of 0.1 deg and the device is useful up to an altitude of 10^6 miles. Power consumption is 3.5 W.
[40] "Gemini-4 Success to Intensify Launch Pace," *Aviation Week*, **82**, 79 (June 14, 1965). Discusses tests of the horizon sensors on the Gemini-4 flight. Purpose of the tests was to see whether the sensors would malfunction when they viewed the sun, moon, and thrust-motor exhaust plumes. No malfunctions were observed, in contrast to the several hundred that were observed during the Gemini-3 flight.
[41] "LES-2 Satellite all Solid State," *Aviation Week*, **82**, 57 (June 21, 1965). The best-oriented antenna on this satellite was to have been selected by an earth sensor operating in the 20 μ region. Troubles with this sensor forced the use of one operating near the end of the visible portion of the spectrum.
[42] "Horizon Sensor," *Aviation Week*, **82**, 81 (June 21, 1965). Describes proposals for the development of an earth or lunar horizon scanner having the following characteristics: weight 8 lb; power consumption 20 W; reliability 0.997 for 90 days of operation; accuracy 0.1 deg; spectral region 20 to 40 μ.
[43] "Gemini-4 Experiments Yield Much Data, Demonstrate Crew Versatility," *Aviation Week*, **82**, 83 (June 21, 1965). Describes a two-color earth-limb photographic experiment as part of a continuing quest for the optimum spectral region for horizon scanners.
[44] "Tacite Sounding Rocket," *Aviation Week*, **83**, 20 (July 5, 1965). Pictures and describes a French sounding rocket developed to investigate the earth-space infrared radiation contrast for horizon scanner design.
[45] "Mockup Illustrates Horizon-Scanning Satellite," *Aviation Week*, **87**, 75 (September 11, 1965). A photograph of the orbital horizon scanner satellite that is described in ref. 19.1.2[57].
[46] A. L. Knoll and M. M. Edelstein, "Estimation of Local Vertical and Orbital Parameters for an Earth Satellite Using Horizon Sensor Measurements," *AIAA J.*, **3**, 338 (1965).
[47] R. J. Orrange, "Autonomous Space Navigation," *Space/Aeronautics*, **45**, 86 (May 1966). A good tutorial article. It characterizes a typical currently available horizon scanner as weighing 7.25 lb, occupying 150 in^3, requiring 4.8 W of power, having an

MTBF of 24,700 hr, and an accuracy of 6 to 10 min of arc in the determination of the local vertical.

[48] "Faltering Observatory Program Could Bring Changes in Space Science," *Space/Aeronautics*, **45**, 23 (June 1966). Analyzes problems with the infrared horizon scanners on the Orbiting Geophysical Observatory vehicles. On OGO-1 a magnetometer boom only partially deployed and stopped directly in front of one horizon scanner. On OGO-2 the horizon scanners vacillated between the horizon and high-altitude clouds, thus exhausting the supply of reaction gas in 10 days. On OGO-3 changes to the horizon scanners included relocation, a reduction in the field of view, and the addition of a filter to reduce the effects of clouds.

[49] P. Alelyunas, "Synergetic Satellites," *Space/Aeronautics*, **46**, 52 (October 1966). Reports that the infrared horizon scanner on the OGO-1 satellite locked onto a metal ball on a partially deployed boom and quickly exhausted its attitude-control gas supply by literally chasing its own tail. For a slightly different explanation of this occurrence, see the previous reference.

[50] "Soviets Pinpoint Luna-12 Equipment," *Aviation Week*, **85**, 27 (November 14, 1966). A photo of the lunar orbiter shows an infrared radiometer and a device that is believed to be either a star tracker or an infrared horizon scanner.

[51] N. H. Hatcher, "Spacecraft Attitude Sensors—Where We Stand Today," *Astronautics and Aeronautics*, **4**, 58 (December 1966). An excellent state-of-the art survey. Gives good descriptions of the operating principles for a variety of horizon scanners.

[52] R. C. Barbera, "An Interesting Infrared Planet Horizon Sensor Design," *Appl. Opt.*, **5**, 471 (1966). Uses a wide-angle lens to image the earth onto 8 pie-shaped detectors. Device works in the 14.8 to 15.5 μ region.

[53] E. P. French, "Effect of Rocket Exhausts on Infrared Planet Sensors," *J. Spacecraft and Rockets*, **3**, 849 (1966). Shows that serious errors can result from tracking through exhaust plumes.

[54] G. Quasius and F. McCanless, *Star Trackers and Systems Design*, London: Macmillan and Co., 1966.

[55] W. J. Normyle, "New Capabilities Broaden Scientific Goals," *Aviation Week*, **86**, 54 (January 23, 1967). Contains a table that shows design objectives for future NASA procurements of horizon scanners.

[56] "Variety of Russian Spacecraft Displayed," *Aviation Week*, **86**, 84 (June 12, 1967). Shows a photograph of what is purported to be a combined star tracker and horizon scanner on the Cosmos-144 meteorological satellite. See also ref. 19.2.3[16].

[57] C. D. LaFond, "Infrared Horizon-Mapping Urged in Space," *Electronic Design*, **15**, 22 (September 27, 1967). A proposal for an unmanned infrared-scanning satellite to make high-resolution measurements of the earth's horizon. Present-day horizon scanners can determine the local vertical to an accuracy of about 0.25 deg. With the knowledge to be gained from this proposed program it appears that the local vertical could be determined to an accuracy of about 0.01 deg. The infrared radiometer would peer through a 26 in. viewport, would have a field of view of 0.01 deg, and would operate in the 15 μ region. The cadmium-doped germanium detectors would be cooled by solid neon. For a photograph that shows a mockup of this satellite, see ref. 19.1.2[45].

[58] M. Bottema, "Guiding of Balloon-Borne Telescopes by Off-Set Sun-Tracking," *Appl. Opt.*, **6**, 213 (1967). A description of the orientation techniques used in the balloon-astronomy program that is described in ref. 19.3.2[35].

[59] R. J. Fitzgerald, "Filtering Horizon-Sensor Measurements For Orbital Navigation," *J. Spacecraft and Rockets*, **4**, 428 (1967).

[60] E. A. Kallett and P. W. Collyer, "Planet Sensor Having a Plurality of Fixed Detectors

About Two Orthogonal Axes and a Partially Attenuating Mask Centrally Located in the Field of View of the Detectors and Smaller Than The Total Field of View," U.S. Patent No. 3,351,756, November 7, 1967.
[61] "Filter Center," *Aviation Week*, **89**, 59 (September 2, 1968). Reports a request for improved horizon scanners capable of maintaining the orientation of communication satellites to within 0.03 deg.

19.2 RADIOMETRIC APPLICATIONS

The applications in this section include the measurement of lunar, stellar, and planetary temperatures; studies of heat transfer in plants, of the efficiency of thermal insulators, and of the solar constant; and the application of infrared sensors to weather satellites.

19.2.1 Measurement of Temperature

Before examining the applications, let us see how the elementary principles of heat transfer can be applied to the prediction of the temperature of planets and satellites. The reader will recall that the transfer of heat occurs by three processes: conduction, convection, and radiation. Since astronomical bodies are isolated in space, once temperature equilibrium has been reached, the only way that these bodies can gain or lose heat is by radiation. The following earth-sun data are necessary elements for a calculation of the temperature of the earth:

Solar radius	6.96×10^{10} cm
Mean solar distance	1.50×10^{13} cm
Effective solar temperature	5900°K
Earth's equatorial radius	6.38×10^8 cm

Starting with the Stefan-Boltzmann law (2-9), and multiplying it by the surface area of the sun, gives the total solar flux

$$P = A_s \sigma T_s^4 \\
= 4\pi (6.96 \times 10^{10})^2 (5.67 \times 10^{-12})(5900)^4 \\
= 4.17 \times 10^{26} \text{ W}. \tag{19-1}$$

This is the number of watts radiated into the 4π steradians surrounding the sun. The radiant intensity of the sun is

$$J = \frac{P}{4\pi} = 3.32 \times 10^{25} \text{ W sr}^{-1}. \tag{19-2}$$

The irradiance at the mean solar distance is

$$H = \frac{J}{d^2} = \frac{3.32 \times 10^{25}}{(1.5 \times 10^{13})^2} = 0.148 \text{ W cm}^{-2}. \tag{19-3}$$

This is the quantity that meteorologists call the solar constant. The generally accepted value of the solar constant is 0.140 W cm^{-2} (Section 3.3). The discrepancy between this value and that calculated above is due to the simple assumption that the temperature of the sun is 5900°K. In Section 3.3 it was noted that the radiation temperature of the sun appears to decrease slowly with wavelength. As a result, an accurate calculation of the solar constant requires some complex procedures that will not be used here because the only purpose of the present calculation is to show the order of magnitude of the quantities involved.

If it is assumed that the earth is in thermal equilibrium, it must be radiating a flux to the universe that is exactly equal to the flux received from the universe. As a practical matter, the sun is the source of essentially all of the flux received by the earth. A simple heat-transfer model for this situation assumes that the earth absorbs the solar flux over an area that is equal to its projected area (a consequence of Lambert's law), while it radiates flux from an area that is equal to its surface area. Hence, for this simple model the earth absorbs as a flat disk and radiates as a sphere. The energy-balance equation for this situation is

$$HA'_e = A_e \sigma T_e^4, \qquad (19\text{-}4)$$

where A'_e is the projected area of the earth and A_e is its surface area. Making the appropriate substitutions gives a value of 284°K for the predicted temperature of the earth.

The simple heat-transfer model errs in assuming that the earth is a blackbody. The major complicating factor is that the earth is sheathed by an atmosphere that is a selective rather than a uniform absorber. In addition, the model assumes that Kirchhoff's law applies. In Section 2.5 it is shown that Kirchhoff's law holds rigorously only when the two bodies exchanging radiation are at the same temperature. Fortunately, many materials show little variation in emissivity (and, hence, absorptance) over moderate temperature intervals, and Kirchhoff's law can be applied even though the two bodies are at a different temperature. However, when the temperature difference is as large as that between the sun and the earth, most materials have an absorptance α for the short-wavelength solar radiation that is different than their emissivity ϵ for the long-wavelength earth radiation. Experience indicates that the effective emissivity of the earth is about 0.90 and its effective absorptance for solar radiation is about 0.84. Thus the energy-balance equation becomes

$$\alpha HA'_e = \epsilon A_e \sigma T_e^4, \qquad (19\text{-}5)$$

and the predicted temperature of the earth is 279°K.

19.2.1 MEASUREMENT OF TEMPERATURE

For a spherical satellite orbiting the earth, the heat-balance equation is

$$P_s + P_{rs} + P_e + P_{int} = P_{sat}, \quad (19\text{-}6)$$

where the first three terms on the left are the fluxes incident on the satellite (P_s is the direct solar flux, P_{rs} is the solar flux reflected from the earth, and P_e is the flux emitted by the earth), P_{int} is the internal heat dissipation (watts) from any systems on board the satellite, and P_{sat} is the flux radiated by the satellite. Approximate expressions for the various terms in (19-6) are

$$P_s = \alpha \pi r^2 H_s = 0.140 \alpha \pi r^2 \quad (19\text{-}7)$$

$$P_{rs} = \alpha \pi r^2 H_{rs} = 0.050 \alpha \pi r^2 \quad (19\text{-}8)$$

$$P_e = \epsilon \pi r^2 H_e = 0.032 \epsilon \pi r^2 \quad (19\text{-}9)$$

$$P_{sat} = 4\pi r^2 \epsilon \sigma T_{sat}^4, \quad (19\text{-}10)$$

where α is the absorptance of the satellite's skin material for solar radiation, ϵ is the emissivity of the skin material at the equilibrium temperature of the satellite T_{sat}, and r is the radius of the (spherical) satellite. Substituting these values and rearranging terms gives

$$T_{sat} = 645 \left[0.0475 \frac{\alpha}{\epsilon} + 0.008 + \frac{0.0796}{\epsilon r^2} P_{int} \right]^{1/4}. \quad (19\text{-}11)$$

Thus the equilibrium temperature of a passive satellite, that is, one dissipating no power internally, depends only on the value of α/ϵ. For an active satellite, the temperature depends on the value of α/ϵ, the internal power dissipation, and the effectiveness of the satellite as a radiator. Table 19.1 compares the equilibrium temperatures of a spherical satellite (50 cm radius) in an orbit 300 miles above the earth for two different skin materials. The satellite with the sandblasted-aluminum skin ($\alpha/\epsilon = 2.00$) is a "hot" satellite, whereas the change to an anodized-aluminum skin ($\alpha/\epsilon = 0.19$) gives a "cold" satellite. For data on other skin materials, see Table 2.4. The satellite temperatures shown in Table 19.1 probably bracket the temperatures that will be found in practice. Therefore, satellites as a class are characterized by relatively low temperatures; thus an infrared system designed to detect them will require a large optical aperture and an 8 to 13 μ detector. If the detection system is above the earth's atmosphere, it can probably take full advantage of the benefits offered by a background-limited detector.

[1] Earl of Rosse, "On the Radiation of Heat from the Moon," *Proc. Roy. Soc. London*, **17**, 436 (1869); **19**, 9 (1870). This appears to be the first reported radiometric determina-

tion of the lunar temperature. Equipment consisted of a 3 ft reflector, a 4-element thermopile, and a Thompson galvanometer.

[2] Earl of Rosse, "On the Radiation of Heat From the Moon, the Law of its Absorption by our Atmosphere and of its Variation in Amount With her Phases," *Phil. Trans. Roy. Soc. London*, **163**, 587 (1873).

[3] S. P. Langley, "The Temperature of the Moon," *Proc. Nat. Acad. Sci.*, **4**, 107 (1889). Measurements made during the eclipse of 1885.

[4] F. W. Very, "Lunar Temperatures," *Astrophys. J.*, **8**, 199, 265 (1898). Temperatures were measured at different phases of the moon. The values reported are somewhat higher than those currently accepted.

[5] E. Pettit and S. B. Nicholson, "Radiometric Observation of Venus," *Popular Astron.* **32**, 614 (1924).

[6] W. W. Coblentz and C. O. Lampland, "Some Measurements of the Spectral Components of Planetary Radiation and Planetary Temperatures," *J. Franklin Inst.*, **199**, 785 (1925); **200**, 103 (1925). Measurements were made by filter radiometry. Typical filters include thin glass, fluorite, and water cells.

[7] W. W. Coblentz and C. O. Lampland, "Further Radiometric Measurements and Temperature Estimates of the Planet Mars, 1926," *J. Research Natl. Bur. Standards*, **22**, 237 (1927).

[8] E. Pettit and S. B. Nicholson, "Lunar Radiation and Temperatures," *Astrophys. J.*, **71**, 102 (1930). The classical paper in this field. Measurements were made with a thermopile detector and the 100 in. Mt. Wilson telescope. The lower detection limit of this combination was between 70 and 100°K. The temperature at the subsolar point (full moon) was found to be 407°K. Measurements made during the lunar eclipse of June 1927 show that the temperature fell from 342°K to 175°K during the first partial phase and dropped to 156°K during totality.

[9] E. Pettit, "Lunar Radiation as Related to Phase," *Astrophys. J.*, **81**, 17 (1935).

[10] C. O. Lampland, "Lowell Observatory Planetary Radiometric Investigations," *Publ. Am. Astron. Soc.*, **9**, 174 (1939).

[11] E. Pettit, "Radiation Measurements on the Eclipsed Moon," *Astrophys. J.*, **91**, 408 (1940).

[12] W. W. Coblentz, "Temperature Estimates of the Planet Mars, 1924 and 1926," *J. Research Natl. Bur. Standards*, **28**, 297 (1942). Contains an excellent discussion of the methods of determining temperature from radiometric data. The author points out that his original calculations of temperature, which were based on Fowle's atmospheric transmission data, are from 50 to 100°C different from the temperatures calculated with the newer transmission data of Adel and Lampland.

[13] J. C. Jaeger and A. F. A. Harper, "Nature of the Surface of the Moon," *Nature*, **166**, 1026 (1950). Author describes an equation that predicts the drop in lunar temperature during an eclipse. The cooling rate is a function of the thermal parameter $(k\rho c)^{-1/2}$, where k is the thermal conductivity, ρ the density, and c the specific heat of the lunar material. Cooling rates determined by radiometric measurements during eclipses allow calculation of the thermal parameter.

[14] J. C. Jaeger, "The Surface Temperature of the Moon," *Australian J. Phys.*, **6**, 10 (1953).

[15] H. Stommel et al., "Rapid Aerial Survey of Gulf Stream with Camera and Radiation Thermometer," *Science*, **117**, 639 (1953). These authors have shown that an airborne infrared radiometer can provide an almost synoptic oceanographic chart of surface thermal gradients over a much larger area, and in a far shorter time, than can be done with surface vessels. The radiometer consists of a Golay cell alternately exposed to the

19.2.1 MEASUREMENT OF TEMPERATURE

TABLE 19.1 EQUILIBRIUM TEMPERATURE FOR A SPHERICAL SATELLITE (50 cm RADIUS) IN AN ORBIT 300 MILES ABOVE THE EARTH'S SURFACE

Skin Material	ϵ	α/ϵ	Internally Dissipated Power (W)	Equilibrium Temperature (°K)
Sandblasted aluminum	0.21	2.00	0	358
Sandblasted aluminum	0.21	2.00	10	414
Anodized aluminum	0.77	0.19	0	234
Anodized aluminum	0.77	0.19	10	277

sea and to a reference blackbody by means of a rotating chopper set at an angle of 45 deg to the optical axis of the radiometer. The temperature of the blackbody is adjusted by admitting hot or cold water until the signals from the sea are equal to those from the blackbody. The authors correctly note that the temperature of the blackbody at the null point is equal to the temperature of the sea surface only if the intervening atmosphere is perfectly transparent. The measurements show that during the winter months a sharp discontinuity of surface temperature of as much as 20°F in 100 yd occurs along the northern edge of the Gulf Stream.

[16] E. Pettit and S. B. Nicholson, "Temperatures on the Bright and Dark Sides of Venus," *Publ. Astron. Soc Pacific*, **67**, 293 (1955). Radiometric measurements made from 1923 to 1928 show that the temperature is 240°K on both the day and the night side of Venus.

[17] F. Gifford, Jr., "The Surface-Temperature Climate of Mars," *Astrophys. J.*, **123**, 154 (1956). Measurements were made with the 42 in. Lowell reflector and a thermocouple detector.

[18] G. P. Kuiper, "Infrared Observations of Planets and Satellites," *Astron. J.*, **62**, 245 (1957).

[19] G. de Vaucouleurs, "Remarks on Mars and Venus," *J. Geophys. Research*, **64**, 1739 (1959). Discusses radiometric temperature determinations and newly discovered absorption bands in the Martian atmosphere.

[20] J. C. Jaeger, "Sub-Surface Temperatures on the Moon," *Nature*, **183**, 1316 (1959). A continuation of the analysis in ref. 19.2.1 [13].

[21] S. M. Greenfield and W. W. Kellogg, "Calculations of Atmospheric Infrared Radiation as Seen From a Meteorological Satellite," *J. Meteorol.*, **17**, 283 (1960). Shows methods of correcting for atmospheric water vapor, carbon dioxide, and ozone in calculations of ground temperature as measured from a satellite.

[22] C. H. Mayer, "The Temperatures of the Planets," *Sci. Am.*, **204**, 58 (May 1961). The primary emphasis in this article is on those measurements made at centimeter wavelengths by standard RF techniques. The author does, however, include a good discussion of the difficulties of making such temperature measurements in the infrared, where planetary atmospheres can seriously affect the results.

[23] "Lunar Radiometer," *Aviation Week*, **75**, 73 (September 4, 1961). Describes a radio-

meter designed to measure the temperature gradient of the moon's crust. The radiometer has a diameter of 1 in. so that it can be lowered into a $1\frac{1}{4}$ in. hole that other equipment will drill in the surface of the moon. The detector is a thermocouple.

[24] V. S. Troitskii, "Radiation of the Moon, Physical Condition and Nature of its Surface," *Akad. Nauk SSSR, Kommissiya po Fizike Planet*, **3**, 16 (1961) (CFSTI: TT-64-19752).

[25] R. A. Hanel, "Low Resolution Unchopped Radiometer for Satellites," *Am. Rocket Soc. J.*, **31**, 246 (1961). The radiometer consists of a polished cone with a thermistor bolometer mounted at its apex. The spectral response can be controlled by the application of selectively absorbing coatings to the thermistor.

[26] G. Ewing, and E. D. McAlister, "On the Thermal Boundary Layer of the Ocean," *Science*, **131**, 1374 (1961). Radiometric measurements of the top 0.1 mm of the ocean surface reveal the existence of a surface layer that is as much as 0.6°C cooler than conventional thermometric measurements of surface temperature. Being very thin, this surface layer can reestablish itself in about 12 sec after being broken by a wave.

[27] D. M. Gates, "Winter Thermal Radiation Studies in Yellowstone Park," *Science*, **134**, 32 (1961). Measurements were made with the Stoll-Hardy radiometer of ref. 18.2.1[3]. These measurements show the radiation temperature of moss, soil, grass, snow, hot springs, and the zenith sky as a function of time of day and cloud cover.

[28] G. P. Kuiper and B. M. Middlehurst, *Planets and Satellites*, Chicago: University of Chicago Press, 1961; see E. Pettit, "Planetary Temperature Measurements," Chapter 10. Discusses the principles of radiometric temperature measurement and summarizes the state of the art, as of 1960, for measurements of lunar and planetary temperatures.

[29] Kuiper and Middlehurst, *op. cit.* (preceding ref., 19.2.1[28]); see W. M. Sinton, "Recent Radiometric Studies of the Planets and the Moon," Chapter 11.

[30] "Solving One of Venus' Big Riddles," *Electronic Design*, **10**, 40 (November 1962). Contains a detailed block diagram of the dual-channel infrared radiometer and its signal processor. Additional optical data, supplementing those given in the next reference, indicate that the thermistor bolometers are germanium immersed so as to give a system focal length of 9.55 mm. The field of view is 0.9 by 0.9 deg.

[31] L. D. Kaplan, G. Neugebauer, and C. Sagan, "The Six Experiments of Mariner 2: Infrared Radiometer," *Western Electronic News*, November 1962, p. 13. Describes a dual-radiometer consisting of an $f/2.4$ optical system of 3 in. focal length, a dichroic filter to define two spectral intervals, a 20 Hz chopper, and thermistor bolometer detectors. A duplicate system aimed 45 deg away from the planetary scanner obtains reference readings from space. The entire device is 6 in. long and 2 in. wide, weighs 2.7 lb, and consumes 2 W. One channel operates in the 8 to 9 μ region. In the event of breaks in the cloud cover it would read surface temperature. The other channel is in the 10 to 10.8 μ region and will measure the temperature at the top of the atmosphere.

[32] H. L. Johnson, "Infrared Stellar Photometry," *Astrophys. J.*, **135**, 69 (1962). Describes a new system of infrared stellar photometry with approximate effective wavelengths of 1.3, 2.2, 3.6, and 5.0 μ. Observational data are given for 50 stars. All measurements were made with a photovoltaic indium antimonide detector cooled to 77°K. Narrowband interference filters are used to define the desired spectral intervals.

[33] D. Q. Wark, G. Yamamoto, and J. H. Lienesch, "Methods of Estimating Infrared Flux and Surface Temperature from Meteorological Satellites," *J. Atmos Sci.*, **19**, 369 (1962). In the five-channel radiometer carried on Tiros 2, two of the channels (8 to 12 μ and 7 to 30 μ) were designed to measure the upward radiation from the earth. This article describes a method of converting such radiometric data to values of the surface temperature and the total upward flux. Corrected radiometric estimates of ground temperature over several cloudless areas are found to range from 3.5°C higher to 5.0°C lower than temperatures measured at the same time on the ground.

19.2.1 MEASUREMENT OF TEMPERATURE

[34] E. A. Burns and R. J. P. Lyon, "Errors in the Measurement of the Temperature of the Moon," *Nature*, **196**, 463 (1962). A discussion of the errors in radiometrically determined values of the lunar temperature that arise from the frequently made assumption that the moon radiates as a blackbody. One interpretation of the authors' argument is that many of the thermal anomalies recently observed on the moon are, in fact, merely changes in the composition of the surface. It is interesting to note that this potential difficulty was discussed as early as 1942 (ref. 19.2.1[12]), but it seems to have escaped the attention of later workers.

[35] E. D. McAlister, "Application of Infrared-Optical Techniques to Oceanography," *J. Opt. Soc. Am.*, **52**, 607 (1962). See ref. 19.2.1[51].

[36] W. M. Sinton, "Temperatures on the Lunar Surface," in Z. Kopal, ed., *Physics and Astronomy of the Moon*, New York: Academic Press, 1962, p. 407.

[37] M. S. Zel'tser, "The Temperature of the Lunar Surface," in A. V. Markov, ed., *The Moon*, Chicago: University of Chicago Press, 1962, p. 175. Originally published by the State Publishing House of Physical-Mathematical Literature, Moscow, 1960. Summarizes Soviet work on the radiometric determination of lunar temperatures. There is a good bibliography, but most of it is devoted to the non-Russian literature.

[38] E. H. Kolcum, "Mariner Reveals 800°F Venus Temperature," *Aviation Week*, **78**, 30 (March 4, 1963). During the fly-by the infrared radiometer measured a temperature of $-54°C$ at the top of the Venusian atmosphere. A note in the same magazine, **78**, 25 (February 25, 1963), reports that data reduction was hampered by crosstalk between the infrared and the microwave radiometers.

[39] H. L. Clark. "100 Inch Radiometer," U.S. Patent No. 3,087,062, April 23, 1963. Describes a radiometer that uses a 100 in. mirror, thermopile detector, 5 Hz chopper, and synchronous detection (rectification). The NEΔT is $5 \times 10^{-4}°C$. This radiometer is designed for studies of the temperature distribution at the surface of the sea.

[40] J. N. James, "The Voyage of Mariner 2," *Sci. Am.*, **209**, 70 (July 1963). Gives a detailed description of the infrared and the microwave radiometers, the measurements they returned, and the interpretation of these measurements.

[41] J. A. Westphal, B. C. Murray, and D. E. Martz, "An 8–14 Micron Infrared Astronomical Photometer," *Appl. Opt.*, **2**, 749 (1963). The photometer consists of a reflective chopper, a cooled mercury-doped germanium detector, a synchronous rectifier and associated signal processing circuitry, and auxiliary mirrors for calibration purposes. The chopper is set at 45 deg to the optical axis and reflects energy from target and background to the detector. As the chopper blade rotates into the open position, it uncovers a small plane mirror that reflects energy from an adjacent portion of the background to the detector. The ac output of the detector represents background plus target minus background or, to a good approximation, target only. The spacing between the reference and measurement beams can be varied from 5 to 20 min of arc. Low-emissivity optical elements were used throughout; these include gold-coated mirrors and a single barium fluoride window over the detector. In addition, the liquid-hydrogen-cooled detector was fitted with a cooled filter and radiation shield. At sea level the resistance of the detector was 1 megohm; this increased to 8 megohms at the 3900 m altitude of the observing site. For calibration, auxiliary mirrors were used to permit the photometer to view blackened metal cavities cooled by ice water and liquid nitrogen. With a 50.8 cm-diameter $f/15$ paraboloidal mirror, sky noise limited minimum detectable temperature to 105°K. Measurements of the effective spectral radiance (8 to 14 μ) of the sky over Mr. Wilson range from 3×10^{-10} W cm^2 sr^{-1} at the horizon to 8×10^{-11} W cm^2 sr^{-1} at the zenith.

[42] V. D. Krotikov and O. B. Shchuko, "Thermal Behavior of the Moon's Surface Layer During a Lunation," *Astron. Zh.*, **40**, 297 (1963).

[43] C. D. Kern, "Desert Soil Temperatures and Infrared Radiation Received by Tiros III," *J. Atmos. Sci.*, **20**, 175 (1963). Analyzes ground-temperature data from a Tiros flight over the Mediterranean Sea and the Libyan and Egyptian deserts.

[44] B. C. Murray, R. L. Wildey, and J. A. Westphal, "Infrared Photometric Mapping of Venus Through the 8 to 14 Micron Atmospheric Window," *J. Geophys. Research*, **68**, 4813 (1963). The 200 in. Mt. Palomar telescope was used to make radiometric maps of Venus that coincided with the Mariner 2 encounter. The radiometer is described in ref. 19.2.1[41]. The angular resolution was equal to about 1/30 of the disk of Venus and the signal-to-noise ratio was over 100. The mean radiation temperature for the center of the disk was 208°K, about 28°K lower than the commonly accepted value. At least some of this difference may be accounted for with more refined instrumental calibrations.

[45] S. C. Chase, L. D. Kaplan, and G. Neugebauer, "The Mariner 2 Infrared Radiometer Experiment," *J. Geophys. Research*, **68**, 6157 (1963). The two-channel radiometer is described. One channel was centered at $8.4\,\mu$ so as to be clear of gaseous absorption bands, and the other channel was centered at $10.4\,\mu$ in a carbon dioxide absorption band. The observed temperatures agree with 8 to $13\,\mu$ earth-based measurements, the temperatures of the light and dark sides are equal, and there is limb darkening. The results indicate that the radiometer measured the temperature at the top of an unbroken cloud structure. Both laboratory and in-flight calibration procedures are described. The average radiation temperature observed in both channels was $235 \pm 10°K$, which is consistent with previous broad-band, earth-based measurements. See also ref. 19.2.1[47].

[46] J. M. Saari and R. W. Shorthill, "Isotherms of Crater Regions on the Illuminated and Eclipsed Moon," *Icarus*, **2**, 115 (1963). Reports measurements that were made in 1960 with a thermistor bolometer and the 60 in. telescope at Mt. Wilson. The field of view was 8 sec of arc. The spectral bandpass was limited by a germanium filter and a KRS-5 window, hence it extended from about 1.8 to $35\,\mu$.

[47] S. C. Chase, L. D. Kaplan, and G. Neugebauer, "Mariner II: Infrared Radiometer," *Science*, **139**, 907 (1963). A preliminary report on measurements made by the infrared radiometer during the Venus flyby. The radiation temperatures measured in the 8 and $10\,\mu$ channels are equal. This indicates that there was little carbon dioxide absorption along the line of sight and implies that the radiometer measured the temperature at the top of high cloud layers and did not observe the surface.

[48] B. C. Murray, R. L. Wildey, and J. A. Westphal, "Venus: A Map of its Brightness Temperature," *Science*, **140**, 391 (1963). Measurements made with the 200 in. Mt. Palomar telescope and a mercury-doped germanium detector cooled to 4.2°K. Field of view was 1.5 sec of arc, or about 1/30 of the planetary diameter. The measurements confirm earlier observations that there is little or no difference in the atmospheric temperature between the daytime and the nightime portions of the planetary disk. An anomalous hot area observed in the southern hemisphere is believed to have been an atmospheric storm.

[49] K. J. K. Buettner and C. D. Kern, "Infrared Emissivity of the Sahara from Tiros Data," *Science*, **142**, 671 (1963). Concludes that the effective emissivity of the Sahara desert is between 0.69 and 0.91 for the 8 to $12\,\mu$ region.

[50] E. M. Wormser, "Radiation Thermometer With In-Line Blackbody Reference," *Instrum. and Control Syst.*, **37**, 101 (December 1964). A comparison radiometer in which the detector and chopper are located within a temperature-controlled blackbody cavity. This radiometer has been mounted in an aircraft and used to map the temperature over large areas of the ocean. Temperatures from -2 to $+35°C$ can be determined to an accuracy of $\pm 0.5°C$.

19.2.1 MEASUREMENT OF TEMPERATURE

[51] E. D. McAlister, "Infrared-Optical Techniques Applied to Oceanography," *Appl. Opt.*, **3**, 609 (1964). Because of the strong absorption of water, radiometrically determined temperatures are characteristic of a very thin surface layer. The effective thickness of this layer is 0.5 mm when the temperature is measured at a wavelength of 2.2 μ, 0.06 mm at 3.8 μ, and 0.02 mm at 10 μ. Hence multiple-wavelength radiometry can be used to measure the heat flow through the surface of the sea. The author notes that a radiometer such as that described in ref. 19.2.1[39] has detected temperature differences of less than 0.001°C between adjacent areas on the ocean.

[52] B. C. Murray and R. L. Wildey, "Surface Temperature Variations During the Lunar Nighttime," *Astrophys. J.*, **139**, 734 (1964). Equipment consisted of a mercury-doped germanium detector cooled to 20°K, a barium fluoride relay lens, an interference filter, and a 19 in. gold-surfaced reflective telescope. With the photometer at the $f/15$ Cassegrain focus, the field of view was 17 sec of arc. The spectral bandpass extended from 8 to 13.5 μ. The photometer switched at a 180 Hz rate so as to alternately view a small portion of the lunar surface and an equal area of sky beside the lunar limb. The NEP of the detector was about 100 times smaller than that of a good thermal detector operated at room temperature. The calibration procedure is fully described; it involves comparisons with an ice-temperature and a liquid-nitrogen-temperature blackbody cavity. A large number of analyzed data are included.

[53] B. C. Murray, R. L. Wildey, and J. A. Westphal, "Observations of Jupiter and the Galilean Satellites at 10 Microns," *Astrophys. J.*, **139**, 986 (1964). Reports the first detailed radiometry of Jupiter in the 8 to 14 μ region. The radiometer consisted of the 200 in. Mt. Palomar telescope and the photometer that is described in ref. 19.2.1[41]. The field of view was 5.6 sec of arc, or about 1/8 of the planetary diameter. The temperature when averaged over about half of the disk was 128°K. The temperature of two of Jupiter's moons was also measured. An unusual enhancement of the radiation from the shadows cast on the surface by the moons was noted. For further information on these hot shadows, see ref. 19.2.1[56].

[54] F. J. Low and H. L. Johnson, "Stellar Photometry at 10 μ," *Astrophys. J.*, **139**, 1130 (1964). Gives values for the effective irradiance from 16 stars in the 8 to 14 μ region.

[55] J. M. Saari, "The Surface Temperature of the Antisolar Point of the Moon," *Icarus*, **3**, 161 (1964). Measurements were made in the 10 to 12 μ region with a mercury-doped germanium detector. The temperature at the antisolar point (lunar midnight) was found to be 99°K, which is significantly lower than the value of 120°K reported in ref. 19.2.1[46]. The authors suggest that the discrepancy can be accounted for by a failure to account for the radiation received through the partial window extending from 16 to 24 μ.

[56] R. Wildey, "On the Infrared Opacity of Jupiter's Outer Atmosphere," *Icarus*, **3**, 332 (1964). Limb-darkening measurements in the 8 to 14 μ region are applied to a determination of the thermal characteristics of Jupiter's atmosphere and to a possible explanation of the hot shadows described in ref. 19.2.1[53].

[57] F. Schwarz and A. Ziolkowski, "2-Channel Infrared Radiometer for Mariner II," *Infrared Phys.*, **4**, 113 (1964). Gives a detailed description of the radiometer that was used during the Venus flyby. For other descriptions, see refs. 19.2.1[38, 40, 45, 47]. Additional information in this paper includes a specification of 2°C for the NEΔT, a minimum survival time of 118 days, and a weight of 2.88 lb.

[58] E. A. Burns and R. J. P. Lyon, "Errors in the Measurement of the Lunar Temperature," *J. Geophys. Research*, **69**, 3771 (1964). Most radiometric measurements of lunar temperature assume that the lunar surface has an emissivity of unity. Studies of minerals suspected to be part of the lunar crust show that the average emissivity is 0.78 to 0.86 (effective in the 2 to 25 μ region at a temperature of 350°K). These

deviations from blackbody conditions result in calculated temperatures that are low by 3.5 to 5.5 per cent.
[59] E. D. McAlister, "Infrared Optics and the Sea," *J. Opt. Soc. Am.*, **54**, 559 (1964).
[60] G. R. Hunt and J. W. Salisbury, "Lunar Surface Features: Mid-Infrared Spectral Observations," *Science*, **146**, 641 (1964). Reports radiometric measurements made with the 42 in. Lowell telescope. This is apparently the first time that such measurements have been made in the 16 to 24 μ partial atmospheric window. The detector was a Golay cell. The measurements contain numerous unexplained thermal anomalies.
[61] Boeing Scientists Find Unexpected Abundance of Hot Spots on the Moon," *Missiles and Rockets*, **16**, 28 (January 18, 1965). Supplements the more complete report that is given in ref. 19.2.1[84]. Information peculiar to this article is that the detector was mercury-doped germanium cooled with liquid neon. The investigators believe that this may have been the first time that liquid neon was transported outside of the United States (to Egypt).
[62] B. C. Murray and J. A. Westphal, "Infrared Astronomy," *Sci. Am.*, **213**, 20 (August 1965). A good survey of the applications of infrared radiometers to the problems of observational astronomy.
[63] D. M. Gates, "Heat Transfer in Plants," *Sci. Am.*, **213**, 76 (December 1965). A basic reference for anyone trying to interpret radiometric measurements of terrain.
[64] E. D. McAlister, "A Two-Wavelength Microwave Radiometer for Measurement of the Heat Exchange at the Air-Sea Interface," *Appl. Opt.*, **4**, 145 (1965). See ref. 19.2.1[51] for an explanation of the reasons for using two wavelengths for this type of measurement. The author states that only 15 per cent of the time does a 30 mile-diameter circle on the earth's surface have a 20 per cent or less cloud cover; that is, infrared equipment will never have an all-weather capability.
[65] J. A. Westphal, R. L. Wildey, and B. C. Murray, "The 8 to 14 Micron Appearance of Venus Before the 1964 Conjunction," *Astrophys. J.*, **142**, 799 (1965).
[66] F. J. Low, "Lunar Nighttime Temperatures Measured at 20 Microns," *Astrophys. J.*, **142**, 806 (1965). Measurements were made in the 17.5 to 22 μ region using a germanium bolometer cooled to 2°K, a potassium bromide window, and an interference filter that cut on at 17.5 μ. This equipment was used with a 28 in. telescope and had a field of view of 18 sec of arc. The mean temperature across the lunar disk was 90°K.
[67] R. V. Annable et al., "Infrared and the Nimbus High Resolution Radiometer," *Elec. Comm.*, **40**, 500 (1965).
[68] K. J. K. Buettner and C. D. Kern, "The Determination of Infrared Emissivities of Terrestrial Surfaces," *J. Geophys. Research*, **70**, 1329 (1965). With the advent of radiation data from satellites it has become increasingly important to know more about terrestrial surface emissivities. This paper describes several means of measuring emissivity and gives an extensive table of effective emissivity (8 to 12 μ) for terrestrial materials.
[69] R. L. Wildey, B. C. Murray and J. A. Westphal, "Thermal Emission of the Jovian Disk," *J. Geophys. Research*, **70**, 3711 (1965). Radiometric measurements made with the 200 in. Mt. Palomar telescope and the photometer unit described in ref. 19.2.1[41]. For these measurements, several minor changes were made to the photometer and a new liquid-hydrogen-cooled mercury-doped germanium detector was used. The angular resolution was approximately 1/7 of the Jovian disk. The radiometer is calibrated with liquid-nitrogen- and water-ice-cooled blackbodies. The temperature at the center of the disk is 128.5°K, the light bands are about 0.5°C cooler than the dark bands, and the Great Red Spot is from 1.5 to 2.0°C cooler than the surrounding disk. The thermal maps show a significant limb darkening, implying an increase of temperature with depth in the atmosphere caused by a green-house mechanism.

19.2.1 MEASUREMENT OF TEMPERATURE

[70] J. B. Pollack and C. Sagan, "The Infrared Limb Darkening of Venus," *J. Geophys. Research*, **70**, 4403 (1965). Radiometric measurements made from the earth and from passing spacecraft indicate a limb darkening of Venus in the 8 to 13 μ region. This paper discusses the experimental measurements and proposes three atmospheric models to explain them.

[71] F. H. Murcray, "The Spectral Dependence of Lunar Emissivity," *J. Geophys. Research*, **70**, 4959 (1965). The spectral radiance of the lunar surface was measured in the 8.0 to 10.4 μ region from an observing site at an altitude of 11,500 ft. Emissivities calculated from these measurements vary from near unity at 8.5 μ, to 0.90 to 0.93 in the region from 10.0 to 10.4 μ. Measurements were made with a small prism spectrometer fitted with an external reflective chopper set at an angle of 45 deg to the optical axis. A reference blackbody was placed so as to fill the field of view of the spectrometer when the chopper blocked the target beam. A thermocouple detector was used. The incident flux was collected by a 12 in., $f/8$ reflecting telescope.

[72] J. M. Saari and R. W. Shorthill, "Thermal Anomalies on the Totally Eclipsed Moon of December 19, 1964," *Nature*, **205**, 964 (1965).

[73] J. H. Veghte and C. F. Herreid, "Radiometric Determination of Feather Insulation and Metabolism of Arctic Birds," *Physiological Zoology*, **38**, 267 (1965).

[74] R. L. Wildey, "Hot Shadows of Jupiter," *Science*, **147**, 1035 (1965). While scanning the Jovian disk for thermal emission measurements in the 8 to 14 μ region, the author noted a large enhancement in the emission from the shadows cast on the surface by two of its moons. An attempt to repeat the observation on another night failed to show any enhancement. The author is convinced that the phenomenon is real and variable with time and is not a peculiarity of the instrumentation. See also refs. 19.2.1 [53] and [56].

[75] R. W. Shorthill and J. M. Saari, "Nonuniform Cooling of the Eclipsed Moon: A Listing of Thirty Prominent Anomalies," *Science*, **150**, 210 (1965).

[76] D. E. Fink, "Sciences Take Brunt of Space Fund Slash," *Aviation Week*, **84**, 27 (January 31, 1966). Among the possible experiments to be carried out on a mid-1967 Venus mission are two infrared experiments to study the heat balance and the constituents of the Venusian atmosphere.

[77] W. C. Wetmore, "Soviets Reveal New Data on Spacecraft," *Aviation Week*, **84**, 26 (May 26, 1966). The Luna 10 spacecraft that was injected into a lunar orbit carried an "infrared detector to determine the moon's thermal radiation." A report in the same magazine at **84**, 33 (April 11, 1966), states that instruments were carried to record the "infrared emissions from the lunar surface." Despite the obscure language, the references are almost certainly to an infrared radiometer.

[78] J. Green, "The Moon's Surface," *Intl. Sci. Tech.*, September 1966, p. 59. An excellent survey article that describes the current theories on the nature of the lunar surface. A considerable portion of the article is devoted to a description and an analysis of the evidence that has been gathered by infrared radiometry.

[79] S. R. Drayson, "Atmospheric Transmission in the CO_2 Band Between 12 μ and 18 μ," *Appl. Opt.*, **5**, 385 (1966). The data presented here are potentially useful for interpreting measurements from satellite-borne radiometers.

[80] W. A. Hovis, Jr., "Optimum Wavelength Intervals for Surface Temperature Radiometry," *Appl. Opt.*, **5**, 815 (1966). Presents total reflectance measurements (0.5 to 23 μ) of a variety of common surface minerals. Many of these minerals show pronounced reflection maxima (restrahlen) at wavelengths that are often included within the spectral bandpass of radiometers that are used to measure lunar and planetary temperatures. If measurements must be made through the earth's atmosphere, the spectral interval from 10 to 12 μ is the most favourable, since most minerals have an emissivity that is near unity in this region.

[81] M. Harwit et al., "A Liquid Nitrogen Cooled, Rocket Borne, Infrared Telescope," *Appl. Opt.*, **5**, 1732 (1966). Describes a wide-field radiometer that was flown in an Aerobee rocket. The entire radiometer was incorporated into a Dewar package so that all of its parts were cooled to the same temperature. Indium arsenide and indium antimonide detectors were used successfully. Gold-doped germanium detectors performed poorly, and lead sulfide detectors gave continuous troubles.

[82] F. J. Low, "The Infrared Brightness Temperature of Uranus," *Astrophys. J.*, **146**, 326 (1966). Observations made through the 17.5 to 25 μ partial atmospheric window show that the brightness temperature of Uranus is $55 \pm 3°K$. Uranus subtends an angle of 3.8 sec of arc and thus did not fill the 10 sec of arc field of view of the instrument. Fifty-eight pairs of 10 sec integrations were required to reduce the sky noise to a level below the planet signal. Observations were made with a telescope having a 5 ft aperture and a germanium bolometer that was cooled to a temperature of 2°K.

[83] J. R. Yoder, J. D. Rehnberg, and G. R. Hunt, "Earth-Based Infrared Lunar Mapper for Thermal and Compositional Studies," *J. Opt. Soc. Am.*, **56**, 1453 (1966). Describes equipment to produce maps of lunar thermal radiance patterns or of lunar surface compositional probability. A liquid-helium-cooled copper-doped germanium detector subtending 5 sec of arc scans the image plane of an astronomical telescope.

[84] J. M. Saari and R. W. Shorthill, "Hot Spots on the Moon," *Sky and Telescope*, **31**, 327 (1966). Contains an unusually detailed report of the results of the authors' measurement program. The basic equipment was a focal-plane scanner that could sweep the entire lunar image in 30 min. It could make simultaneous measurements at 0.445 μ and in the 10 to 12 μ region with an angular resolution of 8 sec of arc. This scanner was used in 1963 with the 60 in. telescope at Mt. Wilson and in 1964 with the 74 in. Kottamia telescope at the Helwan Observatory in Egypt. For the later measurements the scanner was modified to have a resolution of 10 sec of arc and to make 200 traverses of the lunar image in 17 min. For additional information, see ref. 19.2.1[61].

[85] W. N. Hess, D. H. Menzel, and J. A. O'Keefe, eds., *The Nature of the Lunar Surface*, Baltimore: Johns Hopkins Press, 1966. See H. C. Ingrao, A. T. Young, and J. L. Linsky, "A Critical Analysis of Lunar Temperature Measurements in the Infrared," p. 185.

[86] Hess, Menzel, and O'Keefe, *op. cit.* (ref. 19.2.1[85]), see R. W. Shorthill and J. M. Saari, "Recent Discovery of Hot Spots on the Lunar Surface: A Brief Report of Infrared Measurements on the Eclipsed Moon," p. 215.

[87] "Filter Center," *Aviation Week*, **86**, 101 (April 24, 1967). Announces the award of a $500,000 contract for the development of a prototype, two-channel, infrared radiometer that will be used to measure Mars' surface temperatures on flyby missions in 1969.

[88] W. E. Osborne, "Treasure Finders," *Desert*, September 1967, p. 10. It is unfortunate that this article may lure treasure hunters into thinking that simple infrared radiometers can be used to discover buried treasure. The article states, "As any buried object will possess different infrared characteristics to those of the surrounding earth, it is revealed as a reflected pattern on the surface...as the shape of the buried object can be read by an experienced operator, much needless digging is avoided." If these statements were correct, the detection of buried land mines would no longer be a problem for the military (despite claims to this effect by the same author in ref. 16.1.1[31]). The article shows a picture of a small radiometer that is said to use uncooled indium antimonide or lead selenide detectors or, in a more expensive version, an uncooled immersed bolometer. No commercial models are yet available [sic].

[89] H. L. Clark, "Some Problems Associated with Airborne Radiometry of the Sea," *Appl. Opt.*, **6**, 2151 (1967). Notes that measurements of the radiation temperature of

the sea were made from a blimp as early as 1942. The author estimates that current radiometric techniques may be subject to errors of several degrees Centigrade in measurement of ocean temperatures. Among the errors that are discussed are the effects of an intervening atmosphere, reflected sky radiation, radiometric noise, platform instability (angular and altitude), atmospheric inhomogeneities, surface clutter, and time variations.

[90] D. J. Dumin, "Measurement of Film Thickness Using Infrared Interference," *Rev. Sci. Instr.*, **38**, 1107 (1967). Describes the use of a radiometer to monitor the thickness of epitaxially grown silicon films on sapphire. The radiometer measures the emission from the substrate and from the growing film. The relative variation between the two is a function of the thickness of the growing film.

[91] "New IR Radiometer Planned for Tiros M," *Aviation Week*, **88**, 78 (January 22, 1968). Reports a new contract for approximately $750,000 to develop a very-high resolution radiometer for use on future weather spacecraft. The radiometer will provide temperature data plus day and night data on cloud cover over the earth. Satellite motion and a small oscillating mirror give the desired scan pattern. Apparently the detector is to be cooled with a radiative-transfer cooler.

[92] W. A. Hovis, Jr., L. R. Blaine, and W. R. Gallahan, "Infrared Aircraft Spectra Over Desert Terrain 8.5 μ to 16 μ," *Appl Opt.* **7**, 1137 (1968). The measurements reported here were made over desert areas. Surface samples were collected and their optical properties measured to assess the effect of the surface layer on the radiometrically measured temperature. The spectrometer was the filter wedge type described in ref. 19.3.3[26]. It scanned from 8.5 to 16 μ in 30 sec. The detector was a thermistor bolometer maintained at a temperature of 27.5°C. The field of view was 2 × 2 deg. Spectra and emissivity data are given for several dry lakes and a lava field. The measurements show that the accuracy of remotely sensed surface temperature and spectra in the 8.5 to 14 μ region can be seriously affected by gaseous and particulate absorption in the atmosphere. Over dusty areas, such as dry lakes, radiometrically determined temperatures can be in error by as much as 10°C.

[93] P. M. Saunders, "Radiance of Sea and Sky in the Infrared Window 800–1200 cm^{-1}," *J. Opt. Soc. Am.*, **58**, 645 (1968). A study that resulted from the author's attempt to make accurate surface temperature measurements of the ocean from a low-flying aircraft. The radiometer is described in ref. 17.2.1[28]. An interference filter and a barium fluoride window limited the spectral band pass to 8.2 to 12.5 μ. Under a clear sky, the radiance of a calm sea is a maximum when it is viewed normally. The radiance decreases as the angle of view moves away from the normal and it passes through a minimum at about 10 deg below the horizon. The radiance of a rough sea is essentially the same as that of a calm sea for viewing angles within 45 deg of the normal. Beyond 45 deg the surface of a rough sea is blacker than that of a calm sea.

19.2.2 Measurement of Flux

The applications in this section are principally meteorological, that is, measurement of the solar constant, study of the heat balance mechanisms of the earth and its atmosphere, and nighttime mapping of cloud cover and weather fronts. Of particular interest are the special radiometers, such as the Pohl, Black Ball, and Suomi-Kuhn, that are designed for the balloon-borne measurement of the upwelling and the downward-directed radiation fluxes in the atmosphere. Such radiometers are also finding increasing application in macroclimatology and agricultural studies.

19.2.2 MEASUREMENT OF FLUX

[1] G. P. Kuiper, ed., *The Earth as a Planet*, Chicago: University of Chicago Press, 1954; see H. R. Byers, "The Atmosphere up to 30 kilometers," Chapter 7. Contains a good description of the processes that contribute to the heat balance of the earth's atmosphere.

[2] W. Pohl, "Messungen des Ultraroten Strahlungsstromes in der Freien Atmosphäre," *Z. Geophys.*, **22**, 1 (1956). Describes the Pohl infrared net-radiation radiometer. This device has a pair of sensors, each having two horizontally oriented blackened plates. One plate is in radiation exchange with the sky and the atmosphere above the instrument, and the other plate is in radiation exchange with the ground and the atmosphere below. In one sensor the upward-facing plate is electrically heated while the downward-facing plate of the other sensor is heated. To ensure a uniform conductive heat transfer with the surrounding air, the entire radiometer is rotated around a vertical axis by a small wind vane. The temperature of each of the four surfaces is measured by thermistors.

[3] J. L. Gergen, "Black Ball: A Device for Measuring Atmospheric Infrared Radiation," *Rev. Sci. Instr.*, **27**, 453 (1956). The black ball is a nearly spherical polygon made of paper on a balsa-wood frame and surrounded by a thin polyethylene convection shield. The device can be used only at night. An observation consists of measuring the temperature within the black ball and of the surrounding air. These two measurements are then used to compute a measure of the total infrared radiation in the atmosphere.

[4] V. E. Suomi, D. O. Staley, and P. M. Kuhn, "A Direct Measurement of Infrared Radiation Divergence to 160 mb," *Quart. J. Roy. Meteorol. Soc.*, **84**, 134 (1958). Describes the theory, construction, and the errors that can occur in the use of the Suomi-Kuhn radiometer. One of the best descriptions of this radiometer is given in ref. 19.2.2 [13].

[5] R. A. Hanel et al., "The Satellite Vanguard 2: Cloud Cover Experiment," *IRE Trans. Mil. Electronics*, **4**, 245 (1960). Experiment included two radiometric sensors that used lead sulfide detectors. The 1 deg field of view was to be scanned over the earth by the rotation of the vehicle so as to make cloud-cover maps in the 0.7 to 0.8 μ region. Because the satellite tumbled, no data were obtained.

[6] V. E. Suomi and P. Kuhn, "An Economical Net Radiometer," *Tellus*, **10**, 160 (1958). See ref. 19.2.2 [13].

[7] R. W. Fenn and H. K. Weickmann, "Atmospheric Net Radiation Flux During Winter in the Thule Area, Greenland," *J. Geophys. Research*, **65**, 3651 (1960). Describes data received from two balloon flights of the Pohl (ref. 19.2.2 [2]) infrared net-radiation radiometer. These data are applied to a study of the atmospheric heat budget during the arctic winter.

[8] C. B. Tanner, J. A. Businger, and P. M. Kuhn, "The Economical Net Radiometer," *J. Geophys. Research*, **65**, 3657 (1960). The Suomi-Kuhn radiometer (refs. 19.2.2 [4] and [13]) is finding increasing application in meteorological, agricultural, and hydrological research. This paper describes how the use of two or more such radiometers permits the simultaneous measurement of the incoming and the outgoing flux of the total, the solar, and the thermal radiation currents, as well as the net radiation.

[9] P. M. Kuhn and V. E. Suomi, "Infrared Radiometer Soundings on a Synoptic Scale," *J. Geophys. Research*, **65**, 3669 (1960). Reports and analyzes the results of synoptic measurements of infrared flux that were obtained from a simultaneous ascent of 15 Suomi-Kuhn radiometersondes.

[10] R. L. Aagard, "Measurements of Infrared Radiation Divergence in the Atmosphere with the Double Radiometer and the Black Ball," *J. Meteorol.*, **17**, 311 (1960).

[11] "Tiros 2 Radiation Data User's Manual," Goddard Space Flight Center, Greenbelt, Md., May 15, 1961. A supplement, dated May 15, 1962, is also available.

19.2.2 MEASUREMENT OF FLUX

[12] F. Schwarz, "Infrared Circuits in Tiros Satellites," *Electronics*, **34**, 43 (September 22, 1961). Gives a detailed description, complete with circuit diagrams, of the signal processor for the 5-channel meteorological radiometer. The detector is a thermistor bolometer chopped at 46 Hz.

[13] L. S. Bohl, J. L. Gergen, and V. E. Suomi, "Atmosphere Infrared Radiation Detector," U.S. Patent No. 3,014,369, December 26, 1961. Describes a radiometer to measure the total terrestrial and atmospheric infrared radiation for studies of the earth's heat balance. It consists of an outer radiation-transparent shell spaced from an inner radiation-absorbing shell. A thermistor measures the temperature within the inner shell. The function of the outer shell is to prevent any convective cooling of the inner shell. For further discussion of this radiometer, see refs. 19.2.2[4] and [6].

[14] H. T. Mantis, "Observations of Infrared Cooling of a Tropical Air Mass," *J. Geophys. Research*, **66**, 465 (1961). Describes a series of measurements made with black ball radiometers and analyzes their significance.

[15] W. R. Bandeen et al., "Infrared and Reflected Solar Radiation Measurements from the Tiros II Meteorological Satellite," *J. Geophys. Research*, **66**, 3169 (1961). The five-channel medium-resolution scanning radiometer used thermistor bolometer detectors and optical filters to limit the spectral response of each channel. The resolution element at the surface of the earth is about 40 miles square. Channel 1 (6 to 6.5 μ) was to study the radiation emerging from the atmospheric water vapor absorption band; channel 2 (8 to 12 μ), day and night cloud cover; channel 3 (0.25 to 6 μ), albedo; channel 4 (8 to 30 μ), thermal radiation; and channel 5 (0.55 to 0.75 μ), visual maps for comparison with television pictures. The article gives a good description of the entire radiometric system, from the radiometer to the telemetry.

[16] R. W. Astheimer, R. DeWaard, and E. A. Jackson, "Infrared Radiometric Instruments on Tiros 2," *J. Opt. Soc. Am.*, **51**, 1386 (1961). Describes the small-field five-channel radiometer, the wide-field two-channel radiometer, and the horizon pulse generator.

[17] R. A. Hanel and D. Q. Wark, "Tiros 2 Radiation Experiment and its Physical Significance," *J. Opt. Soc. Am.*, **51**, 1393 (1961).

[18] P. M. Kuhn, "Accurary of the Airborne Economical Radiometer," *Monthly Weather Review*, **89**, 285 (1961). See refs. 19.2.2[6] and [13].

[19] W. Nordberg et al., "Preliminary Results of Radiation Measurements from the Tiros 3 Meteorological Satellite," *J. Atmos. Sci.*, **19**, 20 (1962).

[20] A. Liventsov et al., "Experimental Determination of Radiation from the Earth," *Doklady Akad. Nauk SSSR, Geophyzika*, **166**, 344 (1962) (OTS: 63-21109). Used a radiometer that consisted of reflective optics, bolometer, and an 80 Hz chopper. The field of view was a fraction of a degree. The spectral bandpass extended from 2.5 to 40 μ. In order to eliminate errors from radiation emitted by the chopper, the radiation from the earth was alternately compared with the radiation from outer space. The radiometer was carried aboard a high-altitude geophysical rocket. Before the flight the radiometer was calibrated against blackbodies at temperatures of 77 and 195°K. An internal calibration source was viewed at the end of every scan line. At least one set of measurements was made during a solar eclipse. The purpose of these measurements was to study the total radiation from the earth as well as the radiation from selected geographic areas. A table gives the observed fluxes and the effective temperatures.

[21] H. J. Sayer, "The Desert Locust and Tropical Convergence," *Nature*, **194**, 330 (1962). The synoptic weather patterns derived from Tiros infrared measurements have been used on several occasions in conjunction with surface observations to forecast the migration of swarms of the desert locust. When provided with such a warning, existing organizations can attack the swarms with chemicals dispensed from aircraft.

[22] D. B. Clarke, "Radiation Measurements with an Airborne Radiometer over the Ocean East of Trinidad," *J. Geophys. Research*, **68**, 235 (1963). Describes an adaptation of the polyethylene-covered radiometer in ref. 19.2.2[13]. A heat budget for the subcloud layer was constructed from the measurements.

[23] P. M. Kuhn, "Soundings of Observed and Computed Infrared Flux," *J. Geophys. Research*, **68**, 1415 (1963). During some of the early orbits of Tiros 2, a large number of Suomi-Kuhn radiometersonde ascents were made so as to provide comparison data. This paper discusses the results of these comparisons.

[24] R. R. Law, "Proposed Experimental Test of Cosmological Theory by Infrared Measurements," *Proc. Inst. Radio Engrs.*, **51**, 1180 (1963).

[25] R. S. Hawkins, "Infrared Meteorological Satellite Sensors," *Research Rev.*, **4**, 3 (June 1964). Data from the radiometers on Tiros 2 were fed to a computer that automatically plotted maps of cold fronts and cloud heights. These maps showed only marginal agreement with surface observations. The author has examined all of the data available on a particular frontal system and offers explanations for the differences. He concludes that, with proper interpretation, the infrared maps are in agreement with the surface observations.

[26] F. Moller, "Optics of the Lower Atmosphere," *Appl. Opt.*, **3**, 157 (1964). Describes infrared techniques for studying the radiation balance of the earth and its atmosphere.

[27] R. Wexler, "Infrared and Visual Radiation Measurements from Tiros III," *Appl. Opt.*, **3**, 215 (1964).

[28] "Global Map of Emitted Terrestrial Radiation," *Appl. Opt.*, **3**, 302 (1964). Describes a global map of emitted terrestrial radiation in the 8 to 12 μ window as measured with a scanning radiometer on board Tiros III. The map is reproduced in color on the cover of this issue.

[29] S. I. Rasool, "Global Distribution of the Net Energy Balance of the Atmosphere from Tiros Radiation Data," *Science*, **143**, 567 (1964).

[30] "X-15 May Measure Solar Constant," *Missiles and Rockets*, **16**, 23 (February 15, 1965). The need for these measurements arises from the 5 per cent uncertainty in the value of the solar constant and the difficulties that this uncertainty causes in the design of spacecraft. Precision total and filter radiometers would be used.

[31] J. L. Corcoran and L. H. Horn, "The Role of Synoptic Scale Variations of Infrared Radiation in the Generation of Available Potential Energy," *J. Geophys. Research*, **70**, 4521 (1965). Measurements from channel 4 of the five-channel scanning radiometer aboard Tiros 2 are used in investigating the role of infrared radiation in the generation of eddy available potential energy in the atmosphere.

[32] A. Mani, C. R. Sreedhavan, and V. Srinivasan, "Measurements of Infrared Radiative Fluxes Over India," *J. Geophys. Research*, **70**, 4529 (1965). Reports 66 upper-air soundings over India with Suomi-Kuhn radiometersondes (ref. 19.2.2[4, 6, and 13]). These measurements show that, although significant daily and seasonal variations occur in the upward, downward, and net infrared radiative fluxes in the atmosphere, the mean values for the winter, summer, and monsoon months show remarkably consistent features.

[33] "Soviets Show Cosmos 122 Satellite," *Aviation Week*, **85**, 77 (October 10, 1966). The Soviets report that this satellite carries infrared equipment to determine cloud cover over the dark side of the earth. It also carries a simple radiometer for measurements of the earth's heat balance.

[34] P. M. Kuhn, "Use of Radiometers on Balloons for Moisture Determination," *J. Spacecraft and Rockets*, **3**, 754 (1966). The radiometer consists of two thin, circular, blackened aluminum sensing disks separated by an insulating medium. Each disk is

covered with a radiation-transparent window. Bead or rod thermistors are bonded to the disks for temperature sensing. The disks are arranged so that one senses the incoming radiation and the other senses the outgoing radiation. The author shows how the indications of this radiometer can be used to calculate the atmospheric water vapor content.

[35] N. Robinson, ed., *Solar Radiation*, Amsterdam: Elsevier, 1966, Chapters 6 and 7. Combines a theoretical treatment of the radiation balance of the earth and its atmosphere with a good practical discussion of the operating principles and calibration procedures for instruments to measure solar radiation.

[36] U. Fink and P. Ville, "A Balloon-Borne Diffusing System for Infrared Radiation from $1\,\mu$ to $5\,\mu$," *Appl. Opt.*, **6**, 1424 (1967). Most balloon-borne measurements of solar radiation end at an altitude of about 30 km. These authors wanted to reduce instrument weight in order to extend their measurements to 50 km. They replaced the sun-pointing servo with a conical reflector and hemispherical diffusing dome. The dome was made of calcium aluminate, and its surface was ground with 220-grade Carborundum. Initial flight results indicate that the concept is sound.

[37] A. J. Drummond et al., "Multichannel Radiometer Measurement of Solar Irradiance," *J. Spacecraft and Rockets*, **4**, 1200 (1967). Describes a twelve-channel radiometer for airborne measurements of solar irradiance. Two channels measure the total flux, eight channels are fitted with narrow-band spectral filters and two channels are for broad-band spectral filters. The detectors are thermopiles with a lampblack coating that is said to be a nonselective absorber from 0.2 to $6\,\mu$. Early in-flight measurements indicate that the value of the solar constant is 1.95 cal cm^{-2}min^{-1} (0.136 W cm^{-2}) rather than the currently accepted value of 2.00 cal cm^{-2}min^{-1} (0.140 W cm^{-2}).

19.2.3 World Weather Watch

One of the major nonmilitary applications of infrared techniques is the use of infrared instrumentation for meteorological observations from satellites. These applications involve radiometry (Sections 19.2.1 and 19.2.2), spectroradiometry (Section 19.3.1), and thermal imaging (Section 19.4.2); they include determination of the solar constant, measurement of reflected solar radiation, analysis of the earth's heat budget, study of day and night cloud cover, location of weather fronts, determination of atmospheric structure and temperature profiles, measurement of the altitude of cloud tops, determination of land and sea-surface temperatures, and the analysis of ocean circulation patterns.

On the basis of their early successes, it is evident that observational satellites offer benefits transcending national borders. One of the most promising of these is worldwide observation of the weather, popularly called the World Weather Watch. The references in this section might have been placed in one of the sections mentioned above, but they are placed here because each is a tutorial treatment of the role of infrared instrumentation in World Weather Watch.

[1] V. P. Perev and A. A. Sochivko, "Weather and Satellites," *Priroda*, **50**, 25 (July 1961) (OTS: 61-28522). Discusses the principles of infrared instrumentation and their

application to satellite meteorology. The illustrative examples involve only U.S. satellites.

[2] M. Neiburger and H. Wexler, "Weather Satellites," *Sci. Am.*, **205**, 80 (July 1961). Describes the infrared instrumentation carried on Tiros 1 and 2.

[3] *Proceedings of the International Meteorological Satellite Workshop*, National Aeronautics and Space Administration and U.S. Department of Commerce (Weather Bureau), Washington, D.C., November 13 to 22, 1961. Available from the U.S. Government Printing Office.

[4] V. P. Petrov and A. A. Sochivko, *Artificial Earth and Weather Satellites*, Leningrad: Hydrometeorological Publishing House, 1961 (OTS: 63-21665). The translation covers Chapter 3, "Meteorological Satellites," and Chapter 4, "The Upper Atmosphere and Its Investigation with Satellites and Rockets." There is a good description of the principles of infrared sensors and the way in which they can be applied to meteorological measurements. There are no data on Soviet sensors, but the data on U.S. sensors are quite detailed.

[5] T. F. Malone, "Tomorrow's Weather," *Intl. Sci. Tech.*, May 1962, p. 39.

[6] D. Findlay and J. Strasser, "Tiros to Fill Gap as Nimbus Design Hits Snags," *Electronic Design*, **10**, 4 (September 27, 1962). Gives a complete list of the infrared sensors carried on the first six Tiros satellites.

[7] A. Sergeev, "Infrared Technology in Cosmos," *Trud*, March 7, 1963 (OTS: TT-63-23649). A popularized discussion of infrared techniques and their application to satellite meteorology.

[8] W. Nordberg, "Research with Tiros Radiation Measurements," *Astronautics and Aerospace Eng.*, **1**, 76 (April 1963). A comprehensive discussion of the infrared radiometric measurements made from the Tiros vehicles and their application to the study of day and night cloud patterns, atmospheric structure, and global heat balance. The article contains several examples of radiation maps made over various kinds of weather conditions, an interesting discussion of techniques for calibrating infrared sensors in space, and a lengthy bibliography.

[9] R. A. Hanel and D. Q. Wark, "Physical Measurements from Meteorological Satellites," *Astronautics and Aerospace Eng.*, **1**, 85 (April 1963). Discusses infrared instrumentation for the determination of atmospheric temperature profiles, the altitude of cloud tops, and the measurement of the solar constant, and compares infrared and microwave radiometry for satellite use.

[10] S. Singer, "Satellite Meteorology," *Intl. Sci. Tech.*, December 1964, p. 30. An excellent analysis of the uses that are made of the information received from satellite-borne infrared sensors.

[11] M. Tepper and D. S. Johnson, "Toward Operational Weather Satellite Systems," *Astronautics and Aeronautics*, **3**, 16 (June 1965). Notes that measurements from infrared sensors on satellites are being used for studies of the earth-atmosphere heat balance and its relation to planetery circulation, for the determination of cloud-top temperatures and altitudes, for the determination of land and sea-surface temperatures, and for the measurement of reflected solar radiation. See also ref. 19.4.2 [7].

[12] D. E. Fink, "Special Report—NASA Weather Satellite Plans," *Aviation Week*, **84**, 40 (January 3, 1966). Among the infrared sensors that are described are high-, medium-, and low-resolution radiometers, and grating and interferometric spectrometers.

[13] P. Alelyunas, "Synergetic Satellites," *Space/Aeronautics*, **46**, 52 (October 1966). Describes the possibilities of five types of supersatellites: weather, communications, earth resources, surveillance, and scientific. Infrared equipment plays a role in each. Author notes that the infrared sensors aboard Tiros 7 have provided sufficiently

19.3.1 REMOTE SENSING 581

detailed measurements of sea-surface temperatures to be useful to weather forecasters and commercial fishermen.

[14] M. Tepper, "Threshold to the World Weather Watch," *Astronautics and Aeronautics*, **5**, 10 (January 1967). An excellent survey article that gives a good perspective from which to assess the possible uses of infrared instrumentation in weather satellites.

[15] "Weather Satellites," *Space/Aeronautics*, **47**, 84 (January 1967). Summarizes the U.S. weather satellite program well into the 1970's. Describes the various infrared sensors that are now in use and that are planned for the future. Article states that the high-resolution infrared radiometer on Nimbus 2 has provided invaluable data on the path of the Gulf Stream.

[16] "Variety of Russian Spacecraft Displayed," *Aviation Week*, **86**, 84 (June 12, 1967). The photographs show details of infrared and optical sensors on the Cosmos-144 meteorological satellite. These include a mirror system said to be used for infrared photography, a combined star tracker and horizon scanner, and an opening large enough to accomodate a 12 in. optical system.

[17] P. G. Thomas, "Global Weather Forecasting," *Space/Aeronautics*, **48**, 76 (October 1967). An excellent survey of the potential usefulness of meteorological satellites in combination with computers for worldwide weather forecasting. Two especially valuable tables give data on the characteristics of infrared sensors that have already been flown or are planned for future flights on weather satellites.

[18] R. G. Barry and R. J. Chorley, *Atmosphere, Weather and Climate*, London: Methuen, 1968. This book describes the rapid advances that have been made over the last decade in our understanding of atmospheric processes and climatology on a global scale. It describes much of the infrared instrumentation that has been flown in meteorological satellites and provides a sound basis for understanding the significance of their measurements.

19.3 SPECTRORADIOMETRIC APPLICATIONS

Spectroradiometric equipment, with its ability to examine narrow spectral intervals, yields much information about the composition and physical state of planetary atmospheres and offers clues to their surface constituents.

19.3.1 Remote Sensing of the Earth and Its Atmosphere

The applications in this section are limited to observations of the earth and its atmosphere from ground-based, airborne, or space-borne platforms.

[1] S. P. Langley, "Measurements of the Solar Spectrum to 5.5 Microns," *Ann. Astrophys. Obs. Smithsonian Inst.*, **1**, 1 (1900).

[2] N. B. Foster and L. W. Foskett, "A Spectrophotometer for the Determination of the Water Vapor in a Vertical Column of the Atmosphere," *J. Opt. Soc. Am.*, **35**, 601 (1945). See ref. 19.3.1 [5].

[3] J. Yarnell and R. M. Goody, "Infrared Solar Spectroscopy in a High Altitude Aircraft," *J. Sci. Instr.*, **29**, 353 (1952). The spectrometer used a thermistor bolometer, a KRS-5 window, a Perspex chopper, and a silver sulfide–silver chloride filter to

eliminate short wavelength radiation. Spectral resolution was 0.1 μ from 4 to 8.5 μ. Observations were made from an altitude of 36,000 ft.

[4] G. P. Kuiper, ed., *The Earth as a Planet*, Chicago: University of Chicago Press, 1954; see L. Goldberg, "The Absorption Spectrum of the Atmosphere," Chapter 9. Gives the background necessary for interpreting the spectra of celestial objects and the information that such spectra reveal about the composition and physical state of the earth's atmosphere.

[5] L. W. Foskett et al., "Apparatus for Absorption Spectra Analysis," U.S. Patent No. 2,775,160, December 25, 1956. Describes an infrared hygrometer for the determination of atmospheric water vapor content. The instrument measures the transmittance of a sample of the atmosphere in two narrow spectral intervals, one located in an absorption band and the other in a region that is free of absorption. The ratio of the two transmittances is proportional to the water vapor content. The novel feature of this design is the use of a rotating wheel carrying one or more pairs of narrow bandpass filters. See also ref. 19.3.1 [2].

[6] E. E. Bell and L. Eisner, "Infrared Radiation from the White Sands National Monument, New Mexico," *J. Opt. Soc. Am.*, **46**, 303 (1956). These measurements clearly show the 9 μ restrahlen minimum in the spectral radiance of the gypsum sand that is so characteristic of the area.

[7] D. C. Burch, "Infrared Evidence for the Presence of Ozone in the Lower Atmosphere," *J. Opt. Soc. Am.*, **46**, 360 (1956).

[8] D. M. Gates et al., "Near Infrared Solar Radiation Measurements by Balloon to an Altitude of 100,000 Feet," *J. Opt. Soc. Am.*, **48**, 1010 (1958).

[9] L. D. Kaplan, "Inference of Atmospheric Structure from Remote Radiation Measurements," *J. Opt. Soc. Am.*, **49**, 1004 (1959). The classic paper in this field. Shows that the three-dimensional distribution of atmospheric temperature and water vapor can be obtained from measurements of the spectral variation of the atmospheric radiation as viewed from a high-altitude aircraft or satellite.

[10] M. S. Kiseleva, B. S. Neporent, and V. A. Fursinkiv, "Spectral Determination of Water Vapor in the Upper Atmosphere," *Optics and Spectroscopy*, **6**, 522 (1959).

[11] R. W. Buchheim, ed., *Space Handbook*, New York: Modern Library, 1959. A handbook prepared for the Select Committee on Astronautics and Space Exploration, U.S. House of Representatives. Various space applications of infrared techniques are discussed. Although the coverage is necessarily limited, this report does an excellent job of placing the variety of available techniques in a very easily understandable perspective.

[12] D. G. Murcray et al., "Water Vapor Distribution Above 90,000 Feet," *J. Geophys. Research*, **65**, 3641 (1960). Reports observations of the 6.2 μ water vapor absorption band from a balloon-borne spectrometer carried to an eventual altitude of 92,000 ft. The spectrometer used a sodium chloride prism, a thermistor bolometer, an 80 Hz chopper, and a synchronous rectifier. The spectrum from 1 to 10 μ is scanned in 18 sec. The data give new information about the amount of water vapor in the stratosphere.

[13] J. C. Mester, "Upper Atmospheric Research at the Ballistic Research Laboratories," *IRE Trans. Mil. Electronics*, **4**, 222 (1960). Mentions the development of an infrared hygrometer that will be flown in a Nike-Cajun rocket.

[14] N. Ginsburg, W. R. Fredrickson, and R. Paulson, "Measurements with a Spectral Radiometer," *J. Opt. Soc. Am.*, **50**, 1176 (1960). Radiometer uses 10 in., $f/2.5$ telescope optics, a prism monochromator, and cooled lead telluride or gold-doped germanium detectors. The field of view is 0.8 by 3 mrad. Curves in the article show the spectral radiance of a wide variety of terrestrial backgrounds. In one mode of operation, repeated scans were made over the same targets at various times of the day and

night. These sometimes show "washout," wherein targets disappear completely; that is, there is no radiation contrast between the target and its background.

[15] E. E. Bell et al., "Spectral Radiance of Sky and Terrain at Wavelengths Between 1 and 20 Microns. I. Instrumentation. II. Sky Measurements. III. Terrain Measurements," *J. Opt. Soc. Am.*, **50**, 1308, 1313 (1960); **52**, 201 (1962). An excellent reference for anyone who is planning measurements of this type or who is looking for data on backgrounds.

[16] J. Houghton and J. Seeley, "Spectroscopic Observations of the Water Vapour Content of the Stratosphere," *Quart. J. Roy. Meteorol. Soc.*, **86**, 358 (1960). Shows spectra taken from an aircraft at an altitidue of 15 km. Analysis of these spectra gives the mixing ratio for the water vapor above the tropopause.

[17] D. Q. Wark, "On Indirect Temperature Soundings of the Stratosphere from Satellites," *J. Geophys. Research*, **66**, 77 (1961). Proposes a means of data reduction for the measurement that is described in ref. 19.3.1 [9].

[18] R. A. Hanel, "Determination of Cloud Altitude from a Satellite," *J. Geophys. Research*, **66**, 1300 (1961). Proposes a method of determining the altitude of the tops of clouds that does not require assumptions about the relationship between temperature and altitude in the atmosphere. This author suggests use of the 2 μ band of carbon dioxide. For a commentary on this suggestion, see the next reference.

[19] G. Yamamoto and D. Q. Wark. "Discussion of the Letter by R. A. Hanel, Determination of Cloud Altitude from a Satellite." *J. Geophys. Research*, **66** 3596 (1961). These authors propose the use of the 0.76 μ band of oxygen rather than the 2 μ band of carbon dioxide suggested in the previous reference.

[20] D. Fryberger and E. F. Uretz, "Some Considerations Concerning the Measurement of the Atmospheric Temperature Field by Electromagnetic Means," *IRE Trans. Mil. Electronics*, **5**, 279 (1961). Describes a means of inferring atmospheric temperature structure by using an electromagnetic radiometer as a probe. The relationship between atmospheric condition and sensor indication is developed for both infrared and microwave radiometers.

[21] G. Yamamoto, "Numerical Method for Estimating the Stratospheric Temperature Distribution from Satellite Measurements in the Carbon Dioxide Band," *J. Meteorol.*, **18**, 581 (1961). Offers a means of data reduction for the measurement in ref. 19.3.1 [9].

[22] F. Moller, "Atmospheric Water Vapor Measurements at 6–7 Microns from a Satellite," *Planet. Space Sci.*, **5**, 202 (1961).

[23] M. G. Dreyfuss and D. T. Hilleary, "Satellite Infrared Spectrometer, Design and Development," *Aerospace Eng.*, **21**, 42 (February 1962). This spectrometer was originally developed for the determination of the atmospheric temperature profile from a satellite. The breadboard model described here used immersed thermistor bolometers, an Ebert spectrometer, and a 5 in. diffraction grating. Five exit slits isolate narrow spectral intervals, four being in the 15 μ carbon dioxide band and one being at 11.1 μ in the atmospheric window. The device weighs 55 lb (exclusive of electronics) and consumes 10 W. It seems likely that an interferometric spectrometer is a more favorable choice for the final satellite application.

[24] "Satellite Spectrometer," *Aviation Week*, **77**, 59 (December 3, 1962). A 6 lb interferometric spectrometer used to make measurements of earth radiation from a satellite. The spectral region from 1.8 to 16 μ is scanned in 1 sec.

[25] D. G. Murcray, F. H. Murcray, and W. J. Williams, "Distribution of Water Vapor in the Stratosphere as Determined from Infrared Absorption Measurements," *J. Geophys. Research*, **67**, 759 (1962). Reports on a later flight with the equipment that is described in ref. 19.3.1 [12]. The data gathered here permitted the calculation of the vertical distribution of water vapor in the atmosphere. Figure 3 shows values of the

mixing ratio up to an altitude of 31 km and compares them with the work of other investigators.

[26] V. E. Suomi, "Observing the Atmosphere—A Challenge," *Proc. Inst. Radio Engrs.*, **50**, 2191 (1962). A survey paper that contains a good discussion of the determination of atmospheric temperature profiles by the methods proposed in ref. 19.3.1 [9].

[27] D. H. Rank et al., "Abundance of N_2O in the Atmosphere," *J. Opt. Soc. Am.*, **52**, 858 (1962).

[28] R. L. Bowman and J. H. Shaw, "The Abundance of Nitrous Oxide, Methane, and Carbon Monoxide in Ground-Level Air," *Appl. Opt.*, **2**, 176 (1963).

[29] J. I. F. King, "Meteorological Inferences from Satellite Radiometry," *J. Atmos. Sci.*, **20**, 245 (1963). Shows that the atmospheric temperature variation with height, mixing ratio, and total pressure can be deduced from spectroradiometric observations.

[30] "Weather Bureau's Hurricane Trackers," *Aviation Week*, **80**, 83 (October 19, 1964). Among the instrumentation carried on hurricane-tracker aircraft is an infrared hygrometer for continuous monitoring of atmospheric water vapor content.

[31] L. C. Block and A. S. Zachor, "Inflight Satellite Measurements of Infrared Spectral Radiance of the Earth," *Appl. Opt.*, **3**, 209 (1964). Describes measurements in the 6 to 15 μ region with an interferometric spectrometer.

[32] D. G. Murcray, F. H. Murcray, and W. J. Williams, "Variation of the Infrared Solar Spectrum between 2800 and 1500 cm^{-1} with Altitude," *J. Opt. Soc. Am.*, **54**, 23 (1964).

[33] U. Fink, D. H. Rank, and T. A. Wiggins, "Abundance of Methane in the Earth's Atmosphere," *J. Opt. Soc. Am.*, **54**, 472 (1964).

[34] R. J. P. Lyon, "Analysis of Rocks by Spectral Infrared Emission (8 to 25 microns)," *Econ. Geol.*, **60**, 715 (1965). One of the most complete reports available.

[35] D. T. Hilleary, D. Q. Wark, and D. G. James, "An Experimental Determination of the Atmospheric Temperature Profile by Indirect Means," *Nature*, **205**, 489 (1965). An experimental verification of the theory proposed in ref. 19.3.1[9]. The spectrometer is described in ref. 19.3.1[23].

[36] D. E. Fink, "ESSA Enlarges Forecaster's Perspective," *Aviation Week*, **84**, 94 (March 21, 1966). A discussion of the Satellite Infrared Spectrometer (SIRS) that has already been described in ref. 19.3.1[23]. A planned modification would permit determination of atmospheric water vapor at the same time that the temperature profile is being measured.

[37] E. J. Williamson and J. T. Houghton, "A Radiometer-Sonde for Observing Emission from Atmospheric Water Vapor," *Appl. Opt.*, **5**, 377 (1966). Describes the use of a balloon-borne radiometer to measure the downward flux of thermal radiation from atmospheric water vapor in the 6.3 μ region.

[38] "14 Apollo Applications Experiments Picked," *Aviation Week*, **86**, 82 (May 1, 1967). Six of the meteorological experiments involve infrared instrumentation. These include: (1) A spectroradiometer to measure the radiance of the earth in 35 narrow bands between 3.6 and 5.0 μ. These data will be used to compute the vertical temperature profile. (2) A grating spectrometer for measurements in the 15 μ carbon dioxide band to permit the calculation of the vertical temperature profile up to an altitude of 20 miles. (3) A spectroradiometer using six narrow-band filters to cover the carbon dioxide band. (4) A filter-wedge spectroradiometer covering the 1.5 to 6 μ region. (5) A three-channel narrow-band radiometer covering the 0.63 to 0.65 μ, the 0.79 to 0.81 μ. and the 10 to 11 μ regions. (6) An interferometric spectrometer covering the 5 to 20 μ region.

[39] J. N. Howard, "Geophysics Instrumentation," *Phys. Today*, **20**, 49 (July 1967). An excellent, up-to-date survey of improved instrumentation for studies of the earth's

19.3.1 REMOTE SENSING

atmospheric environment from balloons, rockets, and satellites. A considerable portion of the article deals with infrared instrumentation for studies of the aurora, the airglow, the earth's horizon, the lunar surface, and remote sensing in general.

[40] N. Sclater, "Flying Laboratory to Study Atmospheric IR," *Electronic Design*, **15**, 36 (August 16, 1967). Describes a modified KC-135 aircraft outfitted as an atmospheric-infrared research platform. A rapid-scan spectrometer, an infrared sky mapper, several interferometric spectrometers, and two broad-band infrared radiometers are among the equipment complement. One interesting feature is a liquid-nitrogen-cooled optical chopper that is positioned outside one of the aircraft windows so as to permit extremely low-level radiometric measurements.

[41] D. G. Murcray, F. H. Murcray, and W. J. Williams, "A Balloon-Borne Grating Spectrometer," *Appl. Opt.* **6**, 191 (1967). Spectrometer covers the 2 to 14 μ region and is used to measure the way in which the transmission of the atmosphere varies with altitude.

[42] K. Ya. Kondratiev et al., "Direct Solar Radiation up to 30 km and Stratification of Attenuation Components in the Stratosphere," *Appl. Opt.*, **6**, 197 (1967). Describes balloon-borne radiometers and spectroradiometers that have been used over the Soviet Union as high as 30 km. Data are given for the total direct solar radiation and for the radiation in narrow spectral bands extending from 0.35 to 13 μ.

[43] J. H. Shaw, R. A. McClutchey, and P. W. Schaper, "Balloon Observations of the Radiance of the Earth between 2100 cm^{-1} and 2700 cm^{-1}," *Appl. Opt.*, **6**, 227 (1967). A temperature profile of the atmosphere to an altitude of 30 km has been derived from measurements of the 4.3 μ carbon dioxide absorption band. The results are in reasonable agreement with the temperatures obtained from a radiosonde unit operated during the flight. The spectrometer uses a strontium-titanate-immersed lead selenide detector cooled by liquid nitrogen. The chopper blades are mounted on the tines of a 330 Hz tuning fork.

[44] L. W. Chaney, S. R. Drayson, and C. Young, "Fourier Transform Spectrometer— Radiative Measurements and Temperature Inversion," *Appl. Opt.*, **6**, 347 (1967). This satellite-borne spectrometer is designed to measure the earth's thermal radiation in the spectral band extending from 5 to 20 μ, with a radiance error of less than 0.5 per cent (an accuracy that most workers would hesitate to claim, even in the laboratory). Figure 1 of this paper shows the spectral radiance of the earth as viewed from a balloon. Measurements made in the 14 μ carbon dioxide band have been used to calculate the atmospheric temperature profile. Agreement with radiosonde measurements is excellent.

[45] F. Saiedy and D. T. Hilleary, "Remote Sensing of Surface and Cloud Temperatures using the 899 cm^{-1} Interval," *Appl. Opt.*, **6**, 911 (1967). Measurements were made with a balloon-borne spectrometer at a wavelength of 11.1 μ. The results are carefully compared with measurements made by ground-based thermometers and radiometers, and the agreement is quite good. The spectrometer is an improved version of the one that is described in ref. 19.3.1 [23].

[46] W. A. Hovis, Jr., and M. Tobin, "Spectral Measurements from 1.6 μ to 5.4 μ of Natural Surfaces and Clouds," *Appl. Opt.*, **6**, 1399 (1967). Reports a series of airborne measurements that were made to compare with measurements from the Nimbus II meteorological satellite. Authors conclude that within the 3.4 to 4.2 μ interval, variations in surface emissivity cause no appreciable error in radiometric temperature measurements. Terrain measurements show a disappointing lack of information concerning the constituents of the surface layer— that is, cultivated farm land looks essentially the same as desert. The spectrometer that was used for these measurements is described in ref. 19.3.3 [26].

[47] B. S. Neporent et al., "Determination of Moisture in the Atmosphere from Absorption of Solar Radiation," *Appl. Opt.*, **6**, 1845 (1967). Describes a spectrometer that scans a narrow wavelength interval in the 2.6 μ water vapor absorption band. The detector is uncooled lead sulfide. The signal is chopped at 600 Hz and is synchronously rectified. The spectrometer and its sun tracker occupy a cylinder that is 1.5 m long and has a diameter of 0.5 m. It has been flown to an altitude of 28 km. An internal-standard source is included for in-flight calibration.

[48] T. Y. Palmer and W. G. Tank, "Method and Apparatus for Detecting and Measuring Water Vapor and Oxygen in the Ambient Atmosphere," U.S. Patent No. 3,364,351, January 16, 1968. The scheme presented here, measurement in two narrow spectral intervals, one in an absorption band and the other at a wavelength free of absorption, seems to offer little that is new or advantageous when compared with older methods (19.3.1[5]).

[49] S. F. Singer, "Measurement of Atmospheric Surface Pressure with a Satellite – Borne Laser," *Appl. Opt.*, **7**, 1125 (1968). An extension of the method proposed in refs. 19.3.1[18] and [19]. Uses a laser as a source and measures the integrated absorption of the atmosphere. Measurements will be made inside and outside of the oxygen A band at 0.76 μ. By using a laser rather than the sun as a source, measurements can be made at night and it may be possible to determine altitude, pressure, and temperature at specific points in the atmosphere.

19.3.2 Remote Sensing of Astronomical Bodies

One of the most interesting scientific and philosophical questions of the day concerns the possible existence of life on other planets. The answer to this question is vastly complicated by the fact that there is no formal operational definition of life that can serve for its recognition under all possible conditions. Man simply assumes that he will recognize life if he sees it, but how valid is this assumption? One possible approach to an answer is to conduct some sort of meaningful biological experiments on earth that might offer clues to the form that life could take on a distant planet. The essential information needed for the design of such experiments is accurate knowledge of the composition of a planet's atmosphere and of the pressure at its surface. Nearly all such information that is available today has been collected by means of infrared spectroscopy. In principle the procedure is quite simple. The sun serves as a convenient source, and its flux is observed under two sets of conditions – first, after a direct passage through the earth's atmosphere and, second, after passage through the planetary atmosphere and subsequent reflection to the earth-bound observer. Since the solar flux passes through the earth's atmosphere in both measurements, it is, in principle, possible to separate the absorption in the earth's atmosphere from that in the planet's atmosphere. In practice, the separation of the two is not nearly so simple to accomplish. In addition, the atmospheric windows force observations at wavelengths that are not the most favorable for detection of the characteristic absorp-

19.3.2 REMOTE SENSING OF ASTRONOMICAL BODIES

tion bands of some of the molecules thought to be present in planetary atmospheres.

Most of the early detection of planetary atmospheric constituents was done with grating spectrographs and photographic plates, a combination that did not respond to wavelengths beyond 1 μ. Since the fundamental vibration frequencies of most molecules lie at wavelengths beyond 3 μ, such instrumentation could only respond to third or higher overtone bands. Absorption is relatively weak in overtone bands, and only those bands caused by the major atmospheric constituents could be observed. Among the constituents that were identified in this way were methane on most of the planets, ammonia on Jupiter, and carbon dioxide on Mars and Venus. The introduction of photon detectors helped to advance the art, but it was the development of the interferometric spectrometer in the early 1960's that finally made it possible to obtain high-resolution spectra at wavelengths that were sufficiently long so that absorption bands could be detected even from minor constituents of planetary atmospheres. Some of the results of observations with such spectrometers are discussed in this section; descriptions of the theory and construction of these spectrometers are given in the following section (19.3.3).

The reader will find a rather fascinating detective story threading its way through this section. It begins in 1957 when Sinton (ref. 19.3.2[4]) reported the discovery of three absorption bands in spectra of Mars and interpreted them as being due to the presence of hydrocarbons on the planet. Subsequent installments of the story (refs. 19.3.2[5, 10, 11, 21, and 48]), extending over a period of 8 years, include several other seemingly plausible interpretations of the spectra, reexamination of the wavelength calibration and a revision in the wavelength of the observed absorption bands, and finally a detailed reexamination of the evidence and the conclusion that the bands were due to HDO in the earth's atmosphere.

[1] G. P. Kuiper, W. Wilson, and R. J. Cashman, "An Infrared Stellar Spectrometer," *Astrophys. J.*, **106**, 243 (1947). One of the earliest reported astronomical uses of a lead sulfide detector. Complete schematic diagrams are given for the signal processing circuitry.

[2] G. P. Kuiper, "Infrared Spectra of Planets," *Astrophys. J.*, **106**, 251 (1947). Spectra from 0.75 to 2.5 μ were made using an uncooled lead sulfide detector. Author has plans to cool the detector for later work.

[3] R. B. Barnes et al., "Qualitative Organic Analysis and Infrared Spectrometry," *Anal. Chem.*, **20**, 402 (1948). Gives an extremely useful chart that correlates characteristic absorption frequencies and functional atomic groups. Such a chart is essential in the interpretation of spectra.

[4] W. M. Sinton, "Spectroscopic Evidence for Vegetation on Mars," *Astrophys. J.*, **126**, 231 (1957). All organic molecules have absorption bands in the 3.4 μ region. This author's measurements of the reflection spectra of terrestrial plants show that, for most

spectra, a doublet band appears at about 3.46 μ. Spectra of Mars indicate the probable existence of this band. This evidence, when combined with the observed seasonal changes, makes it probable that vegetation exists on Mars. The spectrometer used a lead sulfide detector cooled with liquid nitrogen. It was used with the 200 in. Mt. Palomar telescope. For further commentary on this discovery, see refs. 19.3.2[5, 10, 11, 21 and 48].

[5] W. M. Sinton, "Further Evidence of Vegetation on Mars," *Science*, **130**, 1234 (1959).

[6] W. M. Sinton and J. Strong, "Radiometric Observations of Mars," *Astrophys. J.*, **131**, 459 (1960). Reports measurements made in 1954 with the 200 in. Mt. Palomar telescope. The absorption bands of carbon dioxide at 9.4, 10.4, and 12.6 μ were found in the Martian atmosphere. The absence of characteristic restrahlen indicates the probable scarcity of silicates on the Martian surface. See also the following reference.

[7] W. M. Sinton and J. Strong, "Radiometric Observations of Venus," *Astrophys. J.*, **131**, 470 (1960). A companion paper to the previous reference, but this one contains considerably more information on the methods used for data reduction. Because of the large amount of carbon dioxide in the Venusian atmosphere, these authors were surprised to find only a very weak absorption band at 10.4 μ and no trace of those expected at 9.6 and 12.6 μ. A weak absorption band found at 11.2 μ was tentatively identified as being caused by carbon suboxide.

[8] "IR Device Built to Detect any Vegetation on Mars," *Electronic Design*, **9**, 6 (April 12, 1961). See the following reference.

[9] "Spectrophotometer to Investigate Existence of Vegetation on Mars," *Aviation Week*, **74**, 139 (May 1, 1961). This device is designed to detect absorption bands in the 2 to 4 μ region that are characteristic of certain hydrocarbons found in vegetation. A good schematic drawing of the instrument is included. Among the novel features are a vibrating-reed chopper and a radiative-transfer cooler that keeps a lead selenide detector at a temperature of 195°K.

[10] N. B. Colthup, "Identification of Aldehyde in Mars Vegetation Regions," *Science*, **134**, 529 (1961). Sinton (ref. 19.3.2[4]) originally reported absorption bands at 3.67, 3.56, and 3.43 μ. Many organic materials have absorption bands in the 3.30 to 3.56 μ region, but very few have bands at 3.67 μ. Organic aldehydes, however, have bands at 3.67, 3.56, and 3.93 μ.

[11] W. M. Sinton, (No Title), *Science*, **134**, 529 (1961). Written in answer to the preceding reference. The author has reexamined his wavelength calibration and concludes that more proper values for the wavelengths of the absorption bands in the Mars spectra are 3.69, 3.58, and 3.45 μ.

[12] C. M. Plattner, "Stratoscope Aims at High-Altitude Photos," *Aviation Week*, **77**, 73 (October 15, 1962). An infrared spectrometer covering the 1 to 7 μ region is to be carried. The detector is a cooled germanium bolometer. It is hoped that this instrument will determine whether water vapor, carbon dioxide, carbon monoxide, and methane are present in the Martian atmosphere.

[13] W. M. Sinton, "Infrared Spectroscopy of Planets and Stars," *Appl. Opt.*, **1**, 105 (1962). A short survey article with examples of typical spectra.

[14] D. G. Murcray, "Optical Measurements from High-Altitude Balloons," *Appl. Opt.*, **1**, 121 (1962). Measurements were made with a filter-type radiometer consisting of a 20 cm diameter $f/1.5$ Cassegrain telescope, a scanning mirror, an 80 Hz chopper, and a thermistor bolometer. Filters isolate the following spectral regions: 1 to 2 μ, 2 to 3 μ, 3 to 5 μ, 5 to 8 μ, and 8 to 35 μ.

[15] J. Strong and F. R. Stauffer, "Instrumentation for IR Astrophysics," in W. A. Hiltner, ed., *Astronomical Techniques,* Chicago: University of Chicago Press, 1962, Chapter 12.

19.3.2 REMOTE SENSING OF ASTRONOMICAL BODIES 589

An excellent survey and discussion of the rationale for the selection of infrared instrumentation.

[16] "Stratoscope 2 Aims for Spectral Analyses of Planet's Atmospheres," *Aviation Week*, **78**, 74 (April 1, 1963). A general descriptive article. An item in the same magazine, at **78**, 33 (April 22, 1963), reports that a preliminary analysis of the data from the infrared spectrometer shows no trace of water vapor in the Martian atmosphere. The concentration of carbon dioxide was found to be 21 times that on earth.

[17] I. C. Stone, "Mariner Design Modified for Mars Flyby," *Aviation Week*, **78**, 50 (May 6, 1963). Among the experiments planned for Mariner-C is an infrared spectrometer that will provide data about the reflectance and emission of the Martian surface in the 1.7 to 2.2 μ and the 5.1 to 6.6 μ regions.

[18] R. E. Danielson, "The First Flight of Stratoscope II," *Am. Scientist*, **51**, 375 (1963). Stratoscope II consisted of a 36 in.-diameter $f/33$ Gregorian telescope that was carried to an altitude of 80,000 ft by a balloon. The infrared equipment was designed for a spectroscopic study of Mars in the 1 to 7.5 μ region. It was hoped that analysis of the spectra would yield water vapor and carbon dioxide abundance in the Martian atmosphere and possibly some evidence of organic matter on the surface. Only some of these goals were realized. The infrared equipment consisted of a calcium fluoride prism spectrograph and a germanium bolometer cooled to 2°K, the boiling temperature of liquid helium at the 80,000 ft altitude of the balloon. One interesting observation is that the bolometer action failed below a wavelength of 1.65 μ.

[19] V. I. Moroz, "Evaluation of Carbon Dioxide Gas Content and Total Pressure from Infrared Spectra of Venus and Mars," *Astronomicheskii Tsirkulyar*, **273**, 1 (1963) (SLA: TT-64-18757). Estimates that the surface pressure on Mars is 15 millibars.

[20] V. I. Moroz, "Stellar Photometer and Spectrometer for the 1–2.5 Micron Range," *Novaya Tekhnika v. Astronomii* (monograph), Moscow, 1963 (OTS: TT-64-21262). Photometer uses a lead sulfide detector.

[21] D. G. Rea, T. Belsky, and M. Calvin, "Interpretation of the 3 to 4 Micron Infrared Spectrum of Mars," *Science*, **141**, 923 (1963). This paper examines the evidence presented in ref. 19.3.2[4] and concludes that there are other equally plausible interpretations. No satisfactory explanation is advanced, and the problem remains unresolved.

[22] R. D. Hibben, "Clues to Life Processes Sought in Space," *Aviation Week*, **80**, 91 (February 17, 1964). Describes the various space programs that will utilize infrared spectrometers to gather clues concerning life in space.

[23] W. W. Kellogg, "Mars," *Intl. Sci. Tech.*, February 1964, p. 40. Discusses the contributions that infrared observations have added to the sum total of our knowledge about Mars.

[24] R. A. van Tassel, "Water Vapor in the Venus Atmosphere," *Research Rev.*, **3**, 8 (June 1964). Describes an infrared spectrometer that scans the region around 1.13 μ with the combination of a diffraction grating and a set of 21 exit slits arrayed to match 21 water vapor absorption lines. Analysis of the records shows that the amount of water vapor in the upper atmosphere of Venus is the equivalent of an absolute humidity of 10^{-2} g m^{-3} at the surface of the earth (the paper makes an obvious error, corrected here, in the units used to express the amount of water vapor).

[25] S. Tilson, "Planet Mars," *Space/Aeronautics*, **42**, 46 (July 1964). A discussion of the possibility of life on Mars. Most of the supporting evidence that is discussed here was obtained with infrared sensors.

[26] W. M. Sinton, "Optics and the Atmospheres of the Planets," *Appl. Opt.*, **3**, 175 (1964). Man's knowledge of the atmospheres of the planets has advanced step by step

with the development of optical techniques. Observations made in the infrared have contributed information about the temperature on the planets and the constituents of their atmosphres.

[27] H. Spinrad, "Spectroscopic Research on the Major Planets," *Appl. Opt.*, **3**, 181 (1964). Summarizes recent work in planetary spectroscopy, most of which has been accomplished in the photographic infrared.

[28] F. H. Murcray, D. G. Murcray, and W. J. Williams, "The Spectral Radiance of the Sun from 4 μ to 5 μ," *Appl. Opt.*, **3**, 1373 (1964).

[29] N. J. Woolf, M. Schwarzschild, and W. K. Rose, "Infrared Spectra of Red-Giant Stars," *Astrophys. J.*, **140**, 833 (1964). Reports measurements made at an altitude of 80,000 ft from Stratoscope II. The spectrometer used a pair of indium arsenide detectors cooled by liquid nitrogen. Chopping frequency was 617 Hz. For other measurements made during this same flight, see refs. 19.3.2[18] and [49].

[30] M. Bottema, W. Plummer, and J. Strong, "Composition of the Clouds of Venus," *Astrophys. J.*, **140**, 1640 (1964). Using a modification of the balloon-borne spectrometer that is described in ref. 19.3.2[24], the authors showed that Venusian clouds in the 1.7 to 3.4 μ region are quite similar to the spectra of terrestrial and laboratory ice clouds. This leads to the conclusion that the cloud layer on Venus is composed of particles of frozen water.

[31] W. A. Hovis, Jr., "Infrared Emission Spectra of Organic Solids from 5 to 6.6 Microns," *Science*, **143**, 587 (1964). Author concludes that the prospect of detecting life on Mars by infrared emission (rather than by reflection) spectroscopy is very low.

[32] J. W. Salisbury and P. E. Glaser, *The Lunar Surface Layer*, New York: Academic Press, 1964; see A. van Tassel and I. Simon, "Thermal Emission Characteristics of Mineral Dusts," p. 445. Shows that the band structure evident in the emission spectra of rocks thought to be representative of those comprising the lunar surface, does not show when the rocks are ground to a fine powder. If the particle size used is representative of that found on the lunar surface, spectral analysis of the lunar thermal emission will yield no information on the composition of the surface.

[33] Salisbury and Glaser, *op. cit.* (preceding ref., 19.3.2[32]); see E. A. Burns and R. J. P. Lyon, "Feasibility of Remote Compositional Mapping of the Lunar Surface," p. 469. Concludes that rocks can be identified as to type by the band structure in their thermal emission, provided that the particles on the lunar surface are larger than 10 to 100 μ.

[34] W. R. Corliss, "Detecting Life in Space," *Intl. Sci. Tech.*, January 1965, p. 28.

[35] J. Strong, "Infrared Astronomy by Balloon," *Sci. Am.*, **212**, 28 (January 1965). Contains a very good history of the balloon-borne infrared investigations carried out since 1956 by a group at Johns Hopkins University. The article contains schematic diagrams of the instrumentation. The special spectrometer with its 21 exit slits is discussed in detail in ref. 19.3.2[24]. One of the results of this program is the finding that the clouds of Venus are composed of ice crystals.

[36] "IR Balloon Telescopes to look at Venus, Mars," *Missiles and Rockets*, **16**, 23 (February 15, 1965). A balloon-borne infrared spectroradiometer (ref 19.3.2 [30]) has shown that the Venus cloud cover consists of ice crystals. New plans call for using this equipment to prove planetary surface temperatures.

[37] H. W. Mattson, "Determining Molecular Structure," *Intl. Sci. Tech.*, February 1965, p. 22. A tutorial discussion of the applications of infrared, ultraviolet, nuclear magnetic resonance, and mass spectroscopy.

[38] D. Winston, "Russians May Skip Mars Probe in 1966," *Aviation Week*, **83**, 31 (August 30, 1965). The Russian Zond-3 space vehicle made spectroradiometric measurements of the lunar surface in the 3 μ region.

19.3.2 REMOTE SENSING OF ASTRONOMICAL BODIES

[39] L. Goldberg, "The New Astronomies," *Intl. Sci. Tech.*, August 1965, p. 18. A discussion of the advances that are due to the use of the infrared, ultraviolet, and X-ray portions of the spectrum for astronomical research.

[40] G. Neugebauer, D. E. Martz, and R. B. Leighton, "Observations of Extremely Cool Stars." *Astrophys. J.*, **142**, 399 (1965). Describes a program to survey the sky in two spectral regions, 0.68 to 0.92 μ and 2.01 to 2.41 μ. A number of previously unobserved cool stars have been detected. Their radiometric temperature is about 1000°K. The survey instrument is a 62 in., $f/1$ aluminized plastic mirror that is vibrated at a rate of 20 Hz. The star images move across a detector system that consists of 8 liquid-nitrogen-cooled lead sulfide detectors and a single silicon photodiode.

[41] H. L. Johnson, F. J. Low, and D. Steinmetz, "Infrared Observations of the Neugebauer-Martz-Leighton Infrared Star in Cygnus," *Astrophys. J.*, **142**, 808 (1965). Confirms the 700°K temperature reported in the preceding reference for one of the cooler stars in Cygnus.

[42] W. O. Davies, "Spectroscopic Observations of Mars," *Frontier*, Winter 1965, p. 4. Gives a very readable explanation of the way in which infrared spectroscopic observations can yield information on the constituents of the Martian atmosphere and on the pressure at its surface.

[43] G. R. Hunt, "Infrared Spectral Emission and Its Application to the Detection of Organic Matter on Mars," *J. Geophys. Research*, **70**, 2351 (1965). The author disagrees with the conclusions of ref. 19.3.2 [31] and presents data that he thinks show that it should be possible to obtain emission spectra from Mars and to use them for the identification of organic matter.

[44] W. A. Hovis, Jr., "Discussion of a Paper by Graham R. Hunt, Infrared Spectral Emission and its Application to the Detection of Organic Matter on Mars," *J. Geophys. Research*, **70**, 2359 (1965). A rebuttal to the comments expressed in the preceding reference. This author feels that the experiment described by Hunt is not representative of the problems encountered in the measurement of emission spectra from spacecraft. Hunt's reply appears in the same journal, at **70**, 2361 (1965), and it shows that he still believes his original remarks are correct.

[45] M. Bottema et al., "The Composition of the Venus Clouds and Implications for Model Atmospheres," *J. Geophys. Research*, **70**, 4401 (1965). Describes correction procedures for reflection spectra of Venus.

[46] R. P. Espinola and H. H. Blau, Jr., "Cloud Composition from Infrared Spectra," *J. Geophys. Research*, **70**, 6263 (1965). Presents measurements of the spectral radiance of clouds in the 1.1 to 2.5 μ region. These spectra clearly show the difference between natural-water and ice-crystal clouds, and they should be useful in determining the composition and phase of clouds on other planets. Although the measurements are of reflected sunlight, the authors have treated them as if the clouds were a source, a perfectly valid procedure that often simplifies interpretation and analysis.

[47] G. Herzberg, "Molecular Spectroscopy and Astrophysical Problems," *J. Opt. Soc. Am.*, **55**, 225 (1965). Questions such as the presence of molecular hydrogen on planets or in space and the composition of comets have stimulated much effort in molecular spectroscopy. Conversely, advances in molecular spectroscopy have led to a better understanding of astrophysical phenomena. The author discusses these and other examples of such cross-fertilization.

[48] D. G. Rea, B. T. O'Leary, and W. M. Sinton, "Mars: The Origin of the 3.58- and 3.69-Micron Minima in the Infrared Spectra." *Science*, **147**, 1286 (1965). These authors conclude that the absorption bands that were first reported in ref. 19.3.2 [4] are really due to absorption by HDO in the earth's atmosphere.

[49] R. E. Danielson, "The Infrared Spectrum of Jupiter," *Astrophys. J.*, **143**, 949 (1966). Reports measurements between 0.8 and 3.1 μ from Stratoscope II.

[50] L. Mertz and I. Coleman, "Infrared Spectrum of the Taurus Red Object," *Astrophys. J.*, **143**, 1000 (1966). Spectra made with an interferometric spectrometer of one of the cool stars reported in ref. 19.3.2[40].

[51] M. J. D. Low and I. Coleman, "Measurement of the Spectral Emission of Infrared Radiation of Minerals and Rocks Using Multiple-Scan Interferometry," *Appl. Opt.*, **5**, 1453 (1966). Results are of potential interest to studies of lunar and planetary surfaces. Spectra are given for six samples of rocks and minerals over the 7 to 30 μ spectral range.

[52] G. R. Hunt et al., "Rapid Remote Sensing by Spectrum Matching Technique: I. Description and Discussion of the Method," *J. Geophys. Research*, **71**, 2919 (1966); "II. Application in the Laboratory and in Lunar Observations," **72**, 705 (1967). Describes a method for obtaining information about the composition of surface materials at a remote location. The method consists of instrumentally comparing the emission spectrum from the target with the reflection spectra of polished samples of known composition. Initial observations of the moon indicate that there are small but consistent differences between different areas on the lunar surface.

[53] P. L. Hanst, "Temperatures in the Upper Atmospheres of Venus, as Indicated by the Infrared CO_2 Bands," *J. Opt. Soc. Am.*, **56**, 556 (1966). Based on computations made for the 10.4 μ absorption band of carbon dioxide, measured spectra indicate that the emissivity of the clouds on Venus is at least as high as 0.65 and that the temperature of the clouds is less than 250°K.

[54] E. E. Becklin and J. A. Westphal, "Infrared Observations of Comet 1965f," *Astrophys. J.*, **145**, 445 (1966). Observations were made with the 24 in. reflector at Mt. Wilson and the photometer that is described in ref. 19.2.1[41]. Observations were made in narrow spectral intervals centered about 1.65, 2.2, 3.4 and 10 μ. For the shorter-wavelength observations, the detector was liquid-nitrogen-cooled lead sulfide.

[55] J. Connes and P. Connes, "Near-Infrared Planetary Spectra by Fourier Spectroscopy: I. Instruments and Results," *J. Opt. Soc. Am.*, **56**, 896 (1966). This paper will undoubtedly come to be recognized as a landmark in astronomical spectroscopy. The spectra shown were made in the 1.5 μ region with an interferometric spectrometer. Both Venus and Mars spectra are shown, and they are compared with the best previously available spectra. This indicates that both the resolution and the signal-to-noise ratio are an order of magnitude better in the Connes' spectra. The spectra were recorded through a narrow-bandpass filter (1.43 to 1.67 μ) so that only a fraction of the spectral range covered by the cooled lead sulfide detector was utilized and the full power of the interferometric principle was not exploited. The coverage of instrumental details is outstanding.

[56] L. D. Gray, "New CO_2 Bands in the Connes' Venus Spectra," *J. Opt. Soc. Am.*, **56**, 1455 (1966). The spectra are those reported in the previous reference. Several new absorption bands of carbon dioxide have been found in these spectra, and the relative abundances of several isotopes of carbon and oxygen were determined.

[57] J. T. Houghton and S. D. Smith, *Infrared Physics*, London: Oxford University Press, 1966. Emphasizes the theory and applications of infrared spectroscopy. There is an interesting discussion of the use of these methods for the study of radiative transfer in planetary atmospheres and for determining the constituents of these atmospheres.

[58] "ESRO Satellite Award," *Aviation Week*, **86**, 23 (February 20, 1967). Reports that the first satellite to be built for the European Space Research Organization will carry infrared spectroradiometric equipment for the investigation of stellar spectra.

[59] J. E. Lovelock and D. R. Hitchcock, "Detecting Planetary Life from Earth—Analyzing Planetary Atmospheres," *Sci. J.*, **3**, 56 (April 1967). An excellent survey article. The authors point out that there is no formal operational definition of life that can serve for its recognition in a different form elsewhere. With an earth-based telescope one can look at a planet's atmosphere to see whether its composition is consistent with the coherent chemistry of a living planet or whether it is merely an assembly of substances consistent in composition with the steady state expected of their reactions. The success of such a life-detecting examination depends upon the ability of the instrumentation to analyze the atmospheric composition as completely as possible for all components down to concentrations as low as a few parts in 10^9. It appears that interferometric spectrometers can, indeed, achieve such performance.

[60] L. D. Kaplan, "Interpretations of the Observations," *Sci. J.*, **3**, 64 (April 1967). An interpretation of the Connes' spectra of ref. 19.3.2 [55]. The resolution of these spectra is at least an order of magnitude better than any previously available astronomical spectra, and the noise level is an order of magnitude less. These spectra indicate that substituted methanes are present in the Martian atmosphere. The Venus spectra indicate the presence of hydrogen chloride in concentrations of 1 part per million and hydrogen fluorode in concentrations of 1 part per 100 million.

[61] P. Connes, P. Fellgett, and J. Ring, "Towards a 1000-inch Telescope," *Sci. J.*, **3**, 66 (April 1967). Among other topics is a discussion of the development of astronomical type telescopes especially for use with interferometric spectrometers. These would be used to learn more about the possibility of life on other planets.

[62] "Industry Observer," *Aviation Week*, **87**, 23 (December 11, 1967). Describes a study of the use of a 36 in. airborne infrared telescope to be flown in a CV-990 aircraft. The purpose is to collect spectral information for planetary and solar research.

[63] J. Strong, "Balloon Telescope Optics," *Appl. Opt.*, **6**, 179 (1967). Describes the optical problems associated with making infrared astrophysical observations with a balloon-borne telescope. Topics treated include design of the spectrometer, calibration, shielding, and data handling.

[64] A. F. H. Goetz and J. A. Westphal, "A Method for Obtaining Differential 8–13 μ Spectra of the Moon and Other Extended Objects," *Appl. Opt.*, **6**, 1981 (1967). Describes a method for measuring spectral emissivity differences by comparing spectral scans across two points on the lunar surface. Spectral emissivity differences of 0.5 per cent can be detected. A liquid-hydrogen-cooled mercury-doped germanium detector was used with the photometer unit that is described in ref. 19.2.1 [41].

[65] D. McCammon, G. Munch, and G. Neugebauer, "Infrared Spectra of Low-Temperature Stars," *Astrophys. J.*, **147**, 575 (1967). Spectra are presented for the 1.5 to 1.8 μ and the 2.0 to 2.5 μ regions. The spectrometer used a diffraction grating and a liquid-nitrogen-cooled lead sulfide detector.

[66] P. Connes et al., "Traces of HCL and HF in the Atmosphere of Venus," *Astrophys. J.*, **147**, 1230 (1967). The spectra from an interferometric spectrometer are plotted automatically to give one solar and two planetary traces on a single chart. One of the planetary traces averages all interferograms taken in the early morning (at low elevation angles), and the other averages interferograms taken near the meridian. Comparison of the three traces allows discrimination between telluric absorption, solar Fraunhofer lines, and lines from the absorptions in the atmosphere of Venus. These are the first spectra that unambiguosly show the presence of traces of gaseous HCL and HF in the Venusian atmosphere.

[67] R. M. MacQueen, "A Balloon-Borne Infrared Coronograph," *Appl. Opt.* **7**, 1149 (1968). Prior coronograph observations have been made in the visible part of the spectrum. However, observing in the infrared offers unique opportunities for the examination of the thermal radiation from the outer solar corona. The coronograph described here has a 3.5 cm objective, a 400Hz tuning fork chopper, a narrow bandpass interference filter centered at 2.2 μ, and a dry-ice-cooled lead sulfide detector.

19.3.3 Instrumentation and Miscellaneous Applications

In a conventional spectrograph, a prism or a diffraction grating is used to disperse the light and each resolution element of the spectrum is focused on a separate portion of a photographic plate. Since the photographic plate is an integrating device, each spectral element is observed throughout the total exposure time. Photographic plates do not respond beyond 1.3 μ and therefore such spectrographs cannot be used in the infrared, where most molecules have their characteristic absorption bands. An infrared spectrometer also contains a prism or a grating for dispersion, but it uses an exit slit to isolate a single spectral element for observation by a single detector. The entire spectrum can be observed only by scanning the spectrum past the exit slit. Thus the spectrometer suffers from the disadvantage that the spectrum must be observed element by element, rather than all elements simultaneously as is done in the spectrograph. If, for instance, it takes a time t to observe a spectrum that consists of n spectral elements, each element is observed for a time t/n with the spectrometer and for a time t with the spectrograph. One partial solution to this problem is to use an array of detectors in the spectrometer, but the cost quickly becomes prohibitive.

In 1949 Fellgett (refs. 19.3.3[3] and [20]) studied this problem and conceived the basic principle of what he called multiplex spectrometry. His object was to find a way in which a single detector could simultaneously handle information about many spectral elements. The name he chose is an obvious reference to the multiplex principle that is so commonly used in communication circuits. Individual messages are given distinctive modulations, are combined and transmitted, and upon receipt they are separated by making use of their unique modulations. Later workers have referred to Fellgett's solution as interferometric or Fourier-transform spectroscopy.

The heart of an interferometric spectrometer is a Michelson interferometer. As shown in Figure 19.2, it is a device that divides the incident flux into two beams that can be individually controlled, and then recombines the beams so that they may interfere. If the path length of one of the beams is varied, for example by periodically moving one of the mirrors, it will cause a variation in the intensity of the recombined beams that can be sensed by a detector placed in the emergent beam. In order

19.3.3 INSTRUMENTATION AND MISCELLANEOUS APPLICATIONS

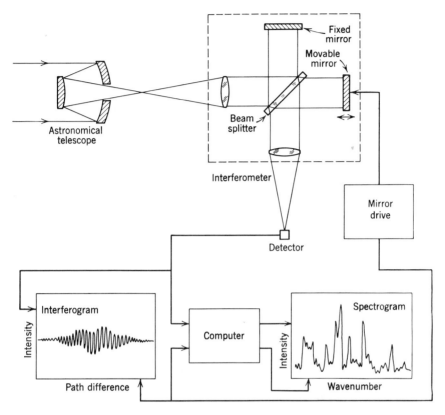

Figure 19.2 The interferometric spectrometer.

to make use of the information contained in the signal from the detector, it is necessary to make simultaneous measurements of the intensity in the emergent beam and of the path difference between the two beams in the interferometer. If this information is plotted, as shown in Figure 19.2, the resulting *interferogram* bears no resemblance to the spectrum, and yet it contains all of the information necessary to reconstruct the spectrum. A mathematical analysis of the process to this point shows that the interferogram is the Fourier transform of the spectrum. Thus the reconstruction process involves the calculation of a transform. The advent of the large digital computer has made it possible to perform the large number of transforms that are required to reconstruct a high-resolution spectrum. Until 1966 (ref. 19.3.3[21]), computer capacity limited the number of interferogram samples that could be transformed to about 3000 samples per hour. In 1966 a new method of computing the transform raised the limit to about 10,000 samples per hour. New-generation computers and improved methods of programming may some day raise this limit to about one-half million samples per hour.

The Connes' Venus spectra (ref. 19.3.2[55] and [60]) show spectral resolution as high as 0.08 cm^{-1}. The full power of the interferometric method becomes evident when it is realized that the best previously available Venus spectra, made with a conventional infrared spectrometer, have a resolution of 7 cm^{-1}. Ultimately the interferometric process is limited by detector noise, a limit that the Connes have nearly reached.

[1] M. Cohu, "Infrared Radiations," *Tech. Mod.*, **28**, 801 (1936). Discusses thermoelectric detectors, reflection, refraction, and polarization of infrared, photographic methods phosphors, selective receivers, infrared spectra, and the applications of infrared spectroscopy to the identification of unknown molecules.

[2] J. Granier and P. Caillon, *L'infrarouge*, Paris: Presses Universitaires de France, 1951.

[3] P. B. Fellgett, "Multi-channel Spectrometry," *J. Opt. Soc. Am.*, **42**, 872 (1952). With the exception of the author's thesis, this appears to be the earliest description of interferometric spectroscopy. This reference is, unfortunately, only an abstract; the full paper was not published at a later date in this journal. For further reminiscences by this author, see ref. 19.3.3[20].

[4] H. Scholze and A. Dietzel, "Investigations on the Water Content of Glasses through Determination of the Infrared Absorption in the Range from 1 to 5 Microns," *Glastechnische Berichte*, **28**, 375 (1955) (SLA: 62-10945). The method consists of monitoring the OH absorption band at 2.95 μ.

[5] G. B. B. M. Sutherland, "Some Comments on the Infrared Spectra of Diamond, Silicon, and Germanium," *J. Opt. Soc. Am.*, **50**, 1201 (1960). Absorption spectra show that there are two quite different types of diamonds, and yet there is difficulty in showing any difference in the physical properties of the two types. This paper offers some possible explanations for the differences in the spectra.

[6] W. G. Spitzer and M. Tanenbaum, "Interference Method for Measuring the Thickness of Epitaxially Grown Films," *J. Appl. Phys.*, **32**, 744 (1961). When epitaxial layers of silicon are grown on a silicon substrate, the transmission spectrum of the combination can be measured with an infrared spectrometer. This spectrum will show interference bands that are characteristic of the thickness of the epitaxial layer.

[7] W. M. Sinton, "Infrared Spectroscopy of Planets," *Appl. Opt.*, **1**, 105 (1962). Contains an excellent discussion of the advantages of using an interferometric spectrometer when a conventional spectrometer is limited by detector noise. The gain in sensitivity is nearly equal to $n^{1/2}$, where n is the number of resolution elements in the spectrum. If, for instance, the resolution in 0.02 μ and the spectrum extends from 2 to 4 μ, the value of n is 100. This means that objects that are 10 times fainter can be observed with an interferometer than can be observed with a spectrometer. Alternatively, if the same object is observed by both methods and n resolution elements are obtained with the spectrometer, then $n^{3/2}$ elements can be obtained with the interferometer for the same signal-to-noise ratio in both spectra. The interferometer does not require narrow entrance and exit slits. All wavelengths of interest are incident continuously on the detector and are observed throughout the entire scan cycle.

[8] W. Wardzynski, "Infrared Radiation," *Problemy* (Poland), **18**, 488 (1962).

19.3.3 INSTRUMENTATION AND MISCELLANEOUS APPLICATIONS 597

[9] E. S. Watson, "Monitoring Spacecraft Atmospheres by Spectrophotometry," *Space/Aeronautics*, **41**, 94 (March 1964). Gives a general description of the principles involved and describes a flight-model carbon dioxide monitor. It works at 4.3 μ and uses a vibrating-reed chopper. The entire unit weighs 1.5 lb, occupies 50 in^3., and consumes 3/4 W.

[10] *Aviation Week*, **80**, 77 (June 8, 1964). Shows a photograph of a neon-immersed infrared interferometric spectrometer that is to be carried on Gemini flights.

[11] R. D. Hibben, "Zero-G Liquid Gage Based on Absorption of Infrared Radiation by Gases," *Aviation Week*, **81**, 56 (December 14, 1964). The unit consists of a trace-gas container, an infrared detector, an output indicator, and a small trace-gas circulating pump. The trace gas must not react with the liquid being gaged or with the gas used to expel the liquid (if an expulsion-type fuel system is to be gaged). Both carbon dioxide and methane have been used as the trace gas. In operation, a small quantity of the trace gas is injected into the volume of the tank that contains the expelling gas. As the fuel is expelled, the trace gas diffuses into an increasingly larger volume. The infrared detector measures the resultant decreased density of the trace gas by monitoring one of its characteristic absorption bands.

[12] G. M. Nikol'skii and S. O. Obashev, "Noneclipse Observations of Coronal Lines in the IR Band," *Akademiya Nauk Kazakhskoi SSR, Vestnik*, **22**, 81 (August 1965). Describes a Soviet image converter tube in which the cathode is cooled to 195°K. The spectral response of the tube extends to 1.2 μ.

[13] R. K. Sloan, "The Scientific Experiments of Mariner IV," *Sci. Am.*, **214**, 62 (May 1966). The onboard magnetometer utilizes the fact that a magnetic field will affect the transmittance at 1.083 μ through a plasma of metastable helium ions. The magnetometer consists of an infrared source, a helium cell, a detector, and a set of coils around the helium cell. The coils produce a rotating magnetic field in the helium that results in a modulation of the output of the detector. The presence of an external magnetic field changes this modulation. This change is sensed as an error signal and is fed back to the coils so as to nullify the effect of the external field.

[14] E. V. Loewenstein, "The History and Current Status of Fourier Transform Spectroscopy," *Appl. Opt.*, **5**, 845 (1966). An excellent tutorial paper.

[15] H. R. Wyss and H. H. Gunthard, "Spectropolarimeter for the Infrared (2–8 μ)," *Appl. Opt.*, **5**, 1736 (1966). Used for the measurement of magneto-optical rotation in the infrared portion of the spectrum.

[16] J. Connes and P. Connes, "Near-Infrared Planetary Spectra by Fourier Spectroscopy: I. Instruments and Results," *J. Opt. Soc. Am.*, **56**, 896 (1966). Gives a detailed analysis of the operating theory and of the errors that can be encountered with interferometric spectrometers. The author's spectrometer is described in great detail. For information about their spectra, see ref. 19.3.2[55]. This paper will undoubtedly be recognized as a landmark in astronomical spectroscopy.

[17] A. E. Martin, *Infrared Instrumentation Techniques*, Amsterdam: Elsevier, 1966. The principal emphasis in this book is on spectroscopic instrumentation. The chapter on interferometric spectrometers is very good.

[18] M. J. D. Low, "Subtler Infrared Spectroscopy," *Intl. Sci. Tech.*, February 1967, p. 52. Contains a somewhat simplified discussion of the interferometric spectrometer and its application to analytical chemistry. One of the illustrations shows such a spectrometer in a housing that is small enough to be held in the palm of one hand. It was made for one of the lunar landings.

[19] "U.S. Chemists Develop Air Pollution Detectors," *Sci. J.*, **3**, 14 (April 1967). A short account of the use of interferometric spectroscopy for detecting atmospheric pollution. A more complete account is to be found in the preceding reference.

19.3.3 INSTRUMENTATION AND MISCELLANEOUS APPLICATIONS

[20] P. Fellgett, "The Genesis of Multiplex Spectrometry," *Sci. J.*, **3**, 58 (April 1967). An excellent account of the theory and development of interferometric spectrometers told by one of their principal developers.

[21] J. Connes and P. Connes, "Multiplex Spectroscopy of Planets," *Sci. J.*, **3**, 61 (April 1967). Contains additional discussion of the principle of the interferometric spectrometer and of the contributions to its development that these authors have made. This paper should be read as a sequel to the previous reference.

[22] "Quartz Gets an Infrared High Q Test," *Sci. Am.*, **217**, 58 (November 1967). Describes a method of measuring the mechanical Q of quartz crystals for oscillators by observing the characteristic absorption bands of the quartz.

[23] R. Beer, "Fourier Spectrometry from Balloons," *Appl. Opt.*, **6**, 209 (1967). Discusses the design and construction of balloon-borne, high-resolution interferometric spectrometers.

[24] S. A. Dolin, H. A. Kruegle, and G. J. Penzias, "A Rapid-Scan Spectrometer that Sweeps Corner Mirrors through the Spectrum," *Appl. Opt.*, **6**, 267 (1967). This device scans the first order of a diffraction grating in 1 msec at rates up to 800 scans per second. By changing detectors, the region from 0.25 to 9 μ can be covered. This paper includes a good bibliography of earlier papers on rapid-scan spectrometers. A feedback-controlled Globar source acts as an internal secondary standard for radiance calibration.

[25] J. C. Camm, R. L. Taylor, and R. Lynch, "Synchronized High Speed Scanning Infrared Spectrometer," *Appl. Opt.*, **6**, 885 (1967). This spectrometer uses an indium antimonide detector and scans a band that is 0.6 μ wide in 30 μsec. The instrument is useful over the range from 2 to 6 μ.

[26] W. A. Hovis, Jr., W. A. Kley, and M. G. Strange, "Filter Wedge Spectrometer for Field Use," *Appl. Opt.*, **6**, 1057 (1967). Gives constructional details of two spectrometers that use circularly wedged interference filters to provide a continuously variable dispersion. One instrument, covering the 1.6 to 5.4 μ range, has been flown in an aircraft and used to study the radiation from the earth. The other instrument, covering the 7.4 to 14.6 μ range, has been flown in a balloon.

[27] A. R. Karoli, J. R. Hickey, and R. E. Nelson, "An Absolute Calibration Source for Laboratory and Satellite Infrared Spectrometers," *Appl. Opt.*, **6**, 1183 (1967). Uses Peltier devices to give temperatures that range from 233 to 343°K. The effective emissivity of the cavity is said to be 0.995.

[28] A. C. S. Van Heel, ed., *Avanced Optical Techniques*, New York: Wiley, 1967; see A. Girard and P. Jacquinot, "Principles of Instrumental Methods in Spectroscopy," Chapter 3. A superb treatment of the latest developments in instrumental spectroscopy. The authors introduce a generalized concept of luminosity in a form that permits useful comparisons between the performance of different spectroscopic instruments.

[29] W. J. Normyle, "Numbus B to Test New Weather Sensors," *Aviation Week*, **88**, 71 (May 6, 1968). Contains an excellent description of the Infrared Interferometric Spectrometer (IRIS) which will make global measurements of the vertical temperature distribution, water vapor content, and ozone distribution of the earth's atmosphere. IRIS weighs 35 lb and will scan from 6 to 20 μ. The field of view is a 100 mile circle on the surface of the earth. Image motion is automatically compensated during the 11 sec scan interval. It is hoped that IRIS will prove that temperature profiles accurate enough for meteorological forecasting can be measured from satellites. For descriptions of the other infrared sensors discussed in this article, see ref. 19.4.2[15].

19.4 THERMAL IMAGING APPLICATIONS.

Thermal imagery is finding increasing application in surveys of earth resources, lunar and planetary studies, and long-range weather forecasting.

19.4.1 Earth Resource Surveys

Thermal imaging equipment is playing an important role in surveys of the world's natural resources. Such equipment, when mounted in aircraft or satellites, provides a survey capability that is both faster and cheaper than the ground methods that have been used in the past. Thermal imaging devices have already proved their usefulness in such diverse fields as geological mapping, petroleum exploration, water pollution studies, forest fire detection and damage assessment, sea-ice reconnaissance, location of underground springs, and the inventory of forest and range lands. One of the most important new tools for this work is multiband spectral analysis, that is, the technique of simultaneously observing an object in several different spectral regions. As experience with this technique grows, it is becoming evident that there are optimum spectral regions for the detection of such things as crop diseases and water pollution.

[1] R. W. Astheimer and E. M. Wormser, "Instrument for Thermal Photography," *J. Opt. Soc. Am.*, **49**, 184 (1959). Describes the use of a thermal imaging device to locate and study the small temperature differentials on the surface of glaciers that are caused by crevasses located at a considerable distance below the surface. For details of the thermal imaging device and its recorder, see refs. 17.4.1 [6] and [7].

[2] R. E. Frost, "The Program of Multiband Sensing Research of the U.S. Army Snow, Ice, and Permafrost Research Establishment," *Photogrammetric Eng.*, **26**, 786 (1960).

[3] R. N. Colwell, "Some Practical Applications of Multiband Spectral Reconnaissance," *Am. Scientist*, **49**, 9 (March 1961). An excellent tutorial treatment by one of the recognized pioneers in the field.

[4] S. F. Singer, "Forest Fire Detection from Satellites," *J. of Forestry*, **60**, 860 (1962). Describes the use of satellite-borne infrared equipment for the detection of forest fires.

[5] D. E. Harris and C. L. Woodbridge, "Terrain Mapping by Use of Infrared Radiation," *Proc. Natl. Electronics Conf.*, 1962, p. 215. Describes the theory and practical design of a line-scan thermal mapper and shows examples of the imagery that could be obtained with such equipment in the 1950's.

[6] "Proceedings of the First Symposium on Remote Sensing of Environment," The University of Michigan, Institute of Science and Technology, Ann Arbor, Michigan, 1962 (CFSTI: AD 274155). See ref. 19.4.1 [33].

[7] "Proceedings of the Second Symposium on Remote Sensing of Environment," The University of Michigan, Institute of Science and Technology, Ann Arbor, Michigan, February 1963 (CFSTI: AD 299841). See ref. 19.4.1 [33].

[8] D. L. Fresh, F. K. Dupre, and J. P. Walker, Jr., "Long Infrared Wavelength Detection with Mercury Doped Germanium and Stirling Cycle Cooling," *Proc. Natl. Electronics Conf.*, 1963, p. 635. Does not contain any description of the thermal mapper, but there is one example of the imagery obtained with it.

[9] L. H. Lattman, "Geological Interpretation of Airborne Infrared Imagery," *Photogrammetric Eng.*, **19**, 83 (1963).

[10] R. N. Colwell et al., "Basic Matter and Energy Relations Involved in Remote Reconnaissance," *Photogrammetric Eng.*, **19**, 761 (1963). An excellent discussion of the physical principles of remote sensing. In this case the "et al." includes seven coauthors.

[11] "Airborne Multi-Sensors Evaluated," *Aviation Week*, **80**, 83 (June 8, 1964). Shows a photograph of a B-25 aircraft that is used to evaluate the usefulness of airborne sensors in geological mapping, petroleum exploration, water pollution studies, and analyses of the thermal efficiency of water sources for power stations. Sensors include intermediate- and far-infrared equipment.

[12] "Infrared Fire Scanner Flown in to Aid Battle," *Los Angeles Times*, September 26, 1964. Describes the use of an infrared thermal mapper to observe the September 1964 forest fire near Santa Barbara, Cal. The article emphasizes that the mapper is able to see through the heavy, visually opaque smoke clouds and clearly indicate the boundaries of the fire. Such a capability is extremely helpful to the fire bosses who must plan the fire-fighting strategy. It is claimed that the device can detect a campfire from an altitude of 20,000 ft.

[13] "News Digest," *Aviation Week*, **81**, 28 (September 28, 1964). Reports that pictures from the high-resolution scanning radiometer on Nimbus show definite indications of volcanic action in the Antarctic Mountains.

[14] "Private Lines," *Aviation Week*, **81**, 103 (October 19, 1964). Reports that the U.S. Army has demonstrated the effectiveness of the infrared sensors on the Mohawk reconnaissance aircraft for spotting incipient forest fires. The system can detect smoldering fires and provide a photographic-type readout showing the location of the fire.

[15] J. O. Morgan, "Recent Developments in Airborne Optical Imaging for Terrestrial Reconnaissance," *J. Opt. Soc. Am.*, **54**, 574 (1964). Describes recent research in the areas of forest-fire surveillance and sea-ice reconnaissance.

[16] D. E. Harris and C. L. Woodbridge, "Terrain Mapping by Use of Infrared Radiation," *Photogrammetric Eng.*, **30**, 135 (1964). The same paper as that listed in ref. 19.4.1[5].

[17] M. R. Holter and F. C. Polcyn, "Comparative Multispectral Sensing," *Photogrammetric Eng.*, **30**, 443 (1964).

[18] J. L. Cantrell, "Infrared Geology," *Photogrammetric Eng.*, **30**, 916 (1964). The primary emphasis in this paper is on the interpretation of the imagery obtained with thermal mapping equipment. Among the applications that are discussed are studies of power plant coolant-discharge systems, drainage patterns, lava flows, and stream pollution.

[19] E. S. Leonardo, "Capabilities and Limitations of Remote Sensors," *Photogrammetric Eng.*, **30**, 1005 (1964). An elementary discussion of modern remote sensors, such as radar, infrared, and photographic, and of the imagery they produce. Figure 11 of this paper shows an excellent thermal map of Dallas.

[20] W. A. Fischer et al., "Infrared Surveys of Hawaiian Volcanoes," *Science*, **146**, 733 (1964). This paper is an outstanding reference because of the many examples of thermal images that it contains. In addition to the studies of volcanoes, a total of 25 fresh-water springs were detected around the periphery of the island of Hawaii.

[21] "Proceedings of the Third Symposium on Remote Sensing of Environment," Report 4864-5-X, The University of Michigan, Institute of Science and Technology, Ann Arbor, Michigan, February 1965. See ref. 19.4.1[33].

19.4.1 EARTH RESOURCE SURVEYS

[22] D. C. Parker and M. F. Wolff, "Remote Sensing," *Intl. Sci. Tech.*, July 1965, p. 20. A good survey article that covers the various techniques for multiband spectral analysis.

[23] "Forest Fires Fought with Electronics," *Electronic Design*, **13**, 11 (September 13, 1965). Describes the use of infrared thermal mappers to pinpoint the location of small forest fires.

[24] S. I. Udall, "Resource Understanding—A Challenge to Aerial Methods," *Photogrammetric Eng.*, **31**, 63 (1965). A general introduction to the rationale for the use of multiband spectral analysis for resource survey and evaluation. The author describes the use of infrared photography for finding water-dependent plants, infrared radiometry for ocean temperature measurements, and thermal mappers for volcanic studies and the location of underground spring outlets. Several good examples of thermal imagery are used to illustrate the article.

[25] C. E. Molineux, "Multiband Spectral System for Reconnaissance," *Photogrammetric Eng.*, **31**, 131 (1965). System operates in the 0.35 to 5 μ regions and consists of a 9-lens multiband camera, color reference cameras, dual spectrometer system, cartographic camera, and a skylight recording camera. System is intended for terrain reconnaissance and especially for detecting manifestations of underground nuclear test activity.

[26] W. B. Foster and T. A. George, "Post-Apollo Experiments for Space," *Astronautics and Aeronautics*, **4**, 50 (March 1966). Discusses the application of multiband spectral techniques to the study of the earth's surface from satellites.

[27] "IR Detection of Diseased Crops," *Space Age News*, April 1966, p. 7. A brief item that tells of the development of an infrared device to study the composition of the lunar surface. It is predicted that similar devices can be used to differentiate between healthy and diseased plants.

[28] W. C. Wetmore, "USSR Sees Gains in Manned Space Tests," *Aviation Week*, **85**, 30 (October 24, 1966). Describes a U.S.-sponsored program in which satellite-borne infrared sensors would be used for worldwide resource studies. The infrared sensors would be particularly effective in the detection of forest fires, the location and identification of cultivated crops, the cataloging of soil fertility, the inventory of forest and range lands, and the detection of insect infestations.

[29] "Infrared Divining Rod," *Time*, November 11, 1966, p. 88. A report that is based on the work reported in ref. 19.4.1 [20]. This article is a good example of the fascination that infrared holds for the layman.

[30] "Earth Resources Satellite Brings Disturbing Questions into Focus," *Space/Aeronautics*, **46**, 22 (November 1966). Shows a photograph and a thermal image of the high-temperature areas in the crater and on the flanks of Kilauea volcano on the island of Hawaii. The text implies that the pictures were made from a Gemini spacecraft. If this is true, these pictures are a fantastic accomplishment. It appears, however, that this implication is not correct because the two pictures are almost certainly Figures 8 and 11 of ref. 19.4.1 [20], which were taken from an airplane at a relatively low altitude.

[31] R. N. Colwell, "Aerial Photography of the Earth's Surface; Its Procurement and Use," *Appl. Opt.*, **5**, 883 (1966). Discusses the concepts of multiband spectral analysis and shows six images of the same area that were made with panchromatic, Aerographic infrared, Aerial Ektachrome, and infrared Ektachrome films, an 8 to 14 μ infrared scanner, and a K-band radar. Another series, taken with the infrared scanner, shows the marked variations in the contrast of a scene that occur during a 24 hr period.

[32] R. A. Wilson, "The Remote Surveillance of Forest Fires," *Appl. Opt.*, **5**, 899 (1966). Discusses the design objectives for a prototype airborne-infrared fire-detection system.

[33] V. L. Prentice and J. O. Morgan, "Third Symposium on Remote Sensing," *Photogrammetric Eng.*, **32**, 98 (1966). This symposium was held in October 1964. For the

official proceedings of this and the two earlier symposia, see refs. 19.4.1[6], [7], and [21]. The papers that were presented cover such fields as planetary studies, earth exploration, meteorology, and oceanography. The lengthy bibliography that is given in this paper contains almost nothing except references to papers that were given at this symposium.

[34] "Earth Survey Satellites," *Space/Aeronautics*, **47**, 85 (January 1967). Mentions the development of an infrared sensor with 20 spectral channels that should give significantly more information than is now gained from sensors having from two to five channels.

[35] J. S. Butz, Jr., "Under the Spaceborne Eye: No Place to Hide," *AF/Space Dig.*, **50**, 93 (May 1967). Describes a commercially available thermal mapper that operates in the 3 to 5 μ region and shows a good example of its imagery.

[36] W. P. Lowry, "The Climate of Cities," *Sci. Am.*, **217**, 15 (August 1967). Shows thermograms of the New York City skyline and of an automibile with its engine idling.

[37] "New Tool to Fight Forest Fires Tested," *Los Angeles Times*, September 7, 1967. Reports that the U.S. Forest Service is currently flying two thermal imaging systems, called Fire Scan, in a continuing patrol of forest lands. The device uses an indium antimonide detector. Various indications of detection range are given in the article—a campfire 8 miles away; a coffee can of burning gasoline in a pine forest; 4 in.^{2} of grassland fire at a distance of 5 miles. Imagery can be obtained through heavy smoke so that the hot spots can be located very accurately.

[38] "Fighting Future Fires," *Time*, **90**, 60 (September 15, 1967). Reports that the most dramatic new aid for forest fire fighters is an infrared thermal mapper. Such a mapper can see through smoke and can detect a fire in a small bucket in a forest from an altitude of 15,000 ft.

[39] R. Blythe and E. Kurath, "Infrared Images—Seeing the World Through New Eyes," *J. Opt. Soc. Am.*, **57**, 1410 (1967). Describes a thermal mapper and some of the characteristics of its imagery. Various applications in the 0.8 to 15 μ region are discussed.

[40] E. Kurath and R. Blythe, "Some Factors Affecting the Interpretation and Uses of IR Imagery," *J. Opt. Soc. Am.*, **57**, 1410 (1967). An introduction to the fundamentals of thermal image evaluation. Applications that are discussed include studies of plant growth, control of agricultural water use, detection of water diffusion and pollution, and studies of microclimatology.

[41] J. H. McLerran, "Infrared Thermal Sensing," *Photogrammetric Eng.*, **33**, 507 (1967). Describes the application of airborne thermal mappers to terrain analysis. The various illustrations in the article indicate that thermal imagery shows major terrain features and that the interpretation of such imagery is more complex than is photographic interpretation. Several examples show the tone reversals and washout effects that occur within a 24-hour period. It is shown that thermal imagery is also useful for studying sea ice in the arctic during daylight and during the arctic night.

[42] F. F. Sabins, Jr., "Infrared Imagery and Geologic Aspects," *Photogrammetric Eng.*, **33**, 743 (1967). Describes structural and stratigraphic interpretations that can be made from 8 to 14 μ thermal imagery made at Indio Hills, Cal. This imagery was obtained in 1961 and was declassified in 1964. No details of the mapper are given.

[43] R. Blythe and E. Kurath, "Infrared and Water Vapor," *Photogrammetric Eng.*, **33**, 772 (1967). Describes the problems of interpreting thermal imagery taken over irrigated areas. The thermal mapper uses a cooled indium antimonide detector and operates in the 3.69 to 5.5 μ region.

[44] R. S. Williams, Jr., and T. R. Ory, "Infrared Imagery Mosaics for Geological Investigations," *Photogrammetric Eng.*, **33**, 1377 (1967).

19.4.2 METEOROLOGICAL APPLICATIONS

[45] R. N. Colwell, "Remote Sensing of Natural Resources," *Sci. Am.*, **218**, 54 (January 1968). A somewhat popularized account of multiband spectral analysis. The illustrations with this article are exceptionally good. One of the thermograms shows how readily campfires can be detected with a thermal mapper. It is claimed that this mapper, working in the 3 to 5 μ region, could detect a single burning charcoal briquette at a distance of 1 mile.

[46] R. N. Einhorn, "Laser-Scanned MOS Detects Infrared," *Electronic Design*, **16**, 34 (February 1, 1968). By scanning a cooled indium antimonide metal-oxide-semiconductor structure with a 1 mW, 0.63 μ laser beam, these workers have been able to detect infrared images focused on the structure.

[47] F. W. Holder, "Electronic Fire and Smoke Detectors," *Electronics World*, **79**, 37 (February 1968). Discusses the use of airborne thermal mappers to detect and map forest fires. Describes and gives a good block diagram of a U.S.-manufactured mapper that uses a cooled indium antimonide detector and operates in the 4.5 to 5.5 μ region. This article also describes a mapper manufactured in Canada for the Ontario Department of Lands and Forests. This mapper is said to be able to detect a single smouldering dead tree stump in moderate to heavily forested areas.

[48] "Technology Application Due Further Study," *Aviation Week*, **88**, 55 (February 26, 1968). Summarizes the 1967 Summer Study of Space Applications that was compiled by committees of The National Science Foundation and The National Research Council.

[49] J. J. Horan, "Spacecraft Infrared Imaging: Principles and Applications," *IEEE Spectrum*, **5**, 71 (June 1968); "Systems Engineering Aspects," **5**, 66 (July 1968); "Erratum," **5**, 121 (August 1968). An excellent tutorial article.

[50] W. H. Gregory, "Several Firms Planning Urban Programs...," *Aviation Week*, **89**, 38 (July 1, 1968). Shows a beautiful example of thermal imagery made over the Connecticut River with a commercially available thermal mapper. Condenser coolant water from a power plant is clearly evident and its course can be followed for several miles downstream from the plant.

[51] P. G. Thomas, "Earth-Resource Survey from Space," *Space/Aeronautics*, **50**, 46 (July 1968). An excellent survey article that gives a good idea of the relative importance of infrared sensors in this program. The discussion of target signatures and their verification with ground-truth is particularly well done.

19.4.2 Meteorological Applications

The weather forecaster has long been hindered by the lack of a global coverage in his meteorological observations. Since less than one-fifth of the total atmosphere is adequately observed by conventional observations, it is possible for large storms to lurk undetected for days in various oceanic, polar, and desert areas. Thermal imaging equipment aboard satellites speeds the early detection of such storms and makes it possible to acquire data over the entire surface of the earth. The applications in this section are limited to those in which thermal imagery is produced at some step in the process; other meteorological applications are listed in Sections 19.2 and 19.3.1.

[1] R. A. Hanel and W. G. Stroud, "Infrared Imaging from Satellites," *J. Soc. Motion Picture Television Engrs.*, **69**, 25 (1960).

[2] "Nimbus IR Sensor to Map Cloud Cover," *Electronic Design*, **12**, 25 (June 22, 1964). Shows a photograph of the radiometer. An unusual feature is the use of a radiative-transfer cooler to maintain the detector at a temperature of 195°K.

[3] H. D. Watkins, "Relay Box Problem Forces Nimbus Delay," *Aviation Week*, **81**, 24 (August 17, 1964). Describes the high-resolution infrared radiometer that will make night cloud-cover pictures and measurements of cloud-top temperatures. The spectral bandpass of the radiometer is from 3.4 to 4.2 μ, and the resolution is about 5 miles on the ground.

[4] "Nimbus IR, TV Photos Show Britain, Russia," *Aviation Week*, **81**, 34 (September 7, 1964). Reports that the television and infrared imagery received from the Nimbus satellite are far better than originally anticipated. A photograph taken over Russia with the high-resolution infrared-scanning radiometer shows such features as the Volga River and the marshy areas surrounding it. Variations in the temperature of the clouds also show clearly in this picture.

[5] "Nimbus I Pictures Cover U.S., Asia, Mediterranean," *Aviation Week*, **81**, 30 (September 14, 1964). Shows two pictures from the high-resolution scanning radiometer. One, taken over Italy, clearly shows the famous Italian boot and an extensive cloud system over the Alps. The other picture shows Madagascar and Southeastern Africa.

[6] S. Singer, "Satellite Meteorology," *Intl. Sci. Tech.*, December 1964, p. 30. One of the illustrations shows the structure of a hurricane as revealed by thermal imagery.

[7] M. Tepper and D. S. Johnson, "Toward Operational Weather Satellite Systems," *Astronautics and Aeronautics*, **3**, 16 (June 1965). Shows a picture from the high-resolution infrared-scanning radiometer on Nimbus I and likens its quality to that of the TV pictures from the Tiros satellites. See also ref. 19.2.3[11].

[8] I. P. Vetlov et al., "Analysis of Infrared Cloud Cover Photographs from the Nimbus I," *Meteorol. i Gidrol.*, September 1965, p. 27 (CFSTI: 66-17737). Two pictures taken at midnight over the Soviet Union are compared with TV cloud-cover photos. Tentative conclusions are that the infrared pictures are superior to synoptic maps, the infrared pictures are inferior to the TV pictures in showing cloud details, and radiation balance data are useful in both infrared and TV pictures.

[9] K. J. Stein, "New Nimbus Avionics Interface Transmits Infrared on APT Link," *Aviation Week*, **84**, 77 (May 2, 1966). Because of the success of the high-resolution radiometer on Nimbus I, it has been decided to use the automatic picture transmission (APT) link on later Nimbus vehicles to transmit the high-resolution infrared pictures as well as the TV pictures to the network of 150 ground stations. The article describes the signal-processing techniques that are used to transmit the infrared pictures. Nimbus II will carry a medium-resolution infrared radiometer that will provide imagery in 5 spectral bands, 0.2 to 4 μ, 6.5 to 7 μ, 10 to 11 μ, 14 to 16 μ, and 7 to 30 μ.

[10] "Nimbus 2 Infrared/APT Picture Shows Southeast U.S.," *Aviation Week*, **84**, 34 (May 23, 1966). In addition to showing a picture from the high-resolution radiometer, this item reports that the medium-resolution infrared radiometer is sending temperature versus altitude information and thereby providing three-dimensional meteorological data.

[11] "Night, Day Weather Shown by Nimbus II," *Aviation Week*, **84**, 88 (May 30, 1966). The pictures in this article provide an excellent comparison between the capabilities of the daytime-only television and the nighttime-only high-resolution infrared radiometer. The infrared picture shows Typhoon Irma. The eye of the storm and the typical spiral-shaped cloud structure are seen clearly.

[12] D. Winston, "New Vigor by Soviets Expected in Manned Space Flight," *Aviation Week*, **86**, 140 (March 6, 1967). Contains an example of the imagery from the infrared

system on the Cosmos 122 satellite. Although this equipment located two typhoon areas, the quality of the imagery is considered to be poor.

[13] K. J. Stein, "Tiros M Design Broadens Capabilities," *Aviation Week*, **86**, 103 (May 8, 1967). Gives a comprehensive description of the instrumentation that is planned for the Tiros M weather satellite. Among this instrumentation are a pair of high-resolution infrared radiometers and associated automatic picture-transmission equipment for studies of nighttime cloud cover. In addition, there is a flat-plate radiometer for studies of the earth's heat balance.

[14] E. C. Barrett, "Satellite Meteorology," *Sci. J.*, **3**, 74 (July 1967). Contains a very comprehensive discussion of infrared instrumentation for meteorological satellites. One of the illustrations shows a nighttime image of the British Isles that was made with the high-resolution infrared radiometer on Nimbus II.

[15] W. J. Normyle, "Nimbus B to Test New Weather Sensors," *Aviation Week*, **88**, 71 (May 6, 1968). Shows pictures of various sensors mounted on the nimbus satellite and gives detailed descriptions of the satellite infrared spectrometer (SIRS) (19.3.1 [23]), the high resolution infrared radiometer (HRIR) (19.4.2[3, 4, 5, 7, 10, 11]), the medium resolution infrared radiometer (MRIR) (19.4.2[10]), and the infrared interferometric spectrometer (IRIS) (19.3.3[29]). The HRIR weighs 18 lb, uses a lead selenide detector in the 3.4 to 4.2 μ region, has a 5 mile ground resolution, can determine land and sea temperatures to $\pm 2°F$. The MRIR weighs 21 lb, has 5 measurement channels, and is used principally for studies of the radiation balance of the earth. Its 5 channels cover 0.2 to 4.0 μ, 10 to 11 μ, 14 to 16 μ, 6.5 to 7 μ, and 20 to 33 μ.

19.4.3 Lunar and Planetary Studies

Thermal imagery is proving useful in the discovery of hot spots on the moon and the study of lunar luminescence. Much of the equipment that is now being developed for earth resource surveys (Section 19.4.1) is also applicable to studies of this type.

[1] "News Digest," *Aviation Week*, **77**, 27 (July 30, 1962). Reports on a series of photographs of the lunar surface that were made with an infrared vidicon sensitive to the 0.9 to 2.3 μ region. For further details on this Russian work, see the following reference.

[2] N. F. Kuprevich, "Television Photography of the Moon," *Priroda*, **52**, 90 (1963) (OTS: 63-31315). Describes the use of an infrared vidicon that is sensitive in the 0.9 to 2.3 μ region for photographs of the lunar surface. Observations are taken on the dark portion of the moon so that the luminescence of the surface can be detected. It is claimed that what appears to be rays emanating from the crater Tycho when viewed in visible light seems to be a long mountain ridge when viewed with the vidicon.

[3] Z. Kopal, "The Luminescence of the Moon," *Sci. Am.*, **212**, 28 (May 1965). Good background information for assessing the report contained in the two preceding references.

[4] P. C. Badgely, "Planetary Exploration from Orbital Altitudes," *Photogrammetric Eng.*, **32**, 250 (1966). Describes new improvements in remote sensors for natural resource studies. Among the functions to be performed by infrared sensors are the determination of terrain and plant composition, the detection of geographically significant thermal anomalies, the location of mineral outcroppings, the tracing of ocean currents and sea ice, and studies of land use. Reports that initial airborne measurements

are already being made with infrared thermal mappers operating in the 8 to 13 μ and the 4.5 to 5.0 μ regions.
[5] J. D. Rehnberg, J. R. Yoder, and G. R. Hunt, "An Earth-Based Infrared Lunar Mapper for Thermal and Composition Studies," *Appl. Opt.*, **6**, 1111 (1967). Describes a focal-plane scanner in which a motor-driven platform scans a rectangular raster in the image plane of an astronomical telescope. A liquid-helium-cooled copper-doped germanium detector subtends 5 sec of arc and scans a total field of view of 15 by 15 min of arc in a time of 15 min.

19.4.4 Miscellaneous

The references in this section describe unique scanning techniques, translation of infrared images into color, and some specific uses of the Evaporograph.

[1] R. C. Hopgood, "Wide-Angle Energy Responsive Scanning System," U.S. Patent No. 2,859,652, November 11, 1958. Describes a rotating-eyeball scanner and how its field of view can be increased by the use of a multielement detector. Displays that are described include an intensity-modulated cathode-ray tube, moving film, and a stylus-operated recorder.
[2] F. G. Willey, "Video Display Device," U.S. Patent No. 2,911,876, November 10, 1959. Presents an improved display for a rotating-eyeball scanner of the type that is described in both the previous and the following references. The key element is a signal-converter or mosaic storage tube to store all scan lines that comprise the full image. The stored information is presented so as to have the appearance of a window aligned with the axis of the scanner, within which the image moves across the display as the aircraft moves across the terrain.
[3] F. G. Willey, "Interlaced Optical Scanning Means," U.S. Patent No. 2,929,293, March 22, 1960 (filed October 19, 1954). A rotating-eyeball scanner in which the optics are tilted so that each scans at a slightly different angle in order to give an interlaced scan. For earlier versions of the same scanner, see the two previous references.
[4] J. L. Woika, "Generating Tangential Sweeps for Infrared Mapping," *Electronics*, **34**, 64 (October 13, 1961). Describes the design of an 8-transistor circuit that generates an approximation of the tangent function that is needed for the slant-range correction of the video signal from an airborne thermal mapper.
[5] C. Hilsum and W. R. Harding, "The Theory of Thermal Imaging and its Application to the Absorption-Edge Image Tube," *Infrared Phys.*, **1**, 67 (1961).
[6] "Announcement of Thermal Vision," Radio Tass, Moscow, October 16, 1962. Claims a thermal imaging device with a frame time of 1 sec and a minimum detectable temperature of 0.1°C. Such performance seems unlikely; hence the report is probably false or else it has been garbled by public relations people.
[7] G. W. McDaniel and D. Z. Robinson, "Thermal Imaging by Means of the Evaporograph," *Appl. Opt.*, **1**, 311 (1962). Mentions the use of an Evaporograph for studies of the temperature distributions in wind tunnel models.
[8] "Soviet Evaporographic Chamber to Represent Remotely the Infrared Zone of the Spectrum," *Jemna Mechanika Optika*, **8**, 325 (1963). Claims that the image of an object having a temperature that is only a few degrees warmer than room temperature can be revealed within several seconds with this particular Evaporograph.

19.5.1 INFRARED PHOTOGRAPHY

[9] R. H. Chase, W. E. Horn, and F. R. Stauffer, "Infrared Circular Scan Aerial Reconnaissance System," U.S. Patent No. 3,134,902, May 26, 1964. Describes a device that uses a circular scan rather than the more common straight-line scan. With the circular scan the area on the ground that is covered by the instantaneous field of view remains constant, which it does not do with a straight-line scanner. A single rotating prism is used to generate the circular scan pattern.

[10] D. S. Lowe and J. G. N. Braithwaite, "A Spectrum Matching Technique for Enhancing Image Contrast," *Appl. Opt.*, **5**, 893 (1966). Discusses a means for implementing multiband sensing.

[11] V. N. Sintsov, "Evaporographic Image Quality," *Appl. Opt.*, **6**, 1851 (1967). Although applications are not discussed, this article is invaluable for its analysis of resolution, frequency response, signal, noise, and signal-to-noise ratio of the Evaporograph.

[12] L. W. Nichols and J. Lamar, "Conversion of Infrared Images to Visible in Color," *J. Opt. Soc. Am.*, **57**, 1411 (1967). Describes a device that scans simultaneously in three spectral regions and produces an image of the scene in the form of a color photograph. The scanner uses a silicon detector for the 0.7 to 1.0 μ region, an indium antimonide detector for the 3.0 to 5.5 μ region, and a mercury-doped germanium detector for the 8.0 to 14.0 μ region. The signal from each detector is used to modulate a distinctively colored light source. The pictures show the effects of both reflected and emitted radiation from the scene. The color of an object in one of these pictures is a function of its temperature as well as its reflective and emissive properties.

19.5 APPLICATIONS INVOLVING REFLECTED FLUX

This section contains references dealing with the techniques of infrared photography and their application to a variety of scientific problems. In addition, there are a number of references to investigations of the reflectance properties of materials and the use of this information for such widely differing tasks as the identification of gems and the determination of surface materials on the moon.

19.5.1 Infrared Photography

To the experienced photointerpreter, infrared film presents an enigma. It has unmistakable advantages, but these are only realized at the expense of some other desirable quality. The developers of multiband spectral analysis have shown that the combination of panchromatic and infrared films usually reveals more information than does either film alone. But if only one type of film can be used on a particular mission, the choice is almost certain to be in favor of the panchromatic type.

Infrared film is rarely used to record the radiation emitted by an object of interest, because this is a task that can usually be done more efficiently by photon or thermal detectors. As a result, in most applications, infrared film is used to record the flux (usually sunlight) that is reflected from an object or a scene. The near-infrared reflectance characteristics of many materials are markedly different from those in the visible portion of the spectrum. It is for this reason that infrared film is often of value. In an aerial photograph made on infrared film, for instance, grass and the leaves of deciduous (hardwood) trees appear very light, as if they were covered with snow. Chlorophyll, which is responsible for the green color of plants, is transparent in the near infrared; hence the incident light passes into the leaf and is reflected back from its tissues. Thus the leaves photograph as if they were a light color, whereas in normal (panchromatic) photographs they appear dark because of the absorption of (visible) light by the chlorophyll. Coniferous (evergreen) trees appear darker than deciduous trees, and trees that are dead appear almost black. Thus rather accurate forest surveys can be made very rapidly with the aid of infrared photography.

Since the scattering of light by atmospheric haze is usually somewhat less in the near infrared than it is in the visible, it is not uncommon to find that infrared film gives a better rendering of distant detail. But do not get trapped by the all-too-common fallacy that infrared has the magical ability to penetrate all kinds of fog and haze.

Infrared films (and image converter tubes as well) are often used to prove or disprove the authenticity of paintings and other works of art. Once again, the key is that pigments that appear identical to the eye often exhibit very different near-infrared reflectances. As a result, it is sometimes possible to detect overpainting and other alterations and to distinguish between a copy and the original. Sometimes paintings that can hardly be seen because of the darkening of the underlying varnish are clearly revealed by infrared photography.

False-color film was developed during World War II for the detection of camouflage from the air. This film takes advantage of the fact that although many camouflage materials may be perfect matches to the eye, their near-infrared reflectance characteristics are usually quite different than those of the natural materials that they are supposed to simulate. False-color film differs from ordinary film in that the three sensitized layers respond to green, red, and near-infrared radiation rather than to the usual combination of blue, green, and red. This film has found important postwar application for multiband spectral analysis. For further information on the interpretation of photographs taken on false-color film, see ref. 19.5.1[14].

[1] N. M. Mohler, "Photographic Penetration of Haze," *J. Opt. Soc. Am.*, **26**, 219 (1936). Photographs that were taken with panchromatic and infrared films show visibility ratios in favor of the infrared that range from 1.1 on days with thick haze or light rain to 1.7 on clear days (visibility 20 miles).

[2] O. Helwich, *Infrared Photography and its Fields of Application*, 2d ed., Harzburg, Germany: W. Heering, 1937.

19.5.1 INFRARED PHOTOGRAPHY

[3] W. Clark, *Photography by Infrared*, 2nd ed., New York: Wiley, 1946. The classic reference on infrared photography. The plates for the first edition were destroyed during the World Wall II bombing of London. The second edition is long since out of print and correspondingly difficult to obtain. It is rumored that a third edition is in preparation.

[4] "False-Color Film for Aerial Photography," *International Photographer*, **19**, 18 (December 1947). Describes the characteristics of Kodak Ektachrome Aero Film (camouflage detection). This film has been replaced by Kodak Ektachrome Infrared Aero Film (ref. 19.5.1[14]).

[5] M. Deribere, *Infrared Photography*, Paris: Paul Montel, 1948.

[6] G. C. Brock, *The Physical Aspects of Aerial Photography*, London: Longmans Green, 1952. Republished in 1967 by Dover, New York. Chapter 14 is probably the best available assessment of the use of infrared film for aerial photography. The author is obviously a fine experimentalist, and he has learned his craft extremely well. An unusual feature of this book is the use of comparison photographs to show the same scenes recorded on both panchromatic and infrared films. Fortunately the photographs are included in the republished edition.

[7] A. Nuernberg, *Infrarot Photographie*, Halle, Germany: Wilhelm Knapp, 1957 (CFSTI: 59-16054).

[8] S. M. Solov'yev, *Infrared Photography*, unidentified monograph, Moscow, 1957 (SLA: 60-17434).

[9] S. M. Solov'yev, "Infrared Light-Sensitive Materials on the Market," *Zhurnal Nauchnoi i Prikladnoi Fotografii i Kinematografii*, **4**, 385 (1959) (SLA: 60-16533). Contains an excellent tabulation of the characteristics of infrared films that are marketed in non-Soviet countries. Suppliers include Agfa (East Germany), Gevaert (Belgium), Guilleminot (France), Ferrania (Italy), Kodak (U.S., Great Britain, France), and Konishiroku (Japan). Soviet-made infrakhrom film is mentioned in the text, but no technical data are given for it. Curiously, all film speed values are given in ASA units.

[10] D. R. Lueder, *Aerial Photographic Interpretation*, New York: McGraw-Hill, 1959.

[11] R. N. Colwell, ed., *Manual of Photographic Interpretation*, Washington: American Society of Photogrammetry, 1960.

[12] I. B. Lewitin, *Fotografia w Infrakrasnych Luczach (Photography with Infrared Rays)*, Moscow: Voenizdat, 1961.

[13] H. K. Meier, "Uses of Infrared Emulsions for Photogrammetric Purposes," *Soc. Phot. Instr. Engrs. J.*, **1**, 4 (October–November 1962). The emphasis is on civil-engineering and geological applications.

[14] "Instructions for Exposing and Processing Kodak Ektachrome Infrared Aero Film," Eastman Kodak Co., Rochester, N.Y., December 1962. This film is a false-color, reversal film designed for use in aerial photography. Its three sensitized layers are sensitive to green, red, and infrared radiation instead of the usual blue, green, and red sensitizings. Since all three layers are sensitive to blue, a yellow filter must always be used. When processed, the green-sensitive layer is developed to a yellow positive image, the red-sensitive layer to a magenta positive image, and the infrared-sensitive layer to a cyan positive image. With this combination the colors are false for most natural objects. Most green camouflage paints that are intended to simulate natural foliage absorb strongly in the near infrared. In the color transparency the areas of natural deciduous foliage appear magenta or red and the painted objects appear purple or blue. More recently, this film has found widespread use in forest surveys. In spring or summer, healthy deciduous trees photograph magenta or red on this film and healthy

evergreen trees photograph bluish purple. Dead or dying deciduous leaves or evergreen needles photograph as a bright green because such leaves and needles have lost their high infrared reflectance. Healthy deciduous trees whose leaves have turned red or yellow in the autumn retain their high infrared reflectance, and the red leaves photograph yellow and the yellow leaves photograph white.

[15] "Fluorescence in the Infrared," *Sci. Am.*, **208**, 79 (April 1963). Shows photographic examples of the near-infrared fluorescence of gallstones, leaves, and greenockite. The fluorescence occurs between 0.74 and 1 μ. The exposure needed to record the fluorescence is about 20,000 times that needed to record the reflected flux.

[16] J. Berchtold, "Infrared Fluorescence of Green Leaves," *J. Opt. Soc. Am.*, **53**, 1006 (1963). Reports that leaves fluoresce in the near infrared and suggests that this fluorescence is partly responsible for the near-white rendering of leaves on infrared film. Apparently this author is unaware that Clark (ref. 19.5.1[3]) discussed this subject in 1946 and referred to French experiments in 1936 that showed that it is doubtful that fluorescence plays any part in the near-white rendering of leaves. The data contained in the previous reference show why this is true.

[17] *Infrared and Ultraviolet Photography*, Publication No. M-3, Eastman Kodak Co., Rochester, N.Y., 1963. A reliable and readily available reference.

[18] R. N. Colwell, "Aerial Photography—A Valuable Sensor for the Scientist," *Am. Scientist*, **52**, 17 (March 1964). One of the best reports available. The 4-page color insert shows some striking examples of the advantages of false-color film.

[19] "Industry Observer," *Aviation Week*, **80**, 19 (May 25, 1964). Announces the possibility of a contract to an observatory for infrared photography of the lunar surface. The objective of this program is to obtain precise data on surface contours and roughness of possible landing sites. It is hoped that the infrared photographs will show surface details that are not apparent in conventional photos.

[20] J. Wolbarst, "Infrared: New Areas Opening Up for Polaroid Camera Users," *Modern Photo.*, **29**, 16 (May 1965). Reports that Polaroid Infrared Film yields a finished print in 15 sec. The illustrations with the article show how this film can be used in aerial photography to detect diseased crops and in microphotography for the study of retinal cones.

[21] N. Rothschild and M. Iger, "Kodak Ektachrome Infrared Aero," *Popular Photo.*, **57**, 158 (December 1965). Reports on the use of false-color film for ordinary pictorial photography. Three pages of illustrations in full color demonstrate the usefulness of the film.

[22] C. J. Robinove, "Infrared Photography and Imagery in Water Resources Research," *Am. Water Works Assn. J.*, **57**, 834 (1965).

[23] T. Owen, "Saturn's Ring and the Satellites of Jupiter: Interpretations of Infrared Spectra," *Science*, **149**, 974 (1965). Interpretation indicates that the absorber in Saturn's ring is water ice. The spectra were recorded out to 1.11 μ by hypersensitized infrared-sensitive photographic plates.

[24] P. Alelyunas, "Synergetic Satellites," *Space/Aeronautics*, **46**, 52 (October 1966). Describes the use of infrared photography through narrow-band filters in order to determine the condition of crops in earth resource studies.

[25] C. H. Strandberg, "Water Quality Analysis," *Photogrammetric Eng.*, **32**, 234 (1966). Discusses the application of aerial photography to the detection of water pollution. Infrared and false-color films are particularly helpful in this work.

[26] W. N. Hess, D. H. Menzel, and J. A. O'Keefe, eds., *The Nature of the Lunar Surface*, Baltimore, Md.: Johns Hopkins Press, 1966, see E. A. Whitaker, "The Surface of the Moon," p. 79. Describes the use of superimposed infrared positive transparencies and

19.5.1 INFRARED PHOTOGRAPHY

ultraviolet negatives to enhance the delicate color differences observed on the moon's surface.

[27] M. M. Thompson, *Manual of Photogrammetry*, 3d ed., Washington, D.C.: American Society of Photogrammetry, 1966, Vols 1 and 2.

[28] J. F. Forkner and D. D. Lowenthal, "A Photographic Recording Medium for $10.6\,\mu$ Laser Radiation," *Appl. Opt.* **6**, 1419 (1967). An unusual application of Kalvar film, which is a thermoplastic resin with an ultraviolet-sensitive compound dispersed within the resin. As normally used, exposure to light decomposes the compound to form a gas within the sensitive layer. Subsequent exposure to heat expands the gas into small vesicles that are trapped in the resin. The vesicles scatter light and thus reveal the image. As used by these authors, the film was pre-exposed to ultraviolet so that subsequent exposure to $10.6\,\mu$ radiation would heat the resin layer and selectively develop the irradiated areas. The characteristic curve for this film indicates that an exposure to 0.6 joules cm^{-2} is required for a density of 0.3 above base fog (pre-exposure of 60 sec to two F15T8-BLB lamps at a distance of 3 cm). The same technique has proved useful for the study of optical images formed in the $10\,\mu$ region.

[29] E. F. Yost and S. Wenderoth, "Multispectral Color Aerial Photography," *Photogrammetric Eng.*, **33**, 1020 (1967). Describes the usefulness of a specially constructed camera that covers four spectral bands in the 0.36 to $0.98\,\mu$ region.

[30] M. P. Meyer and D. W. French, "Detection of Diseased Trees," *Photogrammetric Eng.*, **33**, 1035 (1967). Authors used Kodak Ektachrome Infrared (false-color) film to study Dutch elm disease, oak wilt, and other tree disorders.

[31] N. L. Fritz, "Optimum Methods for Using Infrared-Sensitive Color Films," *Photogrammetric Eng.*, **33**, 1128 (1967). Details the special characteristics of false-color film and how they can be used for reconnaissance and the detection of diseases and pests in agricultural crops. The author deplores the popular notion that infrared-sensitive films are well adapted to detecting objects by their own emission. He states that in the absence of any other source of radiation, an object heated to 650°F (616°K) will just be recorded by the infrared-sensitive layer of false-color film by an exposure of 15 min at $f/2.0$.

[32] C. H. Strandberg, "Photoarchaeology," *Photogrammetric Eng.*, **33**, 1152 (1967). After an extensive study of the films that are available for aerial photography, this author concludes that color and false-color films are the best for photoarchaeological investigations.

[33] C. H. Strandberg, *Aerial Discovery Manual*, New York; Wiley, 1967. This manual is unusually well adapted for the self study of aerial photointerpretation. The advantages and disadvantages of infrared and false-color film are discussed thoroughly. The author describes a 9-lens multiband aerial camera that takes 9 simultaneous photos in various parts of the spectrum. The spectral bands used are 0.400 to $0.460\,\mu$, 0.442 to $0.520\,\mu$, 0.515 to $0.570\,\mu$, 0.525 to $0.605\,\mu$, 0.580 to $0.638\,\mu$ (secondary bandpass at 0.665 to $0.720\,\mu$), 0.662 to $0.720\,\mu$, 0.710 to $0.890\,\mu$, 0.810 to $0.900\,\mu$, (secondary bandpass at 0.350 to $0.400\,\mu$). Many examples of the 9-panel photos that come from this camera are scattered throughout the book. For another description of this camera, see ref. 19.4.1[25].

[34] "High Altitude Photo ... ," *Aviation Week*, **88**, front cover (April 8, 1968). A beautiful example of the use of Kodak Ektachrome Infrared Film. Photo is the southwestern United States and was taken from the X-15 at an altitude of 200,000 ft. The red rendering of vegetation in cultivated desert areas is particularly striking. See also the caption corrections; **88**, 142 (May 13, 1968).

19.5.2 Reflectance Properties of Materials

One of the more useful lessons gained from infrared photography is that the infrared reflectance characteristics of materials can provide valuable clues about the nature of the material. The references in this section pertain to reflectance characteristics of materials in the region beyond the photographic infrared. Since these characteristics can be measured from a distance, this technique offers the chance to learn about the surface materials of lunar and planetary bodies.

[1] A. H. Pfund, "The Identification of Gems," *J. Opt. Soc. Am.*, **35**, 611 (1945). Describes means of identifying gems by their unique infrared reflectance spectra.

[2] H. F. Greener and F. J. O'Neil, "Radiant Energy Reflectance of Men's Wear Colors," *Textile Research J.*, **17**, 63 (1947). The purpose of this study was to find textile dyes that reflected solar radiation so as to keep the wearer cool in hot summer weather.

[3] R. H. Wilhelm and J. B. Smith, "Transmittance, Reflectance, and Absorptance of Near-Infrared Radiation in Textile Materials," *Textile Research J.*, **19**, 73 (1949). The results of this study are of interest for radiation drying of textiles, the analysis of the thermal comfort of clothing, and the suitability of materials for camouflage use.

[4] D. M. Gates and W. Tantraporn, "Reflectivity of Deciduous Trees and Herbaceous Plants in the Infrared to 25 Microns," *Science*, **115**, 613 (1952). For most leaves the reflectance is very small at wavelengths beyond $2\,\mu$. In general, the upper surface reflects more than the lower surface, old leaves reflect more than new leaves, and shade leaves reflect more than sun leaves. For each of these cases, just the opposite is true in the visible.

[5] "Airborne Infrared Monitor Studied," *Aviation Week*, **74**, 84 (March 20, 1961). Describes a program to establish the diffuse reflectance characteristics of various types and conditions of soils and vegetations. An attempt will be made to see if this information will support the use of an airborne monitor to identify terrains and their suitability for vehicular traffic. A commercial spectrophotometer is used, but it has been modified so as to scan the 0.25 to $5\,\mu$ spectral region.

[6] R. J. P. Lyon and E. A. Burns, "Analysis of Rocks and Minerals by Reflected Infrared Radiation," *Econ. Geol.*, **58**, 274 (1963).

[7] D. M. Gates et al., "Spectral Properties of Plants," *Appl. Opt.*, **4**, 11 (1965).

[8] R. B. Wattson and R. E. Danielson, "The Infrared Spectrum of the Moon," *Astrophys. J.*, **142**, 16 (1965). Reports that measurements of the spectral reflectance of the moon between 0.8 and $3.0\,\mu$ were made from Stratoscope 2. In such measurements, the moon's thermal radiation becomes a significant factor beyond $2.5\,\mu$. For other measurements made on this flight and a description of the instrumentation, see refs. 19.3.2 [29] and [49].

[9] W. A. Hovis, Jr., "Infrared Reflectivity of Limonite: Influence on Martian Reflection Spectra," *Icarus*, **4**, 41 (1965). The data reported here extend from 2.5 to $4.5\,\mu$. If Limonite is a constituent of the Martian surface, the low reflectance in the $3\,\mu$ region may cause serious errors in other determinations of water vapor and carbon dioxide in the Martian atmosphere.

[10] C. Sagan, J. P. Phaneuf, and M. Ihnat, "Total Reflection Spectrophotometry and Thermogravimetric Analysis of Simulated Martian Surface Materials," *Icarus*, **4**, 43 (1965). Gives reflection spectra of a variety of minerals that contain ferric oxides and silicates. Samples are in both solid and pulverized form. Pulverized Limonite appears to be a good match to observed reflectance spectra of Mars. Other inferences are drawn concerning the possible origin and evolution of life on Mars.

[11] A. B. Binder, D. P. Cruickshank, and W. K. Hartmann, "Observations of the Moon and of the Terrestrial Rocks in the Infrared," *Icarus*, **4**, 415 (1965). Gives reflectance spectra of various areas on the moon in the wavelength interval from 1 to 2.2 μ. These data support a hypothesis that the unusual reflectance properties of the moon are caused solely by irradiation processes.

[12] W. A. Hovis, Jr., "Infrared Reflectivity of Iron Oxide Materials," *Icarus*, **4**, 425 (1965). These measurements extend from 0.5 to 6 μ.

[13] W. A. Hovis, Jr., "Infrared Spectral Reflectance of Some Common Minerals," *Appl. Opt.*, **5**, 245 (1966). These spectra, taken in the 1 to 6 μ region, might be useful in identifying surface materials on other astronomical bodies.

[14] H. J. Keegan and V. R. Weidner, "Infrared Spectral Reflectance of Frost," *J. Opt. Soc. Am.*, **56**, 523 (1966). Gives reflectance spectra of water and carbon dioxide frosts. The curves are useful in setting up heat budgets for planets, such as Venus, that have water vapor and carbon dioxide in their atmosphere.

[15] W. A. Hovis, Jr. and W. R. Callahan, "Infrared Reflectance Spectra of Igneous Rocks, Tuffs, and Red Sandstone from 0.5 to 22 μ," *J. Opt. Soc. Am.*, **56**, 639 (1966).

[16] J. R. Aronson et al., "Studies of the Middle- and Far-Infrared Spectra of Mineral Surfaces for Application in Remote Compositional Mapping of the Moon and Planets," *J. Geophys. Research*, **72**, 687 (1967).

[17] R. K. Vincent and G. R. Hunt, "Infrared Reflectance from Mat Surfaces," *Appl. Opt.*, **7**, 53 (1968). Much current effort is being expended in the study of reflection spectra in order to learn more about the surface layers of the moon and the planets. Because many of these surfaces may be particulate in nature, it is necessary to better understand the reflection and emission properties of particulate surfaces. This paper provides the basic theory for such an understanding. Spectra are given for calcite and gypsum in the 4 to 14 μ region.

19.6 APPLICATIONS INVOLVING A COOPERATIVE SOURCE

The references in this section pertain to space communications and to some speculations concerning the sensory systems of insects.

19.6.1 Space Communications

The coverage here is limited to optical communications for space use. A review of the discussion of similar systems for terrestrial use (Section 16.6.1) makes one suspicious that optical communication is in the same class as the proverbial bridesmaid, always a contender but never quite able to win the big prize. The requirements posed by astronautical systems may change all of this. The narrow beamwidth of optics and the ease with which high-radiance injection diodes can be modulated make a system that appears to be more efficient than any competitive RF system. Either type of system requires an unobstructed line of sight for its operations. This is no particular disadvantage, since virtually all space communications will be line of sight by their very nature.

[1] J. R. McDermott, "Transmitters and Receivers for Optical Communications," *Space/Aeronautics*, **39**, 98 (June 1963).

[2] K. L. Brinkman, W. K. Pratt, and E. J. Vourgourakis, "Design Analysis for a Deep-Space Laser Communication System," *Microwaves*, **3**, 35 (April 1964).
[3] J. Fusca, "Laser Communications," *Space/Aeronautics*, **41**, 58 (May 1964).
[4] J. S. Greenberg, "On the Narrow Beam Communication System Acquisition Problem," *IRE Trans. Mil. Electronics*, **8**, 28 (1964). Discusses the problem of establishing a communication link between two separated terminals when each is equipped with a narrow-beamwidth transmitter and some prior knowledge of the relative angular location of the other. This paper studies the factors that affect acquisition time, such as uncertainties in the exact direction to the other terminal and the probability of detecting a signal in the presence of noise.
[5] B. Miller, "Gemini Laser Test May Speed Future Use," *Aviation Week*, **83**, 71 (December 13, 1965). Gives a detailed explanation of the laser communication and propagation experiment scheduled for Gemini 7. The astronaut is to aim a 6 lb gallium arsenide laser transmitter at a visible ground-based laser beacon. The voice will be transmitted by pulse modulation of the four gallium arsenide diodes.
[6] E. B. Moss, "Some Aspects of the Pointing Problem for Optical Communication in Space," *J. Spacecraft and Rockets*, **2**, 698 (1965).
[7] S. E. Miller, "Communication by Laser," *Sci. Am.*, **214**, 19 (January 1966).
[8] N. G. Lozins, "Pointing in Space," *Space/Aeronautics*, **46**, 76 (August 1966).
[9] M. Ross, *Laser Receivers*, New York: Wiley, 1966. A treatise on optical communication and information-transfer systems.
[10] E. J. Reinbolt and J. L. Randall, "How Good are Lasers for Deep-Space Communications," *Astronautics and Aeronautics*, **5**, 64 (April 1967). A system-oriented comparison of RF and optical communications.
[11] E. C. Park and L. S. Stokes, "Lasers vs Microwaves for Deep Space Communications," *Microwaves*, **6**, 78 (May 1967). An excellent system analysis that compares a $10.6\,\mu$ laser link with a 3 Gc microwave link.
[12] J. Roschen, "Optoelectronics," *Industrial Research*, August 1967, p. 71. This article stresses the new uses of the infrared portion of the spectrum that will arise from the availability of the injection diode. Among the applications are secure communication, peripheral optics for computers, ranging, terrain illumination for night photography, obstacle detectors for the blind, and various medical applications.
[13] "High Sensitivity Infrared 10.6 Micron Heterodyne Receivers with Large IF Bandwidth," *IEEE Spectrum*, **5**, 5 (July 1968). Designed for space communication experiments, this receiver has an IF bandwidth extending from 10 MHz to 1 GHz. A copper-doped germanium detector serves as the mixer.

19.6.2 Miscellaneous

[1] P. A. Tove and J. Czekajewshi, "Infrared Curtain System Detects and Counts Moving Objects," *Electronics*, **34**, 40 (August 4, 1961). Describes a multiple-beam projector and associated detectors that are used for studying the movements of bats and other nocturnal animals.
[2] E. C. Okress, "Dielectric and Organic Superconducting Waveguide, Resonator, and Antenna Models of Insect's Sensory Organs," *Appl. Opt.*, **4**, 1350 (1965). Discusses the application of waveguide theory to selective sensing of infrared by insects.
[3] W. L. Wolfe and M. J. Lubbers, "Infrared Communication by Nocturnal Insects," *J. Opt. Soc. Am.*, **56**, 540 (1966). These authors investigated the proposal that certain nocturnal insects communicate for mating purposes by infrared radiation. It has been observed that the thorax of the female moth can be as much as 10°C warmer than ambient and that certain organs of the male might function as detectors at a wavelength of $10\,\mu$.

Appendix 1

The Symbols and Abbreviations Used in This Book

Symbols are defined as they are used in the text, and this list is intended only as a guide to the more important usages of each. In particular, no attempt has been made to list the variations that occur when subscripts are used. Whenever possible, the symbols and abbreviations follow the recommendations of the Optical Society of America, the Institute of Electrical and Electronic Engineers, and the United States of America Standards Institute (formerly the American Standards Association). Because of the limited number of characters available in the English and Greek alphabets, many of the symbols must represent more than one quantity. In the few cases where this may cause trouble, a study of the context should resolve the ambiguity.

A. SIMPLE ENGLISH LETTER SYMBOLS

Symbol	Meaning
a	Absorption coefficient; Constant
A	Amplification ratio; Angstrom; Area; Total absorptance; Voltage amplification
b	Constant
B	3 dB bandwidth
c	Constant; Proportionality factor; Radiation constant; Velocity of light; Specific Heat
C	Centigrade
C	Constant; Heat capacity; Number of independent detector elements
d	Depth of focus; Diameter; Distance
D	Constant; Detectivity; Diameter

e		Electronic charge; The number 2.718..., which is the base of natural logarithms
E		Energy
f		Focal length; Frequency
F		Fahrenheit
F		Focal point; Noise factor; Noise figure; Stress; Thrust
g		Transconductance
G		Conductance
h		Altitude; Enthalpy; Planck's constant
H		Irradiance; Principal point
i		Current
I		Current
J		Radiant intensity
k		Boltzmann's constant; Constant; Thermal conductivity
K		Kelvin
K		Reticle rotation constant; Scattering area ratio
L		Distance; Length
m		Mass
M		Mach number; Thermal-stress figure of merit
n		An integer; Index of refraction
N		Noise power; Radiance
p		Partial pressure; Probability density function
P		Pressure; Probability distribution function; Radiant Flux
q		Constant
Q		Constant; Radiant photon emittance
r		Angle of reflection; Number of resolution elements in a search field; Radius; Recovery factor
R		Rankine
R		Range; Resistance
s		Relative spectral response
S		Signal power; Total surface area of a cavity
t		Constant; Time
T		Temperature
u		Final slope angle of a ray; Radiant energy density
U		Radiant energy
v		Velocity; Visibility factor; Voltage
V		Voltage; Visual range
w		Amount of an absorber; Power
W		Radiant emittance; Power (watt)
x		Distance; Length
Y		Young's modulus
z		Thermoelectric figure of merit; Precipitation rate
Z		Zenith angle

APPENDIX 1

B. SIMPLE GREEK LETTER SYMBOLS

Symbol	Meaning
α	Absorptance; Linear expansion coefficient; Plane angle; Seebeck thermoelectric coefficient; Short-circuit common-base current gain of a transistor
β	Angular field of view (rad); Plane angle
γ	Effectiveness factor; Ratio of specific heats; Scattering coefficient
δ	Angular subtense; Phase shift
Δ	Symbol for an increment or small part
ϵ	Emissivity
η	Quantum efficiency
θ	Angular scanning rate; Plane angle
λ	Wavelength
Λ	Rate of energy loss from a thermocouple
μ	Free-carrier mobility; Joule-Thomson coefficient; Micron
ν	Frequency
π	The number 3.14159...
ρ	Absolute humidity; Density; Reflectance; Resistivity
σ	Electrical conductivity; Extinction coefficient; Standard Deviation; Stefan-Boltzmann constant; Wavenumber
τ	Time; Time constant; Transmittance
ϕ	Plane angle; Work function
χ	A factor in chopper calculations
ψ	Scattering exponent
Ψ	A factor in the thermal radiance derivative
ω	Instantaneous field of view (sr); Solid angle
Ω	Search field (sr); Solid angle

C. SPECIAL AND COMPOSITE SYMBOLS

Symbol	Meaning
dB	Decibel
D^*	Detectivity referred to unit bandwidth and unit area
D^{**}	Detectivity referred to unit bandwidth, unit area, and an (effective weighted) solid angle of π steradians for the field of view
Δf	Equivalent noise bandwidth
f/no	f/number
$\mathcal{G}$	Power gain
NA	Numerical aperture
$\mathcal{R}$	Responsivity

S/N Signal-to-noise (power) ratio
$\mathcal{T}$ Frame time; Integration time
$\mathscr{V}$ Ratio of the velocity of an aircraft to its height above the ground (often written as v/h)

D. SELECTED ABBREVIATIONS

Abbreviation	Meaning
AM	Amplitude modulation
atm cm	Atmosphere centimeter
cm	Centimeter
deg	Degree (angular)
°	Degree (temperature)
EFL	Equivalent focal length
EGT	Exhaust gas temperature
FM	Frequency modulation
gr	Generation-recombination (noise)
Hz	hertz (cycle per second)
in.	Inch
ln	Natural logarithm
log	Logarithm to the base 10
mile	Statute mile
mrad	Milliradian
NEI	Noise equivalent irradiance
NEP	Noise equivalent power
NEΔT	Noise equivalent differential temperature
rad	Radian
RF	Radio frequency
sec	Second
sr	steradian

Appendix 2

Symbols and Nomenclature for Radiometry and Photometry

The most definitive statement concerning symbols and nomenclature for radiometry and photometry is the report of the Nomenclature Committee of the Optical Society of America. Short descriptions of the terms are included in this report, but in most cases they are not sufficiently rigorous to serve as definitions and are not to be used as such. The first symbols are recommended by the S.U.N. Commission of IUPAP, ISO Technical Committee 42, and the International Commission on Illumination (CIE). The symbols given in the square brackets do not have international sanction but are widely used in the United States.

This report is reproduced through the courtesy of the Optical Society of America. It was first published in *J. Opt. Soc. Am.*, **57**, 854 (1967). The members of the Nomenclature committee were W. L. Wolfe (chairman); S. S. Ballard; R. D. Hudson, Jr.; R. C. Jones; D. B. Judd; H. W. Nielsen; R. W. Preisendorfer; J. C. Richmond; R. Tousey; K. G. Kessler (chairman, 1964–1965); J. N. Howard, and D. L. MacAdam (ex-officio); H. C. Wolfe (liason with SUN and TC 12).

NOMENCLATURE AND SYMBOLS FOR RADIOMETRY AND PHOTOMETRY

RADIOMETRY

PHOTOMETRY

1. Terms to Describe Processes

radiation. Transfer of energy in the form of electromagnetic waves or photons.

irradiation. The process by which radiant flux is incident on a body.

(radiant) emission. The transformation of a different kind of power into radiant flux.

(radiant) transmission. The passage of radiant flux through a medium.

(radiant) reflection. At an interface between two media, the deviation in direction of travel of incident radiant flux so that the flux remains in the first medium.

(radiant) absorption. The transformation of radiant flux to a different form of power by interaction with matter.

illumination. The process by which luminous flux is incident on a body.

(luminous) emission. The transformation of a different kind of power into luminous flux.

(luminous) transmission. The passage of luminous flux through a medium.

(luminous) reflection. At an interface between two media, the deviation in direction of travel of incident luminous flux so that the flux remains in the first medium.

(luminous) absorption. The transformation of luminous flux to a different form of power by interaction with matter.

2. Terms Used to Describe Devices

radiator. A source of radiant flux.
radiometer. An instrument for measuring radiometric quantities (radiant flux, radiance, radiant intensity, etc.)

light source. A source of luminous flux.
photometer. An instrument for measuring photometric quantities (luminous flux, luminance, luminous intensity, etc.)

3. Terms Used to Describe Quantities

radiant energy. Q, Q_e, [U], Energy in the form of electromagnetic waves or photons.

(radiant) energy density. w, u, Radiant energy per unit volume.

(radiant) flux or **radiant power.** Φ, Φ_e, Time rate of flow of radiant energy.

(radiant) flux density. E or M, [H or W], The radiant flux incident upon or leaving a surface, divided by the area of that surface. A general term including both **exitance** and **irradiance.** May also be a field quantity, referred to an imaginary surface in space.

(radiant) exitance. [formerly called emittance] M, M_e, [W], The radiant flux leaving an infinitesimal element of surface, divided by the area of that surface. (This term can be further refined as **emitted exitance, reflected exitance,** or **transmitted exitance**).

(irradiance) or **(radiant) incidence.** E, E_e, [H], The radiant flux incident per unit area upon a surface.

(radiant) intensity. I, I_e, [J], The radiant flux leaving a point source per unit solid angle.

radiance or **(radiant) sterance.** L, L_e, [N], The radiant flux leaving or arriving at a surface at a given point in a given direction per unit solid angle and per unit of surface area projected orthogonal to that direction.

luminous energy or **quantity of light.** Q, Q_v, The visual aspect of radiant energy.

(luminous) flux. Φ, Φ_v, Time rate of flow of luminous energy.

(luminous) flux density. E or M, The luminous flux incident upon or leaving a surface, divided by the area of that surface. A general term including both **exitance** and **illuminance.** May also be a field quantity, referred to an imaginary surface in space.

(luminous) exitance. M, M_v, The luminous flux per unit time leaving an infinitesimal element of surface, divided by the area of that surface.

illuminance or **(luminous) incidence.** E, E_v, The luminous flux incident per unit area upon a surface.

(luminous) intensity. I, I_v, The luminous flux leaving a point source per unit solid angle.

luminance or **(luminous) sterance.** L, L_v, The luminous flux leaving or arriving at a surface at a given point in a given direction per unit solid angle and per unit of surface area projected orthogonal to that direction.

3A. The Committee Tentatively Endorses These Terms and Offers Them for Consideration

(radiant) sterisent. L^*, L_c^*, The radiant flux generated within a volume (either by emission or by scattering from some other direction into the given direction) in a given direction per unit solid angle and unit volume.

(luminous) sterisent. L^*, L_v^*, The luminous flux generated within a volume (either by emission or by scattering from some other direction into the given direction) per unit solid angle and unit volume.

4. Terms Used to Describe Properties

(radiant) reflectance. ρ, The ratio of the radiant flux reflected from a surface to that incident upon it.
(radiant) transmittance. τ, The ratio of radiant flux transmitted by a body to that incident upon it.
(radiant) absorptance. α, The ratio of radiant flux absorbed by a body to that incident upon it.

(luminous) reflectance. ρ, The ratio of the luminous flux reflected from a surface to that incident upon it.
(luminous) transmittance. τ, The ratio of luminous flux transmitted by a body to that incident upon it.
(luminous) absorptance. α, The ratio of luminous flux absorbed by a body to that incident upon it.

Appendix 3

Conversion Factors

The conversion factors listed here have been limited to those that are most useful to the infrared system engineer.

Multiply	By	To Obtain

Angle
- Degrees .... 60 .... Minutes
- ....3,600 .... Seconds
- ....17.45 .... Milliradians
- Minutes .... 0.2909 .... Milliradians
- Seconds .... 4.848 .... Microradians
- Square degrees .... 3.045×10^{-4} .... Steradian

Area
- Square centimeters .... 0.1550 .... Square inches
- Square inches .... 6.4516 .... Square centimeters

Energy
- Btu .... 252.0 .... Gram calories
- ....1,054.8 .... Joules
- ....0.2928 .... Watt-hours
- Ergs .... 1×10^{-7} .... Joules
- Gram calories .... 4.184 .... Joules
- ....0.239 .... Watt-seconds
- Joules .... 9.480×10^{-4} .... Btu
- ....1×10^{7} .... Ergs
- ....0.239 .... Gram calories
- ....1 .... Watt-seconds

Length

Centimeters	0.3937	Inches
	6.214×10^{-6}	Miles (stat.)
Kilometers	3,281	Feet
	0.6214	Miles (stat.)
Inches	2.54	Centimeters
Feet	30.48	Centimeters
Miles (stat.)	1.609×10^5	Centimeters
	1.609	Kilometers
	5,280	Feet
	0.8684	Miles (naut.)

Pressure

Atmospheres	76.0	Centimeters of mercury
	29.92	Inches of mercury
	14.70	Pounds per square inch

Solar Irradiance

Gram-calories per square centimeter per minute	221.1	Btu per square foot per hour
	6.978×10^{-2}	Watts per square centimeter
Watts per square centimeter	14.33	Gram-calories per square centimeter per minute

Temperature

See footnote on p. 35.

Time

Seconds	1.157×10^{-5}	Days
	3.169×10^{-8}	Years

Velocity

Feet per second	0.6818	Miles (stat.) per hour
	1.097	Kilometers per hour
Miles (stat.) per hour	1.467	Feet per second
	1.609	Kilometers per hour

Volume

Cubic centimeters	6.102×10^{-2}	Cubic inches
Cubic inches	16.39	Cubic centimeters
Gallons (U.S. liquid)	3.785	Liters
Liters	0.2642	Gallons (U.S.)
	61.02	Cubic inches
Ounces (fl.)	2.957×10^{-2}	Liters

Weight

Kilograms	2.205	Pounds (avdp.)
Ounces	28.35	Grams
Pounds	453.6	Grams
	16.0	Ounces

Appendix 4

The Unpublished Literature of the Infrared

In the United States, a large portion of the infrared literature exists in an informal form and is, by the definitions of Section 1.6, properly called unpublished literature. This literature consists of internal company reports, proposals, contractor reports, government reports, proceedings of special conferences, and state-of-the-art reports by special information centers. Unpublished literature is not subject to the formal review procedure normally required of published literature, it is not readily available to all interested readers, and much of it is classified. With a few well-justified exceptions, there are no references to the unpublished literature in this book.

The reader who is working in certain portions of the infrared field may find it necessary to consult the unpublished literature in order to maintain currency in his specialty. This appendix outlines the major sources of such literature. The effective exploitation of these sources is the responsibility of the reader.

Infrared Information Symposia (IRIS)

IRIS is sponsored by the Office of Naval Research and is under joint-service direction[1]. It organizes a continuing series of classified symposia at which the latest developments in military infrared are discussed. A security clearance and a need-to-know are required for attendance. Papers presented at these symposia are published in a classified journal entitled *Proceedings of IRIS*. IRIS also sponsors Specialty Groups on detectors, target measurements, electro-optical information processing, optics and optical materials, image forming sensors, suppression, backgrounds and atmospheric physics, and standards.

APPENDIX 4 627

Infrared Information and Analysis (IRIA) Group

IRIA is a government-sponsored information center for infrared technology, established at the University of Michigan. It receives copies of all contractor reports on government-supported research and development in infrared. None of the reports are available from IRIA, although the qualified inquirer can visit the facility and consult reports in the IRIA library. IRIA also provides bibliographies and state-of-the-art reports on specialized topics in infrared.

Ballistic Missile Radiation Analysis Center (BAMIRAC)

BAMIRAC is another government-sponsored information center at the University of Michigan. It is responsible for the collection, analysis, and dissemination of information about the radiations from ballistic missiles. The files and retrieval system of BAMIRAC are coordinated with those of IRIA.

Defense Documentation Center (DDC)

DDC provides copies of reports generated under Department of Defense contracts. In most cases, inquirers must establish their need-to-know. DDC publishes an announcement bulletin of new acquisitions and provides specialized bibliographies upon request.

Clearinghouse for Federal Scientific and Technical Information (CFSTI)

CFSTI serves as the central source for U.S. government research data in engineering and the physical sciences. It is the mechanism through which unclassified technical reports and translations generated by all government agencies are uniformly indexed and made available to the public. The translations cited in this book, identified by an OTS, SLA, or CFSTI designation, can be ordered by addressing Clearinghouse, U.S. Department of Commerce, 5285 Port Royal Road, Springfield, Va, 22151.

CFSTI has taken over the document distribution program formerly run by the Office of Technical Services (OTS). In 1961, OTS published two Selective Bibliographies on Infrared[2, 3].

The Library of Congress

The Library of Congress holds over 2 million books pertaining to science and technology, over 20,000 technical journals, and the most widely representative collection of the nation's technical reports[4]. Included among these reports are the official service collection of more than 30,000 reports of the World War II Office of Scientific Research and Development (OSRD). Probably the best way to identify OSRD reports on infrared is through the bibliographies in the summary technical reports of Division 16 (refs 16.0. [7] and [8]). Since the U.S. copyright laws of 1865 and 1870 provide for the deposit of copies of all copyrighted publications in the Library of Congress, virtually all of the U.S. published literature of the past century is available here.

The National Referral Center for Science and Technology was established in the Library of Congress in 1962[5]. It compiles a comprehensive register of all information resources having the capability to assist the scientific and technical community, provides referral services to anyone seeking information, and engages in continuous fact-finding and analysis.

The Science and Technology Division provides a variety of reference and bibliographical services[4]. In the mid-1950's, this Division published a series of five bibliographies on infrared[6]–[10]. These cover the period from 1935 to 1952. Since this period predates the establishment of IRIA, these bibliographies complement rather than duplicate the efforts of IRIA.

The Aerospace Technology Division is supported by the Department of Defense. It specializes in Sino-Soviet Bloc activities in the aerospace sciences as evidenced in the foreign published literature. Its reports and analyses are available to the Department of Defense and its contractors through DDC and to the general public through CFTSI.

Miscellaneous

There are two other lesser-known bibliographies that have played an important role in the development of the infrared field. Unfortunately, both were published in limited editions and are probably virtually unobtainable today. A.H. Canada's report[11] is one of the best available guides to the German World War II infrared work and contains many photographs of the equipment. S. S. Ballard's report[12] brought together a tremendous amount of information on reflecting optics and provided the stimulus for many of the optical systems that were designed for infrared equipment during the 1950's.

REFERENCES

[1] A. R. Laufer, "Preface to the Special Issue on Infrared Physics and Technology," *Proc. Inst. Radio Engrs*, 47, 1415, (1959).

[2] *Infrared, Part 1: General Research, Equipment and Materials*, OTS SB-466, U.S. Department of Commerce, Office of Technical Services, Washington, D.C., May 1961. Lists reports and translations acquired by OTS from 1950 to June 1961. Contents include detectors, photography, optics and optical materials, equipment, general research, and government-owned patents. Contains 212 citations.

[3] *Infrared, Part II: Infrared Spectroscopy*, OTS SB-467, U.S. Department of Commerce,

Office of Technical Services, Washington, D.C., May 1961. Lists reports and translations acquired by OTS from 1950 to June 1961. Contains 183 citations.
[4] D. E. Gray, "Science, Technology, and the Library of Congress," *Phys. Today* **18**, 44 (June 1965).
[5] J. F. Stearns, "The Information Maze," *Research/Development*, **15**, 22 (May 1964).
[6] C. R. Brown et al., "Infrared: a Bibliography," Library of Congress, Technical Information Division, Washington, D.C., December 1954 (AD62234). Contains 5541 citations to the published literature from 1935 through 1951.
[7] M. W. Ayton et al., "Infrared: a Bibliography," Part II, Library of Congress, Technical Information Division, Washington, D.C., March 1957 (AD122203). Contains 1600 citations to the unclassified report literature from 1935 through 1952.
[8] M. W. Ayton et al., "Infrared; a Bibliography," Part III, Library of Congress, Technical Information Division, Washington, D.C., June 1957 (AD129600), *Confidential*. Contains 323 citations to reports classified Confidential for the period from 1935 through 1952.
[9] V. H. Boteler and C. R. Brown, "Infrared; a Bibliography," Part IV, Library of Congress, Technical Information Division, Washington, D.C. July 1952 (AD7599), *Confidential*. Contains 832 citations to reports that are classified Confidential for the period from 1935 through 1952.
[10] V. H. Boteler and C. R. Brown, "Infrared: a Bibliography," Part V, Library of Congress, Technical Information Division, Washington, May 1951 (TIP S1741), *Secret*. Contains 758 citations to reports that are classified Secret for the period from 1935 through 1952.
[11] A. H. Canada, "Infrared: Its Military and Peacetime Uses," Data Folder No. 87516, General Electric Co., General Engineering Laboratory, Schenectady, N.Y. First published December 1947, revised March 1951. An extremely valuable report on World War II infrared and its possible peacetime uses. The material was gathered through the author's military service at the Engineer Research and Development Board at Fort Belvoir and his participation with scientific reconnaissance teams that entered Germany immediately after the hostilities ended. In order to avoid security problems, much of the material in this book is taken from German World War II publications. This report also contains an annotated bibliography with 171 citations. It is unfortunate that only 600 copies of this report were printed.
[12] S. S. Ballard, ed., "A Bibliography on Reflecting Optics," Armed Forces–National Research Council Vision Committee, University of Michigan, Ann Arbor, Michigan, November 1950. There is a 2-page supplement dated March 15, 1951. While this report is of interest to all workers in optics, it is particularly valuable to workers in the infrared because of the widespread use of reflecting optics in infrared systems. It contains 228 citations to the published literature and 60 patent citations, for the period from 1925 to 1950. The supplement contains an additional 10 citations. Many of the citations are annotated.

Index

Citations to the annotated references show both the page and the reference number.

Aberrations, 183, 186–195
  chromatic, 187, 192–194
  coma, 187, 189–192
  spherical, 186, 187–189
Abbreviations, 618
Absorptance, 25, 622
  total, 139, 140
Absorption, 47–50
Absorption cell, multiple traversal, 140
Absorption coefficient, 114, 174, 175
Achromatic doublet, 193
Aerodynamic heating, 100–103, 213
Afterburning, 96–98
Aircraft, detection and tracking of, 466[1,3], 467[11], 468[17, 18], 469[25], 470[3, 4, 8, 9], 471[18, 19], 472[24, 33], 475[2, 8], 476[20, 21], 477[2], 478[5–9, 12–14, 16], 479[17–20]
Airy disk, 183, 184
Anab, 480, 485[53], 486[66]
Antireflection coatings, *see* Coatings for optical materials
Aperture stop, 179
Apollo, 584[38]
Apparent radiometric quantities, 24, 432, 433, 513
Applications matrix, 458, 460, 461, end papers
Arsenic trisulfide, 189, 194, 200, 213, 215, 217
Arthritis, diagnostic aid, 543[11], 547[54]
Ash, 480, 485[53], 486[66]
Atmosphere, models of, 117
Atmosphere centimeters, 127
Atmospheric constituents, 118–128
  carbon dioxide, 126
  carbon monoxide, 584[28]

  methane, 584[28, 33]
  nitrous oxide, 584[27, 28]
  ozone, 127, 128, 582[7]
  water vapor, 119, 123, 124, 125, 578[34], 582[9, 10, 12], 583[16, 22, 25], 586[47]
Atmospheric heat balance, measurement of, 576[2–4, 7–10], 577[14–17, 19, 21], 578[22, 23, 26, 28, 29, 31–33], 579[35], 584[37]
Atmospheric structure, pressure, 586[49]
  temperature, 117, 568[33], 570[45], 582[9], 583[17, 20, 21], 584[26, 29, 35], 585[43–45]
Atmospheric transmission, analytical expressions for, 135–142
  field measurements of, 115, 126, 129, 131, 132, 133, 159
  slant path, 158
  standard windows, 158
Atmospheric transmission tables, altitude correction factors, 156
  carbon dioxide, 145, 147, 149, 151, 153, 155
  9.6μ ozone band, 159
  water vapor, 144, 146, 148, 150, 152, 154
Atmospheric windows, 64, 116, 130, 136, 158
Atoll, 480, 482[18], 484[46], 485[50, 53, 54, 56], 486[63, 66]

Backgrounds, 104–110
Ballistic missiles, detection and tracking of, 467[13], 468[22], 469[23], 471[20], 472[22, 27, 29, 30], 473[34, 35, 39, 41, 43], 474[44–46, 49], 475[3, 10], 476[13–15], 477[1], 478[7]

631

signatures of, 473[34], 490[1, 2], 491[3–5, 1–5], 492[6–15], 493[17–23, 25]
BAMIRAC, Ballistic Missile Radiation Analysis Center, 627
Band theory of solids, 278
Bandwidth, AM system, 410
  equivalent noise, 311, 313
  noise frequencies in, 413
  optimum, 408
  3 dB, 312, 313
Beacons, infrared, 499[6, 12], 501[26], 502[37]
Beer's law, 137
Bias, photoconductive detector, 330–336
  photovoltaic detector, 333, 336
  supply requirements, 336
Bismuth telluride, 391
Blackbody, 33, 40
Blackbody radiation laws, 35–38, 59
Blackbody sources, calibration facilities for, 80
  construction of, 72–77, 433, 539[3], 598[27]
  Gouffé analysis of, 68, 70
  molten metal type, 80
Black radiation detector, 274
Blind, obstacle detection for, 538[1], 539[2], 550[1, 2]
Blip detector, 348
Blur circle, 182, 187, 189, 190, 191, 193
Boeing 707 jet transport, 90–96
  possible locations for a search system, 440, 447
  radiometric characteristics, 92, 94, 95
Bohr model of the atom, 46
Bolometer, 273
  carbon, 274, 290
  ferroelectric, 290, 294
  germanium, 274, 290
  superconducting, 274, 290
  thermistor, 265, 273, 290, 294, 360, 371
Boltzmann's constant, 35
Bombs, infrared-guided, 466[4], 479[1], 480[3], 483[30]
Bouguer-Lambert law, 137
Bouwer's optical system, 199
BSC optical glass, 212, 214, 216
Bunsen flame, spectrum of, 52, 88
Burns and frostbite, assessment of, 542[6], 544[22]

Cadmium-doped germanium, 293
Calcium aluminate, 213, 214, 216
Calorimetric detector, 274
Camouflage materials, properties of 501[34], 502[1, 2]
Cancer detection, breast, 540[9, 12, 13, 16], 541, 542[2], 543[12, 15, 18], 545[37, 40, 42], 546[45, 46]
Carbon arc, 83
Carbon dioxide, absorption by, 48–50, 115, 126, 129, 138, 141
  in earth's atmosphere, 126
  emission by, 51, 88, 94, 98
  transmission tables, 145, 147, 149, 151, 153, 155
Carbon monoxide, 584[28]
Carrier, thermally induced, 349
Cassegrainian optics, 196, 198
Catadioptric system, 198
Cathode ray tubes, 411
Cerebrovascular disease, diagnostic assistance, 544[19], 545[41], 547[54]
Chopper, 235
  cooled, 585[40]
  electromagnetically driven, 585[43], 594[67]
  frequency control of, 327
  series operation of, 342
  square wave, 325, 327
Chromatic aberration, 187, 192–194
Circle of least confusion, 182
Claude refrigerator, 388
Clear air turbulence[CAT], detection of, 439, 524, 525[9, 11, 14], 526[15–18, 22, 23], 527[24, 26, 28–30]
CFSTI, Clearinghouse for Federal Scientific and Technical Information, 627
Clouds, altitude determination, 557, 583[18, 19]
  spectral radiance of, 585[46], 590[30, 35, 36], 591[45, 46]
Coatings for optical materials, antireflection, 218–220
  high reflection, 220–222
  surface protective, 222
Collimator, 224–230
  irradiance from, 225, 226, 227, 228, 229
Collision avoidance, aircraft, 466[7], 487[8, 9], 489[34], 535[8], 536[19]
  automobile, 535[14, 15]

INDEX 633

Color translation, 546[53], 607[12]
Coma, 187, 189–192
Command guidance, tactical missiles, 468[17], 508, 508[1], 509[7, 8, 10, 11, 14, 15], 510[16, 18]
 trackers for, 508[3], 509[4, 5]
Committee on Colorimetry, Optical Society of America, 23
Communication systems, modulators for, 503[6], 504[12, 17], 505[24, 29], 506[38–40]
 sources for, 84, 530[8], 504[11, 13, 19], 505[25, 28], 506[34], 614[12]
 space, 613, 613[1], 614[2–13]
 terrestrial, 8, 466[5, 7], 468[18, 20], 503[1, 3, 5–7], 504[9, 10, 14–16, 18, 20], 505[21–23, 26, 29–32], 506[33–37]
Condenser, optical, 203, 204
Conduction band, 278
Conical optical elements, 204, 558[4], 559[10, 20]
Conversion factors, 623
Coolants, cryogenic, 376, 380
 precautions in handling, 377
COP, coefficient of performance, 387
Copper-doped germanium, 265, 293, 294, 361, 370
Countermeasure systems, against infrared-guided missiles, 507[1, 3–5, 7, 10, 11]
 aircraft, 468[22], 473[40, 42], 474[50, 52, 53], 475[55], 507[5, 7, 10, 11]
 reducing aircraft vulnerability, 486[60], 507[6, 8, 9]
Crevasse detection, 599[1, 2]
Critical angle, 175
Crops, diseased, detection of, 601[27], 610[18, 20, 24], 611[30, 31]
 surveys of, 601[28], 602[40], 605[4], 611[34]
Cross-array detector, 255, 428, 470[3], 475[1, 6], 476[12, 17], 479[1], 480[3], 483[30], 487[3], 508[2], 560[32]
Cryostat, Joule-Thomson, 380–384
Cutoff wavelength, optical filter, 223
 photon detector, 277, 280
Cut-on wavelength, optical filter, 223

D^*, 270, 322
 comparisons of for available detectors, 290–294, 360, 361
 increase by radiation shielding, 351, 353, 354, 355, 356, 366, 367, 368, 369, 370
 relation between peak and blackbody values, 344
 theoretical value at spectral peak, 349, 351
D^{**}, 352
Data link, infrared, 506[36], 509[14], 526[20]
DDC, Defense Documentation Center, 627
Decollimation, 229
Dentistry, thermal imaging in, 546[49]
Depth, of field, 181
 of focus, 180
Detection, probability of, 412
Detection range, see Range equations
Detectivity, 269, 322
Detector measurements, area, 330
 data sheet, 322, 323
 frequency response, 339
 noise spectrum, 339
 optimum bias, 335
 response contours, 346
 spectral response, 340–345
  conversion of relative to absolute values, 343
 test set, 322, 324, 340, 341
 time constant, 345
Detectors, distributed capacitance of, 332
 comparison of, 288, 290–294, 360, 361
 cooling, reasons for, 278, 283
 imaging, 264, 296
 integration with refrigerator, 394
 ionizing radiation, effect of, 359
 Lambertian, 352
 linearity of response, 359
 noise spectrum, generalized, 210
 optically immersed, 289, 295
 packaging, 354, 362, 373
 performance descriptions, 266, 322
 photoconductive, 278
 photovoltaic, 284
 point of elemental, 264
 power dissipation, 332
 quantum efficiency, 282, 283, 350
 radiation shielding, 351
 selection of, 358, 440, 441, 442, 446, 449
 spectral response, 286
 typical operating temperatures, 269
 thermal, 266, 271
 ultimate performance limits, 340–358

Dewar, 373
Dew point, 119
Diagnostic assistance, 542–547
Differential radiance, 63, 431
  value of for atmospheric windows, 431
Diffraction, 183
Displays, 411–414
Doping, 280
Dual-mode system, 211
Dwell time, 424

Earth, resource surveys, 599, 601[24, 26, 28], 603[51], 610[22, 24, 25], 611[30]
  surface, radiometry of from space, 568[33], 570[43], 573[38], 577[20, 21], 578[25, 28], 580[8, 9], 584[31], 585[44–46]
Effective radiometric quantities, 82, 84, 436
EGT, exhaust gas temperature, 87
Einstein's photoelectric equation, 277
Electric dipole moment, 48
Electromagnetic radiation, 20
Electronic circuitry, thermal conditions in, 514, 519[63, 67], 520[69], 521[80, 86, 88–90], 522[92, 96, 97], 528[8, 10], 529[12, 15, 16, 19, 20], 530[27–29, 37, 38]
Emissivity, 25, 28, 39–45
  common materials, 43, 45, 570[49], 572[68], 573[71]
Entrance pupil, 179
Episcotister, 235
Equivalent noise bandwidth, 270, 311
Equivalent sea level path, 143
Error response, 256
Ettingshausen effect, 394
Evaporograph, 6, 265, 299, 528[4, 5], 529[13, 14, 17], 531[36], 542[3], 606[7, 8], 607[11]
Exitance, 621
Extinction coefficient, 114
Eye, examination of, 547[1, 2], 548[3–8, 10, 11, 3, 4], 549[5, 6, 8, 17]
  sensing motion of, 549[1], 550[2]

f/number, 179, 180, 197
Falcon, 9, 480, 486[58]
False alarm time, 413, 426
Fiber optics, 204, 518[45], 519[65], 520[78]
Field lens, 203, 427
Field of view, 181
Field stop, 181, 258
Film, epitaxially grown, thickness monitor, 521[85], 575[90], 596[6]
  infrared, black and white, 6, 265, 296, 609[9], 610[17, 20], 611[28]
  false color, 549[19, 20], 608, 609[4, 14], 610[18, 21], 611[30–32, 34]
  photographic, examination during manufacture, 499[12], 532[1]
Filter, gas, for cryostats, 383
  optical, 222–224
Fire control systems, reasons for including infrared equipment, 477
Fires, aircraft fuel tanks, detection of, 512[5], 513[9]
  cooking, detection of, 496[21, 24], 497[26–28], 603[45]
  forest, detection and surveillance of, 511, 512[1, 3, 6, 7], 599[4], 600[12, 14, 15], 601[23, 32], 602[37, 38], 603[47]
Firestreak, 480, 481[16], 484[47], 490[7]
Flares, 508[3], 509[12]
  as a countermeasure technique, 473[40], 507[1–5, 7, 10, 11]
Flesh, human, transmittance of, 61, 547, 548[1]
Flexure of optical elements, 201, 202
Fluorescence of materials, 216, 520[5], 610[15, 16]
Flux, 24, 25, 27, 28, 621
Focal length, 177, 178, 179
Focal plane, 176
Focal point, 176
Focusing, automatic system for slide projector, 533[8]
  motion required, 180
Fog, 160
  droplet sizes in, 161
  photographic penetration of, 162
Forbidden energy gap, 280
Form factor, 317, 318
Fourier transform spectroscopy, *see* Interferometric spectroscopy
Forward looking infrared, (FLIR), 479[22], 496[25]
Frame time, 206, 424
Frost, reflectance of, 613[14]
Fuse, proximity, 466[4], 481[16], 482[25],

485[54], 490[7, 8]
Fused silica (quartz), 189, 194, 212, 214, 216

Gas analyzers, nondispersive, 524[1], 525[6]
Gases, absorption spectra of, 47
Gem identification, 596[5], 612[1]
Gemini, 492[16], 493[18, 19], 560[26], 561[40, 43], 597[10], 601[30], 614[5]
General Electric radiation slide rule, 53, 54, 55
Generation-recombination noise, 309
Geological applications of thermal imagery, 600[9, 11, 13, 18, 20], 601[24, 30], 602[41, 42, 44]
Germanium, bolometer, 274, 290
  cadmium-doped, 293
  copper-doped, 265, 293, 294, 361, 370
  gold-doped, *n*-type, 292
  p-type, 292, 294, 361
  mercury-doped, 265, 293, 294, 361, 369
  as optical material, 200, 212, 215, 218, 220, 221
  zinc-doped, 265, 293, 294
Glass, infrared transmitting, 212, 214, 216
Globar, 83
Golay detector, 274, 290
Gouffé, analysis of blackbody, 68
Graybody, 40
Gregorian mirror system, 196
Gun-flash detection, 473[38], 474, [48, 51]

Haven's limit, 357
Haze, 160
  photographic penetration of, 162, 164, 608[1]
  transmission of, 163, 164
Heating and drying, 534[2–5], 535[10–12]
Herschel, Sir John, 6
Herschel, Sir William, 3, 4, 5
Herschelian optics, 196
Holes, 280
Horizon, analytical models of, 559[21, 22], 560[29, 30, 33], 561[36]
  measurements of, 560[23, 25], 561[37, 38, 43, 44], 562[57]
Horizon scanner, 487[6], 488[25], 558[7], 559[11, 13, 16, 20], 560[24, 26, 28, 32, 34], 561[35, 39, 42, 47], 562[51, 52, 55], 563[61]
  problems with, 557, 559[14, 15, 18], 561[40, 41], 562[48, 49, 53]
Hotboxes (railroad), detection of, 10, 513, 516[19, 22–24], 517[32, 37], 518[43, 46, 48–50, 52, 53, 55], 519[60], 520[77]
Humidity, 119
Hygrometer, spectroscopic type, 581[2], 582[5, 13], 584[30, 36], 586[48]
Hygrometry, 119

Icebergs, detection of, 455, 469[1], 470[2], 520[70]
IFF systems, 466[7], 468[17], 498[2], 499[12]
Image converter tube, 8, 265, 296, 297, 466[1, 2], 467[8, 9], 469[24, 25], 498[1, 2, 3], 499[7, 11–16], 500[21, 25], 501[31, 34], 548[8], 597[12]
  general military uses of, 9, 466[1, 5], 467[8, 9], 468[18, 20, 22], 469[24, 25], 498, 498[2–4], 499[6, 7, 12], 500[18–20, 23, 25], 501[29, 31, 32, 34], 502[35], 532[9], 629[11]
  illuminators and filters for, 8, 297, 499[8, 9], 501[28–30], 502[36]
Incidance, 621
Index of refraction, 171
Indium antimonide, *pc,* 291, 292, 294, 361, 367
  pem, 291, 294, 360
  pv, 292, 294, 361, 368
Indium arsenide, *pem,* 291, 360
  pv, 291, 292, 364
Infrared, applications matrix, 458, 460, 461, end papers
  discovery of, 3
  equipment, market for, 9
  literature of, 16, 626
  origin of term, 4
  Soviet literature of, 457, 465
Infrared system, elements of, 15
Insects, infrared sensing by, 614[2, 3]
Interferogram, 595
Interferometric spectrometers, 492[16], 493[18], 526[17], 583[24], 584[38], 585[40, 44], 592[55], 593[60, 61, 66], 595, 597[10, 16, 18], 598[23, 29], 605[15]
Interferometric spectroscopy, 592[55], 594,

596[3,7], 597[14, 16–19], 598[20, 21, 28]
Intrusion detection, 470[7], 471[14], 472[31], 474[47, 54], 513[10], 533[1], 534[2–8]
Inversion temperature, gas, 383
Irdome, 204, 213, 442, 448
  nonspherical, 481[16], 484[47]
IRIA, Infrared Information and Analysis, 627
IRIS, Infrared Information Symposia, 626
Irradiance, 25, 27, 436, 621
IRTRAN, 200, 212, 214, 217, 218

Johnson noise, 306
Joule-Thomson, cryostat, 380–384
  refrigerators, 380, 387
Jupiter, determination of atmospheric constituents, 592[49], 610[23]
  radiometric measurements of, 571[53, 56], 572[69], 573[74]

Kirchhoff's law, 39, 564
KRS-5, 212, 215, 218

Lambertian surface, 29
Lambert's cosine law, 29
Landing systems, aircraft, 500[17], 501[26], 534[1], 535[9]
Laser, 84
Law enforcement, 512[8], 513[10, 11], 531[3], 532[8, 9, 1, 3], 533[1], 608; see also Intrusion detection
Lead selenide, 265, 291, 292, 294, 336, 339, 340, 341, 344, 353, 354, 356, 360, 366, 375
Lead sulfide, 265, 291, 292, 294, 360, 365
Lead telluride, 292, 294
Leidenfrost flow, 378
Lenses, infrared, typical characteristics, 200
Lichtsprecher, 8, 503[6]
Light, velocity of, 35
Light pipe, 203
Linear expansion coefficient, common optical mounting materials, 213
Line-scan thermal mapper, NEΔT, equation for, 431
Liquid crystals, 530[33], 531[34], 544[26], 546[50]
Liquid-transfer refrigerator, 377
LOPAIR, poison gas detection, 493[1], 494[6]

Mach number, 93
Maksutov optical system, 199
Mangin mirror, 199
Mariner, 568[30, 31], 569[38, 40], 570[45, 47], 571[57], 589[17], 597[13]
Mars, atmospheric constituents, 588[6, 12], 589[16, 18, 19], 591[42], 592[55], 593[60], 612[9, 11]
  radiometric measurements of, 566[7, 12], 567[17, 19], 574[87], 589[17, 23, 25]
  vegetation on, 587, 587[4], 588[5, 8–11], 589[21], 590[31], 591[43, 44, 48]
Matra 530, 480, 482[28], 485[52]
Mercury-cadmium-telluride, 292
Mercury-doped germanium, 265, 293, 294, 361, 369
Metascope, 467[8, 9], 468[18, 20], 498[5], 499[6], 501[33]
Methane, abundance of in atmosphere, 584[28, 33]
Microelectronic circuitry, use in infrared systems, 398, 399, 479[13, 14, 16], 484[45]
  thermal conditions in, 519[58, 64], 520[74], 521[87], 530[25, 26]
Microphonics, 374, 390, 405
Microscope, infrared, 531[2, 4, 5], 532[6, 7, 11], 548[9]
Microvolter, 329, 337
Mil, definitions of, 182
Mine, antipersonnel, detection of, 472[31], 574[88]
Missiles, infrared guided, 466[3, 4], 467[12, 13], 468[17, 18, 22], 469[23], 479, 481[9, 16, 17], 482[18, 19, 23, 28, 29], 483[33, 36], 484[40–42, 44–47], 485[48–57], 486[58, 61, 63, 64, 66, 67], 490[7], 512[3]
  characteristics of, 480
  seeker design, 481[7, 11, 14, 15], 483[35], 486[62]
  seekers for, 480[4], 481[8, 10, 12, 13], 482[20–22, 24, 27], 483[34], 484[43]
  proximity fuses for, 466[4], 481[16], 482[25], 485[54], 490[7, 8]
  tactical, command guidance, 468[17], 508, 508[1,3], 509[4, 5, 7, 8, 10, 11, 14, 15], 510[16, 18]
Mixing ratio, 119

INDEX

Modulation transfer function, 194
Modulators, communication, 503[6], 504[12, 17], 505[24, 29], 506[38–40]
Moisture content of materials, determination of, 532[2], 533[4–7]
Monochromator, 31, 341
  calibration of wave length scales, 222
Moon, surface composition, determination of, 590[32, 33, 38], 592[52], 593[64], 605[1–3], 606[5], 610[19, 26], 612[8], 613[11, 16]
  temperature of, 565[1], 566[2–4, 8, 9, 11, 13, 14], 567[20], 568[24], 569[34, 36, 37, 42], 570[46], 571[52, 55, 58], 572[60, 61, 66], 573[71, 72, 75, 77], 574[83–86]
Multiband spectral analysis, 495[13], 599, 599[3], 600[11, 17], 601[22, 24–26, 31], 602[34], 603[45], 611[29, 33]
Multiple-channel systems, 410

Navigation and flight control systems, 486–489, 554–463
NDRC, National Defense Research Committee, 465
  summary technical reports on infrared, 466[7], 467[8]
Neon lamp displays, 411, 468[17], 472[23]
Nernst glower, 82
Nerve gases, detection of, 493[1], 494[4–6]
Newtonian optics, 196
Night driving systems, 466[1, 5], 467[8, 9], 468[17, 18, 22], 469[24, 25], 498[2, 3], 499[6, 7, 12], 500[18, 19, 23, 25], 501[29, 31, 34]
Nimbus, 559[15], 581[15], 598[29], 604[2–5, 7–11], 605[15]
Noise, detector, summary of, 310
  excess, 309, 348
  flicker, 309, 404
  Gaussian, 313, 318
  generation-recombination, 309, 349
  Johnson, 306
  meters for measurement of, 316, 317, 330
  $1/f$, 309, 349
  peak factor, 314, 315
  photon, 310
  power spectrum of, 307, 339
  preamplifier, measurement of, 338
  Rayleigh, 315
  shot, 308
  thermal, 306
  white, 307
Noise equivalent bandwidth, 311
NE ΔT, noise equivalent differential temperature, 429
  value of, typical equipment, 517[28], 528[7], 530[24, 26, 31], 540[11], 542[8], 569[39], 571[57]
NEI, noise equivalent irradiance, 268, 322, 420
NEP, noise equivalent power, 268, 322
Noise factor, 318
  minimum, matching conditions for, 403–407
Noise figure, 318
Nomenclature Committee, Optical Society of America, 619
Nondestructive test and inspection, 528–531
Numerical aperture, 179
Nutation, 250
  means of accomplishing, 470[10], 481[6], 482[22, 26], 487[3], 488[19], 489[31]

Obscuration, 196
Ocean temperature, measurement of, 566[15], 568[26], 569[35, 39], 570[50], 571[51], 572[59, 64], 574[89], 575[93], 601[24]
Optically immersed detectors, 289
Optical materials, emmisivity of, 218
  physical properties of importance, 210
  preferred, 210, 212
Optical systems, 195–202
Optical thickness, 219
Overtone bands, 49
Oxygen content of semiconductors, monitoring of, 525[4, 5], 527[27]
Ozone, 127, 128, 582[7]
  effective transmittance of 9.6μ band, 159

Paints, reflectance of 501[31, 34], 524[2]
Paraboloid, 188, 226
Paraxial equations, 176
Patent classification system, U.S., 17
Peltier effect, 391
Performance specifications, detector, 266, 322
  system, 435
Personnel, radiometric characteristics of, 103

Phosphors, infrared-sensitive, 6, 299
  system use of, 467[8, 9], 468[18, 20], 471[15], 498[5], 499[6], 501[33], 518[42], 528[3], 542[1], 544[33], 546[47]
Photoconductive detectors, 278
  biasing, 283, 330–336
  cutoff wavelength, 280
  ultimate performance limits, 348
Photodiode, 285
Photoelectric detectors, 276
  cutoff wavelength, 277
Photoelectromagnetic detectors, 286
Photographic plates, transmittance of, 260
Photography, infrared, 607, 608[1], 609 609[3, 5–8, 12], 610[17]
  interpretation of, 502[4], 608, 609[4, 6, 11, 13, 14], 610[18], 611[33]
Photometry, 23
  symbols and nomenclature for, 619
Photon detectors, 266, 275
Photon flux, background, minimizing in system design, 422
Photovoltaic detectors, 284
  biasing, 284, 333, 336
  dynamic impedance, 336
  ultimate performance limits, 348
Pip coils, 242
Placental localization, 544[20, 21], 545[39, 44]
Planck's constant, 35
Planck's law, 34, 35, 36, 38
Planets, atmospheric constituents of, *see* specific planet
  radiometry of, general, 563, 566[6, 10], 567[18, 22], 568[28, 29, 32], 571[54], 572[62], 573[80], 587[2], 588[13], 589[22, 26], 590[24, 27, 35], 591[39, 47], 592[57], 593[59], 596[7], 605[4]
  temperature of, *see* specific planet
Planet sensor, 558[6], 562[60]
Plumes, 86, 87–89, 92, 94, 97, 98, 100, 495[11]
  measurement of, 491[1, 2, 5], 492[6, 7, 9, 10, 14], 493[21, 25]
  radiance, calculation of, 94, 98
  temperature contours, 92, 97
p-n Junction, 284
Pneumatic detector (Golay), 274, 290
Poison gas detection, 493[1], 494[4–6]

Pollution, environmental, detection of, 525[8, 13], 527[25], 541[2], 597[18, 19], 600[11, 18], 602[40], 603[50], 610[22, 25], 611[33]
Position sensing, 522
Pratt and Whitney turbojet and turbofan engines, 90, 92, 94, 97, 98
Preamplifiers, 328, 333, 400–408
  matching for minimum noise factor, 403, 406
  noise, measurement of, 338
Precipitable water, 121
Preferred optical materials, 210–218
Pregnancy, determination of, 545[39], 546[48]
Principal plane and point, 178
Probability of detection, 412, 425, 426
Process control, industrial, general, 515[6, 7, 8, 11, 12, 15], 516[17, 25], 517[29, 35, 39], 518[40, 47, 51], 519[59, 66], 520[72, 73, 75, 79], 521[82, 84, 85], 524[2, 3, 4], 528[9], 529[13, 14, 21], 596[4]
  steel mills, 514[1], 518[54], 519[62], 522[95, 1–9]
Pulse visibility factor, 425, 445, 449

Quantum efficiency, 283, 350
Quartz, *see* Fused silica

Radiance, 25, 26, 621
Radiant emittance, 24, 25, 35
Radiant energy, 24, 25, 621
Radiant flux, 24, 25, 27, 28, 621
Radiant intensity, 25, 26, 432, 621
Radiant photon emittance, 25, 27, 350
Radiant power, 24, 621
Radiation calculations, aids for, 53, 57, 60
Radiation contrast, 63, 237
Radiation production, efficiency of, 60
Radiation shielding, 351–356
Radiation slide rules, 53–57
Radiative-transfer refrigerator, 385, 604[2]
Radiator, selective, 32
Radiometers, black ball, 575, 576[3, 10], 577[14]
  calibration of, 434, 493[21], 517[28], 569[41]
  examples of, 30, 493[21], 514[2, 3], 515[7, 9, 14, 15], 516[16–18, 20, 25, 26], 517[28, 31, 33], 518[44], 519[56],

INDEX 639

Radiometers, examples of (Cont'd),
520[68, 71]. 522[94], 525[11, 13],
526[18], 539[3], 540[16], 562[57],
566[15], 567[23], 568[25, 30, 31],
569[39, 41], 570[45, 47, 50], 571[52,
57], 572[64], 574[81], 575[91],
576[2, 3, 5], 577[13, 15, 16, 20],
578[22, 34], 579[37], 588[15],
604[2, 3, 9, 10], 605[15]
  filter type, 515[4, 10], 516[21], 524[1–
    4], 588[14]
  Hardy dermal, 539[1, 3, 5, 7], 568[27]
  Pohl, 575, 576[2]
  Suomi-Kuhn, 575, 576[4, 6, 8], 577[13,
    18], 578[22, 23]
Radiometry, correcting for nonblackbody
  conditions, 515[5], 517[34], 522[93]
  terminology and symbols, 23, 619
Rain, drop sizes in, 164
  transmission through, 163–165
Ramjets, 98
Range equations, background-limited
  detector, 422
  generalized form, 419
  idealized range, 420
  search system, 425
  sample calculations, 444, 450
  signal-to-noise considerations, 419, 424,
    427
  tracking system, with pulse position modulation, 428
  with reticle, 426
Rangefinding techniques, 466[7], 468[18],
  470[12], 489, 489[2], 490[3–6],
  506[4, 5], 538[1], 539[2]
Ray, 172
Rayleigh criterion, 185
Rayleigh-Jeans law, 34
Rayleigh noise, 315, 318
Reconnaissance, 494–497
Redeye, 399, 480, 483[33], 484[44, 45],
  486[61]
Red Top, 480, 490[7]
Reflectance, 25, 28, 622
  spectral, techniques for measurement of,
    222
Reflection, law of, 173
  loss by, 173–175
  total internal, 175, 204
Reflective coatings, *see* Coatings for optical
  materials

Reflective optical systems, 195
Reflector, diffuse, 29
  specular, 28
Refraction, law of, 173
Refrigerant, solid, temperature and storage
  pressure, 385
Refrigerators for detectors, 377–394
  coefficient of performance, 387
  comparison of available types, 386, 392
Remote sensing, 551, 581, 586, 599[6, 7],
  600[21], 601[22, 33]
Resolution, angular, 186
Responsivity, 267, 322, 363, 434
Restrahlen, 106, 573[80], 582[6]
Reticles, AM, 245, 253, 488[21]
  balanced, 244
  comparison of AM and FM types, 258
  effective transmittance of, 427
  electromagnetic drive for, 256, 488[18],
    489[32]
  fabrication of, 258
  FM, 246, 247, 251, 252, 476[14], 478[7],
    482[20, 22], 484[43], 488[23, 24],
    489[29]
  frequency spectra of signals from, 244
  multiple frequency, 249, 478[9, 10],
    480[4], 488[16, 17], 489[33]
  pulse modulation, 241, 242, 243, 246, 248
  radial transmission gradient for, 253
  rotating, 238, 241–243, 245–249
  stationary, 250, 251, 252, 253
  two-color, 254, 473[36], 521[84]
RF pickup by infrared systems, 400
Rockets, 99
Rotational spectrum, 47

Sapphire, 200, 212, 215, 216
Satellites, detection of, 477[1], 553, 553[1,
  2, 5, 7], 565
  meteorology, 579–581
Scan element, 206
Scanners, typical, 206, 472[25], 477[2],
  478[3, 4], 560[28], 606[1, 2, 3],
  607[9]
Scattering, 159–163
Scattering coefficient, 114
Schmidt optical system, 198, 297
Schmidt trigger, 414, 451
Scintillation, 165, 494[6]
Search field, 181, 206
Search systems, detailed design of, 438–452

range equation, 424
Search/track systems, 469–479, 512–513, 553–554
Secondary mirror, 195, 198
Selective radiator, 40, 46
Semiconductor, extrinsic, 281
  intrinsic, 280
Sex differentiator, 514, 516[27]
Ships, detection of, 466[1], 467[11], 468[17, 18], 470[2, 5, 8, 9, 11], 471[13, 18, 19], 473[37], 475[2], 503[2]
Shot noise, 308
Sidewinder, 9, 480, 481[17], 482[23], 484[42], 485[48, 51, 54, 57], 486[64], 512[3]
Signatures, radiation, 491, 491[1], 492[6, 15], 493[18, 23], 603[51]
Silicon, detector, 291
  monitoring oxygen content, 525[4, 5], 527[27]
  as optical material, 189, 194, 200, 212, 215, 218
Skin, human, effect of smoking on temperature, 545[36], 547[54]
  temperature of, 103, 539[1, 2, 7, 8], 540[11, 12, 15, 17], 542[1, 7], 544[29, 33], 546[47, 50, 53]
Sky radiance of, 108, 569[41], 583[15]
Small angle approximations, 182
Snell's law, 173
Sniperscope, 9, 467[8, 9], 498[4], 499[6], 500[20], 501[34], 502[35], 532[9], 629[11]
Snooperscope, 466[1], 499[6, 7], 501[31, 34], 629[11]
Solar constant, 84, 564, 578[30], 579[37]
Solid angle, 26
Solid-refrigerant refrigerator, 384
Source, point or extended, 26
Sources for equipment calibration, black body, 72–77, 80, 598[27]
  facilities for calibration of, 80
  low temperature, 433
  traceable to NBS, 80, 81
Soviet infrared literature, 457, 465
Space filtering, 238
Spectral response of detectors, 286, 340–345
Spectrometers, 526[19], 581[3], 582[12, 14], 583[23], 584[38], 585[41–43], 586[47], 587[1], 588[9, 12, 15], 589[18, 20], 592[57], 593[63], 594, 598[28], 605[15]
  interferometric, 492[16], 493[18], 526[17], 583[24], 584[38], 585[40, 44], 592[55], 593[60, 61, 66], 595, 597[10, 16, 18], 598[23, 29], 605[15]
  rapid scan, 491[1, 2, 3], 525[3, 7, 14], 589[24], 590[35], 598[24–26]
Spectropolarimeter, infrared, 597[15]
Spectroradiometer, 30
Spectrum, band, 32
  line, 32
Spherical aberration, 186, 187–189
  blur circle due to, 189
Stagnation temperature, 101
Standards, sources, 80, 81
  surface defects on optical elements, 201
Stars, 104, 109, 558[8], 590[29], 591[40, 41], 592[50], 593[65]
Star trackers, 486, 487[3, 5, 7, 10], 488[16–24, 26], 489[28, 29, 31–33], 560[27], 562[54]
Stefan-Boltzmann law, 33, 37, 38
Sterance, 621
Sterisent, 106, 622
Stirling refrigerator, 389
Stratoscope, 588[12], 589[16, 18], 590[29], 592[49], 612[8]
Strontium titanate, 212, 215, 216
Submarines, detection of, 471[18], 494[1, 2], 495[3, 4, 6, 9, 10], 496[15, 20]
Sun, 84, 581[1, 3], 582[8], 584[32], 585[42], 590[28]
Sun tracker, 558[1, 2], 559[12], 562[58]
Surface protective coatings, *see* Coatings for optical materials
Symbols, list of, 615
Symbols and nomenclature for radiometry and photometry, 619
System, 10
  active or passive, 8
  performance specification, 435
System engineer, 13
System engineering, 10
Synchronous rectification, 410

Targets, 85–104
Telluric bands, 129
Temperature, minimum detectable with infrared film, 611[31]

INDEX

Temperature measurement, radiometric equations for, 432, 433
Temperature scales, conversion of, 35
Terrain avoidance systems, 487[4], 488[14], 489[30], 506[1, 2,3]
Terrain materials, emissivity of, 570[49], 572[68]
  radiometric properties of, 568[27], 572[63], 575[92], 582[6, 14], 583[15], 584[31, 34], 585[43, 44, 46], 590[31], 592[51]
  reflectance of, 612[4–7, 9, 10], 613[11–13, 15–17]
Textiles, reflectance of, 502[2], 612[2, 3]
Thermal detectors, 266, 271
  blackening of, 274
  ultimate performance limits, 357
Thermal imagery, color translation in, 546[53], 607[12]
  examples of, 495[4, 11], 496[17, 22], 529[15], 530[27, 32], 543[17, 18], 544[19, 24], 545[36, 39, 41], 546[51, 53], 547[54], 599[5, 6, 7], 600[8, 19–21], 601[24, 30], 602[35, 36, 42], 603[45, 50], 604[4–8, 10–12], 605[14]
  interpretation of, 600[9, 18], 602[39–43]
Thermal imaging equipment, 496[22], 497[28, 30, 33], 528[1, 2, 6, 7], 530[23, 24], 531[38], 542[5, 8], 545[39, 43], 546[52], 547[54], 599[6, 7], 600[21], 602[35, 39], 603[47, 49], 604[2, 3, 10], 606[5, 6], 607[9]
  Evaporograph, 6, 265, 299, 528[4, 5], 529[13, 14, 17], 531[36], 542[3], 606[7, 8], 607[11]
  NEΔT, equation for, 431
  test area for, 497[26]
Thermal noise, 306
Thermal radiance derivative, 63, 431
Thermal radiation laws, 35, 38
Thermal radiator, 32
Thermicon, 299
Thermistor bolometer, 265, 273, 290, 294, 360, 371
Thermocouple, 5, 271, 290, 294, 342
Thermoelectric materials, 391
Thermoelectric refrigerator, 391
Thermograph, 541

Thermomagnetic refrigerator, 394
Thermopile, 6, 272, 290
Thornton radiation slide rule, 53, 56, 57, 350, 423
Threshold detector, 413, 414, 451
Time constant, 267
  measurement of, 345
Time multiplexing, 209, 410
Tiros, 559[11, 13], 560[29, 31], 561[36], 568[33], 570[43, 49], 575[91], 576[11], 577[12, 15–17, 19, 21], 578[23, 25, 27–29, 31], 580[2–4, 6, 8, 11, 13], 605[13]
Traceability to NBS, 78
Tracking systems, 475–479, 533–534
  range equations for, 426, 428
Traffic monitoring, 518[47], 534[6, 7], 535[16]
Translations, non-English references, 17, 457
Transmission of the earth's atmosphere, *see* Atmospheric transmission
Transmittance, 25
  internal, 137
  plane parallel plate, 174
Treasure, location of buried, 574[88]
Tumor, malignant, temperature of, 537, 540[9, 13], 542[2], 543[12, 15], 545[37], 546[46]
Tungsten lamp, 83
Turbofan engines, 89
Turbojet engines, 85
Two-phase flow, 378

Uranus, temperature of, 574[82]

Valence band, 278
Veins, superficial, examination of, 548[2], 549[7, 9, 11]
Venus, atmospheric constituents, 573[76], 588[7], 589[19, 24], 590[30, 36], 591[45], 592[53, 55, 56], 593[60, 66], 613[14]
  temperature of, 566[5], 567[16, 19], 568[30, 31], 569[38, 40], 570[44, 45, 47, 48], 571[57], 572[65], 573[70]
v/h, 430
  sensor for, 487[11, 12], 488[13, 15, 27]
Vibrational spectrum, 48
Vibration-rotation band, 49
Vidicon, 265, 297, 605[1, 2]
Visibility, 159
Visual range, 162

Volcanic activity, detection of, 600[13, 20], 601[30]

War games, hit scoring, 509[6], 510[17]
Washout, 237, 582[14], 602[41]
Water vapor, absorption by, 48–50, 115, 126, 129, 138, 141
  in earth's atmosphere, 119, 123, 124, 125, 578[34], 582[9, 10, 12], 583[16, 22, 25], 586[47]
  mass of in saturated air, 119
  quantities, conversion nomogram for, 122
  transmission tables, 144, 146, 148, 150, 152, 154

Wave analyzer, 329, 330, 339
Waveform, rms conversion factor, 326
Wavelength, 22
Wavenumber, 23
Wiener spectrum, 106
Wien's displacement law, 33, 37, 38, 39
Windows, atmospheric, 64, 116, 130, 136
Working Group on Infrared Backgrounds (WGIRB), 23
World Weather Watch, 579–581

Xenon arc, 84

Zinc-doped germanium, 265, 293, 294

Figure 15.1 INFRARED APPLICATIONS MATRIX.

	Military — Chapter 16	Industrial — Chapter 17
1. Search, track, and range	Intrusion detection Bomber defense Missile guidance Navigation and flight control Proximity fuses Ship, aircraft, ICBM, and mine detection Fire control Aircraft collision warning	Forest fire detection Guidance for fire-fighting missiles Fuel ignition monitor Locating hidden law violators Monitoring parking meters Detect fires in aircraft fuel tanks
2. Radiometry	Target signatures	Hot boxes on railroad cars Noncontact dimensional determination Process control Measurement of the temperature of brake linings, power lines, cutting tools, welding and soldering operations, and ingots
3. Spectro-radiometry	Terrain analysis Poison gas detection Target and background signatures Fuel vapor detection Detection of contaminants in liquid oxygen piping	Detection of clear-air turbulence Analysis of organic chemicals Gas analysis Determination of alcohol in the breath Discovery of leaks in pipelines Detection of oil in water Control of oxygen content in germanium and silicon
4. Thermal imaging	Reconnaissance and surveillance Thermal mapping Submarine detection Detection of underground missile silos, personnel, vehicles, weapons, cooking fires, and encampments Damage assessment	Nondestructive testing Inspection Locating piping hidden in walls and floors Inspection of infrared optical materials Detect and display microwave field patterns Study efficiency of thermal insulators
5. Reflected flux	Night driving Carbine firing Intrusion detection Area surveillance Camouflage detection Station keeping Docking and landing	Industrial surveillance and crime prevention Examination of photographic film during manufacture Detection of diseased trees and crops Travelling matte photography Automatic focusing of projectors
6. Cooperative source	Terrestrial communications Command guidance for weapons Countermeasures for infrared systems Range finding Drone control Intrusion detection	Intrusion detection Automobile collision prevention Traffic counting Radiant heating and drying Data link Intervehicle speed sensing Aircraft landing aid Cable bonding